Hansjörg Hauser, Roland Wagner (Eds.)

Animal Cell Biotechnology

Also of Interest

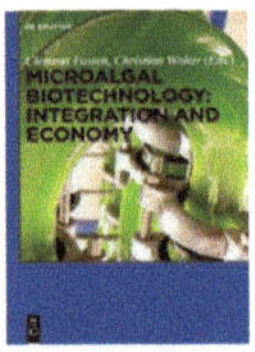

Microalgal Biotechnology: Integration and Economy
Clemens Posten, Christian Walter (Eds.), 2012
ISBN 978-3-11-029827-7, e-ISBN 978-3-11-029832-1

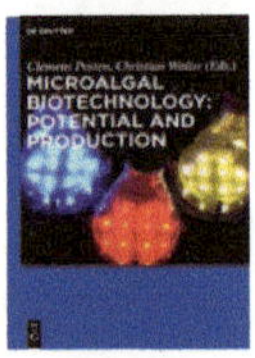

Microalgal Biotechnology: Potential and Production
Clemens Posten, Christian Walter (Eds.), 2012
ISBN 978-3-11-022501-3, e-ISBN 978-3-11-022502-0

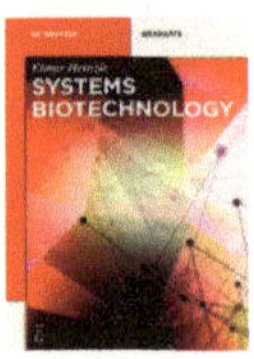

Systems Biotechnology
Elmar Heinzle, Susanne Peifer-Gorges, 2014
ISBN 978-3-11-028924-4, e-ISBN 978-3-11-028926-8

Sustainable Process Engineering
Prospects and Opportunities
Andrzej Benedykt Koltuniewicz, 2014
ISBN 978-3-11-030875-4, e-ISBN 978-3-11-030876-1

Membrane Engineering
Lidietta Giorno, Enrico Drioli, 2015
ISBN 978-3-11-028140-8, e-ISBN 978-3-11-028139-2

Animal Cell Biotechnology

In Biologics Production

Edited by
Hansjörg Hauser and Roland Wagner

DE GRUYTER

Editors
Dr. Hansjörg Hauser
Helmholtz Centre for Infection Research
Gene Regulation and Differentiation
Inhoffenstraße 7
38124 Braunschweig
Germany
E-mail: hansjoerg.hauser@helmholtz-hzi.de

Prof. Dr. Roland Wagner
Rentschler Biotechnologie GmbH
Erwin-Rentschler-Straße 21
88471 Laupheim
Germany
E-mail: roland.wagner@rentschler.de

ISBN 978-3-11-027886-6
e-ISBN (PDF) 978-3-11-027896-5
e-ISBN (EPUB) 978-3-11-038142-9
Set-ISBN 978-3-11-027897-2

Library of Congress Cataloging-in-Publication data
A CIP catalog record for this book has been applied for at the Library of Congress.

Bibliographic information published by the Deutsche Nationalbibliothek
The Deutsche Nationalbibliothek lists this publication in the Deutsche Nationalbibliografie; detailed bibliographic data are available on the Internet at http://dnb.dnb.de.

Typesetting: Werksatz Schmidt & Schulz GmbH, Gräfenhainichen
Printing and binding: CPI books GmbH, Leck
Cover image: Mehau Kulyk/Science Photo Library
∞ Printed on acid-free paper
Printed in Germany

www.degruyter.com

Preface

It is now more than 15 years ago when the first edition of the book 'Mammalian Cell Biotechnology in Protein Production' was published. At that time, realization of the book was driven by the requirements of academia and industry working on cell-based research and production processes to intensify their biological and technological understanding for bringing their inter-disciplinary knowledge to an applicable know-how and for using their technological capacities to an all-embracing capability.

During the last 20 years the biopharmaceutical market showed a substantial growth and can be described as the age of blockbusters. While at the beginning drugs dominated targeting genetic defects (Antithrombin III or Factor VIII), more and more therapeutics for treatment of cancer, cardiovascular, immunological and neurological diseases have been produced. However, treatment of diseases with a broad diversity and physiological dependency on the genetic background of the host is less predictable. As a consequence, not all patients benefit from blockbusters. Current clinical research increasingly reveals information about the molecular basis of these diseases, especially through reliable omics data. Patient stratification methods lead to a facilitated prediction if a certain drug will be effective or if it is better not used. This generates the "virtual patient" and asks for individualized medicines, leading to reduced use of blockbusters and request for a diversification of medications, even if they are only of interest for smaller patient groups. Thus, the next decade will be characterized by more sophisticated direct-to-consumer distribution channels which will diminish the role of wholesalers. The blockbuster sales model will be more and more replaced by products or combinations of drugs that are primarily focused on specialized medication and treatment procedures.

Meanwhile, industry was able to develop new products of the second and third generation and launched the so-called biosimilars that allow companies to rely, at least in part, on the safety and efficacy data of the reference brand product. Increases in knowledge and the more rational approach to drug discovery have contributed to the discovery of many important new classes of biopharmaceuticals but the costs for licensing a product constantly increased and amounts up to more than 1 billion USD.

The regulatory authorities have continuously strengthened their rules and guidelines. This concern not only the quality of the production process and the product, it also influences the average number of patients enrolled in clinical trials. While in 1970 approximately 2,000 individuals were needed for approval of a certain drug, more than 5,000 have been requested in 1990.

Today, the research and development process typically spans more than a decade and still remains subject to considerable risk and uncertainty. This is reflected by the low probability of success. In 2010, Ernst and Young presented an estimation of 5,000 products in phase I/II development and another 20,000 products in preclinical development. This reflects the failure rate in this step. The high number indicates the potential revenues but as well the requirement for new drugs as stated above. Also, the high number requests cheaper, faster and more efficient procedures to obtain clinical trial materials. The capacity to develop manufacturing processes in a competitive time period is often not available at the critical phase and the flexibility of the facility is missing. New cheaper and flexible production facilities are necessary to satisfy the increasing demand. At the same time, the considerable increase in the number of new drug candidates, their use in individualized medicine leads to a reduction of the manufacturing volume forcing companies to leave their established routes and set up flexible facilities.

While the above considerations concern recombinant proteins, mainly antibodies, gene and cell therapies represent a new class of drugs that have the potential for cure. As of 2012, over 2,030 clinical gene therapy trials for human and animal health have to be completed or have been ongoing. Around 200 companies are involved in developing gene therapeutics and their number increased 4-fold during the last decade.

The new book encompasses the major aspects for the development and manufacturing of biopharmaceuticals and cell-based processes ranging from the genetic, molecular and cellular issues up to the final product. It is mainly directed to the expression and production of proteins and builds bridges to other biologics like those based on viral genomes. All these therapeutics have particular requirements on purity, potency and safety with tight specification ranges guaranteed by robust processes of highest efficiency run under cGMP conditions. The book also addresses researchers in industry and academia that need higher amounts of recombinant proteins from animal cells for functional tests and structural investigation.

The chapters have been written by outstanding experts in the respective biotechnological areas. The content is aimed at students interested in the blooming field of protein and virus-based biotechnology for the development and manufacturing of biotherapeutics and at researchers working in this scientific field in academics and industry as well as at all scientists interested in specific aspects of the applied animal cell culture-based biotechnology.

Berlin, in March 2014 *Hansjörg Hauser and Roland Wagner*

Contents

List of Contributing Authors

Claudia Altamirano
Pontificia Universidad Católica de Valparaíso
Facultad de Ingeniería
Escuela de Ingeniería Bioquímica
Av. Brasil 2147
Valparaíso
Chile
E-mail: claudia.altamirano@ucv.cl

David Ausländer
ETH Zurich
Department of Biosystems Science and
Engineering (D-BSSE)
Mattenstrasse 26
4058 Basel
Switzerland
E-mail: david.auslaender@bsse.ethz.ch

Martina Baumann
University of Natural Resources and
Life Sciences
Department of Biotechnology
Muthgasse 18
1190 Wien
Austria
E mail: martina.baumann@boku.ac.at

Sandra V. Bennun
Sandia National Laboratories
Nanobiology Department
1515 Eubank SE
Albuquerque, NM 87185
USA
E-mail: svbennun@gmail.com

Julio Berrios
Pontificia Universidad Católica de Valparaíso
Facultad de Ingeniería
Escuela de Ingeniería Bioquímica
Av. Brasil 2147
Valparaíso
Chile
E-mail: julio.berrios@ucv.cl

Michael J. Betenbaugh
Johns Hopkins University
Department of Chemical and
Biomolecular Engineering
3400 North Charles Street
Baltimore, MD 21218
USA
E-mail: beten@jhu.edu

Emily Blake
Johns Hopkins University
Departmen. of Chemical and
Biomolecular Engineering
3400 North Charles Street
Baltimore, MD 21218
USA
E-mail: eblake5@jhu.edu

Mariela Bollati-Fogolín
Institut Pasteur de Montevideo
Cell Biology Unit
Calle Mataojo s/n
11400 Montevideo
Uruguay
E-mail: mbollati@pasteur.edu.uy

Nicole Borth
University of Natural Resources and
Life Sciences
Department of Biotechnology
Muthgasse 18
1190 Wien
Austria
E-mail: nicole.borth@boku.ac.at

Karina Brinkrolf
Bielefeld University
Center for Biotechnology
Genome Research
of Industrial Microorganisms
Universitätsstraße 25
33615 Bielefeld
Germany
E-mail: kbrinkro@cebitec.uni-bielefeld.de

Michael Butler
University of Manitoba
Department of Microbiology
Animal Cell Technology
Buller Bilding
Winnipeg, MAN R3T 2N2
Canada
E-mail: butler@cc.umanitoba.ca

Natalia Ceaglio
Ciudad Universitaria
Universidad Nacional del Litoral
Faculdade de Bioquímica y Ciencias Biológicas
Laboratório de Cultivos CelularesC. C. 242 -
3000 Santa Fe - Pcia. Santa Fe
Argentina
E-mail: nceaglio@fbcb.unl.edu.ar

Andrew Chung
Johns Hopkins University
Departmen. of Chemical and
Biomolecular Engineering
3400 North Charles Street
Baltimore, MD 21218
USA
E-mail: u921704@gmail.com

Lynn Conley
Biogen Idec Inc.
Cell Culture Development
5000 Davis Drive
Research Triangle Park, NC 27709
USA
E-mail: lynn.conley@biogenidec.com

Wilhelm G. Dirks
Leibniz-Institut DSMZ- German Collection
of Microorganisms and Cell Cultures
Department of Human and Animal Cell Cultures
Inhoffenstraße 7b
38124 Braunschweig
Germany
E-mail: wilhelm.dirks@dsmz.de

Hans G. Drexler
Leibniz-Institut DSMZ-German Collection of
Microorganisms and Cell Cultures
Department of Human and Animal Cell Cultures
Inhoffenstraße 7b
38124 Braunschweig
Germany
E-mail: hans.drexler@dsmz.de

Detlef Eisenkrätzer
Roche Diagnostics GmbH
Pharma Technical Development
Europe Biologicals
Nonnenwald 2
82377 Penzberg
Germany
E-mail: detlef.eisenkraetzer@roche.com

Marina Etcheverrigaray
Ciudad Universitaria
Universidad Nacional del Litoral
Faculdade de Bioquímica y Ciencias Biológicas
Laboratório de Cultivos CelularesC. C. 242 -
3000 Santa Fe - Pcia. Santa Fe
Argentina
E-mail: marina@fbcb.unl.edu.ar

Lore Florin
Boehringer Ingelheim International
Biotherapeutics Research
900 Ridgebury Road
Ridgefield, CT 06877-0368
USA
E-mail: lore.florin@boehringer-ingelheim.com

Nuno Fontes
Boehringer Ingelheim Fremont Inc.
Protein Sciences
6701 Kaiser Drive
Fremont, CA 94555
USA
E-mail: nuno.fontes@boehringer-ingelheim.com

Torben P. Frandsen
CMC Biologics A/S
Vandtaarnsvej 83B
2860 Soeberg, Copenhagen
E-mail: tof@cmcbio.com

Martin Fussenegger
ETH Zurich
Department of Biosystems Science and Engineering (D-BSSE)
Mattenstrasse 26
4058 Basel
Switzerland
E-mail: fussenegger@bsse.ethz.ch

Alexander Goesmann
Justus Liebig University Giessen
Faculty 08 – Biology and Chemistry
Bioinformatics and Systems Biology
Heinrich-Buff-Ring 58
35392 Gießen
Germany
E-mail: alexander.goesmann@computational.bio.uni-giessen.de

Matthias Hackl
University of Natural Resources and Life Sciences
Department of Biotechnology
Muthgasse 18
1190 Wien
Austria
E-mail: matthias.hackl@boku.ac.at

Hansjörg Hauser
Helmholtz Centre for Infection Research
Gene Regulation and Differentiation
Inhoffenstraße 7
38124 Braunschweig
Germany
E-mail: hansjoerg.hauser@helmholtz-hzi.de

Kelley M. Heffner
Johns Hopkins University
Department of Chemical and Biomolecular Engineering
3400 North Charles Street
Baltimore, MD 21218
USA
E-mail: kheff24@gmail.com

Raimund Hoffrogge
Bielefeld University
Faculty of Technology
Cell Culture Technology
Universitätsstraße 25
33615 Bielefeld
Germany
E-mail: raimund.hoffrogge@uni-bielefeld.de

Peter van Hoogevest
Lipoid GmbH
Frigenstraße 4
67065 Ludwigshafen
Germany
E-mail: van.hoogevest@lipoid.com

Hans Henning von Horsten
HTW Berlin – University of Applied Sciences
Campus Wilhelminenhof
Wilhelminenhofstraße 75a
12459 Berlin
Germany
E-mail: hanshenning.vonhorsten@htw-berlin.de

WeiWei Hu
Biogen Idec Inc.
Cell Culture Development
5000 Davis Drive
Research Triangle Park, NC 27709-4627
USA
E-mail: weiwei.hu@biogenidec.com

Vijay Janakiraman
Biogen Idec Inc.
Cell Culture Development
5000 Davis Drive
Research Triangle Park, NC 27709
USA
E-mail: vijay.janakiraman@biogenidec.com

Ingo Jordan
ProBioGen AG
Goethestraße 54
13086 Berlin
Germany
E-mail: ingo.jordan@probiogen.de

Steven Ziyi Kan
Livzon Mabpharm Inc.
Zhuhai, Guangdong
China
E-mail: steve.kan@hotmail.com

Hitto Kaufmann
Sanofi-Aventis Deutschland GmbH
Industrial Affairs, Biologics
Industriepark Hoechst, K703
65926 Frankfurt am Main
E-mail: Hitto.Kaufmann@sanofi.com

Ricardo Kratje
Ciudad Universitaria
Universidad Nacional del Litoral
Faculdade de Bioquímica y Ciencias Biológicas
Laboratório de Cultivos CelularesC. C. 242 -
3000 Santa Fe - Pcia. Santa Fe
Argentina
E-mail: rkratje@fbcb.unl.edu.ar

Natascha Kruse
Helmholtz Centre for Infection Research
RG Model Systems for Infection and Immunity
Inhoffenstraße 7
38124 Braunschweig
Germany
E-mail: natascha.kruse@helmholtz-hzi.de

Kyoungho Lee
University of Minnesota
Chemical Engineering & Material Science
421 Washington Avenue SE
Minneapolis, MN 55455
USA
E-mail: leek@umn.edu

Christoph Lipps
Helmholtz Centre for Infection Research
RG Model Systems for Infection and Immunity
Inhoffenstraße 7
38124 Braunschweig
Germany
E-mail: christoph.lipps@helmholtz-hzi.de

Roderick A. F. MacLeod
Leibniz-Institut DSMZ-German Collection of
Microorganisms and Cell Cultures
Department of Human and Animal Cell Cultures
Inhoffenstraße 7b
38124 Braunschweig
Germany
E-mail: roderick.macleod@dsmz.de

Tobias May
InSCREENeX GmbH
Inhoffenstraße 7
38124 Braunschweig
Germany
E-mail: tobias.may@inscreenex.com

Dethardt Müller
Rentschler Biotechnologie GmbH
Technology Development
Erwin-Rentschler-Straße 21
88471 Laupheim
Gemany
E-mail: dethardt.mueller@rentschler.de

Thomas Noll
Bielefeld University
Faculty of Technology
Cell Culture Technology
Universitätsstraße 25
33615 Bielefeld
Germany
E-mail: thomas.noll@uni-bielefeld.de

Marcos Oggero
Ciudad Universitaria
Universidad Nacional del Litoral
Faculdade de Bioquímica y Ciencias Biológicas
Laboratório de Cultivos Celulares
C. C. 242 - 3000 Santa Fe - Pcia. Santa Fe
Argentinia
E-mail: moggero@fbcb.unl.edu.ar

Takeshi Omasa
Tokushima University
Institute of Technology and Science
Department of Biological Science & Technology,
Biosystems Engineering
Minamijosanjima-cho
Tokushima, 770-8506 2-1
Japan
E-mail: omasa@tokushima-u.ac.jp

Alfred Pühler
Bielefeld University
Center for Biotechnology
Genome Research of Industrial Microorganisms
Universitätsstraße 25
33615 Bielefeld
Germany
E-mail: puehler@cebitec.uni-bielefeld.de

Amy Hong Que
Biologics Process Development
WuXi App Tec
Shanghai
China
E-mail: ahque2010@gmail.com

Søren K. Rasmussen
Symphogen A/S
Pederstrupvej 93
2750 Ballerup
Denmark
E-mail: skr@symphogen.com

Balder Rebmann
Albert-Ludwigs-University Freiburg
Institut für Biology II
Centre for Biological Signalling Studies (BIOSS)
Schänzlestraße 18
79108 Freiburg
Germany
E-mail: balder.rebmann@biologie.
uni-freiburg.de

Thomas Ryll
Biogen Idec Inc.
Cell Culture Development
14 Cambridge Center
Cambridge, MA 02142
USA
E-mail: thomas.ryll@biogenidec.com

Volker Sandig
Cell and Vector Biology
ProBioGen AG
Goethestraße 54
13086 Berlin
Germany
E-mail: volker.sandig@probiogen.de

John Paul Smelko
Biogen Idec Inc.
Cell Culture Development
5000 Davis Drive
Research Triangle Park, NC 27709
USA
E-mail: john.smelko@biogenidec.com

Shawal Spencer
Helmholtz Centre for Infection Research
RG Model Systems for Infection and Immunity
Inhoffenstraße 7
38124 Braunschweig
Germany
E-mail: shawal.spencer@helmholtz-hzi.de

Andreas Tauch
Bielefeld University
Center for Biotechnology
Medical Microbiology and Genomics
Universitätsstraße 25
33615 Bielefeld
Germany
E-mail: tauch@cebitec.uni-bielefeld.de

Venkata Tayi
University of Manitoba
Department of Microbiology
Animal Cell Technology
Buller Bilding
Winnipeg, MAN R3T 2N2
Canada
E-mail: venkata.tayi@gmail.com

Anne B. Tolstrup
Boehringer Ingelheim Pharma GmbH & Co. KG
BP Process Development Germany
Birkendorfer Straße 65
88397 Biberach a. d. Riß
Germany
E-mail: anne.tolstrup@boehringer-ingelheim.com

Cord C. Uphoff
Leibniz-Institut DSMZ- German Collection of Microorganisms and Cell Cultures
Department of Human and Animal Cell Cultures
Inhoffenstraße 7b
38124 Braunschweig
Germany
E-mail: cord.uphoff@dsmz.de

Mauricio Vergara
Pontificia Universidad Católica de Valparaíso
Facultad de Ingeniería
Escuela de Ingeniería Bioquímica
Av. Brasil 2147
Valparaíso
Chile
E-mail: mauricio.vergara@ucv.cl

Jens Vogel
Boehringer Ingelheim Fremont Inc.
Process Science USA
6701 Kaiser Drive
Fremont, CA 94555
USA
E-mail: jens.vogel@boehringer-ingelheim.com

Roland Wagner
Rentschler Biotechnologie GmbH
Erwin-Rentschler-Straße 21
88471 Laupheim
Germany
E-mail: roland.wagner@rentschler.de

Yang Wang
Sino Biological Inc.
Analytical Development and Quality Control
14 Zhong He Street
Beijing 100176
China
E-mail: yang_wang@sinobiological.com

Wilfried Weber
Albert-Ludwigs-University Freiburg
Institute of Biology II
Center for Biological Signalling Studies (BIOSS)
Schänzlestraße 18
79108 Freiburg
Germany
E-mail: wilfried.weber@biologie.uni-freiburg.de

Matthew Westoby
Biogen Idec Inc.
Cell Culture Development
5000 Davis Drive
Research Triangle Park, NC 27709
USA
E-mail: matthew.westoby@biogenidec.com

Karsten Winkler
ProBioGen AG
Pharmaceutical Cell Lines
Goethestraße 54
13086 Berlin
Germany
E-mail: karsten.winkler@probiogen.de

Dagmar Wirth
Helmholtz Centre for Infection Research
RG Model Systems for Infection and Immunity
Inhoffenstraße 7
38124 Braunschweig
Germany
E-mail: dagmar.wirth@helmholtz-hzi.de

Florian M. Wurm
Swiss Federal Institute of Technology Lausanne (EPFL)
Institute of Bioengineering
Laboratory of Cellular Biotechnology (LBTC)
EPFL SV IBI-SV LBTC
1015 Lausanne
Switzerland
E-mail: florian.wurm@epfl.ch

Kelly Wiltberger
Biogen Idec Inc.
Cell Culture Development
5000 Davis Drive
Research Triangle Park, NC 27709-4627
USA
E-mail: kelly.wiltberger@biogenidec.com

Matias Zurbriggen
Albert-Ludwigs-University Freiburg
Institute of Biology II
Center for Biological Signalling Studies (BIOSS)
Schänzlestraße 18
79108 Freiburg
Germany
E-mail: matias.zurbriggen@biologie.
uni-freiburg.de

Ruoheng Zhang †
Boehringer Ingelheim Fremont Inc.
Protein Sciences
6701 Kaiser Drive
Fremont, CA 94555
USA

1 Introduction

1.1 Industrial Use and Perspectives of Animal Cell Culture

Roland Wagner and Hansjörg Hauser

1.1.1 Introduction

In 2013, it was estimated that there were more than 900 products in clinical phases up to submission and several thousand products in preclinical development [1]. The global pharmaceutical sales increased dramatically during the last 2 decades, e.g., it more than doubled from 2000 to 2009. The USA, with around 37 %, is still the world's biggest single market. Growth in this market, which was above average until the first years of the new millennium, has tended to approximate the moderate dynamics of European markets in the past 3 years. Latin American and Asian markets have grown most strongly. Europe's share of the world market declined in 2009 to 31 % (from 32 % in the previous year), due mainly to the Euro's weaker exchange rate against the U.S. dollar. Germany's global market share also fell very slightly in 2009 from 4.5 to 4.3 %. In real terms, assuming a constant exchange rate, Germany's share of the world market decreased from 5.0 to 3.5 % over the past decade. Additionally, the product diversity is broadened by the placement of gene and viral therapeutics as well as cell therapeutics and antisense products to the portfolio. From the overall 907 products in development, 338 count for the largest group of monoclonal antibodies and 250 for vaccines.

In the mid-1980s the average accumulated protein product concentration ranged at ca. 200 mg/L. For producing today's necessary amount of the anti-cancer drug Avastin® of nearly 3 t per year several million liters bioreactor capacity would have been needed based on such a process. Subsequent developments particularly concerned cell culture nutrient media selection of suitable cell lines, cell line engineering, and adapted modular feeding strategies. A combination of achievements led to substantial extension of the cultivation time and hence the available time period for the enrichment of a product such that the amount of producing cells over the prolonged cultivation interval, the cell integral, based on a distinctly higher amount of viable cells, and cell specific productivity is increased. With these methods antibody product titer of more than 5 g/L and above can be achieved within 2 weeks of culture.

1.1.2 Cells as Bioreactors for the Production of Biologics

With the approval of the tissue plasminogen activator (tPA, Activase®) in 1986 mammalian cell culture and particularly the Chinese hamster ovary (CHO) cell line became the most popular production system for the manufacturing of protein therapeutic products. Even 20 years after tPA approval, CHO cells remained as the preferred mammalian cell line for the production of recombinant protein therapeutic for several reasons. CHO cells are easy to handle and can grow in suspension culture, a prerequisite for a homogenous large-scale culture in the industry. Moreover, and very important, CHO cells pose less risk as few human viruses are able to propagate in them [2]. They can grow in serum-free and chemically defined media, which ensures reproducibility between different batches of cell culture and minimizes risk of contamination and impurities. Last, but not least, CHO cells are capable to perform post-translational modifications to recombinant proteins, which are compatible and bioactive in humans [3]. Specifically, glycosylation of glycoproteins produced by CHO cells are more human-like, with the absence of immunogenic α-galactose epitope [4]. Several gene amplification systems are well established to make use of the genome instability of CHO cells to allow for gene amplification, which ultimately result in higher yield of recombinant protein. Currently, recombinant protein titers from CHO cell culture have reached the gram per liter range, which is a 100-fold improvement over similar processes in the 1980s. The significant improvement of titer can be attributed to progress in establishment of stable and high producing clones as well as optimization of culture process. Due to these reasons, CHO cells are established host cell lines for regulatory approvals of therapeutic glycoprotein products [2, 3, 5]. Beside CHO, there are a handful of other cell lines accepted by the regulatory authorities derived from other the Syrian baby hamster, the Muscovy duck, insects and human tissue.

1.1.3 Protein Expression Technologies

The product is expressed in cells by applying appropriate genetic engineering techniques and transfection of the cells with the expression vector bearing the desired genes of interest. The cellular productivity can be modulated by selecting special DNA regulatory elements carried on the vector and by targeting its integration site on the host cell genome. Productivity can also be increased through improving cell culture characteristics via cell line engineering [6].

Antibodies Make up the Biggest Group

Monoclonal antibodies are the fastest growing category starting from 1 % in 1995 to 14 % in 2001 and more than 70 % in 2013. Based on their molecular structure and the resulting binding properties, they have the skills to specifically recognize antigens and cellular markers. An incredible amount of variants and their distinct action is the basis for a nearly inexhaustible therapeutic potential. For the year 2015 the market volume of monoclonal antibodies is estimated to 64 billion U.S. dollars covering at least 38 % of the total biotechnology-based pharmaceutical market which will be shared with other proteins and vaccines then encompassing around 170 billion U.S. dollars. From the 38 billion U.S. dollars in 2009 an approximate double in growth for the antibody market is prognosticated [7].

Personalized Medicine Drives Industry to the Development of new Drugs and More Efficient Manufacturing Processes

Personalized medicine will lead to the renunciation from standard therapy approaches looming today, that has been offered for all patients. At the same time this will require a drastic increase in the diversity of therapeutics. Further, this trend will reduce the market share of so-called blockbuster drugs as the therapeutic antibodies Avastin®, Herceptin®, Rituxan®, or Enbrel®. Industry reacts to this situation by efficiency increase in the development of new drugs and by a substantial reduction of the development period and/or by the application of innovative economic manufacturing processes.

The Biological Potential and the Technological Effort Assign the Process Limit

The rapid and straight increase in the productivity of biologics-producing processes during the last 5 years up to now easily tempts to assume that this progress could keep on with practically endless constancy. While the theoretical end of cellular productivity is not yet reached, from the industrial point of view a final titer will be sufficient when a stable platform process at maximum economy is reached. The maximum cellular productivity might be estimated by the production capacity of a natural high producer cell that has developed throughout biologic evolution. An antibody-producing B lymphocyte is able to produce up to 20,000 antibody molecules per second [8]. Related to a contemporary cell culture in a bioreactor this is equivalent to a production of 6 grams per liter and day at a representative cell concentration of 10 million cells per milliliter culture broth. Translating this capacity to a typical fed-batch process over a cultivation period of 14 days, 80 grams antibody product per liter culture would be produced. This considerable amount is still more than 10-fold higher than a state of the art process today (see above).

Increase in Productivity and Process Realization Have to Form a Unit

In general, product maximization leads to a higher yield. However, this advantage is often accompanied by an increase in rather undesired side products and reduction of product quality that take out the shine from the laudable but unreflected titer numbers. A higher protein concentration imposes additional molecule interactions leading to a higher formation of aggregates and substantially impeding following process steps, which in extreme result in an uneconomic process by a costly and inefficient product purification. Therefore, the economic reason will aim to a balanced situation, which will be adjusted by the competition between the biochemical and cell physiological potency as well as the height of the production capacity and the resulting possibilities. In reality, production bioreactors for cell cultures of today have a working volume of about 10,000 L. Such a volume is integrated in the flow of the total process in a way that the upstream and downstream process form a procedural unit. A significantly higher product concentration will induce new challenges for the purification concept. Chromatographic systems must take enormous dimensions and the purification efficiency has to be substantially increased, such that the higher amount of incompletely or even incorrectly processed, possibly denaturated or aggregated protein will be robustly separated from the desired product in order to guarantee a constant high quality. Primarily, such an additional investment for the manufacturing of new antibody therapeutics would be hardly justifiable. Therefore, the development of cell-based expression systems will preferably focus on the optimization and intensification of the product quality and titers ranging in the one-digit gram range will probably dominate future processes in view of economics. Nevertheless, the request of achieving high product titers will mostly drive the process expectations, especially when competitive biosimilars are developed which shall displace existing products from the market after patent closure.

Biosimilars Require Processes of Highest Robustness

Generics must contain the same active ingredients as the original formulation as they are considered identical in dose, strength, route of administration, safety, efficacy, and intended use (U.S. Food and Drug Administration on generic drugs). This term is only used for small molecule-based drugs produced by simple processes. In contrast, biologics generally exhibit high molecular complexity and are quite sensitive to changes in manufacturing processes since they are made by or derived from a living organism. Differences in impurities and/or breakdown products can have serious health implications. This has created a concern that copies of biologics might perform differently than the original branded version of the product. Consequently, only a few versions of follow-up biologics have been licensed in the regulatory authorities in U.S. (FDA) and Europe (EMA). Examples of those biosimilars are erythropoietin (EPO), human growth hormone (hGH) and granulocyte colony stimulating factor (G-CSF). The development of a biosimilar is only attractive when the following prerequisites are fulfilled:

a) A long ranging market demand.
b) The original product is produced with an inefficient process or organism.
c) No product of a second generation is planned.
d) Patent closure within the next 10 years.

Nevertheless, numerous patents will be closed during the next 5 to 10 years. For proteins of the first generation such as EPO, G-CSF, IFN and FSH the patent protection is already expired or will be expired in the next years. Products of the second generation are blockbuster monoclonal antibodies and/or fusion proteins such as Rituxan®, Herceptin®, Avastin®, and Enbrel® that are most highly attractive for being redesigned as a biosimilar using a more efficient process with a better profitability (see Table 1.1.1). Additionally, products of the third generation are improved mAb/fusion proteins such as Orencia® (Abatacept, extracellular domain of CTLA-4/IgG1-Fc) from Bristol-Myers Squibb or Prolia® (Denosumab) from Amgen and similar new products which are protected by patents beyond 2020. Due to parallel development approaches of different companies the potential of becoming a blockbuster product is comparably lower for this product class. However, the development of a biosimilar is a challenging process. Beside the preclinical data, also the clinical phases, but often in limited extent, have to be passed. Moreover, it takes distinctly more time for market penetration compared to classical generics due to a slower acceptance process of the product by the physicians. Nevertheless, the effort is worthwhile because the development of a new product requires an expenditure amount of 1 billion U.S. dollars and above, which is more than 5 times higher compared to the respective biosimilar.

Table 1.1.1: Blockbuster cell culture-based biologics of the second generation under development as a biosimilar.

Company	**Brand Name**	**Product**	**Cell Line**	**Patent Protection**	**Sales 2011** (in billion USD)
Amgen	Enbrel®	Etanacept, TNFR2-p75/IgG1-Fc	CHO	2012, prolonged to 2022	7.9
AbbVie	Humira®	Adalimumab	CHO	2015	8.2
Genentech/Biogenldec Roche	Rituxan® MapThera®	Rituximab	CHO	2016	6.8
Centocor	Remicade®	Infliximab	SP2/0	2018	7.2
Genentech	Avastin®	Bevacizumab	CHO	2019	6.0
Genentech	Herceptin®	Trastuzumab	CHO	2019	5.9

Antibody-drug Conjugates Force onward the Market

Antibody-drug conjugates (ADCs) make up one of the growing fields in the biopharmaceutical industry. ADCs are a combination of a monoclonal antibody or its fragment (scFv) and a reactive, mostly cytotoxic chemical that has been fragilely fused via a special chemical linker for assembling the final drug vector. This principle allows for the development of highly targeted therapeutic approaches for a wide range of diseases, particularly oncological and hematological indications. By combining the specific targeting capability with the cancer-killing ability of a cytotoxic drug, ADCs allow sensitive discrimination between healthy and diseased tissue. The biochemical reaction between the antibody and the cellular target protein triggers a signal in the tumor cell, which then absorbs or internalizes the antibody together with the cytotoxin (Table 1.1.2). After the ADC is internalized, the cytotoxic drug is released and kills the cancer cell [9].

Table 1.1.2: List of highly potential cytotoxic agents used for the formation of antibody-drug conjugates.

Name	Target	Action
α-Amanitin	RNA polymerase II	Inhibition
Monomethyl auristatin E	Tubulin	Inhibits polymerization
Calicheamicin	DNA (minor groove)	Scission
Duocarmycin	DNA (minor groove)	Alkylate adenine at N3
Doxorubicin	DNA	Intercalation
Maytansinoid	Tubulin	Inhibits polymerization
Pyrrolobenzodiazepines	DNA (minor groove)	Cross-linking

The linker is the crucial part or the ADC and relies usually on chemical synthesis [10]. The complexity in structure, development, and manufacturing impose new challenges to the process. Only three ADCs have received market approval so far. However, after a request from the U.S. Food and Drug Administration (FDA), Pfizer/Wyeth, the developer and marketer of the first ADC to receive marketing approval in 2001 for the treatment of patients with acute myelogenous leukemia (Gemtuzumab ozogamicin, trade name Mylotarg®), withdrew the drug from the market in June 2010. As a result, only two ADCs are marketed (2013), including Brentuximab vedotin (trade name Adcetris®, marketed by Seattle Genetics and Millennium/Takeda) and Trastuzumab emtansine (trade name Kadcyla®, marketed by Genentech/Roche).

Targeting Biologicals to Specific Disease Sites

Cytokines are mediators of cell communication. Their therapeutic use often requires high doses to achieve effective local biological levels. However, the clinical use of some cytokines is limited because of their pleiotropism, which can result in unwanted side effects. Thus, protein engineering technologies that overcome these limitations and enable the targeting of cytokines to specific sites have been developed. One such example uses antibody-based recognition to direct the cytokine to a particular tissue [11]. This method requires the target site-specific cleavage of the fusion protein, thereby exploiting the severity of the pathological process to regulate drug delivery.

Because these technologies are based on the expression of fusion proteins, their application can be extended to diverse biologicals.

1.1.4 Vaccines Are Produced in Mammalian Cells

Prevention is better than cure. The mortality of people infected by the aggressive pox virus *Variola major* was 20 to 60 %. Up to 1967, when the World Health Organization (WHO) started a program to exterminate smallpox, yearly 15 million new infections were registered from which 2 million people died. Thirteen years later, smallpox was nearly eliminated by consequent vaccination campaigns. The word “vaccine” originates from the Latin *Variolae vaccinae* (cow pox), which the British physician Edward Jenner finally demonstrated in 1796 after 6 persons successfully used the principle that could prevent smallpox in humans. Today the term “vaccine” applies to all biological preparations, produced from living organisms, that enhance immunity against disease and either prevent (prophylactic vaccines) or, in some cases, treat disease (therapeutic vaccines). Apart form the adjuvants, vaccines consist of proteins, viruses and bacteria, natural or recombinant, which are either nonpathogenic or inactivated, cells or cocktails from those. Early viral vaccines were produced in primary cells.

The worldwide vaccine business is the fastest growing market in biopharmaceuticals counting for nearly 40 billion U.S. dollars in 2013 and expected to increase over 100 billion U.S. dollars in 2023 with a yearly growth rate of 10 % (according to Exane-PNB Paribas [12]). As stated by the WHO [13] merely the worldwide influenza will increase from 2.9 billion U.S. dollars in 2011 to 3.8 billion U.S. dollars in 2018 (see also Table 1.1.3).

Table 1.1.3: Top product vaccine sales in 2010 [17].

Brand Name	Producer	Type/Composition	2010 Sales (in billion USD)
Prevnar-13®	Pfizer	13-valent pneumococcal conjugate vaccine	2.40
Proquad®	Merck/Sanofi-Aventis	Measles-mumps-rubella and varicella combination vaccine (NMR-V)	1.40
Gardasil®	Merck	HPV	1.35
Prevnar®	Pfizer	7-valent pneumococcal conjugate vaccine	1.20
Fluzone®	Sanofi Pasteur	Influenza (seasonal and H1N1 strains	1.20
Infanrix and Pediarix	Glaxo SmithKline	Infanrix = DTaP, Pediarix = DTap-HepB-IPV (combination DPT-based vaccines with acellular pertussis	1.20

1.1.5 Virotherapy and Gene Therapy Open the Door for Cure

Virotherapy uses specific viruses that have been genetically modified such that they exclusively can propagate in highly active dividing cells, as this is the case for cancer cells. Such oncolytic viruses indicate a new class of high-potency therapeutics. Large amounts of potent viruses have to be produced in animal cells cultivated in large scale bioreactors. The first oncolytic virus candidate was H101, a modified adenovirus, which was licensed in 2005 by China's State Food and Drug Administration (SFDA) for nasopharyngeal carcinoma together with chemotherapy. H101 has been engineered to remove a viral defense mechanism that interacts with a normal human gene P53, which is very frequently dysregulated in cancer cells [14]. Systemic therapy is now marketed under the brand name Oncorine® by Shanghai Sunway Biotech.

Gene therapy is generally defined as the transfer of recombinant nucleic acids to humans as well as animals with the aim to regulate, to repair, to replace, to add, or to eliminate a genetic sequence. Principally, gene therapy is only allowed to perform in somatic cells excluding a passage via the germ line. For the *ex vivo* approach cells (e.g., lymphocytes or hematopoietic stem cells) are taken from the patient, modified outside the body by a therapeutic gene transfer, and the corrected cells are then reinjected. Three virus families are currently covering 90 % of the vector systems used for gene therapy: adenoviruses (AdV), adeno-associated viruses (AAV) and retroviruses (RV) including lentiviruses (LV). All viruses are produced with animal cells in culture. AdVs and AAVs yield at an amount of 10,000 to 100,000 particles per cell, whereas only 1 to 10 RV particles are released per cell. Retrovirus vectors are used for *ex vivo* gene therapy and only relatively low amounts are necessary for the treatment. In contrast, virus doses of up to a quadrillion (1×10^{15}) particles per patient are used for one *in vivo* approach. This is 100 times more than the total amount of human cells in a body (10 trillion = 10^{13}). The production of such a high amount of virus needs highly efficient manufacturing processes. The breakthrough in the acceptance of gene therapy was in 2012 when the European Commission approved Glybera®, an AAV1-LDL^{S447X} gene therapy for the treatment of lipoprotein lipase deficiency (LPLD), as first gene therapeutic in the Western world. In order to treat all 5,000 patients worldwide a minimum amount of 1×10^{17} virus genomes has to be produced using at least a 10,000 L bioreactor capacity, which is comparable to those used for the production of recombinant pharmaproteins [15]. According to Global Industry Analysts Inc. [16] the market for gene therapy is expected to reach 794 million U.S. dollars by 2017. The key factors driving growth in the market include a rising demand for new and efficient therapies for cancers treatment, inability to cure some cancer types and other critical diseases, and prospective launch of gene therapies in major global markets.

1.1.6 Cells as Therapeutics

Cell-based therapies are a viable option for an increasing number of diseases. The transplanted cells are either used as mediators of specific actions or as replacements for diseased or lost cells/tissue. Examples for the first application concerns T cells and mesenchymal stem cells to prevent reactivation of latent cytomegalovirus (CMV) in the transient state of immunodeficiency and graft versus host disease after hematopoietic cell transplantation, respectively. The second application mainly concerns the use of stem cells. A number of animal experiments and a few human studies have demonstrated its usefulness. Further, embryonic stem cells and the recent development of induced pluripotent stem (iPS) cells technology demonstrated the potential of cell-based therapy in rodent models. It was shown that such cells are capable to differentiate *in vitro* and *in situ* and respective cell grafts can improve deficiencies. Altogether, the recent findings have shown great promise for developing the foundation of the cell-based therapy. Unfortunately, there are several challenges faced by researchers that must be overcome before stem cell therapies can become a successful reality for those suffering from disease. While T cell-based immunotherapies are now being included in the clinical practice of transplant recipients to prevent and treat infections and complications associated with CMV, AdV and Epstein-Barr virus (EBV), other cell therapies are still in the development or in clinical trials. Although the projected market for cell therapies and regenerative medicine is estimated to amount several hundred billion U.S. dollars in 2020, significant short-term payoffs are not expected. While the biotechnology of cell therapy applications has not been addressed in this book, it is obvious that the expertise researchers and industrialists have gained from protein production will be essential to further fabrication of such cells.

1.1.7 Protein Production in Pharmaceutical Research

In the development of pharmaceuticals expression of proteins in cells is often required. Medium or large scale production concerns mainly the aspects of structural biology. Despite a multitude of recent technical breakthroughs speeding high-resolution structural analysis of biological macromolecules, production of sufficient quantities of well-behaved, active protein continues to represent the rate-limiting step in many structure determination efforts. To determine protein structure milligram amounts of pure protein are needed. If cells are required for the production of such proteins (due to activity, authentic processing, multisubunit complexes, membrane location), the technologies applied for recombinant pharmaceutical protein production are applied. The same technologies are used for functional assessment of certain proteins. In fact, the trials for the successful production of large quantities of pure proteins for structural biology have significantly contributed to the development of recombinant expression systems as they are used today. While the direct application of these technologies is not directly translated into industrial products, their successful use is often rate-limiting in the development of new drugs.

References

[1] Ernst & Young. 2013. Beyond Borders. Global biotechnology report 2013.

[2] Boeger H, Bushnell DA, Davis R, Griesenbrick J, Lorch Y, Strattan JS, Westoder KD, Kornberg RD, 2005. Structural basis of eukaryotic gene transcription. FEBS Lett 579: 899-903.

[3] Kim JY, Kim YG, Lee GM. 2012. CHO cells in biotechnology for production of recombinant proteins: Current state and further potential. Appl Microbiol Biotechnol 93: 917-930.

[4] Ghadery D, Zhang M, Hurtado-Ziola N, Varki A. 2012. Production platforms for biotherapeutic glycoproteins. Occurrence, impact, and challenges of non-human sialylation. Biotechnol Genet Eng Rev 28: 147-175.

[5] Wiberg FC, Rasmussen SK, Frandsen TP, Rasmussen LK, Tenggbjerg K, Coljee VW, Sharon J, Yang CY, Bregenholt S, Nielsen LS, Haurum JS, Tolstrup AB. 2006. Propduction of target-specific recombinant human polyclonal antibodies in mammalian cells. Biotechnol Bioeng 94: 396-405.

[6] Lai T, Yang Y, Ng SK. 2013. Advances in mammalian cell line development technologies for recombinant protein production. Pharmaceuticals 6: 579-603.

[7] Business Insights 2006. The future of monoclonal antibody therapeutics Business Insights Ltd, London, UK. http://www.globalbusinessinsights.com/content/rbhc0163m.pdf, accessed 15th May 2014.

[8] Kuby J, Kindt TJ, Goldsby RA, Osborne BA. 2007. Kuby immunology. San Francisco: W. H. Freeman.

[9] Chari RVJ, Martell BA., Gross JL, Cook SB, Shah, Sudhir A, Blättler WA, McKenzie SJ, Goldmacher VS. 1992. Immunoconjugates containing novel maytansinoids: promising anticancer drugs. Cancer Res 52: 127-31.

[10] Axup JY, Bajjuri KM, Ritland M, Hutchins BM, Kim CH, Kazane SA, Halder R, Forsyth JS, Santidrian AF, Stafin K, Lu Y, Tran H, Seller AJ, Biroc SL, Szydlik A, Pinkstaff JK, Tian F, Sinha SC, Felding-Habermann B, Smider VV, Schultz PG. 2012. Synthesis of site-specific antibody-drug conjugates using unnatural amino acids. Proc Natl Acad Sci USA 109: 16101-16106.

[11] Gould DJ, Chernajovsky Y. 2007. Novel delivery methods to achieve immunomodulation. Curr Opin Pharmacol 7: 445-450.

[12] Allery C, Berthon S. 2012. Vaccines: Growth boosters. www.exanebnpparisbas-equities.com, accessed 7th May 2014.

[13] Kaddar M. 2013. Economic characteristics of vaccines and vaccination. Workshop on Business Modeling for Sustainable Influenza Vaccine Manufacturing. Washington DC, January 14, 2013.

[14] Garber K. 2006. China approves world's first oncolytic virus therapy for cancer treatment. JNCI J Nat Cancer Inst 98: 298-300.

[15] Brunzell JD. 1999. Familial lipoprotein lipase deficiency. In: Pagon RA, Bird TD, Dolan CR, Stephens K, Adam MP (Eds). GeneReviews™. Seattle: University of Washington.

[16] GIA. 2012. Gene therapy – a global market report. www.strategyr.com/Gene_Therapy_Market_Report.asp, accessed 7th May 2014.

[17] Plotkin SA. 2012. Vaccine development and implementation. Vaccine Fact Book 2012: 52-68.

2 Cell Lines

2.1 Generation of Cell Lines and Biotechnological Applications

Christoph Lipps, Tobias May, and Dagmar Wirth

2.1.1 Introduction

Mammalian organisms contain more than 200 different cell types that are required to fulfill the various functions. Many of these functions depend on the interaction of different cell types. To investigate the molecular basis of diseases, to develop novel treatment options for intervention, and to screen novel drugs, cell based systems are required that reduce the complexity of the organism. For many of such applications the cells need to maintain physiological relevance to mimic the cells *in vivo* as close as possible. For industrial applications, cell based systems should be robust, reliable, and also cost effective.

For many years, cell based systems have been used in basic research as well as in applied disciplines like protein production, drug discovery, and toxicity testing, and they are also used for therapy in regenerative medicine approaches. These systems are based on either primary cells (primary cell cultures, PCCs) or immortalized cell lines, which are also called continuous cell lines, CCLs.

Primary cells can be isolated from various tissues of the organisms and protocols have been established to cultivate many different cell types *in vitro*. These cells closely reflect the properties of cells *in vivo*. Thus, they are considered to be physiologically relevant. However, primary cells are associated with several drawbacks (Figure 2.1.1). As primary cells have a limited proliferation potential they cannot be expanded *in vitro* to larger cell numbers. Thus, to obtain the desired cell numbers, primary cells are collected from many different donors and/or large tissue samples are required. For many cell types, the isolation procedure is a tedious and time-consuming process. Primary cells from humans are characterized by a high donor-to-donor variability. These differences originate, e.g., from (i) the genetic variability between different individuals, (ii) environmental influences, (iii) diseases and medications to which the donor has been exposed, and (iv) technical reasons like variations during the isolation procedure. Although this high degree of variability is scientifically and commercially interesting (e.g., with respect to personalized medicine) it poses a big hurdle for the development of robust and reliable assays. Moreover, primary cells are attributed with a potential risk for adventitious agents. (For a more detailed discussion of the disadvantages of primary cells for production of vaccines see Chapter 2.3.)

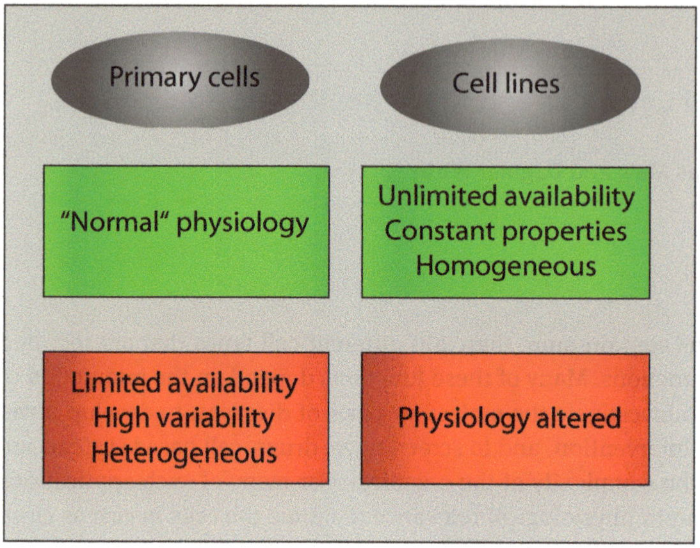

Figure 2.1.1: Comparison of primary cells and cell lines. The major advantages (green) and disadvantages (red) of the respective cell system are given.disadvantages (red) of the respective cell system are given.

As a consequence of the lack of reproducibility of primary cells, many applications rely on unlimitedly proliferating cell lines or CCLs. Cell lines can be either isolated from tumors or generated upon spontaneous or induced immortalization of primary cells. Cell lines by definition show indefinite growth. Thus, they can be easily expanded to the required cell numbers. In addition, if compared to primary cells, they represent a rather homogenous, genetically defined cell population with constant properties. Moreover, they can be genetically modified to display additional features. However, immortalized cells are frequently accompanied with the (partial) loss of relevant physiological properties (Figure 2.1.1). Thus, cell lines lack many features and markers of the tissue they were isolated from.

Since there is no *in vitro* cell system available that meets all the requirements, the selection of a cell system is always a compromise. Accordingly, it strongly depends on the specific application whether physiologically more relevant primary cells or robust cell lines are employed. For example, for the production of recombinant proteins and/or therapeutic viruses and vaccines, cells should show a high productivity, stability, and robustness. Moreover, they should proliferate under defined conditions so that tightly controlled processes can be installed. At the same time, the physiological relevance of the cell system is less important. For such applications, cell lines such as CHO or HEK293 cells proved to be highly instrumental.

In contrast, in drug development the physiological relevance of the employed cell system is of high priority. In this respect, many established cell lines do not meet the expectations. Thus, in the last years emphasis has been laid on the development of new immortalized cell lines with improved properties from specific tissues that have not yet been available previously and that meet the requirements of physiological relevance.

In this chapter, different strategies to establish immortalized cell lines are summarized. This comprises the description of the commonly used immortalizing genes including their proposed molecular action, protocols for cell type specific immortalization, as well as recently emerged conditional immortalization strategies to control cell proliferation. In addition, a noncomprehensive overview is given on biotechnologically relevant cell lines as well as their applications. Emphasis is also given on cell lines used in the drug development process.

2.1.2 Principles of Immortalization

In primary cells, proliferation is tightly controlled. In this surveillance, the tumor suppressor genes p53 and pRb play a pivotal role. These genes control cell cycle progression and contribute to DNA integrity. Moreover, they are involved in the control of cellular senescence and apoptosis, which represent the ultimate mechanisms to eliminate cells that display aberrant proliferation control (for review see, e.g., [1-3]).

In immortal, i.e., infinitively growing cells, this well-balanced regulatory network is perturbed and immortalized cells have lost the original capacity to control cell proliferation. A natural source of immortal cells are tumors in which specific cellular control genes are usually found to be mutated. Tumor derived cell lines are often characterized by genetic instability, which contributes to the generation of additional mutations that lead to immortalization. Tumor derived cell lines display an aberrant growth control and often are tumorigenic. This can be demonstrated by their capacity to form tumors upon injection into immune-compromised mice (see also Chapter 2.3).

As an alternative to tumor derived cell lines, immortal cell lines are generated upon manipulation of primary cells isolated from healthy tissues. For certain cell types like mouse embryo fibroblasts (MEFs), immortalization of cells can simply be achieved upon continuous passaging of primary cells following a regular splitting protocol. In a report by Todaro and Green, MEFs were split every three days and replated with a low density of 3×10^5 per plate (3T3 protocol) [4]. By this protocol, multiple random genetic and epigenetic changes that confer a proliferative advantage are accumulated and finally result in immortalization of cells [5]. Due to the randomness of this process, it is not specified which mutations are induced through this protocol. In later studies the role of the tumor suppressor p53 was investigated, however with nonconsistent results. In one study clear indication was given that p53 is the major driver of the immortalization process [6]. In contrast, another study demonstrated that p53 is only involved in spontaneous immortalization of MEFs if a modified protocol (the 3T12 protocol) but not if the 3T3 protocol is used [7]. This highlights the major disadvantage of spontaneous immortalization protocols: since the genetic alteration that is provoked is not controllable the cells differ in their properties and an extensive screening has to be performed to identify the desired cell clone with the desired properties. Another important drawback of such a random immortalization by cultivation is the fact that it is restricted to certain cell types such as mouse fibroblasts while most cell types cannot become immortalized by this method.

Over the last decades, knowledge has been accumulating on specific genetic manipulations that result in the immortalization of cells. While in the beginning mainly random procedures were employed, in the recent years novel strategies have been emerging that are directed to achieve the rational and producible development of such designer cell lines (Figure 2.1.2). It has to be noted that the various protocols for immortalization are not generally applicable but are usually restricted to certain cell types and/or species. This is exemplified by the recent observation, namely that human cells usually require immortalization strategies that are different to cells from other mammals like, e.g., murine cells [8].

The most important and promising approaches towards a rational development of cell lines are summarized in the following sections.

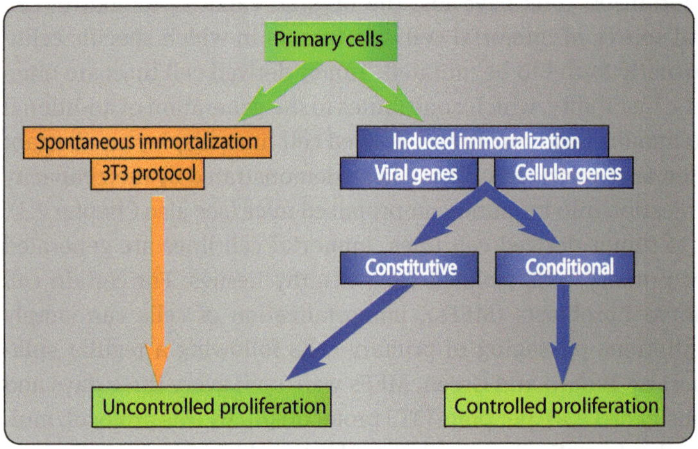

Figure 2.1.2: Schematic representation of the different immortalization approaches. Immortalization can be achieved through spontaneous events like, e.g., outgrowth from tumor material or through defined cultivation regimens (3T3 protocol). A more controlled approach is the introduction of immortalization genes into primary cells of choice. If the immortalization gene is controllable by the exogenous means a growth controlled immortalized cell line is established. In this case proliferation can be controlled by exogenous stimuli.

2.1.2.1 Immortalization of cells based on viral oncogenes

The first cell lines created *in vitro* were derived upon infection of primary cells with viruses that have an intrinsic oncogenic potential (tumor viruses). Various viruses are known to confer infinitive growth of cells by transducing oncogenes that deregulate the cellular growth control. A highly successful strategy for development of cell lines by viral infection represents the use of Epstein-Barr virus (EBV) for the immortalization of B lymphocytes.

Epstein-Barr Virus (EBV) genes

EBV is now in routine use for the establishment of B cell lines that are of particular interest for the development and generation of human monoclonal antibodies (for review see [9]). The gene or the genes from EBV that are responsible for the immortalization of B cells are not yet defined. However, it has been shown that the EBV nuclear antigen 1 (EBNA1) strongly enhances the immortalization process by several orders of magnitude [10]. This indicates that the concerted action of several EBV genes induces immortalization of primary B cells. The relevant pathways of EBNA1 based immortalization include the downregulation of p53 and the induction of reactive oxygen species (ROS). EBNA1 lowers the cellular p53 level indirectly by binding to the herpes virus-associated ubiquitin-specific protease (HAUSP) [11]. This complex binds to the p53-Mdm2 complex and is considered to induce deubiquitination of Mdm2 which in turn leads to the inactivation of p53 [12]. The accumulation of ROS after transduction of EBNA1 is probably induced through the induction of a NADPH oxidase (Nox2) [13]. High ROS levels induce DNA damage and the associated elevated mutation rates are considered to contribute to the immortalization of primary B cells.

Rather than employing replication competent wild type viruses, more controlled immortalization can be achieved by the transfer of the relevant subgenomic viral tumor genes. For this purpose, expression cassettes are generated in which viral or nonviral promoters are used to express the viral genes. These cassettes can be introduced into primary cells by nonviral methods. However, an increase of the efficiency of gene transfer and thus immortalization is often required. Thus, recombinant viruses such as retroviruses or lentiviruses that integrate the immortalizing genes into the cellular genome are of interest. Since lentiviruses can also infect arrested cells, this strategy is currently the method of choice for immortalization of primary cells with limited proliferation capacity.

Human Papillomavirus (HPV) E6 and E7

Out of the 100 human papillomaviruses 15 types are so called high-risk papillomaviruses that are considered to be the causative agents of most cervical cancers (for review see [14]). The major tumor promoting potential of these viruses is based on the viral E6 and E7 proteins. These proteins have been shown to interfere with cell cycle control and the regulation of apoptosis. E7 inhibits the function of various pRb family members by binding and thereby facilitating cell cycle progression (reviewed by [15]). E6 on the other hand is known to promote the degradation of p53 and thereby disrupts the growth control by p53. Another function of E6 is the induction of telomerase activity which supports the immortalization of cells by maintaining telomere length (see below) [16]. For the efficient immortalization of human cells both proteins – E6 and E7 – are required. In certain settings the immortalization of human primary cells can also be achieved through expression of only a single HPV gene. It was demonstrated that both E7 and E6 are able to immortalize primary human foreskin keratinocytes when transferred alone. However, the efficiency was dramatically increased when E6 and E7 were used in combination [17] or when a variant of the E6 gene was employed [18]. It has to be noted that foreskin derived keratinocytes have a high intrinsic proliferation potential which probably contributes to the fact that this cell type is comparably easy to immortalize. Generally, the combination of E6 and E7 has been shown to be particularly efficient for the immortalization of epithelial cells [19].

Adenoviral E1A and E1B

The family of the adenoviruses comprises more than 50 serotypes (which fall into six different groups). Although adenoviruses per se do not induce cancer formation, it has been demonstrated early that the proteins E1A and E1B are able to immortalize primary cells *in vitro* [20]. The E1A gene is the dominant oncogene [21]. E1A is able to immortalize primary cells per se, albeit with a much lower frequency when compared to immortalization regimens which utilize E1A and E1B together [22].

The E1A protein has been shown to interact with retinoblastoma family members (pRb, p107 and p130) (for review see [23]). In the normal setting the tumor suppressor pRb regulates the transition from the G1 to the S phase of the cell cycle. The binding and inactivation of pRb through E1A therefore leads to activation of the S phase genes, which is mainly accomplished through the transcription factor E2F [15, 23]. Another important feature of E1A which contributes to its potential concerning immortalization is the interaction between E1A and p300/CBP. The precise mechanism is not yet completely understood but seems to involve transcriptional deregulation of various genes through epigenetic modulation such as histone deacetylation. One prominent target gene which is activated through this mechanism is the oncogene c-myc [24]. The action of E1B protein is mediated by its interaction and neutralization of p53 [25]. Interestingly, the inactivation of p53 is achieved without degradation of this tumor suppressor protein. As a consequence of this inactivation the initiation of the apoptosis program usually triggered by p53 seems to be prevented [26].

Simian Virus 40 Large T Antigen (TAg)

The simian virus 40 (SV40), a member of the polyomavirus group, is a monkey virus that has been shown to induce various types of tumors in rodents. Whether SV40 large T antigen (TAg) also causes tumors in humans is still under debate. Because of an accidental contamination of poliovirus vaccines in the 1960s, millions of people were exposed to SV40 virus [27]. Fortunately, this did not lead to a significant increase in tumor incidence, which suggests that the tumorigenic potential of the SV40 virus is low. The major protein that is responsible for the immortalization of primary cells is the large T antigen (TAg). For immortalization of cells, a temperature-sensitive mutant of SV40 TAg has gained particular relevance (see below). TAg is known to modulate the activity of a number of cellular proteins but amongst them, p53 and pRb are regarded as the most important ones for immortalization (for review see [28]). The binding of TAg to p53 leads to the inactivation of the tumor suppressor protein, which is essential for the block of its function as a transcription factor and the immortalization capacity of TAg [29]. In addition, TAg mediated immortalization is contributed by the inactivation of pRb. This interaction is mediated through a conserved motif (LXCXE motif) which is also found in "large T proteins" from other polyomaviruses [30].

While TAg alone can be successfully employed for the establishment of rodent cell lines, the immortalization of primary human cells by TAg is a very inefficient process. Transduction of human cells with TAg usually leads to an extension of the cellular life span. However, the TAg-transduced cells still enter into crisis in which most of the cells die and only a very small fraction of cells evade [31]. Importantly, the cells that went through crisis display a grossly altered phenotype when the established cell line is compared to their primary counterparts which is also reflected by an altered karyotype [32]. Therefore, an efficient immortalization of primary human cells is only achieved with a concerted action of TAg with a second immortalizing gene such as c-myc or human telomerase (hTert, see below).

2.1.2.2 Immortalization by hTert and the Impact of Telomeres

A critical part during genome duplication during S phase is the replication of the very ends of chromosomes; the repetitive DNA stretches which are called telomeres. These telomeres consist of repeats of the sequence TTAGGG/CCCTAA. In germ cells, a total of about 15 kb of telomeric repeats are found. In contrast, in differentiated cells the telomere length is significantly shorter. This reduction of telomeric ends is a consequence of the fact that during DNA replication the telomere ends cannot be fully replicated ("end replication problem") [33]. As a consequence the telomeres progressively shorten with every replication cycle.

In human stem cells, telomerase can antagonize this effect via its reverse transcriptase subunit hTert. However, telomerase is not expressed in human somatic cells. Only in tumor cells the activity of hTert is restored leading to the stabilization of the telomeres. It could be shown that ectopic expression of hTert can lead to immortalization of human cells [34]. Since this initial study many human cell types have been successfully immortalized with hTert. In many cases, these cell lines display features of primary cells. This is highlighted by studies that successfully employed the hTert-immortalized cell lines in regenerative medicine approaches, e.g., with bovine adrenocortical cells [35], with human dermal endothelial cells [36], and human mesenchymal stem cells [37]. The use of hTert alone is restricted to certain human cell types, as others need the concerted action of several genes for efficient immortalization [38]. Importantly, prolonged constitutive expression of hTert induces changes in gene expression that lead to a premalignant phenotype [39].

Of note, telomerase is not applicable for the establishment of murine cell lines. This is probably due to the fact that laboratory mouse strains have long telomeres and murine primary cells often exhibit an endogenous telomerase activity [40].

2.1.3 Targeted Immortalization of Specific Cell Types

While the above mentioned genes act in a broader spectrum of cell types, the immortalization capacity of certain genes is restricted to specific cell types (reviewed in [41]). This can be exploited to preferentially immortalize cells in crude preparations of tissue cells that display a mixture of various cell types (e.g., see [42]). Indeed, a number of studies demonstrate that the transfer of certain genes can selectively immortalize specific cell types (Figure 2.1.3).

One example represents the immortalization of murine monocytes from the bone marrow-derived cell populations by v-myc [43, 44]. Another example concerns the specific immortalization of murine embryonic endothelial cells from a crude mixture of embryonic primary mouse cells by polyomavirus middle T antigen (PymT) [45].

Also specific immortalization of various human cell types has been reported. Human keratinocytes were evaluated for specific immortalization by the inhibitor of differentiation 1 (Id1). Cell lines could be generated that show increased telomerase activity and modulations in the two major tumor suppressor pathways pRb and p53 as highlighted by an impaired p53-mediated DNA-damage response. In addition, these cell lines have lost the differentiation potential normally seen in primary keratinocytes [46]. How robust this approach is remains to be shown since another study using this gene could only observe an extended lifespan of primary keratinocytes but not immortalization [47]. Differences in the experimental set-up might contribute to the contradicting observations.

Specific immortalization has also been shown for human hepatocytes. In the study of Ray et al. (2000), the hepatitis C virus (HCV) core protein was applied [48]. However, compared to primary hepatocytes the cell lines displayed an altered cell morphology suggesting that immortalization was accompanied with loss of the differentiated phenotype.

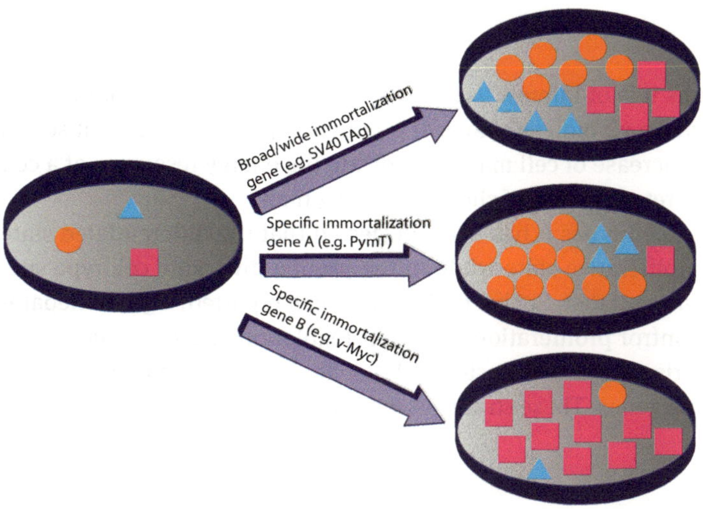

Figure 2.1.3: Schematic representation of cell type specific immortalization. Broadly acting immortalization genes like SV40 large T antigen immortalize primary cells widely independent of the cell type. Therefore if the starting material is a mixture of different primary cell types an immortalization approach with TAg leads to cell lines from the different cell types. In contrast, cell type specific immortalization genes like PymT or v-myc only lead to the immortalization of certain cell types. Therefore, from a mixture of different cell types only those will be immortalized that are amenable to the immortalization of the respective cell type specific immortalization gene.

A promising cell type for specific immortalization are memory B cells which can be immortalized by the ectopic expression of the apoptosis modulating gene Bcl-xl together with Bcl6. Bcl6 is a transcriptional repressor that prevents differentiation of B cells into plasma cells by downregulating p53 and facilitates the expansion of the B cells. Importantly, the resulting cell lines are fully functional with respect to antibody secretion [49].

A specific cell type of interest is the human CD34+ hematopoietic stem cell. It was found that the MLL-AF9 fusion gene, which is predominantly associated with myeloid leukaemias, was able to immortalize this cell type. These cells are still able to differentiate either to the myeloid or to the lymphoid lineage [50].

While these reports represent promising examples for exploitation of the cell type specific activities of immortalizing genes, it has to be emphasized that the potential of the individual genes is hardly predictable and thus has to be evaluated experimentally. A rational exploitation of immortalizing genes is also hampered by the limited knowledge of the signaling pathways that govern cell type specific proliferation.

2.1.4 Growth Controlled Cell Lines

For various applications it is of interest to control the proliferation of immortal cells over time. This concerns, e.g., biotechnological production processes where it seems advisable to block the increase of cell mass to dedicate the energy resources of a cell to the synthesis of a secreted protein of choice over long time.

Moreover, for some cell types it has been suggested that the differentiated state is restricted to a nonproliferative state while in the proliferative state cell type specific pathways are downregulated. Such a finding has been reported, e.g., for hepatocytes *in vivo* [51]. To control proliferation, different regulation strategies have been exploited. These comprise the use of post-translational, transcriptional and genetic control of immortalizing genes. These strategies are briefly described in the following paragraphs.

2.1.4.1 Controlled Proliferation by Post-translational Control of Immortalizing Genes

Growth control can be achieved by the controlled expression of immortalizing genes. Such a conditional immortalization of cells was first achieved with the help of a temperature-sensitive mutant of TAg (tsTAg). This mutant protein is fully active and the cells are cultivated at 33 °C. At this permissive temperature the tsTAg is comparable in its function to the wild type TAg and leads to proliferation and immortalization. Proliferation arrest is achieved by shifting the cells from 33 to 37-39 °C. Exposed to elevated temperatures the mutant tsTAg is misfolded and thereby inactivated and subjected to degradation [52]. tsTAg has proven to be instrumental for establishing various temperature-dependent cell lines. Growth-controlled cell lines have been established by retroviral transfer of tsTAg into primary cells (e.g., [53]) and also by isolation of cells from a tsTAg transgenic mouse, the so-called Immortomouse® [54, 55]. Examples for tsTAg-based conditionally immortalized cell types encompass myogenic cell lines [56], hepatocyte cell lines [57, 58], tissue specific microvascular endothelial cells [59], an astrocyte cell line [60], and more recently an adrenal medullary cell line [61] as well as stroma cell lines [62]. For a more comprehensive overview please see review by Obinata (1997) [63]. tsTAg-immortalized cell lines were used for a multitude of applications (see overview in [64, 65]). However, it has to be noted that the temperature shift itself imposes a significant change in cellular properties [66], which complicates the interpretation of cellular behavior regarding reversion of immortalization. In addition, one report indicates that the regulation control through the tsTAg is not stringent [67]. If tsTAg-immortalized rat hepatocytes were maintained for extended time periods at the nonpermissive temperature outgrowth of cells was observed. Still, in these cells tsTAg protein was temperature-controlled. This indicates that the immortalized cells acquired spontaneous mutations if maintained at the nonpermissive temperature. These mutations in turn induced proliferation independent of tsTAg and consequently growth control was lost [67]. This kind of adaptive mutation might apply for other cell types and systems. However, immortalization systems have been described that retain the conditional phenotype long-term [68, 69].

Since temperature-sensitive mutants are not available for other immortalizing genes, more generic protocols to control activity of these genes have been explored. One example concerns the post-translational control of immortalizing genes upon fusion to the estrogen receptor moiety (ER). In ER fusion proteins the protein activity is structurally impaired. Binding of 4-hydroxy-tamoxifen to the ER domain changes the conformation of the fusion protein and restores the activity. Based on such an approach employing HoxB8-ER, murine macrophage, neutrophil and hematopoietic progenitor cell lines were generated [70].

2.1.4.2 Controlled Proliferation by Transcriptional Control of Immortalizing Genes

Besides post-translational control also transcriptional control of immortalizing genes has been exploited. For this purpose, synthetic, so-called orthogonal regulation systems have been implemented, that allow specific control of a synthetic promoter while not affecting the cellular regulatory network (reviewed by [71]). An extensively explored regulatory system concerns the "tetracycline system" (Tet-system) in which a recombinant transactivator protein specifically binds to a synthetic promoter and thereby induces transcription. The binding of the transactivator is modulated by tetracycline (Tet) or its derivatives such as doxycycline (Dox). Thus, addition or withdrawal of Tet or Dox results in control of transcription.

In the original study from Gossen and Bujard (1992) [72] the Tet-off system was used in which in absence of Tet the transactivator tTA binds to and activates the cognate promoter while in presence of Tet binding is impaired. Further developments in this field comprise the development of a Tet-on system [73], a bidirectional promoter which facilitates the expression of two cistrons [74] and a Tet-dependent promoter with a reduced basal activity [75] (for review see [76]).

While the classical set-up with a constitutive expression of the transactivator results in a gradual, dose-response, an auto-regulated set-up leads to stochastic activation. In such a setting the transactivator is controlled by the synthetic promoter, which allows implementing all components required for regulation and transgene expression in a single expression cassette [77, 78].

Based on an auto-regulated Tet system a highly efficient method for conditional immortalization was established [69]. In these cells, cell proliferation is strictly dependent on Dox. Due to the compactness of the expression cassettes they can be packaged into lentiviral vectors which facilitates an efficient gene transfer of the immortalizing genes in a broad range of cell types [68]. By Tet-controlled expression of SV40 TAg, PymT, hTert, and c-Myc, various types of proliferation-controlled cell lines have been established. Examples of growth-controlled cell lines include mouse fibroblasts [69], lung microvascular derived endothelial cells [68] and liver sinusoidal endothelial cells [79], but also human cell lines such as umbilical vein-derived endothelial cells [68], as well as human mesenchymal stem cells [80, 81].

2.1.4.3 Genetic Control of Immortalization

Finally, also genetic switches have been utilized to control immortalizing genes and thereby revert the immortalization state. For this purpose, site-specific recombinases such as Cre or Flp recombinase were successfully employed to facilitate excision of the immortalizing gene(s) flanked with the cognate recombination target sites (e.g., loxP or FRT sites) [82-85]. However, in contrast to the previously described transcriptional and post-translational control systems, the reversion of the genetic switches is irreversible since the immortalizing gene is eliminated. Efficient transfer of the recombinase by adenoviral transduction combined with an extensive selection process can eliminate nonrecombined cells which would overgrow the population of reverted cells [86]. To further improve this critical step, more recent approaches involve inducible mutants of the recombinase, which are stably integrated into the cells and activated on demand [87]. The use of recombinases implements the risk for unintended genetic rearrangements if multiple copies of the immortalizing genes are integrated at various chromosomal sites. Such rearrangements could also involve cellular chromosomal regions, which can potentially affect cellular properties. Thus, recombinase-based control of proliferation requires additional controls to ensure the cellular phenotype.

2.1.4.4 Controlled Cell Proliferation in Protein Production Processes

Control of cell proliferation is an interesting option to dedicate the cell's capacity to the synthesis of recombinant proteins rather than to the increase of cell mass. Accordingly, controlled proliferation was evaluated with respect to the cellular productivity. To this end, the proliferation of immortalized cell lines was blocked by controlled expression of genes that stop cell cycle progression. One example concerns expression of the cyclin-dependent kinase inhibitors p21 or p27 which stop the cell cycle predominantly in the G1 phase. Interestingly, in the growth arrested state cell lines showed an enhanced production of the protein of interest. For Chinese hamster ovary cell (CHO) a 4- to 30-fold increase in antibody production could be observed in the growth arrested state if compared to proliferating cells [88-90], reviewed by [91]. Also mouse fibroblasts conditionally immortalized with SV40 TAg were analyzed for their productivity in the proliferating and in the nonproliferating state. In this study, the overall increase in production was 2-fold higher in the growth-arrested cells. Importantly, this ratio remained constant throughout the cultivation period [92].

These studies indicate that controlled proliferation enables a higher productivity of producer cells. Moreover, growth-controlled producer cells allow to continuously harvest secreted proteins over extended time periods, which makes this strategy attractive for production. However, to date this technology has not exploited in a certified process of therapeutic protein production.

2.1.5 Cell Lines for Biotechnological Applications

In industrial research cell lines are mainly used either for drug development or for producing biologicals. In the following sections a noncomprehensive list of the most important cell lines for these two industrial applications is provided (Figure 2.1.4).

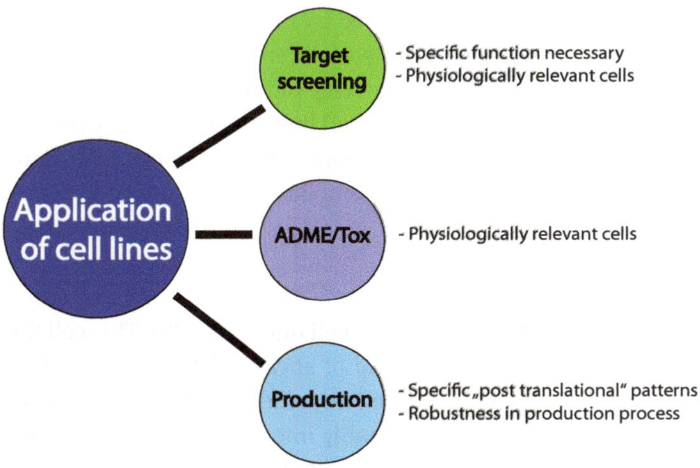

Figure 2.1.4: Applications and requirements of cell lines in biotechnology. The major industrial applications of cell lines are depicted. In addition, the desired properties of the cell lines in these applications are given.

2.1.5.1 Production Cell Lines

One major application of cell lines is the production of biotechnologically relevant proteins, such as antibodies, and other therapeutic proteins, such as human α-antithrombin (hAAT), blood coagulation factors, erythropoietin or follitropin, as well as their use for vaccine production. Also, recombinant viruses for gene therapeutic approaches are produced with the help of continuously growing cell lines that act as helper cells [93-96].

Such cell lines require a number of specific features. Besides a high overall productivity and genetic stability of the cell, a critical property concerns the type of glycosylation that is provided by the producer cell. Generally, a glycosylation pattern is required that confers protein activity and resembles best the glycosylation type in humans. In particular, it should not induce an antigenic response in humans. Further, for protein producer cells also practical issues such as a high efficiency of genetic manipulation of these cell lines is of importance.

Currently, the production of proteins is mainly based on few cell lines that have been shown to provide high productivity, safety, acceptable stability, of recombinant protein production and can be subjected to fermentation processes under serum free conditions. Many production protocols are based on cell lines such as Chinese hamster ovary cells (CHO) [97, 98] and HEK293 cells [99] that have been isolated decades ago. (The application of these cell lines in vaccine production is highlighted in Chapter 2.3). In the recent years, the demand for well-characterized, traceable and rapidly expandable cell lines with defined and improved properties increases. Accordingly, an increasing number of cell lines has been generated by rational transfer of defined immortalizing genes to characterized primary cells of different species and tissues, resulting in the establishment of so-called designer cells. In the following, we give an overview about the history of some of these various cell lines that are or may become relevant for research and biotechnological applications. Since Chapter 2.2 is dedicated to the CHO cell line, it is not included in this compilation.

2.1.5.2 The HEK293 Cell Line and Its Derivatives

One frequently used recombinant protein production cell line is HEK293. This cell line has been generated upon transfection of a mixture of sheared DNA obtained from adenovirus Ad5 into human embryonic cells derived from kidney [99]. In depth characterization revealed that HEK293 cells contain a stably integrated single copy of the left part of the adenoviral genome (nucleotides 1 to 4,344) of Ad5 [100]. This region includes the adenoviral E1A and E1B genes controlled by the viral promoters. Later, it was found that HEK293 cells unexpectedly display a number of neuronal markers suggesting that the immortalized parental cell was of embryonic neuronal origin rather than representing a classical kidney epithelial cell [101]. Indeed, experimental evidence was given that adenoviral Ad5- and also Ad12-based immortalization protocols are more efficient on human embryonic neurons rather than kidney cells [101, 102]. This is probably due to a certain resistance of kidney epithelial cells to immortalization and/or a preference of adenoviral E1A/E1B immortalization for neuronal cells. This example shows that even low numbers of 'contaminating' cells in a cell preparation can give rise to a preferential immortalization while the cell type of interest may not be targeted.

Due to the fact that HEK293 cells express E1A and E1B proteins this cell line has been broadly used as a helper cell line for E1A/E1B-deleted recombinant adenoviral vectors (for review see [103, 104]). Upon transfer of recombinant adenoviral vectors to HEK293 cells, the E1A/E1B function is provided *in trans* and complements the formation of replication-defective viral particles. Further, upon transfection of EBNA1 or SV40 T antigen encoding expression cassettes, derivatives of 293 cells were generated so-called 293E and 293T cells, respectively [105]. These cells support the amplification of episomal plasmids that harbor the respective origin of replication [106]. Moreover, 293T and also 293E cells have been shown to provide high transfection rates. This property makes the cell lines as an efficient tool for various applications and in particular for small and large scale transient expression of proteins. This is also reflected by the fact that a number of pharmaceuticals are produced with this cell type [107]. Moreover, 293 cells and also the SV40 T antigen expressing 293T cells derived thereof are routinely used for the production of recombinant γ-retroviral and lentiviral vectors (reviewed in [108, 109]). Finally, these cells serve as production systems for various viruses such as influenza virus [110] and vaccinia virus [111].

2.1.5.3 The PerC.6 Cell Line

Learned from the lesson of HEK293 cells, human embryonic retinoblasts were used to establish the PerC.6 cell line [112]. For this purpose, the adenoviral genes E1A and E1B (nucleotides 459 to 3,510) controlled by the PGK promoter or the endogenous promoter, respectively, were transferred. In contrast to the HEK293 cell line the employed adenoviral region of the PerC.6 cell line excludes viral sequences that are homologous to E1A/E1B-deleted, recombinant adenoviral vectors. Accordingly, the risk for homologous recombination towards replication competent adenoviruses is significantly reduced in PerC.6 cells. The construction of this cell line was done under well-documented good manufactory practice (GMP) conditions which facilitated its use for pharmaceutical production. Therefore, besides its use for recombinant adenovirus and adeno-associated virus (AAV) production, this cell line proved to be efficient for production of vaccines including influenza (reviewed in [113, 114]), poliovirus [115] and also for protein production [107].

2.1.5.4 Recent Developments

Recent developments of cell lines for biotechnological purposes include the brain-derived cell line AGE1.HN which was generated upon transfer of an expression cassette encoding the E1A and E1B genes driven by the PGK promoter and the E1B-dependent viral promoter, respectively. In addition to these immortalizing genes the adenoviral gene pIX was introduced into this cell line, which alters the cell metabolism and enhances productivity for secreted proteins [116]. The AGE1.HN cell line has been shown to display neuronal but not glial cell markers and was already applied for protein production [117].

Similarly, duck cells were immortalized using the E1A and E1B adenoviral genes with or without pIX (AGE1.CR.pIX and AGE1.CR, respectively). These cell lines have been shown to be suitable for propagation of influenza and vaccinia virus [118, 119].

Another approach was based on primary cells isolated from amnion fluid. Upon immortalization with the abovementioned adenoviral genes E1A/E1B and pIX a cell line was established designated as CAP [120, 121]. Although the specific cell type that is immortalized by this procedure has not yet been defined, these cells have been shown to give rise to high levels of protein [121, 122] as well as influenza virus [123].

2.1.5.5 Cell Lines Used in the Drug Development Process

In the drug development process cell lines are used in high throughput screenings as tools to identify drug candidates. In this early developmental stage, considerable cell numbers (> 1×10^9 cells) are required to perform automated screenings. Due to this fact robust cell systems are mandatory and therefore CHO and HEK293 cells are often used that overexpress the desired drug target [124, 125]. The identified hits are further characterized in more relevant cell systems, e.g., in cell lines displaying a special phenotype or even in primary cells. As this characterization is dependent on the drug target and/or the desired indication there is no golden rule which cell systems are used in this stage of the drug development process.

After narrowing down the number of potential drug candidates they are optimized with respect to efficacy and to pharmacokinetic properties like absorption, distribution, metabolism, excretion, and their toxicity profile (ADMET profiling). For some of these parameters robust cell based *in vitro* test systems have been established.

2.1.5.6 Caco-2 Cells

An important pharmacological property is absorption. Drugs are preferably applied through the oral route. Therefore they have to stand the acidic surrounding of the stomach but also have to be efficiently absorbed in the small intestine where they enter the blood stream and are distributed to the target location. Thus, if the absorption process is inefficient, the dose has to be increased. High drug concentrations are more likely to generate toxic side effects. Accordingly, the optimization of the absorption characteristics is a critical parameter of drug development. For this purpose an excellent *in vitro* model – the Caco-2 cell line – was established. This cell line was derived in the 1970s from a colorectal adenocarcinoma [126]. About ten years later it was recognized that this cell line was able to spontaneously differentiate into a cell type that is similar to small intestinal enterocytes [127]. Today, Caco-2 cells are widespreadly used in various labs working in disciplines like nutrition, pharmacology, and toxicology.

Among the relevant features of this cell line are the expression of specific small intestinal marker proteins and tight junction complexes. Importantly, also the Caco-2 cells form microvilli on the apical side (facing the gut lumen) when cultured on cell culture inserts (e.g., in transwell plates). For drug development purposes the most important property is that Caco-2 cells form a tight barrier. The formation of the barrier is induced when the cells are plated onto cell culture inserts and are maintained for 3 weeks [128]. During this time a tight barrier develops which limits passive transport through the cell layer. It was shown that for such passive transport processes the

Caco-2 cells show a permeability coefficient that is comparable to the *in vivo* situation [129].

Also active transporters play important roles in the absorption of drugs. The Caco-2 cells do express several intestinal transport systems albeit to a lesser extent than in the small intestine *in vivo*. This is probably the reason for the slower transport through the Caco-2 cells compared to the *in vivo* situation. Another difficulty is that the cellular properties change with continuous passaging. Currently, different subclones have arisen from the parental Caco-2 cells. Among the properties that change with continuous passaging are an increase in TEER (trans epithelial electrical resistance) and accordingly a decrease in permeability [130]. Further, the cells change in the cell proliferation rate, in the activity of metabolic enzymes (Cyp3A4), and the cells seem to lose the contact inhibition [131].

2.1.5.7 Keratinocyte Cell Lines (HaCaT, NIKS)

Although not frequently used in the drug development process keratinocyte cell lines play an important role in industrial research. The evaluation of keratinocyte cell lines as a surrogate for skin was promoted by an EU regulation (REACH) which requests that every chemical substance has to be tested for its toxicological profile. So far, this substance testing is mainly performed *in vivo*, which is costly and harbors ethical concerns. As an alternative cell based surrogates for skin already have been developed. *In vitro* three-dimensional skin equivalents can be established and are used for substance testing, e.g., in the cosmetic industry. So far, these skin equivalents are composed of primary keratinocytes and in some cases also other primary cell types like fibroblasts or macrophages [132].

Probably the most famous and still the most frequently used keratinocyte cell line is HaCaT. This cell line was established from adult skin of a donor who came down with melanoma. Cells were isolated from the distant periphery of the tumor and the cell line was established by spontaneous immortalization, which was probably facilitated through cultivation of the cells at elevated temperatures (38.5 °C), a temperature known to increase the proliferation capacity of primary keratinocytes. HaCaT cells are not tumorigenic upon transplantation into nude mice. Importantly, these cells are able to differentiate into stratified epithelia *in vitro*. This, however, happens at a slower rate than with normal primary human keratinocytes [133].

Recently, a novel spontaneously immortalized human keratinocyte cell line, NIKS, was described. NIKS cells are immortal and also nontumorigenic. Importantly, they form a fully stratified squamous epithelium *in vitro* when cultivated in organotypic cultures. NIKS cells were isolated from juvenile foreskin keratinocytes and cultivated for the cell line generation on MEF feeders. The resulting cell line is still fully responsive to growth factor treatment as, e.g., epidermal growth factor (EGF) leads to increased proliferation whereas TGFb1 treatment leads to inhibition of proliferation [134]. NIKS cells even differentiate to the same extent as their primary counterparts when cultivated in organotypic cultures together with dermal fibroblasts. In the meantime companies use this progenitor cell line to create skin equivalents, which can be used for substance testing but are also tested for regenerative purposes (e.g., treating severe burns).

2.1.5.8 Hepatocyte Cell Lines

For drug development hepatocytes are one of the key cell types. This is due the fact that the liver is the major site of drug metabolism. Hepatocytes are highly specialized cells with unique properties including the detoxification of metabolites. Thus, they are of pivotal interest as screening systems for drug development but also for routine toxicology studies.

Human hepatocellular carcinoma tissues represent a source for the establishment of immortal hepatic cell lines. Several cell lines have been established (Mz-Hep.1, KYN-2, BC2) which display some of the hepatic markers (for overview see [135, 136]). The most widely used hepatoma cell line is the human HepG2 cell line which was derived from a well differentiated hepatocellular carcinoma of a 15-year-old Caucasian male [137]. Interestingly, this cell line can synthesize many human plasma proteins including albumin and α-fetoprotein [137]. While albumin is expressed in adult hepatocytes, α-fetoprotein is a marker of the progenitor phenotype and is not expressed in mature hepatocytes. Although HepG2 cells show many liver-specific functions, they lack the functional expression of almost all relevant human cytochrome P450 family members [138]. This limits the use of HepG2 cells as an *in vitro* model for metabolism or toxicology studies. Certain protocols such as 1a,25-dihydroxyvitamin D3-based induction of Cyp3A4 [139] or the cultivation on a nanotechnology-based three-dimensional scaffold [140] provide evidence that this cell line can be used for drug metabolism studies although only a limited number of settings. In addition, the drug-metabolizing potential has to be characterized for each assay before the produced data can be interpreted [141].

Due to the obvious limitations of the HepG2 cell line there is still a big demand for novel hepatic cell lines which was at least partially solved with the establishment of the Huh-7 [142] and HepaRG cell line [143]. The Huh-7 and its subclone Huh7.5 are particularly suited as *in vitro* model systems for studying HCV biology whereas it is not suited for toxicology tests due to its limited hepatocyte functionality (for overview see [144]). The HepaRG cell line on the other hand retains an undifferentiated morphology under normal culture conditions but can be differentiated through modifying the culture conditions (cell density, DMSO addition). This differentiation process induces the typical hepatocyte morphology as well as the expression of various cytochrome P450s, phase II enzymes and drug transporters. Therefore the HepaRG cell line is currently considered as an alternative to primary hepatocytes for toxicology tests.

The obvious drawback associated with the generation of hepatic cell lines from hepatocarcinomas is the unpredictability. Therefore, regimens were developed to standardize the immortalization process of hepatocytes. To this end it was shown that primary human hepatocytes can be immortalized by the transfer of immortalizing genes such as TAg, hTert, E6/E7, or as well the gene encoding the core protein of HCV. Several cell lines like the PH5CH [145], the Fa2N-4 [146], the HepLi5 [147], the HuS-E [148], or the NKNT-3 cell line [149] have been established. For these approaches either TAg alone (PH5CH, Fa2N-4, HepLi5), or hTert in combination with TAg (NKNT-3) or E6/E7 (HuS-E) were employed. However, all these cell lines still show significant differences when compared to primary hepatocytes, which limit their use as predictive models [150, 151].

Another approach was used to establish the cBAL111 cell line. In this case, human fetal liver cells were immortalized by recombinant expression of hTert [152]. Importantly, this cell line has the potential to differentiate to a more mature phenotype so that these cells show a hepatic phenotype which includes albumin expression and urea synthesis. However, also in the case of this cell line, the levels are not comparable to primary mature hepatocytes (overview see [152]).

2.1.6 Outlook

In the last decades, mammalian cell lines gained increasing relevance for biotechnology. This concerns their role as potent production systems for a plethora of biologicals. In addition, recent advances in the development of immortalization protocols allowed the generation of highly specialized cell lines. These cell lines can already partly substitute primary cells for applications that rely on complex properties of specialized cell types. It can be envisioned that the refinement of the protocols, together with the increasing knowledge in the signaling pathways relevant for specialized cells, will allow to further improve the methods for expansion of cells with complex phenotypes.

Functional complementation of immortalized cell lines could be a future option. Based on synthetic biology approaches, relevant properties could be restored. Missing or silenced regulatory cascades could be reactivated or reassembled by implementing genetic circuits, respectively. This might concern the implementation of specific modification patterns (such as genes for distinct glycosylation profiles) but also more complex features such as the detoxification properties of hepatocytes.

Finally, reliable and reproducible methods for immortalizing patient-derived cells might pave the way for their application in regenerative medicine. Certainly, cell lines that reflect the various genetic settings found in the human population will also have a high potential for the establishment of novel *in vitro* screening/test systems.

References

[1] Sherr CJ, McCormick F. The RB and p53 pathways in cancer. Cancer Cell 2002;2:103-12.

[2] Burkhart DL, Sage J. Cellular mechanisms of tumour suppression by the retinoblastoma gene. Nature Reviews Cancer 2008;8:671-82.

[3] Hanahan D, Weinberg RA. Hallmarks of cancer: the next generation. Cell 2011;144:646-74.

[4] Todaro GJ, Green H. Quantitative studies of the growth of mouse embryo cells in culture and their development into established lines. Journal of Cell Biology 1963;17:299-313.

[5] Hahn WC, Weinberg RA. Modelling the molecular circuitry of cancer. Nature Reviews Cancer 2002;2:331-41.

[6] Harvey DM, Levine AJ. p53 alteration is a common event in the spontaneous immortalization of primary BALB/c murine embryo fibroblasts. Genes & Development 1991;5:2375-85.

[7] Rittling SR, Denhardt DT. p53 mutations in spontaneously immortalized 3T12 but not 3T3 mouse embryo cells. Oncogene 1992;7:935-42.

[8] Rangarajan A, Hong SJ, Gifford A, et al. Species- and cell type-specific requirements for cellular transformation. Cancer Cell 2004;6:171-83.

[9] Traggiai E. Immortalization of human B cells: analysis of B cell repertoire and production of human monoclonal antibodies. Methods in Molecular Biology 2012;901:161-70.

[10] Humme S, Reisbach G, Feederle R, et al. The EBV nuclear antigen 1 (EBNA1) enhances B cell immortalization several thousandfold. Proceedings of the National Academy of Sciences of the United States of America 2003;100:10989-94.

[11] Holowaty MN, Zeghouf M, Wu H, et al. Protein profiling with Epstein-Barr nuclear antigen-1 reveals an interaction with the herpesvirus-associated ubiquitin-specific protease HAUSP/USP7. Journal of Biological Chemistry 2003;278:29987-94.

[12] Cummins JM, Vogelstein B. HAUSP is required for p53 destabilization. Cell Cycle 2004;3:689-92.

[13] Gruhne B, Sompallae R, Marescotti D, et al. The Epstein-Barr virus nuclear antigen-1 promotes genomic instability via induction of reactive oxygen species. Proceedings of the National Academy of Sciences of the United States of America 2009;106:2313-8.

[14] Moody CA, Laimins LA. Human papillomavirus oncoproteins: pathways to transformation. Nature Reviews Cancer 2010;10:550-60.

[15] Bellacchio E, Paggi MG. Understanding the targeting of the RB family proteins by viral oncoproteins to defeat their oncogenic machinery. Journal of Cellular Physiology 2013;228:285-91.

[16] Klingelhutz AJ, Foster SA, McDougall JK. Telomerase activation by the E6 gene product of human papillomavirus type 16. Nature 1996;380:79-82.

[17] Halbert CL, Demers GW, Galloway DA. The E7 gene of human papillomavirus type 16 is sufficient for immortalization of human epithelial cells. Journal of Virology 1991;65:473-8.

[18] Niccoli S, Abraham S, Richard C, et al. The Asian-American E6 variant protein of human papillomavirus 16 alone is sufficient to promote immortalization, transformation, and migration of primary human foreskin keratinocytes. Journal of Virology 2012;86:12384-96.

[19] Liu X, Dakic A, Chen R, et al. Cell-restricted immortalization by human papillomavirus correlates with telomerase activation and engagement of the hTERT promoter by Myc. Journal of Virology 2008;82:11568-76.

[20] Freeman AE, Black PH, Vanderpool EA, et al. Transformation of primary rat embryo cells by adenovirus type 2. Proceedings of the National Academy of Sciences of the United States of America 1967;58:1205-12.

[21] Ruley HE. Adenovirus early region 1A enables viral and cellular transforming genes to transform primary cells in culture. Nature 1983;304:602-6.

[22] Barker DD, Berk AJ. Adenovirus proteins from both E1B reading frames are required for transformation of rodent cells by viral infection and DNA transfection. Virology 1987;156:107-21.

[23] Chinnadurai G. Opposing oncogenic activities of small DNA tumor virus transforming proteins. Trends in Microbiology 2011;19:174-83.

[24] Kadeppagari RK, Sankar N, Thimmapaya B. Adenovirus transforming protein E1A induces c-Myc in quiescent cells by a novel mechanism. Journal of Virology 2009;83:4810-22.

[25] Sarnow P, Ho YS, Williams J, et al. Adenovirus E1b-58kd tumor antigen and SV40 large tumor antigen are physically associated with the same 54 kd cellular protein in transformed cells. Cell 1982;28:387-94.

[26] Yew PR, Berk AJ. Inhibition of p53 transactivation required for transformation by adenovirus early 1B protein. Nature 1992;357:82-5.

[27] Sweet BH, Hilleman MR. The vacuolating virus, S.V. 40. Proceedings of the Society for Experimental Biology and Medicine 1960;105:420-7.

[28] Pipas JM. SV40: Cell transformation and tumorigenesis. Virology 2009;384:294-303.

[29] Kierstead TD, Tevethia MJ. Association of p53 binding and immortalization of primary C57BL/6 mouse embryo fibroblasts by using simian virus 40 T-antigen mutants bearing internal overlapping deletion mutations. Journal of Virology 1993;67:1817-29.

[30] Chen S, Paucha E. Identification of a region of simian virus 40 large T antigen required for cell transformation. Journal of Virology 1990;64:3350-7.

[31] Macera-Bloch L, Houghton J, Lenahan M, et al. Termination of lifespan of SV40-transformed human fibroblasts in crisis is due to apoptosis. Journal of Cellular Physiology 2002;190:332-44.

[32] Stewart N, Bacchetti S. Expression of SV40 large T antigen, but not small t antigen, is required for the induction of chromosomal aberrations in transformed human cells. Virology 1991;180:49-57.

[33] Levy MZ, Allsopp RC, Futcher AB, et al. Telomere end-replication problem and cell aging. Journal of Molecular Biology 1992;225:951-60.

[34] Bodnar AG, Ouellette M, Frolkis M, et al. Extension of life-span by introduction of telomerase into normal human cells. Science 1998;279:349-52.

[35] Thomas M, Yang L, Hornsby PJ. Formation of functional tissue from transplanted adrenocortical cells expressing telomerase reverse transcriptase. Nature Biotechnology 2000;18:39-42.

[36] Yang J, Nagavarapu U, Relloma K, et al. Telomerized human microvasculature is functional in vivo. Nature Biotechnology 2001;19:219-24.

[37] Simonsen JL, Rosada C, Serakinci N, et al. Telomerase expression extends the proliferative life-span and maintains the osteogenic potential of human bone marrow stromal cells. Nature Biotechnology 2002;20:592-6.

[38] Kiyono T, Foster SA, Koop JI, et al. Both Rb/p16INK4a inactivation and telomerase activity are required to immortalize human epithelial cells. Nature 1998;396:84-8.

[39] Milyavsky M, Shats I, Erez N, et al. Prolonged culture of telomerase-immortalized human fibroblasts leads to a premalignant phenotype. Cancer Research 2003;63:7147-57.

[40] Blasco MA, Lee HW, Hande MP, et al. Telomere shortening and tumor formation by mouse cells lacking telomerase RNA. Cell 1997;91:25-34.

[41] Lipps C, May T, Hauser H, et al. Eternity and functionality - rational access to physiologically relevant cell lines. Biological Chemistry 2013;394:1637-48.

[42] Schwerk J, Koster M, Hauser H, et al. Generation of mouse small intestinal epithelial cell lines that allow the analysis of specific innate immune functions. PLOS ONE 2013;8:e72700.

[43] Blasi E, Mathieson BJ, Varesio L, et al. Selective immortalization of murine macrophages from fresh bone marrow by a raf/myc recombinant murine retrovirus. Nature 1985;318:667-70.

[44] Blasi E, Radzioch D, Durum SK, et al. A murine macrophage cell line, immortalized by v-raf and v-myc oncogenes, exhibits normal macrophage functions. European Journal of Immunology 1987;17:1491-8.

[45] May T, Mueller PP, Weich H, et al. Establishment of murine cell lines by constitutive and conditional immortalization. Journal of Biotechnology 2005;120:99-110.
[46] Alani RM, Hasskarl J, Grace M, et al. Immortalization of primary human keratinocytes by the helix-loop-helix protein, Id-1. Proceedings of the National Academy of Sciences of the United States of America 1999;96:9637-41.
[47] Nickoloff BJ, Chaturvedi V, Bacon P, et al. Id-1 delays senescence but does not immortalize keratinocytes. Journal of Biological Chemistry 2000;275:27501-4.
[48] Ray RB, Meyer K, Ray R. Hepatitis C virus core protein promotes immortalization of primary human hepatocytes. Virology 2000;271:197-204.
[49] Kwakkenbos MJ, Diehl SA, Yasuda E, et al. Generation of stable monoclonal antibody-producing B cell receptor-positive human memory B cells by genetic programming. Nature Medicine 2010;16:123-8.
[50] Horton SJ, Jaques J, Woolthuis C, et al. MLL-AF9-mediated immortalization of human hematopoietic cells along different lineages changes during ontogeny. Leukemia 2013;27:1116-26.
[51] Klochendler A, Weinberg-Corem N, Moran M, et al. A Transgenic Mouse Marking Live Replicating Cells Reveals In Vivo Transcriptional Program of Proliferation. Developmental Cell 2012;23:681-90.
[52] Tegtmeyer P, Schwartz M, Collins JK, et al. Regulation of tumor antigen synthesis by simain virus 40 gene A. Journal of Virology 1975;16:168-78.
[53] Jat PS, Sharp PA. Cell lines established by a temperature-sensitive simian virus 40 large-T-antigen gene are growth restricted at the nonpermissive temperature. Molecular and Cellular Biology 1989;9:1672-81.
[54] Jat PS, Noble MD, Ataliotis P, et al. Direct derivation of conditionally immortal cell lines from an H-2Kb-tsA58 transgenic mouse. Proceedings of the National Academy of Sciences of the United States of America 1991;88:5096-100.
[55] Oostendorp RA, Medvinsky AJ, Kusadasi N, et al. Embryonal subregion-derived stromal cell lines from novel temperature-sensitive SV40 T antigen transgenic mice support hematopoiesis. Journal of Cell Science 2002;115:2099-108.
[56] Morgan JE, Beauchamp JR, Pagel CN, et al. Myogenic cell lines derived from transgenic mice carrying a thermolabile T antigen: a model system for the derivation of tissue-specific and mutation-specific cell lines. Developmental Biology 1994;162:486-98.
[57] Yanai N, Suzuki M, Obinata M. Hepatocyte cell lines established from transgenic mice harboring temperature-sensitive simian virus 40 large T-antigen gene. Experimental Cell Research 1991;197:50-6.
[58] Allen KJ, Reyes R, Demmler K, et al. Conditionally immortalized mouse hepatocytes for use in liver gene therapy. Journal of Gastroenterology and Hepatology 2000;15:1325-32.
[59] Langley RR, Ramirez KM, Tsan RZ, et al. Tissue-specific microvascular endothelial cell lines from H-2K(b)-tsA58 mice for studies of angiogenesis and metastasis. Cancer Research 2003;63:2971-6.
[60] Langley RR, Fan D, Guo L, et al. Generation of an immortalized astrocyte cell line from H-2Kb-tsA58 mice to study the role of astrocytes in brain metastasis. International Journal of Oncology 2009;35:665-72.
[61] Kohno S, Murata T, Koide N, et al. Establishment and characterization of a noradrenergic adrenal chromaffin cell line, tsAM5NE, immortalized with the temperature-sensitive SV40 T-antigen. Cell Biology International 2011;35:325-34.
[62] Istvanffy R, Oostendorp RA. Generation and establishment of murine adherent cell lines. Methods in Molecular Biology 2013;946:301-14.
[63] Obinata M. Conditionally immortalized cell lines with differentiated functions established from temperature-sensitive T-antigen transgenic mice. Genes to Cells 1997;2:235-44.

[64] Noble M, Groves AK, Ataliotis P, et al. The H-2KbtsA58 transgenic mouse: a new tool for the rapid generation of novel cell lines. Transgenic Research 1995;4:215-25.
[65] Obinata M. Possible applications of conditionally immortalized tissue cell lines with differentiation functions. Biochemical and Biophysical Research Communications 2001;286:667-72.
[66] May T, Wirth D, Hauser H, et al. Transcriptionally regulated immortalization overcomes side effects of temperature-sensitive SV40 large T antigen. Biochemical and Biophysical Research Communications 2005;327:734-41.
[67] Kim BH, Han YS, Choe BK, et al. The escape of temperature-sensitive T antigen immortalized rat hepatocytes from conditional immortalization. Cell Transplantation 2005;14:507-17.
[68] May T, Butueva M, Bantner S, et al. Synthetic gene regulation circuits for control of cell expansion. Tissue Engineering Part A 2010;16:441-52.
[69] May T, Hauser H, Wirth D. Transcriptional control of SV40 T-antigen expression allows a complete reversion of immortalization. Nucleic Acids Research 2004;32:5529-38.
[70] Wang GG, Calvo KR, Pasillas MP, et al. Quantitative production of macrophages or neutrophils ex vivo using conditional Hoxb8. Nature Methods 2006;3:287-93.
[71] Botezatu L, Sievers S, Gama-Norton L, et al. Genetic aspects of cell line development from a synthetic biology perspective. Advances in Biochemical Engineering/Biotechnology 2012;127:251-84.
[72] Gossen M, Bujard H. Tight control of gene expression in mammalian cells by tetracycline-responsive promoters. Proceedings of the National Academy of Sciences of the United States of America 1992;89:5547-51.
[73] Urlinger S, Baron U, Thellmann M, et al. Exploring the sequence space for tetracycline-dependent transcriptional activators: novel mutations yield expanded range and sensitivity. Proceedings of the National Academy of Sciences of the United States of America 2000;97:7963-8.
[74] Baron U, Freundlieb S, Gossen M, et al. Co-regulation of two gene activities by tetracycline via a bidirectional promoter. Nucleic Acids Research 1995;23:3605-6.
[75] Agha-Mohammadi S, O'Malley M, Etemad A, et al. Second-generation tetracycline-regulatable promoter: repositioned tet operator elements optimize transactivator synergy while shorter minimal promoter offers tight basal leakiness. Journal of Gene Medicine 2004;6:817-28.
[76] May T, Hauser H, Wirth D. Current status of transcriptional regulation systems. Cytotechnology 2006;50:109-19.
[77] Unsinger J, Lindenmaier W, May T, et al. Stable and strictly controlled expression of LTR-flanked autoregulated expression cassettes upon adenoviral transfer. Biochemical and Biophysical Research Communications 2004;319:879-87.
[78] May T, Eccleston L, Herrmann S, et al. Bimodal and hysteretic expression in mammalian cells from a synthetic gene circuit. PLOS ONE 2008;3:e2372.
[79] Dag F, Weingartner A, Butueva M, et al. A new reporter mouse cytomegalovirus reveals maintained immediate-early gene expression but poor virus replication in cycling liver sinusoidal endothelial cells. Virology Journal 2013;10:197.
[80] Piper SL, Wang M, Yamamoto A, et al. Inducible immortality in hTERT-human mesenchymal stem cells. Journal of Orthopaedic Research 2012;30:1879-85.
[81] Koch CM, Reck K, Shao K, et al. Pluripotent stem cells escape from senescence-associated DNA methylation changes. Genome Research 2013;23:248-59.
[82] Westerman KA, Leboulch P. Reversible immortalization of mammalian cells mediated by retroviral transfer and site-specific recombination. Proceedings of the National Academy of Sciences of the United States of America 1996;93:8971-6.

[83] Cai J, Ito M, Westerman KA, et al. Construction of a non-tumorigenic rat hepatocyte cell line for transplantation: reversal of hepatocyte immortalization by site-specific excision of the SV40 T antigen. Journal of Hepatology 2000;33:701-8.
[84] Rybkin II, Markham DW, Yan Z, et al. Conditional expression of SV40 T-antigen in mouse cardiomyocytes facilitates an inducible switch from proliferation to differentiation. Journal of Biological Chemistry 2003;278:15927-34.
[85] Zhang Y, Nuglozeh E, Toure F, et al. Controllable expansion of primary cardiomyocytes by reversible immortalization. Human Gene Therapy 2009;20:1687-96.
[86] Narushima M, Kobayashi N, Okitsu T, et al. A human beta-cell line for transplantation therapy to control type 1 diabetes. Nature Biotechnology 2005;23:1274-82.
[87] Wu HL, Wang Y, Zhang P, et al. Reversible immortalization of rat pancreatic beta cells with a novel immortalizing and tamoxifen-mediated self-recombination tricistronic vector. Journal of Biotechnology 2011;151:231-41.
[88] Bi JX, Shuttleworth J, Al-Rubeai M. Uncoupling of cell growth and proliferation results in enhancement of productivity in p21CIP1-arrested CHO cells. Biotechnology and Bioengineering 2004;85:741-9.
[89] Fussenegger M, Schlatter S, Datwyler D, et al. Controlled proliferation by multigene metabolic engineering enhances the productivity of Chinese hamster ovary cells. Nature Biotechnology 1998;16:468-72.
[90] Ibarra N, Watanabe S, Bi JX, et al. Modulation of cell cycle for enhancement of antibody productivity in perfusion culture of NS0 cells. Biotechnology Progress 2003;19:224-8.
[91] Sunley K, Butler M. Strategies for the enhancement of recombinant protein production from mammalian cells by growth arrest. Biotechnology Advances 2010;28:385-94.
[92] May T, Lindenmaier W, Wirth D, et al. Application of a reversible immortalization system for the generation of proliferation-controlled cell lines. Cytotechnology 2004;46:69-78.
[93] Coroadinha AS, Gama-Norton L, Amaral AI, et al. Production of retroviral vectors: review. Current Gene Therapy 2010;10:456-73.
[94] Schucht R, Coroadinha AS, Zanta-Boussif MA, et al. A new generation of retroviral producer cells: predictable and stable virus production by Flp-mediated site-specific integration of retroviral vectors. Molecular Therapy 2006;14:285-92.
[95] Carrondo M, Panet A, Wirth D, et al. Integrated strategy for the production of therapeutic retroviral vectors. Human Gene Therapy 2011;22:370-9.
[96] Wurm FM. Production of recombinant protein therapeutics in cultivated mammalian cells. Nature Biotechnology 2004;22:1393-8.
[97] Tjio JH, Puck TT. Genetics of somatic mammalian cells. II. Chromosomal constitution of cells in tissue culture. Journal of Experimental Medicine 1958;108:259-68.
[98] Walsh G. Biopharmaceutical benchmarks 2010. Nature Biotechnology 2010;28:917-24.
[99] Graham FL, Smiley J, Russell WC, et al. Characteristics of a human cell line transformed by DNA from human adenovirus type 5. Journal of General Virology 1977;36:59-74.
[100] Louis N, Evelegh C, Graham FL. Cloning and sequencing of the cellular-viral junctions from the human adenovirus type 5 transformed 293 cell line. Virology 1997;233:423-9.
[101] Shaw G, Morse S, Ararat M, et al. Preferential transformation of human neuronal cells by human adenoviruses and the origin of HEK 293 cells. FASEB Journal 2002;16:869-71.
[102] Byrd P, Brown KW, Gallimore PH. Malignant transformation of human embryo retinoblasts by cloned adenovirus 12 DNA. Nature 1982;298:69-71.
[103] Silva AC, Peixoto C, Lucas T, et al. Adenovirus vector production and purification. Current Gene Therapy 2010;10:437-55.
[104] Dormond E, Kamen AA. Manufacturing of adenovirus vectors: production and purification of helper dependent adenovirus. Methods in Molecular Biology 2011;737:139-56.

[105] Young JM, Cheadle C, Foulke JS, et al. Utilization of an Epstein-Barr virus replicon as a eukaryotic expression vector. Gene 1988;62:171-85.
[106] Durocher Y, Perret S, Kamen A. High-level and high-throughput recombinant protein production by transient transfection of suspension-growing human 293-EBNA1 cells. Nucleic Acids Research 2002;30:E9.
[107] Durocher Y, Butler M. Expression systems for therapeutic glycoprotein production. Current Opinion in Biotechnology 2009;20:700-7.
[108] Sinn PL, Sauter SL, McCray PB. Gene therapy progress and prospects: development of improved lentiviral and retroviral vectors–design, biosafety, and production. Gene Therapy 2005;12:1089-98.
[109] Gama-Norton L, Botezatu L, Herrmann S, et al. Lentivirus production is influenced by SV40 large T-antigen and chromosomal integration of the vector in HEK293 cells. Human Gene Therapy 2011;22:1269-79.
[110] Le Ru A, Jacob D, Transfiguracion J, et al. Scalable production of influenza virus in HEK-293 cells for efficient vaccine manufacturing. Vaccine 2010;28:3661-71.
[111] Bleckwenn NA, Bentley WE, Shiloach J. Vaccinia virus-based expression of gp120 and EGFP: survey of mammalian host cell lines. Biotechnology Progress 2005;21:186-91.
[112] Fallaux FJ, Bout A, van der Velde I, et al. New helper cells and matched early region 1-deleted adenovirus vectors prevent generation of replication-competent adenoviruses. Human Gene Therapy 1998;9:1909-17.
[113] Genzel Y, Reichl U. Continuous cell lines as a production system for influenza vaccines. Expert Review of Vaccines 2009;8:1681-92.
[114] Barrett PN, Portsmouth D, Ehrlich HJ. Developing cell culture-derived pandemic vaccines. Current Opinion in Molecular Therapeutics 2010;12:21-30.
[115] Sanders BP, Edo-Matas D, Custers JH, et al. PER.C6((R)) cells as a serum-free suspension cell platform for the production of high titer poliovirus: a potential low cost of goods option for world supply of inactivated poliovirus vaccine. Vaccine 2013;31:850-6.
[116] Niklas J, Schrader E, Sandig V, et al. Quantitative characterization of metabolism and metabolic shifts during growth of the new human cell line AGE1.HN using time resolved metabolic flux analysis. Bioprocess and Biosystems Engineering 2011;34:533-45.
[117] Blanchard V, Liu X, Eigel S, et al. N-glycosylation and biological activity of recombinant human alpha1-antitrypsin expressed in a novel human neuronal cell line. Biotechnology and Bioengineering 2011;108:2118-28.
[118] Lohr V, Rath A, Genzel Y, Jet al. New avian suspension cell lines provide production of influenza virus and MVA in serum-free media: studies on growth, metabolism and virus propagation. Vaccine 2009;27:4975-82.
[119] Jordan I, Vos A, Beilfuss S, et al. An avian cell line designed for production of highly attenuated viruses. Vaccine 2009;27:748-56.
[120] Schiedner G, Hertel S, Kochanek S. Efficient transformation of primary human amniocytes by E1 functions of Ad5: generation of new cell lines for adenoviral vector production. Human Gene Therapy 2000;11:2105-16.
[121] Schiedner G, Hertel S, Bialek C, et al. Efficient and reproducible generation of high-expressing, stable human cell lines without need for antibiotic selection. BMC Biotechnology 2008;8:13.
[122] Fischer S, Charara N, Gerber A, et al. Transient recombinant protein expression in a human amniocyte cell line: the CAP-T(R) cell system. Biotechnology and Bioengineering 2012;109:2250-61.
[123] Genzel Y, Behrendt I, Rodig J, et al. CAP, a new human suspension cell line for influenza virus production. Applied Microbiology and Biotechnology 2013;97:111-22.

[124] Lieu PT, Machleidt T, Thyagarajan B, et al. Generation of site-specific retargeting platform cell lines for drug discovery using phiC31 and R4 integrases. Journal of Biomolecular Screening 2009;14:1207-15.
[125] Schucht R, Lydford S, Andzinski L, et al. Rapid establishment of G-protein-coupled receptor-expressing cell lines by site-specific integration. Journal of Biomolecular Screening 2011;16:323-31.
[126] Fogh J, Fogh JM, Orfeo T. One hundred and twenty-seven cultured human tumor cell lines producing tumors in nude mice. Journal of the National Cancer Institute 1977;59:221-6.
[127] Pinto M, Robine Leon S, Appay MD. Enterocyte-like differentiation and polarization of the human colon carcinoma cell line Caco-2 in culture. Biology of the Cell 1983;47:323-30.
[128] Artursson P. Epithelial transport of drugs in cell culture. I: A model for studying the passive diffusion of drugs over intestinal absorptive (Caco-2) cells. Journal of Pharmaceutical Sciences 1990;79:476-82.
[129] Artursson P, Karlsson J. Correlation between oral drug absorption in humans and apparent drug permeability coefficients in human intestinal epithelial (Caco-2) cells. Biochemical and Biophysical Research Communications 1991;175:880-5.
[130] Briske-Anderson MJ, Finley JW, Newman SM. The influence of culture time and passage number on the morphological and physiological development of Caco-2 cells. Proceedings of the Society for Experimental Biology and Medicine 1997;214:248-57.
[131] Lu S, Gough AW, Bobrowski WF, et al. Transport properties are not altered across Caco-2 cells with heightened TEER despite underlying physiological and ultrastructural changes. Journal of Pharmaceutical Sciences 1996;85:270-3.
[132] Lebonvallet N, Jeanmaire C, Danoux L, et al. The evolution and use of skin explants: potential and limitations for dermatological research. European Journal of Dermatology : EJD 2010;20:671-84.
[133] Boukamp P, Petrussevska RT, Breitkreutz D, et al. Normal keratinization in a spontaneously immortalized aneuploid human keratinocyte cell line. Journal of Cell Biology 1988;106:761-71.
[134] Allen-Hoffmann BL, Schlosser SJ, Ivarie CA, et al. Normal growth and differentiation in a spontaneously immortalized near-diploid human keratinocyte cell line, NIKS. Journal of Investigative Dermatology 2000;114:444-55.
[135] Donato MT, Lahoz A, Castell JV, et al. Cell lines: a tool for in vitro drug metabolism studies. Current Drug Metabolism 2008;9:1-11.
[136] Gomez-Lechon MJ, Donato T, Jover R, et al. Expression and induction of a large set of drug-metabolizing enzymes by the highly differentiated human hepatoma cell line BC2. European Journal of Biochemistry 2001;268:1448-59.
[137] Aden DP, Fogel A, Plotkin S, et al. Controlled synthesis of HBsAg in a differentiated human liver carcinoma-derived cell line. Nature 1979;282:615-6.
[138] Rodriguez-Antona C, Donato MT, Boobis A, et al. Cytochrome P450 expression in human hepatocytes and hepatoma cell lines: molecular mechanisms that determine lower expression in cultured cells. Xenobiotica 2002;32:505-20.
[139] Elizondo G, Medina-Diaz IM. Induction of CYP3A4 by 1alpha,25-dyhydroxyvitamin D3 in HepG2 cells. Life Sciences 2003;73:141-9.
[140] Nakamura K, Kato N, Aizawa K, et al. Expression of albumin and cytochrome P450 enzymes in HepG2 cells cultured with a nanotechnology-based culture plate with microfabricated scaffold. Journal of Toxicological Sciences 2011;36:625-33.
[141] Hewitt NJ, Hewitt P. Phase I and II enzyme characterization of two sources of HepG2 cell lines. Xenobiotica 2004;34:243-56.
[142] Nakabayashi H, Taketa K, Miyano K, et al. Growth of human hepatoma cells lines with differentiated functions in chemically defined medium. Cancer Research 1982;42:3858-63.

[143] Gripon P, Rumin S, Urban S, et al. Infection of a human hepatoma cell line by hepatitis B virus. Proceedings of the National Academy of Sciences of the United States of America 2002;99:15655-60.
[144] Scheel TK, Rice CM. Understanding the hepatitis C virus life cycle paves the way for highly effective therapies. Nature Medicine 2013;19:837-49.
[145] Noguchi M, Hirohashi S. Cell lines from non-neoplastic liver and hepatocellular carcinoma tissue from a single patient. In Vitro Cellular & Developmental Biology – Animal 1996;32:135-7.
[146] Mills JB, Rose KA, Sadagopan N, et al. Induction of drug metabolism enzymes and MDR1 using a novel human hepatocyte cell line. Journal of Pharmacology and Experimental Therapeutics 2004;309:303-9.
[147] Pan X, Li J, Du W, et al. Establishment and characterization of immortalized human hepatocyte cell line for applications in bioartificial livers. Biotechnology Letters 2012.
[148] Aly HH, Watashi K, Hijikata M, et al. Serum-derived hepatitis C virus infectivity in interferon regulatory factor-7-suppressed human primary hepatocytes. Journal of Hepatology 2007;46:26-36.
[149] Kobayashi N, Westerman KA, Tanaka N, et al. A reversibly immortalized human hepatocyte cell line as a source of hepatocyte-based biological support. Addiction Biology 2001;6:293-300.
[150] Hariparsad N, Carr BA, Evers R, et al. Comparison of immortalized Fa2N-4 cells and human hepatocytes as in vitro models for cytochrome P450 induction. Drug Metabolism and Disposition 2008;36:1046-55.
[151] Hoekstra R, Deurholt T, ten Bloemendaal L, et al. Assessment of in vitro applicability of reversibly immortalized NKNT-3 cells and clonal derivatives. Cell Transplantation 2006;15:423-33.
[152] Deurholt T, van Til N, Chhatta A, et al. Novel immortalized human fetal liver cell line, cBAL111, has the potential to differentiate into functional hepatocytes. BMC Biotechnology 2009;9:89.

2.2 CHO History, CHO Evolution and CHO Genomics – an Unsolvable Enigma?

Florian M. Wurm

2.2.1 Introduction

Immortalized Chinese hamster ovary (CHO) cells are the source of multiton quantities of protein pharmaceuticals. Such cells are characterized by diversity in genetics and phenotypes. Preexisting diversities in cell populations are further enhanced by selective forces when such populations become separated from each other so that the sharing and mixing of a given gene pool (of a species) is prevented (Darwin). In nature, such spreading of groups of individuals without gene pool sharing results in speciation. The distribution of CHO cell lines to hundreds of laboratories that grow these cells under (very diverse, but practical) conditions represents a strong case of such evolutionary selections.

CHO cells have been used in cell culture for research and for industrial applications for more than 50 years and a bewildering number of differently named cell lines can be found in the literature. This chapter tries to represent, in a cursory and not necessarily comprehensive way, the history of CHO cells, particularly the origin and subsequent fate of key industrially used cell lines. It is, in the opinion of the author, not a radical proposal to suggest that the name CHO represents many truly different cell "species", based on their inherent genetic diversity and their highly dynamic rate of genetic change. The remodeling of the many genomic structures in clonal or nonclonal cell populations, further enhanced by the nonstandardized culture conditions in hundreds of different labs, renders CHO cells a typical case for "groups of quasispecies". This term was coined in the 1970s for families of related (genomic) sequences exposed to high mutation rate environments where a large fraction of offspring is expected to carry one or more mutations. The implications of the quasispecies concept for CHO cells used in protein manufacturing processes are significant. CHO genomics/transcriptomics will provide only limited insights when analysis is performed on a small number of "old" and poorly (historically and otherwise) characterized CHO cell lines. Strictly speaking, any individual cell in a culture appears to carry a peculiar genomic structure. Only screening of many clonal cell lines derived directly and under controlled culture conditions from one relatively well-defined starting material may reveal a narrow diversity of phenotypes with respect to physiological/metabolic activities, and thus, that allows more precise and reliable predictions of the potential of a clone for high-yielding manufacturing processes.

The dominance of Chinese hamster ovary (CHO) cells for manufacture of complex therapeutic proteins is based on a number of characteristics of these cells that had been recognized in the early 1980s by groups of researchers, mostly in early biotechnology start-ups. At Genentech Inc. in South San Francisco, these cells were grown at a very large scale in suspension cultures, resulting in the first recombinant protein from animal cells in bioreactors approved for human therapy, in 1987. Subsequent research and development work in process, science with these cells continued since then to further improve both, protein quality and quantity from CHO cultures in bioreactors. Proteins like etanercept (Enbrel®, a TNF inhibitor for rheumatoid arthritis treatment) and trastuzumab (Herceptin®, an anti-HER-2 breast cancer antibody) are each produced at more than 1 t per year and tens of thousands of patients benefit from these protein drugs. In spite of these very impressive achievements in both, clinical success and manufacturing from cells in large bioreactors, recent publications have expressed hope that a detailed knowledge of DNA sequences and transcription patterns of CHO cell lines could provide urgently needed tools to improve the manufacture of protein pharmaceuticals [1]. However, in the opinion of the author, this research does not take sufficiently into account a profoundly adverse problem with CHO cells that could very much limit "omics" approaches and/or deliver entirely irrelevant data.

CHO cells – whether being referred to as K1, DG44, DX B11, CHO-Toronto, CHOpro3-, or CHO-S – are members of a widely distributed family of related, but profoundly different cell lines, since their individual behaviors/phenotypes (responsiveness to different environmental conditions) in cell culture differ quite significantly. CHO cells are immortalized cells, and the extent of their relatedness in genome structure, genomic sequence composition, and in transcription patterns has not been studied so far. Thus, reasonable conclusions on their similarities are therefore not available. However, in view of the published information on CHO cells and taking into consideration generally available information on fluidity of genomes of immortalized cell populations, it is proposed that each one of the above-mentioned CHO cell lines could be considered a "quasispecies". This term was first coined by Eigen and Schuster in the late 1970s in a series of landmark papers describing a high mutation rate environment where a group or "cloud" of related offspring exists and where one would expect that a large fraction of the offspring carries at least one mutation [2-4]. Since then, the term has been extensively discussed and applied in virology (reviewed in [5]) and, most recently, in the population dynamics of cancer cells [6].

Understanding and appreciating the above reasoning requires a "complete" history of CHO cells prior to their use in modern biotechnology. Fortunately, most of the history of the cells has been recorded in an enormous wealth of publications resulting from research executed with CHO cells from the 1960s to the 1980s. In addition, approximately at the time of the emergence of interest for CHO cells in the biotechnology industry, in 1985 Gottesman [7] edited a 900-page compendium entitled *Molecular Cell Genetics* that contains CHO main work exclusively dedicated to the Chinese hamster and the cells derived from this species. (Unfortunately, this compendium is out of print today but was available to the author for this chapter.)

2.2.2 CHO Cells in Metabolic and Genetic Studies

Today, few users of CHO cells know that they were extremely popular for fundamental research in molecular and classical cell genetics for a 30-year period prior to their use in pharmaceutical biotechnology. Their popularity was based on their practicality in cell culture, their few (11 pairs) and large chromosomes (in comparison to human ones), and the ease with which metabolic and other mutations could be generated or found and subsequently studied in elegant experimental approaches. During this period, the identification of the genes for the encoded enzymes and their expression profiles was a "hot topic". CHO cells became the preferred study object for gene and genome-based research in a mammalian host system. Induced and spontaneous mutations lead with a high success rate to physiological deficiencies (metabolic mutants) that could easily be identified through culture in selective media. Mammalian cell culture technology had matured at the time to an extent that media with partially defined or even serum-free compositions could be generated with ease [8-12]. These procedures permitted the application of concepts of microbial genetics to mammalian cells.

2.2.3 The Early History of CHO Cells

In 1957 CHO cells were established in the laboratory of Puck [13], then at the Eleanor Roosevelt Institute for Cancer Research and later at the Department of Biochemistry of the University of Colorado in Denver, from 0.1 g of ovary tissue from a Chinese hamster. The outbred hamster was provided by Yerganian from the Boston Children's Cancer Research Foundation [14]. It is important to note that out-breeding tries to avoid homozygosity, i.e., reducing the number of identical alleles in the genome, and thus maintains a typically vigorous and diverse genetic background of diploid animals or plants. The issue of allelic diversity in diploid systems will be discussed later again, since it appears that loss of heterozygosity (LOH) is a genetic trend for genomes of immortalized cells: they frequently loose entire chromosomes or fragments of chromosomes and thus increase the degree in homozygosity over time.

From the trypsinized ovary tissue a culture emerged that appeared to be predominantly of a fibroblast type and had a near diploid karyotype with only 1 % of the cell population differing in chromosome number by one less or more from the expected chromosome number of 22 (11 pairs) [13]. However, this small diversion from the strictly diploid character of primary cells is not observed in primary cells of human origin where a fully diploid karyotype is prevalent until senescence occurs ("Hayflick limit") after about 50 population doublings and the cells die [15].

The fibroblast morphology of the primary ovary cells was maintained in culture for more than 10 months, longer than the time limit being established for human fetal cells by Hayflick. At an unknown time thereafter, the morphology of some cells changed, and these cells overgrew the fibroblast cell type. It appears that cells in culture derived from the hamster ovary were liberated from the typical constraint in cell population doublings (i.e., they may have already experienced some type of spontaneous immortalization) while remaining close to a diploid character. However, the mentioning of additional morphological changes after 10 months in culture points towards additional modifications, most likely with a genetic cause whose origin is (still) not understood today. Subsequent recloning of these cells with a modified morphology resulted in the cell line that is now called CHO. The change of morphology from a fibroblast type of culture to a more epitheloid morphology of cells is mentioned in Puck [16, 17]. Unfortunately, no information on the specifics of the cloning step or the potential diversity of CHO cell lines is available. For clarity in this text, the author refers to these original CHO cells as CHO-ori (see Figure 2.2.1).

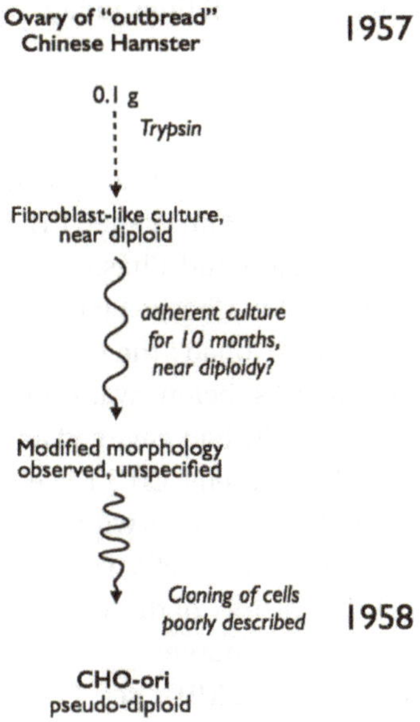

Figure 2.2.1: A scheme representing the generation of the early "original" CHO cell line.

The scheme in Figure 2.2.1 depicts the little knowledge (as available at a time distance of more than 45 years) about the origin, history, and early handling of CHO cells. Obviously, under today's standards of scrutiny in our industry, one can only say that this description is unsatisfying. It is highly unlikely that "cloning" was done at the time in Puck's laboratory with the attention to details we apply today in industry to this step. In retrospect, several pertinent questions arise, unfortunately without a chance to obtain answers. Particularly striking are issues around the immortalization of these cells, their assumed genomic constitution, and their apparent phenotypic change in culture.

Subcultures of these CHO-ori cells, now a CHO cell line (to be distinguished from cell strains with a limited life span, according to Hayflick's terminology), were passed on to many laboratories. They were described as "hardy", growing very well, and fast in adherent culture. They had a high cloning efficiency (always done in an adherent mode in dishes or flasks) even at very low fetal bovine serum concentrations in the culture medium. 10 to 20 % fetal bovine serum (FBS) in commercial media was standard in cell culture at the time. However, Hamilton and Ham [11] already reported in 1977 the growth of these cells in serum-free media. The cells required the addition of proline to the culture medium, and, to the author's knowledge, today all CHO cell lines require it. The first reference to this is from 1963 [16]. Thus, it appears that loss of competence for proline synthesis is an early event in the history of these cells.

The following text does not follow the actual history of passing cells from one laboratory to another, but gives names and pedigree according to the importance for the emergence of the biopharmaceutical industries that uses CHO cell lines today.

2.2.3.1 CHO-DXB11

The very first product made by CHO cells, and thus the starting point of the biotechnology era involving recombinant mammalian cells, employed a cell line called CHO-DXB11. This line was generated at Columbia University by Urlaub and Chasin, who were interested in dihydrofolate reductase (DHFR) [17]. The cells in Chasin's lab were derived from CHO-K1 after a coworker in Puck's lab (Kao) had "cloned" the CHO-ori cells. According to Chasin (personal communication), the CHO-K1 subclone was established in the late 1960s, perhaps almost 10 years after CHO-ori cells had emerged and passed on to others as an immortalized cell line. Similar to CHO-ori cells, CHO-K1 cells were also supplied to many laboratories around the world and experiments with these cells were described in numerous publications. Vials of frozen CHO-K1 cells were also deposited at the American Type Culture Collection (ATCC). The history of these deposited cells was not available to the author at the time of writing this chapter.

Chasin established the cell line CHO-DXB11 (also called DUK-XB11, see Figure 2.2.2). The purpose of the work was to delete DHFR activity. These cells carry a deletion of one locus for DHFR and a missense mutation (T137R) of the second DFHR locus, rendering the cells incapable of reducing folate, a precursor for thymidine and hypoxanthine synthesis [17]. The cell line is not named in the mentioned paper, but it is one of the gamma-ray-induced mutants described. It is interesting to note that this cell line, the first to become a host system for the production of hundreds of kilograms of human tissue plasminogen activator (tPA), was the product of (severe) mutagenesis. The reasons why this and not any other CHO cell line became the pioneering cell line in biotechnology is rapidly explained: the dual inactivation of the DHFR locus rendered this cell line very useful for transgenesis with a functional DHFR gene [18]. Transfer of a functional DHFR gene via plasmid transfection could repair the DHFR deficiency and allow easy selection of recombinant cells in well-defined media. In

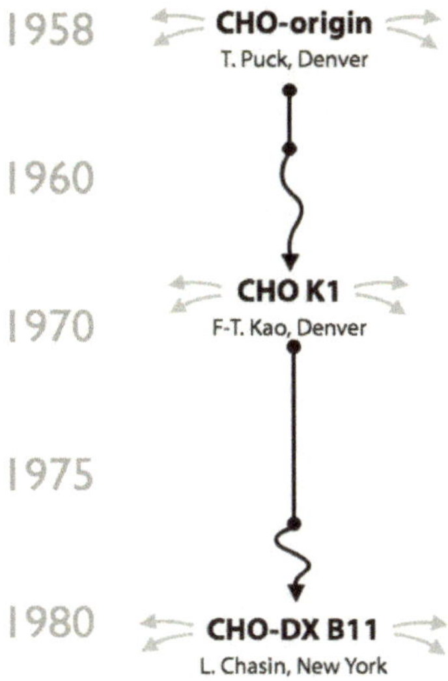

Figure 2.2.2: The origin of CHO-DX B11 cells. A straight line indicates transfers of cells without intentional modification of the cell line. A line with multiple curves indicates intended experimental steps that are expected to change the genotype/phenotype of the cell population. The grey arrows next to the name of the cells indicate the fact that many cells (as frozen vials or as active cultures) were distributed to many labs over extended periods of time. No information on such transfers is available today.

addition, a second, unrelated gene of interest (GOI), encoded by the same plasmid vector, could easily be transferred simultaneously and recombinant clones expressing both the functional transgenic DHFR gene and the desired cotransferred GOI could be recovered [19]. This 1983 publication is the first to describe cotransfer of two genes into CHO cells whereby the two corresponding DNA sequences were provided on two separate plasmids. They were simply cotransfected at different ratios. In the case of the paper quoted, a 1:10 ratio of the DHFR plasmid to the gamma-interferon plasmid gave the highest yielding clones.

The DHFR-negative cells were grown in media containing 5 to 10 % fetal bovine serum (FBS). These cells were then transfected, using calcium phosphate as a reagent, and recombinant clones were identified after exposure of the transfected population in media lacking glycine, hypoxanthine, and thymidine (GHT-minus medium and dialyzed fetal bovine serum). A risk factor mediated by the use of sera from cows, bovine spongiformous encephalopathy (BSE), became an important consideration for pharmaceutical manufacturing in the 1990s. In industry, sera were generally obtained from BSE-free sources (Australia, New Zealand). Whether this practice was followed in academic labs is difficult to assess. Transfection and cloning occurred in an adherent mode, whereby cloning was done by using "cloning rings" or cotton-swabs. It must be assumed that these techniques were also used during the early phases of the CHO history. In all cases, an identified colony, visible to the naked eye, was targeted and many cells from such a colony were transferred into a well of a multiwell plate. Regulatory concerns requires today "single cell cloning", frequently done even twice in order to "prove" clonality. This is the most stringent "population bottleneck" one can imagine – an issue which has important consequences on the genomic constitution(s) of the emerging clonal cell population.

2.2.3.2 CHO-DG44

For the researchers of metabolic studies with these cells, a low rate of reversion to DHFR activity in CHO-DXB11 cells presented a problem. In order to fully eliminate this possibility and to provide also a better DHFR-negative host system for eventual gene transfer, Chasin engaged in another round of DHFR elimination, but not with cells derived from the K1 populations. Figure 2.2.3, established to the best knowledge available to the author, tries to establish the history of CHO-DG44 and other cell lines of relevance to the industry.

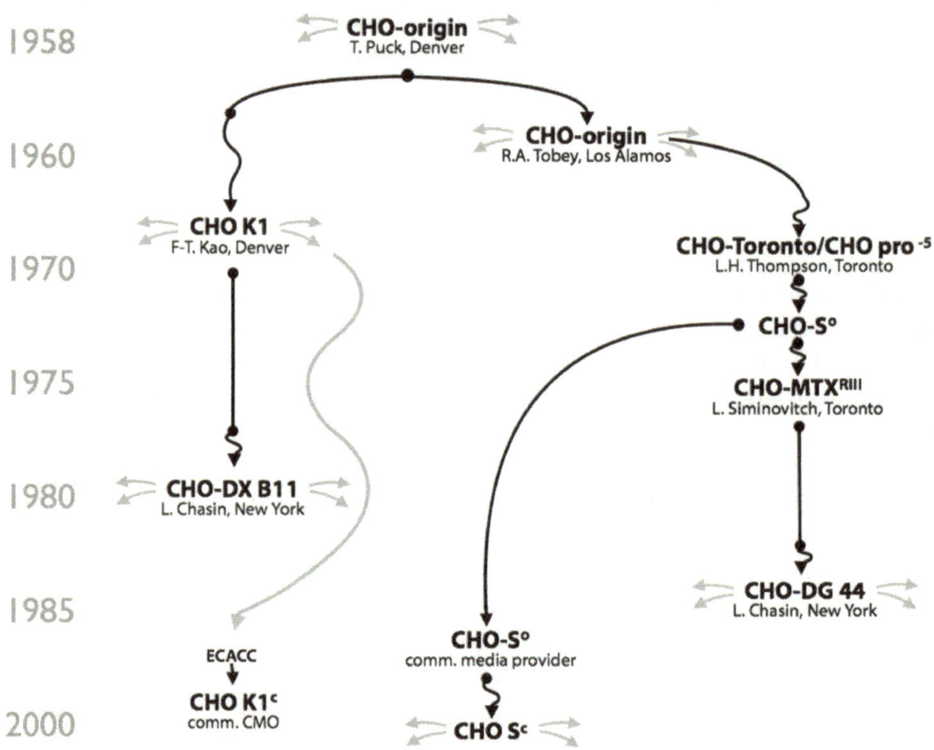

Figure 2.2.3: Origin of CHO-DX B11, CHO-DG 44 cells, commercial CHO S^c cells and commercial CHO $K1^c$ cells. A short multicurved line (black) indicates literature describing experimental steps that are expected to change the genotype/phenotype of the cell population. Year indications are best estimates. Small grey arrows next to a given cell name indicate multiple distributions of cells to different labs and companies. The grey, multicurved line indicates uncertainties or undescribed fates of the CHO $K1^c$ cells.

Flintoff, a coworker of Siminovitch, had generated a useful mutant, named CHO-Mtx^{RIII} derived initially from CHO-ori cells, but other cell lines mentioned in the literature (such as CHO pro5- and CHO-S^o) may have been part of its history as well. This gene-amplified mutant proved to be suitable for deletion of both DHFR alleles [20]. In the same year (1976), Siminovitch published a highly instructive minireview on genetic diversity of cultured somatic (immortalized) cells, based on the many years of insights gained mostly with CHO cells [21].

Chasin and Urlaub used the CHO Mtx^{RIII} cell line to delete the amplified DHFR containing regions of the CHO chromosomes. Their elegant work showing the full deletion of the two DHFR loci on chromosome 2 (actually on chromosome 2 and on a shortened marker chromosome variant Z2) resulted in the now widely-used CHO-DG44 cells, obtained after two "traditional" cloning steps [22, 23, personal information from Chasin].

2.2.3.3 Gene Amplification

The availability of these two DHFR-negative cell lines (CHO-DXB11 and CHO-DG44) allowed an approach to amplify genes with the help of a chemical antagonist of DHFR, methotrexate (MTX). The isolation of "amplified" recombinant cell lines occurred using stepwise increases in the MTX concentration in the culture medium over several subcultivations. Such induced gene amplification usually increases the productivity of cells containing the GOI [24-26]. It is based on selections of mutated cells that have amplified the DHFR containing sections of the genome of the cells, which occurs at the level of large chromosomal sections and thus coamplifies any other DNA as well, in the case of cotransfected DNA the plasmid sequences that contain the GOI. This approach was a key approach for enhancing protein production in clonal subpopulations of transfected CHO cell lines over a 20-year period in the biotechnology industry. During this period, most of the recombinant protein products were derived from CHO cells that had undergone MTX-induced gene amplification. Gene amplification results in large genomic reorganizations, visible to the eye when karyotyping cells [27, 28].

Briefly, new chromosomal structures, known as homogeneously staining regions (HSRs), can be found in metaphases of MTX selected human (cancer) derived cells, as well as in CHO cells. These regions show multiple (up to thousands) repetitions of smaller chromosomal regions (amplicons), all containing DNA containing, at least in part, DHFR genes. In fluorescence *in situ* hybridizations on recombinant CHO cells, large chromosomes were found that contained entire arms and long segments within chromosomal arms, hybridizing with DHFR sequences. Copy-number analysis of such cell lines revealed hundreds and thousands of copies of DHFR in these cells [26-28]. Mechanisms of gene amplification have been studied both in human (cancer) cells and human cell lines [27, 30], as well as in hamster [31, 32] and in mouse cell lines [33, 34], and several mechanisms have been proposed including principles that go back to "fragile chromosomes" and gene transposition as proposed by the work of McClintock in *Zea mays* [35]. The genetic stability of these unusual chromosome structures within a given cell population is poorly understood [29].

2.2.3.4 CHO-K1

In Gottesman [7] we find the following statement: "One subline of the original isolate, called CHO-K1 (ATCC CCL 61) was maintained in Denver by Puck and Kao, whereas another subline was sent to Tobey at Los Alamos. This latter line was adapted to suspension growth by Thompson at the University of Toronto (CHO-S) in 1971 and has given rise to a number of Toronto subclones with similar properties including the line CHO Pro^{-5} used extensively by Siminovitch and numerous colleagues in Toronto, CHO GAT of McBurney and Whitmore, subline 10001 of Gottesman at the NIH, and subline AA8 of Thompson. There are some differences in the karyotypes of the CHO-K1 and CHO-S cell lines, and CHO-S grows well in spinner and suspension culture, whereas CHO-K 1 does not. Both sublines seem to give rise readily to mutant phenotypes." This shows the handling of CHO cells by many laboratories, their diversity in phenotypes (one grows the other not in suspension), and the reason for their popularity at the time. The line "give rise readily to mutant phenotypes" would have to be avoided when presenting cell hosts to regulatory agencies. However, today's popularity of CHO-K1 cells is based on the successful and reliable use of these cells by a well-known contract manufacturing company for the production of high-value pharmaceuticals. Licenses for these cells, now suspension-adapted, in connection with a unique gene transfer system based on the enzyme glutamine synthetase (GS-system), have been taken by a number of their client companies. The GS system was originally designed for NSO cells (a murine myeloma-derived cell line also used for the fusion with B cells in the generation of hybridomas) [36] and was quickly applied to CHO cells as well. The origin of the CHO-K1 cells in the hands of the above mentioned contract manufacturer goes back to a vial of frozen cells derived in November 1989 from the European Collection of Animal Cell Cultures (ECACC). How and when these cells arrived at ECACC is not known to the author. A serum-free, suspension culture was frozen in the year 2000 (11 years later) as a "development bank". Eventually, a subline was generated that gave rise in October 2002 to a "CHO K1 SV" master cell bank under "protein-free" conditions (Metcalfe, personal communication). Worldwide, at the time of writing of this chapter, there are five licensed pharmaceutical products that were made with the help of the GS system in combination with CHO-K1 cells.

Briefly, recombinant CHO-K1 cells can be obtained after cotransfection with a functional glutamine synthetase gene together with the GOI on the same plasmid followed by selection in the absence of glutamine in the medium. In addition, the application of a GS inhibitor (methionine sulphoximine, MSX) allows either an increase of the stringency of selection or the selection for subpopulations of cells with an amplified copy number of the GS gene and the GOI. One has to assume that the principles of gene amplification with the GS system are similar to the ones discussed above for the DHFR system. Unfortunately, no publications with respect to karyotypic characterization of GS/MSX amplified sequences in CHO cells have been published so far.

2.2.3.5 CHO-S

About 50 years ago it was recognized that some (variants of) CHO cell lines have the capacity to grow in singlecell suspension culture [37]. In 1973, Thompson and coworkers described suspension cultures of CHO cells [38] and CHO-S cells were mentioned (see quote from Gottesman above). Thompson's CHO-S cells (in Figure 2.2.3: CHO-S^{o}) were derived from the CHO-Toronto cell line, also referred to as CHO pro^{-5}. Unfortunately, there is confusion about the origin and lack of due credit for CHO-S cells. It appears that the CHO-S cells from Thompson's lab, possibly through Tobey, were provided in the late 1980s to a cell culture media and cell host providing company. The company obtained a vial of CHO cells from Tobey at the Los Alamos National Laboratory. These cells were further adapted to a specific medium provided by that company marketed since 2002 as CHO-S. Here, the author will call the CHO-S from Thompson's lab CHO-S^{o} (o = original) and CHO-S from that company CHO-S^{c} (c = commercial). In view of their culture history it can be assured that these two cell lines will differ with respect to optimal culture conditions and other phenotypic/genetic features.

Neither of these CHO-S cell lines was used at Genentech in the mid 1980s for culture in singlecell suspension. Instead, recombinant clonal subpopulations derived from CHO-DUXB11, first established in adherent cultures with FBS in the medium, were individually adapted to suspension. The suspension adapted, serum-free subpopulations were not recloned prior to generation of master cell banks (personal information provided by the author). This appears surprising but is scientifically defendable, since "stability" and "identity" of a recombinant cell population has a higher chance of being maintained when cloning is avoided (see also the discussion on stability and microevolution below). The approach to suspension-adapted clonally derived cell lines, grown prior and during cloning in serum-containing media, without another recloning step was the basis for the first large-scale (10,000 L) stirred-tank bioreactor-based culture of CHO cells for the production of human recombinant tPA, and it was also used for other products developed by Genentech in the 1990s.

2.2.4 Diversity of Culture Conditions and the Cytogenetics of CHO Cells

CHO cells have been maintained by hundreds of different laboratories under highly diverse conditions. Therefore, the fluidity of genomic structures in immortalized cells will have to be considered. Decades of research into culture of immortalized cells have taught one important lesson: any culture of clonal or nonclonal cell lines will have a dramatic and lasting effect on the diversity of genotypes exhibited by the cell population. Insights into the persistent and continuing fluidity of genomes of immortalized cells go back to the 1960s. In 1961, Hsu, a highly recognized leader in the cytogenetics of mammalian cells, published a landmark paper entitled "Chromosomal Evolution in Cell Populations" [39] that summarized more than a decade of work after the visual analysis of chromosome structures, and their identification had become a standard technique. Chromosomes could be counted and identified and thus provided a suitable means to begin to understand the genomic organization of plants and animals. However, in contrast to the clearly recognizable and stable (in structure) chromosomes of diploid animals and plants, chromosomes of animal-derived immortalized cells showed a strong tendency to be nonidentical in number (from cell to cell) and apparently were able to change their organizational structures. Chromosomes of cell lines were not stable, in contrast to the overall unchanged structures and numbers of chromosomes found in wild-type biological species. The extremely rapid genome modifying impact of immortalization is strikingly visible in the unique and highly unusual chromosomal structures of such cells. For CHO cell lines, nothing different was seen. An example of this is given in Figure 2.2.4, which shows the diploid chromosomes of the Chinese hamster in comparison to those of an ancestral CHO-K1 cell line, based on the karyotyping work of Deaven and Peterson [40].

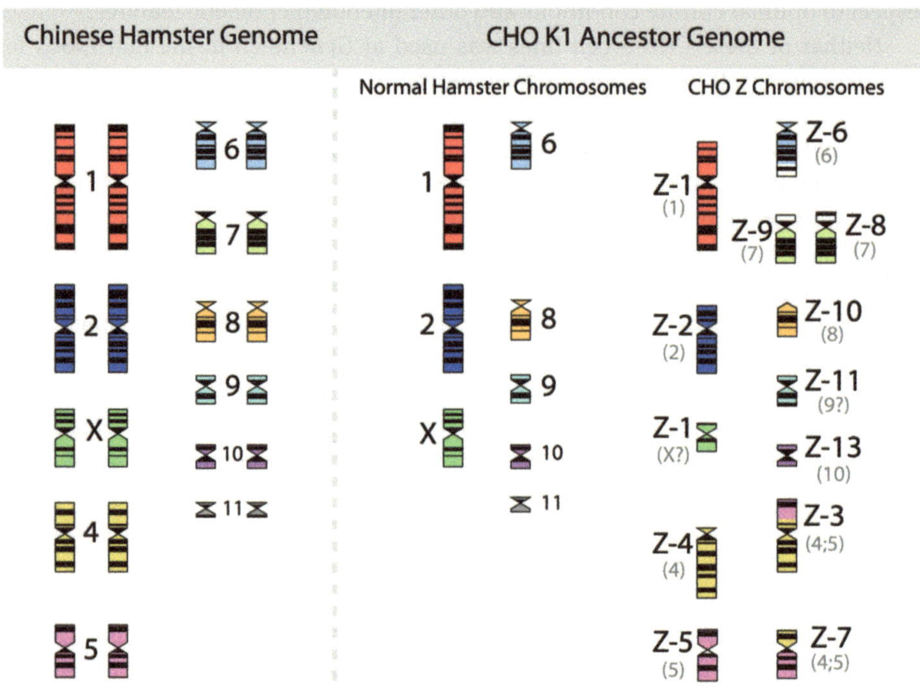

Figure 2.2.4: The 22 chromosomes of the Chinese hamster and the 21 chromosomes of CHO-K1 as identified by G-banding techniques (modified from [40]). A part of this figure was first published in [58] (with permission by Nature Publishing Group).

More recently, Omasa and his group constructed genomic BAC (bacterial artificial chromosome) libraries of available CHO-K1 and CHO-DG44 cell lines in order to establish a map of the hamster chromosomes as fragments of them are distributed in the chromosomes of those cell lines. The BAC-based maps solidify the earlier made karyotyping based findings by Deaven and Peterson: dramatic rearrangement of chromosomal fragments as compared to the diploid (hamster) genome in both cases. Also, only few structures appear "stable" when comparing DG44 and K1 cells [41]. All immortalized cells present similar restructuring of their genomic DNA. What is not shown is the fact that the genomic structure of CHO-K1 shown in Figure 2.2.4 is only one of many different genomic organizations present in a population of CHO-K1 cells. Other cells may show certain similarities to this pattern, but they will rarely (or maybe never) exhibit one that is identical when studying 50 karyotypes of individual cells. Deaven and Peterson observed a distribution of chromosome numbers per cell, ranging from 19 to 23. 60 to 70 % of the cells had 21 chromosomes. Although CHO-K1 cells do not have the 11 pairs of chromosomes of the hamster genome, the majority of the chromosome structures of the hamster genome are present albeit rearranged with only a few elements (G-banding pattern fragments) not clearly accounted for. Much more recently (2006), in the PhD work by Hazelwood [42] done under the guidance of Dickson (University of Manchester), a similarly complex genomic situation of K1 cells as they are/were used by the above mentioned contract manufacturer was revealed: a CHO-K1 cell line showed metaphase spreads over a broad chromosome number range with 16 to 30 chromosomes (100 cells studied), with 18, 23 and 18 % of cells showing 19, 20 or 21 chromosomes, respectively. A CHO-K1 SV cell line, grown under protein-free conditions (mentioned earlier), showed also a very broad chromosome number distribution of 10 to 30 chromosomes. In this instance, 10, 13, 17, 7 and 12 % (total 59 %) of cells showed 16, 17, 18, 19 or 20 chromosomes, respectively. Studies with clonal subpopulations of these cells revealed similar complexities of the karyotypes – none of them matching even approximately the statistics shown for the "parental" cells [42]. The American Type Culture Collection describes the CHO-K1 cell line's genotype as "chromosome frequency distribution 50 cells: 2n = 22. Stemline number is hypodiploid." Thus, the name "CHO-K1" refers to very different cell populations, most clearly represented by their karyotype.

This is not surprising: in cultivated, immortalized animal cells, singlecell cloning, with and without prior gene transfer, but also just modification of cell culture conditions for a given cell population, leads to new and genetically diverse cells as pointed out by Hsu [39]. Each of these populations of cells represent a new quasispecies family in the terminology of Eigen and Schuster [43-45]. If cloning is performed, as is now essential for a manufacturing cell lines, we will never know the genomic composition of **the one** cell that gives rise to the resulting population of cells. For example, we do not even know whether cloning efficiencies of cells with 19 or 20 or 21 chromosomes are different (however, it is not unreasonable to assume that they are different). Whenever we have a chance to do karyotype analysis on a clonal population, we find diversity in chromosome structures. Clonal cell populations analyzed posttransfection and subjected to stringent selective forces show a bewildering genomic restructuring, as judged by simple karyotyping or chromosome counting. Each clonal population analyzed is different. The modal chromosome numbers vary and individual, recognizable chromosomes show rearrangements [46].

Two recent papers on the genomic landscapes of HeLa cells [47, 48], the immortalized cell line [49] that was the foundation of animal cell culture technologies used today, shed a revealing light on the dynamics of genome remodeling under continuous cultivation. A remarkably high level of aneuploidy and numerous large structural variants were found at unprecedented resolution. A fifth of the HeLa cell line genome showed "loss of heterzygosity" (47). The original genome of Henrietta Lacks, the unfortunate woman who developed cervical cancer and whose cells were the source of the many HeLa cell lines being studied over the last 50 years, would be considered near 100 % heterozygote and thus would show significant sequence variations between the allelic DNAs representing the two sets of the 23 chromosomes. In one of the two genome papers on HeLa cells the average chromosome number of these cells is given as 64 and many segments of chromosome have a ploidy status ranging from triploid to octoploid. For example, one large homozygote fragment of chromosome 5 with a size of about 40 million bp is apparently present with 8 copies in HeLa cells. Another fragment, about 90 million bp, essentially the entire q-arm of chromosome 3 is present as 3 copies. These karyotype features result obviously from losses of fragments or arms of chromosomes while the corresponding allelic fragments are duplicated or multiplied. In [47] it is stated: "The extensive genomic rearrangements are indicative of catastrophic chromosome shattering". Up to 2,000 genes in HeLa cells are expressed at higher ranges than those seen in human tissues. More than 700 large deletions and almost 15,000 small deletions (as compared to the human genome) were detected. Most interesting, in view of the major chromosomal rearrangements in CHO, are the results of multiplex fluorescent *in situ* hybridizations (MFISH), a chromosome-painting method. Unfortunately, only 12 metaphase spreads were analyzed in this way. As with CHO, these 12 metaphase spreads show common structural rearrangements of the karyotype but also a number of "single cell events". The latter show unique translocations of chromosomal fragments, not seen in any of the other cells, indicative of a continuing dynamic of restructuring of the HeLa genome. Since HeLa cells usually do not undergo single cell cloning as CHO cells do, the fate of cells with unique rearrangements is difficult to predict. They may be passed on as a constant and small part of the entire population if they do not negatively affect the duplication of a cell. Unfortunately, the term "unique" and "single cell observations" are not to be taken at face value. An analysis of the karyotype of 12 cells does not provide a sufficient basis to make conclusions about populations of hundreds of millions of dividing cells. In the publication by Adey et al. [48] sequence analysis of two widely used HeLa cell lines (CCL-2 and S3, the latter being cloned from the former four years after the establishment of CCL-2) and 8 additional lines were performed. The cell lines were referred to as hypertriploid (without an assessment of chromosome number). Gains and losses of entire chromosome arms were observed, but also more frequent amplifications and deletions. Evidence of "chromothripsis" (chromosome shattering) is provided in both papers [47, 48], indicating an early event, potentially inherited from the original cancer. This phenomenon, poorly understood, results from a fragmentation of one or more chromosomes that subsequently "heal" or become repaired as a new and differently organized chromosome structure. The analysis of these two landmark papers [47, 48] on the genomics of the widely used HeLa cells requires an extensive knowledge of the complexities of sequence analysis techniques and of the likely biology (population dynamics of replicating units, bottlenecks, gene-copy number effects, etc.). In all humbleness, the author of this chapter does not claim to having fully grasped the genomic structures of the populations of HeLa cells studied by the two groups. In discussing these papers with the lead author of one of the two papers [47], many unanswered questions remain on timing of the origin and the dynamics of restructuring of the mammalian genome in such cells and, most importantly, the individuality of cells in terms of chromosome/genome structures can not be resolved with the presently used sequencing techniques when applied to the entire populations.

However, because of the importance of these phenomena apparently linked to cancerogenesis, mechanisms that result in these unusual chromosome structure and ploidy changes within an emergent population are being studied since a short time. Telling, the title of one of these papers is "Cancer genomes evolve by pulverizing single chromosomes" [50-52]. The findings appear to mirror those karyotype-based observations of chromosomal instability seen in immortalized cells over many decades. Comparing the recent findings in HeLa and human cancers (cells) *in vivo* and *in vitro* with so far available CHO data (karyotypes, BAC hybridizations, genome studies) we find a striking similarity.

2.2.5 Stability, Gene Pools and Microevolution

In the context of pharmaceutical production, stability is defined as the reproducible protein yield and quality from a given cell line over extended periods of time, from thawing of the cell line from a master cell bank until a given time point that is considered the longest period allowed for production. The minimal accepted period for requested proof of stability is about 3 months, based on the fact that manufacturing processes at large scale take significant time and will involve many cell population doublings. The expansion of cells towards the large-scale production vessel from a frozen vial can take up to 4 weeks. Subsequently, the production phase in the large-scale reactor can take up to 3 weeks for a fed-batch process and even longer for a perfusion-based manufacturing approach. Since several batches of product are typically being produced sequentially from one thawed vial of cells, a 3-month time window for this work is calculated very tightly. For approved protein products, stability studies cover at least 6 months of culture are standard.

It is difficult to imagine the huge number of cells that can be generated within a 3-month time window. The author needs to elaborate, in order to highlight the occurring genetic “bottlenecking”, CHO cells double their number about once a day (and shorter times have been reported), thus within 3 months about 90 doublings of the initial cell population will occur. Cell banks, the starting population of cells in vials for pharmaceutical manufacturing, are typically made with 4 to 6 × 10^6 cells per vial, corresponding to about 30 μL of cell biomass (compacted cells). If unrestricted for subsequent growth after thaw, this biomass could theoretically multiply within the 3-month time window to a biomass volume of approximately 3 × 10^{22} L. However, a single 10,000 L bioreactor will contain “only” about 10^{13} cells (corresponding to a biomass of about 300 L). Thus, any large-scale production run will only use **a minute fraction of the progeny of cells derived from the starting culture after thawing the cells.** Thus scale-up is, in biological terms, equivalent to the expansion of a single invading species into an unexploited environment (where most progeny die/are selected against). In scale-up and maintenance of cells, many restrictions on the growth of these cells occur and thus a new population of quasispecies will evolve. Due to the genetic diversity of the invading population of quasispecies, the final bioreactor will certainly contain a quasispecies different from the starting one deposited in the master cell bank.

Independent from the diversity in CHO populations, the stability of the transgene(s) within these populations represents another problem that is not sufficiently studied and understood. Due to the lack of control over the site(s) of integration of the GOI within a single CHO cell, the issue of its stability within the genome is an unresolved problem. (In this context, it is noteworthy that regulators and some companies are insisting more and more on “true” clonality and that a single cell cloning exercise is not satisfactory. In view of the discussion above, this level of scrutiny is difficult to justify.) In spite of decades of research in this field, no controllable and reproducible gene transfer system has been developed for CHO cells so far. For this reason, manufacturers screen thousands of clonal cell populations (all of which are to be considered quasispecies populations) and study them in extended subcultivations in order to predict with a reasonable probability that the productivity is maintained (a) over time at small scale and (b) after transfer to large scale for manufacturing. Essentially, each time we clone cells, we generate a founder population that undergoes microevolution while we optimize and scale up our cells into large bioreactors. The diversity of these founder populations must be significant, since cloning efficiencies in CHO cells are high (> 80 %), yet cells differ dramatically in their individual genomic composition. The enormous hardiness of these cells, however, allows rapid expansion in number of cells derived from the **one** genome structure of the founder cell while rapidly restructuring it, as has already been shown by Hsu in 1961 [39]. A recent and important paper in this context on CHO cell populations verifies the expected genetic (heritability) diversity with 199 clonal cell populations derived from a CHO K1 parental host system [53].

Since true clonality cannot be preserved and thus does **not solve the stability problem,** the best approach for maintaining a balanced gene pool in a given quasispecies population is to minimize growth-restricting (selective) conditions. Clearly, for commercial pharmaceutical manufacturing, the maintenance of the gene pool composition of the cells in a master cell bank must be assured by all means. By keeping cell populations under conditions with little environmental changes, one can hope that a trend towards a modified gene pool would be minimized. Unfortunately, many standard cell culture techniques are possibly favoring or selecting for modifications of a given gene pool in a quasispecies population of CHO cells. For example, the shift of cells from adherent culture to suspension cultures represents a major environmental modification and thus will lead to the selection of subpopulations. Also, the composition of media that either prevent or allow cells to grow to high density can be considered a selective condition. Finally, even work with controlled bioreactors may be a cause for a population bottleneck. For example, certain reactors have poor gas exchange capacities and thus need to be stirred or otherwise mixed vigorously and, frequently in addition, need to be sparged with pure oxygen gas to maintain basic metabolic activities for the cells. Such conditions can kill sensitive cells and will select for populations of cells that are adapted to these harsher conditions. Other bioreactor systems, for example OrbShake bioreactors [54, 55] are known to have higher gas transfer rates than stirred-tank bioreactors and thus require less energy (correlated to shear stress and liquid turbulence) in order to distribute oxygen to cells. The milder conditions of such bioreactors would be expected to maintain sensitive cells and thus would not apply a restrictive/selective pressure on a population of cells that is scaled from milliliter cultures to hundreds and thousands of liters.

Awareness of the importance of environmental conditions in cell culture for maintaining stability has only recently been discussed in groups of scientists who deal with manufacturing issues. In this context, however, the complexity of genomic compositions and the diversity of genomes in cell populations have not been discussed.

2.2.6 Concluding Remarks

The dilemma faced by the industry using CHO cells is enormous. CHO cells have been known for decades to be highly robust **and flexible**, one of the main reasons for their popularity in manufacturing. Their gold standard status is rooted in the many successful products that have been made in them without risk to transmit unknown infectious agents. The quality of these products and, more importantly, the volumetric yields obtained for such products have surpassed the productivity of microbial systems. This further encourages the use of these cells.

However, the long and convoluted historic pathway of these cells, resulting undoubtedly in an enormous genetic and physiological diversity, combined with a large variety of culture conditions used in hundreds of labs represent a problem in assessing the genomic composition of any cell line analyzed and its relevance for others. Each population of cells can be considered a **quasispecies.** Each cell line will represent its own unique genome/transcriptome/proteome and, therefore, an evaluation of data in comparison to other cell lines will be difficult if not done within a consortium of most interested parties that are ready to share their data. However, the wish to share data is even more complicated by the fact that companies will have their own, undisclosed ways to engineer their cell lines for more efficient protein manufacturing and are eager to protect them. Each of these efforts will generate population bottlenecks of quasispecies. Also, the work in companies involves different culture media, transfection approaches, reagents, selection steps, bioreactors, freezing and thawing protocols, cloning approaches, and bioprocesses, many of them proprietary.

What we can hope when working with cell lines of CHO origin in a given laboratory over many years is learning to appreciate this diversity and to address it with a large panel of standardized methods. It is a pragmatic approach and continues to be dependent on **screening for favorable phenotypes.** In essence, the genetic diversity problem can be managed by finding the best conditions for a cell population and then keeping the environmental conditions constant in order to minimize subsequent gene pool drifts in the populations.

What role for CHO genomics? A few important papers have been published very recently on sequence compositions and "genomic landscapes" of individual CHO cell lines and of the hamster genome [1, 56, 57]. Data in these papers were generated with cell populations derived from the CHO-K1, DG-44 and CHO-S^{c} lineages. They shed a first light on the complexities of CHO genomes, and the papers show a considerable level of sequence heterogeneity in comparison to the hamster genome and with respect to each other. In the papers by the Palsson group (1, 57) it was found that mutations appear to accumulate rapidly and are unique to a given cell line. For one antibody producing cell line, more than 300,000 single nucleotide polymorphisms (SNPs) arose, representing 9 % of all SNPs in this line. Notably, the authors claim that such nonuniform distribution of mutations have phenotypic relevance. More work has to be done on CHO, since the papers published so far lack quite important information of relevance to the industry:

- What is the degree of individual chromosome ploidy?
- Have entire chromosomes disappeared that generate long segments of homozygosity (as seen in HeLa)?
- Are such losses common among different CHO cell lines?

It appears that such major sequence variations may be shared within a branch of the CHO families, but not across the different branches. These publications have to be considered a beginning of work necessary to unravel the true genetic diversity of the branches and twigs of the CHO tree [58]. To take CHO K1 cells and its genomes: comparison of data sets derived from CHO K1- cells, grown in adherent culture using DMEM and fetal bovine serum, in comparison to K1 suspension cells in chemically defined media, and their respective single cell cloned, but subsequently emerged cell populations, etc., will have to be established. The author could imagine tens if not hundreds of such sequence assessments that could lead to very useful insights eventually. The reader is reminded of the fact that the one female hamster, chosen for the origin of CHO cells, was an "outbred" hamster, whose degree of homozygosity was expected to be low. Did the same happen to CHO cells? Is this the reason why CHO were so useful for mammalian genetics studies in the 1960s to the 1980s? In the process of using CHO cells and mutagenizing them for cytogenetic studies and for the application of microbial genetics to mammalian cells, chromosome losses and rearrangements and copy-number compensations occurred that would favor a higher degree of homozygosity in **each of the descendant cell populations used in our industry**. The representation of sequence data from a CHO cell line in connection with a corresponding karyotype map would be useful, preferably linking Giemsa banding data with gene loci. Identifying those chromosomes that match the typical diploid hamster chromosomes and linking them with sequence data would possibly provide a concept for the more stable parts of the CHO genome whose losses would make the cells nonfunctional. The paper of Borth and colleagues that provides sequences of sorted hamster chromosomes is aligned with this wish [55]. For example, chromosome 1, chromosome Z1, chromosome 2, and chromosome Z2 (see Figure 2.2.2) appear to be relatively stable structures, at least from the perspective of all the karyotype maps available so far. Active gene loci and their structural and functional organization within the heterochromatin on these chromosomes are an important resource. Other smaller chromosomes or fragments of chromosomes, if recognizable, may fall into the same category. This, the author is hopeful, would represent a **CHO core genome** providing a highly useful and readily applicable starting point for more fine-tuned studies.

Eventually, we should hope for insights how a single cell founder genome shapes the resulting quasispecies genomes in clonal subpopulations, all expressing the same gene of interest. Practitioners in industry will assure that profound differences in the metabolic behavior of clones from a single transfection event for the production of one particular protein can be seen frequently. The founder genome, derived from the clonal event, is expected to shape the expression profiles, the physiology, the physical and chemical sensitivity of the resulting cells under production conditions. Possibly, a more profound set of data, for example from a 1,000 CHO genomes project, would further enhance our understanding in this context.

Conflict of Interest

The author is founder and member of the executive team of the company ExcellGene SA, a company that offers services to the industry in the context of recombinant protein production from animal cells in bioreactors.

References

[1] Xu, X., Nagarajan, H., Lewis, N. E., Pan, S., Cai, Z., Liu, X., Chen, W., Xie, M., Wang, W., Hammond, S., Andersen, M. R., Neff, N., Passarelli, B., Koh, W., Fan, H. C., Wang, J., Gui, Y., Lee, K. H, Betenbaugh, M. J., Quake, S. R., Famili, I., Palsson, B. O., Wang, J. The genomic sequence of the Chinese hamster ovary (CHO)-K1 cell line. *Nature Biotechnology* **2011,** 29, 735-41.

[2] Eigen, M., Schuster, P. The Hypercycle. A Principle of Natural Self-Organization. Part A: Emergence of the Hypercycle. *Naturwissenschaften* **1977** 64, 541-65.

[3] Eigen, M., Schuster, P. The Hypercycle. A Principle of Natural Self-Organisation. Part B: The Abstract Hypercycle. *Naturwissenschaften* **1978** 65, 7-41.

[4] Eigen, M., Schuster, P. The Hypercycle. A Principle of Natural Self-Organisation. Part C: The Realistic Hypercycle. *Naturwissenschaften* **1978** 65, 341-69.

[5] Domingo, E. Quasispecies theory in virology. *Journal of Virology* **2002** 76, 1, 463-65.

[6] Napolitani, D., Signore, M., Struppa, D. C. Stem-like adaptive aneuploidy and cancer quasispecies. *arXiv* **2013** arXIV:1303.674v1, 1-16.

[7] Gottesman, M. M., ed. *Molecular Cell Genetics*. **1985** John Wiley and Sons, New York, USA.

[8] Barnes, D., Sato, G. Methods for growth of cultured cells in serum-free medium. *Analytical Biochemistry* **1980** 102, 255-70.

[9] Gasser, F., Mulsant, P., Gillios, M. Long-term multiplication of the Chinese Hamster Ovary (CHO) cell line in a serum-free medium. *In vitro Cellular and Developmental Biology* **1985** 21, 10, 589-92.

[10] Ham, R. G. Clonal growth of mammalian cells in a chemically defined, synthetic medium. *Proceedings of the National Academy of Sciences of the USA* **1965** 53, 288-93.

[11] Hamilton, W. G., Ham, R. G. Clonal growth of Chinese Hamster cell lines in protein free media. *In vitro* **1977** 13, 537-47.

[12] Taub, M., Chuman L., Saier, M. H., Sato, G. Growth of Madin-Darby canine kidney epithelial cell (MDCK) line in hormone-supplemented, serum free medium. *Proceedings of the National Academy of Sciences of the USA* **1979** 76, 7, 3338-42.

[13] Puck, T. T. The genetics of somatic mammalian cells. Adv. Biology. *Medical Physics* **1957** 5, 75-101.

[14] Puck, T. T. Development of the Chinese Hamster Ovary (CHO) Cell for Use in Somatic Cell Genetics. In: Gottesman M. M. *Molecular Cell Genetics* **1985** John Wiley and Sons, New York, USA, 37-64.

[15] Hayflick, L., Moorhead P. S. The serial cultivation of human diploid cell strains. *Experimental Cell Research* **1961** 25, 3, 585-621.

[16] Ham, R. G. An improved nutrient solution for diploid Chinese hamster and human cell lines. *Experimental Cell Research* **1963**, 29, 515-26.

[17] Urlaub, G., Chasin L. A. Isolation of Chinese Hamster Cell Mutants Lacking Dihydrofolate Reductase Activity. *Proceedings of the National Academy of Sciences of the USA* **1980** 77, 4216-20.

[18] Subramani, S., Mulligan, R., Berg P. Expression of the mouse dihydrofolate reductase complementary deoxyribonucleic acid in simian virus 40 vectors. *Molecular and Cellular Biology* **1981** 1, 854-64.

[19] Scahill, S. J., Devos, R., Van der Heyden, J., Fires W. Expression and characterization of the product of a human immune interferon cDNA gene in Chinese Hamster Ovary cells. *Proceedings of the National Academy of Sciences of the USA* **1983** 80, 4654-58.

[20] Flintoff, W. F., Davidson, S. V., Siminovitch, L. Isolation and partial characterization of three methotrexate-resistant phenotypes form Chinese Hamster Ovary cells. *Somatic Cell Genetics* **1976** 2, 3, 245-61.

[21] Siminovitch L. On the Nature of Hereditable Variation in Cultured Somatic Cells. *Cell* **1976** 7, 1-11.
[22] Urlaub, G., Käs, E., Carothers, A. M., Chasin, L A. Deletion of the diploid dihydrofolate reductase locus form cultured mammalian cells. *Cell* **1983**, 33, 405-12.
[23] Urlaub, G., Mitchell, P. J., Kas, E., Chasin, L. A., Fumanage, V. L., Myoda, T. T., Hamlin, J. Effect of gamma rays at the dihydrofolate reductase locus: Deletions and inversions. *Somatic Cell and Molecular Genetics* **1986** 12, 6, 555-66.
[24] Kaufman, R. J., Schimke, R. T. Amplification and loss of dhfr in a CHO cell line. *Molecular and Cellular Biology* **1981** 12, 12, 1069-76.
[25] Kaufman, R. J., Wasley, L. C., Spiliotes, A. J., Gossels S. D., Latt, S. A., Larsen, G. R., Kay, R. M. Coamplification and co-expression of human tissue-type plasminogen activator and murine dihydrofolate reductase sequences in Chinese Hamster Ovary cells. *Molecular and Cellular Biology* **1985** 5, 1750-59.
[26] Weidle, U. H., Buckel, P., Wienberg, J. Amplified expression constructs for human tissue type plasminogen activator in CHO cells: Instability in the absence of selective pressure. *Gene* **1988** 66, 193-203.
[27] Biedler, J. L, Spengler, B. A. A novel chromosome abnormality in human neuroblastoma and antifolate resistant Chinese hamster cell lines in culture. *Journal of the National Cancer Institute* **1976** 57, 683-89.
[28] Pallavicini, M. G., DeTeresa, P. S., Rosette, C., Gray, J. W., Wurm, F. M. Effects of Methotrexate (MTX) on Transfected DNA Stability in Mammalian Cells. *Molecular and Cellular Biology* **1990** 10, 1, 401-04.
[29] Wurm, F. M. Integration, amplification and stability of plasmid sequences in CHO cell cultures. *Biologicals* **1990** 18, 3, 159-64.
[30] Mondelo, C., Giorgi, R. Nuzzo, F. Chromosomal effects of methotrexate on cultured human lymphocytes. *Mutation Research* **1984** 139, 67-70.
[31] Ma, C., Looney, J. E., Leu, T.-H., Hamlin, J. L. Organisation and genesis of dihydrofolate reductase amplicons in the genome of a methotrexate-resistant Chinese Hamster Ovary cell line. *Molecular and Cellular Biology* **1988,** 8, 6, 2316-27.
[32] Nunberg, J. H., Kaufman, R. J., Schimke, R. T., Urlaub, G., Chasin, L. A. Amplified dihydrofolate reductase genes are localized to a homogenously staining region of a single chromo- some in a methotrexate-resistant Chinese hamster ovary cell line. *Proceedings of the National Academy of Sciences of the USA* **1978** 75, 5553-56.
[33] Schimke, R. T., Kaufman, R. J., Alt, F. W., Kellems, R. F. Gene amplification and drug resistance in cultured murine cells. *Science* **1978** 202, 1051-1055.
[34] Schimke, R. T. Gene amplification in cultured animal cells. *Cell* **1984** 37, 705-13.
[35] McClintock, B. The stability of broken ends of chromosomes in Zea Mays *Genetics* **1941** 26, 234-82.
[36] Bebbington, C. R., Renner, G., Thomson, S., King, D., Abrams, D., Yarranton, G. T. High-level expression of a recombinant antibody from myeloma cells using a glutamine syntheses gene as an amplifiable selectable marker. *Biotechnology* **1992** 10, 2, 169-75.
[37] Puck, T. T., Sanders, P., Petersen, D. Life cycle analysis of mammalian cells. II Cells from the Chinese Hamster Ovary grown in suspension culture. *Biophysical Journal* **1964** 4, 441-50.
[38] Thompson, L. H., Baker, R. M. Isolation of mutants of cultured mammalian cells. *Methods in Cell Biology* **1973** 6, 209-81.
[39] Hsu, T. S. Chromosomal Evolution in Cell Populations *International Review of Cytology* **1961** 12, 69-161.
[40] Deaven, L. L., Peterson, D. F. The Chromosomes of CHO, an Aneuploid Chinese Hamster Cell Line: G-Band, C-Band and Autoradiographic Analyses. *Chromosoma* **1973** 41, 129-44.

[41] Cao, Y., Kimura, S., Itoi, T., Honda, K., Ohtake, H., Omasa, T. Construction of BAC-based physical map and analysis of chromosome rearrangement in Chinese Hamster Ovary cell lines. *Biotechnology and Bioengineering* **2011,** 112, 5, 476-84.

[42] Hazelwood, E. Molecular analysis of clonal variation in GS-CHO cell lines. **2006** University of Manchester, UK. Theses Th31711.

[43] Derouazi, M., Martinet, D., Besuchet, N., Schmutz, N., Flaction, R., Wicht, M., Bertschinger, M., Hacker, D. L, Beckmann, J. S. and Wurm, F.M. Genetic characterization of CHO production host DG44 and derivative recombinant cell lines. *Biochemical and Biophysical Research Communications* **2006** 340, 4, 1069-77.

[44] Landry, J., Pyl, P. T., Rausch, T., Zichner, T., Tekkedil, M. M., Stütz, A. M., Jauch, A., Aiyar, R. S., Pau, G., Delhomme, N., Gagneur, J., Korbel, J. O., Huber, Steinmetz, L. M. The genomic and transcriptomic landscape of a HeLa cell line. *G3: Genes, Genomes, Genetics* **2013** 3, 8, 1213-24.

[45] Adey, A., Burton, J. N., Kitzman, J. O., Hiatt, J. B., Lewis, A. P., Martin B. K., Qiu, R., Lee, C., Shendure, J. The haplotype-resolved genome and epigenome of the aneuploid HeLa cancer cell line *Nature* **2013**, 500, 207-211.

[46] Gey, G. O., Coffman, W. D., Kubicek, M. T. Tissue culture studies of the proliferative capacity of cervical carcinoma and normal epithelium. *Cancer Research* **1952** 12, 264-65.

[47] Meyerson, M., Pellman, D. Cancer genomes evolve by pulverizing single chromosomes. *Cell* **2012** 144, 9-10.

[48] Stephens, P. J., Greenman, C. D., Fu, B., Yang, F., Bignell, G. R., Mudie, L. J., Pleasance, E. D., Lau, K. W., Beare, D., Stebbings, L. A., McLaren, S., Lin, M.-L., McBride, D. J., Varela, I. Nik-Zainal, S., Leroy, C., Jia, M., Menzies, A., Butler, A. P., Teague, J. W., Quail, M. A., Burton, J., Swerdlow H., Carter, N. P., Morsberger, L. A., Iacobuzio-Donahue, C., Follows, G. A., Green, A. R., Flanagan, A. M., Stratton, M. R., Futreal, P. A., Campbell, P. J. Massive Genomic Rearrangement Acquired in a Single Catastrophic Event during Cancer Development. *Cell* **2012** 144, 27–40.

[49] Liu, P., Erez, A. Sreenath Nagamani, S. C., Dhar, S. U., Kolodziejska, K. E., Dharmandhikari, A. V., Lance Cooper, M., Wiszniewska, J., Zhang, F., Withers, M. A., Bacino, C. A., Campos-Acevedo, L. D., Delgado, M. R., Freedenberg, D., Garnica, A., Grebe, T. A., Hernandez- Almaguer, D., LaDonna I., Lalani, S. R., McLean, S. D., Northrup, H., Scaglia, F., Strathearn, L., Trapane, P., Kang, S.-H. L., Patel, A., Cheung, S. W., Hastings, P. J., Stankiewicz, P., Lupiski, J. R., Bi, W. Chromosome catastrophes involve replication mechanisms generating complex genomic rearrangements. *Cell* **2013** 146, 889-903.

[50] Davis, S. L., Lovelady, C. S., Grainger, R. K., Racher, A. J., Young, R. J., James, D. C. Functional Heterogeneity and heritability in CHO cell populations. *Biotechnology and Bioengineering* **2012** 110, 260-74.

[51] Zhang, X., Bürki, C.-A., Stettler, M., De Sanctis, D., Perrone, M., Discacciati, M., Parolini, N., De Jesus, M., Hacker, D. L., Quarteroni, A., Wurm, F. M. Efficient oxygen transfer by surface aeration in shaken cylindrical containers for mammalian cell cultivation at volumetric scales up to 1000 L. *Biochemical Engineering* **2009** 45, 41- 47.

[52] Tissot, S., Oberbek, A., Reclari, M. Dreyer, M., Hacker, D. L., Baldi, L., Farhat, M., Wurm, F. M. Efficient and reproducible mammalian cell bioprocesses without probes and controllers? *New Biotechnology* **2011** 29, 3, 387-94.

[53] Lewis, N. E., Liu, X., Li Y., Nagarajan, H., Yerganian, G., O'Brien, G., Bordbar, A., Roth, A. M., Rosenbloom, J., Bian, C., Xie, M., Chen, W., Li, N., Baycin-Hizal, D., Latif, H., Forster, J., Betenbaugh, M. J., Famili, I., Xu, X., Wang, J., Palsson, B. O. Genomic landscapes of Chinese hamster ovary cell lines as revealed by the *Cricetulus griseus* draft genome. *Nature Biotechnology* **2013** 8, 759-65.

[54] Brinkrolf, K., Rupp, O., Laux, H., Kollin, F., Ernst, W., Linke, B., Kofler, R., Romand, S., Hesse, F., Budach, W. E., Galosy, S., Müller, D., Noll, Th., Wienberg, J., Jostock, Th., Leonard, M., Grillari, J., Tauch, A., Goesmann, A., Helk, B., Mott, J. E., Pühler, A., Borth, N. Chinese hamster genome sequenced from sorted chromosomes. *Nature Biotechnology* **2013** 31, 694-95.
[55] Wurm, F. M., Hacker D. L. First CHO genome. *Nature Biotechnology* **2011,** 29, 8, 718-19.

2.3 Cell Lines for Vaccine Production

Ingo Jordan and Volker Sandig

2.3.1 Introduction

„*The cell substrate used in the production of a live vaccine is a critical parameter in the level of attenuation or virulence of live vaccine viruses. If the cells used for production were to be changed, there could be entirely unknown effects on the safety and efficacy of these vaccines.*" WHO, 1998 [148]

The ability to cultivate viruses in cell cultures was decisive for the development of many important vaccines [1, 2]. Initially, primary cells were the only available substrate but today production of vaccines is performed both in finite cell cultures or continuous cell lines (CCLs). The finite cultures can be primary cell cultures (PCCs) that are directly obtained from explanted tissue without any subpassage [3] and usually consist of a mixture of different cell types. If the primary cells have undergone more than 10 or 15 cell doublings they are considered secondary and tertiary cell cultures, respectively. Cells in primary (as well as secondary or tertiary) cultures usually do not proliferate beyond 50 cell doublings. Because a consistent supply with well-characterized finite cultures can be difficult to maintain, continuous cell lines are also being investigated for replication of vaccine strains.

There are several advantages associated with continuous cell lines (compare Chapter 2.1). In addition to an essentially unlimited life-span, bioreactor processes with such cells usually can be better controlled. Continuous cell lines are also amenable to genetic engineering that is required for generation of packaging cell lines that allow replication of some modern vectored vaccines. However, health regulatory properties are not yet fully defined for such substrates. For this reason, adoption of continuous cell lines for vaccine production has been hesitant.

This chapter will focus on historical aspects of cell culture development to highlight the options for vaccine production that are available today.

2.3.2 Derivation and Life-span Defines the Different Cell Substrates

2.3.2.1 Primary Cell Cultures Were Important in the Fight against Poliomyelitis

Embryonated chicken eggs [4-6] and cultures derived from monkey kidneys were historically the most important primary cell substrates for attenuation of viruses and production of vaccines. Especially the development of a vaccine against poliomyelitis depended heavily on the primary monkey kidney cells. Infection with poliovirus was an enormous global burden, associated with frequent epidemics, high morbidity for those that developed clinical symptoms, and often with dramatic neurologic sequelae. Treatment until today is only supportive, a productive infection cannot be cured so that prevention is key to controlling poliomyelitis. To produce a vaccine against poliomyelitis the virus strains had to be produced in appropriate host cells.

"In 1957, at the end of what may be called the first era of the history of live virus vaccine", fresh explants of monkey kidney were more efficient at replication and production of attenuated poliovirus than repeatedly passaged monkey cell lines [7]. Two live vaccines based on attenuated strains were produced on primary monkey kidney cultures and given to human recipients in the 1950s, the preparation developed by Sabin [8] and the preparation developed by Koprowksi and Cox [9, 10].

However, development and investigation of Salk's [11] killed polio vaccine (that used virulent strains of poliovirus) in North America was slightly ahead of the parallel research on the live vaccines. Production at a commercial scale suitable for the ambitious 1955 polio vaccine field trial [12] was performed in minced monkey kidney cells suspended in chemically defined Medium 199 supplemented with calf serum [13]. The field trial was successful and Salk's polio vaccine licensed in the same year. However, "[the] fact that 1,500 monkeys are needed in the production of every million doses of killed-virus vaccine has created great difficulties in the supply of monkeys" [14]. Because live vaccines usually are more efficient at inducing a protective immune response than their killed counterparts, fewer units are required for vaccination. Driven by the increasing production substrate limitation, its greater efficacy, perhaps also partly motivated by the pursuit for regional independence during the cold war and polarization in the polio vaccine research field, Sabin's live vaccine initially was predominantly used to control polio worldwide [15]. Salk's killed vaccine remained the preferred preparation in North America until 1962 [16]. Today, the prevalence of wild poliovirus is considered to be lower as compared with vaccine-derived polioviruses so that killed vaccines are considered to provide a better solution in the fight against the remaining endemic pockets. Primary monkey kidney cells are not an option anymore for any vaccine and other cell substrates are required.

2.3.2.2 Primary Cell Cultures for the Production of Modern Vaccines

A primary substrate that still is in use for currently licensed vaccines are suckling mouse and suckling hamster brains. Some vaccines against Japanese encephalitis and against hemorrhagic fever with renal syndrome (HFRS, currently licensed only in the Republic of Korea and the People's Republic of China) are being produced on this complex substrate. Until recently, some rabies vaccines were also produced in rodent brains.

Today by far, the most important source for primary cells are embryonated chicken eggs. Vaccines currently produced on this substrate protect against pediatric diseases such as measles and mumps and include most of the seasonal doses against influenza. The embryos are extracted from the eggs and mechanically or enzymatically homogenized. This cell suspension, the chicken embryo fibroblasts (CEFs), are transferred to bioreactors for infection with vaccine strains. Alternatively, two small holes are drilled into the shell of embryonated eggs, one for inoculation of virus into the allantoic cavity and the second to allow air to escape so the seed volume is not forced out again. Progeny virus is released into the allantoic fluid which is then harvested.

2.3.2.3 Disadvantages of Primary Cells for Vaccine Production

One disadvantage associated with primary cells as substrate is the requirement that a complex material has to be introduced continuously into the production process. Only eggs of a well-defined regulatory status referred to as specific pathogen free (SPF) are suitable for vaccine production. This status is described in Chapter 5.2.2 of the European Pharmacopoeia. SPF flocks are housed in a controlled facility separated from non-SPF animals and protected against contact with wild birds, rodents, or insects. Food and air is treated to avoid contamination. Personnel should shower and change into protective clothes before entering the controlled facility. Records of the general health of the flock are maintained and no medication that may interfere with the sensitivity of detection of pathogens should be administered. Currently, SPF chicken flocks are screened for 18 pathogens including avian adenoviruses, Marek's disease virus, Newcastle disease virus, and mycoplasma. The number of farms that are equipped to produce such eggs is limited and consequentially the supply with this complex material can also occasionally become limited [17].

Alternative cell substrates for research on and production of vaccine strains have been investigated well before the value of rhesus and cynomolgus macaques was recognized as a limitation. For example, a comprehensive study on the interaction of herpes simplex and pseudorabies viruses with the first continuous "pure cell strain" (the clonal L929 cell line derived from a mouse fibrosarcoma) was published in 1953 [18]. In the same year, replication of poliovirus on the first human continuous cell line (HeLa) was described in great detail [19]. It was already known at the time that the neurotropic poliovirus can replicate in tissue cultures consisting of extraneural cells but this work established that "the cancerous cells of strain HeLa have much in common with the cells previously employed" [19]. Hence, reliable and consistent production of vaccine viruses such as the 3 strains required in a polio vaccine in theory is also possible with an immortal cell line. However, although it was considered highly unlikely that a chemically inactivated vaccine produced on HeLa cells may be carcinogenic, such a property had to be considered [20]. Because it was not possible at the time to "secure in tissue culture a pure line strain of normal mammalian cells [...] proven to be free of any extraneous viral or bacterial contaminants and known to posses no carcinogenic activity" chick embryos were presciently proposed as the most versatile alternative [20].

2.3.2.4 Diploid Finite Cell Lines as Bridge between Primary and Continuous Cell Cultures

The potentially ideal normal cell line was made available in the 1960s with the human diploid cell lines (DCLs) WI-38 [21] and MRC-5 [22]. WI-38 was derived from a female human fetal lung approximately 13 weeks into pregnancy and MRC-5 from a male human fetal lung approximately 14 weeks into pregnancy. Telomere erosion and other molecular mechanism leading to senescence were unknown at the time. It was therefore surprising that "normal" cells turned out to have a very limited life span, usually between 10 and 50 population doublings [23, 24]. This limitation was partly compensated as the refinement of cell culture technology allowed multiple expansion passages of characterized material obtained from cryogenically preserved seeds. Today, the cryogenically preserved characterized materials are known as master cell banks and working cell banks with formalized release testing requirements [149]. Because the properties of primary cultures and cell cultures of cells that have completed 10 or 15 cell doublings (secondary and tertiary cultures) differ, precise passaging records should be maintained [150].

One of the first vaccines that were produced on the WI-38 cell line were vaccines against adenovirus infection. Today, specially designed adenoviruses are well known as transgene carriers for gene therapy and vectored vaccines. But they are also important pathogens that cause acute upper respiratory diseases. One segment of the population that is affected by adenovirus epidemics are newly recruited military personnel. Close crowding in a constantly changing and immunologically heterogenous population together with the physical and psychological stresses appear to promote spread of the virus. Among the more than 50 serotypes of human adenoviruses that have been described, especially serotype 4 and serotype 7 have the potential to severly disrupt military medical services and training commands [25, 26].

The first vaccine that was developed against the febrile respiratory illness was a killed vaccine that consisted of the two most relevant serotypes produced on primary monkey kidney cells [25]. In 1961, contamination with the oncogenic simian virus SV-40 was discovered in many vaccine preparations derived from the primary monkey cultures, including the vaccines against acute respiratory disease and poliomyelitis. Because a vaccine against adenoviruses was effective and significantly reduced morbidity in the military, alternative vaccines were highly desirable. Interestingly, the serotype 4 and 7 adenoviruses cause clinical disease only if they infect the respiratory tract. An infection of the gastrointestinal tract can induce an effective immune response yet usually remains clinically inapparent. Based on this observation research on a live vaccine had initiated already prior to the discovery of the contamination with SV-40 [27, 28]. During the 1960s, a live, nonattenuated vaccine was developed that in its final formulation was produced by Wyeth-Ayerst (Marietta, PA, USA) on WI-38 cells and consisted of the two serotypes in enteric-coated capsules. In 1971, 10 years after the killed adenovirus vaccine was withdrawn, the live vaccine was routinely administered to new male recruits [29, 30].

Difficulties in the production of the new adenovirus vaccine first emerged in 1984, and in 1994 it was decided not to produce this vaccine anymore [29, 31]. The last batch was produced in 1995 and the last vaccine from the stocks was administered in 1999.

The acute febrile adenovirus infections resurged and a new manufacturer was contracted to produce the vaccine. This manufacture appeared to have difficulties with the WI-38 cell line (archived narrative of the Armed Forces Epidemiological Board, Spring 2005 Meeting, page 65) and it was considered to switch production to the younger MRC-5 cell line. Eventually, any problems with the cell line and technical challenges surrounding the enteric coating were resolved and the vaccine (now distributed by Teva Pharmaceuticals) has been approved by the U.S. Food and Drug Administration (FDA) in 2011. The WI-38 as substrate has been maintained and, with the exception of removing antibiotics in the culture media, there have been very few changes to the initial production protocols.

However, supply of the initial human DCLs became a concern not only for the adenovirus vaccine [32, 33]. In 2006, the WHO initiated the generation of a MRC-5 seed bank with cells at a population doubling level (PDL) of 7. The finished bank consists of 450 vials with PDL-12 cells and is expected to provide production substrate for at least 20 years [151].

2.3.2.5 Continuous Cell Lines and Aneuploidy

A molecular counter that leads into a predetermined inhibition of cell proliferation may introduce uncertainty into the sometimes very long development pathways of investigational vaccines. Continuous cell lines are not affected by such reductions in life span but they are usually also not diploid anymore. A functionally unlimited repertoire of population doublings is an enormous biotechnological advantage. However, changes in karyotype are associated with tumors and for this reason cause for regulatory concerns. The discussion of risks that may be associated with aneuploidy is facilitated by the increasing knowledge on the molecular mechanism behind immortalization.

Mechanisms that interfere with the life span of primary cells are senescence and apoptosis. The senescent phenotype is characterized by an essentially irreversible cell cycle arrest, accumulation of the tumor suppressor cyclin-dependent kinase inhibitor 2A (often referred to as $p16^{Ink4A}$), and a persistent global response to DNA damage triggered by the shortening of the telomers [34]. Apoptosis can be induced if cell cycle checkpoints are consistently ignored in the presence of DNA damage and complements senescence in its function of maintaining tissue homeostasis and prevention of cancer [35]. Continuous cell lines have progressed beyond both cellular senescence and apoptosis. Whereas the karyotype of primary cells is diploid and, within natural variations, free of chromosomal translocations or macrodeletions, the price that is usually paid for "immortality" are epigenetic and genomic alternations. Some of the molecular mechanisms behind these changes have been investigated. For example, a central node in the pathways that links mutational stressors to the cell cycle is occupied by the p53 protein [35]. If activated, this tumor suppressor arrests the cell cycle so that DNA damage can be repaired, or induces apoptosis if the damage is irreversible. One of the stressors that induces p53 is constitutive activation of the cell cycle. Because undisturbed cell cycle progression is a prerequisite of immortalization and transformation towards a neoplastic phenotype, the p53 "guardian of the genome" is blocked in many tumors. However, if the sensor for genetic instability is inactivated, the risk of aneuploidy increases with every cell division. For this reason chromosomal aberrations can be observed as a consequence of transformation [36]. Conversely, aneuploidy can also lead into transformation. The mechanism behind this chain of events is connected to telomere erosion, a progressive shortening of the protective caps of the chromosomes with each division of normal cells, and the explanation for replicative senescence observed in 1961 [23, 37, 38]. The risk for chromosomal rearrangements increases if the telomeres are critically short. Although cell viability suffers with genetic instability, some rare escape mutants may survive and progress into a cancerous phenotype [36, 39]. All combined, induced by telomere ablation during the process of replicative senescence, aneuploidy can precede transformation.

2.3.2.6 Spontaneous Immortalization

Immortal cell lines derived by any of the two mechanism just described share aneuploidy. They appear to be similar but one can argue that there are very important differences in the established (final) cell lines. Although cancer often evolves in a multistep process that may be accompanied by a great number of mutations dispersed throughout the genome, maintenance of the transformed phenotype can be mediated by a confined set of few mutated genes. This phenomenon is termed oncogene addiction [40]. For explanted tissues in cell cultures similar effects may apply. If culture conditions are thus that, for example, a feeder layer allows primary cells to survive through replicative senescence [37] then a rare correct combination of dominant mutations in oncogenes or negative mutations in tumor suppressors may lead to spontaneous immortalization. Examples for spontaneously immortalized cell lines are the Vero, MDCK, BHK21, and CHO cell lines.

The Vero cell line was derived from the kidney of a African green monkey (*Cercopithecus aethiops*, today *Chlorocebus aethiops*) by Yasumura and Kawakita in 1962. A wide spectrum of viral vaccine strains that ranges from vaccinia virus against smallpox to poliovirus, rotavirus, Japanese encephalitis virus, and rabies virus is being produced on the Vero cell line (Table 2.3.1). Documentation of this cell line is extensive and the literature is enormous. Vero cells were transferred from Chiba University in Japan to the laboratory of Tropical Virology, National Institute of Allergy and Infectious Diseases, USA at passage 93 in 1964, and submitted to the American Tissue Culture Collection (ATCC) at passage level 113. There, it was propagated to passage level 124 for generation of the ATCC CCL-81 cell bank. The Mérieux Institute (today

Table 2.3.1: Some currently licensed vaccines sorted according to production substrate.

Substrate	Manufacturer	Brand	Indication	Active Ingredient
Resp. bacteria	Glaxo Smith Kline	PEDIARIX/ INFANRIX	Diphteria, tetanus, and pertussis	Purified bacterial antigens
S. cerversiae	Glaxo Smith Kline	PEDIARIX/ ENGERIX-B	Hepatitis B	Purified HBsAg
	Glaxo Smith Kline	ENGERIX-B	Hepatitis B	Purified HBsAg
	Merck	Recombivax HB	Hepatitis B	Purified HBsAg
	Merck	Gardasil	HPV-induced cancers	Adjuvanted HPV VLPs
Trichoplusia ni	Glaxo Smith Kline	CERVARIX	HPV-induced cancers	Adjuvanted HPV VLPs
Mouse brain	Korea Green Cross	Hantavax	Haemorrhagic fever with renal syndrome	Killed HTNV
	Sanofi Pasteur/ BIKEN	JE-VAX	Japanese encephalitis	Partially purified killed JEV strain Nakayama-NIH
Primary hamster kidney	Chengdu Vaccine Plant	SA 14-14-2 JE	Japanese encephalitis	Live attenuated JEV strain SA 14-14-2
Embryonated chicken eggs	Abott	INFLUVAC TC	Influenza	HA antigens
	MedImmune	fluMist	Influenza	Live influenza A and B viruses
	Sanofi Pasteur	YF-VAX	Yellow fever	Live attenuated YFV strain 17D-204
CEF	Merck	M-M-R II	Measles, mumps, and rubella	Live attenuated measles virus strain Edmonston
	Merck	M-M-R II	Measles, mumps, and rubella	Live attenuated mumps virus strain Jeryl Lynn
	Novartis	RabAvert	Rabies	Killed RABV strain Flury LEP
FRhL-2	Wyeth	RotaShield (withdrawn)	Rotavirus gastroenteritis	Live rhesus or rhesus-human reassortant rotaviruses serotypes 1 to 4

MRC-5	Glaxo Smith Kline	HAVRIX	Hepatitis A	Killed HAV
	Merck	VAQTA	Hepatitis A	Killed HAV
	Merck	VARIVAX	Varicella	Live attenuated VZV strain Oka/Merck
	Merck	ZOSTAVAX	Varicella	Live attenuated VZV strain Oka/Merck
	Sanofi Pasteur	IMOVAX RABIES	Rabies	Killed RABV strain PM/WI38 1503-3M
WI-38	Merck	M-M-R II	Measles, mumps, and rubella	Live attenuated rubella virus strain Wistar RA 27/e
	Teva Pharmaceuticals	Adenovirus Type 4 and Type 7 Vaccine, Live, Oral	Acute respiratory disease	Live, not-attenuated HAdV-4 and HAdV-7
Vero	Glaxo Smith Kline	PEDIARIX	Poliomyelitis	Killed Polio virus strains Mahoney, MEF-1, and Saukett
	Glaxo Smith Kline	Rotarix	Rotavirus gastroenteritis	Live attenuated human rotavirus strain G1P[8]
	Intercell Biomedical	IXIARO	Japanese encephalitis	Partially purified killed JEV strain SA14-14-2
	Merck, USA	RotaTeq	Rotavirus gastroenteritis	5 live reassortant rotaviruses
	Sanofi Pasteur	ACAM2000	Smallpox disease	Vaccinia virus strain Dryvax
	Sanofi Pasteur	IPOL	Poliomyelitis	Killed Polio virus strains Mahoney, MEF-1, and Saukett
	Sanofi Pasteur	VERORAB	Rabies	Killed RABV strain PM/WI38 1503-3M
MDCK	Novartis	OPTAFLU	Influenza	Killed influenza A and B viruses

Sanofi Pasteur) obtained one ampoule of the CCL-81 and prepared a primary cell bank (passage 129) and 7 working cell banks. One of the working cell banks (called LS-7) is at passage 137 and was tested according to the WHO Technical Report Series number 745 that was published in 1987 [41-43]. Vero cells at the low passage level used for vaccine production are not tumorigenic but evolve the capacity to form progressing nodules and metastases in immunodepressed newborn rats with increasing passage levels [41]. The 10-87 Vero Reference Cell Bank (RCB) of the WHO is at passage level 134 [44] and related to the cell bank of the Mérieux Institute. Ampoules from the 10-87 RCB are distributed for generation of master cell banks and working cell banks for the production of vaccines. However, the 10-87 RCB is nearing depletion and may be replaced with a bank derived from the CCL-81 at the ATCC to obtain a cell bank with lowest possible passage level [45].

The MDCK cell line was established from primary kidney cells of an apparently healthy adult cocker spaniel by Madin and Darby in 1958. There appears to be no published record for derivation and for the first 49 passages of this cell line [46-48]. There also appears to be some confusion as to the sex of the donor animal that some publications list as female and others as male, and as to the actual strains that are being used in studies on epithelial junctions and cell polarity for which MDCK is especially suitable [47]. It was discovered quite early that MDCK cells are highly permissive for different strains of influenza A viruses [49]. Because they are also suitable for bioreactor processes, several companies have applied for regulatory approval to produce influenza A vaccines in MDCK. The first such approval was granted for production of a trivalent subunit vaccine in 2007 [50, 51].

BHK21 cells were derived from primary Syrian hamster kidney cell cultures prepared in March 1961 [52, 53]. The cells that lead to BHK21 were from 1-day-old animals of litter number 21. Transformation properties of polyomaviruses were investigated with cultures from another litter. Resistance to polyomavirus transformation was observed in the unrelated explants, and to test for possible secreted factors that mediate such a resistance, medium was transferred from the treated to the litter 21 explants at day 41 of culture. On day 65, cells in the litter 21 culture replicated rapidly. However, spontaneous rather than induced immortalization is the most likely explanation for this phenomenon as BHK21 cells had never "been knowingly exposed to polyoma virus during their propagation, and derivatives have since been shown to lack the polyoma specific cell antigen" [53]. Although BHK cells are highly permissive for a wide spectrum of viruses and are frequently used in research on investigational vaccines (for example, [54, 55] no licensed human vaccine is being produced on this cell line.

Contrasting with BHK, the Chinese hamster ovary (CHO) cell line, derived from an inbred (captive) female Chinese hamster [56] is surprisingly refractory against productive infection with various viruses [57-59] and not recognized as substrate for replication of vaccine viruses. However, this cell line is central to the developments in genetic engineering [60] that may have also facilitated regulatory changes in vaccine production. Continuous cell lines were introduced into this field in preparation for clinical trials for tissue plasminogen activator in 1984. One reason is that some recombinant proteins are not processed correctly in *Escherichia coli* and other prokaryotic cells. For such biologicals eukaryotic cell lines are required that can be transfected for stable integration and maintenance of foreign genes. The most versatile cell line for these purposes appears to be the CHO cell line, a continuous cell line with capacity to produce tumors in nude mice and that furthermore releases particles from endogenous proviruses. Improved procedures for tracking and documentation of raw materials, and new methods for purification and filtration to remove mycoplasma and other contaminants, eased concerns about the use of these therapeutics in humans. The first recombinant biological derived from a continuous eukaryotic cell line was licensed in 1986 [60].

2.3.2.7 Designed Immortalization and Designer Cell Lines

Because both, natural tumors and continuous cell lines, are often aneuploid, one paramount concern is that a transforming factor may be carried together with the vaccine into the recipient. If the underlying reason for aneuploidy or immortalization would be known, the potency of transforming factors that may copurify with a vaccine could be better assessed. With increasing knowledge about the involved biochemical pathways, experimentally induced cell immortalization has become accessible to more precise manipulation. One hypothesis of such an approach is that if constitutive activation of the cell cycle initiates prior to the crisis induced by telomere erosion, genomic instability may still induce a number of unrelated mutations but selective pressures towards the neoplastic phenotype are relaxed. The initial immortalizing event may thus have already cleared (and thereby possibly determined) one path for further evolution of the cell line so that activation of multiple and uncharacterized oncogenes may not occur anymore (see also Chapter 2.1).

To differentiate cell lines that are derived from such procedures from cell lines that have been immortalized with chemical mutagens or radiation they are sometimes referred to as designer cell lines. The advantage of designer cell lines is that the oncogenic potential of host cell derived DNA, that may contaminate a vaccine, can be described with greater confidence. This property can be beneficial in defined risks considerations that are used to recommend a required degree of purifcation.

One approach to achieve designed immortalization is to express recombinant viral proteins in primary cells. The underlying biological principle is that turnover of biosynthetic compounds (ATP, NADPH, nucleotides, amino acids and lipids) increases in the acute phase of a virus infection. Because the metabolic activity of an actively dividing cell is greater than that of a resting cell some viruses have evolved factors that induce the cell cycle. If such factors are expressed in a primary cell it may become immortalized. Well-described viral factors are the large T antigen (LT) of SV40 [61], the E6 and E7 functional pair encoded by human papillomaviruses (HPV) [62], and the E1 genes of human adenoviruses [63].

A common biochemical motif of the viruses is to disrupt the association of the cellular pRb protein with E2F family of transcription factors. The targets of E2F proteins include cyclins, cyclin-dependent kinases (CDKs), and proteins important for

DNA replication and repair. Because pRb is a negative regulator, liberation of E2F proteins by pRb is an important step leading to S phase entry and cell cycle progression. Normally, pRb is predominantly controlled by Cdk4 and Cdk6. It is maintained in its inactive form by hyperphosphorylation through S, G2, and M phases and regains its tumor suppressive activity at the exit of M and entry into G1 with the help of the type 1 protein phosphatase (PP1).

Among the small DNA tumor viruses, HPV E7 proteins bind to pRb and induce proteosomal degradation of the tumor suppressor via unscheduled ubiquitination [64]. The adenoviral E1A and the SV40 LT proteins preferentially bind to hypophosphorylated pRB in a fashion that displaces E2F proteins [61, 63].

However, E2F members not only cause cell cycle progression but also induce expression of an alternate reading frame (ARF) of the INK4a/ARF gene locus. Increased levels of ARF lead to increased proteolysis of MDM2. Because the ubiquitin ligase MDM2 marks p53 for destruction, release of E2F proteins is one of several pathways that lead to stabilization of p53 levels.

Increased activity of p53 following the aberrant cell cycle induction is important because this protein is central to events in the nucleus and in the cytoplasm that lead to G1 cell cycle arrest or apoptosis. For this reason the viral factors are complemented by an additional activity, proteins that interact with p53 to block or even usurp the antioncogene's function. In HPV it is the E6 protein that induces proteolysis of p53 by directing a ubiquitin ligase to p53 and by sequestration of the CPB/p300 cofactor [64]. The E1B gene of human adenoviruses encodes two antiapoptotic proteins in overlapping reading frames, 55K and 19K. The 55K protein converts p53 into a suppressor via binding to the chromatin and p53 nucleoprotein complexes [65]. The multifunctional LT antigen of SV40 mimics DNA targets of p53 to associate very stably with p53. Formation of the p53/LT antigen complex prevents that canonical target genes, including MDM2, cannot be transactivated anymore. p53 within this complex is also stabilized because it is not accessible for ubiquitination and degradation. However, the complex does interact with pRB and p300 proteins to activate genes involved in cell cycle progression [66-68].

There are several differences between tumor viruses that are associated with a greater degree of inducing malignancies as opposed to lower-risk viruses. For example, the high-risk HPV E6 protein (in addition to functions shared with adenovirus E1B 55K protein) appears to also induce degradation of proteins implicated in the organization of signaling complexes responsive to cell contact, and may interact with c-myc and other factors to augment telomere maintenance [64]. Furthermore, compared to E7 of low-risk HPV types, the E7 protein of high-risk HPV-16 has a greater affinity for pRb, interferes both with the regulated activity of histone deacetylases and histone acetyl transferases, and can directly (and independently of pRb) bind to and modulate activity of E2F1 and E2F6 [64]. Conversely, the adenovirus E1A protein appears to bridge pRb and p300, causing increased acetylation of pRb and thereby strengthening the interaction of pRb with Mdm2. In this ternary complex, p53 is not tagged for degradation but also cannot function as a transactivator for transcription of cell cycle inhibitors anymore. This would strengthen induction of cell cycle progression by E1A. However, the cytoplasmic p53 still is available for induction of apoptosis via permeabilization of the mitochondrial outer membrane [69]. Such a chain of events may be one explanation for the observation of increased proapoptotic sensitivity of E1A-positive cells [63] and dependence on the complementary protective functions provided by the two coding sequences in E1B. While E1B 55K protein alleviates p53 accumulation, the E1B 19K protein exerts its activity as a homolog to cellular Bcl2 proteins. In this function 19K interferes with the axis that connects release of mitochondrial proteins in adenovirus infected cells to caspase activation and apoptosis [70].

To investigate mechanisms underlying tumor formation induced by infection, early experiments with human primary cells have been performed with live viruses. However, in such a system the induced transformed lesions occur together with cytopathic effects in the susceptible cells or persistence of recoverable infectious units [71-74]. The HeLa cell line was an exception in that normal body cells became immortalized in a tragic infection and that the causative agent, high-risk HPV 18, has suffered genomic deletions rendering the integrated virus noninfectious [75, 76]. It appears that immortalizing viral factors may have to be continuously expressed to prevent the cells from going into senescence or apoptosis [77]. If the viral factors could be expressed without associated infectious activity, immortalization could be performed with greater efficiency and in tissues that normally would be refractory to infection. With the development of efficient transfection methods [78-80] this hypothesis was successfully tested in rodent and human primary cells with the HEK 293 cell as one well-known result in the vaccine community [81]. The experiment was successful, although efficiencies were extremely low and resulted in only two cell colonies, because adenoviruses contain a double-stranded DNA genome that in principle is infectious and does not require viral factors for activation of gene expression [82]. The transfected adenoviral DNA was mechanically sheared to express viral proteins in the cell without concomitant rescue of infectious units.

With the advent of adenoviruses equipped with heterologous expression cassettes as vectors for gene therapy, the significance and requirements for the producer cell line increased as well. The vectors were designed such that viral factors, initially only the E1 region, were deleted from the viral genome and had to be transcomplemented by the host cell. Such vectors are very safe for application in humans because they cannot launch a productive infection anymore while transiently expressing the foreign genes. The HEK 293 cell line is a highly efficient production system for E1-deleted adenoviruses. However, replication competent adenoviruses (RCAs) frequently develop by recombination of vector and the large viral genomic fragment contained in the HEK 293 cell line [83, 84]. Because RCAs pose a significant risk to vector recipients alternative approaches were required. In the generation of the PER.C6 cell line [85] this hurdle has been essentially solved with expression constructs that place an immortalizing E1A and E1B combination under control of distinct heterologous promoters. This design helps to eliminate the overlap of nonessential sequences so that probability of productive recombination is significantly reduced.

Although tightly linked to the development of adenovirus-based applications, HEK 293 and PER.C6 are permissive for many viruses and important cell lines for research on a wide spectrum of indications [86-88]. Regulatory agencies are familiar with this approach of obtaining designed cell lines (for example, FDA briefing document 3750b1_01.pdf, obtainable at http://www.fda.gov/ohrms/dockets/ac/01/briefing/3750b1_01.pdf or by archive search at www.fda.gov). Additional cell lines, including one avian retrovirus-free substitute for CEF, have been generated with E1A and E1B to expand the available host cell spectrum and to complement the existing cell lines (CR.pIX and CAP) [89-92].

In summary, an increasingly wide spectrum of diverse and highly permissive cell lines is available for production of viral vaccines. The diversity is complemented by the enormous biotechnological improvements that allow cultivation and sterile harvesting of large volumes of infected cell suspension in chemically defined media. However, one concern has persisted throughout the development of all variations of primary to continuous cells, from the early phases of cell-based vaccine production in the 1940s and 1950s until today. This concern, the risk that the initial host and derived substrates may add potentially dangerous substances that may be transferred into recipients together with the vaccine preparation, will be reviewed in the next section.

2.3.3 Safety and Acceptance of Cell Substrates

The main concern associated with vaccines is the risk of transferring substances or adventitious agents that may induce infectious disease or cancer in the recipient. In this context, it is important to clearly distinguish between tumorigenicity and oncogenicity, although there may be some overlap between the effects of oncogenic capacity and contamination with certain adventitious agents.

2.3.3.1 Tumorigenicity

Tumorigenicity describes the capacity of viable cells to proliferate in the recipient. There appears to be no marker for this property. For example, one of the hallmarks of cancer is a disturbance in the karyotype but not all cell lines with aberrant karyotype are tumorigenic. Chromosomal abnormalities have been observed to increase with passage level also in diploid cell lines, including the MRC-5 [22]. With modern technology these changes are even more apparent and "possessing a stable karyotype might not be such an important characteristic as was previously thought" [149, 93]. In 1954, however, the American Forces Epidemiological Board decided not only against the HeLa cell line as a substrate for production of a new vaccine against adenoviruses. The board further specified that live and killed vaccines should be produced only on "normal" cells [149, 94, 95] without any subpassaging. This strict definition excluded also the diploid cell lines such as WI-38 that today are known not to be tumorigenic.

For current vaccine substrates, tumorigenicity is determined experimentally in appropriate animal models. To increase sensitivity of the assay the cytotoxic T lymphocyte activity of the test animals is impaired to avoid an immunological reaction against the usually noncognate cell line. Accepted test animals include neonatal mice or rats treated with antithymocyte serum and Nu/Nu genotype adult nude mice. For certain veterinary vaccines that are intended for animals for human consumption the tumorigenicity in the target animal (such as chicken) may also be determined. Important parameters of the assay includes the number of inoculated cells per animal required so that nodules can be detected, the observation period, whether nodules grow progressively or not, and whether nodules are established also distant from the inoculation site. Cell lines that produce nodules that fail to grow progressively, or below a threshold of at least 2 out of 10 animals, may be considered not to be tumorigenic.

Tumorigenicity is not a distinct property but is evident in varying degrees. For example, HEK 293 is a weakly tumorigenic cell line that produces tumors only at a dose of 10^7 cells per animal. The MDCK cell line is associated with a variable tumorigenic phenotype [48]. An MDCK cell strain adapted to proliferation in suspension may form tumors in athymic nude mice already at a dose of 10 cells per animal whereas a strain of MDCK that still grows in adherence and exhibits contact inhibition is positive only if at least 10^5 cells are in inoculated per animal. The Vero cell line acquires the capacity to form tumors starting with a passage level of 140 to 250, depending on prior cultivation intervals [96].

Cytopathic effect and vaccine production processes are expected to be highly effective in removing all viable cells from injectible vaccine preparations (for example, page 19 of the Solvay Pharmaceuticals briefing document for the Vaccines and Related Biological Products Advisory Committee (VRBPAC), November 2005, obtainable at http://www.fda.gov/ohrms/dockets/ac/05/briefing/5-4188B1_19a.pdf). For this reason the tumorigenic risk in a final vaccine should be extremely remote.

2.3.3.2 Oncogenicity

Oncogenicity describes the capacity of acellular components in a vaccine to induce tumors in the recipient. Transfer of oncogenic potential does not require intact cells and is more difficult to assess than tumorigenic potential. The risk is usually considered to be associated with residual host cell-derived DNA that may encode oncogenes or allow expression of infectious oncoviruses. Oncogenic potential in coding or regulatory cellular RNA or protein complexes within a cell lysate is a formal possibility. However, the inadvertent presence of these agents is probably too transient to carry a cell through the multiple stages that are postulated to be important for induction of tumors [36]. Oncogenic infectious particles capable of self-replication within a cell (such as SV40 in primary monkey kidney cultures) are often discussed in the context of adventitious agents rather than oncogenic potential of a cell substrate.

Today, the admissible level of host cell-derived DNA is 10 ng per dose of parenteral vaccine that is being produced on a continuous cell line [97]. This value also applies to the Vero cell line that is not tumorigenic at the passage level used for vaccine production. However, because Vero is an aneuploid continuous cell line a limit of less than 10 ng DNA per vaccine dose applies for such preparations as well [44].

The requirement for less than 10 ng of DNA derived from a continuous host cell appears to be high compared to the early threshold that in 1986 allowed only 100 pg of DNA, but it is low compared to the 75 to 450 µg of DNA contained in each unit of transfused blood. The current value was derived by defined risk calculations that estimate the probability for an oncogene to be encoded in a given segment of DNA in relation to the size of the host's genome and mass of DNA being transferred. With such a formula a transformational event is expected once in 10^9 recipients if 1 ng of cellular DNA containing 100 copies of an activated oncogene would be transferred. The calculation can be further refined by incorporating enzymatic fragmentation of the DNA to below 1,000 bp [98]. The results are expected to overestimate the risk of transferring oncogenic activity. For example, biological activity is further reduced if β-propiolactone is used in the production of killed vaccines [97], and human cells appear not to incorporate injected DNA as efficiently as the rodent test animals [99]. Consistent with the theoretical considerations, no tumors were observed in a longitudinal study with monkeys that were injected with up to 1,000 µg of DNA derived from the continuous human PER.C6 cell line [100].

While improved purification is one important approach to reducing the risk of adverse events, the choice of immortalizing transgenes for designer cell substrates may further contribute to the safety of vaccine production processes. For E1-positive cell lines [81, 85, 89, 90] several beneficial properties combine: Human adenoviruses are both endemic and pandemic pathogens responsible for 2 to 7 % of upper respiratory infections in children. Although they have also the capacity to cause severe disease especially in immunocompromized patients [101] they are not associated with human tumors [63]. Because the E1A protein, the immortalizing agent, sensitizes cells for apoptotic stimuli and suppresses proliferation of some tumors [63, 102], virotherapy [103] and E1A gene therapy is investigated as therapeutic option for cancer patients [102, 104, 105]. Finally, activation of proapoptotic pathways caused by E1A is neutralized only by coexpression expression of E1B (that by itself has no oncogenic functions), necessitating two concurrent transfers of contaminating E1A and E1B functions into the target cell.

2.3.3.3 Adventitious Agents

Adventitious agents are microorgansims that are introduced inadvertently together with the cell substrate or during production. Contamination of diphteria antitoxin with *Clostridium tetanii* in St. Louis, USA, is one example demonstrating that adventitious agents have been a cause for concern well prior to the dawn of cell culture based preparations. This tetanus tragedy of 1901 led to the death of 13 children and implementation of the Biologics Control Act (or Virus-Toxin Law) that authorized the Hygienic Laboratory of the Public Health and Marine Hospital Service (today known as Center for Biologics Evaluation and Research (CBER)) to inspect and certify vaccine producer. Another example is the "Lübeck Disaster" in Germany that was caused by accidental inoculation of attenuated BCG strain with virulent *Mycobacterium tuberculosis* [106, 107]. Appropriate safety precautions were ignored with the effect that 77 of the vaccinated 256 neonates died of tuberculosis, introduction of the BCG vaccine was delayed in Germany, and codification of the German Health Law was initiated.

One early example of adventitious contamination in cell culture-derived vaccines was caused by SV40 that was naturally present in the primary monkey cultures used for production of the vaccines against poliomyelitis and adenoviruses [30, 108, 109]. It is estimated that between 10 and almost 100 million children and thousands of army recruits in the USA were exposed to the contaminated vaccine between 1955 and 1963. That infectious SV40 appears to have survived the formalin inactivation step used for preparation of killed vaccines has contributed to the exposure rates. The situation is further complicated because more than one strain of SV40 may have contaminated the vaccines [110] and because tainted vaccines may have been administered in Eastern Europe until the 1980s [109].

SV40 is an oncogenic polyomavirus that can induce tumors in laboratory animals and that can be used to immortalize primary cells. There still is a controversial discussion on whether SV40 infection can also induce tumors in humans [30, 108, 109, 111, 112]. Exposure and productive or abortive infection cannot be reliably determined because not all who were exposed to SV40 produced antibodies (although virus was detectable in the stool) and because antibodies induced by human polyomaviruses cross-react with SV40. PCR for detection of SV40 sequences in human tumors also yielded contradictory results. Epidemiological studies do not suggest that tumor incidence may have increased because of the SV40 contamination.

One consequence of the SV40 contamination in monkey primary cells was that the human diploid cell line WI-38 was approved for live polio vaccine production in 1972 although it is derived by serial passage [95]. Another consequence was generation of the Vero cell line. This cell line was initially intended as substrate for the investigation of the oncogenic properties of SV40 [41] but today is better known as a versatile and efficient host for production of various vaccines.

Another vaccine that was widely distributed due to urgent demand until a serious contamination was detected is the yellow fever vaccine of the 1941/1942 lots. Human serum was added to stabilize the live 17D vaccine strain. The serum donors were all presumed healthy as the concept of a serum-associated infectious jaundice had not been fully developed [113]. After recognition, that serum contaminated with a transmissible agent (today known to be hepatitis B virus) had been used, the formulation of the 17D vaccine was changed to serum-free preparations, and distribution was restricted to military personnel scheduled for deployment in regions with endemic yellow fever [114].

More recent contamination events include a calicivirus in CHO cell cultures [115] used for production of biologicals and probably introduced together with a nutrient additive and presence of *Serratia marcescens* in influenza vaccine produced on embryonated chicken eggs [116]. The latter contamination was detected prior to release of the affected vaccine lots.

There are also instances where some contaminants may pose an extremely remote risk for recipients whereas the measurable benefit provided by the vaccine may vastly outweigh any concerns. Under these exceptional circumstances, especially if the contamination cannot be avoided with the current technology, the presence of the adventious agent may be accepted in a final vaccine preparation. One example are the endogenous retroviruses that are dispersed throughout the genome of chickens. Some of these proviruses can become activated during the production process and be expressed into secreted viral particles. The viral particles may contain reverse transcriptase and activity of this enzyme can be detected in the vaccine with a sensitive variation of the PCR method [117]. However, there is no virological and epidemiological indication that avian retroviruses are dangerous to humans [118]. Because production of vaccines on primary chicken material is an extremely well characterized procedure for several important vaccines the WHO does not recommend changes to the current procedure [119]. The endogenous retroviruses cannot be bred out of the chicken genome but recently avian cell lines free of particle-associated reverse transcriptase have been provided [90].

Another adventitious agent that currently is accepted in a live vaccine preparation are porcine circoviruses (PCV1 and PCV2) in the human rotavirus vaccine [120]. Signs of PCV1 and PCV2 infection can be detected worldwide in virtually all pig populations. PCV1 appears to cause no disease in pigs, PCV2 predisposes pigs to a number of clinical conditions and can be cause for more severe complications of postweaning multisystemic wasting syndrome and porcine reproductive and respiratory syndrome. Both types of the tiny viruses appear to be unable to infect humans. They were probably introduced into the vaccines together with trypsin isolated from porcine pancreas and required for cultivation of the adherent Vero cell line, the production substrate for virus seeds and final vaccine preparations. Rotavirus infection causes significant morbidity in industrialized countries and is a major cause of death in developing countries. The benefits of vaccination against rotavirus by far outweigh the risks and use of the vaccines should be continued until pure versions can be provided [121].

Finally, cell lines have also been central in descriptions of contaminations that later have not been found to be accurate. One example is the OPV/AIDS theory or "river hypothesis" that suggests that also chimpanzee primary cells may have been used in the production of the 1957 to 1960 experimental oral polio vaccine (OPV) that was investigated in Central Africa, and that this additional production substrate has contaminated some OPV preparations with the SIVcpz primate immunodeficiency virus. If this chain of events had been the case then a precursor of the group M human immunodeficiency virus would have had the opportunity to be iatrogenically introduced into the population [122]. Careful phylogenetic analysis and search for genomic traces related to chimpanzees or SIVcpz strongly favor the alternative theory of multiple independent zoonotic transmissions to bush-meat hunters decades prior to the OPV trials [123-126].

Today, extensive regulatory dossiers suggest a number of procedures for effective screening of adventitious agents [127]. The suggested procedures are designed to identify a wide spectrum of known pathogens via PCR and serology in test animals. They are also designed to detect previously uncharacterized pathogens via assays such as electron microscopy, induction of cytopathic effects by cocultivation of the cell substrate with sentinel cell lines, and induction of disease by inoculation of cell lysate into embryonated eggs, suckling, and adult mice and guinea pigs.

2.3.3.4 Vaccine Controversy and Cell Lines

Ethical and political aspects also influence adoption of a cell substrate for vaccine production purposes. The ideal vaccine is suitable and accepted for application in a heterogeneous population. It is highly reactogenic so that the number of booster shots can be reduced for immunization of the very young, a property that may facilitate vaccine compliance and coverage because fewer visits to health care centers are required. The highly reactogenic ideal vaccine will also protect recipients with a weaker immune system, the elderly in the industrialized nations and those in developing countries that may be malnourished or suffer from chronic infections. At the same time, the ideal vaccine is very safe, essentially free of adverse events, and purified to a degree that no undesired components remain. Finally, it will be administered transdermally or by the mucosal route and not via injection.

These properties may not be reconciled in the near future. A negative correlation has already been described between very high degree of purification and reactogenicity for vaccines against influenza [128]. For some applications, a combined stimulation of the adaptive and innate immune pathways is required to achieve efficient priming of an immune response. Activation of innate immune system may be insufficient with highly purified vaccine preparations that lack cellular danger signals such as ruptured membranes and pathogen associated patterns such as viral genomic fragments and double stranded RNA. Adjuvants are being investigated as a means to reconstitute the "missing contamination" with a characterized component [129, 130]. However, the mechanism of stimulation by adjuvants is not completely understood and some formulations appear to be associated with actual [131, 132] and perhaps perceived risks [133]. Life viruses that replicate in the host are efficient stimulators of immunity but they also bear the risk of reversion to the pathogenic origin or can be dangerous to immunocompromised recipients or contact persons [134]. Highly attenuated or host-restricted viral vectors may help to bridge the gap between a desire for highly purified and characterized vaccines and the concomitant decreased capacity to stimulate cellular and innate immunity. However, with the exception of host-restricted poxviruses (modified vaccinia virus Ankara, fowlpox virus and canarypox virus), the host-restricted vectors usually are recombinant viruses that depend on transcomplementation of structural or transactive factors by the recombinant helper or packaging cell line (for example [55, 85]). All summarized, effective immunization against targets that also require induction of cellular immunity may require researchers to choose between addition of adjuvant or use of a vectored vaccine. If the choice is on a viral vector then continuous cell lines or helper viruses may be necessary.

Another aspect of cell cultures for vaccine production concerns conflict with strong views in religious communities. Parents may decline vaccination of their children for reasons that should be discussed to avoid lapses in vaccine coverage and subsequent outbreaks of vaccine-preventable diseases [135-138]. One concern appears to be that porcine and human excipients may contaminate a vaccine. This concern can be addressed with adjusted production processes and information that is also conveyed in a simplified format.

That human diploid cell lines and some designed cell lines are derived from tissue from aborted human fetuses is another cause for ethical concern. The dilemma is especially strong for parents of children that are to be vaccinated, and appears to be especially pronounced in Western industrialized countries. This field has been also the topic of extensive bioethical considerations [139, 140]. The concerns can be partially alleviated by clearing a misconception about further and continuing abortions also for present vaccine production [139]. It may also be possible to use alternative cell lines of nonhuman origin [141, 142] or from sources without ethical connotations [92] for certain future applications to relieve vaccine scepticism.

2.3.4 Conclusion

Animal cell technology for the production of viral vaccines is an incredibly complex and diverse field. Many topics have not been reviewed here, including details on health regulatory requirements; technical aspects of improved bioreactor technologies and novel purification options; certain cell lines such as the diploid FRhL-2 cell line derived from the lung of a fetal rhesus monkey and used in the production of the (now withdrawn) RotaShield vaccine; cell lines used for the production of some veterinary viral vaccines, including vaccines to protect companion animals or animals intended for human consumption; and the fascinating field of stem cell lines, where at least one (EB66) [143] is on the regulatory track for approval of vaccine production.

The wide spectrum of and possible designs for helper and packaging cell lines for production of host-restricted vectors has only been briefly visited in discussion of the HEK 293 and PER.C6 cell lines.

Vaccines are one of the most efficacious and effective health interventions available [144]. Among the remaining challenges are, that yet safer vaccines need to be developed and that risks and benefits need to be communicated more effectively. For example, some adverse reactions that merely reflect the desired activation of the immune system may never be controlled. Regulatory assessment will be facilitated and adoption of vaccines among health care providers and the recipients may be improved if the knowledge about a vaccine preparation could be further expanded. A complete characterization would include the biochemical properties and requirements of the cell lines used in vaccine production and the genetic composition of the vaccine strains adapted to that substrate and that may evolve within the few generations required to provide a final product. Advanced technologies such as deep sequencing are expected to contribute to such a level of detail [145].

A new challenge in animal cell technology is to provide some of the novel recombinant viruses that have been developed as therapeutic vaccines against certain cancers and chronic infectious diseases [146, 147]. Attenuated viruses are stimulators also of cellular immunity but their tropism, efficacy, and sensitivity to preexisting immunity varies. Furthermore, for safety reasons they should be host-restricted in a way that prevents replication in the intended recipient, including immunocompromised patients. This can be accomplished by adaptation of the vaccine strain to novel cell lines unrelated to the natural host or by recombinant technologies where essential transcomplementing factors are provided by the producer cell line but cannot be found in the recipient. This increase in vaccine diversity and process demands will require improved and novel approaches for the provision of a wide spectrum of suitable and acceptable cell lines, biochemical and recombinant manipulations of host cells, and biotechnological advances for production and purification. It does not require prescience to predict exciting times for all those who are working in this field.

References

[1] Enders JF, Weller TH, Robbins FC. Cultivation of the Lansing Strain of Poliomyelitis Virus in Cultures of Various Human Embryonic Tissues. Science 1949;109:85–7.

[2] Enders JF, Robbins FC, Weller TH. Classics in infectious diseases. The cultivation of the poliomyelitis viruses in tissue culture by John F. Enders, Frederick C. Robbins, and Thomas H. Weller. Rev Infect Dis 1980;2:493–504.

[3] Schaeffer WI. Terminology associated with cell, tissue, and organ culture, molecular biology, and molecular genetics. Tissue Culture Association Terminology Committee. In Vitro Cell Dev Biol 1990;26:97–101.

[4] Dawson JR. A study of chick-embryo-adapted rabies virus. Am J Pathol 1941;17:177–188.7.

[5] Koprowski H. Biological modification of rabies virus as a result of its adaptation to chicks and developing chick embryos. Bull World Health Organ 1954;10:709–24.

[6] Mayr A, Munz E. Changes in the vaccinia virus through continuing passages in chick embryo fibroblast cultures. Zentralbl Bakteriol Orig 1964;195:24–35.
[7] Koprowski H. Historical aspects of the development of live virus vaccine in poliomyelitis. Br Med J 1960;2:85–91.
[8] Sabin AB, Ramos-Alvarez M, Alvarez-Amezquita J, et al. Live, orally given poliovirus vaccine. Effects of rapid mass immunization on population under conditions of massive enteric infection with other viruses. JAMA 1960;173:1521–6.
[9] Courtois G, Flack A, Jervis GA, et al. Preliminary report on mass vaccination of man with live attenuated poliomyelitis virus in the Belgian Congo and Ruanda-Urundi. Br Med J 1958;2:187–90.
[10] Plotkin SA, Lebrun A, Courtois G, et al. Vaccination with the CHAT strain of type 1 attenuated poliomyelitis virus in Leopoldville, Congo. 3. Safety and efficacy during the first 21 months of study. Bull World Health Organ 1961;24:785–92.
[11] Sald JE. Studies in human subjects on active immunization against poliomyelitis. I. A preliminary report of experiments in progress. J Am Med Assoc 1953;151:1081–98.
[12] Lambert SM, Markel H. Making history: Thomas Francis, Jr, MD, and the 1954 Salk Poliomyelitis Vaccine Field Trial. Arch Pediatr Adolesc Med 2000;154:512–7.
[13] Rhodes AJ. Research on the development of a poliomyelitis vaccine: Toronto, 1950-1953. Can Med Assoc J 1956;75:48–9.
[14] Smorodintsev AA, Drobyshevskaya AI, Bulychev NP, et al. Immunological and epidemiological effectiveness of live poliomyelitis vaccine in the USSR. Bull World Health Organ 1960;23:705–25.
[15] WHO. Inactivated poliovirus vaccine following oral poliovirus vaccine cessation. Wkly Epidemiol Rec 2006;81:137–44.
[16] Pearce JMS. Salk and Sabin: poliomyelitis immunisation. J Neurol Neurosurg Psychiatr 2004;75:1552.
[17] Uscher-Pines L, Barnett DJ, Sapsin JW, et al. A systematic analysis of influenza vaccine shortage policies. Public Health 2008;122:183–91.
[18] Scherer SF. The utilization of a pure strain of mammalian cells (Earle) for the cultivation of viruses in vitro. I. Multiplication of pseudorabies and herpes simplex viruses. Am J Pathol 1953;29:113–37.
[19] Scherer WF, Syverton JT, Gey GO. Studies on the propagation in vitro of poliomyelitis viruses. IV. Viral multiplication in a stable strain of human malignant epithelial cells (strain HeLa) derived from an epidermoid carcinoma of the cervix. J Exp Med 1953;97:695–710.
[20] Cox HR. Active immunization against poliomyelitis. Bull N Y Acad Med 1953;29:943–60.
[21] Hayflick L. The linited in vitro lifetime of human diploid cell strains. Exp Cell Res 1965;37:614–36.
[22] Jacobs JP, Jones CM, Baille JP. Characteristics of a human diploid cell designated MRC-5. Nature 1970;227:168–70.
[23] Hayflick L, Moorehead PS. The serial cultivation of human diploid cell strains. Exp Cell Res 1961;25:585–621.
[24] Hayflick L. The illusion of cell immortality. Br J Cancer 2000;83:841–6.
[25] Hilleman MR. Efficacy of and indications for use of adenovirus vaccine. Am J Public Health Nations Health 1958;48:153–8.
[26] Russell KL, Hawksworth AW, Ryan MAK, et al. Vaccine-preventable adenoviral respiratory illness in US military recruits, 1999-2004. Vaccine 2006;24:2835–42.
[27] Couch RB, Chanock RM, Cate TR, et al. Immunization with types 4 and 7 adenovirus by selective infection of the intestinal tract. Am Rev Respir Dis 1963;88:394–403.

[28] Top FH Jr. Control of adenovirus acute respiratory disease in U.S. Army trainees. Yale J Biol Med 1975;48:185–95.
[29] Barraza EM, Ludwig SL, Gaydos JC, et al. Reemergence of adenovirus type 4 acute respiratory disease in military trainees: report of an outbreak during a lapse in vaccination. J Infect Dis 1999;179:1531–3.
[30] Rollison DEM, Page WF, Crawford H, et al. Case-control study of cancer among US Army veterans exposed to simian virus 40-contaminated adenovirus vaccine. Am J Epidemiol 2004;160:317–24.
[31] Gray GC, Goswami PR, Malasig MD, et al. Adult adenovirus infections: loss of orphaned vaccines precipitates military respiratory disease epidemics. For the Adenovirus Surveillance Group. Clin Infect Dis 2000;31:663–70.
[32] Nichols WW, Murphy DG, Cristofalo VJ, et al. Characterization of a new human diploid cell strain, IMR-90. Science 1977;196:60–3.
[33] Takahashi M. Development of a live varicella vaccine-past and future. Jpn J Infect Dis 2000;53:47–55.
[34] Rodier F, Campisi J. Four faces of cellular senescence. J Cell Biol 2011;192:547–56.
[35] Rich T, Allen RL, Wyllie AH. Defying death after DNA damage. Nature 2000;407:777–83.
[36] Hahn WC, Weinberg RA. Modelling the molecular circuitry of cancer. Nat Rev Cancer 2002;2:331–41.
[37] Ramirez RD, Morales CP, Herbert BS, et al. Putative telomere-independent mechanisms of replicative aging reflect inadequate growth conditions. Genes Dev 2001;15:398–403.
[38] Shay JW, Zou Y, Hiyama E, et al. Telomerase and cancer. Hum Mol Genet 2001;10:677–85.
[39] Rudolph KL, Chang S, Lee HW, et al. Longevity, stress response, and cancer in aging telomerase-deficient mice. Cell 1999;96:701–12.
[40] Weinstein IB, Joe AK. Mechanisms of disease: Oncogene addiction--a rationale for molecular targeting in cancer therapy. Nat Clin Pract Oncol 2006;3:448–57.
[41] Montagnon BJ. Polio and rabies vaccines produced in continuous cell lines: a reality for Vero cell line. Dev Biol Stand 1989;70:27–47.
[42] Vincent-Falquet JC, Peyron L, Souvras M, et al. Qualification of working cell banks for the Vero cell line to produce licensed human vaccines. Dev Biol Stand 1989;70:153–6.
[43] Montagnon BJ, Vincent-Falquet JC. Experience with the Vero cell line. Dev Biol Stand 1998;93:119–23.
[44] Trent D, Shin J, Hombach J, et al. WHO Working Group on technical specifications for manufacture and evaluation of dengue vaccines, Geneva, Switzerland, 11-12 May 2009. Vaccine 2010;28:8246–55.
[45] Knezevic I, Stacey G, Petricciani J, et al. Evaluation of cell substrates for the production of biologicals: Revision of WHO recommendations. Report of the WHO Study Group on Cell Substrates for the Production of Biologicals, 22-23 April 2009, Bethesda, USA. Biologicals 2010;38:162–9.
[46] Gaush CR, Hard WL, Smith TF. Characterization of an established line of canine kidney cells (MDCK). Proc Soc Exp Biol Med 1966;122:931–5.
[47] Dukes JD, Whitley P, Chalmers AD. The MDCK variety pack: choosing the right strain. BMC Cell Biol 2011;12:43.
[48] Omeir RL, Teferedegne B, Foseh GS, et al. Heterogeneity of the tumorigenic phenotype expressed by Madin-Darby canine kidney cells. Comp Med 2011;61:243–50.
[49] Gaush CR, Smith TF. Replication and plaque assay of influenza virus in an established line of canine kidney cells. Appl Microbiol 1968;16:588–94.
[50] Doroshenko A, Halperin SA. Trivalent MDCK cell culture-derived influenza vaccine Optaflu (Novartis Vaccines). Expert Rev Vaccines 2009;8:679–88.

[51] Gregersen J-P, Schmitt H-J, Trusheim H, et al. Safety of MDCK cell culture-based influenza vaccines. Future Microbiol 2011;6:143–52.
[52] MacPherson I, Stoker M. Polyoma transformation of hamster cell clones--an investigation of genetic factors affecting cell competence. Virology 1962;16:147–51.
[53] Stoker M, MacPherson I. Syrian hamster fibroblast cell line BHK21 and its derivates. Nature 1964;203:1355–7.
[54] Albariño CG, Ghiringhelli PD, Posik DM, et al. Molecular characterization of attenuated Junin virus strains. J Gen Virol 1997;78:1605–10.
[55] Polo JM, Belli BA, Driver DA, et al. Stable alphavirus packaging cell lines for Sindbis virus and Semliki Forest virus-derived vectors. Proc Natl Acad Sci USA 1999;96:4598–603.
[56] Tjio JH, Puck TT. Genetics of somatic mammalian cells. II. Chromosomal constitution of cells in tissue culture. J Exp Med 1958;108:259–68.
[57] Marsh M, Bron R. SFV infection in CHO cells: cell-type specific restrictions to productive virus entry at the cell surface. J Cell Sci 1997;110:95–103.
[58] Berting A, Farcet MR, Kreil TR. Virus susceptibility of Chinese hamster ovary (CHO) cells and detection of viral contaminations by adventitious agent testing. Biotechnol Bioeng 2010;106:598–607.
[59] Xu X, Nagarajan H, Lewis NE, et al. The genomic sequence of the Chinese hamster ovary (CHO)-K1 cell line. Nat Biotechnol 2011;29:735–41.
[60] Lubiniecki AS. Historical reflections on cell culture engineering. Cytotechnology 1998;28:139–45.
[61] Sáenz-Robles MT, Sullivan CS, Pipas JM. Transforming functions of Simian Virus 40. Oncogene 2001;20:7899–907.
[62] McLaughlin-Drubin ME, Münger K. Oncogenic activities of human papillomaviruses. Virus Res 2009;143:195–208.
[63] Frisch SM, Mymryk JS. Adenovirus-5 E1A: paradox and paradigm. Nat Rev Mol Cell Biol 2002;3:441–52.
[64] Korzeniewski N, Spardy N, Duensing A, et al. Genomic instability and cancer: lessons learned from human papillomaviruses. Cancer Lett 2011;305:113–22.
[65] Berk AJ. Recent lessons in gene expression, cell cycle control, and cell biology from adenovirus. Oncogene 2005;24:7673–85.
[66] Bocchetta M, Eliasz S, De Marco MA, et al. The SV40 large T antigen-p53 complexes bind and activate the insulin-like growth factor-I promoter stimulating cell growth. Cancer Res 2008;68:1022–9.
[67] Cheng J, DeCaprio JA, Fluck MM, et al. Cellular transformation by Simian Virus 40 and Murine Polyoma Virus T antigens. Semin Cancer Biol 2009;19:218–28.
[68] Topalis D, Andrei G, Snoeck R. The large tumor antigen: a "Swiss Army knife" protein possessing the functions required for the polyomavirus life cycle. Antiviral Res 2013;97:122–36.
[69] Green DR, Kroemer G. Cytoplasmic functions of the tumour suppressor p53. Nature 2009;458:1127–30.
[70] Cuconati A, Mukherjee C, Perez D, et al. DNA damage response and MCL-1 destruction initiate apoptosis in adenovirus-infected cells. Genes Dev 2003;17:2922–32.
[71] Jensen F, Koprowski H, Ponten JA. Rapid transformation of human fibroblast cultures by simian virus. Proc Natl Acad Sci USA 1963;50:343–8.
[72] Jensen FC, Girardi AJ, Gilden RV, et al. Infection of human and simian tissue cultures with Rous sarcoma virus. Proc Natl Acad Sci USA 1964;52:53–9.
[73] Aaronson SA, Todaro GJ. SV40 T antigen induction and transformation in human fibroblast cell strains. Virology 1968;36:254–61.

[74] Todaro GJ, Aaronson SA. Human cell strains susceptible to focus formation by human adenovirus type 12. Proc Natl Acad Sci USA 1968;61:1272–8.
[75] Schwarz E, Freese UK, Gissmann L, et al. Structure and transcription of human papillomavirus sequences in cervical carcinoma cells. Nature 1985;314:111–4.
[76] Meissner JD. Nucleotide sequences and further characterization of human papillomavirus DNA present in the CaSki, SiHa and HeLa cervical carcinoma cell lines. J Gen Virol 1999;80:1725–33.
[77] Hall AHS, Alexander KA. RNA interference of human papillomavirus type 18 E6 and E7 induces senescence in HeLa cells. J Virol 2003;77:6066–9.
[78] McCutchan JH, Pagano JS. Enchancement of the infectivity of simian virus 40 deoxyribonucleic acid with diethylaminoethyl-dextran. J Natl Cancer Inst 1968;41:351–7.
[79] Graham FL, van der Eb AJ. Transformation of rat cells by DNA of human adenovirus 5. Virology 1973;54:536–9.
[80] Abrahams PJ, van der Eb AJ. In vitro transformation of rat and mouse cells by DNA from simian virus 40. J Virol 1975;16:206–9.
[81] Graham FL, Smiley J, Russell WC, et al. Characteristics of a human cell line transformed by DNA from human adenovirus type 5. J Gen Virol 1977;36:59–74.
[82] Graham FL. Covalently closed circles of human adenovirus DNA are infectious. EMBO J 1984;3:2917–22.
[83] Spector DJ. The pattern of integration of viral DNA sequences in the adenovirus 5-transformed human cell line 293. Virology 1983;130:533–8.
[84] Louis N, Evelegh C, Graham FL. Cloning and sequencing of the cellular-viral junctions from the human adenovirus type 5 transformed 293 cell line. Virology 1997;233:423–9.
[85] Fallaux FJ, Bout A, van der Velde I, et al. New helper cells and matched early region 1-deleted adenovirus vectors prevent generation of replication-competent adenoviruses. Hum Gene Ther 1998;9:1909–17.
[86] Samina I, Havenga M, Koudstaal W, et al. Safety and efficacy in geese of a PER.C6-based inactivated West Nile virus vaccine. Vaccine 2007;25:8338–45.
[87] Le Ru A, Jacob D, Transfiguracion J, et al. Scalable production of influenza virus in HEK-293 cells for efficient vaccine manufacturing. Vaccine 2010;28:3661–71.
[88] Sanders BP, Edo-Matas D, Custers JHHV, et al. PER.C6(®) cells as a serum-free suspension cell platform for the production of high titer poliovirus: a potential low cost of goods option for world supply of inactivated poliovirus vaccine. Vaccine 2013;31:850–6.
[89] Schiedner G, Hertel S, Kochanek S. Efficient transformation of primary human amniocytes by E1 functions of Ad5: generation of new cell lines for adenoviral vector production. Hum Gene Ther 2000;11:2105–16.
[90] Jordan I, Vos A, Beilfuss S, et al. An avian cell line designed for production of highly attenuated viruses. Vaccine 2009;27:748–56.
[91] Jordan I, Northoff S, Thiele M, et al. A chemically defined production process for highly attenuated poxviruses. Biologicals 2011;39:50–8.
[92] Genzel Y, Behrendt I, Rödig J, et al. CAP, a new human suspension cell line for influenza virus production. Appl Microbiol Biotechnol 2013;97:111–22.
[93] Rosolowsky M, McKee R, Nichols W, et al. Chromosomal characterization of MRC-5 cell banks utilizing G-banding technique. Dev Biol Stand 1998;93:109–17.
[94] Hilleman MR. Cells, vaccines, and the pursuit of precedent. Natl Cancer Inst Monogr 1968;29:463–9.
[95] Hilleman MR. History, precedent, and progress in the development of mammalian cell culture systems for preparing vaccines: safety considerations revisited. J Med Virol 1990;31:5–12.
[96] Manohar M, Orrison B, Peden K, et al. Assessing the tumorigenic phenotype of VERO cells in adult and newborn nude mice. Biologicals 2008;36:65–72.

[97] Grachev V, Magrath D, Griffiths E. WHO requirements for the use of animal cells as in vitro substrates for the production of biologicals (Requirements for biological susbstances no. 50). Biologicals 1998;26:175–93.
[98] Yang H, Zhang L, Galinski M. A probabilistic model for risk assessment of residual host cell DNA in biological products. Vaccine 2010;28:3308–11.
[99] Bolhassani A, Safaiyan S, Rafati S. Improvement of different vaccine delivery systems for cancer therapy. Mol Cancer 2011;10:3.
[100] Wierenga DE, Cogan J, Petricciani JC. Administration of tumor cell chromatin to immunosuppressed and non-immunosuppressed non-human primates. Biologicals 1995;23:221–4.
[101] Munoz FM, Piedra PA, Demmler GJ. Disseminated adenovirus disease in immunocompromised and immunocompetent children. Clin Infect Dis 1998;27:1194–200.
[102] Frisch SM. E1A as a tumor suppressor gene: commentary re S. Madhusudan et al. A multicenter Phase I gene therapy clinical trial involving intraperitoneal administration of E1A-lipid complex in patients with recurrent epithelial ovarian cancer overexpressing HER-2/neu oncogene. Clin Cancer Res 2004;10:2905–7.
[103] Choi I-K, Yun C-O. Recent developments in oncolytic adenovirus-based immunotherapeutic agents for use against metastatic cancers. Cancer Gene Ther 2013;20:70–6.
[104] Madhusudan S, Tamir A, Bates N, et al. A multicenter Phase I gene therapy clinical trial involving intraperitoneal administration of E1A-lipid complex in patients with recurrent epithelial ovarian cancer overexpressing HER-2/neu oncogene. Clin Cancer Res 2004;10:2986–96.
[105] Yamaguchi H, Chen C-T, Chou C-K, et al. Adenovirus 5 E1A enhances histone deacetylase inhibitors-induced apoptosis through Egr-1-mediated Bim upregulation. Oncogene 2010;29:5619–29.
[106] Unknown Author. The Lübeck disaster. Am J Public Health Nations Health 1931;21:282.
[107] Unknown Author. Topics of Current Interest. Can Med Assoc J 1932;26:362–5.
[108] Vilchez RA, Butel JS. Emergent human pathogen simian virus 40 and its role in cancer. Clin Microbiol Rev 2004;17:495–508.
[109] Poulin DL, DeCaprio JA. Is there a role for SV40 in human cancer? J Clin Oncol 2006;24:4356–65.
[110] Rizzo P, Di Resta I, Powers A, et al. Unique strains of SV40 in commercial poliovaccines from 1955 not readily identifiable with current testing for SV40 infection. Cancer Res 1999;59:6103–8.
[111] Lundstig A, Dejmek A, Eklund C, et al. No detection of SV40 DNA in mesothelioma tissues from a high incidence area in Sweden. Anticancer Res 2007;27:4159–61.
[112] Qi F, Carbone M, Yang H, et al. Simian virus 40 transformation, malignant mesothelioma and brain tumors. Expert Rev Respir Med 2011;5:683–97.
[113] Senior JR, London WT, Sutnick AI. The Australia antigen and role of the late Philadelphia General Hospital in reducing post-transfusion hepatitis and sequelae. Hepatology 2011;54:753–6.
[114] Frierson JG. The yellow fever vaccine: a history. Yale J Biol Med 2010;83:77–85.
[115] Oehmig A, Büttner M, Weiland F, et al. Identification of a calicivirus isolate of unknown origin. J Gen Virol 2003;84:2837–45.
[116] Enserink M. Influenza. Crisis underscores fragility of vaccine production system. Science 2004;306:385.
[117] Weissmahr RN, Schüpbach J, Böni J. Reverse transcriptase activity in chicken embryo fibroblast culture supernatants is associated with particles containing endogenous avian retrovirus EAV-0 RNA. J Virol 1997;71:3005–12.

[118] Hussain AI, Shanmugam V, Switzer WM, et al. Lack of evidence of endogenous avian leukosis virus and endogenous avian retrovirus transmission to measles, mumps, and rubella vaccine recipients. Emerging Infect Dis 2001;7:66–72.
[119] Unknown Author. Reverse transcriptase activity in chicken-cell derived vaccine. Wkly Epidemiol Rec 1998;73:209–12.
[120] Victoria JG, Wang C, Jones MS, et al. Viral nucleic acids in live-attenuated vaccines: detection of minority variants and an adventitious virus. J Virol 2010;84:6033–40.
[121] Kuehn BM. FDA: Benefits of rotavirus vaccination outweigh potential contamination risk. JAMA 2010;304:30–1.
[122] Hooper E. Experimental oral polio vaccines and acquired immune deficiency syndrome. Philos Trans R Soc Lond B Biol Sci 2001;356:803–14.
[123] Gao F, Bailes E, Robertson DL, et al. Origin of HIV-1 in the chimpanzee Pan troglodytes troglodytes. Nature 1999;397:436–41.
[124] Weiss RA. Polio vaccines exonerated. Nature 2001;410:1035–6.
[125] Vartanian J-P, Wain-Hobson S. Analysis of a library of macaque nuclear mitochondrial sequences confirms macaque origin of divergent sequences from old oral polio vaccine samples. Proc Natl Acad Sci USA 2002;99:7566–9.
[126] Worobey M, Santiago ML, Keele BF, et al. Origin of AIDS: contaminated polio vaccine theory refuted. Nature 2004;428:820.
[127] Hess RD, Weber F, Watson K, et al. Regulatory, biosafety and safety challenges for novel cells as substrates for human vaccines. Vaccine 2012;30:2715–27.
[128] Koyama S, Aoshi T, Tanimoto T, et al. Plasmacytoid dendritic cells delineate immunogenicity of influenza vaccine subtypes. Sci Transl Med 2010;2:25ra24.
[129] Akira S. Innate immunity and adjuvants. Philos Trans R Soc Lond, B, Biol Sci 2011;366:2748–55.
[130] Rueckert C, Guzmán CA. Vaccines: from empirical development to rational design. PLoS Pathog 2012;8:e1003001.
[131] Nohynek H, Jokinen J, Partinen M, et al. AS03 adjuvanted AH1N1 vaccine associated with an abrupt increase in the incidence of childhood narcolepsy in Finland. PLoS ONE 2012;7:e33536.
[132] Miller E, Andrews N, Stellitano L, et al. Risk of narcolepsy in children and young people receiving AS03 adjuvanted pandemic A/H1N1 2009 influenza vaccine: retrospective analysis. BMJ 2013;346:f794.
[133] Lippi G, Targher G, Franchini M. Vaccination, squalene and anti-squalene antibodies: facts or fiction? Eur J Intern Med 2010;21:70–3.
[134] Kemper AR, Davis MM, Freed GL. Expected adverse events in a mass smallpox vaccination campaign. Eff Clin Pract 2002;5:84–90.
[135] Henderson L, Millett C, Thorogood N. Perceptions of childhood immunization in a minority community: qualitative study. J R Soc Med 2008;101:244–51.
[136] Kaufmann JR, Feldbaum H. Diplomacy and the polio immunization boycott in Northern Nigeria. Health Aff (Millwood) 2009;28:1091–101.
[137] Ling WY, Razali SM, Ren CK, et al. Does the success of a school-based HPV vaccine programme depend on teachers' knowledge and religion? – A survey in a multicultural society. Asian Pac J Cancer Prev 2012;13:4651–4.
[138] Grabenstein JD. What the World's religions teach, applied to vaccines and immune globulins. Vaccine 2013;31:2011–23.
[139] Furton EJ. Vaccines and the right of conscience. Natl Cathol Bioeth Q 2004;4:53–62.
[140] Leiva R. A brief history of human diploid cell strains. Natl Cathol Bioeth Q 2006;6:443–51.

[141] Lohr V, Rath A, Genzel Y, et al. New avian suspension cell lines provide production of influenza virus and MVA in serum-free media: studies on growth, metabolism and virus propagation. Vaccine 2009;27:4975–82.
[142] Genzel Y, Dietzsch C, Rapp E, et al. MDCK and Vero cells for influenza virus vaccine production: a one-to-one comparison up to lab-scale bioreactor cultivation. Appl Microbiol Biotechnol 2010;88:461–75.
[143] Olivier S, Jacoby M, Brillon C, et al. EB66 cell line, a duck embryonic stem cell-derived substrate for the industrial production of therapeutic monoclonal antibodies with enhanced ADCC activity. MAbs 2010;2.
[144] Ehreth J. The value of vaccination: a global perspective. Vaccine 2003;21:4105–17.
[145] Prachi P, Donati C, Masciopinto F, et al. Deep sequencing in pre- and clinical vaccine research. Public Health Genomics 2013;16:62–8.
[146] Rappuoli R. From Pasteur to genomics: progress and challenges in infectious diseases. Nat Med 2004;10:1177–85.
[147] Plotkin SA. Vaccines: the fourth century. Clin Vaccine Immunol 2009;16:1709–19.
[148] Unknown Author Reverse transcriptase activity in chicken-cell derived vaccine. Relevé Épidémiologique Hebd. Wkly Epidemiol Rec Health Sect Secr Leag Nations 1998;73:209–12.
[149] World Health Organization TRS 878, Annex 1: Recommendations for the evaluation of animal cell cultures as substrates for the manufacture of biological medicinal products and for the characterization of cell banks. WHO Technical Report Series 2010.
[150] World Health Organization WHO Expert Committee on Biological Standardization, Fifty-first report. TRS 910. WHO Technical Report Series 2002.
[151] World Health Organization WHO Expert Committee on Biological Standardization, Fifty-seventh report. TRS 962. WHO Technical Report Series 2011.

2.4 Cell Line Monitoring: Molecular Cytogenetic Characterization

Roderick A. F. MacLeod and Hans G. Drexler

2.4.1 Background

Although the total number of continuous cell lines established has yet to be counted accurately, by extrapolating from one well documented subgroup – human leukemia-lymphoma, where some 1,500 cell lines have been described [1, 2] – a grand total in the order of 10^4 is reached. It follows that today the life sciences would look very different if a resource of this magnitude were no longer available. Cancer cell lines permit preclinical investigations to be performed in the absence of contaminating normal cells, while production cell lines, such as Chinese hamster ovary (CHO), provide a platform for expressing recombinant proteins which is acceptable to regulatory agencies worldwide. It is scarcely imaginable to conceive biomedicine without the resource and bedrock of knowledge provided by cell lines, whether serving as protein factories, oncogenomic resources, or disease models enabling worldwide experimental replication and data accrual. Cell lines cover many species, tissues and, in humans, most main types of tumors. Yet, despite this seeming cornucopia, several cancers and their subtypes – including common and biologically important entities – are unrepresented by cell lines [3].

More than any other single technique, cytogenetic analysis has come to play a central role in characterizing and authenticating cell lines. Cytogenetics informs species of origin, state of neoplastic transformation, and identity. While biochemical methods necessarily yield homogeneous data covering whole cell populations, cytogenetics in common with other microscopic techniques enables observations to be performed on individual cells, thus allowing cell-to-cell differences to be discerned. This is a key asset, given that heterogeneity both in normal and tumor cell populations has become an emerging issue. The classical cytogenetic karyogram (Figure 2.4.1a) presents both, a genomic picture of an individual cell and a portal to whole genome sequencing data. Karyotypes provide the same information in a written symbolic notation systematized by rules laid down by an international convention which is continually updated to take technical and scientific advances into account [4]. For both, tumor and production cell lines, at least karyotypes are effectively unique and thus constitute descriptors independent of DNA profiles. For those working with animal cell lines, cytogenetics provides a rapid means of verifying identity from their species of origin since the chromosome morphologies of most species represented by cell lines are readily distinguishable by those with a modicum of experience.

Cytogenetics has been widely applied to the identification and characterization of cell lines, notably those of tumor origin, almost right from their inception in the second half of the last century. The first major breakthrough in cytogenetics, identification of the human chromosome number at 46, was made in the mid 1950s shortly after the establishment of the first human cell line HeLa [5]. This was followed by the first clue that a specific chromosomal alteration, the Philadelphia chromosome, was closely associated with a specific type of cancer, chronic myeloid leukemia [6]. This rearrangement was resolved to a reciprocal translocation between chromosomes 9 (band q34) and 22 (band q11) following the introduction of chromosome banding techniques, chiefly: Q(uinacrine)-banding [7], C(entromere)-banding [8], and G(iemsa)-banding [9] – the last still widely used, albeit in simplified form [10]. Cytogenetic studies performed on cell lines also first revealed the bugbear of cross-contamination [11].

Molecular cytogenetics bridges the physical and conceptual gap between cells and molecules and owes its widespread popularity to informational and clone resources provided in the wake of the various genome-sequencing projects. The collaboration between molecular cytogenetics and array-based gene expression profiling has proved particularly fruitful when characterizing cell lines.

In the following assessment of cytogenetics for characterizing cell lines, a brief description of the techniques used will be followed by how these may be applied to address current challenges.

2.4.2 The Importance of Harvesting

It is sometimes assumed that obtaining good cytogenetic preparations from continuous cell lines must be simpler than with primary cells due to the unlimited quantities of cells on offer. Alas, this optimistic supposition is not backed up by our experience. Unlike the primary cells normally harvested for cytogenetic diagnosis, notably white blood cells, epidermis, and chorionic villus, human and animal cell lines derive from a plethora of cell types and developmental stages each displaying idiosyncratic responses to culture and cytogenetic harvesting conditions. Population doubling time is a key characteristic of cells in culture. While for the ubiquitous and well-trained CHO cells this time may be as short as 12 h, most human cell lines have population doubling times of 24 h or more. Indeed, some need a week or more to double, a factor hampering cytogenetic harvest. Because mammalian cells transit rather quickly (approx. 30 to 60 min) through the mitotic metaphase when chromosomes are visible, it is necessary to enrich the dividing fraction with arrest enforced by the mitotic poison, colcemid.

To maximize the chances of obtaining optimal chromosome preparations from uncooperative cell lines, we have adopted an evidence based trial-and-error approach which starts by exposing new candidate cell lines to a standard variety of hypotonic buffers, noting their effectiveness according to 3 criteria, mitotic frequency, spreading, and chromosomal quality. If unsuccessful, this information may be used to assess the extent to which alternative hypotonic regimes, imposed by varying buffers, strengths, temperatures, etc., lead to significant improvement. Following hypotonic treatment, cells are fixed in methanol/acetic acid, and preparations assessed microscopically under phase contrast. A detailed account is given in [12]. Once successfully harvested, metaphase chromosome preparations may be cryoconserved for future use: either in the form of microscope slides (at −80 °C), or as cell suspensions in fixative (at −25 °C).

2.4.3 G-banding

G-banding (for details see [12]), though a classical method, is the essential first stage needed to plan additional molecular investigations and to assist their interpretation. For G-banding, slides must be first aged by heat treatment, either at 90 °C for 1 to 2 h, or overnight at 60 °C. After short immersion in trypsin solution, slides are stained in buffered Giemsa solution for bright-field analysis at the highest possible magnification (with a 63 to 100 x objective). For general analysis, about 300 bands per haploid genome are sufficient, and, while for clinical analysis of genetic defects higher resolutions (500 to 1,000 bands) are needed, nowadays this application really demands parallel array comparative genomic hybridization (discussed below) which is far more sensitive for detecting minute copy number alterations. Karyotyping is performed using on-screen image analysis to produce a standardized karyogram showing any chromosomal alterations present arranged numerically (Figure 2.4.1a). However, for cryptic rearrangements, formed by translocations involving morphologically similar segments from different chromosomes, and complex rearrangements, it is necessary to resort to fluorescence *in situ* hybridization (FISH) based methods (see below).

2.4.4 Telling Species Apart

Unbanded metaphase chromosomes may be used as substrates for banding studies or FISH, or used directly for rapid species verification. After training, nonspecialist operators can quickly recognize whether chromosomes are consistent with supposed species-of-origin, most often encountered among cell lines in common circulation, namely, Chinese hamster (Figure 2.4.1b), human, mouse (Figure 2.4.1c), and rat. Indeed a “quick and dirty” cytogenetic analysis may be performed within a few hours enabling rapid detection of cross-contaminations involving most species represented as cell lines. In such cases chromosome morphologies provide strong clues to species of origin. Thus for example, human chromosomes are generally (numbers 6-12 and X) middle-sized submetacentrics (centromeres lying away from the middle) but include two larger (1 and 3) and two smaller (19 and 20) metacentrics (centromeres located near the middle), together with 5 acrocentric (centromeres located near the ends) pairs (13-15, 21, 22) depicted in Figure 2.4.1a (upper panel). Hamster chromosome complements are fewer than in man and require G-banding for positive discrimination from the former (Figure 2.4.1b). Mouse chromosomes, on the other hand, are uniformly telocentric (Figure 2.4.1c), except where these undergo centric fusion. Of course, cell lines, notably those from cancer patients or transformed *in vitro* by viral immortalization (Figure 2.4.2a), often carry multiple chromosomal rearrangements, which may blur discrimination in extreme cases.

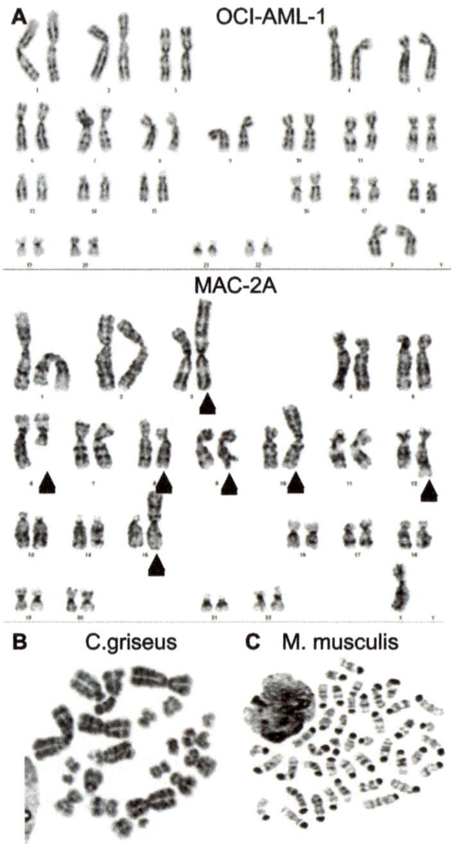

Figure 2.4.1: Cytogenetic analysis. Trypsin G-banding of human **a)**, Chinese hamster **b)**, and mouse **c)** chromosomes. Note dense centromeres and telocentric organization of mouse chromosomes, while human and hamster chromosomes have variable centromeric indices combined with less contrasting banding. The chromosomes of acute myeloid leukemia cell line OCI-AML-1 (upper) and anaplastic large cell lymphoma cell line MAC-2A are organized into karyograms. Note absence of visible rearrangements, unusual in tumor cells, in OCI-AML-1, while MAC-2A carries 7 rearranged chromosomes (arrows).

2.4.5 Fluorescence *in situ* Hybridization (FISH)

A major technical advance was the advent of FISH (for details see [13]) during the 1990s [14, 15]. This widely used technique both informs chromosomal content and enables the gene targets of cancer rearrangements to be identified. FISH exploits the stability and specificity of DNA-DNA hybrids formed after exposure of chromosomes to homologous DNA under renaturing conditions, and utilizes nonisotopically labeled deoxynucleotides and a straightforward method for their efficient incorporation into

DNA by nick translation. These advances prompted the commercial development of chromosome library ("painting") probes, specific for each of the 24 different human chromosomes. Pairwise or three-way combinations of painting probes together with counterstaining the remaining chromosomes is used to resolve chromosome translocations in combination with G-banding (Figure 2.4.2b).

For more complex genomes found in cell lines, such as those of tumor origin, m(ultiplex)-FISH and spectral karyotyping (SKY) are ideal [16]. To tackle the problem

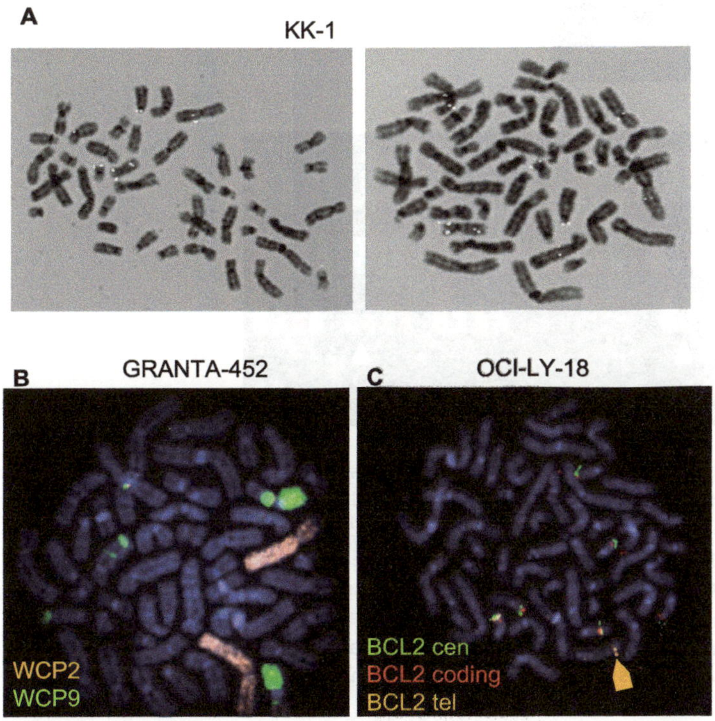

Figure 2.4.2: FISH-based analysis. **a)** FISH analysis of a cell line (KK-1) from a patient with T cell leukemia with a plasmid encoding the HTLV-I genome. Images (left and right) show multiple virus insertions in the same homologs implying stability in vitro. Plasmid DNA was labelled with Cy3-dUTP. G-banding was generated electronically by reversal of DAPI staining. **b)** FISH analysis by chromosome painting to reveal multiple rearrangements of chromosome 9 in a diffuse large B cell lymphoma (DLBCL) cell line GRANTA-452. **c)** Breakpoint analysis of BCL2 with BAC clones in another DLBCL cell line OCI-LY-18. Note multiple rearrangements of BCL2 associated with activational translocation to IGH locus on chromosome 14 (arrow).

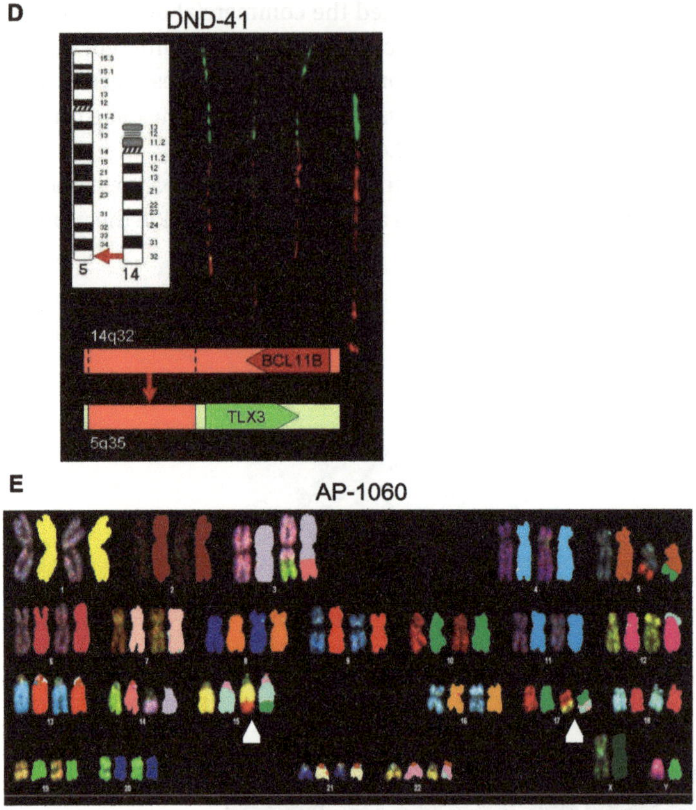

Figure 2.4.2: FISH-based analysis. **d)** Shows fiber FISH analysis of the junction region of fusion between TLX3 (green) and BCL11B (red) genes in a T cell acute lymphoblastic leukemia (T-ALL) cell line, DND-41, effected by t(5;14)(q35;q32). All our detailed knowledge about the configuration of t(5;14) rearrangements, which seem to arise by microinsertion, is based on experiments performed on T-ALL cell lines. **e)** SKY analysis of an acute promyelocytic leukemia cell line AP-1060. Note t(15;17)(q22;q21) which is typical of this entity (arrows). SKY image shows raw (left) and processed (right) images of each chromosomal homolog.

of distinguishing complex chromosomal rearrangements comprised of 24 different homologs in cell lines, given the availability of 6 or 7 reliably separable colors, both methods employ a grid of 24 unique color mixtures which are distinguished either by band-pass filters (m-FISH) or via an interferometer and Fourier transformation (SKY). While neither method allows unambiguous color separation, that of SKY has proved the more robust.

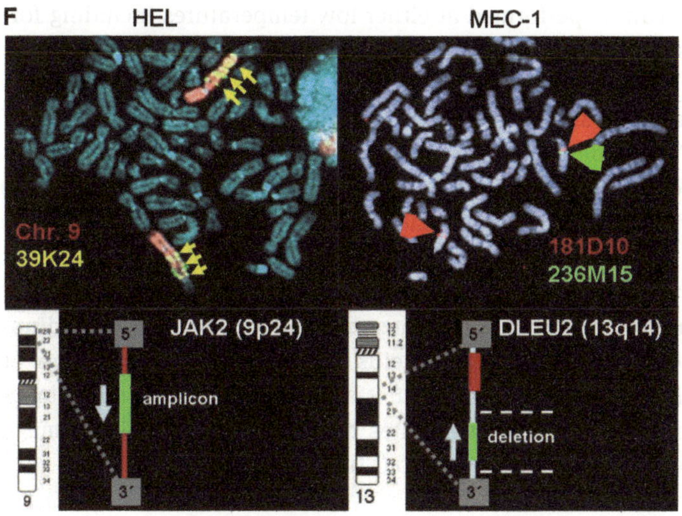

Figure 2.4.2: FISH-based analysis. f) Mutated JAK2 is rendered chromosomally visible by selection in AML cell line HEL (left panel). Arrows show JAK2 amplicons in marker chromosomes. Deletion of DLEU2 present in a B cell chronic leukemia cell line, MEC-1 (right), is visible due to loss of the encompassing 100 kb (green labelled bar and corresponding arrow) genomic region, while the contiguous genomic region (red bar and arrow) is left intact (right panel).

In the wake of the Human Genome Project (HGP) sequence mapped large insert clones, notably fosmid (approx. 40 kb) and bacterial artificial chromosome (BAC) clones (approx. 50 to 300 kb), were made commercially available. This resource enables genes targeted by chromosomal rearrangements in cell lines to be identified using first BACs, then fosmids, to identify flanking or straddling clones. Hence, FISH-using mapped clones bridges the gap between microscopic and molecular biological worlds. And while single locus probes may also be purchased commercially, those wishing to investigate new genetic or neoplastic diseases using cell lines may prefer their own "tailor-made" probes fine-tuned to specific breakpoints (Figure 2.4.2c). The highest FISH resolutions are obtained by the more challenging method known as "fiber FISH" whereby DNA from cell lines is released onto slides for hybridization (Figure 2.4.2d). Using this methodology it was possible to demonstrate that t(5;14)(q35;q32) rearrangements in T cell acute lymphoblastic leukemia (T-ALL) operate to produce microinsertions rather than classical translocations, probably occasioned by the need to avoid disrupting a powerful enhancer region present in the distal 3´-BCL11B region [17].

BAC clones used to prepare these probes contain repeat sequences which require suppression by prehybridization with unlabeled Cot-1 DNA. The posthybridization stringency wash, which can be performed at either low temperatures including formamide to lower the stability of the DNA double helix or at higher temperatures using low salt concentrations alone, is critical to success. Stringency washing allows the operator to balance probe signal intensity against background. The stability of DNA-DNA hybrids on FISH slides allows repeated cycles of stringency washing. For those starting with untested FISH probes, it is feasible to start off using a less stringent wash, which, if yielding unacceptable background levels, can be repeated at higher stringencies (i.e., at lower salt concentrations). The highest stringency washes are performed in water alone. Regardless of the probe combination chosen, counterstaining is usually essential. The standard chromosomal counterstain is 6-diamidino-2-phenylindole dihydrochloride (DAPI), which yields deep blue color, most intense at the centromeres, notably those of chromosomes 1, 9, and 16, and in the terminal long-arm region of the Y chromosome. DAPI generates negative G-bands, which image analysis programs can electronically convert into G-bands (Figure 2.4.2a).

2.4.5.1 Spectral Karyotyping (SKY)

For analysis by SKY, 5 types of fluorescent dye, Spectrum Orange, Texas Red, Cy5, Spectrum Green, and Cy5.5, are used singly and in combination to yield the 24 distinct hues needed to distinguish the human chromosome complement (chromosomes 1-22, X and Y), plus DAPI for counterstaining and reverse G-banding (Figure 2.4.2a, e). SKY allows simultaneous visualization of all chromosomal rearrangements present in a complex karyotype with a resolution approaching 1 Mb. However, analysis of intrachromosomal structural abnormalities, such as inversions, deletions, insertions, and duplications, is limited to gross changes and, hence, demands recourse to single locus probes flanking the breakpoint regions involved. Therefore, in order to identify the site of chromosomal breakage, it is usually necessary to use SKY beforehand. This may then be followed up with G-banding to locate the breakpoints macroscopically, followed by rounds of FISH using single locus probes.

2.4.5.2 Recurrent Chromosome Rearrangements

A key aspect of cancer is the occurrence of recurrent chromosome translocations, e.g., (15;17)(q22;q21). This rearrangement occurs exclusively in acute promyelocytic leukemia (APL) and its retention in candidate APL cell lines confirms their presumptive tumor origin (Figure 2.4.2e). In addition to translocations, point mutations may be observed cytogenetically, albeit indirectly after genomic segmental amplification (Figure 2.4.2f, left panel) or deletion (Figure 2.4.2f, right panel). High throughput sequencing of cancer patients and cognate cell lines have shown that activating mutations, e.g., in several kinase genes, are common in cancer to add to inactivating mutations in tumor suppressor genes that were already known. Mutations may also be introduced by viral infection or transfection. Selective pressure acting on these mutations may be applied by a variety of mechanisms, whether indirect following tumor chemotherapy in humans or directed by selective agents present in the culture medium which favor the upregulation by (epi)genetic means of specific, often metabolic, genes. Mutations may be conceptually divided into those affecting untransformed versus transformed cells. Since somatic cells cease to divide after their Hayflick limits are reached, in all types of cell line genes both innate (e.g., CDKN2B, hTERT, MYC, RB, TP53) and viral in origin (e.g., EBV, HPV6/7, HTLV1, SV40) must be altered or introduced (Figure 2.4.2a) to maintain proliferation. Since most, if not all, of these genes or viruses are implicated in cancer, it is quite difficult to partition genes underlying cell immortalization from those of carcinogenesis. For reasons that remain poorly understood, gene and virus transformed cells initially acquire additional chromosome rearrangements, or undergo ploidy alterations. Hence, untransformed cells are often (incorrectly) simply termed diploid as if this alone is a sufficient hallmark of normality. However, several highly tumorigenic cell lines are now known which bear normal karyotypes (Figure 2.4.1a, upper panel). Nevertheless, cell lines with triploid or tetraploid complements are almost invariably transformed.

2.4.6 Comparative Genomic Hybridization (CGH)

CGH is a cytogenetic method originally based on two-color FISH, and detects genomic imbalances with a resolution approximating 100 kb [18]. Briefly, contrastingly labeled test DNA and normal reference DNA are competitively hybridized along with cold competitor Cot-1 DNA (to reduce unspecific cross-hybridization) onto a normal human metaphase spread. The relative chromatic signal intensities at any given locus reflect the relative abundance of the sequences present at that locus in the 2 DNA samples. Using a sensitive camera, FISH imaging yields visual and quantitative graphic depictions of the actual genomic copy numbers present in the original sample. Over conventional FISH, CGH confers one great advantage: it is largely independent of sample quality enabling direct quantitative evaluation of intractable samples.

The method termed, a(rray) CGH is a derivate of the foregoing technique. aCGH is continually evolving and now provides detailed precise copy number estimates at a high degree of resolution [19]. Moreover, incorporation of single nucleotide polymorphism (SNP) probes yields valuable information on loss of heterozygosity (LOH) which is a feature of tumor-derived cells. aCGH relinquishes metaphase chromosomes in favor of defined genomic segments as targets. Instead, absolute signal ratios are calculated using pooled human DNAs as reference to control for copy number polymorphisms (Figure 2.4.3a, b). Genomic segments derived from tiling arrays are spotted at random (to control field effects) onto glass slides. At first, arrays were comprised of DNA from whole large insert clones, notably golden path BACs. Nowadays, oligonucleotide-based arrays are often used, based on oligomer SNP probes (reviewed in [20, 21]).

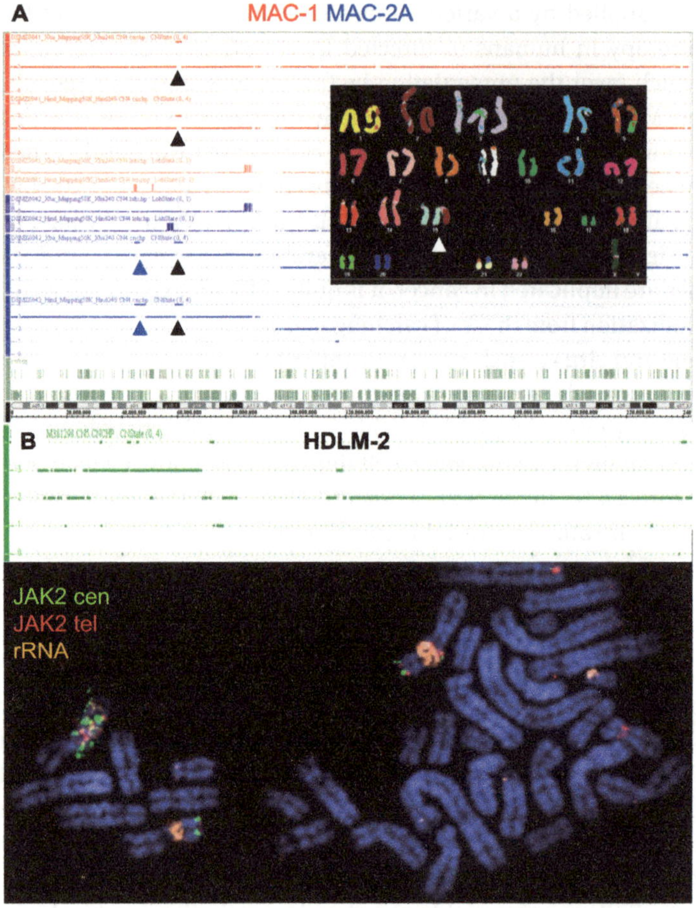

Figure 2.4.3: Array CGH. **a)** Copy number alterations and loss of heterozygosity in sister T cell lymphoma cell lines visualized using an Affymetrix array (50K). Arrows show amplification of MAC-1 (red), MAC-2A (blue), and both cell lines (black). Congruence of XbaI and HindIII restriction enzyme digests serves to validate data. Note copy number increase affecting whole short arm region of chromosome 2 in MAC-2A undetected in the XbaI digest but confirmed by parallel SKY analysis (white arrow shows translocation involving chromosome 15). **b)** Amplification of the JAK2 region in a Hodgkin lymphoma cell line (HDLM-2). Upper figure shows analysis via Affymetrix SNP6 (http://www.sanger.ac.uk/cgi-bin/genetics/CGP/cghviewer/CghHome.cgi), while lower figure shows FISH analysis of JAK2 (red/green) combined with a ribosomal RNA gene clone (orange). Although JAK2 amplification is evident using both methods, detecting the rRNA amplification (and rearrangement) is currently only feasible using FISH due to the high levels of repeat DNA present which resist sequence based analysis.

Useful in its own right, or in combination with cytogenetic or transcriptional microarray data, aCGH measurements yield global genomic high-resolution quantification of copy number alterations which enable small deletions or amplifications to be detected (Figure 2.4.3b). Genomic microarrays vary widely in density which may be tailored to investigators' specific needs, whether single locus or genome wide. By including SNP probes in the array, it is also possible to detect regions undergoing LOH. Since both copy number variations and LOH play critical roles in gene dysregulation and cancer, aCGH has become an essential implement in the oncogenomics "batterie de cuisine". Thanks to their greater sensitivities, high density arrays which combine SNP and copy-number tags enable detection of genetic heterogeneity, since cryptic subclones serve to modify allelic representation. By combining aCGH with cytogenetic data, it is possible to circumvent its main deficiency – its blindness to balanced chromosome rearrangements. Since certain polymorphisms may be pathologically significant, provision of these data alone may be useful. aCGH has revolutionized our picture of the genome which we must now consider a much more fluid entity scattered with regions of variable copy number than hitherto realized. As for chromosome translocations, application of high-density arrays shows that textbook-balanced exchange of genomic material may now turn out the exception rather than the rule. That aside, the corresponding gains in resolution conferred by high density arrays are offset by the increased risk of false positive associations. In oncogenomics precise mapping of chromosomal breakpoints using tiling arrays is now feasible for unbalanced rearrangements [21].

2.4.7 Mutations and Aneuploidy

Gains or losses of single chromosomes (aneuploidy) may occur sporadically in non-transformed cells (often associated with ageing). This tendency is offset by that of homogenization due to subclonal selection (see below). Mutations may serve to silence genes inimical to cell proliferation or favoring apoptosis (tumor suppressor genes), or, on the other hand, to activate genes favoring cell growth and division. These mutations fall into two categories: the widespread and unspecific, e.g., those affecting loss of the TP53 gene, or those specific to narrowly defined entities, e.g., activation of MYCN. Both types of aberration are stably maintained in cell lines derived from the entities involved, again confirming their fitness as models.

2.4.8 Authentication and Cross-contamination

Cell line identification has emerged in recent years as a key issue. About 17 % so-called "new" cell lines supplied to the DSMZ repository by their "originators" turned out after cytogenetic investigation to bear untoward similarities to older classic cell lines, raising the specter of cross-contamination. In almost all cases, this supposition was confirmed by DNA profiling [22, 23]. The current method of choice for cell line authentication, "short tandem repeat" (STR) profiling, which is discussed in detail in the next section, provides positive cell line authentication only in the minority of cases where parallel donor profiling data are also available. For most cell lines, authenticity rests on the "uniqueness" of their STR profiles, leaving the possibility that their supposed identities may be claimed one day by even older "dormant" cell lines open. Moreover, many STR cell line profiles undercut the 80 % similarity bar due to LOH and the microsatellite instability of certain cancers (as discussed in the next section). This is where cytogenetics can come to the rescue, at least where karyotypic data have been published, since most cell line karyotypes are effectively unique. That of the anaplastic large cell lymphoma cell line MAC-2A (Figure 2.4.1a, lower panel), although closely related to those of its sister cell lines MAC-1 and MAC-2B, is still readily distinguishable from its siblings [24].

2.4.9 The Stability Question

The extent to which mutations found in cell lines occur during *in vitro* culture is a matter of ongoing debate. Indeed, there may be no general law. To settle this vexed question there have been, alas, vanishingly few longitudinal studies reported which chart the appearance of gene alterations from primary tissue/tumor through early and late passage cell lines using cloned material to exclude the emergence of weak or dormant clones. Whatever the case, once immortalized cell lines despite high levels of chromosomal rearrangement are usually remarkably stable: thus, the chromosomal signatures of HeLa cells, now in its 7th decade and K-562 well into its 40s, are instantly recognizable, regardless of history or provenance. A recent whole genome sequencing study, which addressed the question at the molecular level in 10 HeLa subclones, concluded that, despite *in vitro* culture over 6 decades, single nucleotide differences were minor and that while genomic differences were present these were mainly early events, such as might have occurred during establishment [25]. Another recent genomic resequencing study has compared seven CHO cell lines derived from the CHO-K1, CHO-S, and DG44 lineages with primary cells from animals taken from the donor colony. This study, in contrast to that on HeLa, uncovered substantial differences which the authors attributed to the various mutagenic and selection treatments to which production cells are subjected unlike cancer cell lines, say [26]. Notwithstanding evidence to the contrary, whenever genomic discrepancies between cultures of the same cell line in different places or at different times do arise, these are all too often attributed to "genomic instability" when other mechanisms may be responsible, such as cross-contamination or genomic heterogeneity as discussed next.

2.4.10 Genetic Heterogeneity

Genetic heterogeneity has recently emerged as a major contributor to cell-to-cell variation within both primary tumors and derived cell lines. Thus, for example, it has been reported that in a brain tumor 3 different subpopulations respectively carried activating mutations in receptor kinases EGFR, MET, and PDGFRA, consistent with independently evolving subclones acquiring diverse second or subsequent hits [27]. Tumor heterogeneity has also been recently observed in tumor cell lines. The cell line in question, U-2932, was established from a patient with diffuse large cell lymphoma (DLBCL) and bears distinct patterns of chromosomal rearrangement closely associated with this entity [28]. In the case of cell lines established from nonneoplastic tissue heterogeneity is less easy to identify. Peripheral blood lymphocytes immortalized by exposure to Epstein-Barr virus (EBV) are initially polyclonal but may homogenize *in vitro* as revealed by uniform integration sites observed in the minority of cell lines where genomic integration occurs (MacLeod, unpublished). In the case of CHO, homogenization of early subclonal disparities has also been documented [26]. It is likely that younger cell lines retain genetic heterogeneity conservation of which depends on rigorous adherence to good cell culture practice, namely establishing separate seeding and experimental banks for each cell line, minimizing over-dilution which merely promotes bottlenecking selection.

2.4.11 Further Cytogenetic Applications

Cytogenetic analysis of induced chromosome breakage is sometimes used to estimate radiation exposure. Similar procedures may be applied to cells in culture when attempting to measure the effects of radiation or clastogenicity of chemical or environmental agents. Both classical and FISH-based methods may be used. Because radiation sensitivity may vary during the cell cycle, times elapsed between treatment and harvesting are critical. One test, the so-called "G2 assay" used to identify cancer susceptibility in primary cell cultures, has also been applied to cell lines [29]. Direct measurement of breakage induced in cytogenetically invisible G1 cells has been achieved by premature chromosome condensation whereby interphase cells are fused to mitotic cells by exposure to polyethylene glycol or Sendai virus [30, 31].

Finally, cytogenetic analysis is normally required to monitor pluripotent stem cell lines in which the question of stability *in vitro* has yet to be resolved one way or the other. (For conflicting views see [32, 33].) Stem cell lines will of course retain genomic aberrations present in donor patients.

2.4.12 Conclusion

The resource afforded by continuous cell lines has profited from the application of cytogenetics. Their analysis using clones provided in the wake of the Human Genome Project and more recently when applied to other species has illuminated the pathology and evolution of increasing numbers of cell lines. Moreover, cytogenetics was instrumental in alerting to the scourge of false cell lines. Now, thanks to an effective purging and continual monitoring by first cytogenetics and STR profiling (see below), the cell line resource is now probably less affected by the presence of undetected false cell lines than ever before. Cytogenetics is also well-positioned to partner array-based analyses, both genomic using aCGH to map breakpoints and transcriptional to identify translocation targets. Finally, cytogenetic data are needed to validate deletions, amplifications, and translocations identified by whole genome sequencing.

References

[1] Drexler HG. The Leukemia-Lymphoma Cell Line FactsBook. New York, USA: Academic Press, 2001.

[2] Drexler HG. Guide to Leukemia-Lymphoma Cell Lines, 2nd ed. Braunschweig, Germany: DSMZ, 2010.

[3] MacLeod RA, Dirks WG, Drexler HG. Where have all the cell lines gone? Int J Cancer 2013;132:1232-4.

[4] Shaffer LG, McGowan-Jordan J, Schmid M. ISCN: An International System for Human Cytogenetic Nomenclature. Basel, Switzerland: S Karger, 2013.

[5] Tijo JH, Levan A. The chromosome number of man. Hereditas 1956,42:1-6.

[6] Nowell PC, Hungerford DA. A minute chromosome in human chronic granulocytic leukemia. Science 1960;132:1497.

[7] Caspersson T, Zech L, Johansson C. Analysis of human metaphase chromosome set by aid of DNA-binding fluorescent agents. Exp Cell Res 1970;62:490-2.

[8] Sumner AT. A simple technique for demonstrating centromeric heterochromatin. Exp Cell Res 1972;75:304-6.

[9] Sumner AT, Evans HJ, Buckland RA. New technique for distinguishing between human chromosomes. Nat New Biol 1971;232:31-2.

[10] Seabright M. A rapid banding technique for human chromosomes. Lancet 1971;2:971-2.

[11] Masters JR. HeLa cells 50 years on: the good, the bad and the ugly. Nat Rev Cancer 2002; 2:315–9.

[12] MacLeod RA, Kaufmann M, Drexler HG. Cytogenetic harvesting of commonly used tumor cell lines. Nat Protoc 2007;2:372-82.

[13] MacLeod RA, Drexler HG. Classical and molecular cytogenetic analysis. Methods Mol Biol 2013;946:39-60.

[14] Cremer T, Lichter P, Borden J, et al. Detection of chromosome aberrations in metaphase and interphase tumor cells by in situ hybridization using chromosome-specific library probes. Hum Genet 1988;80:235-46.

[15] Lichter P, Cremer T, Borden J, et al. Delineation of individual human chromosomes in metaphase and interphase cells by in situ suppression hybridization using recombinant DNA libraries. Hum Genet 1988;80:224-34.

[16] Schrock E, du Manoir S, Veldman T, et al. Multicolor spectral karyotyping of human chromosomes. Science 1996;273:494–97.
[17] Nagel S, Scherr M, Kel A, et al. Activation of TLX3 and NKX2-5 in t(5;14)(q35;q32) T-cell acute lymphoblastic leukemia by remote 3'-BCL11B enhancers and coregulation by PU.1 and HMGA1 Cancer Res 2007;67:1461-71.
[18] Kallioniemi A, Kallioniemi OP, Sudar D, et al. Comparative genomic hybridization for molecular cytogenetic analysis of solid tumors. Science 1992;258:818-21.
[19] Pinkel D, Segraves R, Sudar D, et al. High resolution analysis of DNA copy number variation using comparative genomic hybridization to microarrays. Nat Genet 1998;20:207-11.
[20] Lockwood WW, Chari R, Chi B, et al. Recent advances in array comparative genomic hybridization technologies and their applications in human genetics. Eur J Hum Genet 2006;14:139-48.
[21] Le Scouarnec S, Gribble SM. Characterising chromosome rearrangements: recent technical advances in molecular cytogenetics. Heredity 2012;108:75-85.
[22] MacLeod RA, Dirks WG, Matsuo Y, et al. Widespread intraspecies cross-contamination of human tumor cell lines arising at source. Int J Cancer 1999;83:555-63.
[23] American Type Culture Collection Standards Development Organization Workgroup ASN-0002. Cell line misidentification: the beginning of the end. Nat Rev Cancer 2010;10:441-8.
[24] Ehrentraut S., Nagel S, Scherr ME, et al. t(8;9)(p22;p24)/PCM1-JAK2 activates SOCS2 and SOCS3 via STAT5. PLoS One 2013;8:e53767.
[25] Adey A, Burton JN, Kitzman JO, et al. The haplotype-resolved genome and epigenome of the aneuploid HeLa cancer cell line. Nature 2013;500:207-11.
[26] Lewis NE, Liu X, Li Y, et al. Genomic landscapes of Chinese hamster ovary cell lines as revealed by the Cricetulus griseus draft genome. Nat Biotechnol 2013;31:759-65.
[27] Snuderl M, Fazlollahi L, Le LP, et al. Mosaic amplification of multiple receptor tyrosine kinase genes in glioblastoma. Cancer Cell 2011;20:810-17.
[28] Quentmeier H, Amini RM, Berglund M, et al. U-2932: two clones in one cell line, a tool for the study of clonal evolution. Leukemia 2013;27:1155-64.
[29] Bryant PE, Gray L, Riches AC, et al. The G2 chromosomal radiosensitivity assay. Int J Radiat Biol 2002;78:863-6.
[30] Johnson RT, Rao PN. Mammalian cell fusion: Induction of premature chromosome condensation in interphase nuclei. Nature 1970;26:717–22.
[31] Darroudi F, Nararajan AT. Premature chromosome condensation, a novel method for biological dosimetry. In: High Levels of Natural Radiation. Proc 10th Int Congr Vienna.1993:479–85.
[32] Dekel-Naftali M, Aviram-Goldring A, Litmanovitch T, et al. Screening of human pluripotent stem cells using CGH and FISH reveals low-grade mosaic aneuploidy and a recurrent amplification of chromosome 1q. Eur J Hum Genet 2012;20:1248-55.
[33] Funk WD, Labat I, Sampathkumar J, et al. Evaluating the genomic and sequence integrity of human ES cell lines; comparison to normal genomes. Stem Cell Res 2012;8:154-64.

2.5 Quality Control Essentials in Human Cell Culture: Cell Line Cross-contamination and Microbiological Infections

Wilhelm G. Dirks and Cord C. Uphoff

2.5.1 Cell Line Cross-Contamination

The biomedical sciences are regularly startled by reports that many cell lines used in research are false and therefore not suitable in the service of any research. Consequences of using cross-contaminated cell lines have been recently reported by Boonstra of the University Medical Center Rotterdam [1], who had asked laboratories from around the world to send in cell lines that had originally been established of certain tumors (adenocarcinomas) of the esophagus. Within a recurrently used panel of 13 cell lines modeling tumorigenesis of the esophagus, they found that the cell lines SEG-1, BIC-1, and SK-GT-5 did in reality originate from lung, stomach, and intestine, respectively. Experiments including the three obviously false cell lines had led to eleven patents in the United States and more than 100 professional publications worldwide on clinical studies with patients suffering from esophageal cancer have been published. This case and other cross-contamination incidents are a legitimate reason to increase the fear for errors in respective translational research within the development of new therapies. Here, the waste of research funds on the basis of misleading results and records due to incorrect model systems would be reason enough to conduct a rigorous quality control of cell lines. Since the number of false cell lines in circulation is unacceptably high, authentication of cell line model systems should be carried out prior to commencement of a project.

2.5.1.1 Resistance to Accept False Cell Lines in Case of ECV-304

When false cell lines are used as a source of basic science or to study a common cellular process, there are few if any negative consequences. However, when a study depends on a cell line being representative of a specific genetic background, mistakenly using a cell line from a different background can invalidate the entire study and put to waste substantial investments of time and money. Originally, the first human endothelial cell line, ECV-304, was described by the originator as a spontaneously transformed cell line derived from human umbilical vein endothelial cells (HUVEC) of a new born Japanese female [2]. Using new PCR-based single locus fingerprint technique, the German bioresource center DSMZ discovered that this cell line is identical to T-24, a urinary bladder carcinoma cell line [3]. Since T-24 was established in 1970 compared to 1984 for ECV-304 it was assumed that ECV-304 is the false cell line, which subsequently could be confirmed by the cell banks ATCC (USA), ECACC (UK),

RIKEN (Japan), JCRB (Japan) and KCLB (Korea). The unmasking of ECV-304 as a model for *in vitro* studies of endothelial functions has been a sensitive setback for vascular research and a return to the usage of freshly isolated primary HUVECs, which are difficult to isolate and of short lifespan in culture, appeared inevitable. Perhaps this is the reason why, despite the overwhelming facts of the nonexistence of ECV-304, some groups still publish studies using ECV-304 as a valuable model for the study of cellular processes in the endothelium and angiogenesis [4, 5].

2.5.1.2 History of Methods for Human Cell Line Authentication

The application of specific species markers including cell surface antigens and chromosomes showed that interspecies misidentification has been even in the 1970s a widespread problem [6, 7]. Subsequently, it was shown that intraspecies contamination of human cell cultures was also a serious problem, which could be monitored by the innovation of isoenzymatic analysis [8]. After extending this approach to multiple polymorphic isoenzymes, the persistence of specific marker chromosomes in long-term-passaged cell lines demonstrated the unique power of cytogenetics [9]. Based on the detection of chromosomal markers it was convincingly demonstrated that multiple cell lines under active investigations were actually derived from one source, namely the HeLa cell line [10]. Furthermore, wrong identities among established cell lines occurred at frequencies as high as 16 to 35 % in the late 1970s [11]. Recently, the Department of Human and Animal Cell Cultures of DSMZ could demonstrate an incidence of 14 % of false human cell lines [12-14] indicating intraspecies cross-contaminations as a chronic problem and highlighting the badly neglected need for intensive quality controls regarding cell line authenticity.

Compared to polymorphic isoenzymes or marker chromosomes, a much higher resolution in discrimination among human cell lines was achieved using restriction fragment length polymorphism (RFLP) of simple repetitive sequences [15], which lead subsequently to the concept of "DNA fingerprinting" [16]. The principle of the method is based on the phenomenon that genomes of higher organisms harbor many variable number of tandem repeat (VNTR) regions, which show multiallelic variation among individuals [17]. Sequence analysis demonstrated that the structural basis for polymorphism of these regions is the presence of tandem-repetitive, nearly identical DNA elements, which are inherited in a Mendelian way. Depending on the length of the repeats, VNTRs are classified into minisatellites consisting of 9 to > 70 bp core sequences and microsatellites which include all short tandem repeats (STRs) with core sizes from 1 to 6 bp. STR typing of tetrameric repeats is meanwhile the gold standard and an international reference technique for authentication of human cell lines [18].

2.5.1.3 STR DNA Profiling as a Global Standard in Human Cell Line Authentication

A more informative but also more complex technique for authentication of human cell lines is based on STR DNA typing. The multi allelic variation of the sequences is so great that combination of a few loci allows distinguishing individuals except identical twins. A STR is a microsatellite, consisting of a unit of 2 to 7 nucleotides repeated generally 10 to 40 fold in a head to tail orientation on the DNA strand. This method differs from restriction fragment length polymorphism analysis (RFLP) since STR analysis does not cut the DNA with restriction enzymes. STR analysis measures the exact number of repeating units by using fluorescence primers in a PCR (polymerase chain reaction) to amplify the respective STR loci followed by capillary electrophoresis for a high-resolution fragment size determination. STR analysis is therefore becoming the prevalent method for the DNA analyses. Masters et al. [18] have established STR based typing systems for cell lines, which exclude accidental identity with a probability of less than 1.14×10^{8}. It represents a robust and quick method for the generation of DNA profiles. In less than 8 h the uniqueness of the DNA profile can be checked and due

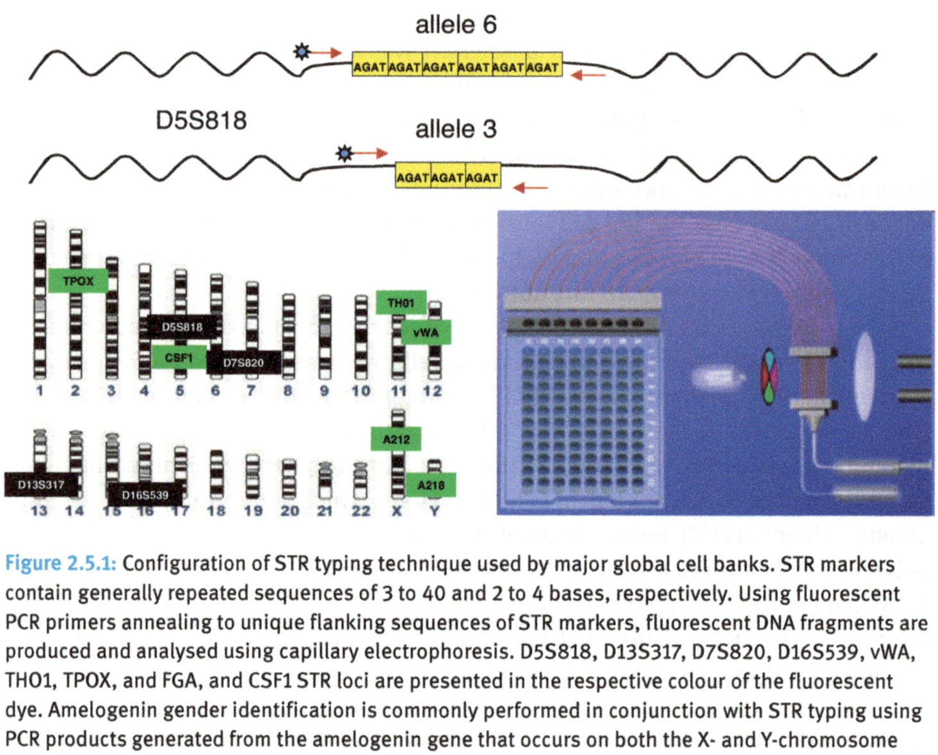

Figure 2.5.1: Configuration of STR typing technique used by major global cell banks. STR markers contain generally repeated sequences of 3 to 40 and 2 to 4 bases, respectively. Using fluorescent PCR primers annealing to unique flanking sequences of STR markers, fluorescent DNA fragments are produced and analysed using capillary electrophoresis. D5S818, D13S317, D7S820, D16S539, vWA, THO1, TPOX, and FGA, and CSF1 STR loci are presented in the respective colour of the fluorescent dye. Amelogenin gender identification is commonly performed in conjunction with STR typing using PCR products generated from the amelogenin gene that occurs on both the X- and Y-chromosome (209 bp and 215 bp, respectively).

to the international standardization the STR reference databases of the DSMZ, the ATCC, the JCRB, and RIKEN can easily be performed [19]. Under the direction of an international working group made up of subject matter experts possessing relevant experience in DNA profiling technologies, the ATCC Standards Development Organization (SDO) has published the standardized procedure for unambiguous authentication and identification of human cell lines using STR profiling [18]. Based on the set of STR loci as shown in Figure 2.5.1, the major global cell banks ATCC, DSMZ, HPACC, JCRB, and RIKEN have built STR databases that are continuously updated.

2.5.1.4 DNA STR Typing as Quality Control in Routine Cell Culture Combined to an Online Tool

The quality control of cell lines in research is still a voluntary step, although cells with false identity can influence results of research significantly. Furthermore, the use of cell line models is impaired by reliance on misidentified examples representing entities with biological characteristics different from those supposed. In recent years, the major cell banks and a confusing number of small companies have started offering cell line authentications. Because the use of an official authentication service is not anonymous and finding false cell lines could be embarrassing for many scientists, DSMZ has launched an online tool for STR analysis [19].

After converting an STR profile to a table of alleles of the respective STR loci of a human cell line (see Table 2.5.1), the suspected identity can be proven by online verification of customer-made STR data sets on the homepage of the DSMZ institute (http://www.dsmz.de/services/services-human-and-animal-cell-lines/online-str-analysis.html).

Registered users simply login at the online STR analysis site on the DSMZ homepage and will be guided. Aided by simple prompts, users can input their own cell line STR data to retrieve best matches with authenticated cell lines listed on the database. Inevitably, reference profiles remain subject to revision until all commonly held cell lines have been STR-typed across participating repositories. At present, about 2,342 such cell lines have been STR-typed and are represented as reference sets on the database. Armed with this tool, online verification of cell line identity should prove a vital weapon to combat the havoc of cell line cross-contamination which has dogged cancer research since inception. The combination of easy and rapidly generated STR profiles and their authentication by an online screen at the reference database constitute a major and novel progress in decreasing the use of false cell lines.

Table 2.5.1: Allele organization and sizes of amplified human STR Loci. Nucleotide range and the number of known alleles of each STR loci are summarized. Further information is available at http://www.cstl.nist.gov/div831/strbase. Regular fragment sizes in base pairs of alleles are printed plain, variant alleles are printed in italics.

Allele	D5S818	D13S317	D7S820	D16S539	vWA	TH01	TPOX	CSF1PO	Amelo-genin
3						169			
4						173			209 = X
5		164	212	266		177	220	287	215 = Y
6	114	168	216	270		181	224	291	
7	118	172	220	274		185	228	295	
8	122	176	224	278		189	232	299	
9	126	180	228	282		193	236	303	
9.3						196			
10	130	184	232	286	118	197	240	307	
11	134	188	236	290	122	201	244	311	
12	138	192	240	294	126	205	248	315	
13	142	196	244	298	130		252	319	
14	146	200	248	302	134		256	323	
15	150	204	252	306	138			327	
16	154				142			331	
17	158				146				
18					150				
19					154				
20					158				
21					162				
22					166				
23					170				

2.5.2 Microbiological Contaminations of Cell Cultures

Microbiological contaminations are certainly the most common and aggravating incidents in cell culture technology. They can occur during manipulation of the cell cultures in consequence of unsterile handling of cell culture tools, media, or supplements. This usually leads to the contamination with airborne bacteria, yeasts, or fungi that rapidly overgrows the eukaryotic cells and can be detected shortly after introduction. On the other hand, a microbial contamination can already be present in the primary cells originating from infected tissue or in cell culture media and supplements due to contaminated raw material [20]. Another source of infection can be the transmission of bacteria or viruses from one cell culture to another cell culture (cross-contamination). In contrast to the accidental intrusion of microorganisms in the laboratory the chronic residual contaminants and those arising from cross-contamination are much more difficult to detect as they are often tightly connected to the eukaryotic cells and do not overgrow them to visible amounts. Unfortunately, until today no general methods exist which are suitable to detect bacterial and/or viral contaminations in general. Although some attempts were made to generally detect contaminating organisms by nucleic acid amplification procedures (e.g., highly conserved ribosomal RNA sequences of eubacteria or degenerate-oligonucleotide primed PCR for DNA and RNA viruses [21]) or protein determination applying matrix-assisted laser desorption/ionization, time of flight (MALDI-TOF) mass spectrometry for bacteria and viruses [22], none of the methods are developed enough to be conveniently employed in cell culture technology. Thus, methods for the detection of the specific unwanted adventitious organisms are necessary to uncover or to exclude their presence. In this regard, it is of utmost importance to assess the probability and the consequence of a contamination of a given cell culture or any compound in the cell culture system which cannot be sterilized both for the safety of employees handling the cell cultures or patients treated with cell culture products and for the reliability of experimental results.

To assess the potential presence of additional organisms in a cell culture it is advantageous to have a well characterized and documented origin of the primary material. Concerning human material, it should be verified that the donor tissue is free of infectious diseases and animal material should preferably be explanted from pathogen free breeding [23]. Potentially infected material, e.g., cervix carcinoma cells from papillomavirus infected patients or B cells from Epstein-Barr virus (EBV) infected persons, should be tested for the presence and expression of the respective virus. Beside these relatively clear cases of infections originating from the primary material a few microorganisms emerged to become more or less widespread contaminants in cell cultures with little species or tissue specificity: the most common bacteria found in cell cultures are mycoplasmas, but also mycobacteria and several other bacteria appear once in a while undetected permanently in cell cultures [24]. To make matters worse, some of these bacteria penetrate the eukaryotic cells and exist intracellularly (e.g., *Mycoplasma fermentans*, *Mycobacterium avium* complex [25]). Furthermore, several strains turned out to be resistant to multiple antibiotics. Concerning viruses, xeno- and polytropic mouse leukemia viruses (X-/P-MLV) exhibit the highest prevalence in established human and animal cell cultures [26]. X-/P-MLV affect numerous cell types expressing the XPR1 surface protein and can – similarly to bacteria – be transmitted to already established cell lines. Other virus infections usually belong to the primary infections. Taken together, more than 30 % of the cell lines circulating in the cell culture laboratories are contaminated either with mycoplasmas and/or with X-/P-MLV (unpublished data). In contrast to the virus infections, mycoplasmas can be eliminated applying antibiotics.

2.5.2.1 Mycoplasma Detection

Mycoplasmas are characterized by their lack of a rigid cell wall, the small size regarding morphology as well as genetic equipment, and their restricted metabolic capabilities. They constitute a separate bacterial class designated mollicutes. Nowadays, mycoplasmas are most likely transmitted to mycoplasma-free cell cultures by cross-contamination from an already infected culture. Thus, the most effective way to prevent mycoplasma contamination is to strictly adhere to the good microbiological practices with special emphasis to cell culture. A protocol for the prevention of contaminations should be designed to minimize dissemination of any microbes and eukaryotic cells from one culture to another [27]. Additionally, one or – preferably – two independent methods should be established in the laboratory to test all cell cultures for mycoplasma contamination at introduction into the laboratory and regularly during cell culturing. According to the regulations for the preparation of medicinal or pharmacological products from cell cultures (United States, European and Japanese pharmacopeia), two methods are still considered as standard methods: the microbiological culture method and the indirect fluorochrome staining method. Both show high sensitivities, but are time consuming (almost 3 weeks and 1 week, respectively) and require a certain degree of experience to evaluate the results. However, the recent editions of the pharmacopeia also permit the application of nucleic acid amplification testing (NAT) when the method detects verifiably at least 10 colony forming units (CFU) per mL. Additionally, several validation parameters need to be addressed to replace the standard methods: specificity, detection limit, robustness, repeatability, and ruggedness [28]. A number of PCR assays are described or commercially available which can be established in the laboratory and the sensitivity of the assay can be adapted to the claimed parameters.

For research laboratories, which are not directly linked to medical or pharmacologic applications or product quality control with the obligatory high evaluation and documentation standards, the key aspects of the detection assays are reliability in sensitivity and specificity, rapidness and convenience. The titer of mycoplasmas in chronically infected cell cultures is usually high and the detection limit of an assay is not the crucial factor, but the detection of the diverse mycoplasma species among the eukaryotic cells is important. By now there are several detection assays available which fulfill these criteria and enables the user of cell cultures to check the active cultures for contamination frequently and with little effort. Although the reliabilities of the tests are generally very high, it is still advisable to apply two different and independent assays to achieve highest certainty. The most common molecular biological detection assays are based on the PCR technology [29] and a multitude of ready-to-use kits are commercially available. Several variants were described in the literature and can easily be established in the laboratory. The selection of a PCR method requires some preconditions depending on the presence of Taq-polymerase inhibitors in cell culture supernatants:

1) Use cell culture supernatant containing few eukaryotic cells to reduce the excess of eukaryotic DNA;
2) perform a DNA extraction to get rid of most of the Taq-polymerase inhibitors;
3) apply an internal control DNA at a limiting concentration for each sample to determine any loss in sensitivity;
4) perform the necessary control reactions with positive, negative, and water control to avoid false positive or false negative results [30].

Additional efficient methods for the detection of mycoplasmas are DNA-RNA-hybridization with labeled (radioactively or with acridinium ester for luminometric measurement) single stranded DNA probes (Mycoplasma Tissue Culture Non-Isotopic Rapid Detection by Merck Millipore, Billerica, MA, USA), fluorescence *in situ* hybridization (FISH) with mycoplasma specific fluorescently labeled probes and biochemical assays for the determination of ATP production by mycoplasma-specific enzymes (MycoAlert Mycoplasma Detection Kit by Lonza, Amboise, France).

2.5.2.2 Mycoplasma Elimination

Due to the absence of a rigid cell wall and their reduced metabolic abilities, mycoplasmas are resistant to many commonly applied antibiotics, such as penicillin or streptomycin. However, several antibiotics, particularly from the antibiotic groups of fluoroquinolones, macrolides, pleuromutilins, and tetracyclins are effective in killing the mycoplasmas. The recommendations of some companies to add antibiotics prophylactically to the culture medium at low concentrations has to be rejected strictly and should be subjected to special applications only. The permanent use of antibiotics in cell cultures causes the development and propagation of resistant strains. Furthermore, antibiotic free cell cultures compel cell culture technologists to strictly adhere to good cell culture practices and possible lapses can be determined.

As experience has shown, a single antibiotic treatment has a success rate of 65 to 85 %, depending on the antibiotic applied [31]. However, not only resistances account for the failures, but also the loss of cell cultures during or shortly after the treatment period caused by an intensive impairment of the mycoplasma contaminated cells. To avoid resistances, two or more treatments in parallel with different antibiotics or combinations of antibiotics should be applied. The authors were successful using a single fluoroquinolone (e.g., ciprofloxacin or enrofloxacin), a combination treatment with a fluoroquinolone and a macrolide (Plasmocin by Cayla-InvivoGen, Toulouse, France), and the alternating administration of a tetracyclin (minocyclin) and a pleuromutilin (tiamulin) (BM-Cyclin by Roche, Mannheim, Germany). At least one of the treatments is usually effective. The treatment is not only harmful for the mycoplasmas, but certainly also stressful for the eukaryotic cells due to effects on the mitochondria. Thus, the cells should be cultured under optimal conditions at high cell densities and serum concentrations. The antibiotic concentrations should be kept constant by frequent medium exchange as far as the cells tolerate this. After the treatment period the cells have to be cultured at least for a further 2 weeks without any antibiotics to get rid of residual dead mycoplasmas which might produce false positive results in the subsequent mycoplasma test or to allow residual viable mycoplasmas to grow to detectable titers to prevent false negative mycoplasma test results [32].

2.5.2.3 Viral Contamination of Cell Cultures

In contrast to bacterial, fungi, or yeast contaminations, viral infections are usually species and tissue specific and are already present in the primary cells prior to the cell culture set-up. The exclusion of serum antibodies in the blood of the donor individual qualifies the excised cells as free for the tested virus because later infections are unlikely. But as the infection status of the donors is unknown in most cases, cell lines should be tested for potential viral contaminations to assess possible impacts on the experimental results obtained with an infected cell culture and the risk for the user handling the cells. Special attention should be given to viruses, which are prevalent in cell cultures and/or associated with specific diseases, e.g., the Epstein-Barr viruses (EBV, human herpesvirus type 4) in human B cells and Burkitt lymphoma cell lines, human herpesvirus type 8 (HHV-8) in Kaposi sarcoma cell lines, or hepatitis B viruses (HBV) in hepatocellular carcinoma cell lines. All these viruses can be detected by PCR. However, the confirmation of the presence of viral DNA by PCR is not identical to the production of the virus. Thus, further investigations are necessary to determine the eventual propagation of the virus in PCR-positive samples. This can be performed by the detection of active viruses in the supernatant of cell cultures or the detection of virus propagation-associated proteins [33]. For example, the immediate early protein BZLF1 of EBV indicates the lytic state of an infection.

As mentioned above, some mouse and monkey retroviruses (X-/P-MLV) and squirrel monkey retrovirus (SMRV) were found to infect human cell cultures and are detected in an increasing number of human cell lines [26]. The source of the infections is currently unknown, but several transmission ways are conceivable:

1) many primary cell cultures are grown on feeder layer originating from X-/P-MLV infected mice;
2) a number of human cell lines were transplanted into immunocompromised mice infected with murine retroviruses for growth support or for the verification of tumor formation and subsequent recovered from the animal;

3) contaminated cell culture supplements of murine origin (e.g., growth factors) might have infected the cells
4) the viruses could have been transmitted (similarly to mycoplasma or eukaryotic cell contaminations) via cross-contamination from infected cell cultures.

Recombination of two related viruses of a double infected cell line with murine leukemia viruses was described for the prostate cancer cell line 22RV1 [34]. The newly discovered virus was designated "xenotropic murine leukemia virus-related virus" (XMRV). The detection of X-/P-MLV in cell lines and other samples raises concerns regarding the production of biologically active substances by infected mice, murine, or even human cell lines. The products can contain residual DNA or infectious particles that either affect the detection of the same virus when contaminated detection kits

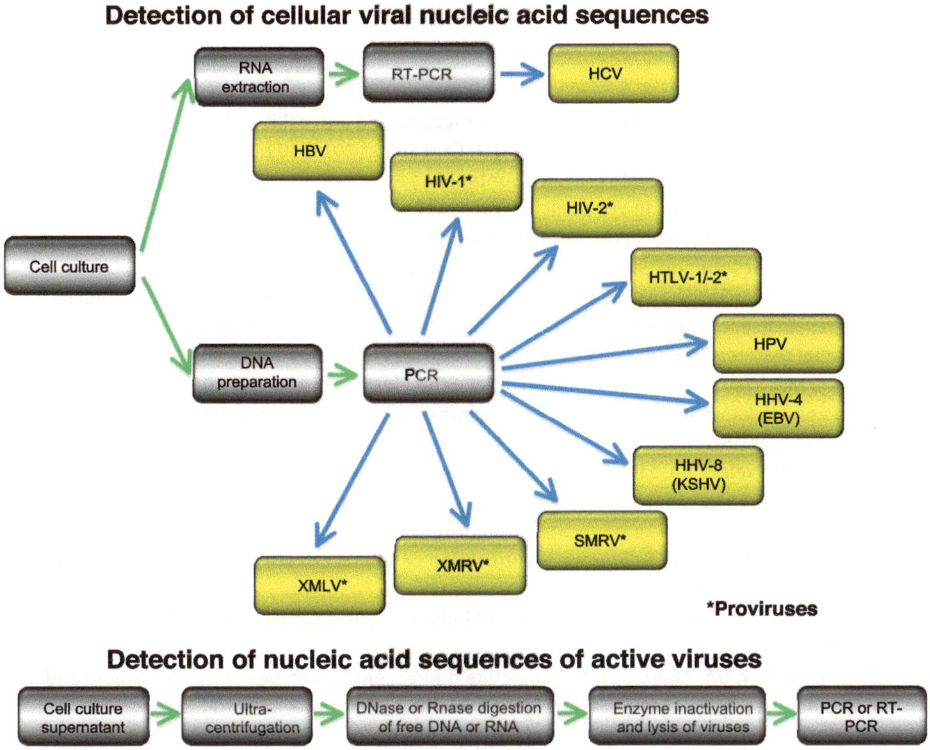

Figure 2.5.2: Schematic illustration of the detection of human pathogenic and cell culture contaminating viruses. Note that the human pathogenic viruses are tissue specific. EBV = Epstein-Barr virus, HBV = hepatitis B virus, HCV = hepatitis C virus, HHV-4 = human herpesvirus type 4, HHV-8 = human herpesvirus type 8, HIV-1 = human immunodeficiency virus type 1, HIV-2 = human immunodeficiency virus type 2, HPV = human papillomavirus, SMRV = squirrel monkey retrovirus, XMLV = xenotropic mouse leukemia retrovirus, XMRV = xenotropic mouse leukemia virus-related virus.

are produced or might contaminate the sample [35]. An example of multiple virus contamination is the cell line NAMALWA of which some subclones (NAMALWA.CSN/70 and NAMALWA.KN2) are demonstrably infected with EBV, SMRV, and XMLV, whereas other subclones (e.g., NAMALWA, NAMALWA.PNT) only contain the first two viruses. The cell line has been extensively used for the production of interferon alpha (IFN-α).

Many human cell lines have been tested for the presence and absence of a variety of human pathogenic and other viruses [33]. A schematic illustration of the detection of different viruses in cell cultures is shown in Figure 2.5.2. Except for the X-MLV infections, which can be diverse depending on the history of the given cell culture, most results are cell line specific and do not differ from one culture to another of the same cell line.

In summary, the quality control of cell cultures for microbiological contaminations is important as to the purity of cell culture products, the reliability of experimental results, and the risk assessment for cell line handling.

References

[1] Boonstra JJ, van Marion R, Beer DG, et al. Verification and unmasking of widely used human esophageal adenocarcinoma cell lines. J Natl Cancer Inst 2010; 102:271-4.
[2] Takahashi K, Sawasaki Y, Hata J, et al. Spontaneous transformation and immortalization of human endothelial cells. In Vitro Cell Dev Biol 1990; 26:265-74.
[3] Dirks WG, Drexler HG, MacLeod RAF. ECV304 (endothelial) is really T24 (bladder carcinoma): cell line cross-contamination at source. In Vitro Cell Dev Biol 1999; 35:558-9.
[4] Suda K., Rothen-Rutishauser B., Gunthert M., et al. Phenotypic characterization of human umbilical vein endothelial (ECV304) and urinary carcinoma (T24) cells: endothelial versus epithelial features. In Vitro Cell Dev Biol Anim 2001; 37:505-14.
[5] Andozia MB, Vieira CS, Franceschini SA, et al. Ethinylestradiol and estradiol have different effects on oxidative stress and nitric oxide synthesis in human endothelial cell cultures. Fertil Steril 2010; 94:1578-82.
[6] Rothfels FH, Axelrad AA, Simonovitch L. The origin of altered cell lines from mouse, monkey, and man as indicated by chromosomes and transplantation studies. Proc Can Cancer Res Conf 1959; 3:189-214.
[7] Simpson WF, Stuhlberg CS. Species identification of animal cell strains by immunofluorescence. Nature 1963; 199:616-7.
[8] Gartler SM. Apparent HeLa contamination of human heterodiploid cell lines. Nature 1968; 217:750-1.
[9] Miller OJ, Miller, DA, Allerdice PW. Quinacrine fluorescent karyotypes of human diploid and heteroploid cell lines. Cytogenetics 1971; 10:338-41.
[10] Nelson-Rees WA, Flandermeyer RR. HeLa cultures defined. Science 1976; 191:96-8.
[11] Nelson-Rees WA. The identification and monitoring of cell line specificity. Prog Clin Biol Res 1978; 26:25-79.
[12] MacLeod RAF, Dirks WG, Matsuo Y, et al. Widespread intraspecies cross-contamination of human tumor cell lines arising at source. Int J Cancer 1999; 83:555-63.
[13] Drexler HG, Dirks WG, MacLeod RAF. False human hematopoietic cell lines: cross-contaminations and misinterpretations. Leukemia 1999; 13:1601-7.

[14] Dirks WG, MacLeod RAF, Jaeger K, et al. First searchable database for DNA profiles of human cell lines: sequential use of fingerprint techniques for authentication. Cell Mol Biol 1999; 5:841-53.
[15] Epplen JT, McCarrey JR, Sutou S, et al. Base sequence of a cloned snake chromosome DNA fragment and identification of a male-specific putative mRNA in the mouse. Proc Natl Acad Sci USA 1982; 793:798-802.
[16] Jeffreys AJ, Wilson V, Thein SL. Hypervariable minisatellite regions in human DNA. Nature 1985; 314:67-73.
[17] Nakamura Y, Leppert M, O´Connell P, et al. Variable number of tandem repeat [VNTR] markers for human gene mapping. Science 1987; 235:1616-22.
[18] Masters JR, Thompson JA, Daly-Burns B, et al. Short tandem repeat profiling provides an international reference standard for human cell lines. Proc Natl Acad Sci 2001; 98:8012-7.
[19] Dirks WG, MacLeod RAF, Nakamura Y, et al. Cell line cross-contamination initiative: An interactive reference database of STR profiles covering common cancer cell lines. Int J Cancer 2010; 126:302-4.
[20] Uryvaev LV, Dedova AV, Dedova LV, et al. Contamination of cell cultures with bovine viral diarrhea virus (BVDV). Bull Exp Biol Med 2012; 153:77-80.
[21] Nanda S, Jayan G, Voulgaropoulou F, et al. Universal virus detection by degenerate-oligonucleotide primed polymerase chain reaction of purified viral nucleic acids. J Virol Methods 2008; 152:18-24.
[22] Karger A, Bettin B, Lenk M, et al Rapid characterisation of cell cultures by matrix-assisted laser desorption/ionisation mass spectrometric typing. J Virol Methods 2010; 164:116-21.
[23] WHO. Recommendations for the evaluation of animal cell cultures as substrates for the manufacture of biological medicinal products and for the characterization of cell banks. 2010.
[24] Delong-Rebel IH, Piemont Y, Fabre M, et al. Mycobacterium avium-intracellulare contamination of mammalian cell cultures. In Vitro Cell Dev Biol Anim 2009; 45:75-90.
[25] Lelong-Rebel IH, Piemont Y, Fabre M, et al. Mycobacterium avium-intracellulare contamination of mammalian cell cultures. In Vitro Cell Dev Biol Anim 2009; 45:75-90.
[26] Hué S, Gray ER, Gall A, et al. Disease-associated XMRV sequences are consistent with laboratory contamination. Retrovirol 2010; 7:111.
[27] Uphoff CC, Drexler HG. Prevention of mycoplasma contamination in leukemia-lymphoma cell lines. Human Cell 2001; 14: 244-7.
[28] Deutschmann SM, Kavermann H, Knack Y. Validation of a NAT-based Mycoplasma assay according European Pharmacopoiea. Biologicals 2010; 38:238-48.
[29] Uphoff CC, Drexler HG. Detection of mycoplasma contaminations in cell cultures by PCR analysis. Human Cell 1999; 12:229-36.
[30] Uphoff CC, Drexler HG. Detection of Mycoplasma contaminations. Methods Mol Biol 2013; 946:1-13.
[31] Uphoff CC, Drexler HG. Comparative antibiotic eradication of mycoplasma infections from continuous cell lines. In Vitro Cell Dev Biol Anim 2002; 38:86-9.
[32] Uphoff CC, Drexler HG. Eradication of Mycoplasma contaminations. Methods Mol Biol 2013; 946:15-26.
[33] Uphoff CC, Denkmann SA, Steube KG, et al. Detection of EBV, HBV, HCV, HIV-1, HTLV-I and -II, and SMRV in human and other primate cell lines. J Biomed Biotechnol 2010; 904767.
[34] Paprotka T, Delviks-Frankenberry KA, Cingöz O, et al. Recombinant origin of the retrovirus XMRV. Science 2011; 333:97-101.
[35] Kearney MF, Spindler J, Wiegand A, et al. Multiple sources of contamination in samples from patients reported to have XMRV infection. PLoS One 2012; 7:e30889.

3 Genetic Engineering of Cells

3.1 Cell Line Evolution and Engineering

Hitto Kaufmann

3.1.1 Guiding Principles

Immortalized mammalian cells have become the basis for the majority of processes to manufacture therapeutic proteins for use in humans [1]. Almost since the advent of recombinant cell culture processes a field of cell science has emerged aiming to improve such manufacturing processes and the respective protein therapeutic by improving the host cell lines. Interestingly, only a small number of immortalized host cell types serve as hosts for protein manufacturing. Chinese hamster ovary (CHO) cells dominate the portfolio due to the fact that they robustly grow in suspension in large-scale bioreactors. During the early days of industrial mammalian cell culture some products were also expressed in Sp20 and NSO cells. More recently, some proteins manufactured from a human cell line called PER.C6 were introduced into clinical studies.

This chapter discusses approaches to improve mammalian hosts either through evolution protocols or genetic engineering aiming to create a phenotype that either produces more product, displays a higher viability in large-scale high density cell cultures, or generates a product of superior efficacy or prolonged half-life. It will be described how the emerging understanding of complex regulatory networks in mammalian cells can be utilized to refine and improve such engineering strategies. Furthermore it will be outlined what needs to be considered to successfully apply such engineering concepts to establish new robust manufacturing processes.

3.1.2 Host Cell Selection and Evolution

3.1.2.1 Mutagenesis and Selection

A largely hemizygos genome and its susceptibility to genotypic instability enabled the generation of various loss-of-function CHO cell derivatives, that were subsequently widely used in industrial applications (see also Chapter 2.2 and Chapter 2.4). A prominent CHO subfamily of cell lines lacks the gene for the enzyme dehydrofolate reductase (DHFR) [2]. Selection procedures are typically carried out with and without prior mutagenesis using agents such as 5-azacytidine (5-azaC) and M-methyl-N'-nitro-N-nitrosoguanidine (MNNG). A number of CHO cell lines were selected for resistance to toxic plant lectins. These cells obtained their resistance through impaired binding of lectin to the cell surface conferred by alterations in N- and O-glycan synthesis pathways [3].

3.1.2.2 Modern Evolution Concepts

Another approach to improve cell culture performance is based on subjecting a host cell line to directed evolution experiments to generate novel phenotypes such as hybridoma cells that have been adapted to growth in the presence of high concentrations of ammonia [4]. An important step in advancing mammalian cell culture-based bioprocesses was the introduction of serum-free cell culture media. Existing host cell lines had to undergo iterative adaptation to those conditions prior to use in a bioreactor [5]. The adaptation approach was taken further in 2007 when a study demonstrated that a series of bioreactor evolution experiments enabled the selection of a novel DG-44 strain that can grow to higher viable cell densities in fed-batch cultures [6]. Interestingly, the authors used DG-44 cells that had gone through adaption to serum-free media prior to the bioreactor evolution without a subcloning step so the initial population was most likely genetically heterogeneous. This shows that a mutagenesis step using chemical agents is not generally necessary when selecting novel CHO phenotypes. Fluorescence activated cell sorting (FACS) and magnetic activated cell separation (MACS) are both techniques that allow rapid sorting of heterogeneous cell populations according to key cellular parameters on a cell-by-cell basis. Recently it was demonstrated that iterative reduction of glutamine levels in cell culture media combined with the selective sorting of viable cells after short batch cultivations led to CHO-K1-derived populations that display comparable or even better viable cell counts than their ancestors and grow in glutamine free media [7]. The basis for the efficient evolution protocols in CHO cells is thought to be strongly supported by the inherent genetic instability of this cell line as outlined in Chapter 2.2.

3.1.3 The Engineering Toolbox

3.1.3.1 Overexpression

The most straight-forward genetic engineering approach is the heterologous expression of a single gene to increase the protein level within the host cell beyond that of the endogenous protein. It could either be the gene encoding the protein from the respective host cell species (e.g., hamster for CHO cells) or a gene homolog from another organism. The strength of the promoter is one of the key factors influencing the new phenotype and the engineering step is usually followed by clonal selection procedures to obtain a genetically determined phenotype.

3.1.3.2 Chromosomal Engineering

The standard approach to generate a stable mammalian production cell line involves random integration of plasmid DNA encoding the product gene followed by a thorough selection procedure. Since this method potentially generates a new genotype and phenotype, the novel cellular entity needs careful thus time-consuming characterization prior to use for biologic manufacturing. The development of mammalian artificial chromosomes enabled a different route that avoids modifying the host cell genome by integrating heterologous genetic sequences encoding the product gene. Artificial chromosome expression (ACE) is based on large repetitive satellite DNA sequences that do not integrate into the host genome and nevertheless have the ability to self-replicate [8]. Another feature of the ACE technology is the possibility to rapidly extract loaded artificial chromosomes from a given cell line and subsequently transfer it to different host cells. This can be used for comparative studies across several immortalized cell lines without uncontrolled modifications to their genome. Using this approach several different CHO backgrounds were recently audited regarding their capability to produce large amounts of an antibody stably over time in culture [9].

3.1.3.3 Knockout

Deleting or inactivating a single gene is a very powerful tool for modifying the phenotype of a cell. By homologous recombination a number of cellular phenotypes have been generated to establish mammalian cells with improved properties for manufacturing proteins [10]. Another possibility to rapidly generate a phenotype with a significantly reduced activity for a single gene of interest is gene knockdown by RNAi. However, this inactivation is often incomplete, can be accompanied with unpredictable off-target effects, and is transient in nature. Recently, gene knockouts have been introduced by editing approaches (see Section 3.1.3.4).

3.1.3.4 Genome Editing

Genome editing employs engineered nucleases composed of sequence-specific DNA-binding domains fused to a nonspecific DNA cleavage module.

Recently, a novel targeted gene knockout approach was developed making use of imperfect cellular repair mechanisms triggered by double strand breaks in mammalian cells. The cellular response to double stand breaks consists of two distinct cellular repair mechanisms homology directed repair (HDR) that accurately repairs the site of damage, and nonhomologous end joining (NHEJ) that often results in changes to the genetic sequence [11]. In CHO cells more than half of the double strand breaks are repaired by the imperfect NHEJ, in some cases leading to loss of function of a specific gene. Engineered site-specific zinc-finger nucleases (ZFNs) can be used to carry out targeted gene knockouts in mammalian cells [12]. These customized nucleases consist of a heterologous zinc-finger protein DNA-binding domain fused to the catalytic domain of the endonuclease FokI with the zinc-finger domain being composed of 3 to 6 fingers which specifically bind the target sequence [13, 14]. Various methodologies have been developed to design zinc-finger nucleases that bind a given target sequence including context dependent assembly, oligomerized pool engineering, and modular assembly. Overall, the current state-of-the-art targeted gene knockout using zinc-finger nucleases has drastically accelerated the path to novel cellular phenotypes. For this technology and respective follow-up developments that are based on double strand breaks are detailed in Chapter 3.4.

3.1.4 Engineering Key Cellular Networks

3.1.4.1 Proliferation Control to Generate Bi-phasic Manufacturing Processes

Mammalian cells in culture typically divide every 20 to 24 h, and this doubling rate is relatively constant throughout the standard phases of a manufacturing process including inoculum cultures, seed train expansion, and final stage bioreactor. It was already hypothesized early on that the specific productivity of a mammalian cell in suspension culture may be higher at lower growth rates. Indeed process formats that decoupled the initial growth phase from a phase with controlled proliferation through lowering the culture temperature or adding substances such as sodium-butyrate did result in higher titers [15, 16]. The molecular understanding of cell cycle regulation in mammalian cells furthermore enabled engineering production cells by overexpression of cyclin-dependent kinase inhibitors such as p27 and p21, leading to the accumulation of cells in the G1 phase of the cell cycle along with a significantly increase in the productivity of several therapeutic proteins [17, 18]. However, inducible expression of these growth inhibitors is mandatory to achieve high cell densities in the production bioreactor prior to a phase of slow growth and high productivity.

3.1.4.2 Anti-apoptotic Engineering to Generate Robust Cell Hosts

Fed-batch cultivation currently represents the state-of-the-art process format for producing biologics. The integrated viable cell density directly translates into product yield at harvest and onset of apoptosis highly determines this parameter. Especially since the introduction of serum- and animal component-free culture media throughout the bioprocess, preventing apoptosis in cell culture as long as possible became the focus of many cell engineering studies [19, 20]. Similar to the progress in proliferation engineering described above, the emerging understanding of how cells integrate survival and cell death across a complex molecular network paved the way for multiple possibilities to generate superior cell phenotypes for production purposes [20]. Key proteins that influence the survival versus death balance inside a mammalian cell are the sensors such as cell death receptors that detect apoptotic triggers and signal them to the core machinery that executes the cell death program – mainly the caspase molecules. On the other hand, there are a number of antiapoptotic molecules that intervene with this cascade of events at multiple levels such as Bcl-2 and Bcl-xL or the inhibitor of apoptosis proteins (IAPs) [21]. Various studies have investigated the effect of overexpression of Bcl-2 in hybridoma, NS0, and CHO cells and could demonstrate improved cell culture robustness towards different stresses [22-25]. A similar positive effect on culture viability was described for BcL-xL [26-28]. Some studies successfully demonstrated that these engineering strategies could translate into bioprocesses yielding higher amounts of product [25, 29]. Caspase-9 is central to programmed cell death in mammalian cells via activation of the executioner caspase-3 [21]. Interestingly, a caspase-9 mutant was described that lacked catalytic activity due to the replacement of the catalytic cysteine by an alanine [30]. In various adherent cell cultures it could be shown that when overexpressed, this mutant acts as a dominant negative caspase-9 and inhibits cell death triggered by various stimuli [31]. When expressed in CHO production cells, Caspase-9DN protected these cells from programmed cell death during freeze/thaw, inoculum cultures and most importantly when grown in bioreactors [32]. Viral proteins often act as survival factors by interacting with the regulatory circuit inside the cells of their mammalian host. Expression of the adenoviral gene encoding the Bcl-2 homolog E1B-19k alone and in combination with the anti-apoptotic protein Aven prolonged culture duration of batch and fed-batch CHO cultures. Expression of both genes simultaneously led to higher titers of a monoclonal antibody in fed-batch bioreactor cultures and, interestingly, these engineered cells also grew at higher rates in these experiments [33]. For apoptosis engineering see also Chapter 3.5.

3.1.4.3 Improving the Secretory Machinery

Over the last decades an evolving toolbox for expressing therapeutic proteins in mammalian cells became available. This includes strong promoters, transcription-enhancing elements, various methods for gene amplification, and procedures that allow selection for integration into highly transcribed regions [34, 35]. Especially for large glycosylated molecules, such as antibodies, it was hypothesized that these successes created another bottleneck transcriptionally or even posttranslationally. Indeed, it could be demonstrated that transient or stable overexpression of XBP-1 can increase therapeutic protein productivity in CHO cells expressing high amounts of a heterologous protein [36-38]. XBP-1 is a key regulator that drives the differentiation of B cells into plasma cells by regulating the expression of up to 100 genes, and in CHO cells an enlarged endoplasmatic reticulum (ER) a typical hallmark of XBP-1 triggered differentiation was observed. The cellular secretion process requires targeted shuttling of protein cargo from the ER to the Golgi and from the Golgi to the plasma membrane. The fusion of the transport vesicles with their target membrane requires functional assembly of soluble N-ethylmaleimide-sensitive factor attachment protein receptor (SNARE) complexes [39]. In addition, this process requires the presence of the SM (Sec1/Munc18) family of proteins, and some studies suggest that they act by enhancing the activity of a SNARE complex. Knockdown studies in SEAP expressing HEK-293 cells elucidated the central role of Sly1, known to facilitate the ER-Golgi transition and Munc18c, known to facilitate the Golgi plasma membrane transition in protein secretion. Expressing both genes, either alone or in combination, increased the secretion of model proteins such as SEAP and VEGF in CHO cells [40]. The recent discovery of the importance of the lipid transfer protein CERT for constitutive Golgi secretory activity provided a novel candidate for secretion engineering. Expressing CERT, and even to a larger extent a gain of function mutant of (CERT S132A) resulted in higher Immunoglobulin G (IgG) productivity in fed-batch processes [41]. All these studies clearly demonstrated that currently achievable product gene mRNA levels can create bottlenecks during the transport and secretion process in mammalian production cells, thus creating a need for improvement of these processes.

3.1.4.4 MiRNA Engineering

Initially discovered in the worm *Caenorhabditis elegans* miRNAs, approximately 22 nucleotides long noncoding RNAs, regulate gene expression in invertebrates and vertebrates by base-pairing with the 3' untranslated regions of target mRNAs. Inhibition of the expression of these target genes mechanistically occurs either through inhibition of translation or degradation of the mRNA templates with the specific silencing mechanism depending on the distinct nature of the mRNA target [42]. With more than a thousand human miRNAs known to date and experimental and computational data showing that each individual miRNA regulates up to hundreds of different mRNAs it is currently assumed that about 20 to 30 % of the mammalian transcriptome is subject to regulation through miRNAs [43]. MiRNAs regulate various physiological processes such as cell growth and division, cell death and differentiation, and core metabolic pathways. It is therefore not surprising that recent studies revealed that disruptions in miRNA gene regulation are part of the molecular characteristics of various human diseases including autoimmune disorders and cancer [44]. Accumulating evidence describes the role of endogenous miRNAs in regulating growth, survival, and metabolism of CHO cells in culture. A microarray study in CHO cells grown in nutrient-depleted media described the upregulation of the mouse miR-297-669 cluster and linked the expression of mmu-miR-446h to inhibition of several antiapoptotic genes as part of the stress response [45]. In a qRT-PCR-based expression profiling study several miRNAs where found to be either up- or downregulated in CHO cells 24 h after reduction of culture temperature from 37 to 31 °C [46]. Another transcriptome-profiling approach demonstrated that as many as 100 miRNAs are differentially regulated when a CHO batch culture progresses through lag phase, exponential growth phase, and stationary phase towards the final decline in culture viability prior to harvest [47]. The fact that miRNAs are involved in many processes that are key to heterologous protein production in immortalized mammalian cells grown in bioreactors and regulate multiple targets simultaneously make them an attractive target for cell engineering. Indeed it could be shown that stable inhibition of the previously described mmu-miRNA-466h-5p by shRNA expression improves CHO cell survival and enhances secretion of a model protein in batch culture [48]. Moreover, a recent functional miRNA screen in IgG producing CHO cells identified several miRNAs that enhanced productivity upon transient expression. Stable coexpression of 2 of those miRNAs lead to higher IgG titers in fed-batch cultures [49], suggesting that miRNA engineering may be an encouraging approach to fundamentally reprogram production cells.

3.1.4.5 Improving Product Quality

Key parameters of biologics such as *in vivo* efficacy, half-life, and safety are not only determined by the amino acid sequence but also by posttranslational modifications. Mammalian cells modify the amino acid chain of secreted proteins along with the folding and transport process. Since not every molecule that is processed through this machinery located within the ER and Golgi is modified and trimmed in the exact same manner, the "protein drug substance" that is obtained from these cells is a chemically heterologous product. (For detailed information about protein glycosylation see Chapter 4.1). For some of the key modifications structure-function relationships have been established. A prominent example is the influence of the N-linked biantennary complex-type oligossacharides bound to the constant region (Fc) of antibodies on their effector function of. It could be demonstrated *in vitro* and *in vivo* that the removal of the core fucose from that structure enhances the binding of the Fc region of the antibody to the FcgRIIIa receptor that is present on specialized immune cells such as natural killer cells, macrophages, and neutrophils. This binding reaction triggers antibody dependent cytotoxicity (ADCC) that is considered a key element of the mode of action of many antibodies developed for cancer therapy [50]. In a key study it could be shown that regulated expression of heterologous 1,4-Nβ-acetylglucosaminylt ransferase III (GnT-III) in antibody-producing cells yields molecules with significantly enhanced ADCC in cell-based assays [51]. The addition of bisecting GlcNAc molecules to an oligosacharide by GnT-II blocks central reactions of the biosynthetic pathway such as core-fucosylation. A further refined engineering approach described how fusing the catalytic domain of GnT-III to localization domains of Golgi proteins further enhanced the competition of the modified GnT-III against α1,6-fucosyltransferase and α-mannosidase II (ManII) [52]. Another genetic setup leading to high ADCC activities was the coexpression of GnT-III and ManII. CHO cells engineered with wild type GnT-III and MnII overexpression were then used to generate a new form of a type II anti-CD20 antibody with high direct B cell cytotoxicity as well as enhanced ADCC [53]. This molecule termed GA101 was preclinically compared to rituximab and ofatumumab and showed superiority *in vitro* and in xenograft models [54]. The recent approval of the

FDA for this molecule underlines the impact of glycoengineering of mammalian production cell lines. In a more direct approach Yamane-Ohnuki and coworkers deleted the gene encoding the α1,6-fucosyltransferase FUT8 in CHO cells and could show that antibodies produced from these cells lack fucose, demonstrate stronger binding to the FcγRIIIa receptor and higher ADCC [10]. Also this engineering approach led to the marketing approval for the novel drug mogamulizumab, targeting relapsed or refractory CCR4-positive adult T cell leukemia-lymphoma [55]. Recently, another approach demonstrated the generation of highly active antibody molecules by expressing the microbial enzyme GDP-6-deoxy-D-lyxo-4-hexulose reductase (RMD) [56]. It could be shown that RMD-expressing CHO cells metabolically deflect a precursor of the fucose *de novo* pathway towards a metabolic dead end thereby generating antibodies that lack fucose and show high ADCC. This approach could also be used to modify existing antibody production cell lines and thereby generating 2' generation antibodies in one transfection step. For glycosylation engineering see also Chapter 3.5.

3.1.5 Adapting to Complexity – Advanced Engineering Concepts

Any engineering approach in mammalian cells means intervening with highly complex regulatory networks that are often linked across the different functional areas. As an example, a growth arrest that is mediated through molecular switches that regulate the cell cycle can easily tip the survival-death balance within a cell and trigger apoptosis. Achieving a pronounced improved phenotype may therefore require altering gene expression patterns of more than one pathway. The first study that demonstrated the power of multigene engineering improved controlled proliferation strategies that were based on overexpressing cyclin-dependent kinase inhibitors by combining p21 expression with C/EBPα and p27 with the antiapoptotic Bcl-xL [57]. Along the same line, overexpression of XBP-1 to debottleneck the secretion pathway of recombinant IgGs in CHO cells could only be sustained towards a genetically stable phenotype by preventing apoptosis induction through coexpressing the caspase inhibitor XIAP (x-linked inhibitor of apoptosis) [58]. While the straight-forwardest approach towards a novel production cell phenotype involves constitutive expression of a specific gene that is linked to improving the targeted pathway, it may not be beneficial to the system to always aim for the highest expression level of this gene at all moments throughout the entire cell culture process. Any growth inhibition should only occur once a high cell density has been reached and some enzymes require a specific intracellular concentration to generate the desired glycopattern of the protein product. This can be achieved by making use of inducible systems such as the tetracycline inducible or repressable gene expression system [57, 51]. Even though these approaches do enable fine-tuning of gene regulation, the use of the inducing agents at larger manufacturing scales is often not feasible. Novel inducible systems have been developed that translate signals into transgene expression levels that are intrinsic to a bioprocess such as temperature or components of cell culture media such as gaseous acetaldehyde [59]. Recently, another concept leading to more refined cell engineering reflecting and integrating the specific temporal needs for metabolic alterations was described. Here, pooled expression patterns of endogenous mammalian cell promoters were quantified and time profiles pointed to the thioredoxin interacting protein (Txnip) promoter for its capability to drive transgene expression in concert with cell growth [60]. Expressing the mouse GLUT5 fructose transporter in CHO cells under the control of the Txnip promoter synchronized sugar metabolism to cell growth and resulted in improved process characteristics and higher titers [60].

3.1.6 From Prediction to Production – Improving Input Parameters and Applicability

The field of production cell engineering is rapidly advancing and some of this success is due to the rapidly progressing 'omic' characterization of the total molecular composition of cells under varying conditions on all levels of cell regulation. Studies of the genome, transcriptome, proteome, metabolome, and glycome of CHO cells in particular have resulted in vast profiling data that offer hints for possible molecular alterations to existing host cell systems [61]. However, it still takes thorough studies to investigate each one of those hypotheses by generating and assessing novel phenotypes. In the future it will be particularly important to directly screen for novel phenotypes in cell culture environments that resemble the actual bioprocesses as closely as possible [49]. Miniaturized automated bioreactor systems that allow running multiple fed-batch cultures in parallel are one approach that seems useful to screen broadly and in relevant systems. Whatever genetic alterations are introduced to give rise to an improved mammalian host cell for protein production, the heterogeneity profile of protein products secreted from these cells needs to be carefully monitored as these alterations can change product attributes that are key to *in vivo* safety, efficacy, and half-life [15, 38]. For the CHO cell world a fully sequenced organism and rapidly increasing functional genomic characterization are building the required platform to develop sophisticated mathematical models that quantitatively describe the molecular pathways for key cellular processes (see also Chapter 5.2). This could be the basis for predicting engineering phenotypes or the applicability of genetic interventions across the diverse world of CHO substrains cultured in laboratories all over the world [62].

References

[1] Durocher Y, Butler M. 2009. Expression systems for therapeutic glycoprotein production. Curr Opin Biotechnol. 20: 700-707.

[2] Urlaub G, Chasin LA. 1980. Isolation of Chinese hamster cell mutants deficient in dihydrofolate reductase activity. Proc Natl Acad Sci USA. 77: 4216-4220.

[3] Patnaik SK, Stanley P. 2006. Lectin-resistant CHO glycosylation mutants. Methods Enzymol. 416: 159-182.
[4] Matsumura M, Shimoda M, Arii T, et al. 1991. Adaptation of hybridoma cells to higher ammonia concentration. Cytotechnology. 7: 103-112.
[5] Sinacore MS, Drapeau D, Adamson SR. 2000. Adaptation of mammalian cells to growth in serum-free media. Mol Biotechnol. 15: 249-257.
[6] Prentice HL, Ehrenfels BN, Sisk WP. 2007. Improving performance of mammalian cells in fed-batch processes through "bioreactor evolution". Biotechnol Prog. 23: 458-464.
[7] Bort JA, Stern B, Borth N. 2010. CHO-K1 host cells adapted to growth in glutamine-free medium by FACS-assisted evolution. Biotechnol J. 5: 1090-1097.
[8] Lindenbaum M, Perkins E, Csonka E, et al. 2004. A mammalian artificial chromosome engineering system (ACE System) applicable to biopharmaceutical protein production, transgenesis and gene-based cell therapy. Nucleic Acids Res. 32: e172.
[9] Kennard ML, Goosney DL, Monteith D, et al. 2009. Auditioning of CHO Host Cell Lines Using the Artificial Chromosome Expression (ACE) Technology. Biotechnol Bioeng. 104: 526-539.
[10] Yamane-Ohnuki N, Kinoshita S, Inoue-Urakubo M, et al. 2004. Establishment of FUT8 knockout Chinese hamster ovary cells: an ideal host cell line for producing completely defucosylated antibodies with enhanced antibody-dependent cellular cytotoxicity. Biotechnol Bioeng. 87: 614-622.
[11] Sonoda E, Hochegger H, Saberi A, et al. 2006. Differential usage of non-homologous end-joining and homologous recombination in double strand break repair. DNA Repair (Amst). 5: 1021-1029.
[12] Santiago Y, Chan E, Liu PQ, et al. 2008. Targeted gene knockout in mammalian cells by using engineered zinc-finger nucleases. Proc Natl Acad Sci USA. 105: 5809-5814.
[13] Durai S, Mani M, Kandavelou K, et al. 2005. Zinc finger nucleases: custom-designed molecular scissors for genome engineering of plant and mammalian cells. Nucleic Acids Res. 33: 5978-5990.
[14] Urnov FD, Miller JC, Lee YL, et al. 2005. Highly efficient endogenous human gene correction using designed zinc-finger nucleases. Nature. 435: 646-651.
[15] Kaufmann H, Mazur X, Fussenegger M, et al. 1999. Influence of low temperature on productivity, proteome and protein phosphorylation of CHO cells. Biotechnol Bioeng. 63: 573-582.
[16] Gorman CM, Howard BH, Reeves R. 1983. Expression of recombinant plasmids in mammalian cells is enhanced by sodium butyrate. Nucleic Acids Res. 11: 7631-7648.
[17] Mazur X, Fussenegger M, Renner WA, et al. 1998. Higher productivity of growth-arrested Chinese hamster ovary cells expressing the cyclin-dependent kinase inhibitor p27. Biotechnol Prog. 14: 705-713.
[18] Meents H, Enenkel B, Eppenberger HM, Werner RG, and Fussenegger M. 2002. Impact of coexpression and coamplification of sICAM and antiapoptosis determinants bcl-2/bcl-x(L) on productivity, cell survival, and mitochondria number in CHO-DG44 grown in suspension and serum-free media. Biotechnol Bioeng. 80: 706-716.
[19] Zanghi JA, Fussenegger M, Bailey JE. 1999. Serum protects protein-free competent Chinese hamster ovary cells against apoptosis induced by nutrient deprivation in batch culture. Biotechnol Bioeng. 64: 108-119.
[20] Arden N, Betenbaugh MJ. 2004. Life and death in mammalian cell culture: strategies for apoptosis inhibition. Trends Biotechnol. 22: 174-180.
[21] Kaufmann H, Fussenegger M. 2004. Caspase regulation at the molecular level. In: Al-Rubeai M, Fussenegger M (eds). Cell Engineering. pp 1-23.
[22] Mastrangelo AJ, Hardwick JM, Bex F, et al. 2000. Part I. Bcl-2 and Bcl-x(L) limit apoptosis upon infection with alphavirus vectors. Biotechnol Bioeng. 67: 544-554.

[23] Mastrangelo AJ, Hardwick JM, Zou S, et al. 2000. Part II. Overexpression of bcl-2 family members enhances survival of mammalian cells in response to various culture insults. Biotechnol Bioeng. 67: 555-564.
[24] Kim NS, Lee GM. 2002. Response of recombinant Chinese hamster ovary cells to hyperosmotic pressure: effect of Bcl-2 overexpression. J Biotechnol. 95: 237-248.
[25] Tey BT, Singh RP, Piredda L, et al. 2000. Bcl-2 mediated suppression of apoptosis in myeloma NS0 cultures. J Biotechnol. 79: 147-159.
[26] Mastrangelo AJ, Hardwick JM, Bex F, et al. 2000. Part I. Bcl-2 and Bcl-x(L) limit apoptosis upon infection with alphavirus vectors. Biotechnol Bioeng. 67: 544-554.
[27] Meents H, Enenkel B, Eppenberger HM, et al. 2002. Impact of coexpression and coamplification of sICAM and antiapoptosis determinants bcl-2/bcl-x(L) on productivity, cell survival, and mitochondria number in CHO-DG44 grown in suspension and serum-free media. Biotechnol Bioeng. 80: 706-716.
[28] Jung D, Cote S, Drouin M, et al. 2002. Inducible expression of Bcl-XL restricts apoptosis resistance to the antibody secretion phase in hybridoma cultures. Biotechnol Bioeng. 79: 180-187.
[29] Tey BT, Singh RP, Piredda L, Pet al. 2000. Influence of bcl-2 on cell death during the cultivation of a Chinese hamster ovary cell line expressing a chimeric antibody. Biotechnol Bioeng. 68: 31-43.
[30] Pan G, O'Rourke K, Dixit VM. 1998. Caspase-9, Bcl-XL, and Apaf-1 form a ternary complex. J Biol Chem. 273: 5841-5845.
[31] Sitailo LA, Tibudan SS, Denning MF. 2002. Activation of caspase-9 is required for UV-induced apoptosis of human keratinocytes. J Biol Chem. 277: 19346-19352.
[32] van de Goor J. 2004. In: Al-Rubeai M, Fussenegger M (eds). Cell Engineering. pp 211-221.
[33] Figueroa B, Jr., Ailor E, Osborne D, et al. 2007. Enhanced cell culture performance using inducible anti-apoptotic genes E1B-19K and Aven in the production of a monoclonal antibody with Chinese hamster ovary cells. Biotechnol Bioeng. 97: 877-892.
[34] Sautter K, Enenkel B. 2005. Selection of high-producing CHO cells using NPT selection marker with reduced enzyme activity. Biotechnol Bioeng. 89: 530-538.
[35] Pearson MJ, Khazaipoul S, Optun A, et al. 2012. Albumin 3'untranslated region facilitates increased recombinant protein production from Chinese hamster ovary cells. Biotechnol J. 7: 1405-1411.
[36] Tigges M, Fussenegger M. 2006. Xbp1-based engineering of secretory capacity enhances the productivity of Chinese hamster ovary cells. Metab Eng. 8: 264-272.
[37] Ku SC, Toh PC, Lee YY, et al. 2010. Regulation of XBP-1 signaling during transient and stable recombinant protein production in CHO cells. Biotechnol Prog. 26: 517-526.
[38] Becker E, Florin L, Pfizenmaier K, et al. 2008. An XBP-1 dependent bottle-neck in production of IgG subtype antibodies in chemically defined serum-free Chinese hamster ovary (CHO) fed-batch processes. J Biotechnol. 135: 217-223.
[39] Hong W, Lev S. 2013. Tethering the assembly of SNARE complexes. Trends Cell Biol. 24: 35-43.
[40] Peng RW, Fussenegger M. 2009. Molecular engineering of exocytic vesicle traffic enhances the productivity of Chinese hamster ovary cells. Biotechnol Bioeng. 102: 1170-1181.
[41] Florin L, Pegel A, Becker E, et al. 2009. Heterologous expression of the lipid transfer protein CERT increases therapeutic protein productivity of mammalian cells. J Biotechnol. 141: 84-90.
[42] Eulalio A, Huntzinger E, Izaurralde E. 2008. Getting to the root of miRNA-mediated gene silencing. Cell. 132: 9-14.
[43] Bushati N, Cohen SM. 2007. miRNA functions. Annu Rev Cell Dev Biol. 23: 175-205.
[44] O'Connell RM, Rao DS, Chaudhuri AA, et al. 2010. Physiological and pathological roles for miRNAs in the immune system. Nat Rev Immunol. 10: 111-122.

[45] Druz A, Chu C, Majors B, er al. 2011. A novel miRNA mmu-miR-466h affects apoptosis regulation in mammalian cells. Biotechnol Bioeng. 108: 1651-1661.
[46] Barron N, Kumar N, Sanchez N, et al. 2011. Engineering CHO cell growth and recombinant protein productivity by overexpression of miR-7. J Biotechnol. 151: 204-211.
[47] Hernandez Bort JA, Hackl M, Hoflmayer H, et al. 2012. Dynamic mRNA and miRNA profiling of CHO-K1 suspension cell cultures. Biotechnol J. 7: 500-515.
[48] Druz A, Son YJ, Betenbaugh M, et al. 2013. Stable inhibition of mmu-miR-466h-5p improves apoptosis resistance and protein production in CHO cells. Metab Eng 16: 87-94.
[49] Strotbek M, Florin L, Koenitzer J, et al. 2013. Stable miRNA expression enhances therapeutic antibody productivity of Chinese hamster ovary cells. Metab Eng 20C: 157-166.
[50] Cartron G, Dacheux L, Salles G, et al. 2002. Therapeutic activity of humanized anti-CD20 monoclonal antibody and polymorphism in IgG Fc receptor FcgammaRIIIa gene. Blood. 99: 754-758.
[51] Umana P, Jean-Mairet J, Moudry R, et al. 1999. Engineered glycoforms of an antineuroblastoma IgG1 with optimized antibody-dependent cellular cytotoxic activity. Nat Biotechnol. 17: 176-180.
[52] Ferrara C, Brunker P, Suter T, et al. 2006. Modulation of therapeutic antibody effector functions by glycosylation engineering: influence of Golgi enzyme localization domain and co-expression of heterologous beta1, 4-N-acetylglucosaminyltransferase III and Golgi alpha-mannosidase II. Biotechnol Bioeng. 93: 851-861.
[53] Mossner E, Brunker P, Moser S, et al. 2010. Increasing the efficacy of CD20 antibody therapy through the engineering of a new type II anti-CD20 antibody with enhanced direct and immune effector cell-mediated B-cell cytotoxicity. Blood. 115: 4393-4402.
[54] Herter S, Herting F, Mundigl O, et al. 2013. Preclinical Activity of the Type II CD20 Antibody GA101 (Obinutuzumab) Compared with Rituximab and Ofatumumab In Vitro and in Xenograft Models. Mol Cancer Ther. 12: 2031-2042.
[55] Beck A, Reichert JM. 2012. Marketing approval of mogamulizumab: a triumph for glyco-engineering. MAbs. 4: 419-425.
[56] von Horsten HH, Ogorek C, Blanchard V, et al. 2010. Production of non-fucosylated antibodies by co-expression of heterologous GDP-6-deoxy-D-lyxo-4-hexulose reductase. Glycobiology 20: 1607-1618.
[57] Fussenegger M, Schlatter S, Datwyler D, et al. 1998. Controlled proliferation by multigene metabolic engineering enhances the productivity of Chinese hamster ovary cells. Nat Biotechnol. 16: 468-472.
[58] Becker E, Florin L, Pfizenmaier K, et al. 2010. Evaluation of a combinatorial cell engineering approach to overcome apoptotic effects in XBP-1(s) expressing cells. J Biotechnol. 146: 198-206.
[59] Weber W, Fussenegger M. 2007. Inducible product gene expression technology tailored to bioprocess engineering. Curr Opin Biotechnol. 18: 399-410.
[60] Le H, Vishwanathan N, Kantardjieff A, et al. 2013. Dynamic gene expression for metabolic engineering of mammalian cells in culture. Metab Eng 20: 212-220.
[61] Datta P, Linhardt RJ, Sharfstein ST. 2013. An 'omics approach towards CHO cell engineering. Biotechnol Bioeng. 110: 1255-1271.
[62] Kildegaard HF, Baycin-Hizal D, Lewis NE, et al. 2013. The emerging CHO systems biology era: harnessing the 'omics revolution for biotechnology. Curr Opin Biotechnol. 24:1102-1107.

3.2 Chromosome Rearrangements and Gene Amplification

Takeshi Omasa and Kyoungho Lee

3.2.1 Introduction

The genome era began at the beginning of the 21st century. The human genome sequence is about 3,200,000,000 (3.2 G) bp containing more than 23,000 genes. The mouse and hamster genomes are of similar size as that of human genome. DNA is a polymer of nucleotides and is composed of a double helix. One turn of the helix contains about 10 bp and the height of the helix is about 3.5 nm [1]. In other words, the theoretical size of the human DNA polymer is estimated to be about $3{,}200{,}000{,}000/10 \times 3.5 \times 10^{-9}$ m = 1.12 m. This very long polymer is packed into one nucleus that is smaller than a 10 μm-diameter cell. The figure below shows the overview of the structure of DNA relative to the nucleus (Figure 3.2.1).

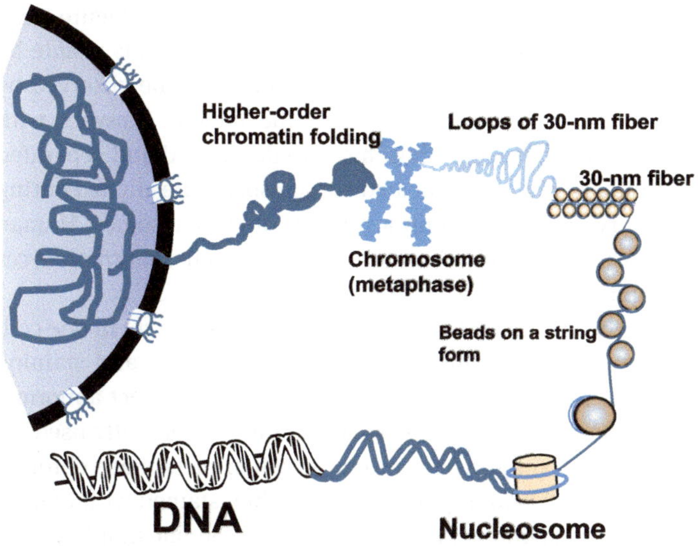

Figure 3.2.1: Overview of the structure of genes and chromosomes.

In order to package this very long DNA thread into a very small cell, an exact packaging mechanism is necessary. The key protein that is involved in packaging in eukaryotic cells is a group of highly alkaline proteins, the histones. Histones bound to the DNA polymer are known as a nucleosome. Nucleosomes are the basic building blocks for chromatin and chromosomes, and they form the tight packaging of the DNA. However, this tight package needs to be loosened during mitosis. When the genetic information is copied in mitosis, the enzymes related to DNA duplication must access the DNA polymers. If this tight package is not loosened, these enzymes cannot access the genetic information. The same phenomenon occurs in the gene expression process.

In this chapter, we introduce the chromosome structure and analysis. Further, we introduce the application of this knowledge and the related technology for gene amplification in industrial protein production using mammalian cells, especially Chinese hamster ovary (CHO) cells. In particular, the chapter describes the basics of chromosomal DNA, its chromatin packaging and impact on gene expression, chromosomal arrangements and re-arrangements in the example of CHO cells, mechanisms of gene amplification, its selection, and manipulation of efficiency as well as technical aspects including FISH and BAC libraries.

3.2.2 Chromosome Structure in Mammalian Cells

The typical chromosome structure in metaphase of mitosis is shown in Figure 3.2.2. Chromosome means colored body, which is easily stained by dye and is visible by microscope. In eukaryotic cells, the chromosome cannot ordinarily be observed by microscope because the chromosome is tightly packed in the nucleus. However, with the progression of the cell cycle phase, chromosomes are duplicated and separated for cell division in the M (mitosis) phase. Duplicated chromosomes are aligned during metaphase of mitosis. Eukaryotic cells contain a number of chromosomes. Human cells have 22 paired chromosomes and sex chromosomes. Yeast cells, *Saccharomyces cerevisiae*, have 16 paired chromosomes.

The karyotype refers to the visual appearance of the chromosome set in one cell, and it contains information about the chromosome number, length, and staining pattern (Figure 3.2.3, see also Chapter 2.4). Karyotypes are used to detect abnormal chromosomes related to human hereditary disease. G-banding is generally used for analytical methods of karyotype. G-bands are produced when metaphase chromosomes are subjected briefly to mild heat or proteolysis and then stained with Giemsa reagent (DNA dye). G bands correspond to large regions of the genome that have low G+C contents. G-banding is a particularly useful procedure for the initial screening for chromosomal aberrations because the entire genome can be evaluated in a single experiment. G-banding analysis is routinely used to identify human disease. According to the obtained chromosome-banding patterns, chromosomal aberrations are determined. In some cases, chromosome-banding patterns are difficult to interpret by observation.

Hybridization-based karyotype analysis was recently developed. A technique termed spectral karyotyping (SKY), which is based on the simultaneous hybridization of 24 chromosome-specific painting probes labeled with different fluorochromes or

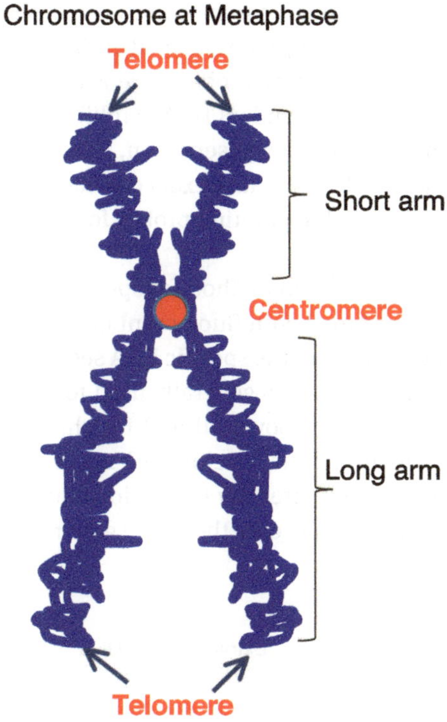

Figure 3.2.2: Diagram of a replicated and condensed chromosome at metaphase.

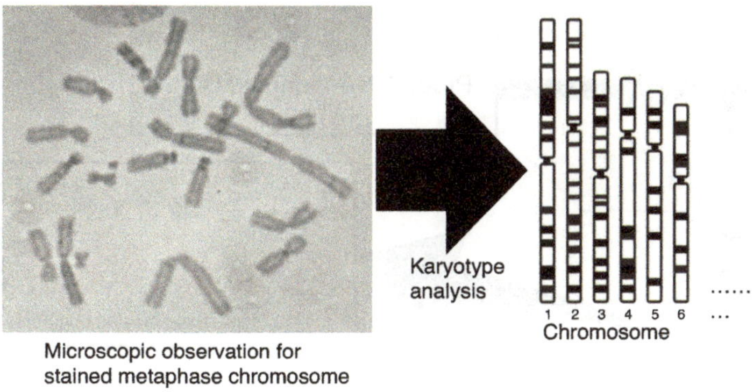

Figure 3.2.3: The concept for analysis of banding patterns from a metaphase spread.

fluorochrome combinations, was developed [2]. All human chromosomes can easily be visualized by SKY in 24 different colors using painting probes labeled with different combinations of 5 fluorochromes. Chromosome painting probes were generated from flow-sorted human chromosomes by sequence-independent DNA amplification. This SKY technology attains user-friendly operation and easy observation, but sorted chromosomes are necessary for the hybridization probes. Because exact chromosome sorting is not an easy technique, the number of available painting probes for SKY is limited for humans, mice, and rats.

Fluorescence *in situ* hybridization (FISH) is a detection method for specific DNA on chromosomes. Specific DNA sequences are combined with fluorescent chemicals and hybridized with a chromosome spread. The location of the specific DNA sequence can be observed by fluorescence microscopy. This method is generally used to identify chromosomal aberrations. Multicolor FISH is also a powerful tool for chromosome analysis. To detect the hybridization, the length of the specific DNA sequences is important. Long DNA sequences give a strong signal intensity of hybridization. A bacterial artificial chromosome (BAC) library, which represents the entire genome of

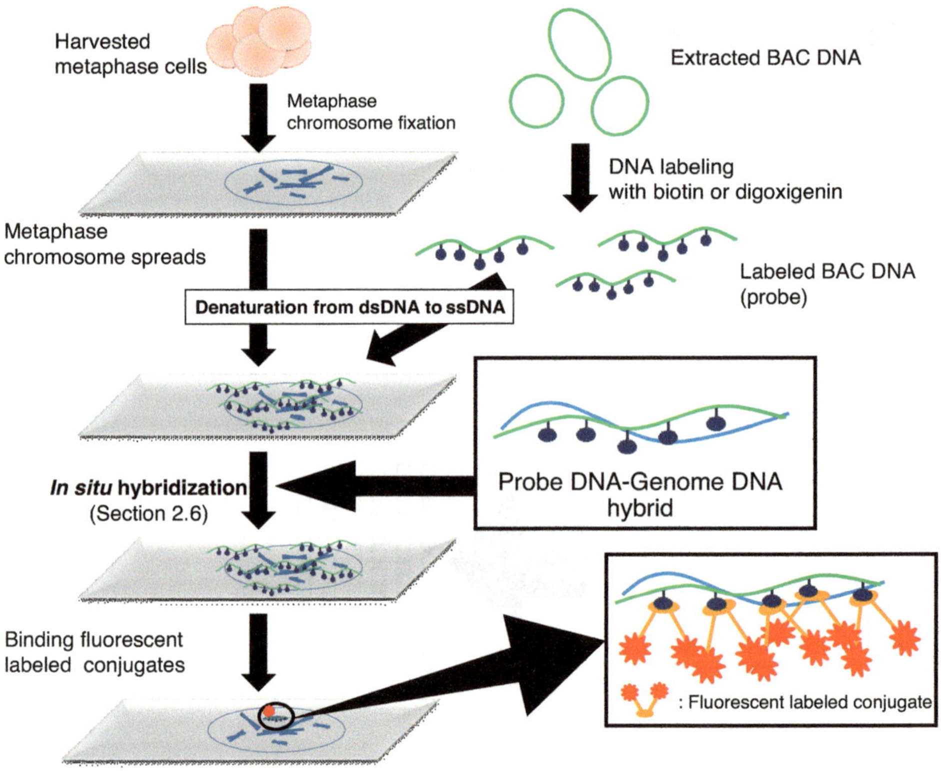

Figure 3.2.4: Experimental procedure of fluorescence *in situ* hybridization using bacterial artificial chromosome DNA as a probe (BAC-FISH; modified from [5]).

an organism without cloning artifacts or rearrangements, can reduce the complexity of the genome and provide clones that are physically separated in an addressable format [3]. BAC libraries also provide scaffolding information for mapping sequence contigs to localized genomic regions using a direct genomic shotgun sequencing approach. Each BAC clone contains quite a long DNA sequence (several hundred kilo-base pairs) and is suitable to use as a hybridization probe. Consequently, FISH imaging using BAC clones as hybridization probes (BAC-FISH) is used to construct a physical map of a genome [4] (Figure 3.2.4).

With the development of genome projects, microarray-based comparative genomic hybridization (aCGH) has become a powerful method for the genome-wide detection of chromosomal imbalances [6]. aCGH enables the evaluation of chromosomal imbalances in tumor tissues without the preparation of metaphase chromosomes from tumor cells. Furthermore, this method permits the genome-wide screening of DNA copy number changes. The recent development of CGH for arrays of mapped genomic DNA segments, i.e., BAC clones, has greatly improved the resolution of the analysis.

3.2.3 Chromosome Rearrangements and Detection in CHO Cells

The CHO cell line was established from the ovary of Chinese hamster (*Cricetulus griseus*), which was used as a laboratory animal before rat or mouse usage (see also Chapter 2.2). CHO cells were easily cultivated and were used for molecular cytogenetic materials during the early 1980s. Today, CHO cells are not used for basic science purposes. These cells play a very important role as a host cell for the commercial-scale production of therapeutic proteins [7-9] (Figure 3.2.5). Two subclones of CHO cells, proline-requiring CHO K1 [10] and dihydrofolate reductase (DHFR) gene-deficient CHO DG44 [11], are the most widely used clones for industrial applications [12, 13].

The reasons for the wide usage of CHO cells are as follows [8].

1. The empirical method to establish highly productive cell lines is well known, and the production of more than 10 g/L of antibody has been reported.

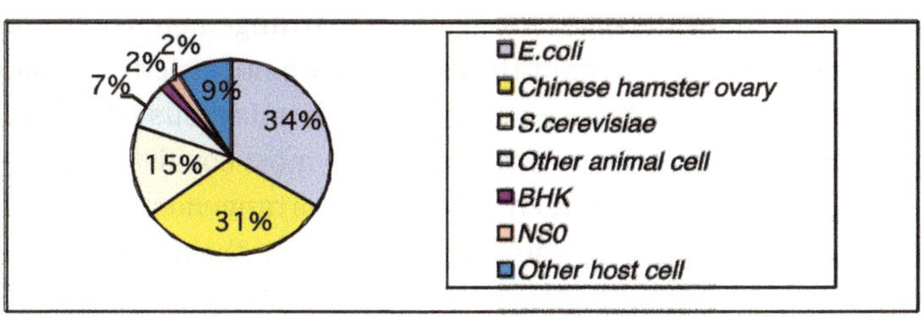

Figure 3.2.5: Good manufacturing practice (GMP)-certified host cells for biopharmaceutical production. Among 191 biopharmaceuticals, 60 products are produced by CHO cells [14].

2. Industrial serum-free media have been developed.
3. High-density large-scale fed-batch cultivations have been developed and scale-up technology is well established.
4. The *de facto* standard of good manufacturing practice (GMP)-certified production has been established.

Genome-wide analysis is of considerable importance for the establishment and improvement of productive CHO cell lines. Wlaschin and Hu constructed a scaffold for the Chinese hamster genome [15]. With the progress of next-generation sequencing techniques, the CHO cell and Chinese hamster genomic analyses were accelerated. Recently, Xu et al. determined the first draft genome sequence of the CHO K1 cell line [16]. This group decoded and compared 6 different CHO cell line genome sequences and revealed the advantage of CHO cells for biopharmaceutical production [17]. Other groups reported the draft sequence of Chinese hamster using sorted chromosomes [18]. These approaches could greatly contribute to the genome-scale reconstruction of industrial mammalian cells.

Previously, the authors constructed the first genomic BAC library from the gene-amplified CHO cell (CHO DR1000L-4N) genome for genome-wide analysis [19]. The BAC clones of this library could be landmarks for a physical map of the CHO cell genome, which would be essential for the basic research and industrial applications of the CHO cell genome. This BAC library was constructed from the whole genome of the CHO DR1000L-4N cell line and consisted of 122,281 clones. The CHO DR1000L-4N cell line was established from the CHO DG44 cell line using the mouse DHFR gene amplification system. It contains a stable high copy number of the exogenous gene of the human granulocyte-macrophage colony-stimulating factor (hGM-CSF) and mouse DHFR genes [20, 21].

The applications of this constructed CHO genomic BAC library were classified into 3 categories: cell engineering, chromosome identification, and BAC microarray for genome analysis [5]. Chromosome identification using CHO BAC clones have the potential to contribute to the chromosomal instability of CHO cells and to improve our understanding of chromosome organization during the establishment of recombinant CHO cells. Deaven and Petersen analyzed the karyotype of CHO cell lines on the basis of banding and concluded that extensive chromosome rearrangements occur in CHO cell lines [22]. Siminovitch also noted that CHO cell lines are not strictly diploid in the functional sense and that many genes are present in the hemizygous state [23]. Genome instability during cell line establishment is one of the most consistent characteristics of CHO cells [24]. The karyotype and chromosome rearrangements in CHO cell lines have been analyzed by banding techniques [22, 24, 25]. In the authors' previous study, a detailed physical chromosomal map of the CHO DG44 cell line was constructed by BAC-FISH using 303 randomly selected BAC clones as hybridization probes [26]. On the basis of this constructed physical map and end sequences, the chromosome rearrangements of CHO DG44, CHO K1, and primary Chinese hamster cells were investigated (Figure 3.2.6). Among 20 CHO chromosomes, 8 were conserved without large rearrangements in CHO DG44, CHO K1, and primary Chinese hamster cells. Chromosome rearrangements are also important aspects to consider in cell line construction. The combination of next-generation sequencing analysis and these BAC-FISH approaches could greatly contribute to the next-generation construction of industrial workhorse cells.

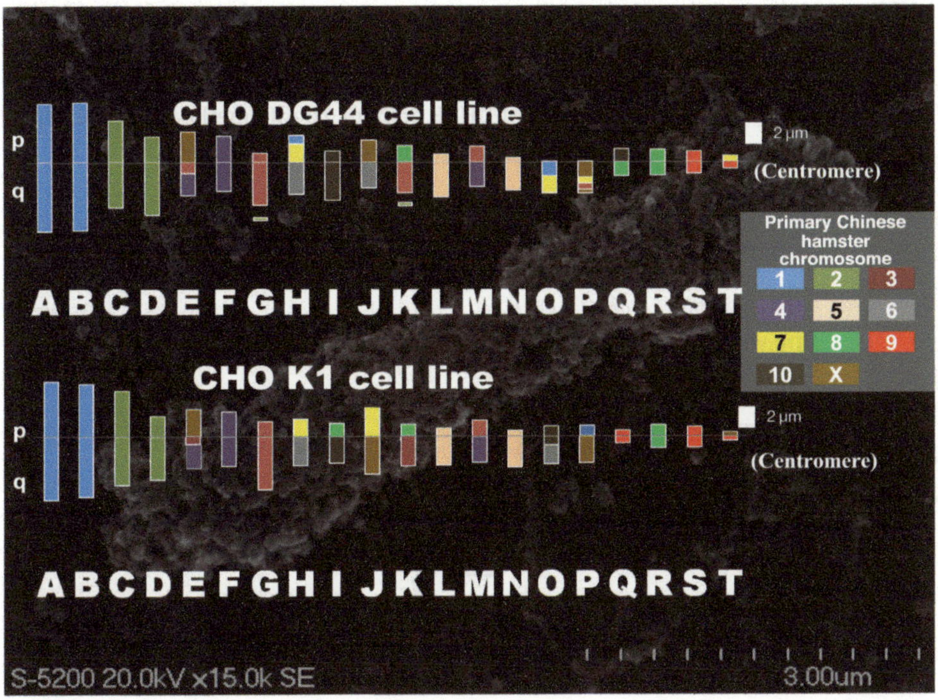

Figure 3.2.6: Karyotypic comparison of CHO DG44, K1, and primary Chinese hamster cells based on BAC-FISH results. The homologous regions of chromosomes from CHO DG44, CHO K1, and primary Chinese hamster cells are colored according to the Chinese hamster chromosomes. The scale bar corresponds to 2 µm. (Background photo: scanning electron microscope image of a CHO cell chromosome [26].)

3.2.4 Gene Amplification

Gene amplification means "the selective, repeated replication of a certain gene or genes without a proportional increase in other genes in the genome" [27]. Gene amplification is a widespread phenomenon in eukaryotes. It is an important process in the development of many organisms, the emergence of drug resistance in tumor cells and some human parasites, and the maturation of tumors.

Gene amplification events can be classified into 2 categories [28, 29], developmentally regulated gene amplification and spontaneously occurring gene amplification. Spontaneously occurring gene amplification is a widespread genetic phenomenon that generates cells and organisms with new genetic potential. The well-known gene amplification in mammalian cells is DHFR gene amplification. In the late 1950s, Hakala investigated the biochemical basis of amethopterin (methotrexate (MTX)) resistance in mammalian cells [30]. MTX is an analogue of dihydrofolate, inhibits DHFR, and kills rapidly growing cells. It is used as a therapeutic agent for leukemias. MTX inhibits not only DHFR but also other folic acid and nucleic acid metabolizing enzymes (Figure 3.2.7). Depending on selective conditions, the resulting cells exhibited up to 155 times the original level of DHFR activity. Highly methotrexate-resistant cells were selected by increasing the concentration of the drug.

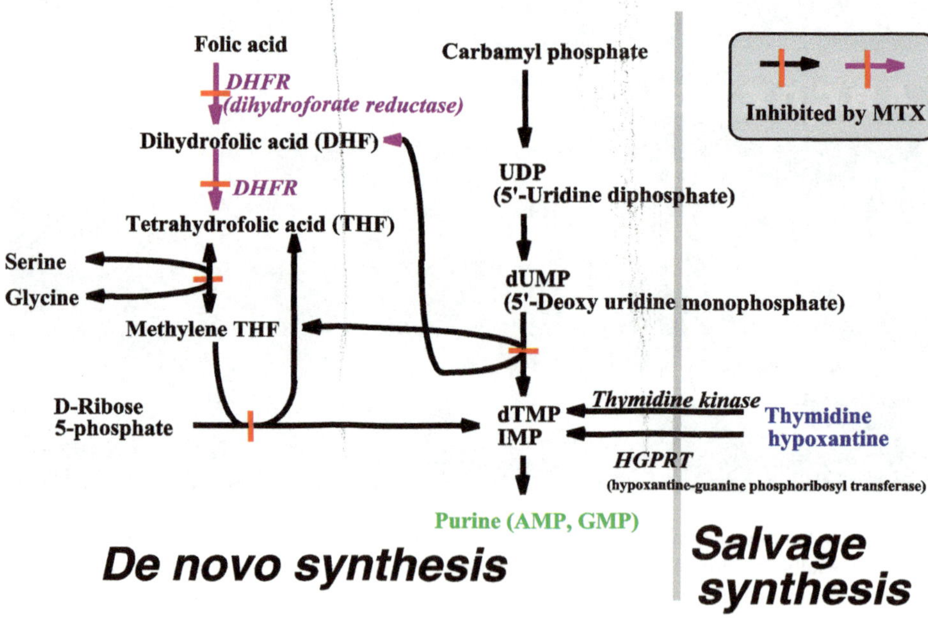

Figure 3.2.7: *De novo* and salvage biosynthesis pathways for purines [29].

From the early 1980s, this gene amplification procedure is used in the industrial production of recombinant proteins. To increase productivity, gene amplification (cell engineering) techniques for the transfected gene are extensively applied, whereby the amplification-promoting gene is used as a selectable marker for the transfected vector. First, the vector, which contains the cDNA of the objective protein (gene of interest (GOI)) and amplification-promoting gene, is introduced into the host cell line. When the selectable and amplification promoting gene is amplified, genetically linked sequences containing the objective cDNA are coamplified. Finally, the copy number of the GOI dramatically increases in the gene-amplified cell line (Figure 3.2.8).

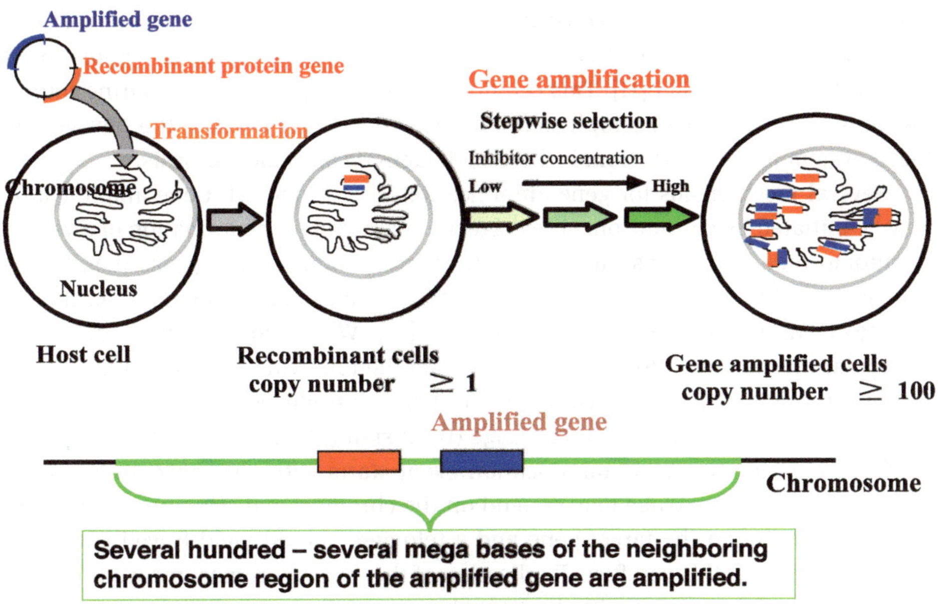

Figure 3.2.8: General procedure for the construction of a gene-amplified cell line for high-level expression [29].

With the increased copy number of the GOI, a high expression level of the GOI can be easily attained. CHO is the most conventional host cell line for gene amplification techniques, and it is easier to construct a gene-amplified CHO cell line than from other cell lines [31]. Siminovitch commented that CHO cells are not strictly diploid in the functional sense and that many genes are present in the hemizygous state [23]. According to the authors' results, the CHO cell line contains only one paired chromosome [26]. The hemizygous state may result in these unpaired chromosomes because of chromosome rearrangements.

The general procedure for the construction of gene-amplified cell lines for the high-level expression of exogenous genes is shown in Figure 3.2.8. The gene amplification procedure consists of 2 steps. The first step is the introduction of an amplifiable vector that contains not only the amplification promoting and selectable marker gene but also the objective GOI. The second step is the gene amplification itself. The transfected cell line ordinarily contains one copy or a low copy number of the transfected gene. In the second step, the transfected gene is selectively amplified by the stepwise increase of selection pressure.

In the coamplification of a heterologous gene in CHO cells, Weidle et al. reported that human t-PA and mouse DHFR cDNAs, which were under the control of the SV40 early promoter and the major late promoter of adenovirus 2 (AMLP), respectively, were amplified to 300 to 1,100 copies per genome at an MTX concentration of 5.0 μM [32]. The most convenient way to construct a gene-amplified cell line is to use a DHFR-deficient cell line. The most frequently used cell line for heterologous expression is the DHRF-deficient DG44 [11]. This cell line requires the presence of a purine source, thymidine, and glycine to complement the DHFR deficiency because the salvage synthesis system remains functional when a cell lacks *de novo* biosynthesis.

Several mechanisms for gene amplification were proposed. Gene amplification may be initiated by chromosome breakage and may lead to the formation of a variety of chromosomal structures, including expanded chromosomal regions (i.e., homogeneously staining regions (HSRs)) and extra chromosomal elements (i.e., double minutes (DMs)), and dicentric chromosomes [33]. Windle and Wahl classified the molecular mechanism into 3 categories [34]. In the following section, typical breakage-fusion-bridge (B-F-B) molecular models will be introduced [29].

The B-F-B model consists of breakage (B), fusion (F), and bridging (B) steps [35, 36] (Figure 3.2.9). The gene amplification event starts with chromosome breakage, and the site of the breakage may depend on the chromosomal structure. A breakage that occurs between the target gene and a telomere should yield frayed ends, and these sister chromatids can fuse. Replication of this fused chromosome would generate a dicentric chromosome with duplicated copies of the target gene that are bridged by the fusion point. A new breakage at a different site produces a chromosome with two copies of the target gene, and the B-F-B cycle is then repeated. According to this B-F-B model, the inverted repeat should be formed in the amplicon.

In DHFR gene amplification in the CHO cell line, the DHFR amplicon and its locus were defined in the endogenous CHO DHFR-amplified CHOC 400 cell line [37, 38]. The head-to-head DHFR amplicon in CHOC 400 has been reported, and its structure generally supports the idea of sister chromatid fusion [39, 40]. The authors also investigated the genomic structure of the exogenous DHFR amplicon in the CHO DR1000L-4N cell line from the construction of the CHO BAC library [19, 41] (Figure 3.2.10). They determined the complete nucleotide sequence of the insert in the Cg0031N14 BAC clone, which contains the amplicon, by shotgun sequencing. Two identical CHO genomic sequences were located in opposite directions (inverted repeat), and the exogenous pSV2-DHFR/hGM-CSF vector sequence was found at the junction between the 2 inverted repeat units. Consequently, the determined sequence had a palindromic structure containing a small inverted repeat in the junction region. The junction contains 2 copies of mouse DHFR and 1 copy of human GM-CSF. The structure of the junction is considerably different from the structure of the original pSV2-DHFR/hGM-CSF vector. The obtained structure cannot be explained by a simple B-F-B cycle or other proposed mechanisms. Further investigation is necessary to explain the gene amplification mechanism in the exogenous DHFR amplicon.

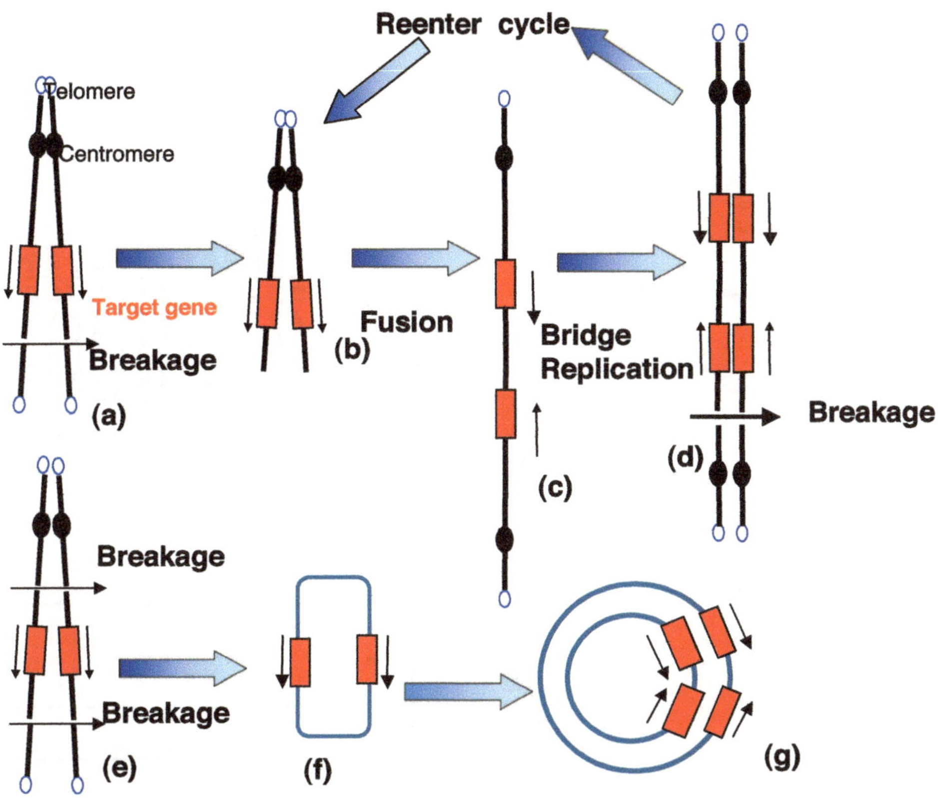

Figure 3.2.9: Breakage fusion bridge (B-F-B) cycle model in gene amplification [29].

3.2.5 Recent Developments of Recombinant CHO Cells using Gene Amplification and Chromosome Engineering

Gene amplification is a conventional procedure for the construction of highly productive cell lines [42]. The most time-consuming and critical steps are cell line screening and subcloning during the gene amplification procedure. The transfected vector is normally randomly integrated into the CHO genome by nonhomologous recombination. The cells containing the stable transgene can survive under selective conditions. The surviving CHO cells have a wide variety of productivity. Among these heterogeneous cell pools, stable and highly productive CHO cell lines should be contained. Consequently, the probability of obtaining a highly productive cell line increases with increasing numbers of screened clones. Because of the recent development of a high-throughput automated cell screening system, various types of clone-picking machines are available [43]. These robotic systems could greatly contribute to the effective construction of highly productive cell lines.

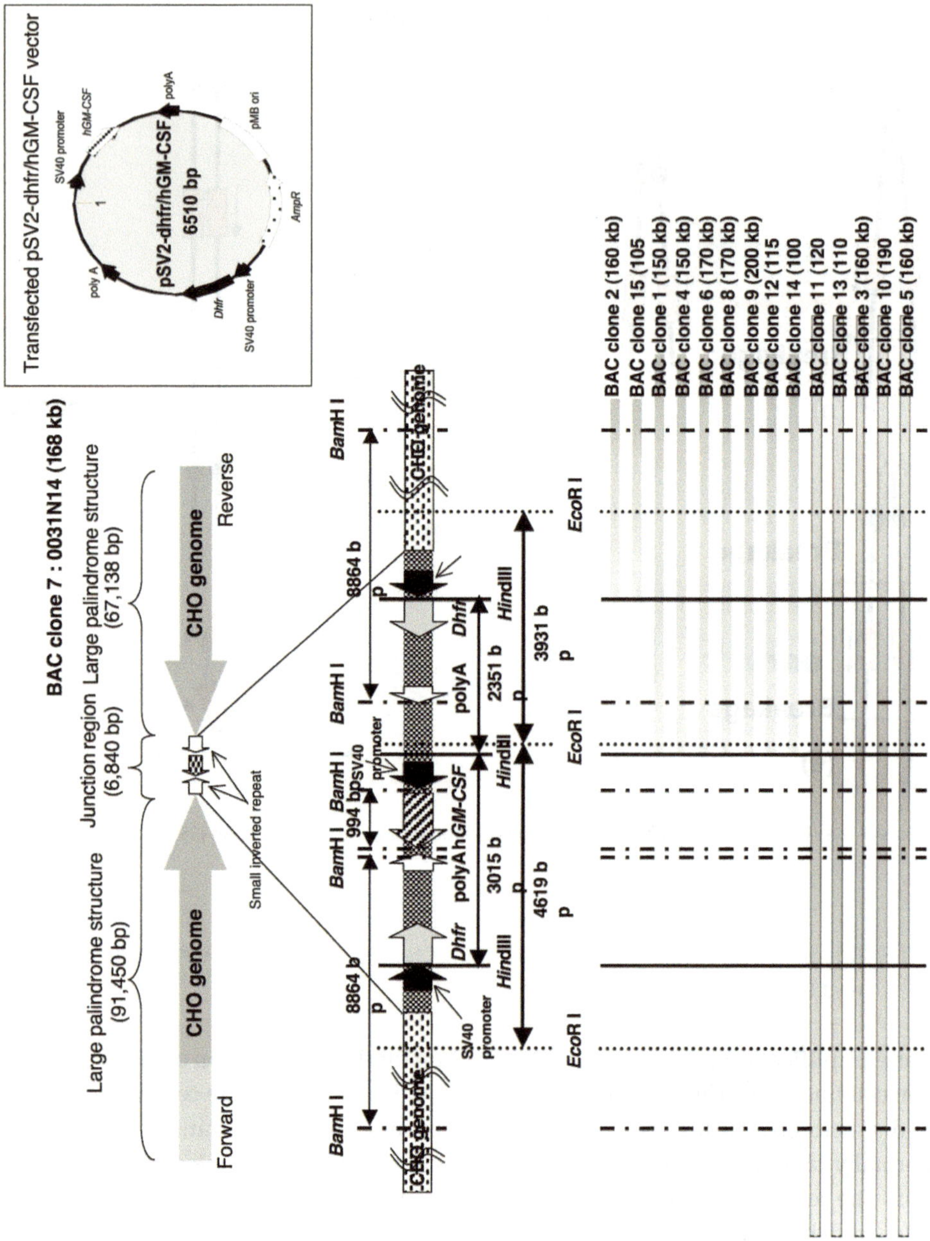

Figure 3.2.10: Proposed structure of the DHFR-amplified chromosomal region of the CHO DR1000L-4N genome. Relationships between the core region of the proposed structure of the CHO DR1000L-4N genome and 15 BAC clones that contain exogenous DHFR are shown. The positions of BAC clones were determined by BAC end sequencing and Southern blot analysis [41].

In contrast, several researchers developed an effective construction method for highly productive cell lines that is based on gene amplification mechanisms and chromosome engineering. For example, it was reported that the mammalian replication initiation region (IR) and a matrix attachment region (MAR) are spontaneously amplified in transfected human cells and efficiently generate chromosomal HSRs and/or extra-chromosomal DMs [44]. On the basis of this phenomenon, a new type of gene amplification vector was constructed, and high productivity was achieved in CHO cells using this novel gene amplification system [45].

Another approach to increase the transgene copy number in the host genome is to use an artificial chromosome system. Artificial chromosomes are composed of an artificial chromosomal element, and they can contain very long exogenous sequences. The artificial chromosome expression system was developed based on preengineered artificial chromosomes with multiple recombination acceptor sites to improve CHO cell line generation [44-48]. A similar artificial chromosome system using human artificial chromosomes was also used in human factor VIII production in CHO cells [49].

The authors recently developed a new approach to accelerate gene amplification using cell cycle engineering. Cell cycle engineering is a useful strategy for achieving suitable production of therapeutic proteins using cell cycle control. Previous researchers have focused on cell cycle modulation to increase the proliferation rate and viable cell density [50, 51] and on cell cycle arrest to achieve productivity enhancement in CHO cells [52, 53]. The authors' approach is to use cell cycle engineering to generate highly productive cells in the cell line development process using gene amplification. Gene amplification is caused by genomic instability, which is closely related to the cell cycle control mechanism. The authors focused on cell cycle engineering to enhance the efficiency of the conventional gene amplification system by escaping cell cycle control (Figure 3.2.11).

The authors down-regulated a cell cycle checkpoint kinase, ataxia-telangiectasia and Rad3-related (ATR), to accelerate gene amplification [54]. ATR is activated in response to DNA damaging agents, and it activates checkpoint kinases 1 and 2 (CHK1 and CHK2) to initiate signal cascades. The authors also focused on the targets of these signaling pathways, cell cycle division 25 (CDC25) phosphatases. CDC25 phosphatases activate complexes of cyclins and cyclin-dependent kinases (CDKs), which in turn regulate cell cycle transitions [55]. Among the 3 isoforms, CDC25A, B, and C, CDC25A is considered to be the key component [56] and is involved in both G1–S and G2–M phase transitions [57-59]. The authors investigated the effects of CDC25A and mutated CDC25A overexpression on gene amplification and productivity during the construction process of productive cell lines [60]. Both wild-type and mutated CDC25A-overexpressing CHO cells showed a rapid increase in transgene copy number compared with mock cells during the gene amplification process in both cell pools and individual clones. The authors' results show the improvement of the conventional gene amplification systems via cell cycle engineering at an early stage of cell line development.

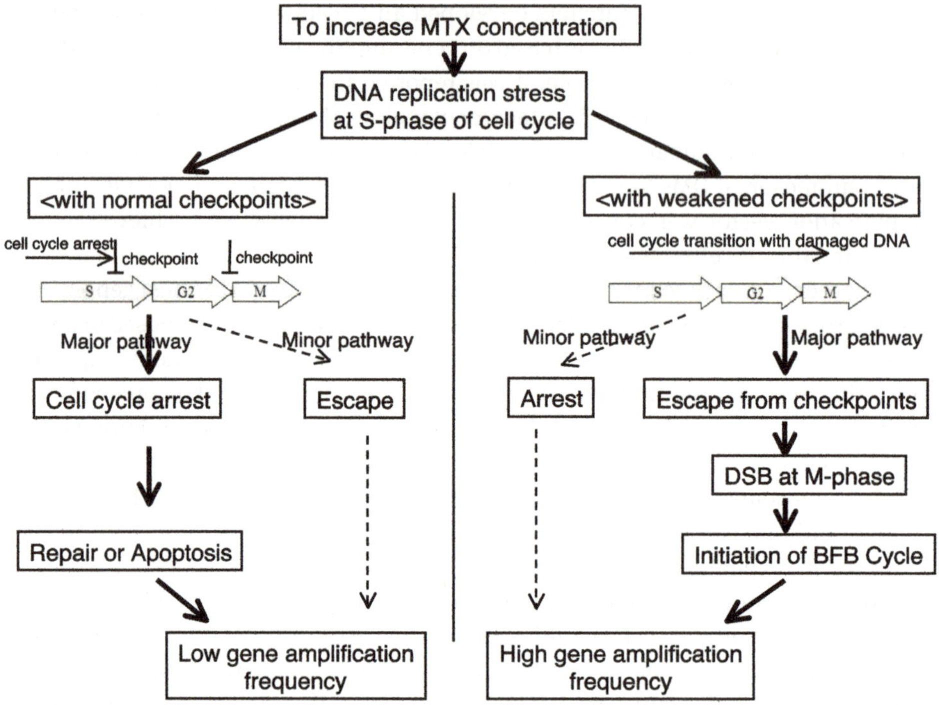

Figure 3.2.11: Model of the relationship between cell cycle checkpoints and gene amplification frequency [54]. DSB: double-strand break, BFB: breakage-fusion-bridge.

3.2.6 Conclusion

In this chapter, the authors introduced the chromosome structure and recent developments in chromosome identification and gene amplification and chromosome-related technologies in CHO cells. For industrial production, the basic scientific investigation could greatly contribute to cell line engineering. Further progress in this area will allow the construction of artificial industrial workhorse mammalian cell lines in the near future.

References

[1] Lodish H, Berk A, Kaiser CA, et al. Molecular Cell Biology (7th edition). New York, USA: WH Freeman & Co Ltd; 2012.
[2] Schrock E, Veldman T, Padilla-Nash H, et al. Spectral karyotyping refines cytogenetic diagnostics of constitutional chromosomal abnormalities. Hum Genet 1997;101:255-62.
[3] Asakawa S, Abe I, Kudoh Y, et al. Human BAC library: construction and rapid screening. Gene 1997;191:69-79.

[4] McPherson JD, Marra M, Hillier L, et al. A physical map of the human genome. Nature 2001;409:934-41.
[5] Cao Y, Kimura S, Itoi T, et al. Fluorescence in situ hybridization using bacterial artificial chromosome (BAC) clones for the analysis of chromosome rearrangement in Chinese hamster ovary cells. Methods 2012;56:418-23.
[6] Chung YJ, Jonkers J, Kitson H, et al. A whole-genome mouse BAC microarray with 1-Mb resolution for analysis of DNA copy number changes by array comparative genomic hybridization. Genome Res 2004;14:188-96.
[7] Greber D, Fussenegger M. Mammalian synthetic biology: engineering of sophisticated gene networks. J Biotechnol 2007;130:329-45.
[8] Omasa T, Onitsuka M, Kim WD. Cell engineering and cultivation of Chinese hamster ovary (CHO) cells. Curr Pharm Biotechnol 2010;11:233-40.
[9] Hacker DL, De Jesus M, Wurm FM. 25 years of recombinant proteins from reactor-grown cells – where do we go from here? Biotechnol Adv 2009;27:1023-7.
[10] Kao FT, Puck TT. Genetics of somatic mammalian cells, VII. Induction and isolation of nutritional mutants in Chinese hamster cells. Proc Natl Acad Sci USA 1968;60:1275-81.
[11] Urlaub G, Chasin LA. Isolation of Chinese hamster cell mutants deficient in dihydrofolate reductase activity. Proc Natl Acad Sci USA 1980;77:4216-20.
[12] Griffin TJ, Seth G, Xie HW, Bet al. Advancing mammalian cell culture engineering using genome-scale technologies. Trends Biotechnol 2007;25:401-8.
[13] Wurm FM. Production of recombinant protein therapeutics in cultivated mammalian cells. Nat Biotechnol 2004;22:1393-8.
[14] Walsh G. Biopharmaceutical benchmarks 2010. Nat Biotechnol 2010;28:917-24.
[15] Wlaschin KF, Hu WS. A scaffold for the Chinese hamster genome. Biotechnol Bioeng 2007;98:429-39.
[16] Xu X, Nagarajan H, Lewis NE, et al. The genomic sequence of the Chinese hamster ovary (CHO)-K1 cell line. Nat Biotechnol 2011;29:735-41.
[17] Lewis NE, Liu X, Li Y, et al. Genomic landscapes of Chinese hamster ovary cell lines as revealed by the Cricetulus griseus draft genome. Nat Biotechnol 2013;31:759-65.
[18] Brinkrolf K, Rupp O, Laux H, et al. Chinese hamster genome sequenced from sorted chromosomes. Nat Biotechnol 2013;31:694-5.
[19] Omasa T, Cao Y, Park JY, et al. Bacterial artificial chromosome library for genome-wide analysis of Chinese hamster ovary cells. Biotechnol Bioeng 2009;104:986-94.
[20] Yoshikawa T, Nakanishi F, Ogura Y, et al. Amplified gene location in chromosomal DNA affected recombinant protein production and stability of amplified genes. Biotechnol Prog 2000;16:710-5.
[21] Yoshikawa T, Nakanishi F, Itami S, et al. Evaluation of stable and highly productive gene amplified CHO cell line based on the location of amplified genes. Cytotechnology 2000;33:37-46.
[22] Deaven LL, Petersen DF. The chromosomes of CHO, an aneuploid Chinese hamster cell line: G-band, C-band, and autoradiographic analyses. Chromosoma 1973;41:129-44.
[23] Siminovitch L. On the nature of hereditable variation in cultured somatic cells. Cell 1976;7:1-11.
[24] Worton RG, Ho CC, Duff C. Chromosome stability in CHO cells. Somatic Cell Genet 1977;3:27-45.
[25] Derouazi M, Martinet D, Besuchet Schmutz N, et al. Genetic characterization of CHO production host DG44 and derivative recombinant cell lines. Biochem Biophys Res Commun 2006;340:1069-77.
[26] Cao Y, Kimura S, Itoi T, et al. Construction of BAC-based physical map and analysis of chromosome rearrangement in Chinese hamster ovary cell lines. Biotechnol Bioeng 2012;109:1357-67.

[27] Smith AD, Datta SP, Smith GH, et al. (eds). Oxford dictionary of biochemistry and molecular biology. Oxford, UK, New York, USA, Tokyo, Japan: Oxford University Press; 1997.
[28] Kellems RE (ed). Gene amplification in mammalian cells - A comprehensive guide. New York, USA: Marcel Dekker; 1993.
[29] Omasa T. Gene amplification and its application in cell and tissue engineering. J Biosci Bioeng 2002;94:600-5.
[30] Hakala MT. Prevention of toxicity of amethopterin for sarcoma-180 cells in tissue culture. Science 1957;126:255.
[31] Omasa T, Yamanaka M, Tanimura N, et al. Expression and amplification of glutamine synthetase gene endows HepG2 cells with ammonia-metabolizing activity for bioartificial liver support system. Enzyme Microb Technol 2004;35:519-24.
[32] Weidle UH, Buckel P, Wienberg J. Amplified expression constructs for human tissue-type plasminogen activator in Chinese hamster ovary cells: instability in the absence of selective pressure. Gene 1988;66:193-203.
[33] Paulson TG, Almasan A, Brody LL, et al. Gene amplification in a p53-deficient cell line requires cell cycle progression under conditions that generate DNA breakage. Mol Cell Biol 1998;18:3089-100.
[34] Windle BE, Wahl GM. Molecular dissection of mammalian gene amplification: new mechanistic insights revealed by analyses of very early events. Mutat Res 1992;276:199-224.
[35] Ruiz JC, Wahl GM. Chromosomal destabilization during gene amplification. Mol Cell Biol 1990;10:3056-66.
[36] Kuo MT, Sen S, Hittelman WN, et al. Chromosomal fragile sites and DNA amplification in drug-resistant cells. Biochem Pharmacol 1998;56:7-13.
[37] Looney JE, Hamlin JL. Isolation of the amplified dihydrofolate reductase domain from methotrexate-resistant Chinese hamster ovary cells. Mol Cell Biol 1987;7:569-77.
[38] Ma C, Looney JE, Leu TH, et al. Organization and genesis of dihydrofolate reductase amplicons in the genome of a methotrexate-resistant Chinese hamster ovary cell line. Mol Cell Biol 1988;8:2316-27.
[39] Okuno Y, Hahn PJ, Gilbert DM. Structure of a palindromic amplicon junction implicates microhomology-mediated end joining as a mechanism of sister chromatid fusion during gene amplification. Nucleic Acids Res 2004;32:749-56.
[40] Sasaki T, Ramanathan S, Okuno Y, et al. The Chinese hamster dihydrofolate reductase replication origin decision point follows activation of transcription and suppresses initiation of replication within transcription units. Mol Cell Biol 2006;26:1051-62.
[41] Park JY, Takagi Y, Yamatani M, et al. Identification and analysis of specific chromosomal region adjacent to exogenous Dhfr-amplified region in Chinese hamster ovary cell genome. J Biosci Bioeng 2010;109:504-11.
[42] Omasa T. Chinese hamster ovary cell genome: impact on cell engineering. In: The 23rd Annual and International Meeting of the Japanese Association for Animal Cell Technology (JAACT2010). Sapporo, Japan: JAACT; 2010:58.
[43] Lindgren K, Salmen A, Lundgren M, et al. Automation of cell line development. Cytotechnology 2009;59:1-10.
[44] Shimizu N, Hashizume T, Shingaki K, et al. Amplification of plasmids containing a mammalian replication initiation region is mediated by controllable conflict between replication and transcription. Cancer Res 2003;63:5281-90.
[45] Noguchi C, Araki Y, Miki D, et al. Fusion of the Dhfr/Mtx and IR/MAR gene amplification methods produces a rapid and efficient method for stable recombinant protein production. PLoS One 2012;7:e52990.

[46] Kennard ML, Goosney DL, Monteith D, et al. The generation of stable, high MAb expressing CHO cell lines based on the artificial chromosome expression (ACE) technology. Biotechnol Bioeng 2009;104:540-53.
[47] Kennard ML, Goosney DL, Monteith D, et al. Auditioning of CHO host cell lines using the artificial chromosome expression (ACE) technology. Biotechnol Bioeng 2009;104:526-39.
[48] Combs RG, Yu E, Roe S, et al. Fed-batch bioreactor performance and cell line stability evaluation of the artificial chromosome expression technology expressing an IgG1 in Chinese hamster ovary cells. Biotechnol Prog 2011;27:201-8.
[49] Kurosaki H, Hiratsuka M, Imaoka N, et al. Integration-free and stable expression of FVIII using a human artificial chromosome. J Hum Genet 2011;56:727-33.
[50] Renner WA, Lee KH, Hatzimanikatis V, et al. Recombinant cyclin E expression activates proliferation and obviates surface attachment of chinese hamster ovary (CHO) cells in protein-free medium. Biotechnol Bioeng 1995;47:476-82.
[51] Kuystermans D, Al-Rubeai M. cMyc increases cell number through uncoupling of cell division from cell size in CHO cells. BMC Biotechnol 2009;9:76.
[52] Fussenegger M, Schlatter S, Datwyler D, et al. Controlled proliferation by multigene metabolic engineering enhances the productivity of Chinese hamster ovary cells. Nat Biotechnol 1998;16:468-72.
[53] Sunley K, Butler M. Strategies for the enhancement of recombinant protein production from mammalian cells by growth arrest. Biotechnol Adv 2010;28:385-94.
[54] Lee KH, Onitsuka M, Honda K, et al. Rapid construction of transgene-amplified CHO cell lines by cell cycle checkpoint engineering. Appl Microbiol Biotechnol 2013;97:5731-41.
[55] Boutros R, Lobjois V, Ducommun B. CDC25 phosphatases in cancer cells: key players? Good targets? Nat Rev Cancer 2007;7:495-507.
[56] Ferguson AM, White LS, Donovan PJ, et al. Normal cell cycle and checkpoint responses in mice and cells lacking Cdc25B and Cdc25C protein phosphatases. Mol Cell Biol 2005;25:2853-60.
[57] Mailand N, Podtelejnikov AV, Groth A, et al. Regulation of G(2)/M events by Cdc25A through phosphorylation-dependent modulation of its stability. EMBO 2002;21:5911-20.
[58] Blomberg I, Hoffmann I. Ectopic expression of Cdc25A accelerates the G(1)/S transition and leads to premature activation of cyclin E- and cyclin A-dependent kinases. Mol Cell Biol 1999;19:6183-94.
[59] Hoffmann I, Draetta G, Karsenti E. Activation of the phosphatase activity of human cdc25A by a cdk2-cyclin E dependent phosphorylation at the G1/S transition. EMBO 1994;13:4302-10.
[60] Lee KH, Tsutsui T, Honda K, et al. Generation of high-producing cell lines by overexpression of cell division cycle 25 homolog A in Chinese hamster ovary cells. J Biosci Bioeng 2013;116:754-60.

3.3 Synthetic Biology Principles for Engineering Mammalian Designer Cells

David Ausländer and Martin Fussenegger

3.3.1 Introduction

In the past century, recombinant DNA technology enabled simple reshuffling of genetic components and provided the basis for molecular biology that started to unravel the molecular basis of biological activities. Subsequent contributions of the "-omics" disciplines and the systems biology era have led to encyclopedic knowledge on the molecular components and their reaction and interaction dynamics that orchestrate essential activities within and among living systems. Today, synthetic biology merges engineering principles with molecular biology [1-3] and has become the science of reassembling cataloged and standardized biological components in a rational manner to create and engineer functional biological designer devices, systems, and organisms with novel, useful, and preferably therapeutic functions. Mammalian synthetic biology focuses on engineering increasingly complex gene networks in mammalian cells. This research can foster advances in gene-function analysis, functional genomic research, drug discovery, biopharmaceutical manufacturing, and prototype gene- and cell-based therapies [4-7].

Molecular switches represent fundamental components of artificial gene networks. They are based on specialized biomolecules that show trigger-controlled activities modulated by conditional protein-protein, protein-DNA, protein-RNA or protein-small-molecule interactions, enzymatic reactions and stability [8-10]. Here the authors provide an overview of the latest developments in molecular switch assembly and how they can be used to engineer synthetic networks in mammalian designer cells to perform biomedically relevant tasks.

3.3.2 Inducible Gene Switches

An important class of molecular switches is the trigger-controlled gene switch, which fine-tunes product-protein levels in mammalian cells in response to molecular or physical cues. This can be achieved by either inducing or repressing gene expression at the transcriptional or post-transcriptional level [6, 11]. Many gene switches have a modular design and are assembled from standardized biological parts with specific function such as DNA-binding domains (DBD), RNA-binding proteins and transactivation, transsilencing, dimerization or RNA-based domains.

A distinctive feature of a switching mechanism is its capability to bind a specific ligand by a molecular domain, called a sensor domain. This sensor domain transmits its status to a second domain by either conformational changes or enabling molecular interactions. Actuator domains (ADs) can be subsequently recruited to the point of action, thereby affecting gene expression. Gene switches take advantage of downstream reporter gene expression or activity that enables simple quantification of protein concentrations in living cells or organisms. Gene switches can be characterized in terms of their kinetic, dynamic behavior. In particular, secreted reporter proteins, such as the human placental secreted alkaline phosphatase (SEAP) or secreted versions of luciferase, can be used to monitor the performance of gene switches at different time-points *in vitro* as well as in peripheral circulation of animals. Also, recent advances in fluorescent protein engineering and microscopy have enabled precise, multicolor, parallel observation of different gene switches in living cells in real-time with accurate dynamic and kinetic characteristics [12, 13]. There are several important parameters that characterize gene switches:

- **Dose response:** The product-gene expression is adjustable by changing the inducer concentration, resulting in fine-tuning of the designer cell response (adjustability).
- **Kinetics:** The protein-product concentration should stay in its appropriate state (either OFF or ON) without the trigger molecule (robustness) and should efficiently turn ON or OFF when adding the trigger compound. Induction or repression of gene expression requires a distinct time frame of presence of the trigger molecule (pulse width).
- **Dynamics:** The induction ratio between ON and OFF states should be as high as possible (induction ratio). The gene switch should be exclusively responsive to a specific trigger molecule (specificity) and the expression switches should be completely reversible (reversibility).

Additional optimization of individual functional parts, or their combination with other modules, may enhance the performance of the above-mentioned parameters. For example, the addition of localization domains such as the nuclear localization signal (NLS) to transcription factors may enhance induction fold. Likewise, mutagenesis of transcription factors and their operator modules may alter specificity, kinetics, or robustness [14-16].

An essential aspect for the utility of a gene switch in mammalian cells is the nature of its inducer molecule. These inducer molecules can range from small molecules, to gaseous compounds, to physical cues such as light and temperature. Thus, the complexity of gene networks in mammalian cells could be significantly advanced through the interplay of well-characterized molecular switching systems, leading to multi-input/-output biocomputing devices [17]. The suitability of gene switches has been proven by a multitude of applications in biotechnology and biomedicine. In particular, mammalian designer cells can perform specific therapeutic tasks when plugged into mouse metabolism, especially when equipped with autonomous sensor-effector devices, called prosthetic gene networks [18].

This chapter presents the fast-growing diversity of gene switches for mammalian designer cells, their benefits for complex, synthetic gene networks, and useful applications in the promising field of cell-based treatment strategies.

3.3.2.1 Synthetic Mammalian Transcription-Control Switches

Synthetic mammalian transcription-control switches share several common features. They consist of a synthetic transcription factor including a DNA-binding domain (DBD) targeting a specific operator sequence (O), a sensor domain (SP) that switches the operator-binding capacity in a trigger-controlled manner and an actuator domain (AD) that is either directly fused or conditionally linked to the DBD via chemically-induced dimerization (CID) and modulates the transcription machinery to induce or repress target promoters. These target promoters contain transcription factor-specific

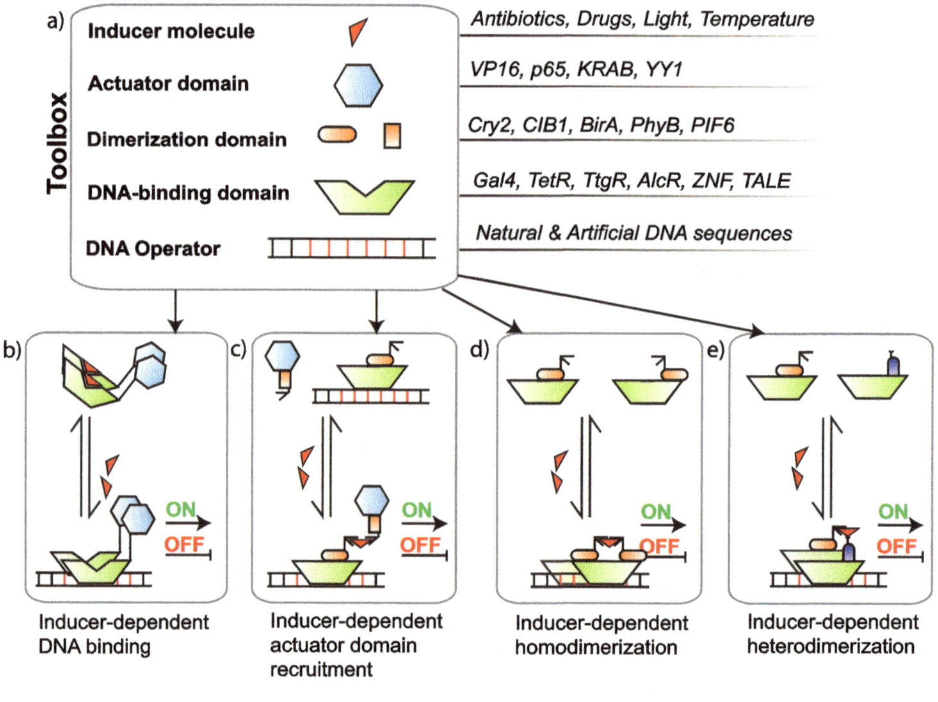

Figure 3.3.1: Schematic illustration of inducer-dependent mechanisms underlying synthetic, mammalian transcription controllers. **a)** Various tools derived from all kingdoms of life can be combined to engineer synthetic, mammalian transcription controllers. The ability to bind DNA is an essential feature of synthetic transcription controllers. **b)** The interaction between DNA and a transcription factor can be controlled by triggers that induce transgene expression or repression. **c)** The interaction between an actuator domain and the DNA-binding domain can be sensitive to an inducer. **d, e)** DNA-binding domains (DBDs) often must dimerize before binding to DNA, which can be accomplished by fusing ligand-responsive homo- or heterodimerization domains to the DBD.

operator modules linked to either constitutive or minimal promoter variants (Figure 3.3.1a). In the following sections, we outline the configuration of the main synthetic mammalian transcription-control switches along with a brief description of some exemplary studies.

Bacteria-derived Transcriptional Repressor Proteins

The survival of bacteria under environmental stress requires adaptive responses partially triggered by transcriptional repressor proteins that contain a sensor domain detecting signal molecules, leading to a conformational change that will affect the interaction between the DBD and its cognate DNA operator site. The Tet system is a pioneering example, in which the homodimerized Tet repressor protein (TetR) binds to its cognate DNA operator sequence (tetO) and switches its conformation upon binding of its ligand (tetracycline and its derivatives), thereby detaching from tetO [19]. Gossen and Bujard (1992) reported the transfer of the Tet system into mammalian cells by covalently fusing the transactivation domain of the virion protein 16 (VP16) from the herpes simplex virus as an actuator domain. This resulted in the chimeric construct TetR-VP16, called tetracycline-dependent transactivator (tTA) (see Figure 3.3.1b) [20]. In the absence of tetracycline, tTA binds and activates a tetracycline-responsive promoter ($P_{hCMV*-1}$) consisting of a heptameric operator module ($tetO_7$) 5' of a minimal version of the human cytomegalovirus immediate early promoter. However, in the presence of tetracycline tTA is released from $P_{hCMV*-1}$ and transgene expression is dose-dependently repressed. Therefore, this configuration became known as the TET_{OFF} system, since the gene is shut down in the presence of the trigger compound. VP16 was successfully exchanged by p65 or E2F4 to create transactivator variants with the identical trigger input but modified control dynamics [21, 22]. In contrast, replacing the transactivation with transsilencing domains, such as the Krüppel-associated box (KRAB) or Yin Yang-1 (YY1) domains, resulted in tetracycline-dependent transsilencers (tTS) that repressed constitutive promoters when bound to adjacent tetO modules in the absence of tetracycline. Since the target gene is induced in the presence of tetracycline as tTS is released from tetO and the constitutive promoter is dose dependently activated, this gene switch configuration is known as the TET_{ON} system (see Figure 3.3.1b) [23]. Mutational studies revealed a TetR mutant with a reverse DNA-binding capability, which was successfully used to design the corresponding reverse tetracycline-dependent transactivator (rtTA) or transsilencer (rtTS) [16, 24]. Further studies developed TetR mutants in which the DNA operator sequence could be altered, or TetR was modified to bind different small-molecule ligands as well as peptides [14, 15, 25].

Capitalizing on the same design concept other prokaryotic TetR family repressors with different trigger specificities and operator sequences have been successfully converted for use as trigger-controlled mammalian transgene switches. This significantly increased the availability of functionally compatible transcription controllers and enabled the design of synthetic gene networks for application in biotechnology and biomedicine [26] (see Figure 3.3.1b). For example, other antibiotics such as macrolides [27, 28] (e.g., erythromycin) or streptogramins (e.g., pristinamycin I) [26, 29] controlled transgene expression in complex synthetic gene networks in bioreactors and mice. An exogenous inducer molecule, such as the apple-metabolite phloretin supplemented in skin cream was applied to regulate transgene expression of microencapsulated cells subcutaneously implanted into mice [30]. Meanwhile, 2-phenylethyl-butyrate revealed novel antituberculosis compounds in cell culture [31] and 6-hydroxy-nicotine enabled multilevel transgene control in mammalian cells [32]. Physical cues, such as temperature, could be converted to induce transgene expression in a chicken B cell line [33]. Endogenous inducer molecules such as NADH [34], biotin [35, 36], L-arginine [37], urate [18], L-tryptophan [38], acetaldehyde [39, 40], or vanillic acid [41] were successfully used to trigger product-gene expression in bioreactors, to induce transgene expression in mice, for the development of the first prosthetic gene network in mice [18], and for the design of synthetic intercellular communication devices of mammalian cells [38].

Daisy chain assembly of well-characterized prokaryotic repressor proteins such as ScbR and TetR and fusion to VP16 resulted in a bipartite transcription factor (ScbR-TetR-VP16) which, at low intracellular concentrations, could reversibly program their individual chimeric or hybrid promoters for trigger-adjustable transgene expression using γ-butyrolactones and tetracycline, respectively [42]. Interestingly, ScbR-TetR-VP16 exhibited double-pole double-throw (DPDT) relay switch characteristics at high intracellular concentrations. In the genetic DPDT version, tetracycline programs ScbR-TetR-VP16 to progressively switch from TetR-specific promoter-driven expression of transgene 1 to ScbR-specific promoter-driven transcription of transgene 2 while ScbR-TetR-VP16 flips back to exclusive transgene 1 expression in the absence of the trigger antibiotic.

Altogether, the first generation of conditional gene expression systems is a stunning example of the successful implementation of bacterial modules into mammalian cells. The research significantly advances the diversity and application portfolio of heterologous gene switches.

Chemically-induced Dimerization Mechanisms

In contrast to bacteria-derived transcriptional repressor, proteins in which an actuator domain is covalently fused to a DBD of interest, the chemically induced dimerization (CID) mechanism triggers recruitment of the AD to a desired DBD or bacterial repressor protein (see Figure 3.3.1c). The basis of this mechanism is the yeast two-hybrid system, which was initially developed for screening unidentified protein-protein interactions. A DBD was covalently fused to a bait molecule that potentially bound a prey domain fused to an actuator domain (for example, transactivation domain VP16). Subsequent interaction recruited the AD into proximity of the transcription start site, thereby affecting reporter-gene expression [43]. This principle introduced ligand-dependency into a commonly ligand-independent transcriptional system. Moreover, ligand-dependent dimerization mechanisms found in all life kingdoms were implemented into ligand-insensitive DBDs and ADs. This turned their fusion constructs into regulatable transcription devices for mammalian cells [44].

The widely used yeast Gal4 transcription factor requires dimerization before binding its cognate DNA operator sequence, or upstream activator sequence (UAS), and inducing a downstream gene of interest. However, lacking the original dimerization domain, Gal4's DBD fused to alternative dimerization domains that triggered homo- or heterodimerization with an equally fused eukaryotic AD, pioneered regulatable gene switches (see Figures 3.3.1d and 3.3.1e). For example, the human estrogen receptor (hER) [44], the human progesterone receptor (hPR) [45], or the thyroid hormone receptor [46] was fused to Gal4 in the same fashion as a transactivation domain. This led to its homodimerization upon ligand binding, thereby inducing transgene expression. Furthermore, shuffling the DBD (for example, lacI, TetR) or the dimerization domains led to heterodimerizing synthetic transcription controllers subject to a variety of triggers, including molecules like abscisic acid [47], biotin [35], ecdysone [48], rapamycin [49-51], and coumermycin [52] or physical cues such as ultraviolet [53], blue [54-56] or red light [57].

Programmable DNA Binders

In the aforementioned mechanisms, the DNA sequence of the operator site is usually predetermined by the transcription factor DBD and, if unknown, must be identified by biochemical and bioinformatical methods. However, this limitation can be overcome by designing novel artificial DBDs, such as the zinc-finger (ZF) domains or transcription activator-like effectors (TALE) that can be rationally redesigned to target any designated DNA sequence. Both proteins are suitable transcription factors when fused to ADs, but lack any ligand-dependence [58-60]. In principle, both proteins can benefit from dimerization mechanisms to build a synthetic ligand-dependent transcription controller. However, only ZF has been engineered to respond to rapamycin or hormones [51, 61]. In future, programmable DNA binders with nuclease activity might have a large effect upon synthetic biology as suitable tools for trigger-controlled genome editing [62-64].

3.3.2.2 RNA-derived Gene Expression Mediators

Aptamers are the most important sensor domains for RNA-based, ligand-dependent switching mechanisms. They are short, highly structured RNA molecules that bind a trigger molecule with superior affinity and specificity (see Figure 3.3.2a) [65, 66]. The development of a powerful selection technique, called SELEX (systematic evolution of ligands by exponential enrichment), enabled researchers to select novel RNA aptamers to virtually any molecule [67, 68].

Many of these artificial RNA aptamers were selected to specifically bind a wide variety of compounds, including small molecules such as antibiotics (e.g., tetracycline, kanamycin [69, 70]), other drugs (e.g., theophylline [71]), proteins (e.g., TetR [72]), MS2 coat protein [73], L7Ae [74]), nucleotides (for example, ATP [75]), amino acids (e.g., L-arginine [76]), whole cells (e.g., African trypanosomes [77]), and Burkitt's lymphoma cells [78]. The theophylline aptamer is a widely used RNA aptamer. It induces a conformational shift upon ligand binding and discriminates the methylated analogue of theophylline, called caffeine, with a thousand-fold weaker binding affinity [71, 79]. Capitalizing on the modular construction of RNA controllers, researchers engineered RNA-based gene switches by interconnecting aptamers' molecular recognition with RNA actuator domains derived from interference RNAs, hammerhead ribozymes, splicing domains, or actuating aptamers (see Figure 3.3.2a).

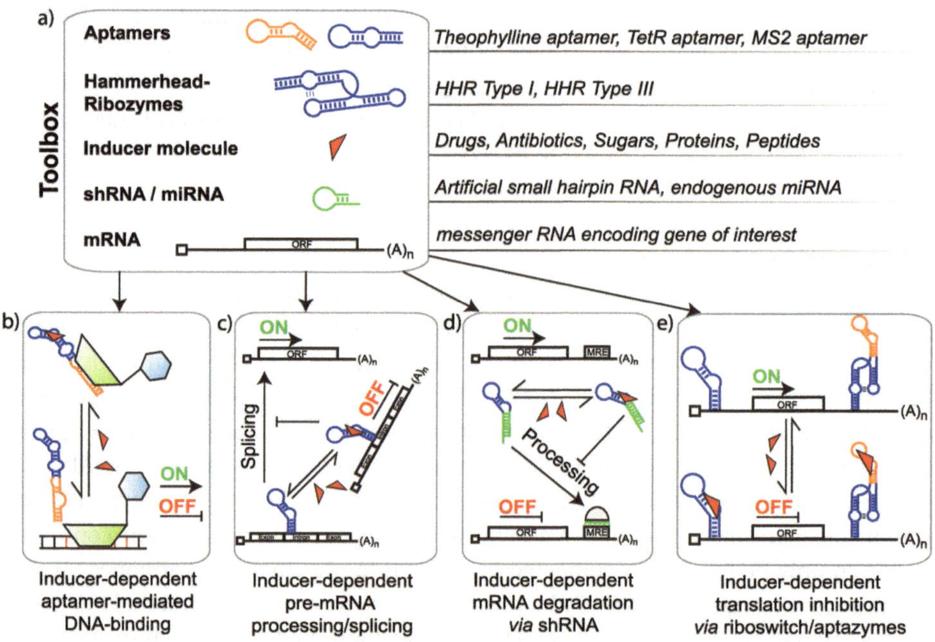

Figure 3.3.2: Schematic illustration of RNA-based, synthetic gene control in mammalian cells. **a)** RNA-based biomolecules can be modularly combined to build ligand-sensitive RNA devices that control transgene expression. Aptamers are important sensor domains that tightly bind their cognate ligand molecules. **b)** Fusing a sensor aptamer with an actuator aptamer led to theophylline-dependent binding of the RNA device to a transcription factor, thereby influencing gene expression. **c)** Splicing efficiency could be regulated by ligand-binding aptamers fused to important splicing domains, such as the branching point. **d)** Aptamers interconnected to shRNA or miRNA domains control transgene expression in a ligand-responsive manner. **e)** Aptamer domains inserted into the 5'-UTR of mRNA molecules can block transgene translation upon binding to their cognate ligand molecules. Also, aptazymes introduced into the 5'-or 3'-UTR can regulate mRNA integrity with a trigger.

Allosteric Ribozymes (Aptazymes)

Hammerhead ribozymes are well-characterized RNA structures that catalyze site-specific cleavage of RNA molecules [80]. When fused to an aptamer domain, hammerhead ribozymes can be converted into allosteric ribozymes (called aptazymes), whose activity is subsequently controlled by the concentration of the aptamer's ligand [81]. Bioengineers implemented such aptazymes into the 5'-untranslated region (5'-UTR) or 3'-UTR of reporter-gene messenger RNA (mRNA), whose self-cleavage activity leads either to the removal of the 5'-cap or the 3'-poly A tail. These are both essential structural components for mRNA integrity and stability (see Figure 3.3.2e). Hence, reporter-gene expression was controlled by adding the ligand of the aptamer that activates the hammerhead ribozyme. This prevented translation or induced mRNA degradation [82-85].

RNA Interference

RNA interference (RNAi) is an essential eukaryotic RNA-based gene regulation mechanism that capitalizes on the RNA-induced silencing complex (RISC). It can be initiated by short dsRNA molecules targeted to any desired mRNA molecule, thereby inducing degradation or inhibiting translation [86]. The specific RNA molecule can be genetically encoded using short-hairpin RNA (shRNA) or miRNA, whose transcripts are driven by constitutive or inducible promoters, or introduced into intronic regions [87-92]. The transcripts are extensively processed, which is mediated by nucleases (including Drosha and Dicer) loading the guide strand of the siRNA on RISC that serves as a template for the complementary target mRNA molecules with miRNA recognition elements (MRE) [93]. Artificial ligand-dependent control over RNA interference mechanisms has been achieved using aptamer domains. Incorporating the theophylline aptamer domain into an essential loop region of a shRNA that is usually the target sequence for Dicer processing led to efficiently inhibiting shRNA maturation by adding theophylline (see Figure 3.3.2d) [94]. Similarly, ligand-responsive miRNAs enabled ligand control of Drosha processing and eventual gene downregulation [95].

A sophisticated design strategy included fusing an aptazyme to a miRNA precursor molecule and an inhibitory strand hybridizing with the miRNA sequence. By adding the aptamer ligand, the aptazyme was activated and deliberated the miRNA analogue, which can be further processed to downregulate a gene of interest [96]. Furthermore, the combination of translational (RNAi) and transcriptional (transrepressor protein) mechanisms was a very effective gene-regulation switch in mammalian cells [97].

Artificial Riboswitches

Natural RNA controllers, called riboswitches, are RNA sequences found in messenger RNA (mRNA) that can bind molecules by an aptamer domain that transmits the information to an adjacent AD, affecting gene expression [98]. Riboswitches have been found in mRNA molecules of prokaryotes, where they regulate transcription termination or translation initiation [98]. Interestingly, in eukaroytes as single riboswitch (TPP riboswitch) has been found so far, which alter gene expression by influencing alternative splicing [99]. Artificial riboswitches for ligand-dependent control of gene expression in mammalian cells were engineered by introducing aptamer domains into the 5'-UTR of gene transcripts (see Figure 3.3.2e). For example, aptamers that bound Hoechst dyes, the bacteriophage MS2 coat protein, or the archael ribosomal protein L7Ae were successfully used for ligand-dependent gene control [100-102].

Splicing Efficiency

Alternative splicing is an ingenious gene-regulation mechanism in eukaryotes in which introns are removed from the pre-mRNA, thereby joining exonic regions and forming the mature mRNA. Ligand-dependent control of splicing efficiency in mammalian cells was achieved by sequestering the branchpoint sequence with a theophylline aptamer (see Figure 3.3.2b) [103]. Another sophisticated design strategy rewired endogenous proteins to regulate splicing via aptamer domains inserted at 5' or 3' of an intron. Aptamers that bind the MS2 coat protein, subunits of NF-κB (p50 and p65), or b-catenin were introduced into human cells, detecting endogenous signaling pathway activation and rewiring these signals to efficiently alter splicing patterns of reporter transcripts [104].

Aptamer-regulated Aptamers

Aptamers that bind transcriptional repressor proteins such as TetR can potentially compete with their cognate DNA operator sequence and control gene expression in mammalian cells [72, 105]. By replacing the loop sequence of the TetR aptamer with the theophylline aptamer, the resulting RNA device binds and inhibits tTA in a theophylline-dependent manner in human cells transfected with the Tet_{OFF} system (see Figure 3.3.2a) [105].

3.3.2.3 Protein stability

The protein concentration can also be controlled by its degradation rate, which involves the ubiquitin-proteasome pathway in mammalian cells [106]. The main strategy is to covalently or temporarily mark a protein of interest with a degradation signal (called degron), enabling recognition by the ubiquitin system and facilitating proteasomal degradation. The interkingdom transfer of the plant-derived auxin-inducible degron (AID) system into mammalian cells is a sophisticated example of this strategy. Fusing AID to a protein of interest facilitates an auxin-dependent interaction between AID and the transport inhibitor response 1 (TIR1) protein. This subsequently recruits an E2 ligase responsible for polyubiquitylation of AID, marking the entire fusion protein for proteasomal degradation [107, 108]. Similar systems use proteins fused to destabilizing domains (DD) or ligand-induced degradation (LID) domains. Upon ligand binding, the stability of the fusion protein can be positively (DD) or negatively (LID) affected [109-113]. In another system, proteins were fused to a bacterial dehalogenase domain (HaloTag) that formed a covalent bond upon adding a haloalkane reactive linker [114, 115]. By fusing the linker sequence to hydrophobic moieties, the tagged fusion protein was efficiently degraded in mammalian cells [116].

3.3.2.4 Surface Receptor-based Gene Control

In contrast to intracellular protein detection of molecules, cells evolved the capability to sense many important molecules on their surface by receptors that transmit information about extracellular ligand concentrations via signal-transduction pathways. An important family of these receptors builds the G-protein coupled receptor (GPCR) class, which is one of the largest gene families in humans.

Its molecular structure includes seven transmembrane domains and a C-terminal intracellular domain. The C-terminal domain binds to heterotrimeric G proteins transmitting the signal into a cellular response by activating distinct signaling pathways [117]. The key endogenous signaling pathways include important second-messenger molecules, such as cyclic AMP (cAMP) and calcium (Ca^{2+}). For example, activating the $G\alpha_s$ protein leads to the adenylyl-cyclase-dependent accumulation of cAMP that subsequently activates protein kinase A. This then phosphorylates a series of target proteins, including cAMP-response element binding protein 1 (CREB1) [118]. Finally, nuclear-located CREB1 binds DNA-encoded cAMP-response elements and transcribes a series of response genes. Similarly, activating the $G\alpha_q$ protein mobilizes phospholipase C (PLC) and protein kinase C (PKC), which cause an intracellular influx of Ca^{2+}. Calmodulin, a calcium-binding protein, then associates with calcineurin, which dephosphorylates the transcription factor nuclear factor of activated T cells (NFAT), and subsequently translocates into the nucleus and drives response gene transcription [118].

Synthetic biology takes advantage of GPCRs by engineering the sensory and signaling capabilities of GPCRs and constructing gene switches for therapeutic applications. For example, receptor chimeras increase GPCR signal-transduction versatility. One approach used the sensory part of light-sensing rhodopsin, a GPCR that requires the chromophore retinal for light sensitivity. A chimeric receptor with novel transduction functionality was engineered by replacing its cytoplasmic domain by its cognate β-adrenergic receptor domain [119].

This principle has been further extended by developing so-called OptoXRs, which are based on rhodopsin's extracellular and membranous part. However, its cytoplasmic domain was shuffled to light-dependently decrease cyclic GMP amounts, increase cAMP levels, or activate PLC in mammalian cells [120]. To prevent potential crosstalk between endogenous signaling pathways in designer circuits, engineered orthogonal signaling pathways might provide useful for therapeutic applications. Such pathways were constructed by capitalizing on the ligand-dependent recruitment of β-arrestin to the cytoplasmic domain of GPCRs. A specific protease domain from the tobacco etch virus (TEV) was fused to β-arrestin, while its cognate cleavage site functioned as a linker sequence between the C-terminal domain of a GPCR and a transcription factor [121]. Upon ligand binding, β-arrestin-TEV was recruited to the engineered GPCR, cleaved, and released the transcription factor, which transcribed a transgene. Another approach aimed to develop mutated receptors only activated by specific small molecules. These receptors, called receptors activated solely by synthetic ligands (RASSLs), lost their original capability to bind their cognate endogenous ligand but bind a novel, pharmacologically inert, synthetic molecule and transduce its signal to their endogenous signaling pathway [122, 123].

GPCRs are also gene switches for therapeutic applications by sensing an endogenous trigger molecule or an externally applied signal. For example, an artificial insemination device was engineered in cellulose microencapsulated mammalian cells by expressing a luteinizing hormone receptor, which induced secretion of an engineered cellulase in a lutropin-dependent manner [124]. This sophisticated approach led to ovulation-triggered sperm release that successfully fertilized Swiss dairy cows. As mentioned above, nature evolved a series of GPCRs that can detect light and transmit such physical signals into living cells. This enables transformation into a biological response. For example, expression of the light-sensitive protein melanopsin enabled blue-light-dependent secretion of glucagon-like peptide 1 [125]. This optogenetic device also enabled blue-light-controlled glucose homeostasis in a type 2 diabetes mouse model.

3.3.3 Design Principles of Complex Gene Networks

Complex gene networks efficiently process environmental input signals into distinct cellular responses (outputs). Most complex gene circuits are modularly built by sensors, such as gene switches, and are interconnected by processing parts that end up in a final physiological response, such as oscillatory gene-transcription induction. Basic research in cell and systems biology aims to dissect and comprehend such endogenous gene networks that nature evolved over millions of years. These advances enabled biologists to tackle engineering and designing novel, tailored, synthetic gene networks with desired functions from the ground-up. The following sections focus on a series of engineering strategies for complex gene networks in mammalian designer cells performing customized functions.

3.3.3.1 Functional Mammalian Gene Circuits

Toggle Switches

A synthetic, bistable, toggle switch describes 2 states that alternate by adding 2 different input signals. One of the signals induces the first state (gene expression ON), while another input toggles the second state (gene expression OFF). Neither input signals is required to maintain the state, but only to interchange into the reversed state (see Figure 3.3.3a).

The first synthetic toggle switch described in mammalian cells consisted of 2 constructs whose expressions were controlled by each other. In the first construct, a macrolide-inducible promoter transcribed a bicistronic construct, including the streptogramin-dependent transsilencer (PIP-KRAB) and an IRES-driven reporter output (SEAP). In the second construct a streptogramin-inducible promoter expressed a macrolide-dependent trans-silencer (E-KRAB). Upon adding erythromycin (EM, input 1), E-KRAB detached from DNA, thereby inducing transcription of PIP-KRAB and the reporter protein SEAP. This represents state 1. State 2 is induced by adding pristinamycin I (PI, input 2), which represses PIP-KRAB and SEAP transcription. This synthetic circuit demonstrated a bistable toggle switch alternated by two different antibiotics [126]. Inserting intronically encoded shRNAs that targeted the opposite construct enhanced toggle performance significantly [90].

Müller et al. illustrated another toggle switch design in mammalian cells whose cellular states could be alternated by illuminating light with different wavelengths (see Figure 3.3.3b). The system is based on the light-controllable and chromophore-dependent interaction of phytochrome B (PhyB) and the phytochrome-interacting factor 6 (PIF6) of *Arabidopsis thaliana*. Upon illuminating light with a wavelength of 660 nm (input 1) the split-transcription factors TetR-PIF6 and PhyB-VP16 heterodimerize and drive transcription of an output protein (state 1). Conversely, when exposing cells to far-red light (740 nm, input 2) heterodimerization is efficiently diminished, leading to reporter-output downregulation (state 2) [57].

Synthetic Memory Devices

Synthetic, cellular memory devices are engineered to maintain a distinct cellular state after a transient trigger but unlike toggle switches they cannot be reset to the prior state (see Figure 3.3.3c) [127]. This principle tracks human cell fate in response to ultraviolet light or hypoxia [128]. The network topology includes a synthetic transcription factor coupled to a trigger-dependent positive-feedback loop and the expression of a reporter protein. Strongly induced subpopulations emerge, exceeding a specific reporter-threshold level over multiple generations. They exhibit changes in multiple cellular processes, including gene expression, growth rate, and viability [128].

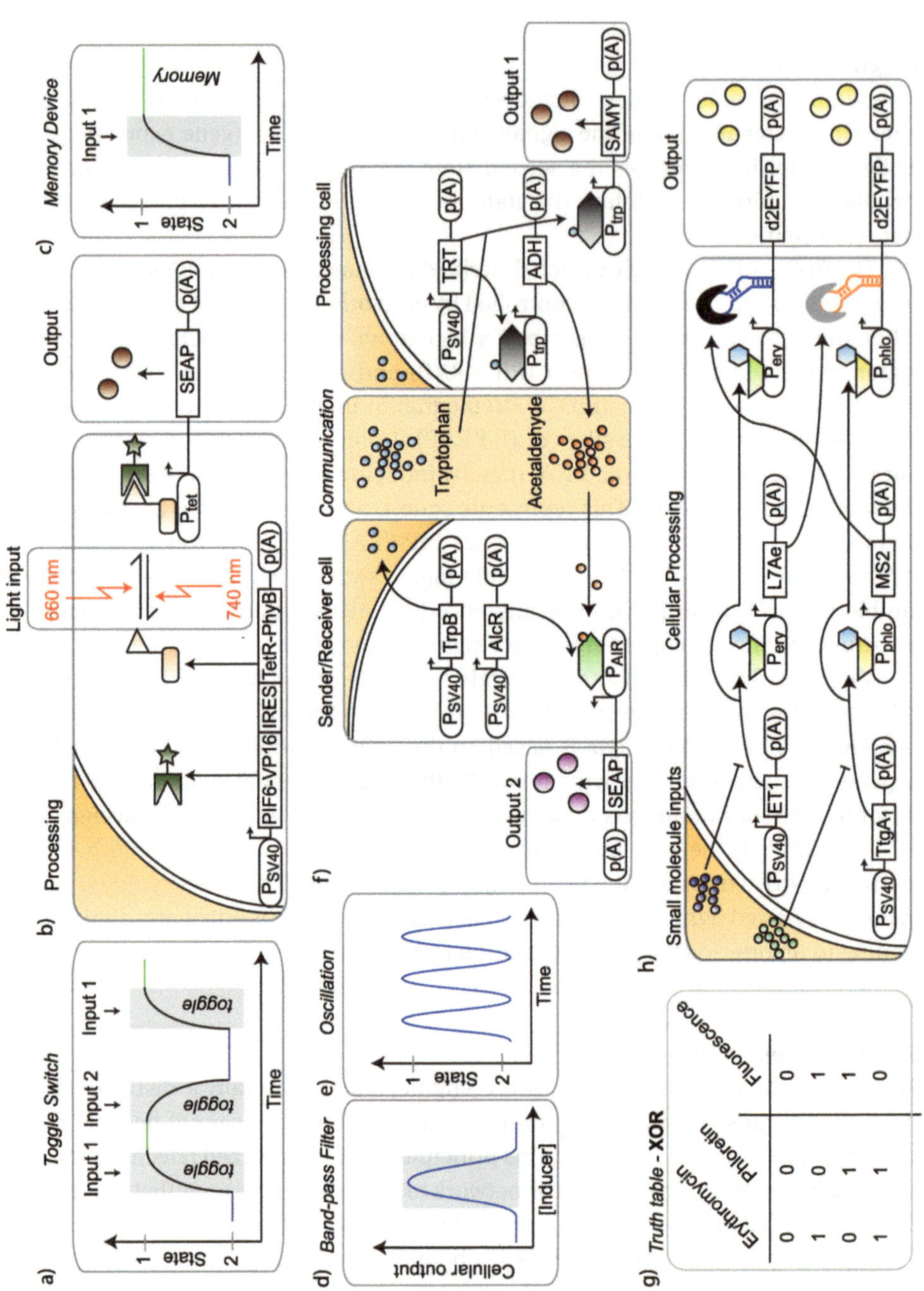

Figure 3.3.3: Schematic illustration of complex, synthetic gene network behavior and design. **a)** A bistable, synthetic toggle switch interchanges between two discrete cellular states by adding corresponding triggers. **b)** A light-dependent toggle switch based on the red-/far-red-sensitive heterodimerization domains PIF6 and PhyB. PhyB fused to TetR heterodimerized with PIF6-VP16 upon

Genetic Band-pass Filters

Genetically encoded band-pass filters enable a physiological response that depends on a precise concentration window of an inducer molecule. In contrast to a gene switch, in which the inducer molecule alters gene expression depending on its concentration, the genetic band-pass system filters a distinct inducer concentration in which the physiological response takes place (see Figure 3.3.3d). A band-pass filter controlling transgene expression in mammalian cells was designed by adding a genetic inverter into the gene network that represses the expression of a transgene by higher inducer-molecule concentrations [129]. In brief, tTA regulates the expression of 3 constructs containing tetracycline-dependent promoters driving the expression of PIP-KRAB, E-KRAB, and by a positive-feedback loop, tTA itself. An erythromycin-dependent promoter controls the expression of the reporter protein and is repressed by E-KRAB. Therefore, low tetracycline concentrations lead to a strong expression of E-KRAB, inhibiting reporter expression. However, adding tetracycline lowers E-KRAB levels, which induces reporter expression. The genetic inverter is the main component of the system, which is realized by another construct encoding the pristinamycin PIP-KRAB repression and enabling E-KRAB transcription, which then inhibits reporter expression [129].

Oscillatory Control of Gene Expression

Oscillatory control of diverse physiological functions plays a large role in biological systems. For example, the circadian rhythm in many organisms maintains robust, 24-hour rhythms of gene expression and other biological functions [130]. Oscillatory control of mammalian cell transgene expression is characterized by the autonomous and repeated interchange between two expression states of a gene of interest, such as

illumination of 660 nm wavelength light, inducing transgene expression. However, heterodimerization and transgene expression could be abolished by far-red light. **c)** Synthetic cellular memory devices maintain a distinct cellular state upon a brief trigger. **d)** Biological genetic band-pass filters enable a physiological response that depends on a precise concentration window of an inducer molecule. **e)** A genetically encoded circuit that oscillates a transgene by switching gene expression ON and OFF. **f)** An intercellular communication system based on sender and receiver cells. The sender cells convert the molecule indole into tryptophan that the receiver cells detect and then convert ethanol into acetaldehyde and express the reporter protein SAMY. Acetaldehyde then diffuses back to the sender cell, which responds by expressing SEAP. **g)** An illustration of a truth table of an XOR gate with erythromycin and phloretin as inputs and a fluorescent output. **h)** A synthetic circuit that programs for XOR behavior in mammalian cells. Two transcriptional controllers, whose activation can be controlled by erythromycin and phloretin, combine to regulate fluorescent-output and RNA-binding protein expression. The reporter transcript harbors an RNA motif in its 5'-UTR that regulates translation by its cognate RNA-binding protein. The absence or presence of both input molecules leads to a cellular OFF state, while the presence of only one input efficiently expresses the fluorescent-output protein (ON state).

a destabilized, fluorescent reporter protein (see Figure 3.3.3e) (ON and OFF). A time-delayed, negative-feedback loop is a major component of these synthetic oscillatory gene circuits. For example, a gene network that consists of tTA-dependent expression of green fluorescent protein (GFP) and tTA-autoregulated positive-feedback loop activates its own promoter and drives the synthetic circuit into the ON state. However, a time-delayed, tTA-dependent expression of the pristinamycin I-dependent transactivator (PIT) initiates a negative-feedback loop, comprising an antisense RNA targeting the tTA transcript. This represses tTA and GFP expression (cellular OFF state). Another cycle is initiated when tTA-controlled PIT levels decrease and tTA expression dominates again [131].

A similar approach comprises a tTA-autoregulated positive-feedback loop and a tTA-dependent expression of the erythromycin-dependent transactivator (ET1), whose exons are separated by an intronically encoded siRNA targeting the tTA-encoding transcript. Additionally, ET1 controls the transcription of a destabilized variant of the yellow fluorescent protein. However, the siRNA expression unit is a negative-feedback loop that downregulates tTA and expression of a fluorescent protein [132].

Fusing TetR to a fluorescent reporter protein resulted in a gene circuit that drives oscillatory fluorescent readouts, which is encoded only on one plasmid. This fusion construct represses its own transcription, which is controlled by a time-delayed, negative-feedback loop introduced by long intronic regions upstream of the fusion construct. This also allows for oscillation frequency adjustments [133].

Intercellular Communication Systems

Multicellular biological systems are based on sophisticated, intercellular communication systems to ensure cell functionality in different tissues and organs and guarantee tissue-wide distribution of information and organism survival. For example, the hormone epinephrine gets secreted into the bloodstream from specific tissues in the central nervous system in response to stress, modulating heart rate and smooth muscle cells. Synthetic, intercellular communication systems allow specialized cells to notify other cells about a distinct cellular state, which can respond to the specific information. Such interactions can include direct interaction by cell surface membrane proteins with receptor proteins on the surface of a neighboring cell (called juxtacrine cell signaling). It can also include paracrine interaction, where molecules synthesized by a cell are secreted and diffused to distant cells that can detect the molecules.

For example, notch signaling mediates juxtacrine cell signaling in mammalian cells. Sender cells express a ligand protein (such as the delta protein) on the cell surface that binds the notch receptor of the receiving cell. This transactivates and induces a transcriptional cascade. Adjustable expression of the key signaling components notch/delta followed by sophisticated time-lapse microscopy revealed the precise dynamics and spatial distributions [134].

In another approach, a multicellular consortium consisting of sender/receiver and processor cells resulted in a two-way communication system based on paracrine, synthetic signaling pathways [38]. Tryptophan synthase-expressing sender/receiver cells convert the molecule indole into tryptophan, which is then secreted into the medium (see Figure 3.3.3f). Tryptophan is a signaling molecule that diffuses to a processing cell unit, which senses elevated tryptophan concentrations. This results in the expression of a reporter molecule and alcohol dehydrogenase (ADH). ADH converts ethanol into volatile acetaldehyde that can diffuse back to the sender/receiver cell which responds by expressing another reporter protein [38].

In principle, such multicellular consortia enable the construction of sophisticated, higher-order networks, eliminating limitations of single-cell engineering by dividing individual workloads into specialized cells that are logically interconnected into an artificial microtissue [135].

3.3.3.2 Cellular Mammalian Biocomputing Devices

Despite their extraordinary complexity, decision-making biological processes share basic features with electronic devices, which switch a universal signal (electricity) between two discrete states (ON and OFF) [6]. By using electronic circuits as blueprints, biologists seek to implement synthetic gene networks into mammalian cells. These networks perform computational operations, resulting in Boolean logic or fundamental arithmetic operations. Living cells can be programmed to convert information about two input molecules into one output by performing logic calculations.

Logic Gates

Boolean logic gates including NAND, NOT IF, and NOR signal integrations were engineered by a series of synthetic transcription controllers regulated by small molecule inputs [136]. Artificial hybrid promoters encoding binding sites for up to 3 different transcription controllers enable transgene expression logics depending on various small molecules. In a circuit that programs mammalian cell NOR logic behavior, E-VP16 allows PIP-VP16 expression in an erythromycin-dependent manner. PIP-VP16 expresses a reporter gene that is only expressed in the absence of both inducer molecules. In contrast, a hybrid promoter expressing a reporter gene and consisting of binding sites for 2-(1'-hydroxy-6-methylheptyl)-3-(hydroxymethyl)-butanolide (ScbR-VP16)- and pristinamycin- (PIP-KRAB-)dependent transcriptional controllers results in a NOT IF behavior of the synthetic circuit in mammalian cells [136].

Similarly, certain combinations of synthetic transcription controllers that transcribe miRNAs control the quantity of mRNA molecules encoding a fluorescent output protein enabling a series of logic gates [137]. Another sophisticated circuit was based on transcriptional and posttranscriptional modules and programmed to logically detect intracellular cancer cell states by taking advantage of endogenous miRNA profiles [138]. The program successfully discriminates between noncancer and cancer cells and activates a killer switch upon detection of tumor cells. Dimerizable transcription factors also show a dual-promoter integrator behaving as an AND gate if signaling pathways are activated, such as in malignant cells, expressing a proapoptotic protein [139]. Furthermore, the combination of two CID systems led to an AND logic gate with superior kinetic parameters [140].

Biocomputing Devices

Half-adders and half-subtractors are the basic circuits that perform any arithmetic calculation in digital electronics. In a combinatorial plug-and-play manner, synthetic transcriptional and posttranscriptional controllers can be rewired to program mammalian cells to run XOR, half-adder, or half-subtractor operations [17]. For example, combining the phloretin- and erythromycin-dependent transcription controllers that regulate the expression of a reporter-gene transcript (which harbors an RNA motif in its 5'-UTR) with an RNA controller-inhibiting translation of the complementary reporter transcript led to the first XOR gate in mammalian cells (see Figures 3.3.3g and 3.3.3h). The circuit activation completely depends on the presence of one of the 2 input molecules. It is turned off if no input or both inputs are present (see Figures 3.3.3g and 3.3.3h). Implementing another N-IMPLY or AND gate into the XOR background enables a logic rewiring of an additional output signal, resulting in half-subtractor or half-adder operations, respectively. This study illustrates the highly complex, multicomponent, synthetic circuits in mammalian designer cells that are the basis for biocomputers, which may provide novel prospects for future therapeutics.

3.3.4 Applications of Gene Switches in Cell-based Therapies

Cells possess many obvious advantages for self-regulating treatment of chronic diseases. First, they autonomously produce or secrete desired therapeutic molecules, such as biologics or bioactive small molecules, replacing potentially lifelong, high-dose drugs. Second, cells can be engineered with sensory modules to detect potential disease states much earlier than when patients notice symptoms, which often occurs in a late stage of a disease. Third, by taking advantage of the complex signal-processing machinery of cells, smart designer cells can integrate a series of metabolic molecules into a customized therapeutic response. In contrast to small-molecule drugs or biologics, therapeutics using the extraordinary complexity of mammalian cells provide novel opportunities, including intelligent and autonomous treatment strategies mimicking biocomputing organs [1, 6, 7].

There are several ways to introduce or graft engineered cells into the patient's body. For example, cells harvested from the host can be cultured and engineered *ex vivo* before they are reinjected into the host (autotransplantation). This procedure was effective in a human clinical trial, where host-derived T cells were modified *ex vivo* with a chimeric antigen receptor that targeted malignant B cells to treat chronic lymphoid leukemia [141]. Stem-cell therapy and regenerative medicine are based on autotransplanting reprogrammed cells or whole tissues. Research has also focused on microencapsulating engineered allogeneic cells. This has several advantages (see Figure 3.3.4a): The cells are protected from the host immune system, the body is shielded from the engineered designer cells, and the encapsulation-material pore size can be tuned for tailor-made responses [142, 143]. The selectively permeable polymer-gel matrix can be built up with sodium alginate-generating microbeads of a desired diameter when it comes in contact with positively charged calcium ions [144]. Cell microencapsulation is suitable for proving functionality of synthetic therapeutic gene networks in mice. This provides the opportunity for cells to tap into the host's metabolism and intervene with customized therapeutic responses (see Figure 3.3.4a) [1, 18, 142]. Cell therapy holds great promise as the next generation of therapeutic treatment strategies and could potentially be a huge market for the pharmaceutical industry [145].

3.3.4.1 Exogenously-controllable Cell Implants (Open-loop Systems)

Cells engineered with sensor modules that detect exogenous ligands can secrete a protein of interest by exogenously administering the ligand representing so-called open-loop systems (see Figure 3.3.4b). For example, human T cells were engineered with an RNA-based synthetic gene circuit that included aptazymes, which responded to theophylline [82]. Administering theophylline into the bloodstream inactivated self-cleavage of theophylline-responsive aptazymes that were integrated into the mRNA of a potent proliferation-enhancer protein, interleukin-2 (IL-2). As a result, IL-2 expression induced the proliferation of the engineered T cell population. However, upon theophylline withdrawal, the activated aptazymes reduced IL-2 protein levels, leading to apoptosis and a reduced T cell population [82]. Another remote-controlled system based on the temperature-sensitive, transient-receptor, potential cation-channel subfamily V member 1 (TrpV1) and iron-oxide nanoparticles allows radio wave-induced secretion of engineered insulin of tumor xenografts, lowering blood glucose levels in mice [146]. By decorating an engineered TrpV1 protein with antibody-coated, iron-oxide nanoparticles, transgene-expression induction is remotely controlled by a low-frequency magnetic field heating the nanoparticle microenvironment, resulting in TrpV1 activation.

A series of microencapsulated cells harboring gene switches sensitive for various exogenous ligand molecules were implanted into mice to secrete a protein of interest (see Figure 3.3.4b). In addition to the tetracycline/doxycycline-inducible transgene-expression systems, other antibiotics have been used in mice to control protein secretion into the bloodstream. For example, an erythromycin-dependent gene switch was engineered into microencapsulated cells that were implanted intraperitoneally into

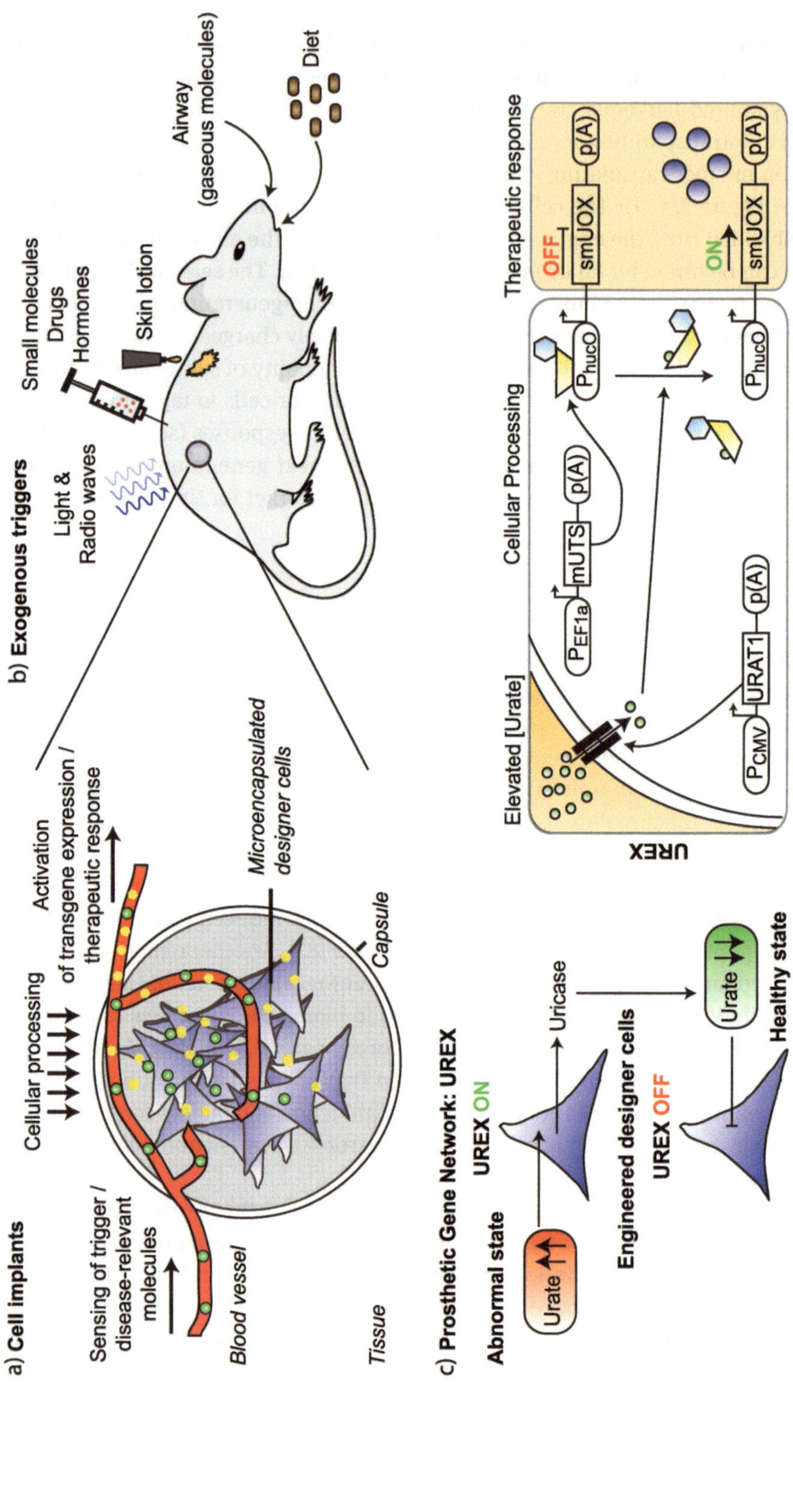
a) Cell implants
Sensing of trigger / disease-relevant molecules
Cellular processing
Activation of transgene expression / therapeutic response
Blood vessel
Microencapsulated designer cells
Capsule
Tissue
b) Exogenous triggers
Light & Radio waves
Small molecules
Drugs
Hormones
Skin lotion
Airway (gaseous molecules)
Diet
c) Prosthetic Gene Network: UREX
Abnormal state
Urate
UREX ON
Uricase
Engineered designer cells
UREX OFF
Urate
Healthy state
UREX
Elevated [Urate]
Cellular Processing
Therapeutic response
P_CMV
URAT1
p(A)
P_EF1a
mUTS
p(A)
P_hucO
OFF
smUOX
p(A)
ON
smUOX
p(A)
P_hucO

mice [28]. Upon administration of the macrolide, SEAP expression was induced and detected in mouse plasma. Other molecules, including drugs and vitamins, have been used to control transgene expression in mice [35, 147]. For example, the hypertensive drug guanabenz binds and dose-dependently activates the trace-amine-associated receptor 1 (cTAAR1), which triggers a synthetic signaling cascade that induces the expression of the peptide hormones GLP-1 and leptin. In mice developing symptoms of the metabolic syndrome this three-in-one treatment strategy was able to simultaneously attenuate hypertension (guanabenz), hyperglycemia (GLP-1) as well as obesity and dyslipidemia (GLP-1, leptin) [147].

However, other ligand-administration routes, including the airway, skin creams or food additives, work well in mice (see Figure 3.3.4b). Microencapsulated cells engineered with an acetaldehyde-inducible regulation (AIR) system and implanted into mice were sensitive to increased acetaldehyde concentration in the atmosphere, inducing SEAP expression and secretion into the mouse bloodstream [39]. Another sophisticated approach used cells modified to sense the apple metabolite phloretin, which was mixed into a skin cream. Applying the cream onto the skin of mice that harbored subcutaneous phloretin-sensitive cell implants revealed SEAP secretion into the blood [30]. The licensed food additive vanillic acid (VA) induced SEAP expression when added to the food of mice [41]. The amino acid tryptophan also induced transgene expression in mice when added to food [38].

Recently, optogenetics made impressive advances in the field of neuroscience by using light as a convenient trigger. Dosage can be precisely and noninvasively controlled. Wavelength can be altered to reach maximum skin penetration, and the spatiotemporal resolution is outstanding. Biologists took advantage of optogenetic-derived tools to reprogram light-insensitive mammalian cells to become light-sensitive [142, 148]. For example, a blue-light-dependent dimerization mechanism, called light-on, resulted in the homodimerization of a fusion protein consisting of Gal4 and the light-oxygen-voltage (LOV) domain vivid. This induced transgene expression in

Figure 3.3.4: Mammalian designer-cell implants trigger transgene expression in vivo. a) Cell implants consist of engineered cells that are microencapsulated into alginate-poly(L)-lysine-alginate beads, which protect the designer cells from the host immune system. Such cell implants can be plugged into the host metabolism, enabling engineered cells to sense the appropriate exogenous or endogenous trigger molecule. A synthetic circuit processes the information into a customized (therapeutic) output signal. b) Exogenous triggers for cell implants can be applied by electromagnetic waves, injections of small molecules or proteins, skin creams harboring inducer molecules, the airway, or specific dietary additives. c) The first prosthetic gene network, called uric-acid-responsive transgene expression (UREX). A urate-sensitive transporter protein translocates extracellular urate into the cytosol of engineered cells. A urate-dependent transcription controller (mUTS) represses the expression of a secreted variant of a urate oxidase (smUox), but detaches from its cognate operator sequence upon urate binding. This autonomously activates the prosthetic gene network. The therapeutic response lowers uric-acid levels in the blood therefore rebalancing homeostasis.

mammalian cells [55]. Transfection of type 1 diabetic mice with LightOn components modified for blue-light-dependent insulin secretion revealed a drop in blood glucose. A GPCR approach based on melanopsin-engineered microencapsulated cells was used for the blue-light-dependent control of a glucagon-like peptide 1 known to regulate blood-glucose homeostasis in diabetic mice [125]. This circuit is specially suited for therapeutic applications as it is based on fully human-derived components and uses vitamin A as cofactor.

3.3.4.2 Prosthetic Gene Networks as Therapeutic Cell Implants (Closed-loop Systems)

To exhaust the maximum potential of cell-based therapies, engineered designer cells should operate completely independently when plugged into the host organism. This enables autonomous detection of the disease state, logic intracellular processing, and self-regulated therapeutic intervention by fine-tuned secretion of functional molecules equilibrating homeostasis. Such synthetic sensor-effector gene networks are called prosthetic gene networks, as they are engineered to correct metabolic disturbances in the body. Their autonomous capability to rebalance metabolic homeostasis makes them closed-loop systems (see Figure 3.3.4a).

The first prosthetic gene network was constructed to autonomously treat gouty arthritis [18]. Abnormally high uric-acid levels in the bloodstream characterize this metabolic disorder, leading to hyperuricemia and deposition of uric-acid crystals in the kidney and other organs (see Figure 3.3.4c). The synthetic circuit consists of a transmembrane protein, called URAT1, which catalyzes the transport of extracellular urate into the cytosol of mammalian cells. A sensor protein derived from *Deinococcus radiodurans*, HucR, fused to the transsilencer domain KRAB, resulted in a mammalian urate-dependent transsilencer (mUTS). This introduced urate-sensitive, transgene expression into mammalian cells. Upon urate binding, mUTS detaches from its cognate DNA operator sequence, enabling transgene expression (see Figure 3.3.4d). Since urate sensitivity is linked to the expression and secretion of a urate oxidase (Uox) that converts urate into a renally secretable molecule, called allantoin, the synthetic circuit represents a closed-loop system. Microencapsulated cells engineered with this prosthetic gene network and implanted into mice reduced urate levels in serum and urine [18].

In summary, microencapsulated, mammalian designer cells are an efficient and autonomous treatment strategy for metabolic disorders. In future, even more complex circuits, such as biocomputers that sense a series of biomarkers and process such information into a customized therapeutic response, could pave the way for the next generation of cell-based therapeutics.

3.3.5 Conclusion and Outlook

During the past decade the number, diversity, and complexity of synthetic mammalian gene switches increased considerably. This large portfolio of functionally compatible gene switches enables their rational assembly to complex synthetic networks whose topologies, dynamic behavior, and processing capacity are starting to reach the complexity of metabolic networks. It is therefore conceivable to engineer designer cells containing sophisticated synthetic gene networks that upon implantation interface with host metabolism and sense and correct metabolic disturbances in a seamless and self-sufficient manner. The first pioneering studies reporting the use of designer cells incorporating prosthetic networks for the treatment of metabolic disorders in animal models are very promising. However, despite applying a rigorous design cycle alternating model-based predictions with molecular implementation, it remains challenging to engineer network topologies that match the dynamics and sensitivity of endogenous metabolic networks and so provide a functional physiologic interface. Also, the functional interdependence and spatial distribution of many pathologies will require multilayered networks that coordinate multiple inputs with several therapeutic outputs across different cell types and organs. Although the first intercellular communication networks and biocomputers have successfully passed the proof-of-concept phase the robustness, reliability, predictability and processing capacity of these devices will need to be significantly improved. However, synthetic networks and designer cells have all it takes to complement and replace drug-based therapies and become the treatment strategies of the future.

References

[1] Weber W, Fussenegger M. Emerging biomedical applications of synthetic biology. Nat Rev Genet. 2012;13(1):21-35.

[2] Khalil AS, Collins JJ. Synthetic biology: applications come of age. Nat Rev Genet. 2010;11(5):367-79.

[3] Andrianantoandro E, Basu S, Karig DK, et al. Synthetic biology: new engineering rules for an emerging discipline. Mol Syst Biol. 2006;2:2006-28.

[4] Slusarczyk AL, Lin A, Weiss R. Foundations for the design and implementation of synthetic genetic circuits. Nat Rev Genet. 2012;13(6):406-20.

[5] Nandagopal N, Elowitz MB. Synthetic biology: integrated gene circuits. Science. 2011;333(6047):1244-8.

[6] Auslander S, Fussenegger M. From gene switches to mammalian designer cells: present and future prospects. Trends Biotechnol. 2013;31(3):155-68.

[7] Fischbach MA, Bluestone JA, Lim WA. Cell-based therapeutics: the next pillar of medicine. Sci Transl Med. 2013;5(179):179ps7.

[8] Chen YY, Smolke CD. From DNA to targeted therapeutics: bringing synthetic biology to the clinic. Sci Transl Med. 2011;3(106):106ps42.

[9] Weber W, Fussenegger M. Molecular diversity--the toolbox for synthetic gene switches and networks. Curr Opin Chem Biol. 2011;15(3):414-20.

[10] Muller M, Auslander S, Auslander D, et al. A novel reporter system for bacterial and mammalian cells based on the non-ribosomal peptide indigoidine. Metab Eng. 2012;14(4):325-35.
[11] Wieland M, Fussenegger M. Engineering molecular circuits using synthetic biology in mammalian cells. Annu Rev Chem Biomol Eng. 2012;3:209-34.
[12] Zrazhevskiy P, Gao X. Quantum dot imaging platform for single-cell molecular profiling. Nat Commun. 2013;4:1619.
[13] Zhang J, Campbell RE, Ting AY, et al. Creating new fluorescent probes for cell biology. Nat Rev Mol Cell Biol. 2002;3(12):906-18.
[14] Henssler EM, Bertram R, Wisshak S, et al. Tet repressor mutants with altered effector binding and allostery. FEBS J. 2005;272(17):4487-96.
[15] Krueger M, Scholz O, Wisshak S, et al. Engineered Tet repressors with recognition specificity for the tetO-4C5G operator variant. Gene. 2007;404(1-2):93-100.
[16] Urlinger S, Baron U, Thellmann M, et al. Exploring the sequence space for tetracycline-dependent transcriptional activators: novel mutations yield expanded range and sensitivity. Proc Natl Acad Sci USA. 2000;97(14):7963-8.
[17] Auslander S, Auslander D, Muller M, et al. Programmable single-cell mammalian biocomputers. Nature. 2012;487(7405):123-7.
[18] Kemmer C, Gitzinger M, Daoud-El Baba M, et al. Self-sufficient control of urate homeostasis in mice by a synthetic circuit. Nat Biotechnol. 2010;28(4):355-60.
[19] Meier I, Wray LV, Hillen W. Differential regulation of the Tn10-encoded tetracycline resistance genes tetA and tetR by the tandem tet operators O1 and O2. EMBO J. 1988;7(2):567-72.
[20] Gossen M, Bujard H. Tight control of gene expression in mammalian cells by tetracycline-responsive promoters. Proc Natl Acad Sci USA. 1992;89(12):5547-51.
[21] Akagi K, Kanai M, Saya H, et al. A novel tetracycline-dependent transactivator with E2F4 transcriptional activation domain. Nucleic Acids Res. 2001;29(4):E23.
[22] Urlinger S, Helbl V, Guthmann J, et al. The p65 domain from NF-kappa B is an efficient human activator in the tetracycline-regulatable gene expression system. Gene. 2000;247(1-2):103-10.
[23] Deuschle U, Meyer WK, Thiesen HJ. Tetracycline-reversible silencing of eukaryotic promoters. Mol Cell Biol. 1995;15(4):1907-14.
[24] Hayakawa T, Yusa K, Kouno M, et al. Bloom's syndrome gene-deficient phenotype in mouse primary cells induced by a modified tetracycline-controlled trans-silencer. Gene. 2006;369:80-9.
[25] Klotzsche M, Berens C, Hillen W. A peptide triggers allostery in tet repressor by binding to a unique site. J Biol Chem. 2005;280(26):24591-9.
[26] Fussenegger M, Morris RP, Fux C, et al. Streptogramin-based gene regulation systems for mammalian cells. Nat Biotechnol. 2000;18(11):1203-8.
[27] Weber W, Fux C, Daoud-el Baba M, et al. Macrolide-based transgene control in mammalian cells and mice. Nat Biotechnol. 2002;20(9):901-7.
[28] Weber W, Marty RR, Keller B, et al. Versatile macrolide-responsive mammalian expression vectors for multiregulated multigene metabolic engineering. Biotechnol Bioeng. 2002;80(6):691-705.
[29] Weber W, Kramer BP, Fux C, et al. Novel promoter/transactivator configurations for macrolide- and streptogramin-responsive transgene expression in mammalian cells. J Gene Med. 2002;4(6):676-86.
[30] Gitzinger M, Kemmer C, El-Baba MD, et al. Controlling transgene expression in subcutaneous implants using a skin lotion containing the apple metabolite phloretin. Proc Natl Acad Sci USA. 2009;106(26):10638-43.
[31] Weber W, Schoenmakers R, Keller B, et al. A synthetic mammalian gene circuit reveals antituberculosis compounds. Proc Natl Acad Sci USA. 2008;105(29):9994-8.

[32] Malphettes L, Weber CC, El-Baba MD, et al. A novel mammalian expression system derived from components coordinating nicotine degradation in arthrobacter nicotinovorans pAO1. Nucleic Acids Res. 2005;33(12):e107.
[33] Weber W, Marty RR, Link N, et al. Conditional human VEGF-mediated vascularization in chicken embryos using a novel temperature-inducible gene regulation (TIGR) system. Nucleic Acids Res. 2003;31(12):e69.
[34] Weber W, Link N, Fussenegger M. A genetic redox sensor for mammalian cells. Metab Eng. 2006;8(3):273-80.
[35] Weber W, Lienhart C, Baba MD, et al. A biotin-triggered genetic switch in mammalian cells and mice. Metab Eng. 2009;11(2):117-24.
[36] Weber W, Bacchus W, Gruber F, et al. A novel vector platform for vitamin H-inducible transgene expression in mammalian cells. J Biotechnol. 2007;131(2):150-8.
[37] Hartenbach S, Daoud-El Baba M, Weber W, et al. An engineered L-arginine sensor of Chlamydia pneumoniae enables arginine-adjustable transcription control in mammalian cells and mice. Nucleic Acids Res. 2007;35(20):e136.
[38] Bacchus W, Lang M, El-Baba MD, et al. Synthetic two-way communication between mammalian cells. Nat Biotechnol. 2012;30(10):991-6.
[39] Weber W, Rimann M, Spielmann M, et al. Gas-inducible transgene expression in mammalian cells and mice. Nat Biotechnol. 2004;22(11):1440-4.
[40] Weber W, Rimann M, de Glutz FN, et al. Gas-inducible product gene expression in bioreactors. Metab Eng. 2005;7(3):174-81.
[41] Gitzinger M, Kemmer C, Fluri DA, et al. The food additive vanillic acid controls transgene expression in mammalian cells and mice. Nucleic Acids Res. 2012;40(5):e37.
[42] Folcher M, Xie M, Spinnler A, et al. Synthetic mammalian trigger-controlled bipartite transcription factors. Nucleic Acids Res. 2013;41(13):e134.
[43] Fields S, Song O. A novel genetic system to detect protein-protein interactions. Nature. 1989;340(6230):245-6.
[44] Braselmann S, Graninger P, Busslinger M. A selective transcriptional induction system for mammalian cells based on Gal4-estrogen receptor fusion proteins. Proc Natl Acad Sci USA. 1993;90(5):1657-61.
[45] Wang Y, Xu J, Pierson T, et al. Positive and negative regulation of gene expression in eukaryotic cells with an inducible transcriptional regulator. Gene Ther. 1997;4(5):432-41.
[46] Collingwood TN, Butler A, Tone Y, et al. Thyroid hormone-mediated enhancement of heterodimer formation between thyroid hormone receptor beta and retinoid X receptor. J Biol Chem. 1997;272(20):13060-5.
[47] Liang FS, Ho WQ, Crabtree GR. Engineering the ABA plant stress pathway for regulation of induced proximity. Sci Signal. 2011;4(164):rs2.
[48] No D, Yao TP, Evans RM. Ecdysone-inducible gene expression in mammalian cells and transgenic mice. Proc Natl Acad Sci USA. 1996;93(8):3346-51.
[49] Ho SN, Biggar SR, Spencer DM, et al. Dimeric ligands define a role for transcriptional activation domains in reinitiation. Nature. 1996;382(6594):822-6.
[50] Liberles SD, Diver ST, Austin DJ, et al. Inducible gene expression and protein translocation using nontoxic ligands identified by a mammalian three-hybrid screen. Proc Natl Acad Sci USA. 1997;94(15):7825-30.
[51] Rivera VM, Clackson T, Natesan S, et al. A humanized system for pharmacologic control of gene expression. Nat Med. 1996;2(9):1028-32.
[52] Zhao HF, Boyd J, Jolicoeur N, et al. A coumermycin/novobiocin-regulated gene expression system. Hum Gene Ther. 2003;14(17):1619-29.

[53] Muller K, Engesser R, Schulz S, et al. Multi-chromatic control of mammalian gene expression and signaling. Nucleic Acids Res. 2013;41(12):e124.
[54] Kennedy MJ, Hughes RM, Peteya LA, et al. Rapid blue-light-mediated induction of protein interactions in living cells. Nat Methods. 2010;7(12):973-5.
[55] Wang X, Chen X, Yang Y. Spatiotemporal control of gene expression by a light-switchable transgene system. Nat Methods. 2012;9(3):266-9.
[56] Yazawa M, Sadaghiani AM, Hsueh B, et al. Induction of protein-protein interactions in live cells using light. Nature Biotechnology. 2009;27(10):941-U105.
[57] Muller K, Engesser R, Metzger S, et al. A red/far-red light-responsive bi-stable toggle switch to control gene expression in mammalian cells. Nucleic Acids Res. 2013;41(7):e77.
[58] Bae KH, Do Kwon Y, Shin HC, et al. Human zinc fingers as building blocks in the construction of artificial transcription factors. Nature Biotechnology. 2003;21(3):275-80.
[59] Cong L, Zhou R, Kuo YC, et al. Comprehensive interrogation of natural TALE DNA-binding modules and transcriptional repressor domains. Nat Commun. 2012;3:968.
[60] Zhang F, Cong L, Lodato S, et al. Efficient construction of sequence-specific TAL effectors for modulating mammalian transcription. Nat Biotechnol. 2011;29(2):149-53.
[61] Beerli RR, Schopfer U, Dreier B, et al. Chemically regulated zinc finger transcription factors. J Biol Chem. 2000;275(42):32617-27.
[62] Boch J. TALEs of genome targeting. Nature Biotechnology. 2011;29(2):135-6.
[63] Miller JC, Tan S, Qiao G, et al. A TALE nuclease architecture for efficient genome editing. Nat Biotechnol. 2011;29(2):143-8.
[64] Perez EE, Wang JB, Miller JC, et al. Establishment of HIV-1 resistance in CD4(+) T cells by genome editing using zinc-finger nucleases. Nat Biotechnol. 2008;26(7):808-16.
[65] Stoltenburg R, Reinemann C, Strehlitz B. SELEX--a (r)evolutionary method to generate high-affinity nucleic acid ligands. Biomol Eng. 2007;24(4):381-403.
[66] Tuerk C, Gold L. Systematic evolution of ligands by exponential enrichment: RNA ligands to bacteriophage T4 DNA polymerase. Science. 1990;249(4968):505-10.
[67] Ellington AD, Szostak JW. In vitro selection of RNA molecules that bind specific ligands. Nature. 1990;346(6287):818-22.
[68] Green R, Ellington AD, Szostak JW. In vitro genetic analysis of the Tetrahymena self-splicing intron. Nature. 1990;347(6291):406-8.
[69] Berens C, Thain A, Schroeder R. A tetracycline-binding RNA aptamer. Bioorgan Med Chem. 2001;9(10):2549-56.
[70] Kwon M, Chun SM, Jeong S, et al. In vitro selection of RNA against kanamycin B. Mol Cells. 2001;11(3):303-11.
[71] Jenison RD, Gill SC, Pardi A, et al. High-resolution molecular discrimination by RNA. Science. 1994;263(5152):1425-9.
[72] Hunsicker A, Steber M, Mayer G, et al. An RNA aptamer that induces transcription. Chem Biol. 2009;16(2):173-80.
[73] Witherell GW, Gott JM, Uhlenbeck OC. Specific interaction between RNA phage coat proteins and RNA. Prog Nucleic Acid Re. 1991;40:185-220.
[74] Turner B, Melcher SE, Wilson TJ, et al. Induced fit of RNA on binding the L7Ae protein to the kink-turn motif. RNA. 2005;11(8):1192-200.
[75] Sassanfar M, Szostak JW. An RNA motif that binds ATP. Nature. 1993;364(6437):550-3.
[76] Famulok M. Molecular recognition of amino-acids by RNA-aptamers – an L-citrulline binding RNA motif and its evolution into an L-arginine binder. J Am Chem Soc. 1994;116(5):1698-706.
[77] Homann M, Goringer HU. Combinatorial selection of high affinity RNA ligands to live African trypanosomes. Nucleic Acids Res. 1999;27(9):2006-14.

[78] Mayer G, Ahmed MS, Dolf A, et al. Fluorescence-activated cell sorting for aptamer SELEX with cell mixtures. Nat Protoc. 2010;5(12):1993-2004.
[79] Lee SW, Zhao L, Pardi A, et al. Ultrafast dynamics show that the theophylline and 3-methylxanthine aptamers employ a conformational capture mechanism for binding their ligands. Biochemistry-Us. 2010;49(13):2943-51.
[80] Prody GA, Bakos JT, Buzayan JM, et al. Autolytic processing of dimeric plant-virus satellite RNA. Science. 1986;231(4745):1577-80.
[81] Tang J, Breaker RR. Rational design of allosteric ribozymes. Chem Biol. 1997;4(6):453-9.
[82] Chen YY, Jensen MC, Smolke CD. Genetic control of mammalian T-cell proliferation with synthetic RNA regulatory systems. Proc Natl Acad Sci USA. 2010;107(19):8531-6.
[83] Auslander S, Ketzer P, Hartig JS. A ligand-dependent hammerhead ribozyme switch for controlling mammalian gene expression. Mol Biosyst. 2010;6(5):807-14.
[84] Ketzer P, Haas SF, Engelhardt S, et al. Synthetic riboswitches for external regulation of genes transferred by replication-deficient and oncolytic adenoviruses. Nucleic Acids Res. 2012;40(21):e167.
[85] Wieland M, Auslander D, Fussenegger M. Engineering of ribozyme-based riboswitches for mammalian cells. Methods. 2012;56(3):351-7.
[86] Fire A, Xu S, Montgomery MK, et al. Potent and specific genetic interference by double-stranded RNA in Caenorhabditis elegans. Nature. 1998;391(6669):806-11.
[87] Dickins RA, Hemann MT, Zilfou JT, et al. Probing tumor phenotypes using stable and regulated synthetic miRNA precursors. Nat Genet. 2005;37(11):1289-95.
[88] Malphettes L, Fussenegger M. Macrolide- and tetracycline-adjustable siRNA-mediated gene silencing in mammalian cells using polymerase II-dependent promoter derivatives. Biotech Bioeng. 2004;88(4):417-25.
[89] Wu RH, Cheng TL, Lo SR, et al. A tightly regulated and reversibly inducible siRNA expression system for conditional RNAi-mediated gene silencing in mammalian cells. Journal of Gene Medicine. 2007;9(7):620-34.
[90] Greber D, El-Baba MD, Fussenegger M. Intronically encoded siRNAs improve dynamic range of mammalian gene regulation systems and toggle switch. Nucleic Acids Res. 2008;36(16):e101.
[91] Li X, Liu S, Wang D, et al. Adenoviral delivered eGFP-intron splicing system for multiple gene RNAi. Biotechnol Lett. 2011;33(9):1723-8.
[92] Ying SY, Chang CP, Lin SL. Intron-mediated RNA interference, intronic miRNAs, and applications. Methods Mol Biol. 2010;629:205-37.
[93] Macrae IJ, Zhou K, Li F, et al. Structural basis for double-stranded RNA processing by Dicer. Science. 2006;311(5758):195-8.
[94] An CI, Trinh VB, Yokobayashi Y. Artificial control of gene expression in mammalian cells by modulating RNA interference through aptamer-small molecule interaction. Rna-a Publication of the Rna Society. 2006;12(5):710-6.
[95] Beisel CL, Chen YY, Culler SJ, et al. Design of small molecule-responsive miRNAs based on structural requirements for Drosha processing. Nucleic Acids Res. 2011;39(7):2981-94.
[96] Kumar D, An CI, Yokobayashi Y. Conditional RNA interference mediated by allosteric ribozyme. J Am Chem Soc. 2009;131(39):13906-7.
[97] Deans TL, Cantor CR, Collins JJ. A tunable genetic switch based on RNAi and repressor proteins for regulating gene expression in mammalian cells. Cell. 2007;130(2):363-72.
[98] Mandal M, Breaker RR. Gene regulation by riboswitches. Nat Rev Mol Cell Biol. 2004;5(6):451-63.
[99] Sudarsan N, Barrick JE, Breaker RR. Metabolite-binding RNA domains are present in the genes of eukaryotes. RNA. 2003;9(6):644-7.

[100] Stripecke R, Oliveira CC, McCarthy JE, et al. Proteins binding to 5' untranslated region sites: a general mechanism for translational regulation of mRNAs in human and yeast cells. Mol Cell Biol. 1994;14(9):5898-909.

[101] Saito H, Kobayashi T, Hara T, et al. Synthetic translational regulation by an L7Ae-kink-turn RNP switch. Nat Chem Biol. 2010;6(1):71-8.

[102] Werstuck G, Green MR. Controlling gene expression in living cells through small molecule-RNA interactions. Science. 1998;282(5387):296-8.

[103] Kim DS, Gusti V, Dery KJ, et al. Ligand-induced sequestering of branchpoint sequence allows conditional control of splicing. BMC Mol Biol. 2008;9:23.

[104] Culler SJ, Hoff KG, Smolke CD. Reprogramming cellular behavior with RNA controllers responsive to endogenous proteins. Science. 2010;330(6008):1251-5.

[105] Auslander D, Wieland M, Auslander S, et al. Rational design of a small molecule-responsive intramer controlling transgene expression in mammalian cells. Nucleic Acids Res. 2011;39(22):e155.

[106] Ravid T, Hochstrasser M. Diversity of degradation signals in the ubiquitin-proteasome system. Nat Rev Mol Cell Biol. 2008;9(9):679-90.

[107] Nishimura K, Fukagawa T, Takisawa H, et al. An auxin-based degron system for the rapid depletion of proteins in nonplant cells. Nat Methods. 2009;6(12):917-22.

[108] Kampf MM, Engesser R, Busacker M, et al. Rewiring and dosing of systems modules as a design approach for synthetic mammalian signaling networks. Mol Biosyst. 2012;8(6):1824-32.

[109] Schneekloth JS, Jr., Fonseca FN, Koldobskiy M, et al. Chemical genetic control of protein levels: selective in vivo targeted degradation. J Am Chem Soc. 2004;126(12):3748-54.

[110] Banaszynski LA, Chen LC, Maynard-Smith LA,. A rapid, reversible, and tunable method to regulate protein function in living cells using synthetic small molecules. Cell. 2006;126(5):995-1004.

[111] Bonger KM, Chen LC, Liu CW, et al. Small-molecule displacement of a cryptic degron causes conditional protein degradation. Nature Chemical Biology. 2011;7(8):531-7.

[112] Miyazaki Y, Imoto H, Chen LC, et al. Destabilizing domains derived from the human estrogen receptor. J Am Chem Soc. 2012;134(9):3942-5.

[113] Iwamoto M, Bjorklund T, Lundberg C, et al. A general chemical method to regulate protein stability in the mammalian central nervous system. Chem Biol. 2010;17(9):981-8.

[114] Los GV, Wood K. The HaloTag: a novel technology for cell imaging and protein analysis. Methods Mol Biol. 2007;356:195-208.

[115] Los GV, Encell LP, McDougall MG, et al. HaloTag: a novel protein labeling technology for cell imaging and protein analysis. ACS Chem Biol. 200820;3(6):373-82.

[116] Neklesa TK, Tae HS, Schneekloth AR, et al. Small-molecule hydrophobic tagging-induced degradation of HaloTag fusion proteins. Nat Chem Biol. 2011;7(8):538-43.

[117] Lagerstrom MC, Schioth HB. Structural diversity of G protein-coupled receptors and significance for drug discovery. Nat Rev Drug Discov. 2008;7(4):339-57.

[118] Ritter SL, Hall RA. Fine-tuning of GPCR activity by receptor-interacting proteins. Nat Rev Mol Cell Biol. 2009;10(12):819-30.

[119] Kim JM, Hwa J, Garriga P, et al. Light-driven activation of beta 2-adrenergic receptor signaling by a chimeric rhodopsin containing the beta 2-adrenergic receptor cytoplasmic loops. Biochemistry. 2005;44(7):2284-92.

[120] Airan RD, Thompson KR, Fenno LE, et al. Temporally precise in vivo control of intracellular signalling. Nature. 2009;458(7241):1025-9.

[121] Barnea G, Strapps W, Herrada G, et al. The genetic design of signaling cascades to record receptor activation. Proc Natl Acad Sci USA. 2008;105(1):64-9.

[122] Armbruster BN, Li X, Pausch MH, et al. Evolving the lock to fit the key to create a family of G protein-coupled receptors potently activated by an inert ligand. Proc Natl Acad Sci USA. 2007;104(12):5163-8.
[123] Conklin BR, Hsiao EC, Claeysen S, et al. Engineering GPCR signaling pathways with RASSLs. Nat Methods. 2008;5(8):673-8.
[124] Kemmer C, Fluri DA, Witschi U, et al. A designer network coordinating bovine artificial insemination by ovulation-triggered release of implanted sperms. J Control Release. 2011;150(1):23-9.
[125] Ye H, Daoud-El Baba M, Peng RW, et al. A synthetic optogenetic transcription device enhances blood-glucose homeostasis in mice. Science. 2011;332(6037):1565-8.
[126] Kramer BP, Viretta AU, Daoud-El-Baba M, et al. An engineered epigenetic transgene switch in mammalian cells. Nat Biotechnol. 2004;22(7):867-70.
[127] Burrill DR, Silver PA. Making cellular memories. Cell. 2010;140(1):13-8.
[128] Burrill DR, Inniss MC, Boyle PM, et al. Synthetic memory circuits for tracking human cell fate. Genes Dev. 2012;26(13):1486-97.
[129] Greber D, Fussenegger M. An engineered mammalian band-pass network. Nucleic Acids Res. 2010;38(18):e174.
[130] Bell-Pedersen D, Cassone VM, Earnest DJ, et al. Circadian rhythms from multiple oscillators: lessons from diverse organisms. Nat Rev Genet. 2005;6(7):544-56.
[131] Tigges M, Marquez-Lago TT, Stelling J, et al. A tunable synthetic mammalian oscillator. Nature. 2009;457(7227):309-12.
[132] Tigges M, Denervaud N, Greber D, et al. A synthetic low-frequency mammalian oscillator. Nucleic Acids Res. 2010;38(8):2702-11.
[133] Swinburne IA, Miguez DG, Landgraf D, et al. Intron length increases oscillatory periods of gene expression in animal cells. Genes Dev. 2008;22(17):2342-6.
[134] Sprinzak D, Lakhanpal A, Lebon L, et al. Cis-interactions between notch and delta generate mutually exclusive signalling states. Nature. 2010;465(7294):86-90.
[135] Bacchus W, Fussenegger M. Engineering of synthetic intercellular communication systems. Metab Eng. 2012;16C:33-41.
[136] Kramer BP, Fischer C, Fussenegger M. BioLogic gates enable logical transcription control in mammalian cells. Biotechnol Bioeng. 2004;87(4):478-84.
[137] Leisner M, Bleris L, Lohmueller J, et al. Rationally designed logic integration of regulatory signals in mammalian cells. Nat Nanotechnol. 2010;5(9):666-70.
[138] Xie Z, Wroblewska L, Prochazka L, et al. Multi-input RNAi-based logic circuit for identification of specific cancer cells. Science. 2011;333(6047):1307-11.
[139] Nissim L, Bar-Ziv RH. A tunable dual-promoter integrator for targeting of cancer cells. Mol Syst Biol. 2010;6:444.
[140] Miyamoto T, DeRose R, Suarez A, et al. Rapid and orthogonal logic gating with a gibberellin-induced dimerization system. Nat Chem Biol. 2012;8(5):465-70.
[141] Porter DL, Levine BL, Kalos M, et al. Chimeric antigen receptor-modified T cells in chronic lymphoid leukemia. N Engl J Med. 2011;365(8):725-33.
[142] Auslander S, Wieland M, Fussenegger M. Smart medication through combination of synthetic biology and cell microencapsulation. Metab Eng. 2012;14(3):252-60.
[143] Krishnamurthy NV, Gimi B. Encapsulated cell grafts to treat cellular deficiencies and dysfunction. Crit Rev Biomed Eng. 2011;39(6):473-91.
[144] Zimmermann H, Shirley SG, Zimmermann U. Alginate-based encapsulation of cells: past, present, and future. Curr Diab Rep. 2007;7(4):314-20.
[145] Mason C, Brindley DA, Culme-Seymour EJ, et al. Cell therapy industry: billion dollar global business with unlimited potential. Regen Med. 2011;6(3):265-72.

[146] Stanley SA, Gagner JE, Damanpour S, et al. Radio-wave heating of iron oxide nanoparticles can regulate plasma glucose in mice. Science. 2012;336(6081):604-8.
[147] Ye H, Charpin-El Hamri G, Zwicky K, et al. Pharmaceutically controlled designer circuit for the treatment of the metabolic syndrome. Proc Natl Acad Sci USA. 2013;110(1):141-6.
[148] Bacchus W, Fussenegger M. The use of light for engineered control and reprogramming of cellular functions. Curr Opin Biotechnol. 2012;23(5):695-702.

3.4 Rational Approaches for Transgene Expression: Targeted Integration and Episomal Maintenance

Natascha Kruse, Shawal Spencer, and Dagmar Wirth

Summary/Overview

Genetic modification of mammalian cells is a convenient method to specifically alter cellular properties. Applications range from the investigation of gene function and regulation, the utilization of cells as cell factories, e.g., for the production of recombinant proteins, but also for designing and adapting the cells for defined purposes like drug screening.

In the last 30 years until today the strategies to genetically modify cells relied on random integration of expression cassettes into the genome. As such, they are accompanied by chromosomal position effects that dramatically modulate the transcriptional performance of incoming vectors. Accordingly, screening is required to isolate cell clones with the expression properties of interest. Meanwhile, methods emerged that allow introducing specific changes at defined positions in the genome. This increases the predictability of modifications and paves the way for a rational design of cells.

In this chapter, we give an overview on the chromosomal elements that have a dominant influence on the transcriptional activity of transgenes upon integration. (For other cellular processes that affect transgene expression such as posttranscriptional and posttranslational modifications the reader is referred to Chapters 3.3, 3.5 and 4.1.) Further, we describe various strategies for exploiting defined chromosomal sites for targeted integration of expression cassettes. This comprises methods based on site-specific recombinases such as Flp, Cre and ΦC31, supporting the targeted integration of transgenes or recombinase-mediated cassette exchange (RMCE) in a highly efficient manner. As well, homologous recombination is described. This method gained importance for the modification of somatic cells and cell lines with the recently emerged strategies to induce double strand breaks (based on meganucleases, zinc finger nucleases, TALE nucleases and the CRISPR/Cas9 system). We discuss the exploitation of bacterial artificial chromosome (BAC) vectors to avoid chromosomal position effects. Finally, we briefly describe the utilization of episomal vectors to avoid integration-associated variability of gene expression.

3.4.1 Introduction

The generation of genetically modified cells involves the transfer of recombinant DNA (e.g., plasmids) encompassing the regulatory elements (promoters and enhancers) as well as effector sequences encoding proteins or regulatory RNAs, which are usually combined with selection markers or reporters. Upon physico/chemical transfer of the plasmids into the cells (e.g., by DNA/calcium phosphate coprecipitation, electroporation, lipofection, or nucleofection) cellular repair mechanisms sense the incoming recombinant DNA. The activation of these repair mechanisms usually results in illegitimate recombination and integration of the recombinant DNA into the cellular chromosomes. A significantly more efficient method to achieve stable chromosomal integration of transgenes is provided by retro- or lentiviral gene transfer by which viral integrases ensure the efficient insertion of the reverse transcribed recombinant vector DNA [1-3]. Also, transposase mediated integration of expression cassettes has emerged as a powerful tool for efficient integration of transgenes (see [4] for review).

Independent of the method used to transfer the recombinant DNA: once integrated into the genome, the nature of the integration site turns out to be a crucial factor that affects the level and stability of expression over time and eventually also the regulation of transgene expression. In random transfer protocols, the chromosomal site of transgene integration is an unpredictable parameter and is the main cause of heterogeneity of expression in individual cell clones. This is usually summarized as the "position effect" [5-7] (see Figure 3.4.1a). Transfer of DNA using the above mentioned methods results in nondirected integration of the DNA without any specific preferences to particular chromosomal sites. Still, integration is 'semirandom' since the various transfer protocols exert a certain bias with respect to the 'nature' (e.g., the accessibility or specific properties) of the chromosomal integration sites. As such, the methods differ with respect to the probability with which transgenes integrate, e.g., into transcriptionally active sites, into promoter regions, into coding sequences, or into intergenic regions. This affects the probability with which transgene integration can cause deregulation or accidental disruption of endogenous genes. For many applications the deregulation of endogenous genes by the integration event is not a relevant issue. However, the nature of the integration site is of particular concern in gene therapy applications, and can have deleterious consequences.

The pattern and distribution of integration sites obtained with the different protocols has been investigated. Transposase mediated gene transfer protocols show equal integrations into exons, introns, and intergenic regions and less preference for promoter-near regions [8, 9], summarized by [10]. In contrast, γ-retroviral transfer is characterized by a preference for promoter regions, while lentiviral vectors more frequently integrate into coding sequences [11, 12].

3.4.2 Chromosomal Elements that Modulate Transgene Expression upon Integration into the Cellular Chromosomes

In the last decades, a number of different regulatory DNA elements have been identified. These elements contribute to the expression of gene cassettes once integrated into the host's chromosomal DNA by mechanisms that are different from classical enhancer/silencer effects. Although in some cases distinct sequence motifs have been identified, these elements are largely operationally defined. The characterization of such elements is challenging and in the focus of ongoing research activities. However the characterization is not always straightforward since their function is often manifold and the functions of differently designated elements can overlap. In many cases, a pronounced context dependency suggests that the underlying principle is of complex nature.

In the following a few of the better characterized transcriptionally active elements are described.

3.4.2.1 Scaffold-/Matrix-Attachment Regions (S/MARs)

Matrix-associated regions or matrix-attachment elements (MARs) were initially identified by their specific (re)association with the nuclear matrix (i.e., the remnants of a salt-extraction protocol), whereas scaffold-attachment elements (SARs) were mostly characterized by their (re)association with nuclear scaffolds (i.e., the remnants of a lithium 3,5-diiodosalicylate (LIS, a mild detergent) extraction procedure [13]). The observation that the elements recovered by the reassociation methods are identical or closely related has coined the consensus-term "S/MAR" and is nowadays used interchangeably for these elements.

S/MARs are AT rich sequences of about 300 to 3,000 bp and are spaced about every 30 kb in mammalian genomes [13, 14]. They associate with the nuclear scaffold proteins and form chromatin loops, thereby establishing separate chromatin domains. S/MARs are cis-acting elements. They are responsible for the arrangement of chromatin in domains especially at the time of nucleosome formation [15]. This separate domain thereby insulates the polynucleosomes from the topographical effects, i.e., effects from the neighboring regulator elements that might have some kind of influence on the integrated cassette. The classical SAR elements were reported to be present at the putative borders of numerous chromatin domains adjacent to enhancer elements [16]. Interestingly, genomic S/MARs were also found close to retroviral integration sites, suggesting that the specific DNA structures favors viral integration [17]. Mechanistically, SARs are considered to be responsible for exposing single strands of DNA, be it by unwinding, by the formation of cruciform, slippage, or triple helical structures which may all follow the buildup of negative super helical tension [17]. It has been shown that bending and/or torsional stress is necessary for the unpairing to occur [18]. Such kind of unpairing is implicated to be of major importance in creating an open complex formation in transcription [19]. Based on the sequence motifs and structural properties, several *in silico* tools have been designed to predict the S/MAR elements [20-22].

It could be shown that S/MARs can increase transgene expression upon random transfer of expression cassettes [23]. By integrating S/MAR elements into expression vectors transgene expression is more stable for longer time periods [24]. S/MARs have also been integrated in retroviral vectors, thereby giving rise to higher titers [25] and higher levels of expression in infected cells which were found to be more resistant to silencing [26] and in plants at the site of a T-DNA integration event [27].

Accordingly, S/MARs were also exploited for protein production in CHO cells. Several studies show that S/MARs can increase the levels of protein production and an overall proportion of high-producer cells. This not only reduces the number of cell clones to be screened but also decreases the time of production [28-30].

3.4.2.2 Insulators

The organization of the eukaryotic genome in distinct functional domains necessitates the presence of elements that form discrete interdomain boundaries and help in maintaining the identity of such domains. One such class of cis-acting boundary elements is called chromatin insulators [31-33]. Insulators do not have a transcriptional activity of their own and must be physically present between the promoter and enhancer/heterochromatin in order to provide the insulation from the respective transcriptional effects (see Figure 3.4.1b). While the impact of proteins for insulation is established, recent studies suggest also RNA-based mechanisms to play a role in the insulation function [34].

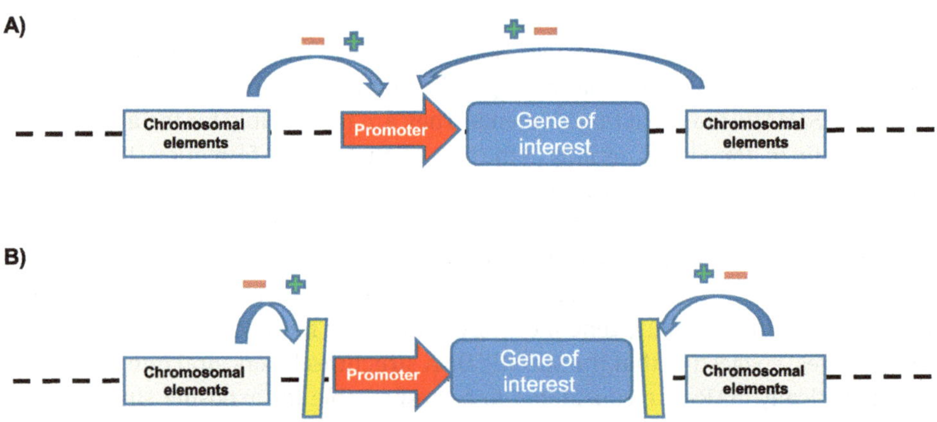

Figure 3.4.1: **a)** Unshielded expression. Upon integration into the genome, the expression cassette is exposed to chromosomal influences from various distal regulatory elements that might have negative or positive effects on transgene expression. **b)** Shielded expression. Certain chromosomal elements (e.g., insulator, UCOE, indicated as yellow boxes) can shield the transgene thereby eliminating not only the negative influences but also positive ones.

The CTCF proteins are reported to play a central role in the appropriate functioning of insulator elements. These proteins bind to insulator elements and are thought to be involved in establishing higher-order chromatin structures, which affect gene expression [32, 35]. In addition to their structural function, CTCF proteins are thought to affect local genomic DNA methylation by binding to and excluding DNA methyltransferases from the region, thus regulating gene expression [36]. CTCF binding itself appears to be sensitive to methylation and the lack of CTCF binding is associated with default methylation.

Two kinds of functional properties have been attributed to insulator elements that arise from their positional activity [37, 38] "enhancer blocking insulation" and "barrier insulation". Insulators that prevent interaction between the enhancer element and a promoter are called "enhancer blockers". For this kind of insulation it is proposed that 2 insulators interact with each other to form a chromatin loop that physically separates the enhancer and promoter in 2 separate chromatin domains. "Barrier insulators" protect promoters from silencers if integrated in between the two. The underlying model for this functional aspect is based on the observation that insulator elements usually have an open chromatin structure. They show a defined pattern of histone acetylation and methylation marks representing an active histone state. Together, this is thought to stop the spread of heterochromatin.

Some native insulators elements possess both functional attributes. This includes mouse short interspersed elements (SINE) [39], the chicken hypersensitive HS4 (cHS4) element and the *Drosophila melanogaster gypsy* SF1 insulator. To shield transgene cassettes from the neighboring chromosomal environment, insulators have been integrated into expression cassettes. In particular, they were employed in lentiviral and γ-retroviral vectors aiming at improvement of safety and performance (e.g., [40-43]. In general, the insulators seem to increase the likelihood and/or consistency of transgene expression ([44, 45] for review). However, the performance is affected by the nature, size and the orientation of the insulator element as well as the promoter and cell types used. Moreover, it has been shown that the incorporation of insulators in lentiviral vectors can also cause additional effects that contribute to overall expression properties: e.g., it has been shown that they alter the preference of viral integration sites [42] or the transcriptional termination [40].

A major drawback with using the insulators for vector design is their massive size (2.5 kb in case of cHS4 with 1.25 monomeric units). This makes construction of expression cassettes a tedious task and limits the application in retroviral vectors. Moreover, a significant reduction in the viral titers has been observed especially when used in double copies [46].

3.4.2.3 Ubiquitously Acting Chromatin-Opening Elements

Another type of chromosomal elements concerns the so-called ubiquitously acting chromatin-opening elements (UCOEs). UCOEs comprise of 2 divergent promoter elements, which flank a CpG island and are found upstream of several housekeeping genes. These elements are supposed to have an open chromatin conformation in most cell types. The 2 best investigated UCOEs are derived from the murine ribosomal protein S3 (RPS3) locus and the human HNRPA2B1-CBX3 locus. They have been investigated in various studies in a number of different configurations. They have been shown to confer stable transgene expression when integrated in expression vectors [47]. Moreover, they can prevent epigenetic silencing in lentiviral vectors [48, 49]. Recently, it has been shown that the intrinsic splicing signals can be eliminated without compromising the function of UCOEs, thereby increasing the safety of these elements [50].

3.4.2.4 Locus Control Regions (LCRs)

LCRs have been operationally defined as chromosomal DNA elements that provide copy number dependent and position independent properties. They are involved in regulation of entire loci or clusters of, e.g., developmentally regulated genes. LCRs are constituted by different moieties that include typical cis-acting elements like insulators, silencers, enhancer, S/MARs elements etc. where each of these cis elements has a differential affect on gene expression. It is their collective activity that justifies the functional definition as LCR to a particular locus. The components LCRs commonly colocalize to sites of DNAse I hypersensitivity (HS) in the chromatin of target gene expressing cells. The most prominent activity of LCRs is their strong transcription enhancing activity, as was studied in detail in case of the human developmentally regulated β-globin locus [51]. The functional role of LCRs became evident with studies in transgenic mice where it was determined that the LCR was responsible for developmental expression of the ε-globin genes [52, 53]. LCRs are mostly located upstream of their target gene. Less frequently, they are found in introns, as is the case for the human adenosine deaminase (ADA) LCR, or downstream of the gene they regulate, like the CD2 LCR [54] or TH2 LCRs [55, 56].

The most widely accepted model of LCR action is the looping model [53, 57]. According to this model the 5' hypersensitive site of the LCR forms a folded structure where the hypersensitive sites are in the core and act as docking site for the transcription factors. This folded structure of the LCR physically forms a loop that brings the enhancer with bound transcription factors in close proximity to the gene promoter allowing an interaction between the distant enhancer with the promoter. This interaction between promoter and enhancer allows the formation of a functional transcriptional apparatus by sharing LCR-bound coactivators and transcription factors with the basal transcription apparatus present at the promoter [51].

LCRs have been employed for gene therapy of sickle cell disease and β-thalassemias by inserting them into retroviral vectors [58-60]. Of note, different reports are found concerning the ability of LCRs to confer position independent expression: while in some studies position independent expression was confirmed [61, 62] other studies do not observe this property [58]. In this respect, a report is of interest that shows that the particular design of the expression cassette can contribute to this property [63].

Several studies concern the question if LCRs can prevent deregulation of cellular genes flanking the integration site of transgenes. While some assays showed no influence on deregulation (mutagenic potential) [64], reduced risk for insertional deregulation was observed in other settings [65] suggesting that this function is dependent on the setting/constellation of LCR, transgene, and endogenous genes.

3.4.2.5 Potential of Chromosomal Elements for Biotechnology

The application of chromosomal elements for design of expression vectors is of particular interest for the development of therapeutic vectors since these elements could shield the transgenic expression cassettes and thereby avoid interference with the host cell's genes through cis-effects [66]. If successful, this could improve the safety of gene therapy strategies. However, despite some promising reports, the predictability of all such genetic elements in routine vector application remains moderate. This is because our knowledge of the underlying mechanisms is still rather incomplete. In particular, the question how an element will work in a different genomic environment, in different organisms and at different stages of development still remains largely unanswered. This currently impairs the predictability in employing these elements as 'building blocks' in gene expression vectors.

3.4.3 Exploitation of Chromosomal Hotspots by Targeted Integration

To deal with the heterogeneity of gene expression upon random integration of cassettes into the host's chromosomes, standard protocols employ screening procedures to identify those integration events that meet the requirements for transgene expression (e.g., stable or regulated expression). This imposes a limitation, in particular when resources for screening are restricted and/or if the monitoring of transgene expression is not straightforward. Thus, efforts have been undertaken to develop novel strategies for predictable, reliable, and, most importantly, effective manipulation of mammalian cells. The recently emerged methods are designed to target expression cassettes to preselected chromosomal loci, in particular chromosomal sites, which support high level of expression (so-called 'hotspots'). Following such strategies, the unpredictability associated with the random integration of expression cassettes is overcome or at least reduced since the modulation mediated by the particular chromosomal site is known. In particular, when repeatedly the same integration site is used the requirement for large-scale screens for every new transgene that needs to be expressed is obviated (Figure 3.4.2).

In the following, we describe the use of strategies based on site-specific recombinases as well as strategies that rely on the cellular ability to integrate DNA by homologous recombination. These strategies are summarized in Figure 3.4.3.

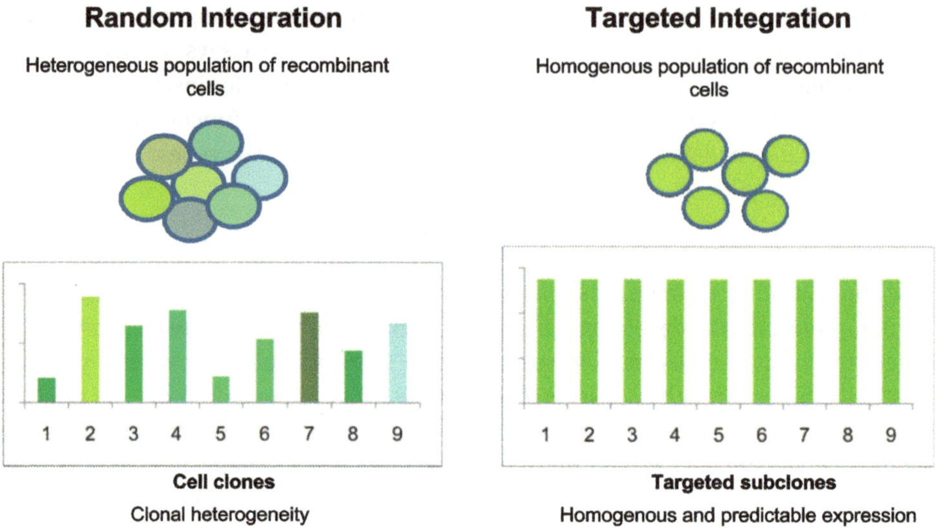

Figure 3.4.2: Random versus targeted integration. Random integration of transgene cassettes leads to high levels of clonal heterogeneity due to the nature of the respective chromosomal integration site. If a particular site that supports expression is targeted (a chromosomal hotspot), predictable expression pattern are obtained in the subclones thereby eliminating large-scale screening efforts.

3.4.3.1 Site-specific Recombination for Genetic Modification of Chromosomal Sites in Mammalian Cells

In some species, site-specific recombinases have been identified that exclusively bind to and recombine short DNA motifs, so-called recombination targets (RTs). These recombinases play a pivotal role in the life cycles of many microorganisms including yeast, bacteria, and bacteriophages [67]. The recombinases can catalyze intra- and intermolecular reactions, namely DNA integration, excision, or inversion (see below). The observation that site-specific recombinase systems can be transferred to heterologous organisms paved the way to exploit these systems for highly efficient and precise targeted manipulation and reengineering of many different genomes including mammalian cells.

Recombinases and Mode of Action

Most of the recombinases fall in 2 main families: serine recombinases and tyrosine recombinases. The names indicate the amino acid present at the active site of the enzyme that catalyzes the recombination reaction. The recombinases that are routinely used for genetic engineering purposes are described in the following.

The 2 major family members of tyrosine recombinases are Cre and Flp. Cre ("causes recombination") recombinase was derived from the P1 bacteriophage. The

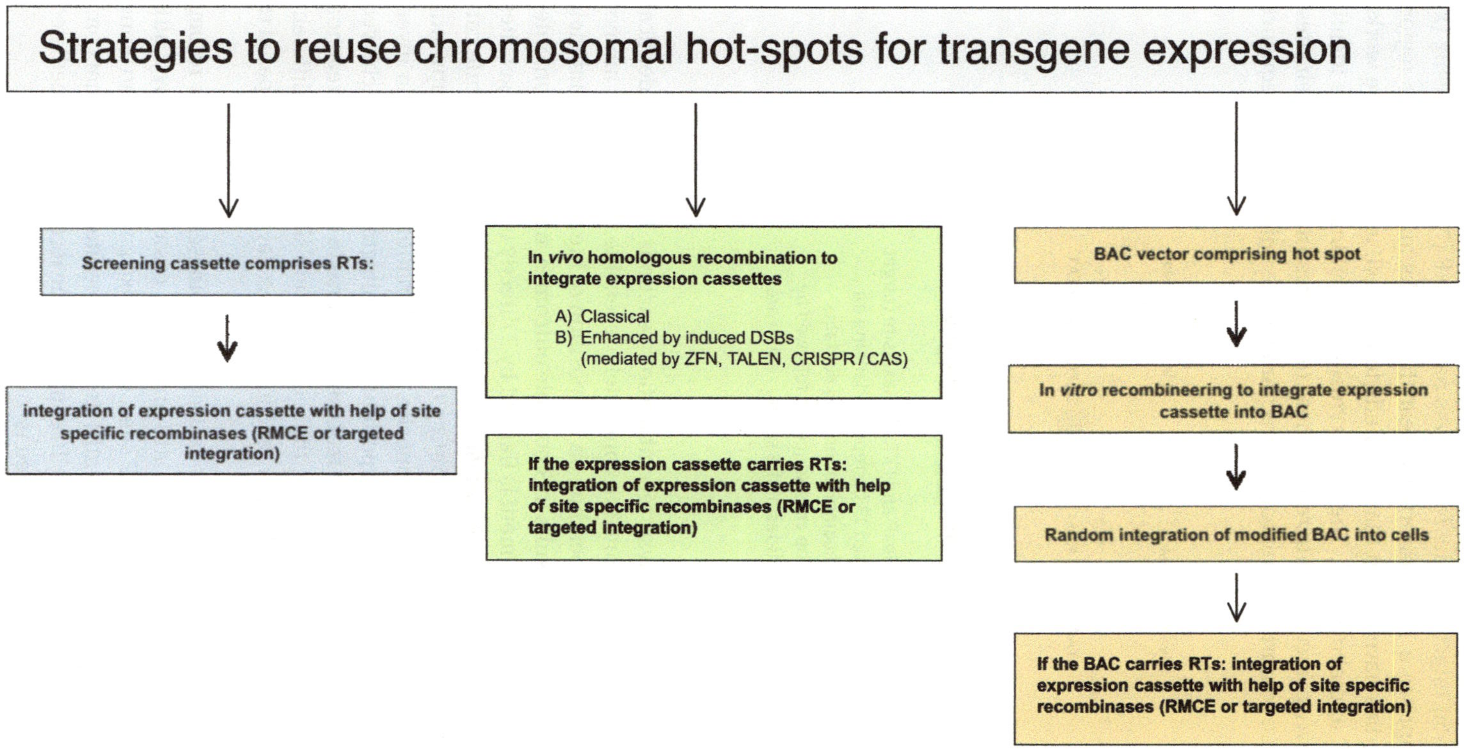

Figure 3.4.3: Strategies to reuse chromosomal hotspots for transgene expression. **left**) Chromosomal hot-spots identified by random integration of RT site(s) together with an expression cassette can be reused by integrating a cassette of interest that encodes the same RT sites either by targeted integration or by RMCE. **middle**) Direct integration of expression cassettes into chromosomal hotspots via homologous recombination. Efficiency of homologous recombination is increased by double-strand breaks. **right**) A BAC vector that contains a hotspot engineered with the expression cassette can be integrated randomly into the genome of the host cell since transgene expression will be controlled by the BAC encoded sequences.

34 bp recombination target (RT) site for Cre was named loxP ("locus of crossing (x) over, P1"). It comprises a 13 bp inverted repeat that flanks an 8 bp spacer region (Figure 3.4.4). Studies have shown that Cre binds to the LoxP as a dimer. Cre works optimally at 37 °C and above [68]. The other tyrosine recombinase, Flp, was identified in *Saccharomyces Cerevisiae*. Flp recombinase recognizes and binds to their sites, called the FRT sites, comprising 3 repeat regions flanking the unique spacer sequence (Figure 3.4.4).

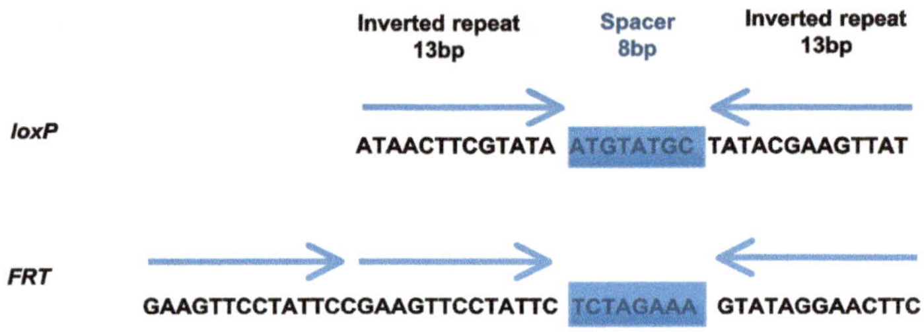

Figure 3.4.4: Recombination target sites loxP and FRT. The loxP site (target of Cre recombinase) comprises two inverted 13 bp repeats flanking an 8 bp spacer sequence. The FRT site (Flp recombination site) contains two inverted 13 bp repeats flanking an 8 bp core sequence followed by an additional 13 bp direct repeat. Nucleotides of the spacer region are marked in blue.

Tyrosine recombinases can catalyse recombination reactions between 2 identical recombination targets. The reactions are completely reversible, i.e., the RTs are not altered during recombination. Depending on the position of the target sites and their orientation, the recombination results in 3 possible outcomes, namely excision, integration and inversion. If a DNA element is flanked by 2 directly orientated target sites it will be excised by the recombinase, providing a small circular DNA molecule carrying one RT and the parental DNA maintaining the other RT (Figure 3.4.5a, upwards reaction). If 2 RT sites are orientated in opposite, recombination will result in inversion of the fragment between both sites. Importantly, also integration can be achieved by tyrosine recombinases. The integration reaction is reverse to the excision reaction (Figure 3.4.5a, downwards reaction). Here, the recombination sites are on 2 different DNA molecules. Since this is a bimolecular reaction, it is less favourable compared to excision and inversion.

ΦC31 integrase belongs to the group of serine recombinases. It mediates recombination between the heterotypic RT sites attB and attP. A cassette flanked by 2 attB sequences will recombine with a set of attP sites, thereby exchanging the intervening DNA sequence. As a result, hybrid attL and attR sites are generated which are not compatible for any further recombination event. In contrast to the recombination mediated by tyrosine recombinases, the ΦC31-mediated recombination is not reversible.

Targeted Integration and Recombinase-mediated Cassette Exchange

Site-specific recombinases were exploited for highly efficient modification of chromosomal loci in mammalian cells. In a first step, a single RT or a set of specific RT(s) are integrated into the genome of a cell line (so-called tagging). This can be achieved by transfection or infection of an expression vector that carries the RT(s). Upon random integration of this vector and screening, cell clones are isolated that provide the appropriate expression pattern (e.g., high, stable, or regulated transgene expression). Alternatively, the RT site(s) can be specifically integrated into defined chromosomal loci by homologous recombination (see Section 3.4.3.2). In a second step, such cell clones are then targeted with a gene of interest by simple cotransfection of a vector harboring the RT(s) and a vector encoding the recombinase. As a result of the site-specific recombination, integration of the cassette is achieved ('targeting'). Two strategies have been followed, "targeted integration" and "recombinase-mediated cassette exchange" (RMCE) (Figure 3.4.5, for review see [69, 70]).

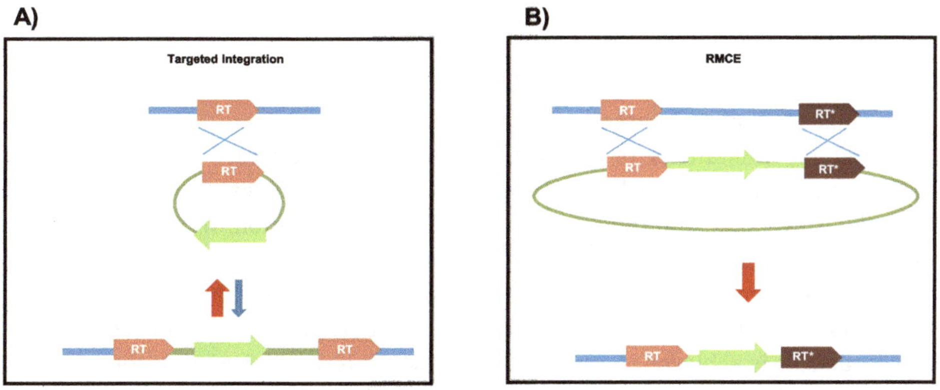

Figure 3.4.5: **a)** Targeted integration. The integration of one RT site into the genomic DNA (indicated by the blue line) allows the integration of a plasmid (green) carrying the identical RT site upon recombination. However, the targeted state can be readily reverted by the highly favored excision reaction. **b)** RMCE. Two heterospecific, noninteracting RT sites integrated in the genome will recombine with identical RT sites delivered by an external plasmid resulting in recombinase-mediated cassette exchange (RMCE). The reverse reaction is kinetically and thermodynamically unfavored and virtually excluded.

Targeted Integration

Targeted integration is mediated by enzymes such as Cre and Flp. In a first tagging step, a plasmid carrying a single RT (loxP, FRT) is integrated into the desired genomic location (by random integration or homologous recombination). For targeting, an expression vector containing a homologous RT is transferred to the tagged cell line in presence of the respective recombinase. Intermolecular recombination between the

2 identical RTs will result in integration of the whole plasmid into the tagged chromosomal site (Figure 3.4.5a, downwards).

To facilitate isolation of recombined cells, usually, selection strategies are employed (see below). One limitation of targeted integration is the reversibility of the integration reaction since the targeted state represents a substrate for the subsequent excision, the thermodynamically and kinetically favoured reaction. This renders the modification instable. Thus, transient expression of the recombinase has to be assured. Moreover, strict selection for the integrated product can help to eliminate cells that have undergone the back reaction.

Another limitation of this simple type of targeted integration is the fact that not only the gene of interest is integrated but also the bacterial backbone. Since the prokaryotic sequences are frequently associated to DNA methylation, expression can be affected by silencing.

Meanwhile, targeted integration has been used for the development of various cell lines for constitutive or regulated protein expression [71, 72]. Some of them are commerically available (e.g., from Life Technologies).

Recombinase-mediated Cassette Exchange (RMCE)

To avoid the limitations of targeted integration, recombinase-mediated cassette exchange (RMCE) was developed. In contrast to targeted integration, this method is virtually irreversible and thus is not assocciated with excision. In this method, targeting is a "cut and paste" mechanism that involves recombinase-mediated excision of a DNA segment flanked by a set of 2 heterologous RTs and integration of a cassette flanked with the very same set of RTs. Thereby, an exchange of the intervening sequences is achieved. A prerequisite for RMCE is a set of heterologous RTs that cannot recombine with each other (Figure 3.4.5b). The RTs of ΦC31 derived from *Streptomyces phage* ΦC31 naturally provide the requirements for RMCE.

Also Cre and Flp have been adapted for this purpose, mainly based on the finding that these enzymes can also recombine mutant RTs. Mutant FRT and also loxP sites are now available that show no significant cross-reactivity and thus represent ideal RTs for RMCE [70,73]. Flp-based RMCE protocols turned out to be highly efficient. If combined to a stringent selection procedure 100 % targeting efficiency can be achieved [74-76].

The availability of such mutants also allows to even expand the concept of cassette exchange to more than one integration site in the genome, thereby 'multiplexing' targeted integration. By flanking the various integration sites with different independent sets of RTs, specific targeting can be achieved if expression cassettes are provided that carry the corresponding sets of RTs. A prerequisite for multiplexing is that the heterospecific sites cannot interact among themselves; otherwise this would lead to excision reactions or chromosomal rearrangements. Various combinations of mutants make such an approach possible (reviewed in [77]).

Predictability of Transgene Expression upon Targeting Preselected Chromosomal Sites

Due to its high efficiency and fast and straightforward protocols, recombinase-based genome modification has gained momentum over the last years and is now the method of choice for many applications. It has been used in various different fields in basic science and also biotechnology including the generation of recombinant protein-expressing cell lines and also in the generation of a wide variety of transgenic mice. Basically, the concept allows to screen integration sites using reporters that facilitate easy detection (e.g., GFP) and to establish these clones as master cell lines for targeting cassettes of choice. Due to the fact that targeting is fast, efficient, and precise and the genetic manipulation is defined, the expression properties of the final producer clones are predictable.

Meanwhile, ample evidence is given that targeted integration of expression cassettes allows to reproducibly generate cell lines that express a gene or vector of interest. This is in particular true if the targeting of chromosomal integration sites is, e.g., restricted to coding sequences only. However, the prediction of gene expression is not straightforward anymore once major changes are done in the design of the targeted cassettes and in particular once also regulatory elements are exchanged or added. The following examples should illustrate this (see Figure 3.4.6 for a schematic illustration).

Gama-Norton et al. investigated the impact of the orientation of the targeting cassette including a promoter in 2 distinct loci in 293 cells that were identified because of their capability to support high expression of retroviral vectors. Expression cassettes based on 3 different promoter designs were targeted in sense and antisense orientation by RMCE. It was observed that only one of the loci could support expression of either orientation. In contrast, in the other locus the high expression levels were only achieved when the targeting cassettes were integrated in the same orientation as the initial tagging vector while expression droped up to hundredfold for the other orientation [78]. Similarly, an orientation-dependent expression of the synthetic Tet promoter was observed in the ubiquitously expressed mouse Rosa26 locus [79]. This suggests that the influence of the chromosomal surroundings can be affected by the position of the elements within an expression cassette.

Another study compared 2 loci, identified to support high expression with an SV40 and MSCV promoter, respectively. These sites were targeted using 3 different promoters, including the promoters used for screening. Interestingly, high expression could only be restored if the cells were targeted with the promoter used for screening. In contrast, the other promoters failed to support high expression in these sites [80].

In another study by Nehlsen and coworkers various antibody expression cassettes controlled by potent promoters were targeted into preselected high expression loci of HEK293 and CHO cells. Interestingly, for a given integration site, good antibody expression levels were only achieved with particular promoters, while other potent promoters failed. However, in another chromosomal context, the relative performance

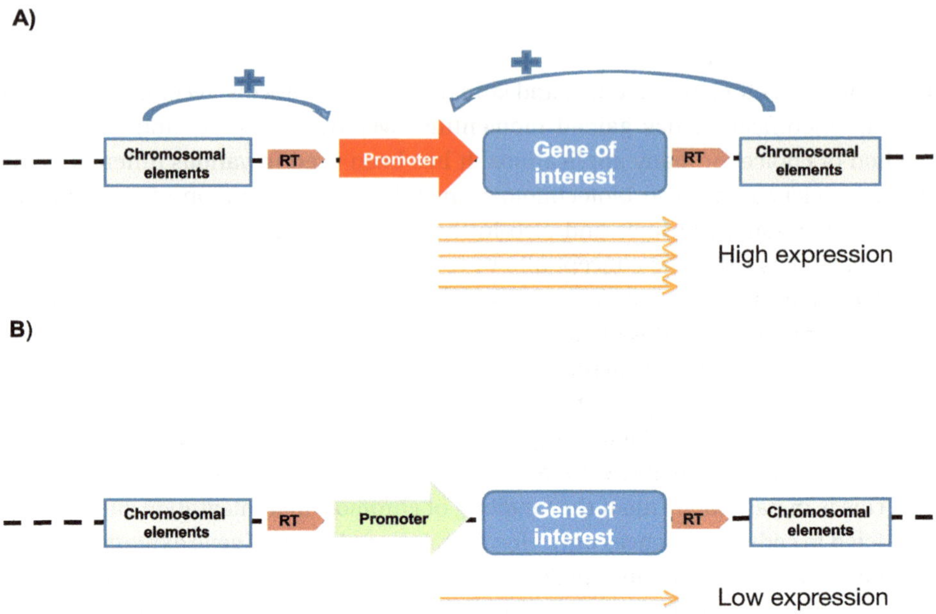

Figure 3.4.6: Expression of transgenes after RMCE. **a)** High susceptibility of the promoter to positive factors provided by the chromosomal environment. A chromosomal site supports high level of transcription of an expression cassette in which transgene is controlled by the promoter (red promoter). Transcription is indicated by orange arrows. **b)** Integration of an unsusceptible promoter (green promoter) will result in reduction of expression. Upon RMCE via the RT sites, a new cassette is introduced that is controlled by this promoter. Since promoter B is not susceptible to the positive influence mediated by the chromosomal surroundings, transcription will not be supported to the same level.

of the promoters completely changed [81]. From this study it can be concluded that it is not only the nature of the integration site that critically defines the expression level. Rather, it is the performance of a regulatory cassette embedded in the chromosomal context. Similar observations were obtained for various promoters integrated into the Rosa26 locus of mouse embryonic stem (ES) cells [82, 83]. This shows that the sensitivity and susceptibility of regulatory elements towards the environmental influence of a particular chromosomal site can vary dramatically.

From these studies it can be concluded that it is not only the nature of the integration site that critically defines the expression level. Rather, it is the performance of a regulatory cassette embedded in the chromosomal context. Together, these studies indicate that the high expression potential of chromosomal loci is restricted to a specific cassette design, i.e., a specific orientation and/or a specific promoter. Thus, to exploit the potential of a chromosomal site a suitable cassette design has to be provided. Since the perfomance of an expression cassette in a chromosomal context cannot be predicted, for each integration site the design of the targeted cassette has to be experimentally verified. Alternatively, the targeting vector should reflect as close as possible the design of the screening vector.

Recombinant Protein Expression

The repeated use of genome integration sites that allow high expression of recombinant proteins is of particular benefit for the generation of production cell lines. Once having tagged and expressing a hotspot with recombination target sites, integration of any transgene of interest can be achieved in a short time. Since screening is not required anymore, the cell line development is significantly speeded up [84, 85].

A number of studies have been performed to investigate the application of targeted integration strategies for generation of cell lines for production of recombinant proteins. They demonstrated that applying RMCE-based strategy in CHO cells allows to reuse genomic sites supporting high and stable protein expression but also ensures the homogeneity in growth and productivity [81, 86, 87].

To avoid unwanted recombinations, recombinase-based targeted integration relies on single copy tagged sites. This seems to be conflicting with the fact that high expression of most recombinant CHO producer cell lines relies on multiple copy numbers of the transgene. On the other hand, single copy sites have been shown to provide high levels of recombinant protein expression. Single copy integrates have the advantage of a higher genomic stability and are less prone to rearrangements if compared to multicopy integrations. The RMCE concept is still compatible with high copy number mediated expression. Once a high producer locus is tagged with RTs, RMCE can be used to integrate preformed multicopy plasmids [88, 89]. Further, it has been shown that targeted integration can be combined with methotrexate- (MTX-) based gene ampflication. Upon recombinase-based targeting of dhfr-expressing vectors and subsequent MTX-based amplification, an increase of protein expression could be achieved [90-92].

While many applications show the benefit of this approach for generating antibody expressing cell lines, the strategy can also be applied to other proteins. Schucht and coworkers [84] used high expressing master cell lines to express G protein-coupled receptors (GPCRs). Wilke et al. used RMCE in mutant CHO Lec3.2.8.1 cells to produce glycoproteins with the well-established glycosylation pattern in a homogenous form [93].

A highlight application of the RMCE technology represents the production of a human polyclonal anti-RhD antibody in CHO cells [94]. Individual antibody expression cassettes were targeted into the same chromosomal site. These cell clones expressing individual antibodies were mixed and cell pools were generated that gave rise to the production of a polyclonal antibody in a highly reproducible manner. Since the individual cell clones have comparable growth properties, the pool can be stably propagated without compromising the composition of the polyclonal antibody under conditions allowing a direct industrial application.

Recombinant Viral Vector Production

RMCE has been succesfully applied for the generation of γ-retroviral helper cell lines. By screening and tagging chromosomal integration sites in HEK293 cells that support high level of recombinant vector production, high titer retroviral helper cell lines have been generated in which the recombinant vector can be exchanged via RMCE [75, 95]. This strategy not only allows the site-specific targeting of the well characterized integration site but obviates the need for selectable genes within the retroviral vector. By targeting different retroviral vector constructions into tagged loci, the vector design could be optimized resulting in titers of 10^7 ip/10^6 cells [96]. Further, based on this principle an integrated, GMP-compatible strategy for production of therapeutic retroviral vectors was established [85].

Recombinase-based Manipulation of Primary Cells and Mice

Since the efficiency of recombinase-based manipulation is high, the manipulation of primary cells becomes feasible. Targeted integration has been achieved in primary fibroblasts of mice [97]. Ortiz-Urda et al. used the ΦC31 integrase-mediated targeting of primary epidermal progenitor cells from patients suffering from the blistering skin disorder recessive dystrophic epidermolysis bullosa. This disease is caused by mutations in the large COL7A1 gene. ΦC31-mediated stable integration of the COL7A1 cDNA into pseudosites in the genome was reported to correct the gene defect [98].

Targeted integration could even be realized in the tissue of mice. In one study, the ΦC31 integrase was used for targeted integration in mice, thereby exploiting pseudosites in the genome that serve as recombination targets. A plasmid encoding the human Factor IX flanked by attB was coapplied along with integrase expressing plasmid via high pressure tail vein injection. By this protocol, transduction and targeted integration of hepatocytes was successfull. Interestingly, integration occurred both in 2 endogenous pseudosites (see Section 3.4.3.1.6) or sites with partial similarity to attP. Importantly, hFIX serum levels were reported to be similar to normal FIX levels [99]. In another study the integrase system was used to correct hereditary tyrosinemia type 1 in a mouse model by transfer of the fumarylacetoacetate hydrolase (FAH) gene into the liver. Seven targeted integration sites were identified [100]. The frequency of integration was low (1/1,000 or less). Still, these study show proof of principle for potential application *in vivo*.

Recombinase-based strategies are of particular benefit to generate transgenic mice. Upon integration of RTs to defined chromosomal loci of embryonic stem (ES) cells fast and efficient targeting of various cassettes is achieved. This allows to significantly speed up the generation of transgenic mice. Meanwhile, ES cell platforms lines have been established in which defined loci are tagged with RTs. This includes the ubiquitously expressed collagen A1 locus [101], the Rosa26 locus [76,102], the ß-actin locus [103] and the p53 locus [104].

Recombinase-mediated Genome Replacement (RMGR)

While most studies employed recombinase-based strategies for targeted integration of plasmid vectors, the potential of this method is even higher. Wallace et al. introduced the method of recombinase-mediated genome replacement (RMGR), a method in which chromosomal domains are exchanged [105]. In this study, a 2 step homologous recombination was used to insert heterologous FRT sites into the α-globin locus of the mouse genome. In a second step, a bacterial artificial chromosome vector comprising the human globulin locus was flanked with corresponding FRT sites. Flp-mediated targeting allowed to replace the 117 kb spanning FRT-flanked mouse globin locus with the human counterpart. This study gave evidence that even the exchange of these extended regions is feasible although the efficiency is severely compromised.

Novel Developments in Recombinase Systems

Since the wild type Flp recombinase works at an optimum temperature of 30 °C but has low efficiency at 37 °C [106], an enhanced Flp variant (Flpe) was developed [107]. Flpe was shown to have fivefold more efficiency then the Flp at 37 °C although it is still not as efficient as the Cre. However, for most in vitro settings this difference does not seem to be a particular limitation. Raymond and Soriano [108] introduced "Flpo", a mouse codon optimized variant that was shown to be equally efficient in comparison to Cre. This was validated by several studies that followed based on this variant [109-112]. Similar to Flpo a codon-improved Cre recombinase was designed to reduce high CpG content in the coding sequence [113]. It was shown that this prevents silencing of Cre by DNA methylation and allows efficient and stable Cre expression. For *in vitro* targeting approaches this variant is of less interest since only a transient expression of the recombinase is required. However, it has an impact for applications in which long-term expression is crucial, i.e., in transgenic animals.

Some new recombinase systems have been recently identified. These include the Dre/Rox [114] the Vcre/VloxP and Scre/SloxP from *Vibrio* species and *Shewanella* species, respectively [115, 116], and the Vica/vox system [117]. So far, they are not yet used for routine genetic manipulation but might represent interesting options for further exploration of this field.

Pitfalls in the Site-specific Recombination Technology

Accidental Random Integrations Accompanied with Targeted Integration

The use of recombinase-based targeting involves the transfection of the targeting cassette together with the recombinase gene. Beside the expected targeting reaction, it is possible that a few cells integrate the transfected plasmids randomly. If the recombinase expressing plasmid integrates it could be detrimental in case the targeted state represents a substrate for the recombinase. Then, continous expression of the recombinase would result in excision of the cassette. For RMCE, reversion of the targeted state is highly unfavourable and chromosomal integration of the recombinase expression vector does not compromise the stability of the targeted cassette. One study investigated the frequency of random vector integrations that occur in addition to the intended targeted integration. The study shows that the efficiency of the unwanted events depends both on the transfection method but also on the cell line employed. While in mouse HEK293 cells less than 3 % of clones with a targeted integration event showed random integration, in CHO cells the frequency was around 15 % [81]. Since random integration along with the appropriate targeted ones will be coselected with most of the selection procedures, they will remain undiscovered by simple characterization protocols. In this regard, a method employing negative selection based on the HSV thymidine kinase as a suicide gene might be of benefit [118]. It should be mentioned that for most applications an additional integration of vectors is not a matter of concern.

Genotoxicity in Recombinase-based Approaches

RMCE and targeted integration enable efficient gene swapping by precisely replacing a genomic target sequence by any donor construct with compatible heterospecific RT sites. However to be effective, the recombinase enzymes have to be applied at an adequate concentration. In case of Cre, it has been shown that an excess of the recombinase might trigger endonucleolytic activities and therefore can induce cellular toxicity [119, 120]. On the other hand, many transgenic mouse lines have been established that constitutively express Cre without deleterious consequences to the maintenance of the animals. Nevertheless, to overcome genotoxicity attributed with Cre optimization of time and level of expression is advised.

Endogenous RT sites have been identified that are recognized by the recombinase without pronounced homology (so-called pseudosites). Such sites have been identified for the ΦC31 system in the genome of human cells [121]. In this study it is speculated that the cryptic ΦC31 attachment sites flanking the transgene and cryptic ΦC31 attachment sites in the host genome recombine with each other leading to genomic rearrangements. Also pseudosites for Cre have been identified [122, 123]. No pseudosites have been reported so far for the Flp recombinase.

3.4.3.2 Homologous Recombination

Homologous recombination is a natural process that occurs during meiosis or mitosis. On one hand, it creates genetic diversity; on the other hand it is a mechanism to repair deleterious double strand breaks that can arise from cell stress (e.g., irradiation, reactive oxygen species). Homologous recombination is exploited for targeted integration of foreign DNA upon transfer to cells. It was first employed for site-specific genome engineering of mouse (ES) cells in the 1980s [124, 125] (reviewed in [126]). Since then, homologous recombination has been mainly applied for specific modification of genes in ES cells for establishment of genetically modified mice: knock-out mice in which a specific gene is deleted and knock-in mice in which a reporter mice mimics the expression of an endogenous gene [126, 127].

The efficiency of homologous recombination is defined by several parameters, the transfer protocols, the length of homologous arms, the type of cells, the nature of the targeting locus, and the extent of homology (i.e., isogenicity) of DNA [128-132]. If all parameters are optimized, high efficiencies can be achieved. This is highlighted by a report showing that upon optimization of the protocol, frequencies of recombination as high as 10^{-1} could be obtained in mouse ES cells [130].

Also the optimal length of homologous arms was defined. To target a gene of interest into a defined genomic locus by homologous recombination the gene of interest needs to be flanked by DNA sequences, which are homologous to that specific site. Generally, the targeting efficiency correlates with the length of homologous arms. A twentyfold higher targeting efficiency was reported in ES cells when increasing the length of homology arms by twofold [124]. The critical length of homology arms to obtain efficient recombination was also investigated. While 1,7 kb of homology is insufficient for targeted recombination [129], a length above 14 kb leads to a saturation of the recombination system [133].

The major drawback of homologous recombination for targeted genome modification is its low efficiency. It was reported that the ratio of targeted integration (homologous recombination) compared to random integration approaches varies within the species and cell types. While in yeast homologous recombination is the preferential mechanism to integrate foreign DNA, in mammalian cells the frequency is much lower [134]. In mammalian stem cells such as ES cells the frequency of recombination was 10^{-5} [126, 130]. In somatic cells, the frequency of homologous recombination is significantly lower (ten- to hundredfold lower, reviewed in [134]), in some cell types not detectable at all. Indeed, somatic cells preferentially undergo nonhomologous recombination [128] resulting in random integration of DNA [135].

Accordingly, the genetic modification of somatic cells by homologous recombination is challenging. A few approaches successfully adapted this technique to modify the genome of mammalian cells. This is exemplified by the inactivation of the p53 or p21 gene in HCT-1116 cells or human fibroblasts to study the cell cycle or by targeting the human ß-globin locus [136-138].

The DT40 chicken immortalized B cell line represents a somatic cell line with remarkably low frequencies of random integration. In this cell line, homologous recombination efficiencies of 50 to 90 % can be achieved [134, 139]. Due to the high frequency of targeted integration this cell line is of interest to study not only different knock-outs but also to elucidate the mechanism of homologous recombination [140-142].

Another determinant for homologous recombination is the nature of the chromosomal locus to be targeted, i.e., whether it is accessible for recombination or not. Studies show that the recombination is favored in transcriptionally active sites [143].

To increase the transfer rate adenoassociated viral vectors (AAV) have been used. By incorporating the respective homology arms into an AAV vector, homologous recombination was achieved in up to 1 % of mammalian cells as exemplified for the HPRT locus [144]. Other loci, like collagen A1 and A2 and the Rosa26 locus, were also shown to be successfully targeted by the AAV vectors [145-147].

Double-strand Break-induced Homologous Recombination

Although some examples cited in the previous section show the potential to increase the efficiency of homologous recombination, this does not facilitate the establishment of efficient routine protocols. Thus, homologous recombination is usually a laborious and time-consuming procedure in mammalian cells. Recently, homologous recombination gained relevance as a tool for modification of mammalian genomes with the development of methods to specifically induce double-strand breaks (DSBs). The breakthrough observation was made 20 years ago when it was found that the generation of DSBs at the site of interest could dramatically increase the frequency of homologous recombination up to 1,000-fold [148-153]. It took another 10 to 15 years to exploit methods for efficient and specific generation of double-strand breaks for routine protocols. DSB induced homologous recombination further allows to reduce the size of homology arms (in one study only 50 to 100 bp were used [154]) which simplifies both, the construction of targeting vectors and the detection of homologous recombination events.

The first reports for DSB-induced homologous recombination were based on rare-cutting endonucleases (meganucleases). For this purpose, the respective restriction sites were first introduced into the genome by classical homologous recombination, in a second step these loci were then targeted by DSB induced homologous recombination. This procedure is comparable to the use of site-specific recombinases for targeted integration of cassettes described above.

Even more advanced methods emerged that allow to specifically introduce DSBs at nearly any position in the genome without previous modification of the locus. This opened a new era of genetic engineering and paved the way for homologous recombination in somatic cells and cell lines.

In the following, the currently used methods to induce DSBs (meganucleases, zinc finger nucleases, TALE nuclease and CRISPR/CAS9) are summarized (see also Figure 3.4.7a and Figure 3.4.7b).

Meganucleases

Sequence-specific endonucleases that recognize sequences of more than 12 bp are classified as meganucleases. Due to the large restriction site, these endonucleases are highly specific. They are found in bacteria, yeast, algae, fungi, phages, and some plant organelles. Meganuclease recognition sites are usually associated with mobile

A)

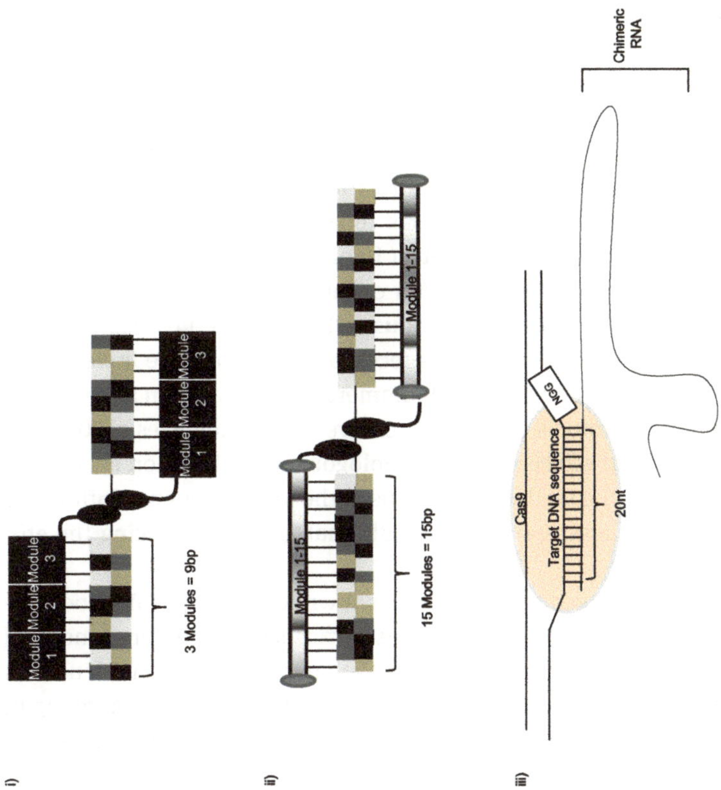

B)

	ZFNs	TALENs	CRISPR/CAS
Homology recognizing module	30 aa	33-35 aa	Chimeric guide RNA
Catalytic enzyme	e.g. Fok nucleaseI, modified FokI nuclease		e.g. Cas9, modified Cas9
Size of recognition	9 -12 per monomer	15-18 per monomer	20 nucleotides per RNA
Frequency of targetable sequences in the genome	Ca. every 600 bp	Ca. every 35 bp	Ca. every 12 bp

Figure 3.4.7: Strategies to induce double-strand breaks at the site of interest. **a)** ZFNs consist of 3 to 4 modules as schematically depicted in (i). While one module recognizes 3 bp, a total 9 or 12 bp per ZFN will be targeted. Binding of two ZFN to adjacent sites will lead to heterodimerization and activation of the nuclease domain (ii). The TALENs are composed of individual modules consisting of 33 to 35 amino acids which recognize 1 bp each (iii). The CRISPR/Cas system is based on the design of an RNA carrying a specific targeting sequence of 20 nucleotides. The Cas9 enzyme recognizes the RNA/DNA hybrid sequence and induces double-strand breaks. The features of the 3 systems are summarized in **b)**.

genetic elements. Upon the expression of the meganucleases DSBs are induced and homologous recombination is initiated, leading to the homing of these genetic elements [155]. The I-SceI endonuclease has been frequently used to modify mammalian genomes [148, 149, 156]. This endonuclease recognizes a 18 bp motif which is much larger than any other known restriction site [157]. Of note, there are no endogenous I-SceI recognition sites in the genome of mouse and man. To exploit this tool for efficient homologous recombination, the I-SceI restriction site has to be first introduced into the genome by classical homologous recombination. In a second step, the targeting vector encoding the gene of interest flanked by homology arms is then cotransfected together with an expression vector encoding the I-SceI nuclease. This procedure has been successfully applied in various cell types [148, 149, 152, 153, 158]. In these studies, the frequency of homologous recombination could be increased at least hundredfold. In case of the dopachrome tautomerase (Dct) gene, an I-SceI-based homologous recombination protocol still could achieve only frequencies of 10^{-7}. This suggests that locus dependent effects impair homologous recombination even upon induction of DSBs [159]. Also the methylation status of the target sequence influences the activity of such meganucleases and is an additional factor influencing the efficiency of recombinations [160].

Interestingly, I-SceI-based recombination was shown to allow specific modification in differentiated tissues *in vivo*. For this purpose, transgenic mice were generated in which an I-SceI target site was integrated into the LagoZ gene. Upon adenoviral transfer of the I-SceI meganuclease gene via tail vein injections, specific targeting of the LagoZ gene was achieved in up to 1.3 % of hepatocytes [161]. However, high lethality was observed which might be an effect of the meganuclease and/or the adenovirus infection per se.

A follow-up concerns the modification of meganucleases to redirect them to specific endogenous DNA sequences of interest. Since for DNA binding multiple amino acid contacts and a defined tertiary structure are needed, such an endeavor is highly complex. Nevertheless, redesigning of the I-Cre meganuclease for other target sequences was demonstrated to be successful [162, 163].

Zinc **F**inger protein-linked **N**ucleases (ZFNs)

An alternative approach to induce double-strand breaks is based on zinc finger protein-linked nucleases (ZFN). ZFN are artificially designed proteins in which a nuclease domain is targeted to a specific DNA sequence. For this purpose, the DNA-binding moiety of zinc finger proteins is used. Zinc finger proteins represent a group of DNA-binding proteins (such as eukaryotic transcription factors) comprising a DNA-binding domain, which is stabilized by one or more zinc ions. A zinc finger consists of approximately 30 amino acids and recognizes a sequence of 3 to 4 bp (called a module, see Figure 3.4.7a, (i)) [164]. Already in the early 1990s it was shown that zinc finger proteins could be modified in its DNA-binding specificity [165, 166].

To induce DSBs, zinc finger domains are fused to a nuclease to form the zinc finger nucleases (ZFN). For this purpose, the FokI nuclease has been broadly used. Since FokI is active as a dimer, binding of 2 ZFN monomers to adjacent DNA sites is required to achieve dimerization of the FokI domains and subsequent generation of DSBs (Figure 3.4.7). Three to 4 zinc finger modules constitute one ZFN monomer that facilitates binding to a DNA sequence of 9 to 12 bp [167]. Binding of the 2 ZFN monomers (left ZFN arm and right ZFN arm) thus is mediated by a 18 to 24 bp sequence. The 2 ZFN monomer-binding sites are not directly adjacent but are usually separated by a spacer. The linker domain connecting the nuclease with the DNA binding moiety and the DNA spacer sequence between the adjacent binding sites have as well an impact on the cleavage activity of the ZFNs [168]. Synthetic zinc finger nucleases have been successfully used to modify the genome of different organisms including zebrafish, pig, *Caenorhabditis elegans*, *Arabidopsis*, hamster, *Drosophila*, humans, and mouse.

Two repair mechanisms of DSBs are exploited for genetic manipulation: (i) nonhomologous end joining (NHEJ) repair that is characterized by random addition or elimination of nucleotides and (ii) homologous recombination for accurate integration of cassettes resulting in the correction and addition of genes, respectively (reviewed in [169]).

A significant lack of specificity has been described for many ZFNs. The specificity and precision of ZFN-based gene modification is determined by the frequency of cleavage at nonhomologous sites ('off-target sites'). Theoretically, a 9 bp recognition target occurs many times in the genome while a 18 bp recognition sequence is expected to be unique in the genome of mammalian cells [170, 171]. To increase the sequence specificity, ZFN proteins that target even 18 bp sequences were developed by fusing 2 × 3 or 3 × 2 segments [170].

Unintended binding of the zinc finger moieties to DNA sequences that have only partial homology to the target sequence [172, 173] contribute to the lack of precision. To lower these off-target effects experiments were carried out that increased the DNA-binding specificity by optimizing the zinc finger binding moieties [174].

Moreover, a significant context dependency in the binding of the zinc finger modules is observed, i.e., a particular finger will provide affinity and specificity for a given triplet in one sequence, but not in others. Complex methods are employed to predict binding (reviewed in [167]). The numbers of putative binding sites for zinc fingers are dramatically varying. Depending on the methods being used to predict such target sites sometimes more than 600 bp need to be searched to find one ZFN site (reviewed [164]). Professional service for the design and the construction of ZFNs are offered by companies (e.g., Sangamo Bioscience or Sigma Aldrich). *In vitro* pretesting is recommended to validate the ZFN pairs. Even with these preselections, the prediction of ZFN pairs is not satisfactory and a substantial proportion of ZFN pairs fail to act properly *in vivo* [169, 175]. To overcome the failure rate, usually multiple sequence targets have to be generated and extensively tested in vitro.

Another factor of concern is the nuclease itself. As discussed before, FokI cleavage depends on dimerization. However, if 2 identical ZFNs bind to adjacent sites rather than the intended heterodimers, an unspecific cleavage can occur. To overcome such problems a modified FokI domain was developed by iterative structure-based design that allowed to decrease the unspecific cleavage initiated by the formation of homodimers [176, 177]. In an effort to increase the nuclease activity, a hyperactive variant of the FokI cleavage domain, called Sharkey, was employed and that resulted in a fifteenfold increase in cleavage activity [178].

In simple protocols, the ZFN encoding vectors are cotransfected together with the homology vectors (targeting plasmids). To limit the activity of the ZFN and to avoid side reactions, protocols have been established for transient expression, thereby using nonintegrating DNA vectors and the direct delivery of zinc finger proteins [179]. Further, the stability of ZFN was modulated by controlled proteosomal degradation [180]. As discussed above, the efficiency of DSB-induced homologous recombination is reduced by a side reaction, the error prone nonhomologous end joining (NHEJ) repair mechanism. NHEJ results in the repair of DSBs by processing of DNA ends and relegation, usually giving rise to mutations (insertions, deletions). It was found that nicked DNA does not activate the NHEJ repair mechanism but still stimulates homologous recombination [181]. By replacing or modifying the FokI nuclease with a nicking enzyme, less off-target events were observed. However, the efficiency of homologous recombination events was lower compared to the DSB inducing endonucleases [182-184].

TALEN

Transcription activator-like effectors (TALEs) are DNA-binding proteins [185, 186] that are secreted by the plant pathogenic bacteria genus *Xanthomonas*. These proteins comprise a modular DNA-binding domain that provides binding to plant promoter sequences and thereby activation of plant genes that support the bacterial infection. TALE proteins form the basis of a technology that can be regarded as an alternative to the ZFN technology.

The TALE DNA binding-domains comprise of the individual modules which each consist of approximately 33 to35 amino acids ([187] see Figure 3.4.7a, (ii)). The 2 amino acids at position 12 and 13 (so-called repeat variable diresidue or RVD) of each module can vary and define the binding of the module to a specific nucleotide. Since for all nucleotides TALE modules are known these TALE modules can be combined in a way that new DNA-binding proteins that specifically recognize a short stretch of DNA are built. Upon fusion of this DNA-binding domain to a nuclease such as FokI, resulting in a transcription activator like effector nuclease "TALEN", specific DNA cleavage can be achieved.

Although theoretically TALENs can be designed for any target site, it has been shown that many of such custom TALENs can fail. This suggests that yet unknown rules govern the assembly of functional repeat domains. Software programs to predict optimal TALEN binding sites were developed, e.g., [188, 189]. Since by these programs, TALEN sites are found every 35 bp in average of mammalian genomes [189], the TALENs show a largely extended flexibility in the selection of target sites if compared to the above discussed ZFNs. Also, the design of TALENs is characterized by less restrictions than ZFNs. The fact that defined modifications such as the translational start site of genes could be mutated underlines the potential power of TALENs [190].

TALENs have already been used for gene disruption and gene addition for genomes such as pigs, bovines, humans, *Caenorhabditis elegans*, rice, and zebrafish (reviewed in [169, 191]). While previously much time and effort was required to assemble the TALENs, a number of new protocols and reagents were developed to construct TALENs by rapid modular assembly or fast ligation-based automatable solid phase high throughput platform (FLASH) resulting in a 88 % efficiency of creating active TALENs. These methods facilitate the construction and require much less time [192-196]. Since the nature of the chromosomal site significantly contributes to the efficiency of homologous recombination, a comparison of TALENs with ZFN is difficult. However, 2 studies show for defined sites in human cells that the efficiency of homologous recombination mediated by ZFN and TALEN induced DSBs is comparable [197, 198].

The reduced restrictions in the design of TALE-based DNA-binding proteins makes this technology a highly interesting method not only for introducing DSBs but also for targeting other functional activities such as transcription factors, histone modifiers and DNA modifiers to specific sites in the DNA [167, 199]. Of note, due to the highly repetitive sequence elements in TALE proteins, they are prone to rearrangements if transduced via γ-retroviral or lentiviral vectors [200].

The CRISPR/Cas9 System

The ZFN- and TALEN-based strategies for inducing DSBs require the *de novo* design of DNA-binding protein(s). As such, however, the development of the modular DNA-binding proteins can be difficult and time consuming.

Recent developments have opened the door for an alternative strategy that relies on a RNA-guided mechanism to introduce DSBs. This strategy is based on the type II clustered regularly interspaced short palindromic repeats (CRISPR/CRISPR-associate systems (Cas)). The CRISPR/Cas systems belong to bacterial and archae defense systems that protect the organism from invading foreign DNA. In the natural situation, the foreign DNA is excised in a three-step process. First, short fragments of the foreign invading DNA are integrated in the CRISPR loci of the organism. In the second step, a precursor RNA, which is complementary to the foreign DNA, is synthetized (pre-cr RNA). This precursor RNA is activated by a RNA (trac-RNA) complementary to the 3' end of the pre-cr RNA. This RNA complex then binds to the DNA template and activates an endonuclease (e.g., Cas9). Cas9 induces a DSB to eliminate the foreign DNA sequence [201, 202].

It was shown that the system can be simplified for application in mammalian cells. The 2 RNA moieties can be combined in a short synthetic 'guide RNA', comprising both, the cr-RNA and the trac-RNA. This guide RNA facilitates both, the binding to the bacterial endonuclease Cas9 and the binding to the DNA motif, and directs cleavage (Figure 3.4.7a, (iii)). From the 100 nt spanning guide RNA 20 nt confer the binding to a complementary DNA target site. Upon expression of this synthetic guide RNA together with an expression vector for Cas9, binding of the RNA to the site of interests activates the Cas9 endonuclease which efficiently induces cleavage [201]. Successful cleavage by Cas9 requires that the target site contains a NGG motif located 1 bp downstream of the complementary RNA target sequence. Thus this method allows the targeting of DNA sequences occurring every 8 to 12 bp in the genome [203] which makes it to a powerful system compared to TALENS and ZFNs which show less frequent potential target sites (see above). Meanwhile, vectors have been introduced which facilitate coexpression of the short RNA by a polymerase III promoter (e.g., the U6 promoter) and the humanized Cas9 protein controlled by a mammalian Pol II promoter [204]. Alternative protocols use cotransfer of *in vitro* synthetized guide RNA and the Cas9 RNA [205, 206].

Although the method of CRISPR/Cas-mediated genome engineering is comparably new, it is rapidly involving and was successfully applied in various species including human [201, 207, 208], zebrafish [209], mouse [206], yeast [210], and in bacterial cells [211]. Due to the comparably simple design and generation of targeting RNAs, CRISPR/Cas-based induced homologous recombination turns out to be a comparably robust method that supports routine applications. Of particular notice is the fact that this system has been even used to achieve simultaneous targeting of various loci: such multiplexing was achieved by specific targeting of 5 different loci based on such an RNA guided approach [206]. Meanwhile, a number of modifications were explored to increase the efficiency of homologous recombination. Among them, the Cas9 was converted into a nickase [204].

So far, a few restrictions of the CRISPR/Cas system have been reported. On the one hand this concerns the large size of the Cas9 protein, which could be problematic when being delivered by recombinant viruses that own limited cargo size. More importantly, a comparably high frequency of off-target events was detected in some studies showing that the specificity of RNA guided nucleases varies from target site to target site tolerating some mismatches [212, 213] (reviewed in [164]). One way to reduce this limitation could be achieved by transient expression, e.g., by cytoplasmic transfer of a Cas9 encoding, *in vitro* generated RNA [205, 206]. Another option is to increase the target sequence by using paired guide RNAs, which was successful in improving specificity if combined with a nicking [214].

Application of DSB-induced Modification for Biotechnology and Biomedicine

Improving the efficiency of homologous recombination by DSBs facilitates site-specific modification not only in stem cells but also in somatic cells and cell lines. Thus, these methods gained importance also in the field of biomedicine and biotechnology.

The technologies are of particular interest for gene therapies. ZFN are already in first clinical trials aiming at mutating the CCR5 gene, the coreceptor HIV. In this respect it is also of note that ZFN-, TALENs-, and CRISPR/Cas-based approaches have been successfully used for correcting disease-causing point mutations in iPSCs [215-217].

A number of chromosomal loci in producer cell lines such as CHO and HEK293 are now accessible by meganuclease, ZFN- or TALEN-based targeting protocols. An example is the knock-out of the dihydrofolate reductase gene in CHO cells by transient delivery of ZFNs achieving a disruption frequency of up to 15 % of alleles in the cell population [218]. Also, glycosylation has been modified in CHO cells [219, 220]. Further, multiple gene knockouts were achieved using a sequential ZFNs-based approaches in CHO and K562 cells [221, 222].

In 293 cells efficiencies of up to 10^{-1} for a meganuclease approach and 10^{-2} for ZFN-mediated gene targeting approaches were achieved (reviewed in [223]). However, as discussed for the site-specific recombinases, for high titer protein expression relevant integration sites that give rise to high and stable expression levels need to be identified.

3.4.4 'Safe Harbors' in the Mammalian Genome

Previous and ongoing discussions tackled the question of the right chromosomal locus to integrate a transgene. Apart from the functional considerations, i.e., that the transgenes should integrate into transcriptionally active sites, safety aspects are of importance, in particular for biomedical applications. In this respect, the idea of a so-called 'safe harbor' emerged. Theoretically, such "safe harbors" would be sites in the genome in which transgene can be integrated without disturbing the activity of endogenous genes and thus without deleterious consequences (such as the promotion of cell growth that could give rise to cancer). More so, the transgenes should be expressed in such sites in a predictable manner. From these theoretical considerations, attempts have been made to define such safe harbor sites by criteria such as a minimal distance to standard genes and to tumor promoting genes [224, 225]. However, such definitions do not consider the topology of DNA, the possibility of interchromosomal crosstalk [226], and also the eventual perturbation induced by the integrated expression cassette. Thus, the theoretical prediction of safe harbors based on these simple criteria does not seem to be feasible and might even be misleading. In any case, rigorous testing would be needed to validate such a concept experimentally.

Another issue that concerns the predictability of transgene expression is the site of integration. Our understanding of chromosomal crosstalk has been substantiated by studies employing site-specific recombination to defined loci (this is discussed in more detail in Section 3.4.3.1.2). Indeed, evidence accumulates that a predefined site, shown to support high and efficient transgene expression for one specific cassette design, not necessarily supports other cassette designs. Rather, the combination of a specific chromosomal site with a specific regulatory element from the transgene will define the final expression level and stability (compare Figure 3.4.6 and [78, 81, 227]). As long as methods that predict the interaction of the integrated cassettes with the genome are not available, experimental validation of hotspots or safe harbor sites is required.

3.4.5 The Use of Bacterial Artificial Chromosomes (BACs) and *in vitro* Recombineering

Bacterial artificial chromosome (BAC) vectors represent an interesting option to overcome or at least reduce the effects that are caused by random integration of transgene expression cassettes into cell lines. In this paragraph, the potential of BAC vectors for the genetic manipulation of mammalian cells is discussed.

Since BAC vectors comprise large genomic regions (up to 300 kb) they encompass not only the coding sequences but also regulatory elements located in cis. Thus, upon random transfer of a BAC with a transgene positioned in its center, transgene expression is modulated by the BAC elements and is not affected by the nature and the features of the chromosomal integration site.

BAC libraries have been generated for various species and can accommodate 150 to 300 kb of genomic DNA. For the construction of BAC libraries from CHO cells the reader is referred to Chapter 3.2. BAC vectors are propagated and can be manipulated in *Escherichia coli*. In particular, methods have been developed that facilitate the modification of BACs by homologous recombination in *Escherichia coli* (recombination mediated genetic engineering, so-called 'recombineering'; reviewed in [228-230]). Recombination in *Escherichia coli* requires only short homology regions of 50 to 70 bp, which can be easily attached to fragments of choice by PCR. These PCR fragments are introduced into bacteria carrying a defined BAC together with vectors encoding the needed recombinases for pursuing the recombination reaction.

To facilitate the screening for successful recombination, the target cassette usually encompasses a selection marker. If this marker is flanked by recombination target sites, such as FRT or loxP, this cassette can be later excised by the site-specific recombinase, leaving the recombination target site. Even more accurate protocols have been developed for BAC manipulation without any remnants (so-called seamless manipulation [231-234]).

One limitation of the exploitation of BACs lies in its fragility that can result in fragmentation of the BAC DNA upon manipulation of the cells. This is in particular true for linear BAC vectors which are even more sensitive to shear stress. As a consequence, the integrity of the BAC vector can be compromised upon transfer to the cells and only fragments are integrated into the genome. In such a situation, chromosomal position effects can come into account and the benefit of the large chromosomal regions is obviously reduced. To overcome this limitation, BAC vectors have been equipped with ITRs, the target sites for transposases. By cotransfering ITR-BACs in presence of the transposase, precise integration of full length BAC vectors could be achieved [235, 236]. Interestingly, using this approach the copy number was shown to be low and most clones showed only a single integration event.

While BAC vectors have been mainly used for studying the gene regulation in an authentic context *in vitro* and *in vivo* (see [237, 238] for review) only few examples exist for their exploitation for biotechnological purposes. One of the reasons might be that the nature of the preferred locus of transgene expression is not yet identified. However, proof of principle has been shown for the well-characterized Rosa26 locus identified in the mouse genome [239, 240] or the dhfr locus [241].

3.4.6 Episomal Vectors

Stable integration of plasmid vectors into the chromosomal DNA of the host cell provides long-term expression of transgenes. Once integrated, the transgenic DNA is embedded within the host's chromosomal DNA. Accordingly, the replication and maintenance of the transgenes completely relies on the cellular replication machinery and the controlled partition of chromosomes upon cell division. Generally, plasmid vectors allow transgene expression also without chromosomal integration. However, in the nonintegrated state, they are lost due to continuous division of cells and concomitant dilution of plasmid vectors. Various strategies have been figured out to ensure extrachromosomal maintenance of foreign plasmid DNA.

One approach relies on the design of artificial chromosomes, either by chopping natural chromosomes to manageable sizes (top-down) or by implementing the relevant components of chromosomes that are required for segregation into artificial chromosomes, reviewed in [242]. Mammalian artificial chromosomes of some (6 to 10) megabases have been constructed which comprise both centromers (the sites of kinetochore formation and spindle attachment) and telomeric ends. This allowed the construction of 'synthetic' chromosomes that are maintained in proliferating cells. So far, however, the design of such artificial chromosomes is complex and the modification is challenging due to the size of these artificial chromosomes. In this respect, the implementation of site-specific recombinase-based approaches can improve this technology [243, 244].

Loss of episomal vectors in replicating cells can be overcome by including elements that provide amplification of episomes (reviewed in [245]). Such elements are found in viruses that have evolved strategies that maintain their genomes extrachromosomally. Basically, viral episomes harbor origins of replication that facilitate replication in presence of certain viral proteins. One prominent example concerns vectors based on EBV elements. EBV-based episomal vectors which have been shown to provide long term gene expression in B cell lines but also in cell lines such as HEK293 and some rodent cells [246]. Replication of these vectors was shown to be facilitated by expression of EBNA-1. EBNA-1 binds to the cellular chromosomes at newly replicated regions, which is considered to be crucial to ensure controlled partition of the episome during cell division [247]. However, without applying selection pressure episomes can be lost in particular in highly replicating cells.

Episomal vectors have been constructed that replicate in absence of viral proteins [248]. In these vectors, nuclear scaffold-/matrix-attached regions (S/MARs) mediate the association with the nuclear matrix, which is suggested to provide mitotic stability ([249], for description of S/MARs see 3.4.2.1). Indeed, S/MAR-harboring episomal vectors were shown to cosegregate with the host chromosome during mitosis [250]. Interestingly, the maintenance of the episomes upon cell proliferation is critically dependent on the integration of S/MAR elements within the transcribed regions [251, 252]. This has important implications for the design of the vectors. S/MAR-based episomal vectors have been shown to provide transgene expression in various cell lines such as CHO cells [253], but also primary cells [254], in stem cells [255], and transgenic pigs [256]. While the episomal maintenance of such vectors has been proven in various studies, this does not necessarily rule out that aberrant chromosomal integration of individual vector copies can occur. If important, this has to be confirmed by thorough analysis.

As episomal vectors do not need to integrate into the cellular chromosomes for maintenance, per definition, they are not affected by chromosomal position effects. If integration indeed does not occur, they are not associated with the risk for insertion-mediated deregulation of cellular genes (insertional mutagenesis). In this sense, they represent an interesting option for therapeutic applications [257].

In biotechnology the most frequent application of episomal vectors lies in the extended time span of transient expression. Protocols for CHO cells, HEK293 cells, but also other cell types were developed that allow short-term efficient production of proteins using appropriate vectors, DNA transfer methods, and cultivation protocols. In these situations long-term maintenance of the vectors are not required and aberrant integration does not play a significant role (for review see [258]).

3.4.7 Concluding Remarks and Perspectives

Rational modification of mammalian cells has successively moved into the focus of research, biotechnology and biomedicine. While episomal and/or transient expression obviates many aspects discussed in this chapter, long-term and stable expression still requires the integration of the expression cassettes into the chromosomal DNA. Systematic design of "stable" and predictable expression is on the one hand supported by the increasing knowledge on the various chromosomal elements that modulate transgene expression upon chromosomal integration. On the other hand, novel methods have been developed that allow to efficiently introduce genetic alterations at defined chromosomal sites with considerable accuracy. It can be expected that these technologies will pave the way for future applications. They will become an integral part of applications in the emerging field of synthetic biology. Due to the complexity of the mammalian genome the construction of completely artificial cells is currently not possible. However, the predictable modification of cellular properties based on precise genetic engineering seems to be realistic in the next years.

References

[1] Blesch A. Lentiviral and MLV based retroviral vectors for ex vivo and in vivo gene transfer. Methods 2004;33:164-72.
[2] Delenda C. Lentiviral vectors: optimization of packaging, transduction and gene expression. J Gene Med 2004;6(1):S125-38.
[3] Matrai J, Chuah MK, VandenDriessche T. Recent advances in lentiviral vector development and applications. Mol Ther 2010;18:477-90.
[4] Ivics Z, Izsvak Z. Nonviral gene delivery with the sleeping beauty transposon system. Hum Gene Ther 2011;22:1043-51.
[5] Batenchuk C, St-Pierre S, Tepliakova L, et al. Chromosomal position effects are linked to sir2-mediated variation in transcriptional burst size. Biophys J 2011;100:L56-8.
[6] Daboussi F, Zaslavskiy M, Poirot L, et al. Chromosomal context and epigenetic mechanisms control the efficacy of genome editing by rare-cutting designer endonucleases. Nucleic Acids Res 2012;40:6367-79.
[7] Moreno R, Martinez I, Petriz J, et al. Boundary sequences stabilize transgene expression from subtle position effects in retroviral vectors. Blood Cells Mol Dis 2009;43:214-20.
[8] Carlson CM, Dupuy AJ, Fritz S, et al. Transposon mutagenesis of the mouse germline. Genetics 2003;165:243-56.
[9] Yant SR, Wu X, Huang Y, et al. High-resolution genome-wide mapping of transposon integration in mammals. Mol Cell Biol 2005;25:2085-94.
[10] Baum C, Kustikova O, Modlich U, et al. Mutagenesis and oncogenesis by chromosomal insertion of gene transfer vectors. Hum Gene Ther 2006;17:253-63.
[11] Wu X, Li Y, Crise B, et al. Transcription start regions in the human genome are favored targets for MLV integration. Science 2003;300:1749-51.
[12] Mitchell RS, Beitzel BF, Schroder AR, et al. Retroviral DNA integration: ASLV, HIV, and MLV show distinct target site preferences. PLoS biology 2004;2:E234.
[13] Bode J, Schlake T, Rios-Ramirez M, et al. Scaffold/matrix-attached regions: structural properties creating transcriptionally active loci. Int Rev Cytol 1995;162(A):389-454.

[14] Grandjean M, Girod PA, Calabrese D, et al. High-level transgene expression by homologous recombination-mediated gene transfer. Nucleic Acids Res 2011;39:e104.
[15] Schubeler D, Mielke C, Maass K, et al. Scaffold/matrix-attached regions act upon transcription in a context-dependent manner. Biochemistry 1996;35:11160-9.
[16] Laemmli UK, Kas E, Poljak L, et al. Scaffold-associated regions: cis-acting determinants of chromatin structural loops and functional domains. Curr Opin Genet Dev 1992;2:275-85.
[17] Mielke C, Maass K, Tummler M, et al. Anatomy of highly expressing chromosomal sites targeted by retroviral vectors. Biochemistry 1996;35:2239-52.
[18] Khan E, Mack JP, Katz RA, et al. Retroviral integrase domains: DNA binding and the recognition of LTR sequences. Nucleic Acids Res 1991;19:851-60.
[19] Eckdahl TT, Anderson JN. Conserved DNA structures in origins of replication. Nucleic Acids Res 1990;18:1609-12.
[20] Evans K, Ott S, Hansen A, et al. A comparative study of S/MAR prediction tools. BMC Bioinformatics 2007;8:71.
[21] Liebich I, Bode J, Frisch M, et al. S/MARt DB: a database on scaffold/matrix attached regions. Nucleic Acids Res 2002;30:372-4.
[22] Girod PA, Nguyen DQ, Calabrese D, et al. Genome-wide prediction of matrix attachment regions that increase gene expression in mammalian cells. Nat Meth 2007;4:747-53.
[23] Klehr D, Maass K, Bode J. Scaffold-attached regions from the human interferon beta domain can be used to enhance the stable expression of genes under the control of various promoters. Biochemistry 1991;30:1264-70.
[24] Harraghy N, Gaussin A, Mermod N. Sustained transgene expression using MAR elements. Curr Gene Ther 2008;8:353-66.
[25] Buceta M, Galbete JL, Kostic C, et al. Use of human MAR elements to improve retroviral vector production. Gene Ther 2011;18:7-13.
[26] Bell AC, Felsenfeld G. Stopped at the border: boundaries and insulators. Curr Opin Genet Dev 1999;9:191-8.
[27] Dietz A, Kay V, Schlake T, et al. A plant scaffold attached region detected close to a T-DNA integration site is active in mammalian cells. Nucleic Acids Res 1994;22:2744-51.
[28] Harraghy N, Buceta M, Regamey A, et al. Using matrix attachment regions to improve recombinant protein production. Methods Mol Biol 2012;801:93-110.
[29] Grabundzija I, Wang J, Sebe A, et al. Sleeping Beauty transposon-based system for cellular reprogramming and targeted gene insertion in induced pluripotent stem cells. Nucleic Acids Res 2013;41:1829-47.
[30] Girod PA, Zahn-Zabal M, Mermod N. Use of the chicken lysozyme 5' matrix attachment region to generate high producer CHO cell lines. Biotechnol Bioeng 2005;91:1-11.
[31] Murrell A, Heeson S, Reik W. Interaction between differentially methylated regions partitions the imprinted genes Igf2 and H19 into parent-specific chromatin loops. Nat Genet 2004;36:889-93.
[32] Reik W, Murrell A, Lewis A, et al. Chromosome loops, insulators, and histone methylation: new insights into regulation of imprinting in clusters. Cold Spring Harb Symp Quant Biol 2004;69:29-37.
[33] West AG, Gaszner M, Felsenfeld G. Insulators: many functions, many mechanisms. Genes Dev 2002;16:271-88.
[34] Matzat LH, Dale RK, Lei EP. Messenger RNA is a functional component of a chromatin insulator complex. EMBO Rep 2013;14(10):916-22.
[35] Lewis A, Murrell A. Genomic imprinting: CTCF protects the boundaries. Curr Biol 2004;14:R284-6.

[36] Herold M, Bartkuhn M, Renkawitz R. CTCF: insights into insulator function during development. Development;139:1045-57.
[37] Valadez-Graham V, Razin SV, Recillas-Targa F. CTCF-dependent enhancer blockers at the upstream region of the chicken alpha-globin gene domain. Nucleic Acids Res 2004;32:1354-62.
[38] Rincon-Arano H, Furlan-Magaril M, Recillas-Targa F. Protection against telomeric position effects by the chicken cHS4 beta-globin insulator. Proc Natl Acad Sci USA 2007;104:14044-9.
[39] Roman AC, Gonzalez-Rico FJ, Molto E, et al. Dioxin receptor and SLUG transcription factors regulate the insulator activity of B1 SINE retrotransposons via an RNA polymerase switch. Genome Res 2011;21:422-32.
[40] Hanawa H, Yamamoto M, Zhao H, et al. Optimized lentiviral vector design improves titer and transgene expression of vectors containing the chicken beta-globin locus HS4 insulator element. Mol Ther 2009;17:667-74.
[41] Uchida N, Hanawa H, Yamamoto M, et al. The chicken hypersensitivity site 4 core insulator blocks promoter interference in lentiviral vectors. Hum Gene Ther Meth 2013;24:117-24.
[42] Evans-Galea MV, Wielgosz MM, Hanawa H, et al. Suppression of clonal dominance in cultured human lymphoid cells by addition of the cHS4 insulator to a lentiviral vector. Mol Ther 2007;15:801-9.
[43] Aker M, Tubb J, Groth AC, et al. Extended core sequences from the cHS4 insulator are necessary for protecting retroviral vectors from silencing position effects. Hum Gene Ther 2007;18:333-43.
[44] Recillas-Targa F, Valadez-Graham V, Farrell CM. Prospects and implications of using chromatin insulators in gene therapy and transgenesis. Bioessays 2004;26:796-807.
[45] Emery DW. The use of chromatin insulators to improve the expression and safety of integrating gene transfer vectors. Hum Gene Ther 2011;22:761-74.
[46] Nielsen TT, Jakobsson J, Rosenqvist N, et al. Incorporating double copies of a chromatin insulator into lentiviral vectors results in less viral integrants. BMC Biotechnology 2009;9:13.
[47] Dharshanan S, Chong H, Cheah SH, et al. Stable expression of H1C2 monoclonal antibody in NSO and CHO cells using pFUSE and UCOE expression system. Cytotechnology 2013 (Epub ahead of print).
[48] Zhang F, Frost AR, Blundell MP,et al. A ubiquitous chromatin opening element (UCOE) confers resistance to DNA methylation-mediated silencing of lentiviral vectors. Mol Ther 2010;18:1640-9.
[49] Pfaff N, Lachmann N, Ackermann M, et al. A ubiquitous chromatin opening element prevents transgene silencing in pluripotent stem cells and their differentiated progeny. Stem Cells 2013;31:488-99.
[50] Knight S, Zhang F, Mueller-Kuller U, et al. Safer, silencing-resistant lentiviral vectors: optimization of the ubiquitous chromatin-opening element through elimination of aberrant splicing. J Virol 2012;86:9088-95.
[51] Kim S, Kim YW, Shim SH, et al. Chromatin structure of the LCR in the human beta-globin locus transcribing the adult delta- and beta-globin genes. Int J Biochem Cell Biol 2012;44:505-13.
[52] Raich N, Enver T, Nakamoto B, et al. Autonomous developmental control of human embryonic globin gene switching in transgenic mice. Science 1990;250:1147-9.
[53] Kim A, Dean A. Chromatin loop formation in the beta-globin locus and its role in globin gene transcription. Mol Cells 2012;34:1-5.
[54] Indraccolo S, Minuzzo S, Roccaforte F, et al. Effects of CD2 locus control region sequences on gene expression by retroviral and lentiviral vectors. Blood 2001;98:3607-17.
[55] Koh BH, Hwang SS, Kim JY, et al. Th2 LCR is essential for regulation of Th2 cytokine genes and for pathogenesis of allergic asthma. Proc Natl Acad Sci USA 2010;107:10614-9.
[56] Smale ST. A Th2 cytokine LCR. Adding a new piece to the regulatory puzzle. Immunity 2003;19:1-2.

[57] Palstra RJ, de Laat W, Grosveld F. Beta-globin regulation and long-range interactions. Adv Genet 2008;61:107-42.
[58] Pawliuk R, Bachelot T, Raftopoulos H, et al. Retroviral vectors aimed at the gene therapy of human beta-globin gene disorders. Ann N Y Acad Sci 1998;850:151-62.
[59] Ellis J, Pannell D. The beta-globin locus control region versus gene therapy vectors: a struggle for expression. Clin Genet 2001;59:17-24.
[60] Morianos I, Siapati EK, Pongas G, et al. Comparative analysis of FV vectors with human alpha- or beta-globin gene regulatory elements for the correction of beta-thalassemia. Gene Ther 2012;19:303-11.
[61] Kowolik CM, Hu J, Yee JK. Locus control region of the human CD2 gene in a lentivirus vector confers position-independent transgene expression. J Virol 2001;75:4641-8.
[62] Su Y, Liebhaber SA, Cooke NE. The human growth hormone gene cluster locus control region supports position-independent pituitary- and placenta-specific expression in the transgenic mouse. J Biol Chem 2000;275:7902-9.
[63] Mei Q, Kothary R, Wall L. The tkNeo gene, but not the pgkPuro gene, can influence the ability of the beta-globin LCR to enhance and confer position-independent expression onto the beta-globin gene. Exp Cell Res 2000;260:304-12.
[64] Montiel-Equihua CA, Zhang L, Knight S, et al. The beta-globin locus control region in combination with the EF1alpha short promoter allows enhanced lentiviral vector-mediated erythroid gene expression with conserved multilineage activity. Mol Ther 2012;20:1400-9.
[65] Arumugam PI, Higashimoto T, Urbinati F, et al. Genotoxic potential of lineage-specific lentivirus vectors carrying the beta-globin locus control region. Mol Ther 2009;17:1929-37.
[66] Antoniou MN, Skipper KA, Anakok O. Optimizing retroviral gene expression for effective therapies. Hum Gene Ther 2013;24:363-74.
[67] Hirano N, Muroi T, Takahashi H, et al. Site-specific recombinases as tools for heterologous gene integration. Appl Microbiol Biotechnol 2011;92:227-39.
[68] Buchholz F, Ringrose L, Angrand PO, et al. Different thermostabilities of FLP and Cre recombinases: implications for applied site-specific recombination. Nucleic Acids Res 1996;24:4256-62.
[69] Wirth D, Gama-Norton L, Riemer P, et al. Road to precision: recombinase-based targeting technologies for genome engineering. Curr Opin Biotechnol 2007;18:411-9.
[70] Turan S, Zehe C, Kuehle J, et al. Recombinase-mediated cassette exchange (RMCE) – a rapidly-expanding toolbox for targeted genomic modifications. Gene 2013;515:1-27.
[71] Spitzer J, Landthaler M, Tuschl T. Rapid Creation of Stable Mammalian Cell Lines for Regulated Expression of Proteins Using the Gateway(R) Recombination Cloning Technology and Flp-In T-REx(R) Lines. Methods Enzymol 2013;529:99-124.
[72] Wu H, Mao C, Duenstl G, et al. Assay development of inducible human renal phosphate transporter Npt2A (SLC34A1) in Flp-In-Trex-HEK293 cells. Eur J Pharmacol 2013;721(1-3):332-40.
[73] Wong ET, Kolman JL, Li YC, et al. Reproducible doxycycline-inducible transgene expression at specific loci generated by Cre-recombinase mediated cassette exchange. Nucleic Acids Res 2005;33:e147.
[74] Verhoeyen E, Hauser H, Wirth D. Evaluation of retroviral vector design in defined chromosomal loci by Flp-mediated cassette replacement. Hum Gene Ther 2001;12:933-44.
[75] Schucht R, Coroadinha AS, Zanta-Boussif MA, et al. A new generation of retroviral producer cells: predictable and stable virus production by Flp-mediated site-specific integration of retroviral vectors. Mol Ther 2006;14:285-92.
[76] Sandhu U, Cebula M, Behme S, et al. Strict control of transgene expression in a mouse model for sensitive biological applications based on RMCE compatible ES cells. Nucleic Acids Res 2011;39:e1.

[77] Turan S, Galla M, Ernst E, et al. Recombinase-mediated cassette exchange (RMCE): traditional concepts and current challenges. J Mol Biol 2011;407:193-221.
[78] Gama-Norton L, Botezatu L, Herrmann S, et al. Lentivirus production is influenced by SV40 large T-antigen and chromosomal integration of the vector in HEK293 cells. Hum Gene Ther 2011;22:1269-79.
[79] Strathdee D, Ibbotson H, Grant SG. Expression of transgenes targeted to the Gt(ROSA)26Sor locus is orientation dependent. PLoS One 2006;1:e4.
[80] Nehlsen K, da Gama-Norton L, Schucht R, et al. Towards rational engineering of cells: Recombinant gene expression in defined chromosomal loci. BMC Proc 2011;5 Suppl 8:O6.
[81] Nehlsen K, Schucht R, da Gama-Norton L, et al. Recombinant protein expression by targeting pre-selected chromosomal loci. BMC Biotechnology 2009;9:100.
[82] Tchorz JS, Suply T, Ksiazek I, et al. A modified RMCE-compatible Rosa26 locus for the expression of transgenes from exogenous promoters. PLoS One 2012;7:e30011.
[83] Chen CM, Krohn J, Bhattacharya S, et al. A comparison of exogenous promoter activity at the ROSA26 locus using a PhiiC31 integrase mediated cassette exchange approach in mouse ES cells. PLoS One 2011;6:e23376.
[84] Schucht R, Lydford S, Andzinski L, et al. Rapid establishment of G-protein-coupled receptor-expressing cell lines by site-specific integration. J Biomol Screen 2011;16:323-31.
[85] Carrondo M, Panet A, Wirth D, et al. Integrated strategy for the production of therapeutic retroviral vectors. Hum Gene Ther 2011;22:370-9.
[86] Zhou H, Liu ZG, Sun ZW, et al. Generation of stable cell lines by site-specific integration of transgenes into engineered Chinese hamster ovary strains using an FLP-FRT system. J Biotechnol 2010;147:122-9.
[87] Kaufman WL, Kocman I, Agrawal V, et al. Homogeneity and persistence of transgene expression by omitting antibiotic selection in cell line isolation. Nucleic Acids Res 2008;36:e111.
[88] Kawabe Y, Makitsubo H, Kameyama Y, et al. Repeated integration of antibody genes into a pre-selected chromosomal locus of CHO cells using an accumulative site-specific gene integration system. Cytotechnology 2012;64:267-79.
[89] Crawford Y, Zhou M, Hu Z, et al. Fast identification of reliable hosts for targeted cell line development from a limited-genome screening using combined phiC31 integrase and CRE-Lox technologies. Biotechnology Prog 2013;29(5):1307-15.
[90] Kito M, Itami S, Fukano Y, et al. Construction of engineered CHO strains for high-level production of recombinant proteins. Appl Microbiol Biotechnol 2002;60:442-8.
[91] Huang Y, Li Y, Wang YG, et al. An efficient and targeted gene integration system for high-level antibody expression. J Immunol Methods 2007;322(1-2):28-39.
[92] Kim MS, Lee GM. Use of Flp-mediated cassette exchange in the development of a CHO cell line stably producing erythropoietin. J Microbiol Biotechnol 2008;18:1342-51.
[93] Wilke S, Groebe L, Maffenbeier V, et al. Streamlining homogeneous glycoprotein production for biophysical and structural applications by targeted cell line development. PLoS One 2011;6:e27829.
[94] Wiberg FC, Rasmussen SK, Frandsen TP, et al. Production of target-specific recombinant human polyclonal antibodies in mammalian cells. Biotechnol Bioeng 2006;94:396-405.
[95] Coroadinha AS, Schucht R, Gama-Norton L, et al. The use of recombinase mediated cassette exchange in retroviral vector producer cell lines: predictability and efficiency by transgene exchange. J Biotechnol 2006;124:457-68.
[96] Gama-Norton L, Herrmann S, Schucht R, et al. Retroviral vector performance in defined chromosomal Loci of modular packaging cell lines. Hum Gene Ther 2010;21:979-91.
[97] Mao J, Barrow J, McMahon J, et al. An ES cell system for rapid, spatial and temporal analysis of gene function in vitro and in vivo. Nucleic Acids Res 2005;33:e155.

[98] Ortiz-Urda S, Thyagarajan B, Keene DR, et al. Stable nonviral genetic correction of inherited human skin disease. Nat Med 2002;8:1166-70.
[99] Olivares EC, Hollis RP, Chalberg TW, et al. Site-specific genomic integration produces therapeutic Factor IX levels in mice. Nat Biotechnol 2002;20:1124-8.
[100] Held PK, Olivares EC, Aguilar CP, et al. In vivo correction of murine hereditary tyrosinemia type I by phiC31 integrase-mediated gene delivery. Mol Ther 2005;11:399-408.
[101] Beard C, Hochedlinger K, Plath K, et al. Efficient method to generate single-copy transgenic mice by site-specific integration in embryonic stem cells. Genesis 2006;44:23-8.
[102] Perez-Pinera P, Ousterout DG, Brown MT, et al. Gene targeting to the ROSA26 locus directed by engineered zinc finger nucleases. Nucleic Acids Res 2012;40:3741-52.
[103] Shmerling D, Danzer CP, Mao X, et al. Strong and ubiquitous expression of transgenes targeted into the beta-actin locus by Cre/lox cassette replacement. Genesis 2005;42:229-35.
[104] Wei QX, Odell AF, van der Hoeven F, et al. Rapid derivation of genetically related mutants from embryonic cells harboring a recombinase-specific Trp53 platform. Cell Cycle 2011;10:1261-70.
[105] Wallace HA, Marques-Kranc F, Richardson M, et al. Manipulating the mouse genome to engineer precise functional syntenic replacements with human sequence. Cell 2007;128:197-209.
[106] Buchholz F, Angrand PO, Stewart AF. A simple assay to determine the functionality of Cre or FLP recombination targets in genomic manipulation constructs. Nucleic Acids Res 1996;24:3118-9.
[107] Buchholz F, Angrand PO, Stewart AF. Improved properties of FLP recombinase evolved by cycling mutagenesis. Nat Biotechnol 1998;16:657-62.
[108] Raymond CS, Soriano P. High-efficiency FLP and PhiC31 site-specific recombination in mammalian cells. PLoS One 2007;2:e162.
[109] Hirsch MR, d'Autreaux F, Dymecki SM, et al. A Phox2b::FLPo transgenic mouse line suitable for intersectional genetics. Genesis 2013;51:506-14.
[110] Kranz A, Fu J, Duerschke K, et al. An improved Flp deleter mouse in C57Bl/6 based on Flpo recombinase. Genesis 2010;48:512-20.
[111] Wu Y, Wang C, Sun H, et al. High-efficient FLPo deleter mice in C57BL/6J background. PLoS One 2009;4:e8054.
[112] Turan S, Galla M, Ernst E, et al. Recombinase-mediated cassette exchange (RMCE): traditional concepts and current challenges. J Mol Biol 2011;407:193-221.
[113] Shimshek DR, Kim J, Hubner MR, et al. Codon-improved Cre recombinase (iCre) expression in the mouse. Genesis 2002;32:19-26.
[114] Sauer B, McDermott J. DNA recombination with a heterospecific Cre homolog identified from comparison of the pac-c1 regions of P1-related phages. Nucleic Acids Res 2004;32:6086-95.
[115] Minorikawa S, Nakayama M. Recombinase-mediated cassette exchange (RMCE) and BAC engineering via VCre/VloxP and SCre/SloxP systems. Biotechniques 2011;50:235-46.
[116] Suzuki E, Nakayama M. VCre/VloxP and SCre/SloxP: new site-specific recombination systems for genome engineering. Nucleic Acids Res 2011;39:e49.
[117] Karimova M, Abi-Ghanem J, Berger N, et al. Vika/vox, a novel efficient and specific Cre/loxP-like site-specific recombination system. Nucleic Acids Res 2013;41:e37.
[118] Preuss E, Treschow A, Newrzela S, et al. TK.007: A novel, codon-optimized HSVtk(A168H) mutant for suicide gene therapy. Hum Gene Ther 2010;21:929-41.
[119] Loonstra A, Vooijs M, Beverloo HB, et al. Growth inhibition and DNA damage induced by Cre recombinase in mammalian cells. Proc Natl Acad Sci USA 2001;98:9209-14.
[120] Baba Y, Nakano M, Yamada Y, et al. Practical range of effective dose for Cre recombinase-expressing recombinant adenovirus without cell toxicity in mammalian cells. Microbiol Immunol 2005;49:559-70.

[121] Ehrhardt A, Engler JA, Xu H, et al. Molecular analysis of chromosomal rearrangements in mammalian cells after phiC31-mediated integration. Hum Gene Ther 2006;17:1077-94.
[122] Thyagarajan B, Guimaraes MJ, Groth AC, et al. Mammalian genomes contain active recombinase recognition sites. Gene 2000;244:47-54.
[123] Semprini S, Troup TJ, Kotelevtseva N, et al. Cryptic loxP sites in mammalian genomes: genome-wide distribution and relevance for the efficiency of BAC/PAC recombineering techniques. Nucleic Acids Res 2007;35:1402-10.
[124] Thomas KR, Capecchi MR. Site-directed mutagenesis by gene targeting in mouse embryo-derived stem cells. Cell 1987;51:503-12.
[125] Doetschman T, Gregg RG, Maeda N, et al. Targetted correction of a mutant HPRT gene in mouse embryonic stem cells. Nature 1987;330:576-8.
[126] Capecchi MR. Gene targeting in mice: functional analysis of the mammalian genome for the twenty-first century. Nat Rev Genet 2005;6:507-12.
[127] Babinet C, Cohen-Tannoudji M. Genome engineering via homologous recombination in mouse embryonic stem (ES) cells: an amazingly versatile tool for the study of mammalian biology. An Acad Bras de Cienc 2001;73:365-83.
[128] Cervantes RB, Stringer JR, Shao C, et al. Embryonic stem cells and somatic cells differ in mutation frequency and type. Proc Natl Acad Sci USA 2002;99:3586-90.
[129] Hasty P, Rivera-Perez J, Bradley A. The length of homology required for gene targeting in embryonic stem cells. Mol Cell Biol 1991;11:5586-91.
[130] Templeton NS, Roberts DD, Safer B. Efficient gene targeting in mouse embryonic stem cells. Gene Ther 1997;4:700-9.
[131] Sedivy JM, Sharp PA. Positive genetic selection for gene disruption in mammalian cells by homologous recombination. Proc Natl Acad Sci USA 1989;86:227-31.
[132] te Riele H, Maandag ER, Berns A. Highly efficient gene targeting in embryonic stem cells through homologous recombination with isogenic DNA constructs. Proc Natl Acad Sci USA 1992;89:5128-32.
[133] Lu ZH, Books JT, Kaufman RM, et al. Long targeting arms do not increase the efficiency of homologous recombination in the beta-globin locus of murine embryonic stem cells. Blood 2003;102:1531-3.
[134] Porteus M. Using homologous recombination to manipulate the genome of human somatic cells. Biotechnol Gent Eng Rev 2007;24:195-212.
[135] Tichy ED, Pillai R, Deng L, et al. Mouse embryonic stem cells, but not somatic cells, predominantly use homologous recombination to repair double-strand DNA breaks. Stem Cells Dev 2010;19:1699-711.
[136] Smithies O, Gregg RG, Boggs SS, et al. Insertion of DNA sequences into the human chromosomal beta-globin locus by homologous recombination. Nature 1985;317:230-4.
[137] Brown JP, Wei W, Sedivy JM. Bypass of senescence after disruption of p21CIP1/WAF1 gene in normal diploid human fibroblasts. Science 1997;277:831-4.
[138] Bunz F, Dutriaux A, Lengauer C, et al. Requirement for p53 and p21 to sustain G2 arrest after DNA damage. Science 1998;282:1497-501.
[139] Buerstedde JM, Takeda S. Increased ratio of targeted to random integration after transfection of chicken B cell lines. Cell 1991;67:179-88.
[140] Storck S, Thiry M, Bouvet P. Conditional knockout of nucleolin in DT40 cells reveals the functional redundancy of its RNA-binding domains. Biol Cell 2009;101:153-67.
[141] Lee S, Liu X, Takeda S, et al. Genotoxic potentials and related mechanisms of bisphenol A and other bisphenol compounds: a comparison study employing chicken DT40 cells. Chemosphere 2013;93:434-40.

[142] Qing Y, Yamazoe M, Hirota K, et al. The epistatic relationship between BRCA2 and the other RAD51 mediators in homologous recombination. PLoS genetics 2011;7:e1002148.
[143] Thomas BJ, Rothstein R. Elevated recombination rates in transcriptionally active DNA. Cell 1989;56:619-30.
[144] Russell DW, Hirata RK. Human gene targeting by viral vectors. Nat Genet 1998;18:325-30.
[145] Chamberlain JR, Deyle DR, Schwarze U, et al. Gene targeting of mutant COL1A2 alleles in mesenchymal stem cells from individuals with osteogenesis imperfecta. Mol Ther 2008;16:187-93.
[146] Miller DG, Wang PR, Petek LM, et al. Gene targeting in vivo by adeno-associated virus vectors. Nat Biotechnol 2006;24:1022-6.
[147] Chamberlain JR, Schwarze U, Wang PR, et al. Gene targeting in stem cells from individuals with osteogenesis imperfecta. Science 2004;303:1198-201.
[148] Rouet P, Smih F, Jasin M. Introduction of double-strand breaks into the genome of mouse cells by expression of a rare-cutting endonuclease. Mol Cell Biol 1994;14:8096-106.
[149] Choulika A, Perrin A, Dujon B, et al. Induction of homologous recombination in mammalian chromosomes by using the I-SceI system of Saccharomyces cerevisiae. Mol Cell Biol 1995;15:1968-73.
[150] Elliott B, Richardson C, Winderbaum J, et al. Gene conversion tracts from double-strand break repair in mammalian cells. Mol Cell Biol 1998;18:93-101.
[151] Cohen-Tannoudji M, Robine S, Choulika A, et al. I-SceI-induced gene replacement at a natural locus in embryonic stem cells. Mol Cell Biol 1998;18:1444-8.
[152] Donoho G, Jasin M, Berg P. Analysis of gene targeting and intrachromosomal homologous recombination stimulated by genomic double-strand breaks in mouse embryonic stem cells. Mol Cell Biol 1998;18:4070-8.
[153] Porteus MH, Baltimore D. Chimeric nucleases stimulate gene targeting in human cells. Science 2003;300:763.
[154] Orlando SJ, Santiago Y, DeKelver RC, et al. Zinc-finger nuclease-driven targeted integration into mammalian genomes using donors with limited chromosomal homology. Nucleic Acids Res 2010;38:e152.
[155] Burt A, Koufopanou V. Homing endonuclease genes: the rise and fall and rise again of a selfish element. Curr Opin Genet Dev 2004;14:609-15.
[156] Hartlerode A, Odate S, Shim I, et al. Cell cycle-dependent induction of homologous recombination by a tightly regulated I-SceI fusion protein. PLoS One 2011;6:e16501.
[157] Bellaiche Y, Mogila V, Perrimon N. I-SceI endonuclease, a new tool for studying DNA double-strand break repair mechanisms in Drosophila. Genetics 1999;152:1037-44.
[158] Szczepek M, Brondani V, Buchel J, et al. Structure-based redesign of the dimerization interface reduces the toxicity of zinc-finger nucleases. Nat Biotechnol 2007;25:786-93.
[159] Fenina M, Simon-Chazottes D, Vandormael-Pournin S, et al. I-SceI-mediated double-strand break does not increase the frequency of homologous recombination at the Dct locus in mouse embryonic stem cells. PLoS One 2012;7:e39895.
[160] Valton J, Daboussi F, Leduc S, et al. 5'-Cytosine-phosphoguanine (CpG) methylation impacts the activity of natural and engineered meganucleases. J Biol Chem 2012;287:30139-50.
[161] Gouble A, Smith J, Bruneau S, et al. Efficient in toto targeted recombination in mouse liver by meganuclease-induced double-strand break. J Gene Med 2006;8:616-22.
[162] Seligman LM, Chisholm KM, Chevalier BS, et al. Mutations altering the cleavage specificity of a homing endonuclease. Nucleic Acids Res 2002;30:3870-9.
[163] Rosen LE, Morrison HA, Masri S, et al. Homing endonuclease I-CreI derivatives with novel DNA target specificities. Nucleic Acids Res 2006;34:4791-800.

[164] Segal DJ, Meckler JF. Genome engineering at the dawn of the golden age. Annu Rev Genomics Hum Genet 2013;14:135-58.

[165] Choo Y, Klug A. Selection of DNA binding sites for zinc fingers using rationally randomized DNA reveals coded interactions. Proc Natl Acad Sci USA 1994;91:11168-72.

[166] Rebar EJ, Pabo CO. Zinc finger phage: affinity selection of fingers with new DNA-binding specificities. Science 1994;263:671-3.

[167] Davis D, Stokoe D. Zinc finger nucleases as tools to understand and treat human diseases. BMC Medicine 2010;8:42.

[168] Wilson KA, McEwen AE, Pruett-Miller SM, et al. Expanding the Repertoire of Target Sites for Zinc Finger Nuclease-mediated Genome Modification. Mol Ther Nucleic Acids 2013;2:e88.

[169] Urnov FD, Rebar EJ, Holmes MC, et al. Genome editing with engineered zinc finger nucleases. Nat Rev Genet 2010;11:636-46.

[170] Klug A. The discovery of zinc fingers and their applications in gene regulation and genome manipulation. Annu Rev Biochem 2010;79:213-31.

[171] Wirt SE, Porteus MH. Development of nuclease-mediated site-specific genome modification. Curr Opin Immunol 2012;24:609-16.

[172] Gabriel R, Lombardo A, Arens A, et al. An unbiased genome-wide analysis of zinc-finger nuclease specificity. Nat Biotechnol 2011;29:816-23.

[173] Pattanayak V, Ramirez CL, Joung JK, et al. Revealing off-target cleavage specificities of zinc-finger nucleases by in vitro selection. Nat Meth 2011;8:765-70.

[174] Gupta A, Christensen RG, Rayla AL, et al. An optimized two-finger archive for ZFN-mediated gene targeting. Nat Meth 2012;9:588-90.

[175] Kim JS, Lee HJ, Carroll D. Genome editing with modularly assembled zinc-finger nucleases. Nat Meth 2010;7:91.

[176] Miller JC, Holmes MC, Wang J, et al. An improved zinc-finger nuclease architecture for highly specific genome editing. Nat Biotechnol 2007;25:778-85.

[177] Doyon Y, Vo TD, Mendel MC, et al. Enhancing zinc-finger-nuclease activity with improved obligate heterodimeric architectures. Nat Meth 2011;8:74-9.

[178] Guo J, Gaj T, Barbas CF, 3rd. Directed evolution of an enhanced and highly efficient FokI cleavage domain for zinc finger nucleases. J Mol Biol 2010;400:96-107.

[179] Gaj T, Guo J, Kato Y, et al. Targeted gene knockout by direct delivery of zinc-finger nuclease proteins. Nat Meth 2012;9:805-7.

[180] Pruett-Miller SM, Reading DW, Porter SN, et al. Attenuation of zinc finger nuclease toxicity by small-molecule regulation of protein levels. PLoS Genet 2009;5:e1000376.

[181] McConnell Smith A, Takeuchi R, Pellenz S, et al. Generation of a nicking enzyme that stimulates site-specific gene conversion from the I-AniI LAGLIDADG homing endonuclease. Proc Natl Acad Sci USA 2009;106:5099-104.

[182] Kim E, Kim S, Kim DH, et al. Precision genome engineering with programmable DNA-nicking enzymes. Genome Res 2012;22:1327-33.

[183] Wang J, Friedman G, Doyon Y, et al. Targeted gene addition to a predetermined site in the human genome using a ZFN-based nicking enzyme. Genome Res 2012;22:1316-26.

[184] Ramirez CL, Certo MT, Mussolino C, et al. Engineered zinc finger nickases induce homology-directed repair with reduced mutagenic effects. Nucleic Acids Res 2012;40:5560-8.

[185] Boch J, Scholze H, Schornack S, et al. Breaking the code of DNA binding specificity of TAL-type III effectors. Science 2009;326:1509-12.

[186] Moscou MJ, Bogdanove AJ. A simple cipher governs DNA recognition by TAL effectors. Science 2009;326:1501.

[187] Deng D, Yan C, Pan X, et al. Structural basis for sequence-specific recognition of DNA by TAL effectors. Science 2012;335:720-3.

[188] Doyle EL, Booher NJ, Standage DS, et al. TAL Effector-Nucleotide Targeter (TALE-NT) 2.0: tools for TAL effector design and target prediction. Nucleic Acids Res 2012;40:W117-22.
[189] Cermak T, Doyle EL, Christian M, et al. Efficient design and assembly of custom TALEN and other TAL effector-based constructs for DNA targeting. Nucleic Acids Res 2011;39:e82.
[190] Moore FE, Reyon D, Sander JD, et al. Improved somatic mutagenesis in zebrafish using transcription activator-like effector nucleases (TALENs). PLoS One 2012;7:e37877.
[191] Sun N, Zhao H. Transcription activator-like effector nucleases (TALENs): a highly efficient and versatile tool for genome editing. Biotechnol Bioeng 2013;110:1811-21.
[192] Schmid-Burgk JL, Schmidt T, Kaiser V,et al. A ligation-independent cloning technique for high-throughput assembly of transcription activator-like effector genes. Nat Biotechnol 2013;31:76-81.
[193] Reyon D, Maeder ML, Khayter C, et al. Engineering customized TALE nucleases (TALENs) and TALE transcription factors by fast ligation-based automatable solid-phase high-throughput (FLASH) assembly. Curr Protoc Mol Bio 2013;Chapter 12:Unit 12.16.
[194] Briggs AW, Rios X, Chari R, et al. Iterative capped assembly: rapid and scalable synthesis of repeat-module DNA such as TAL effectors from individual monomers. Nucleic Acids Res 2012;40:e117.
[195] Sanjana NE, Cong L, Zhou Y, et al. A transcription activator-like effector toolbox for genome engineering. Nat Protoc 2012;7:171-92.
[196] Morbitzer R, Elsaesser J, Hausner J, et al. Assembly of custom TALE-type DNA binding domains by modular cloning. Nucleic Acids Res 2011;39:5790-9.
[197] Hockemeyer D, Soldner F, Beard C, et al. Efficient targeting of expressed and silent genes in human ESCs and iPSCs using zinc-finger nucleases. Nat Biotechnol 2009;27:851-7.
[198] Mahfouz MM, Li L, Shamimuzzaman M, et al. De novo-engineered transcription activator-like effector (TALE) hybrid nuclease with novel DNA binding specificity creates double-strand breaks. Proc Natl Acad Sci USA 2011;108:2623-8.
[199] Li Y, Moore R, Guinn M, et al. Transcription activator-like effector hybrids for conditional control and rewiring of chromosomal transgene expression. Sci Rep 2012;2:897.
[200] Holkers M, Maggio I, Liu J, et al. Differential integrity of TALE nuclease genes following adenoviral and lentiviral vector gene transfer into human cells. Nucleic Acids Res 2013;41:e63.
[201] Jinek M, Chylinski K, Fonfara I, et al. A programmable dual-RNA-guided DNA endonuclease in adaptive bacterial immunity. Science 2012;337:816-21.
[202] Horvath P, Barrangou R. CRISPR/Cas, the immune system of bacteria and archaea. Science 2010;327:167-70.
[203] Ran FA, Hsu PD, Wright J, et al. Genome engineering using the CRISPR-Cas9 system. Nat Protoc 2013;8:2281-308.
[204] Cong L, Ran FA, Cox D, et al. Multiplex genome engineering using CRISPR/Cas systems. Science 2013;339:819-23.
[205] Yang H, Wang H, Shivalila CS, et al. One-Step Generation of Mice Carrying Reporter and Conditional Alleles by CRISPR/Cas-Mediated Genome Engineering. Cell 2013;154:1370-9.
[206] Wang H, Yang H, Shivalila CS, et al. One-step generation of mice carrying mutations in multiple genes by CRISPR/Cas-mediated genome engineering. Cell 2013;153:910-8.
[207] Cho SW, Kim S, Kim JM, et al. Targeted genome engineering in human cells with the Cas9 RNA-guided endonuclease. Nat Biotechnol 2013;31:230-2.
[208] Mali P, Yang L, Esvelt KM, et al. RNA-guided human genome engineering via Cas9. Science 2013;339:823-6.
[209] Hwang WY, Fu Y, Reyon D, et al. Efficient genome editing in zebrafish using a CRISPR-Cas system. Nat Biotechnol 2013;31:227-9.

[210] DiCarlo JE, Norville JE, Mali P, et al. Genome engineering in Saccharomyces cerevisiae using CRISPR-Cas systems. Nucleic Acids Res 2013;41:4336-43.

[211] Jiang W, Bikard D, Cox D, et al. RNA-guided editing of bacterial genomes using CRISPR-Cas systems. Nat Biotechnol 2013;31:233-9.

[212] Fu Y, Foden JA, Khayter C, et al. High-frequency off-target mutagenesis induced by CRISPR-Cas nucleases in human cells. Nat Biotechnol 2013;31:822-6.

[213] Cradick TJ, Fine EJ, Antico CJ, et al. CRISPR/Cas9 systems targeting beta-globin and CCR5 genes have substantial off-target activity. Nucleic Acids Res 2013;41:9584-92.

[214] Ran FA, Hsu PD, Lin CY, et al. Double nicking by RNA-guided CRISPR Cas9 for enhanced genome editing specificity. Cell 2013;154:1380-9.

[215] Soldner F, Laganiere J, Cheng AW, et al. Generation of isogenic pluripotent stem cells differing exclusively at two early onset Parkinson point mutations. Cell 2011;146:318-31.

[216] Yusa K, Rashid ST, Strick-Marchand H, et al. Targeted gene correction of alpha1-antitrypsin deficiency in induced pluripotent stem cells. Nature 2011;478:391-4.

[217] Zou J, Mali P, Huang X, et al. Site-specific gene correction of a point mutation in human iPS cells derived from an adult patient with sickle cell disease. Blood 2011;118:4599-608.

[218] Santiago Y, Chan E, Liu PQ, et al. Targeted gene knockout in mammalian cells by using engineered zinc-finger nucleases. Proc Natl Acad Sci USA 2008;105:5809-14.

[219] Malphettes L, Freyvert Y, Chang J, et al. Highly efficient deletion of FUT8 in CHO cell lines using zinc-finger nucleases yields cells that produce completely nonfucosylated antibodies. Biotechnol Bioeng 2010;106:774-83.

[220] Sealover NR, Davis AM, Brooks JK, et al. Engineering Chinese hamster ovary (CHO) cells for producing recombinant proteins with simple glycoforms by zinc-finger nuclease (ZFN)-mediated gene knockout of mannosyl (alpha-1,3-)-glycoprotein beta-1,2-N-acetylglucosaminyltransferase (Mgat1). J Biotechnol 2013;167:24-32.

[221] Liu PQ, Chan EM, Cost GJ, et al. Generation of a triple-gene knockout mammalian cell line using engineered zinc-finger nucleases. Biotechnol Bioeng 2010;106:97-105.

[222] Cost GJ, Freyvert Y, Vafiadis A, et al. BAK and BAX deletion using zinc-finger nucleases yields apoptosis-resistant CHO cells. Biotechnol Bioeng 2010;105:330-40.

[223] Paques F, Duchateau P. Meganucleases and DNA double-strand break-induced recombination: perspectives for gene therapy. Curr Gene Rher 2007;7:49-66.

[224] Papapetrou EP, Lee G, Malani N, et al. Genomic safe harbors permit high beta-globin transgene expression in thalassemia induced pluripotent stem cells. Nat Biotechnol 2011;29:73-8.

[225] Sadelain M, Papapetrou EP, Bushman FD. Safe harbours for the integration of new DNA in the human genome. Nat Rev Cancer 2012;12:51-8.

[226] Nunez E, Fu XD, Rosenfeld MG. Nuclear organization in the 3D space of the nucleus – cause or consequence? Curr Opin Genet Dev 2009;19:424-36.

[227] Nehlsen K, da Gama-Norton L, Schucht R, et al. Towards rational engineering of cells: Recombinant gene expression in defined chromosomal loci. BMC Proc 2011;5 Suppl 8:O6.

[228] Thomason L, Court DL, Bubunenko M, et al. Recombineering: genetic engineering in bacteria using homologous recombination. Curr Protoc Mol Biol 2007;Chapter 1:Unit 1 16.

[229] Copeland NG, Jenkins NA, Court DL. Recombineering: a powerful new tool for mouse functional genomics. Nat Rev Genet 2001;2:769-79.

[230] Narayanan K, Chen Q. Bacterial artificial chromosome mutagenesis using recombineering. J Biomed Biotechnol 2011;2011:971296.

[231] Muyrers JP, Zhang Y, Benes V, et al. Point mutation of bacterial artificial chromosomes by ET recombination. EMBO Rep 2000;1:239-43.

[232] Warming S, Costantino N, Court DL, et al. Simple and highly efficient BAC recombineering using galK selection. Nucleic Acids Res 2005;33:e36.
[233] Zhang Y, Muyrers JP, Rientjes J, et al. Phage annealing proteins promote oligonucleotide-directed mutagenesis in Escherichia coli and mouse ES cells. BMC Mol Biol 2003;4:1.
[234] Bird AW, Erler A, Fu J, et al. High-efficiency counterselection recombineering for site-directed mutagenesis in bacterial artificial chromosomes. Nat Meth 2012;9:103-9.
[235] Li MA, Turner DJ, Ning Z, et al. Mobilization of giant piggyBac transposons in the mouse genome. Nucleic Acids Res 2011;39:e148.
[236] Rostovskaya M, Fu J, Obst M, et al. Transposon-mediated BAC transgenesis in human ES cells. Nucleic Acids Res 2012;40:e150.
[237] Pulverer JE, Rand U, Lienenklaus S, et al. Temporal and spatial resolution of type I and III interferon responses in vivo. J Virol 2010;84:8626-38.
[238] Poser I, Sarov M, Hutchins JR, et al. BAC TransgeneOmics: a high-throughput method for exploration of protein function in mammals. Nat Meth 2008;5:409-15.
[239] Blaas L, Musteanu M, Grabner B, et al. The Use of Bacterial Artificial Chromosomes for Recombinant Protein Production in Mammalian Cell Lines Recombinant Gene Expression. In: Lorence A (ed): Springer New York; 2012:581-93.
[240] Blaas L, Musteanu M, Eferl R, et al. Bacterial artificial chromosomes improve recombinant protein production in mammalian cells. BMC Biotechnol 2009;9:3.
[241] Bian Q, Belmont AS. BAC TG-EMBED: one-step method for high-level, copy-number-dependent, position-independent transgene expression. Nucleic Acids Res 2010;38:e127.
[242] Kazuki Y, Oshimura M. Human artificial chromosomes for gene delivery and the development of animal models. Mol Ther 2011;19:1591-601.
[243] Kennard ML. Engineered mammalian chromosomes in cellular protein production: future prospects. Meth Mol Biol 2011;738:217-38.
[244] Kouprina N, Earnshaw WC, Masumoto H, et al. A new generation of human artificial chromosomes for functional genomics and gene therapy. Cell Mol Life Sci 2013;70:1135-48.
[245] Lufino MM, Edser PA, Wade-Martins R. Advances in high-capacity extrachromosomal vector technology: episomal maintenance, vector delivery, and transgene expression. Mol Ther 2008;16:1525-38.
[246] Mizuguchi H, Hosono T, Hayakawa T. Long-term replication of Epstein-Barr virus-derived episomal vectors in the rodent cells. FEBS Lett 2000;472:173-8.
[247] Ito S, Gotoh E, Ozawa S, et al. Epstein-Barr virus nuclear antigen-1 is highly colocalized with interphase chromatin and its newly replicated regions in particular. J Gen Virol 2002;83:2377-83.
[248] Hagedorn C, Wong SP, Harbottle R, Lipps HJ. Scaffold/matrix attached region-based nonviral episomal vectors. Hum Gene Ther 2011;22:915-23.
[249] Baiker A, Maercker C, Piechaczek C, et al. Mitotic stability of an episomal vector containing a human scaffold/matrix-attached region is provided by association with nuclear matrix. Nat Cell Biol 2000;2:182-4.
[250] Stehle IM, Postberg J, Rupprecht S, et al. Establishment and mitotic stability of an extra-chromosomal mammalian replicon. BMC Cell Biol 2007;8:33.
[251] Stehle IM, Scinteie MF, Baiker A, et al. Exploiting a minimal system to study the epigenetic control of DNA replication: the interplay between transcription and replication. Chromosome Res 2003;11:413-21.
[252] Jenke AC, Stehle IM, Herrmann F, et al. Nuclear scaffold/matrix attached region modules linked to a transcription unit are sufficient for replication and maintenance of a mammalian episome. Proc Natl Acad Sci USA 2004;101:11322-7.

[253] Piechaczek C, Fetzer C, Baiker A, et al. A vector based on the SV40 origin of replication and chromosomal S/MARs replicates episomally in CHO cells. Nucleic Acids Res 1999;27:426-8.
[254] Papapetrou EP, Ziros PG, Micheva ID, et al. Gene transfer into human hematopoietic progenitor cells with an episomal vector carrying an S/MAR element. Gene Ther 2006;13:40-51.
[255] Lufino MM, Poppleston AR, Cowley SA, et al. Episomal transgene expression in pluripotent stem cells. Methods Mol Biol 2011;767:369-87.
[256] Manzini SA, Vargiolu IM, Stehle ML, et al. Genetically modified pigs produced with a nonviral episomal vector. Proc Natl Acad Sci USA 2006;103(47):17672-7.
[257] Argyros O, Wong SP, Harbottle RP. Non-viral episomal modification of cells using S/MAR elements. Expert Opin Biol Ther 2011;11:1177-91.
[258] Hacker DL, Kiseljak D, Rajendra Y, et al. Polyethyleneimine-based transient gene expression processes for suspension-adapted HEK-293E and CHO-DG44 cells. Protein Expr Purif 2013;92:67-76.

3.5 Manipulation of Cell Growth, Metabolism and Product Quality Attributes

Hans Henning von Horsten, Karsten Winkler, and Volker Sandig

3.5.1 Introduction

Biopharmaceutical upstream product yield for a commercial manufacturing process correlates up to a limit of 5 g/L with the upstream-associated cost of goods per gram product [1]. Low yield upstream processes require large overall volumes for fermentation media and bioreactor capacity and therefore present a high capital burden that has triggered intensive research and development activities to overcome limitations of upstream processing technology [1] and the intrinsic limitations of pharmaceutical producer clones. The achievement of maximum product titres has long been the center point of cell line and upstream process engineering [2]. Since upstream product concentration per unit volume is a function of average specific productivity and the integral of viable cell density (IVCD), these 2 process-determining parameters have become an area of significant focus for process developers in the biomanufacturing arena. By 2010, advances in mammalian cell culture technology have already increased product titers for certain therapeutic antibodies well above the 5 g/L benchmark with some companies reporting 10 to 13 g/L for extended culture durations. [1, 3]. The dramatic successes in product titre improvement have shifted the attention of bioprocess development towards monitoring and achieving desired quality attributes of the manufactured product. Therefore, the improvements in cell specific productivity per unit time have also triggered a renewed interest in continuous bioprocessing [4, 5]. While continuous bioprocessing has long been applied to the manufacturing of degradation prone products, high cell-specific productivities that can now be achieved with top level producer cell lines enable commercially viable upstream processes even for those proteins that are stable enough to survive standard fed-batch processing. Since product quality is a reciprocal function of residence time in the bioreactor, it is highly desirable to minimize this latter parameter by switching to a continuous mode of upstream operations. Another driving force behind this shift of focus towards process impact on product quality attributes is the revision of the quality guideline ICH Q8 in the wake of the US Food and Drug Administration's Quality-by-Design (QbD) initiative. The QbD initiative, a joint pilot program between the US FDA and the biopharma industry, is aimed at achieving an improved understanding of the physical or chemical changes known to affect safety or efficacy of biotherapeutics and how these critical quality attributes (CQAs) are affected by the parameters of the production process [6]. The ultimate goal of a QbD-based manufacturing approach is to design production processes in such a way that product CQAs can be consistently controlled within predefined acceptance limits [7-9].

Aside from the relevant process parameters, the intrinsic properties of a pharmaceutical producer clone are the dominant actuating variables that affect product titre and quality. This chapter introduces and summarizes the different cell line engineering approaches aiming at high titres and improved product quality.

3.5.2 Cell Line Properties affecting Product Titer and Quality

Average specific productivity (Q_p) and the integral of viable cell density (IVCD) over process duration are the 2 parameters that determine the upstream product concentration per unit volume in a bioreactor run. Both, IVCD and average specific productivity observed for a given producer clone, depend not only on the upstream environmental conditions in the bioreactor but also on the intrinsic features of the particular producer clone deployed for production.

3.5.2.1 IVCD

Maintaining the viability of cell cultures over the period of time required for upstream processing remains one of the major challenges in biomanufacturing. IVCD represents the cumulative area under the growth curve that results from plotting viable cell density against time in culture [10, 11]. Under the presumption of constant cell specific productivities, IVCD becomes the determining factor for the finally achieved volumetric yield. Cell death during batch and fed-batch culture can occur in response to a wide spectrum of stimuli, ranging from agitation induced insults to perturbations of the bioreactor environment that result in suboptimal growth and viability conditions. The latter changes could be nutrient limitation and depletion, accumulation of toxic metabolites and byproducts, pH-fluctuations, and perturbations of the oxygen supply [12]. One of the parameters directly related to IVCD is peak viable cell density. In a given environment this parameter is unique to a specific producer clone and several intrinsic features of the clone determine its performance in terms of its maximum achievable viable cell density and IVCD. This clone intrinsic growth performance is likely to be associated with the cells energy metabolism and ability to efficiently utilize available nutrients.

3.5.2.2 Specific Productivity (Q_p)

Average specific productivity is calculated by dividing the arithmetic mean of daily product concentrations per production run by the cumulative number of viable cells per run. For clone ranking purposes, it is common practice to determine average specific productivity only for the exponential phase of the growth curve where cell growth and viability are most robust and where the secreted product is not yet exposed to degrading enzymes that originate from dying cells. Specific productivity is cell cycle-

and growth phase-associated [13, 14]. This is not surprising since the protein synthesis and secretion machinery is regulated at multiple stages and competes with mitosis and cell growth processes for the limited intracellular energy. Pharmaceutical producer clones that have been selected to produce high levels of heterologous mRNA experience bottlenecks at several rate-limiting steps within the protein expression and secretion machinery [15] some of which may be controlled by the cell cycle [14, 16]. The G1 resting phase of the cell cycle has been linked to the highest specific productivity. Increasing the percentage of cells in the G1 phase typically leads to a concomitant increase of Q_p [13, 17]. Induction of G1 growth arrest can be achieved by mild hypothermic temperature shift, addition of chemicals such as DMSO, or by overexpression of the cell cycle inhibitory proteins p21Cip1 or p27Kip1 [17]. The underlying mechanism of all these strategies is the up-regulation of cyclin-dependent kinase inhibitors [17]. Recently, Clarke et al. have used gene expression profiles to build a predictive model of CHO productivity in bioprocess cell culture [18].

Engineered mammalian cells for industrial bioprocessing can suffer from cell line instability and a concomitant loss of specific productivity. This phenomenon appears to be rather common for Chinese hamster ovary (CHO) clones regardless of the gene expression system used [19]. A producer cell line that maintains stability of production is one of the most important prerequisites for the successful commercial manufacture of a therapeutic protein [20, 21]. Contract manufacturers define a production cell line as unstable if its product titer declines by more than 30 % over a period of time that covers at least 55 cumulative population doublings [21, 22]. Epigenetic methylation-induced transcriptional silencing of the CMV promoter driving transgene transcription and progressive loss of recombinant transgene copies in a proliferating CHO cell population as well as other yet unidentified mechanisms are thought to cause Q_p loss during extended maintenance culture [19]. The link between loss of transgene copy number, mRNA level, and Q_p-decline has not always been confirmed [20, 23] which emphasizes the notion that Q_p is a function of diverse cellular parameters affecting the complex expression and secretion machinery.

3.5.2.3 Interplay of Energy and Sugar Nucleotide Metabolism

Over the course of their life cells perform mechanical, transport, and chemical work, and they manage their energy resources to complete all these tasks by energy coupling. This means that they use the exergonic process energy gained from cellular respiration and central metabolism to drive endergonic chemical reactions such as the synthesis of glycoproteins from aminoacyl-t-RNA and sugar nucleotide monomers. The 2 dominant energy nutrients of mammalian cells in culture are glucose and glutamine [24], both of which ultimately feed into the TCA cycle and electron transport reactions. Under ideal standard conditions, each mole of glucose that is utilized by respiration can yield about 686 kcal of energy, which can then be used for endergonic biosynthesis processes [25]. But the conditions in cancer cells and many immortal cell lines differ from this ideal situation as these cells suffer from the Warburg effect [26, 27]. Even if oxygen supply is unlimited, these rapidly proliferating cells typically utilize glucose via glycolysis and secrete massive levels of lactate. In contrast to normal cells that can efficiently utilize glucose, most cancer cells and immortal cells including CHO cells show a blocked metabolic flux at the level of the pyruvate dehydrogenase complex, which effectively results in a decoupling of glycolysis and TCA cycle and electron transfer chain [28, 29].

Of the 3 enzymes that are major regulators of glycolytic flux (hexokinase, phosphofructokinase, and pyruvate kinase), hexokinase is also contributing to the Warburg effect from yet another angle: in rapidly growing cancer cells, mitochondrially-bound hexokinase II is highly overexpressed and has been identified as the driving force for the extremely high level of glycolysis observed in Warburg-affected tumor cells [30-32]. The dependence of hexokinase on mitochondrial ATP indicates that glycolysis can only work properly if the cells have access to other sources of energy besides glucose. These alternative energy molecules must feed into anaplerotic pathways which generate ATP within the mitochondrial TCA cycle. Indeed, glutamine feeding is required to maintain the TCA cycle activity while the combined effects of pyruvate dehydrogenase block and excessive hexokinase activity in Warburg-affected cancer cells results in an accumulation of pyruvate which is then directly converted into lactate [33]. Since aerobic glycolysis is a very inefficient way to generate ATP, CHO cells consume glucose very rapidly as long as this source of energy is available [34]. The activity of the glycolytic key enzyme phosphofructokinase 1 (PFK-1) is the last step of the energetically expensive preparatory phase and initiates the pay-off phase of glycolysis where an excess of ATP energy equivalents is generated. Metabolic block of PFK-1 can cause an accumulation of glucose-6-phosphate which in turn can be utilized by the pentose phosphate pathway to provide cellular NADPH. A metabolic block of PFK-1 activity is mediated by its allosteric inhibitor fructose 2,6-bisphosphate, a metabolite synthesized by the bifunctional enzyme phosphofructokinase 2/fructose-2,6-bisphosphatase [35, 36]. Overexpression of TIGAR (TP53-induced glycolysis and apoptosis regulator), a protein carrying such a phosphofructokinase/fructose-2,6-bisphosphatase activity has been observed in several cancer cells [37, 38]. TIGAR activity causes an accumulation of the PFK-1 allosteric inhibitor fructose-2,6-bisphosphate and in turn causes an accumulation of glucose-6-phosphate and increases the activity of the pentose phosphate cycle [39-41]. This is beneficial for cancer cell survival, since NADPH derived from the pentose phosphate cycle is the major source of reducing power needed for quenching reactive oxygen species (ROS) produced during rapid cell proliferation [42]. Cytosolic sugar nucleotide biosynthesis is intrinsically linked to the cellular energy metabolism and strongly dependent on availability of glycolytic pathway intermediates [43]. As a consequence of increased glycolytic flux and pentose phosphate cycle activity, nutrient limitation causes important key intermediates of the glycolytic pathway to become less available as substrates for anabolic biosynthesis reactions including sugar nucleotide and oligosaccharide synthesis.

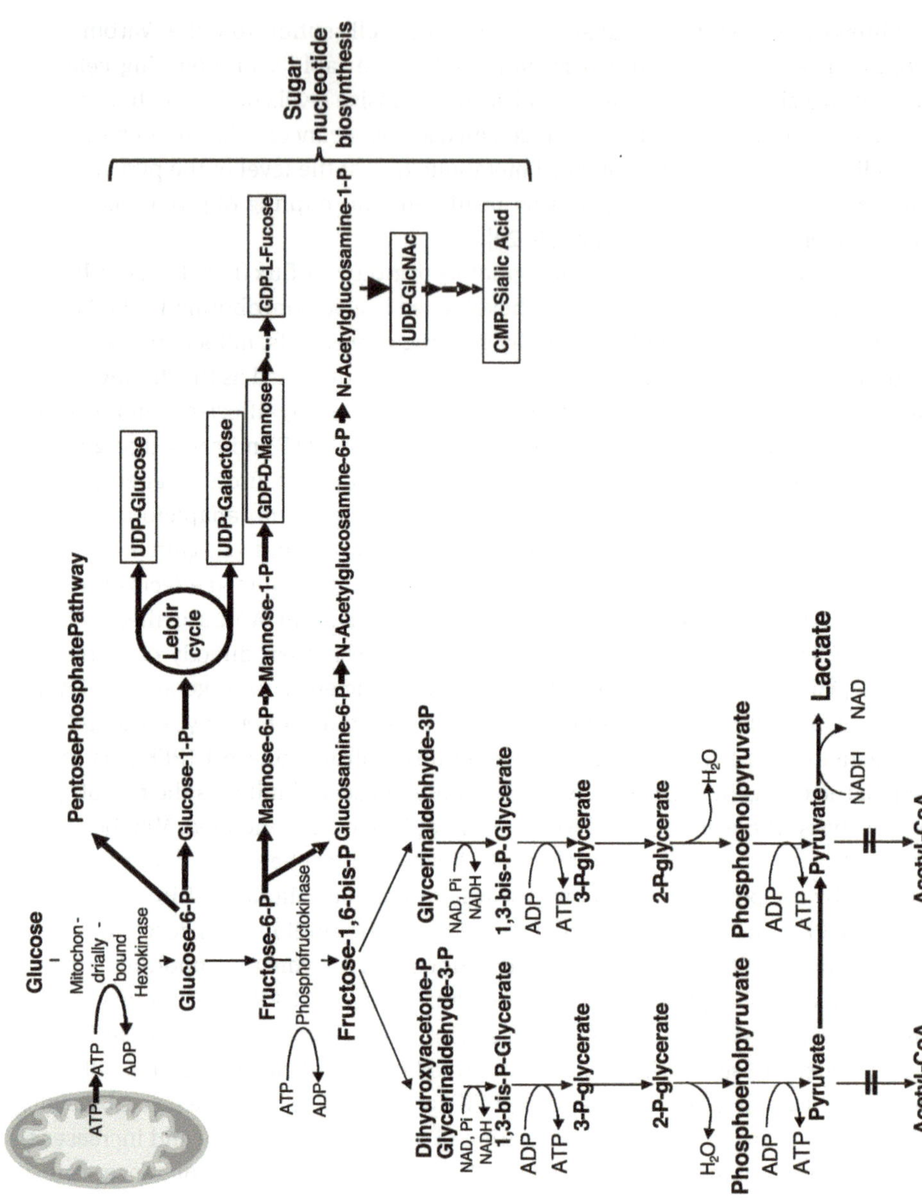

Figure 3.5.1: Sugar nucleotide metabolism as an interdependent metabolic sink for energy and redox metabolism intermediates.

Both, fructose-6-phosphate and glucose-6-phosphate, are glycolytic intermediates that contribute to the accumulation of the UDP-GlcNAc-, UDP-Gal- and GDP-mannose primary sugar nucleotide pools (Figure 3.5.1) Therefore, modulation of these pathways in particular a reduction of the glycolytic flux would be desirable.

The intracellular pools of these primary sugar nucleotides have a direct effect on the proportion of fucose and sialic acid precursors [43]. GDP-L-fucose and CMP-sialic acid are synthesized from these primary sugar nucleotides, and their efficient biosynthesis is therefore even more reliant on a sufficient supply of glycolytic pathway intermediates. The cytosolic UDP-GlcNAc-pool has been suggested to affect the antennarity of N-glycans [43]. Most recently, UDP-GlcNAc has been identified as a key substrate for the posttranslational attachment of an O-linked β-N-acetylglucosamine to serine 529 of PFK-1 in response to hypoxia [44]. O-GlcNAcylation at serine 529 blocks PFK-1 activity [44] and boosts pentose phosphate cycle and sugar nucleotide metabolism at the expense of downstream glycolytic flux. An elevated cytosolic pool of UDP-GlcNAc and efficient O-GlcNAcylation of PFK-1 may therefore cause an interruption of glycolytic lactate production. The cellular pool of GDP-mannose is likely to be a rate limiting substrate for the synthesis of the high-mannose dolichol pyrophosphate-linked oligosaccharide donor that is assembled on the cytoplasmic face of the rough endoplasmic reticulum. Considering that the oligosaccharyltransferase complex catalyses the initial step in the biosynthesis of N-linked glycoproteins and that it acts as a gatekeeper for the secretory pathway [45], it becomes obvious that a well-filled cytosolic GDP-mannose pool is an important prerequisite for the efficient assembly and secretion of glycoproteins. The high glycolytic flux and pentose phosphate cycle activity in Warburg-affected cancer cells are able to deplete the cytosolic GDP-mannose pool and therefore run counter to a seamlessly efficient glycoprotein synthesis and secretion. The observed rate limiting activity of dolichol pyrophosphate oligosacharyltransferase during glycoprotein secretion has led to the generation of a cell line modified to express lipid-linked oligosaccharide (LLO) flippase [46]. Aside from this, engineering of the sugar nucleotide metabolism has been used to achieve certain desired product quality attributes pertaining to glycosylation. These will be discussed in Section 3.5.3.3.

3.5.2.4 ER Processing and Quality Control

It has been observed in many cases that the mRNA level does not correlate well with the expression level of the respective protein. This also holds true for the expression of therapeutic proteins from pharmaceutical producer clones [23]. Bottlenecks downstream of translation have been discussed as a cause for this correlation mismatch [47, 48]. One of the well-known bottlenecks for achieving high product titers in cell culture supernatant is the ER processing and quality control machinery. Since the endoplasmic reticulum is a luminal compartment of the cell involved in folding and sorting of newly synthesized secretory cargo proteins, several quality control mechanisms are in place within the ER to ensure proper folding and secretion of expressed proteins. The lines of defense against secretion of misfolded proteins include the calnexin-calreticulin cycle, the unfolded protein response (UPR) [49], the ER-associated protein degradation (ERAD) [50], and the ER overload response [51]. Many engineering attempts have been made to interfere with ER quality control in order to improve volumetric yield for target proteins.

3.5.2.5 Golgi Processing and the Glycosylation Machinery

The Golgi processing and glycosylation machinery is not only relevant as a bottleneck to optimal secretion efficiency but also as the subcellular compartment that supports proper posttranslational modifications, in particular glycosylation.

All of these subcellular functional units have been subject to cell line engineering with the aim of improving either the total product output per unit time or the overall quality of the secreted product. These cell line engineering approaches will be discussed in the following section.

3.5.3 Cell Line Engineering Approaches

3.5.3.1 Cell Line Engineering Approaches Aimed at Achieving an Improved IVCD

Because of the tremendous impact of cellular growth and viability on volumetric product titer, several approaches have been pursued to engineer cells for improved bioreactor growth parameters. The majority of these approaches have focused on the suppression of apoptotic cell death and on improving metabolic utilization of chemical energy.

Apoptosis Blocker

Mammalian cells are very sensitive to environmental changes of their growth conditions. Cell death not only causes a decrease in overall viable cell density but the resulting loss of compartmental integrity can liberate degrading enzymes from the dying cells to accumulate in the cell culture supernatant. These leached degrading enzymes put the precious product at risk of being fragmented and modified in unfavourable terms. Blocking cell death during upstream fermentation is therefore critical for optimal bioprocessing. The 2 prominent causes of cell death in a bioreactor are necrosis and apoptosis. While necrosis is a form of cell death that is caused by factors external to the cell, e.g., mechanical trauma, toxin exposure etc., apoptosis is a programmed form of cell death that occurs in response to a number of different stimuli [52, 53]. In fed-batch cell culture processes, apoptosis is considered to be the prevalent form of cell death, accounting for about 80 % of CHO cell death in standard serum-free batch cultures [52-54]. Proapoptotic factors that may cause cells to undergo apoptosis include reactive oxygen species (ROS) that originate from mitochondrial activity and/or from the bioreactor oxygen sparger and autocrine factors that are released by the cells in response to increasing cell density [55]. Serum apparently has a protective effect on the cells. Mammalian cells grown in serum-free media are more prone to die by apoptosis [12, 55]. It has been suggested that serum may contain some effective antiapoptotic factors that protect cells from apoptosis during all stages of cell growth following nutrient depletion [52]. In order to limit the extent of apoptotic cell death during upstream fermentation runs, engineers have focused on improving the cell culture media and feed compositions and on genetic engineering approaches [52, 55, 56]. Approaches to engineer cells to become more resistant to proapoptotic triggers include coexpression of the antiapoptotic proteins Bcl-2, which was the first antiapoptic gene used for this purpose, and Bcl-xL, another antiapoptotic regulator [52]. Inducible overexpression of the antiapoptotic genes E1B-19K and Aven has resulted in increased culture performance and mAb titers for mammalian CHO cell cultures [57]. Since the specific productivities of cells expressing E1B-19K and Aven genes were comparable or slightly lower compared to the specific productivities of the parental hosts, the observed increases in volumetric productivity were primarily attributed to gains in IVCD over fed-batch duration [57]. Reports about the protective effect of silkworm hemolymph against apoptotic cell death have led to the discovery of the responsible 30K protein, which was found to be encoded by the Bombyx mori 30Kc6 gene [58]. Transient expression of 30Kc6 was reported to significantly suppress apoptotic cell death induced by serum deprivation and to increase IVCD and product titer by five- to tenfold [58]. Originally, the positive effects of 30Kc6 expression on IVCD and productivity in CHO cells were attributed to stable maintenance of mitochondrial activity [58]. Wang et al. demonstrated in 2012 that stable coexpression of 30Kc6 increased cell viability by 46.7 % and overall productivity of a monoclonal antibody by 3.4-fold [59]. The authors suggest that 30Kc6 expression inhibits apoptosis by repressing the Bax translocation from cytosol to mitochondria, which in turn downregulates apoptotic cascade responses including cytochrome c release and caspase-3 activation [59]. Majors et al. [60] demonstrated that overexpression of the E2F-1 cell cycle transcription factor in CHO DG44 antibody producer lines postpones entry into stationary phase in mammalian cells and causes an average 20 % increase of viable cell density over nontransfected control cell lines in batch culture. Unfortunately, this positive effect was no longer evident under commercially relevant fed-batch conditions [60]. Majors et al. employed the Ramos B cell line as a mutagenesis and selection tool for evolution of a novel gain-of-function mutant of the apoptosis-protective bcl-xL gene [61]. CHO cells engineered to overexpress the strongly active Bcl-x(L) Asp29Asn variant demonstrated a stronger resistance against apoptosis when compared with CHO cells expressing the wild-type Bcl-x(L) protein [61]. Provided that the elevated protection is maintained during fed-batch and scale-up, stable coexpression of Bcl-x(L) Asp29Asn seems to be a very promising route for antiapoptotic cell line engineering.

The most important application for overexpressing antiapoptotic genes is to suppress cell death during bioreactor processing and to improve IVCD over process duration. In spite of all the successes of antiapoptotic cell line engineering, a cell line-adapted approach to bioprocessing still remains the best means to achieve an optimal and sustained cell growth in upstream cell culture.

In the future, miRNA-based cell line engineering may turn out as an additional approach towards regulation of cellular growth and productivity of biopharmaceutical producer cell lines [62-64].

Metabolic Engineering

One of the major triggers of cell death and declining cell growth are nutrient limitation at the end of batch or fed-batch processing and accumulation of toxic metabolites. The 2 most common growth-inhibiting metabolites are lactate and ammonia. The strong lactate accumulation by certain producer clones that suffer from Warburg effect is undesirable, not only as it impacts cell growth but also as it causes a need for frequent alkaline titration during bioprocessing. These frequent temporary shifts in pH are likely to trigger deamidation of solvent-accessible asparagine and glutamine residues within the manufactured protein products. Elevated levels of accumulated ammonia also simultaneously affect cell viability and product quality. Elevated ammonia above the critical threshold level of ~5 mM must be avoided in order to maintain seamless glycoprocessing [65-67]. The negative impact of ammonia correlates with the presence of the unprotonated ammonia species, which accounts for the pH-dependency of ammonia toxicity. Local pH-maxima generated during alkaline titration required to compensate for strong lactate production also have an undesirable impact on product glycosylation. They contribute to the formation of unprotonated ammonia species, which can freely dissociate across membranes and are thus far more toxic than the ionized form of ammonia. Ammonia has been shown to collapse the Golgi-cytosol pH gradient and cause an alkaline shift within the acidic Golgi compartments. This changes both, pH-optima of glycosylation enzymes and also their compartmentally localized membrane association. One well-known consequence of ammonia accumulation within the Golgi compartment is the interruption of the N-glycan processing cascade at the stage of the Man(5GlcNAc(2)-structure. Ammonia apparently blocks the transfer of N-acetylglucosamine (GlcNAc) residues to Man(5GlcNAc(2), an enzymatic activity catalyzed by α-1,3-mannosyl-glycoprotein 2-β-N-acetylglucosaminyltransferase (GnT-I) [68]. In line with this, the GnTI defective CHO cell mutant Lec-1 produces homogeneous N-linked Man(5)GlcNAc(2) glycan structures [69]. However, ammonia poisoning does not seem to be the only root cause for the occurrence of elevated levels of undesirable Man(5)GlcNAc(2)-structures. Increases in undesirable Man(5)GlcNAc(2)-structures have also been linked to a combination of high basal and feed media osmolality and increased run duration [70].

Given the negative impact of ammonia on cell growth and product quality, cell line engineering approaches have focused on ammonia reduction for a long time. Mammalian cells contain only 2 major enzymes capable of utilizing free ammonia as a substrate – one is glutamine synthetase (GS) and the other is the urea-cycle-enzyme carbamoylphoshate synthetase I [71, 72]. Both of these enzymes have been used to engineer cells for increased resistance towards ammonia [71-74]. Glutamine synthetase (GS) has widely been used as an amplifiable selection marker in NSO- and CHO-based expression systems that were developed by scientists at Celltech Ltd., a company that was later acquired by Lonza [71, 75]. Recently, a CHO starter cell line carrying a biallelic knock-out of the endogenous glutamine synthetase gene was generated in order to improve the use of GS as a selection marker with additional ammonia reduction activity [76]. The other major metabolic engineering approach for pharmaceutical CHO cell lines focuses on carbon distribution between glycolysis and the TCA cycle and particularly on minimizing lactate accumulation. During fed-batch production the producer cells progress through multiple metabolic stages whereby the changing cellular metabolism during fed-batch growth is characterized by an inverse relationship between cell growth and target protein productivity [77]. The stationary phase that follows the exponential cellular growth typically shows an increased productivity for the heterologous protein product [77]. Unfortunately, the Warburg effect is most dominant during exponential growth and therefore this phase is typically accompanied by strong lactate production and minimal TCA-cycling [77, 78]. Nutrients in media and feeds seem to have a strong impact on lactate production. While glutamine has been shown to be effective in replenishing the anaplerotic flux, it has also been shown to contribute stronger than glucose to the overall lactate accumulation during exponential growth [77]. Interestingly, it was reported that CHO cells utilize glucose more effectively for anaplerotic replenishment during the stationary phase of cell growth as the Warburg block seems to be less prominent at this stage [77]. Dean and Reddy suggested that the magnitude of this metabolic shift at the stationary phase is important for achieving high viable cell densities and product titers. They also found that the 2 metabolic enzymes phosphoenolpyruvate carboxykinase (PEPCK1) and pyruvate kinase (PK) are differentially regulated during exponential and stationary phases [77]. Moreover, it had been shown that certain CHO cell clones are able to consume lactate, particularly as they enter the stationary phase [79]. Those lactate-consuming clones apparently also have a much better energy efficiency compared to lactate producing clones as they show a 6 times greater ATP production per total C-mol substrate consumed [80]. Zagari et al. did a comparative analysis of the mitochondrial membrane potential and oxygen consumption of lactate consumers and lactate producers and found a positive correlation between a strong mitochondrial oxidative metabolism and lactate consumption [79]. Templeton et al. found that energy was primarily generated through oxidative phosphorylation upon the switch from exponential growth to peak protein production in the stationary phase [78]. This switch was found to be associated with an elevated oxidative pentose phosphate pathway activity that alleviated the reliance of the cellular energy metabolism on lactate-promoting glycolytic flux [78]. Differential expression of metabolic enzymes and mitochondrial redox regulators in lactate consuming clones may provide new tools for metabolic engineering aimed at reducing lactate accumulation and improving flux into the TCA cycle. An early approach to engineer CHO cells for restricted lactate production was the coexpression of human pyruvate carboxylase (hPC) [81]. Compared to control cells, the specific lactate production rate of clones coexpressing hPC was decreased by 21 to 39 % [81]. Dorai et al. showed that apoptosis-resistant CHO cell lines coexpressing the antiapoptotic genes E1B-19K, Aven (EA167) and a mutant of XIAP (EAX197) were also lactate-consuming clones while their parental clones accumulated significant levels of lactate [82]. The benefit of combining lactate engineering approaches with simultaneous coexpression of antiapoptotic transgenes was demonstrated by Jeon et al. who developed a BCL-2-overexpressing dhfr(-) CHO cell line with a downregulated lactate dehydrogenase A gene [83].

3.5.3.2 Engineering Approaches Aimed at Improving Protein Secretion and Q_p

Specific productivity, the other contributing factor to overall process volumetric productivity, is of course dependent on all compartments of the cell engaged in protein synthesis, transport, processing, and export. Since approaches for designing efficient expression cassettes for stable high level production of transgene mRNA and codon optimization for efficient translation have been extensively discussed elsewhere, this section will keep the focus on the more recent cell line engineering approaches that have targeted the secretory capacity of pharmaceutical producer cell lines. The amount of product that is released from each cell into the culture supernatant during upstream processing is not only dependent on the efficiency of the expression cassette in providing high levels of stable mRNA but also on the efficiency of intracellular processing of the translated protein. As discussed in the introductive section, the translated protein encounters several bottlenecks en route to the plasma membrane from where it is finally secreted. Heterologous overexpression of therapeutic proteins from strong promoters can contribute to protein overload and intracellular aggregate accumulation if the cells folding machinery and ER capacity become overwhelmed. Accumulation of misfolded protein within the ER compartment triggers the unfolded protein and ER overload responses and thereby causes an undesirable decline in cell specific productivity. Coexpression of the UPR-associated transcription factor GADD34 (gene encoding growth arrest and DNA damage inducible protein 34) has been shown to increase production of recombinant human antithrombin III from CHO cells [84]. A very interesting approach to overcome bottlenecks in intracellular protein processing is based on artificial expansion of the ER capacity. Heterologous coexpression of the active, spliced form of human X-box binding protein 1 (XBP-1(s)) transcription factor has been shown to cause ER expansion and a concomitant increase in specific therapeutic antibody productivity of CHO-DG44 cell cultures [85, 86]. However, in order to maintain this type of enhanced secretion it was necessary to coexpress an additional antiapoptotic transgene [86]. Other approaches have focused on engineering the chaperone repertoire, e.g., by overexpressing BIP, the ER quality control mechanism including the calnexin calreticulin cycle as well as other enzymes such as protein disulfide isomerase (PDI) and prolyl-cis-transisomerase that are involved in the protein folding capacity of the ER. Overexpression of protein disulfide-isomerase even decreased target protein secretion [87]. Later it was found that the oxidative state of PDI is important for activity and an expression system for inducible coexpression of PDI was devised [88].

But the ER is not the only bottleneck. Vesicle transport between the ER and Golgi compartments and the plasma membrane is as critical for efficient secretion. Just like other cells that grow in suspension CHO cells have lost their polarity and it has been suggested that CHO suspension cells retain the capacity for both apical and basolateral secretion [89]. Heterologous expression of the ceramide transfer protein (CERT), a protein involved in protein kinase D-dependent protein transport from the Golgi to the plasma membrane, resulted in higher specific productivities of the target therapeutic protein by engineered cells grown in suspension cultures [90]. These data suggested that a bottleneck in recombinant protein secretion at the Golgi complex has a strong impact on specific productivity in pharmaceutical producer cell lines grown in suspension [90]. The final stage of Golgi vesicle maturation, fusion with the plasma membrane and exocytosis, has been targeted as well: Sec1/Munc18 proteins stabilize a specific conformation of syntaxin as part of the SNARE complex bringing the 2 membranes together and initiate fusion [91]. Overexpression of components of the fusion complex, either Munc18 alone [92] or combined expression of SNAP23 and VAMP8 [93] has also been shown to increase the production of coexpressed therapeutic proteins in mammalian cells.

As posttranslational modifications – in particular glycosylation – commonly suffer at higher expression levels, a deficit should not only be expected in the final stages of Golgi maturation. During the transport of cargo proteins from the ER through Golgi cisternae towards the plasma membrane, ER and Golgi resident proteins would gradually get lost without an efficient sorting mechanism. Cargo protein transport is accomplished by COP complexes. COPI is responsible for retrograde transport of ER and Golgi resident functional proteins while COPII moves cargo proteins in the anterograde direction. These protein complexes form a cytoplasmatic coat for the membrane system directing vesicles through interaction with respective tails of functional and cargo membrane proteins [94] and could represent another potential target for enhancement.

When boosting a single point in the "assembly line" other elements are likely to become rate limiting in a producer cell. Therefore, it would be intriguing to target the whole secretion apparatus at once with one single regulator. As the driver for locomotion of the various vesicles the dynamic actin cytoskeleton could represent such unifying factor. Its assembly is regulated by the Rho GTPAse CDC42 which recruits the actual motor dynein, interacts with the γ-subunit of COPI as well as with syntaxin I [95-97]. However, as a central regulator of cellular processes CDC42 not only controls secretion but also affects transcription, translation, stress response, and apoptosis. Moreover, simple overexpression is not useful because conversion of CDC42 into its active GTP-bound form is strongly regulated itself by guanine nucleotide exchange factors (GEFs) and GTPase-activating proteins (GAPs). Even overexpression of a permanently active GTPase-defective cdc42 mutant is not beneficial: while generally promoting secretion it also inhibits transport of some proteins and causes cell cycle arrest [98, 99].

We have recently shown that in contrast to the constitutively active variant a mutant shuttling between GDP and GTP bound stages in the absence of GEFs is able to boost productivity of various antibodies in CHO cells up to 2.6-fold. The approach is applicable to empty starter and preexisting producer cell lines and requires a well-adjusted expression of CDC42 [100].

3.5.3.3 Engineering Approaches Aimed at Improving Product Quality

Given the past successes in improving upstream product titer, the focus of manufacturers in the biopharma industry has started to shift towards achieving an improved product quality. The physicochemical complexity of large macromolecular therapeutics complicates the assessment of product quality and requires a practically feasible and meaningful approach. A rational approach towards assessing product quality is to focus on measurable parameters of therapeutic activity such as pharmacokinetics and pharmacodynamics and on currently known product safety attributes.

Engineering of the Glycosylation Machinery

One of the most common, yet complex critical quality attributes of biotherapeutics produced by mammalian cell culture is product glycosylation [101]. Currently, 3 distinct types of glycosylation have been described to occur on proteins: N-linked, O-linked, and C-type [102]. While the latter is less common and has a rather simple chemistry, N- and O-linked glycosylation show a much higher degree of molecular complexity and have therefore been a greater focus of biopharmaceutical analytics. N- and O-linked glycans contribute significantly to the physicochemical microheterogeneity of therapeutic proteins. Each molecule of a protein drug can potentially contain N- and O-glycans that differ in complexity from those glycans attached to another protein molecule. To make things worse, glycan sites may only partially be occupied [103]. This however is less of an issue for proteins made in mammalian cell culture but more so for those products that are made in yeast [103].

The terminal sugars of N- and O-glycans have been linked to pharmacokinetic and pharmacodynamic properties of biopharmaceuticals [104, 105]. The degree of sialylation has been shown to impact the pharmacokinetic behaviour including circulation half-life of glycoproteins [105, 106]. Terminal sialylation is now regarded as such an important quality attribute of biopharmaceuticals that high throughput methods have been developed by industry to monitor intraclonal variability of glycoprotein sialylation at early stages of upstream development [107]. Terminal mannose has been linked to increased clearance receptor binding [108-110] although this effect of terminal high mannose on antibody pharmacokinetics still seems to be an issue of debate [111]. Terminal fucose has been implicated in affecting antibody pharmacodynamics [104].

In order to achieve the production of therapeutic glycoproteins with desired glycosylation quality attributes and quality target glycoforms there are 2 major routes of cell line engineering: the cytosolic sugar nucleotide metabolism and the compartmentalized glycan-processing machinery (see also Chapter 4.1).

Glycoengineering Methods Targeting the Sugar Nucleotide Metabolism

Sugar nucleotides provide the building blocks from which complex oligosaccharide O- and N-glycans are assembled. As mentioned earlier, the biosynthesis of sugar nucleotides requires primary metabolites of the central energy metabolism and therefore operates in competition to cellular growth – particularly under nutrient limitation. Specifically adapted feed regimes have already been devised to support an improved sugar nucleotide biosynthesis and ultimately a better product glycosylation [112]. For historic and economic reasons, glucose and glutamine are still the dominant sources of energy in cell culture media and feeds. This implies that under standard bioreactor upstream conditions the synthesis of other sugars is typically starting from the primary metabolites of these original energy sources. The important sugar nucleotide substrate UDP-galactose is synthesized by an interconversion of glucose and galactose, which is catalyzed by the enzymes of the Leloir cycle, in particular UDP-galactose-4-epimerase [113]. This central galactose-glucose-interconversion enzyme utilizes not the sugars themselves but their UDP-sugar nucleotides. As a consequence, an efficient supply of the nucleoside diphosphates such as UDP is therefore also critical for the efficient biosynthesis of UDP-galactose and other sugar nucleotides. Researchers at Genentech were among the first to recognize this and devised a cell line engineering approach where the pharmaceutical host cell line is selected for resistance to N-phosphonacetyl-L-aspartate, an inhibitor of the central enzyme complex responsible for pyrimidin *de novo* synthesis [114]. These host cell lines selected for overexpression of the CAD multienzyme complex display an increased activity of carbamoyl-phosphate synthetase 2, aspartate transcarbamylase and dihydroorotase and allegedly support an improved level of target product galactosylation. This original work is in line with the data of Gramer et al. who demonstrated the effect of uridine feeding on target product galactosylation [112].

Cytidine monophosphate-sialic acid (CMP-sialic acid) is another sugar nucleotide that has been targeted by metabolic cell line engineering approaches. The first 2 committed, rate-limiting steps of the biosynthesis of the sialic acid sugar nucleotide CMP-5-N-acetylneuraminic acid (CMP-Neu5Ac) is regulated by the bifunctional key enzyme uridine diphospho-N-acetylglucosamine (UDP-GlcNAc) 2-epimerase/N-acetylmannosamine (ManNAc) kinase (GNE/MNK) [115]. In its natural wild type form the epimerase activity of this enzyme is subject to allosteric feedback inhibition by CMP-Neu5Ac [115]. Point mutations within the allosteric domain of GNE/MNK abolish the feedback inhibition and result in an excessive overproduction of CMP-sialic acid [116]. Such feedback-abolishing mutations are known to be associated with codons 263 and 266 of the GNE/MNK gene [117]. Heterologous coexpression of this sialuria mutant of GNE/MNK has been claimed to enhance the steady state concentration of CMP-sialic acid in the cytosol and to achieve a concomitant increase in the degree of target protein terminal sialylation [118]. Later, Bork et al. confirmed these results for EPO expression in CHO [119] and Son et al. demonstrated that the CMP-sialic acid concentration of CHO cells engineered for sialuria-mutated GNE/MNK (R263L-R266Q) expression was increased by more than tenfold and that the sialic acid content of rhEPO produced from these engineered cells was 43 % higher compared to the rhEPO sialylation of control cells [120].

In most animal cells, and in particular rodent cells including CHO, the synthesis of sialic acid sugar nucleotides proceeds to N-glycolyl-neuraminic acid (Neu5Gc) because these cells typically possess CMP-N-acetylneuraminic acid hydroxylase (CMP-Neu5Ac hydroxylase) activity [121]. However, human cells lack this enzyme and therefore glycostructures containing terminal Neu5Gc-sialic acid can be considered a foreign epitope to the human immune system. Since all humans have Neu5Gc-specific antibodies, the presence of covalently bound Neu5Gc can lead to the formation of immune complexes and accelerated drug clearance [122, 123]. Therefore, the absence of Neu5Gc-sialic acid is an important quality attribute of biotherapeutics. In order to mitigate this potential problem Chenu et al. have engineered CHO lines to achieve a downregulated CMP-Neu5Ac hydroxylase activity [121].

Two other cell line engineering approaches to enhance recombinant glycoprotein galactosylation and sialylation focused on enhancing the transport of sugar nucleotides across membranes of the luminal compartments. Wong et al. reported that overexpression of the CMP-sialic acid transporter (CMP-SAT) in CHO cells has led to an increased CMP-sialic acid intraluminal pool and a concomitant increase of target protein sialylation [124]. Later Jeong et al. and Son et al. confirmed this beneficial effect [120, 125].

The other approach focuses on the antiport activity concerning the import of UDP-sugar nucleotides into the Golgi. Upon transfer of the monosaccharide sugars from the activated sugar nucleotide precursor to the nascent oligosaccharide N- and O-glycans the respective nucleoside diphosphate is released. These nucleoside diphosphates however work as inhibitors for the glycosyltransferases [126]. Moreover, the required UDP-sugar nucleotides are imported into the Golgi lumen via antiport against uridin-monophosphate (UMP) [127]. To accommodate these 2 reactions, the Golgi lumen contains nucleoside diphosphatase activity whereby the Golgi resident UDPase activity is most prominent [128]. It was found that Golgi UDPase is upregulated in response to ER stress and ER capacity expansion [129]. In line with this context, a patent application filed by Momenta Pharmaceuticals claims that overexepression of a nucleoside diphosphatase in particular uridin diphosphatase (apyrase) improves galactosylation and overall glycosylation quality [130].

A desirable quality target glycoform concerning monoclonal antibodies is the absence of α-1,6-linked core-fucose from the Fc-Asn297-linked N-glycans. Antibodies lacking core-fucose show a much higher antibody dependent cell-mediated cytotoxicity (ADCC) and related antitumor activity. The ADCC effector function of IgG1, -2 and -3 antibodies is the dominant mechanism by which therapeutic antibodies can kill tumor cells. ADCC is mediated by the binding of antigen-antibody complexes to Fcγ receptors expressed on immune cells whereby the mechanism underlying ADCC is the binding of the FcγRIIIa (CD16a) receptor on natural killer (NK) cells to the antibody Fc domain of the IgG molecule. Upon binding between IgG-Fc and the FcγRIIIa receptor afucosylated Fc-N-glycan structures enable additional carbohydrate-carbohydrate interactions between the N-glycans of the FcgRIIIa receptor and the afucosylated Fc, which are absent in the complex structure with fucosylated Fc [131]. This explains the observed fortified binding activity of afucosylated antibodies.

In April 2012 Kyowa Hakko Kirin's mogamulizumab (trade name Poteligeo®), a humanized monoclonal antibody targeting CC chemokine receptor 4 (CCR4), became the first afucosylated therapeutic mAb that was approved for market authorization [132, 133]. A less well-known quality attribute concerning the terminal sugar fucose is the absence of antennary fucosylation. Spontaneous gain of function mutations in CHO can lead to the addition of fucose moieties to the C-3 or C-4 antennary GlcNAc or galactose residues resulting in antennary fucosylated Lewis X structures [134, 135]. This type of antennary fucosylation has been reported to target such molecules to sites of inflammation and will therefore adversely affect biodistribution, pharmacokinetics and *in vivo* activity of therapeutic glycoprotein drugs [135-138].

Heterologous coexpression of the pseudomonadal enzyme GDP-4-keto-6-deoxy-D-lyxohexulose reductase (RMD, synonyms GDP-4-keto-6-deoxy-D-mannose reductase, GDP-4-keto-D-rhamnose reductase) in the cytosol of eukaryotic cells leads to an immediate shut down of the fucose *de novo* synthesis pathway and efficiently depletes the cytosolic sugar nucleotide pool of GDP-L-fucose [139, 140]. In consequence, antibodies secreted from RMD-coexpressing cells are almost completely devoid of both, core- and antennary fucose [139, 140]. The GlymaxX® technology, developed by ProBioGen, is based on heterologous stable coexpression of this RMD enzyme.

Fucose is either taken up by the cells from the culture medium and utilized via the salvage pathway or it is *de novo* synthesized inside the cell from GDP-mannose via short lived intermediates before it is transported to the Golgi apparatus for attachment to the nascent glycan [141]. Quantitative assessments of fucose metabolism in mammalian cells suggest that more than 90 % of the cytosolic GDP-L-Fucose sugar nucleotide pool originate from *de novo* synthesis, even if fucose is present in the surrounding culture medium [141-143]. Omission of fucose from the culture medium can shut off the salvage pathway source completely so that GDP-L-fucose is only generated from *de novo* synthesis. Coexpressed RMD interrupts the fucose *de novo* synthesis pathway at the stage right after GDP-mannose has been converted by GDP-mannose dehydratase (GMD) into GDP-4-keto-6-deoxy mannose. The latter intermediate is a substrate for the Fx-epimerase-reductase enzyme complex that converts it into GDP-L-fucose. The intermediate GDP-4-keto-6-deoxy mannose is also a substrate for RMD, which converts this molecule into GDP-D-rhamnose, a rare 6-deoxy-D-sugar nucleotide that cannot be utilized by mammalian enzymes and would therefore constitute a dead end product within the mammalian cellular context [139]. Surprisingly, HPAEC-PAD monosaccharide analysis of hydrolysed cytosolic extract did not detect any D-rhamnose that would be indicative of an accumulated GDP-D-rhamnose pool within the cytosolic compartment of RMD coexpressing cells [139]. In addition, the extremely powerful effect of RMD coexpression on the depletion of the cytosolic GDP-L-fucose pool can hardly be explained by competitive substrate removal from the *de novo* synthesis pathway alone.

Mammalian-type GMD is a member of the short chain dehydrogenase/reductase (SDR) family of NADP-dependent oxidoreductases [144]. GMD folds into an N-terminal Rossmann fold domain, which binds the cofactor NADP and a second domain, which binds the substrate GDP-D-mannose [144]. Evidence suggests that both mammalian and pseudomonadal GMD are active as a homotetramer [144, 145]. Although there are suggestions that the mammalian GMD enzyme is active as a dimer, tetramers of the human GMD were already discovered in the cross-linking

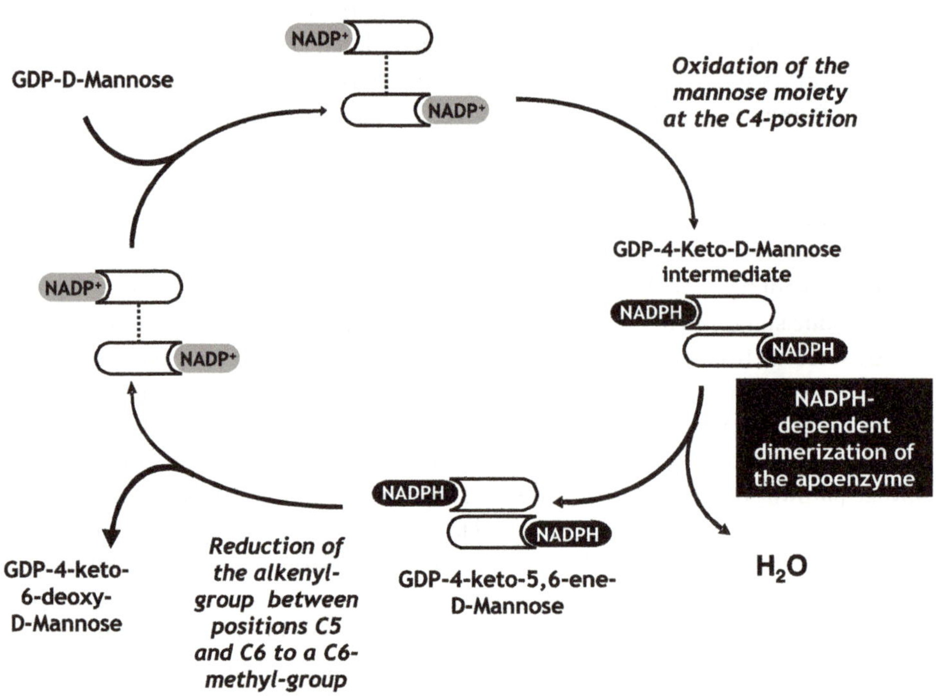

Figure 3.5.2: NADP-NADPH-shuttling and apoenzyme dimerization during the GMD catalytic cycle.

study done by Bisso et al. (1999), suggesting that the mechanism of action is similar for the mammalian and bacterial GMD enzymes [145]. The GMD enzymatic activity involves $NADP^+$-NADPH-shuttling and apoenzyme dimerization (Figure 3.5.2). In the homotetrameric state the cofactor $NADP^+$ becomes entrapped within the structure and has only limited solvent access [144]. Most importantly, the cofactor $NADP^+$ is regenerated *in situ* from the entrapped NADPH within the dimerized apoenzyme at the last catalytic step where the mannose moiety of the GDP-mannose substrate is reduced at the C6-position (Figure 3.5.2). Since the 6-deoxysugar D-rhamnose is stereochemically derived from D-mannose and therefore still shares most of its conformational features with mannose (Figure 3.5.3), it is easy to see that GDP-D-rhamnose would fit just as easily into the GMD substrate pocket as the natural substrate GDP-D-mannose. However, there is a critical difference between D-rhamnose and D-mannose: the C6-position of D-rhamnose differs from that of D-mannose and can no longer be reduced by NADPH. Therefore, we hypothesize that the mammalian GMD

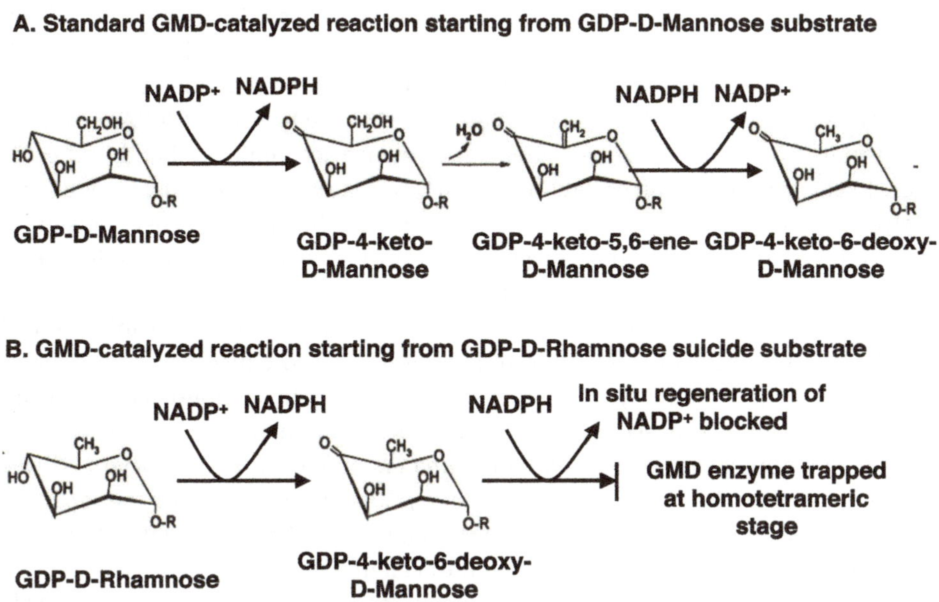

Figure 3.5.3: Overview of GMD suicide substrate hypothesis. **a)** Standard GMD-catalyzed reaction starting from GDP-D-mannose substrate **b)** GDP-D-rhamnose fits into the substrate binding pocket of mammalian GMD and gets converted directly to GDP-4-keto-6-deoxy-D-rhamnose (synonym GDP-4-keto-D-rhamnose). Simultaneously, NADPH is generated *in situ* and entraps the GMD complex in the homotetrameric state. The second reduction step needed for $NADP^+$ regeneration and release of the tightly bound complex, however, is no longer possible because the C6-position in rhamnose is already at the deoxydized methyl-group stage and therefore does not support this reaction. As a consequence, the GMD complex becomes entrapped in an NADPH-bound, GDP-4-keto-6-deoxy-D-mannose-bound, homotetrameric state.

complex gets arrested in the NADPH- bound state with a GDP-4-keto-D-rhamnose intermediate entrapped within the GMD catalytic site when mammalian GMD comes in contact with GDP-D-rhamnose (Figure 3.5.3). The $NADP^+$ cofactor can no longer be regenerated *in situ*, factually entrapping the GMD in an inactive catalytic state (Figure 3.5.3).

Based on all these facts we hypothesize that GDP-D-rhamnose is a suicide substrate for the mammalian GDP-D-mannose-4,6-dehydratase (GMD). Coexpression of RMD therefore is a very efficient and powerful tool for achieving the manufacturing of non-fucosylated biotherapeutics [139].

Glycoengineering Methods Targeting the N-glycan Processing Machinery

Since the absence of core-fucose is such an important quality attribute of many therapeutic antibodies, much effort has been put into the development of engineered cell lines that enable the production of such afucosylated glycoproteins. The majority of these methods have targeted the N-glycan processing machinery. The first approach to achieve production of ADCC-enhanced antibodies was developed by Umana and colleagues [146]. Overexpression of N-acetylglucosaminyltransferase-III (GnT-III) in antibody producer cells lead to the production of ADCC-enhanced antibodies from these cells [146]. Back in 1999 this effect was attributed to the bisecting GlcNAc structure that results from GnTIII activity [146]. Later, evidence accumulated that this effect is more likely to be due to afucosylation [131, 147]. This is also supported by the idea that a glycan carrying a bisecting GlcNAc is no longer a substrate for fucosyltransferase 8 [148]. Recently published data suggest that GntIII expression is not as effective as other methods to achieve afucosylation [149].

The most prominent approach for engineering an industrially applicable host cell line for the production of therapeutic antibodies lacking core-fucose was the deliberate disruption of both fucosyltransferase 8 (FUT8) alleles in a Chinese hamster ovary (CHO)/DG44 cell line by sequential homologous recombination [150]. This cell line engineering approach was devised by researchers at Kyowa Hakko Kogyo and was later marketed as the Potelligent®-technology [150]. On April 30, 2012 the Japanese Ministry of Health granted approval to mogamulizumab (Poteligeo®), the first drug manufactured by a Potelligent®-modified host cell line [133].

While the level of N- and O-glycan galactosylation has not yet been conclusively linked to major effects on pharmaceutical efficacy of glycoprotein drug substance, it still has become a quality attribute of therapeutic glycoproteins that is closely monitored during bioprocessing [151-154]. The future will tell whether major deviations from the galactosylation quality target product profile will be tolerated for lot release and which galactosylation profile of production lots will be considered to be out of compliance with approved process specifications. Manufacturers have already started to address this issue proactively. The industries interest in achieving improved galactosylation profiles is indicated by the fact that at least one patent application for supplementing production media with galactose with the aim of improving N-glycan structure has recently been filed [155].

Therapeutic antibodies expressed in CHO cells are known to be generally less galactosylated compared to antibodies expressed in mouse myeloma cells such as NS0 and SP2/0 [154]. Therefore, efforts have been made to engineer CHO cells for improved galactosylation. The first approach to achieve improved galactosylation in CHO cells was made by Ryll and colleagues at Genentech who selected CHO cells for enhanced expression of the CAD complex (see Chapter 3.4). Later approaches focused on engineering the N-glycan processing machinery to improve target protein galactosylation. Overexpression of β-1,4-galactosyltransferase in CHO cells that were expressing interferon-γ led to an unexpectedly widespread effect on target protein galactosylation [156]. Surprisingly, Fukuta et al. reported that the contents of hybrid-type and high-mannose-type sugar chains increased in these clones [156]. The authors also suggested that β-1,4-galactosyltransferase competes with α-mannosidase II in cells and that overexpression of β-1,4-galactosyltransferase could suppress the conversion of high-mannose-type sugar chains to the hybrid type [156]. However, Weikert et al. did not find this effect and have shown that N-linked oligosaccharide structures synthesized by cells overexpressing the galactosyltransferases showed an increased level of oligosaccharides terminating with galactose [157].

The extent of glycan terminal sialylation has strongly been linked to pharmacokinetic properties of many glycoprotein drugs [105, 106]. While terminal sialylation appears to have little impact on the circulatory half-life of Fc-glycosylated therapeutic antibodies, it has been shown that this modification of IgG-Fc-glycans reduces Fc-γ receptor binding and converts IgG antibodies to antiinflammatory mediators by increasing their affinity for the lectin receptor DC-SIGN [158, 159]. Because of these benefits for drug substance activity, cell line engineering approaches targeting the N-glycan processing machinery have also focused on achieving higher levels of glycoprotein terminal sialylation.

In 1996 sialidase activity was discovered in cell-free supernatant of batch-cultivated CHO cells and it was shown that this sialidase could cause product desialylation *in situ* [160]. To suppress this negative effect of endogenous sialidase, CHO cells were engineered to express sialidase antisense RNA [161]. These cells, however, were never used in large scale manufacturing since progress in cell culture media design and upstream process engineering seem to have alleviated this problem (for example [162]).

Weikert et al. demonstrated that overexpression of α-2,3-sialyltransferase resulted in sialylation of about 90 % of available branches on the target protein N-glycan structures [157]. Zhang et al. conducted a systematic functional analysis of 31 N-glycosylation-related genes on sialylation of recombinant EPO in 6 different mammalian cell lines [163]. Of the 6 cell lines analyzed CHO and BHK cells were found to sialylate recombinant EPO most effectively [163]. Interestingly, Zhang et al. found that coexpression of additional genes upstream in the glycosylation cascade failed to synergize with these sialyltransferases to further enhance target protein sialylation [163].

Rodent cell lines such as CHO are not able to attach terminal sialic acid to N-glycans via an α-2,6-linked glycosidic bond [164]. In order to achieve an efficient formation of α-2,6-linked sialylation, CHO host cells were transfected with ST6Gal [164].

3.5.4 Engineering of Cell Lines to Achieve Specific Critical Quality Attributes of Therapeutic Glycoproteins

Several pharmaceutical glycoproteins require specialized posttranslational processing to display optimal therapeutic activity. Some proteins require proper proteolytic processing which can be ineffective in standard pharmaceutical host cell lines. One example is bone morphogenetic protein-7 (BMP-7), a protein involved in cartilage, bone, and kidney development. The functionally active mature dimer of BMP-7 is generated from a larger precursor via proteolytic cleavage by subtilisin-like proprotein convertase [165]. Unmodified CHO host cells were found to be unable to properly process the BMP-7 precursor and consequently, about 69 % of the total rhBMP-7 was secreted as undesirable uncleaved precursor forms. In order to overcome this problem, Sathayamurthy et al. overexpressed a soluble form of the paired basic amino acid cleaving enzyme (PACEsol) in the rhBMP-7-expressing CHO cell line [165]. This turned out to be an effective strategy as PACEsol is responsible for the majority of the processing events occurring in the constitutive secretory pathway in mammalian cells. As consequence, operations for downstream purification of mature rhBMP-7 were significantly facilitated by this dramatic reduction of product related impurities from the harvest bulk [165].

Another example for a therapeutic protein requiring proteolytic processing is the blood-clotting factor FVII. Like several of the vitamin K-dependent coagulation proteins required for normal blood coagulation, factor VII is a highly complex protein with multiple structural domains each being associated with a specific functional property essential for its overall effectiveness in controlling hemostasis. For its normal function factor VII must undergo extensive posttranslational modification. Not surprisingly, achieving high levels of functional factor VII by recombinant technology has been limited by its structural complexity and requirement for highly specific posttranslational processing that is not adequately supported by standard production host cell lines. In fact, multiple intracellular bottlenecks have been described to delay proper secretion of factor VII [166]. Moreover, Sutkevikiute et al. found that CHO-derived factor VII showed a lower autoactivation rate compared to BHK cells and that CHO cells were less effective in secreting fully activated factor VII [167]. To overcome this disadvantage of CHO-based factor VII production, Halabian et al. cotransfected factor VII and hepsin, a membrane-associated serine protease, into a CHO cell line and were able to obtain increased amounts of biologically active rFVIIa [168].

Factor IX, another member of the vitamin K-dependent clotting factors, has also been reported to be poorly processed in CHO cells [169]. Coexpression of the paired basic amino acid cleaving enzyme (PACE) improved proteolytic processing of profactor IX to the mature form and led to a two- to threefold increase in specific activity of the secreted fraction [169].

In analogy to the above, Himmelspach et al. had reported for the coagulant protein factor X that propeptide removal, single chain precursor processing and γ-carboxylation became impaired at high expression levels [170].

Currently used mammalian host cell lines for recombinant bioprocessing have only a limited capacity for post translational γ-carboxylation, an attribute of crucial importance for biological activity of the vitamin K-dependent clotting factors factor VII, factor IX, and protein C. [171]. The amino-terminal Gla domain, a stretch of amino acid sequence rich in glutamic acid, is the site where these proteins require γ -carboxylation. Factor IX has 12 and factor VII has 10 glutamic acid residues with their respective Gla domains, a majority of which must be γ-carboxylated for maximum level bioactivity [172]. As a consequence of insufficient γ-carboxylation capacity, the recovery of fully γ-carboxylated and functional proteins expressed from standard host cell lines has been low [171]. To achieve production of properly γ-carboxylated clotting factors from baby hamster kidney cells (BHK), Wajih et al. engineered stably transfected BHK cells to coexpress γ-carboxylase and the vitamin K-2,3-epoxide reductase complex subunit VKORC1 from a bicistronic vector [173]. Interestingly, they found that coexpressed γ-carboxylase did not have much of an effect on target protein γ-carboxylation and that VKOR complex activity is in fact the rate-limiting step in the γ-carboxylation system [173]. Coexpression of the VKORC1 subunit enabled the successful engineering of cells containing a recombinant vitamin K-dependent γ-carboxylation system with enhanced capacity for γ-carboxyglutamic acid modification [171, 173-175]. Further improvement in target protein γ-carboxylation was achieved by stable coexpression of VKORC1 and an additional stable silencing of the γ-carboxylase inhibitory protein calumenin [171, 176].

3.5.5 Conclusion

The pharmaceutical producer cell line remains one of the most important factors in biomanufacturing. Not only does it influence volumetric yield and upstream cost of goods but also efficacy and safety of the manufactured drug substance. The growing pipeline of therapeutic proteins in development as well as requirements for controlled posttranslational modifications of the manufactured proteins will likely lead to an increasing demand for specifically engineered host cell lines.

References

[1] Kelley B. Industrialization of mAb production technology. The bioprocessing industry at a crossroads. MAbs. 2009;1(5):443-52.

[2] O'Callaghan PM, McLeod J, Pybus LP, et al. Cell line-specific control of recombinant monoclonal antibody production by CHO cells. Biotechnol Bioeng. 2010;106(6):938-51.
[3] Shukla A, Thömmes J. Recent advances in large-scale production of monoclonal antibodies and related proteins. Trends Biotechnol, 2010;28(5): 253-61.
[4] Jungbauer A. Continous bioprocessing: An interview with Konstantin Konstantinov from Genzyme. Biotechnol J. 2011:6:1431-4.
[5] Warikoo V, Godawat R, Brower K, et al. Integrated continuous production of recombinant therapeutic proteins. Biotechnol Bioeng. 2012;109(12):3018-29.
[6] US Food and Drug Administration. Docket number FDA-2008-N-03551. Notice of pilot program for submission of quality information for biotechnology products in the Office of Biotechnology Products. FDA, USA, 2008.
[7] Rathore AS, Branning R, Cecchini D. Quality: Design space for biotech products. BioPharm Int. 2007:22(11):36-40.
[8] Rathore AS, Mhatre R (eds). Quality by Design for Biopharmaceuticals. John Wiley & Sons, USA, 2009.
[9] Goetze AM, Schenauer MR, Flynn GC. Assessing monoclonal antibody product quality attribute criticality through clinical studies. MAbs. 2010;2(5):500-7.
[10] Chu L, Blumentals I, Maheshwari G. Production of Recombinant Therapeutic Proteins by Mammalian Cells in Suspension Culture. Methods Mol Biol. 2005;308:107-21.
[11] Adams D, Korke R, Hu W. Application of Stoichiometric and Kinetic Analyses to Characterize Cell Growth and Product Formation. Anim Cell Biotechnol. 2007;24:269-84.
[12] Butler M. Animal cell cultures: recent achievements and perspectives in the production of biopharmaceuticals. Appl Microbiol Biotechnol. 2005;68(3):283-91.
[13] al-Rubeai M, Emery AN, Chalder S, et al. Specific monoclonal antibody productivity and the cell cycle-comparisons of batch, continuous and perfusion cultures. Cytotechnology. 1992;9 (1-3):85-97.
[14] Dutton RL, Scharer J, Moo-Young M. Cell cycle phase dependent productivity of a recombinant Chinese hamster ovary cell line. Cytotechnology. 2006;52(1):55-69.
[15] Benham AM. Protein secretion and the endoplasmic reticulum. Cold Spring Harb Perspect Biol. 2012;4(8):a012872.
[16] Gu MB, Todd P, Kompala DS. Foreign gene expression (β-galactosidase) during the cell cycle phases in recombinant CHO cells. Biotechnol Bioeng. 1993;42:1113-23.
[17] Sunley K, Butler M. Strategies for the enhancement of recombinant protein production from mammalian cells by growth arrest. Biotechnol Adv. 2010;28(3): 385-94.
[18] Clarke C, Doolan P, Barron N,. Predicting cell-specific productivity from CHO gene expression. J Biotechnol. 2011;151:159-65.
[19] Kim M, O'Callaghan PM, Droms K, et al. A mechanistic understanding of production instability in CHO cell lines expressing recombinant monoclonal antibodies. Biotechnol Bioeng. 2011;108(10): 2434-46.
[20] Chusainow J, Yang YS, Yeo JH, et al. A study of monoclonal antibody-producing CHO cell lines: what makes a stable high producer? Biotechnol Bioeng. 2009;102(4):1182-96.
[21] Dorai H, Corisdeo S, Ellis D, et al. Early prediction of instability of Chinese hamster ovary cell lines expressing recombinant antibodies and antibody-fusion proteins. Biotechnol Bioeng. 2012;109(4):1016-30.
[22] Seitz S, von Horsten H, Rose T, et al. Toolbox approach for fast generation of stable CHO production cell lines from different hosts. BMC Proc. 2011;5(8):P28.
[23] Reisinger H, Steinfellner W, Stern B, et al. The absence of effect of gene copy number and mRNA level on the amount of mAb secretion from mammalian cells. Appl Microbiol Biotechnol. 2008;81(4):701-10.

[24] Wellen KE, Lu C, Mancuso A, et al. The hexosamine biosynthetic pathway couples growth factor-induced glutamine uptake to glucose metabolism. Genes Dev. 2010;24(24):2784-99.
[25] Lehninger A, Nelson DL, Cox MM. Lehninger Principles of Biochemistry. W H Freeman, USA, 2008.
[26] Warburg O. On the origin of cancer cells. Science 1956;123(3191):309-14.
[27] Lopez-Lazaro M. The Warburg effect: why and how do cancer cells activate glycolysis in the presence of oxygen?. Anticancer Agents Med Chem. 2008;8(3):305-12.
[28] Hitosugi T, Fan J, Chung TW, et al. Tyrosine phosphorylation of mitochondrial pyruvate dehydrogenase kinase 1 is important for cancer metabolism. J Mol Cell. 2011;44(6):864-77.
[29] Papandreou I, Goliasova T, Denko NC. Anticancer drugs that target metabolism: Is dichloroacetate the new paradigm? Int J Cancer. 2011;128(5):1001-8.
[30] Bustamante E, Pedersen P. High aerobic glycolysis of rat hepatoma cells in culture: role of mitochondrial hexokinase. Proc Natl Acad Sci USA. 1977;74(9):3735-9.
[31] Adams V, Griffin L, Towbin J, et al. Porin interaction with hexokinase and glycerol kinase: metabolic microcompartmentation at the outer mitochondrial membrane. Biochem Med Metab Biol. 1991;45(3):271-91.
[32] Dakubo GD. Mitochondrial Genetics and Cancer. Chapter 2: The Warburg Phenomenon and other metabolic alterations of cancer cells. Springer, USA, 2010.
[33] DeBerardinis RJ, Mancuso A, Daikhin E, et al. Beyond aerobic glycolysis: Transformed cells can engage in glutamine metabolism that exceeds the requirement for protein and nucleotide. PNAS. 2007;104(49):19345-50.
[34] Ahn WS, Antoniewicz MR. Towards dynamic metabolic flux analysis in CHO cell cultures. Biotechnol J. 2012;7(1):61-74.
[35] Kurland IJ, Pilkis SJ. Covalent control of 6-phosphofructo-2-kinase/fructose-2,6-bisphosphatase: insights into autoregulation of a bifunctional enzyme. Protein Sci. 1995;4(6):1023-37.
[36] Wu C, Khan SA, Peng LJ, et al. Roles for fructose-2,6-bisphosphate in the control of fuel metabolism: beyond its allosteric effects on glycolytic and gluconeogenic enzymes. Adv Enzyme Regul. 2006;46(1):72-88.
[37] Wanka C, Steinbach JP, Rieger J. Tp53-induced glycolysis and apoptosis regulator (TIGAR) protects glioma cells from starvation-induced cell death by up-regulating respiration and improving cellular redox homeostasis. J Biol Chem. 2012;287(40):33436-46.
[38] Won KY, Lim SJ, Kim GY. Regulatory role of p53 in cancer metabolism via SCO2 and TIGAR in human breast cancer. Human Pathology. 2012;43:221-8.
[39] Bensaad K, Tsuruta A, Selak MA, et al. TIGAR, a p53-inducible regulator of glycolysis and apoptosis. Cell. 2006;126(1):107-20.
[40] Madan E, Gogna R, Bhatt M, et al. Regulation of glucose metabolism by p53: emerging new roles for the tumor suppressor. Oncotarget. 2011;2(12):948-57.
[41] Mor I, Cheung EC, Vousden KH. Control of glycolysis through regulation of PFK1: old friends and recent additions. Cold Spring Harb Symp Quant Biol. 2011;76:211-6.
[42] Cairns RA, Harris IS, Mak TW. Regulation of cancer cell metabolism. Nat Rev Cancer. 2011;11:85-95.
[43] Burleigh SC, van de Laar T, Stroop CJM, et al. Synergizing metabolic flux analysis and nucleotide sugar metabolism to understand the control of glycosylation of recombinant protein in CHO cells. BMC Biotechnol. 2011;11:95.
[44] Yi W, Clark PM, Mason DE, et al. Phosphofructokinase 1 glycosylation regulates cell growth and metabolism. Science. 2012;337(6097):975-80.
[45] Dempski RE, Imperiali B. Oligosaccharyl transferase: gatekeeper to the secretory pathway Curr Opin Chem Biol. 2002; 6(6):844-50.

[46] Helenius J, Neupert C, Aebi M, et al. Novel tools for the production of glycosylated proteins in host cells. Patent Application WO2010049177, filed 10/2009.
[47] Greenbaum D, Colangelo C, Williams K, et al. Comparing protein abundance and mRNA expression levels on a genomic scale. Genome Biol. 2003;4(9):117.
[48] Maier T, Güell M, Serrano L. Correlation of mRNA and protein in complex biological samples. FEBS Lett. 2009;583: 3966-73.
[49] Cao SS, Kaufman RJ. Unfolded protein response. Curr Biol. 2012;22(16):R622-6.
[50] Brodsky JL. Cleaning up: ER-associated degradation to the rescue. Cell. 2012;151(6):1163-7.
[51] Pahl HL, Baeuerle PA. The ER-overload response: activation of NF-kappa B. Trends Biochem Sci. 1997;22(2):63-7.
[52] Arden N, Betenbaugh MJ. Life and death in mammalian cell culture: strategies for apoptosis inhibition. Trends Biotechnol. 2004;22(4):174-80.
[53] Arden N, Betenbaugh MJ. Regulating apoptosis in mammalian cell cultures. Cytotechnology. 2006;50(1-3):77-92.
[54] Goswami J, Sinskey AJ, Steller H, et al. Apoptosis in batch cultures of Chinese hamster ovary cells. Biotechnol Bioeng. 1999;62(6):632-40.
[55] Laken HA, Leonard MW. Understanding and modulating apoptosis in industrial cell culture. Curr Opin Biotechnol. 2001;12(2):175-9.
[56] Krampe B, Al-Rubeai M. Cell death in mammalian cell culture: Molecular mechanisms and cell line engineering strategies. Cytotechnology. 2010;62(3):175-88.
[57] Figueroa B Jr, Ailor E, Osborne D, et al. Enhanced cell culture performance using inducible anti-apoptotic genes E1B-19K and Aven in the production of a monoclonal antibody with Chinese hamster ovary cells. Biotechnol Bioeng. 2007;97(4):877-92.
[58] Choi SS, Rhee WJ, Kim EJ, et al. Enhancement of recombinant protein production in Chinese hamster ovary cells through anti-apoptosis engineering using 30Kc6 gene. Biotechnol Bioeng. 2006;95(3):459-67.
[59] Wang Z, Ma X, Fan L, et al. Understanding the mechanistic roles of 30Kc6 gene in apoptosis and specific productivity in antibody-producing Chinese hamster ovary cells. Appl Microbiol Biotechnol. 2012;94(5):1243-53.
[60] Majors BS, Arden N, Oyler GA, et al. E2F-1 overexpression increases viable cell density in batch cultures of Chinese hamster ovary cells. J Biotechnol. 2008;138(3-4):103-6.
[61] Majors BS, Chiang GG, Pederson NE, et al. Directed evolution of mammalian anti-apoptosis proteins by somatic hypermutation. Protein Eng Des Sel. 2012;25(1):27-38.
[62] Barron N, Kumar N, Sanchez N, et al. Engineering CHO cell growth and recombinant protein productivity by overexpression of miR-7. J Biotechnol. 2011;151(2):204-11.
[63] Barron N, Sanchez N, Kelly P, et al. MiRNAs: tiny targets for engineering CHO cell phenotypes? Biotechnol Lett. 2011;33(1):11-21.
[64] Sanchez N, Lao N, Gallagher C, et al. Engineering CHO cell growth by stable manipulation of miRNA expression. BMC Proc. 2011;5(8):22.
[65] Hansen HA, Emborg C. Influence of ammonium on growth, metabolism, and productivity of a continuous suspension Chinese hamster ovary cell culture. Biotechnol Prog. 1994;10:121-4.
[66] Yang M, Butler M. Effect of ammonia on the glycosylation of human recombinant erythropoietin in culture. Biotechnol Prog. 2000;16(5):751-9.
[67] Chen P, Harcum SW. Effects of elevated ammonium on glycosylation gene expression in CHO cells. Metab Eng. 2006;8(2):123-32.
[68] von Schaewen A, Sturm A, O'Neill J, et al. Isolation of a mutant Arabidopsis plant that lacks N-acetyl glucosaminyl transferase I and is unable to synthesize Golgi-mediated complex N-linked glycans. Plant Physiol. 1993;102:1109-18.

[69] Zhong X, Cooley C, Seth N, et al. Engineering novel Lec1 glycosylation mutants in CHO-DUKX cells: molecular insights and effector modulation of N-acetylglucosaminyltransferase I. Biotechnol Bioeng. 2012;109(7):1723-34.
[70] Pacis E, Yu M, Autsen J, et al. Effects of cell culture conditions on antibody N-linked glycosylation-what affects high mannose 5 glycoform. Biotechnol Bioeng. 2011 (Epub ahead of print).
[71] Cockett MI, Bebbington CR, Yarranton GT. High Level Expression of Tissue Inhibitor of Metalloproteinases in Chinese Hamster Ovary Cells Using Glutamine Synthetase Gene Amplification. Nat Biotechnol. 1990;8:662-7.
[72] Park H, Kim IH, Kim IY, et al. Expression of carbamoyl phosphate synthetase I and ornithine transcarbamoylase genes in Chinese hamster ovary dhfr-cells decreases accumulation of ammonium ion in culture media. J Biotechnol. 2000;81(2-3):129-40.
[73] Kim HJ, Kim HJ. Glycosylation variant analysis of recombinant human tissue plasminogen activator produced in urea-cycle-enzyme-expressing Chinese hamster ovary (CHO) cell line. J Biosci Bioeng. 2006;102(5):447-51.
[74] Zhang F, Sun X, Yi X, et al. Metabolic characteristics of recombinant Chinese hamster ovary cells expressing glutamine synthetase in presence and absence of glutamine. Cytotechnology. 2006;51(1):21-8.
[75] Barnes LM, Bentley CM, Dickson AJ. Advances in animal cell recombinant protein production: GS-NS0 expression system. Cytotechnology. 2000;32:109-123.
[76] Fan L, Kadura I, Krebs LE, et al. Improving the efficiency of CHO cell line generation using glutamine synthetase gene knockout cells. Biotechnol Bioeng. 2012;109(4):1007-15.
[77] Dean J, Reddy P. Metabolic analysis of antibody producing CHO cells in fed-batch production. Biotechnol Bioeng. 2013;110(6):1735-47.
[78] Templeton N, Dean J, Reddy P, et al. Peak antibody production is associated with increased oxidative metabolism in an industrially relevant fed-batch CHO cell culture. Biotechnol Bioeng. 2013;110(7):2013-24.
[79] Zagari F, Jordan M, Stettler M, et al. Lactate metabolism shift in CHO cell culture: the role of mitochondrial oxidative activity. N Biotechnol. 2013;30(2):238-45
[80] Martínez VS, Dietmair S, Quek LE, et al. Flux balance analysis of CHO cells before and after a metabolic switch from lactate production to consumption. Biotechnol Bioeng. 2013;110(2):660-6.
[81] Kim SH, Lee GM. Functional expression of human pyruvate carboxylase for reduced lactic acid formation of Chinese hamster ovary cells (DG44). Appl Microbiol Biotechnol. 2007;76(3):659-65.
[82] Dorai H, Kyung YS, Ellis D, et al. Expression of anti-apoptosis genes alters lactate metabolism of Chinese Hamster Ovary cells in culture. Biotechnol Bioeng. 2009;103(3):592-608.
[83] Jeon MK, Yu DY, Lee GM. Combinatorial engineering of ldh-a and bcl-2 for reducing lactate production and improving cell growth in dihydrofolate reductase-deficient Chinese hamster ovary cells. Appl Microbiol Biotechnol. 2011;92(4):779-90.
[84] Omasa T, Takami T, Ohya T, et al. Overexpression of GADD34 enhances production of recombinant human antithrombin III in Chinese hamster ovary cells. J Biosci Bioeng. 2008;106(6):568-73.
[85] Becker E, Florin L, Pfizenmaier K, et al. An XBP-1 dependent bottle-neck in production of IgG subtype antibodies in chemically defined serum-free Chinese hamster ovary (CHO) fed-batch processes. J Biotechnol. 2008;135(2):217-23.
[86] Becker E, Florin L, Pfizenmaier K, et al. Evaluation of a combinatorial cell engineering approach to overcome apoptotic effects in XBP-1(s) expressing cells. J Biotechnol. 2010;146(4):198-206.

[87] Davis R, Schooley K, Rasmussen B, et al. Effect of PDI overexpression on recombinant protein secretion in CHO cells. Biotechnol Prog. 2000;16(5):736-43.

[88] Mohan C, Lee GM. Effect of inducible co-overexpression of protein disulfide isomerase and endoplasmic reticulum oxidoreductase on the specific antibody productivity of recombinant Chinese hamster ovary cells. Biotechnol Bioeng. 2010;107(2):337-46.

[89] Yoshimori T, Keller P, Roth MG, et al. Different biosynthetic transport routes to the plasma membrane in BHK and CHO cells. J Cell Biol. 1996;133(2):247-56.

[90] Florin L, Pegel A, Becker E, et al. Heterologous expression of the lipid transfer protein CERT increases therapeutic protein productivity of mammalian cells. J Biotechnol. 2009;141 (1-2):84-90.

[91] Hou J, Tyo K, Liu Z, et al. Engineering of vesicle trafficking improves heterologous protein secretion in Saccharomyces cerevisiae. Metab Eng. 2012;14(2):120-7.

[92] Peng RW, Guetg C, Tigges M, et al. The vesicle-trafficking protein munc18b increases the secretory capacity of mammalian cells. Metab Eng. 2010;12(1):18-25.

[93] Peng RW, Abellan E, Fussenegger M. Differential effect of exocytic SNAREs on the production of recombinant proteins in mammalian cells. Biotechnol Bioeng. 2011;108(3):611-20.

[94] Duden R. ER-to-Golgi transport: COP I and COP II function. Mol Membr Biol. 2003;20(3):197-207.

[95] Minin AA. Dispersal of Golgi apparatus in nocodazole-treated fibroblasts is a kinesin-driven process. J Cell Sci 1997;110:2495–2505.

[96] Daniel S, Sharp GW. A link between Cdc42 and syntaxin is involved in mastoparan-stimulated insulin release. Biochemistry. 2002;41(30):9663-71.

[97] Chen JL, Fucini RV, Lacomis L, et al. Coatomer-bound Cdc42 regulates dynein recruitment to COPI vesicles. J Cell Biol. 2005;169:383-9.

[98] Wu WJ, Erickson JW, Lin R, et al. The gamma-subunit of the coatomer complex binds Cdc42 to mediate transformation. Nature. 2000;405:800-4.

[99] Musch A, Cohen D, Kreitzer G, et al. cdc42 regulates the exit of apical and basolateral proteins from the trans-Golgi network. EMBO J. 2001;20:2171-9.

[100] Sandig V, Winkler K, von Horsten HH, et al. Enhancement of Protein Production Yield Mediated by a fast shuttling cdc42 GTPase. Patent Application WO2012152945, filed 05/2012.

[101] Lingg N, Zhang P, Song Z, et al. The sweet tooth of biopharmaceuticals: importance of recombinant protein glycosylation analysis. Biotechnol J. 2012;7(12):1462-72.

[102] Spiro RG. Protein glycosylation: nature, distribution, enzymatic formation, and disease implications of glycopeptide bonds. Glycobiology. 2002;12(4):43R-56R.

[103] Choi BK, Warburton S, Lin H, et al. Improvement of N-glycan site occupancy of therapeutic glycoproteins produced in Pichia pastoris. Appl Microbiol Biotechnol. 2012;95(3):671-82.

[104] Raju TS. Terminal sugars of Fc glycans influence antibody effector functions of IgGs. Curr Opin Immunol. 2008;20(4):471-8.

[105] Solá RJ, Griebenow K. Glycosylation of therapeutic proteins: an effective strategy to optimize efficacy. BioDrugs. 2010;24(1):9-21.

[106] Bork K, Horstkorte R, Weidemann W. Increasing the sialylation of therapeutic glycoproteins: the potential of the sialic acid biosynthetic pathway. J Pharm Sci. 2009;98(10):3499-508.

[107] Markely LR, Wang DI. High-throughput analysis of intraclonal variability of glycoprotein sialylation. Biotechnol Prog. 2012;28(2):591-4.

[108] Goetze AM, Liu YD, Zhang Z, et al. High-mannose glycans on the Fc region of therapeutic IgG antibodies increase serum clearance in humans. Glycobiology. 2011;21(7):949-59.

[109] Alessandri L, Ouellette D, Acquah A, et al. Increased serum clearance of oligomannose species present on a human IgG1 molecule. MAbs. 2012;4(4):509-20.

[110] Yu M, Brown D, Reed C, et al. Production, characterization, and pharmacokinetic properties of antibodies with N-linked mannose-5 glycans. MAbs. 2012;4(4):475-87.
[111] Bumbaca D, Boswell CA, Fielder PJ, et al. Physiochemical and biochemical factors influencing the pharmacokinetics of antibody therapeutics. AAPS J. 2012;14(3):554-8.
[112] Gramer MJ, Eckblad JJ, Donahue R, et al. Modulation of antibody galactosylation through feeding of uridine, manganese chloride, and galactose. Biotechnol Bioeng. 2011;108(7):1591-602.
[113] Holden HM, Rayment I, Thoden JB. Structure and function of enzymes of the Leloir pathway for galactose metabolism. J Biol Chem. 2003;278(45):43885-8.
[114] Ryll T. Improved galactosylation of recombinant glycoproteins. Patent Application WO 2001059089, filed 02/2000.
[115] Keppler OT, Hinderlich S, Langner J, et al. UDP-GlcNAc 2-epimerase: a regulator of cell surface sialylation. Science. 1999;284(5418):1372-6.
[116] Weiss P, Tietze F, Gahl WA, et al. Identification of the metabolic defect in sialuria. J Biol Chem. 1989;264(30):17635-6.
[117] Leroy J G, Seppala R, Huizing M, et al. Dominant inheritance of sialuria, an inborn error of feedback inhibition. Am J Hum Genet. 2001;68:1419-27.
[118] Ryll T. Improved sialylation of glycoproteins. Patent Application WO 2001059075, filed 02/2000.
[119] Bork K, Reutter W, Weidemann W, et al. Enhanced sialylation of EPO by overexpression of UDP-GlcNAc 2-epimerase/ManAc kinase containing a sialuria mutation in CHO cells. FEBS Lett. 2007;581(22):4195-8.
[120] Son YD, Jeong YT, Park SY, et al. Enhanced sialylation of recombinant human erythropoietin in Chinese hamster ovary cells by combinatorial engineering of selected genes. Glycobiology. 2011;21(8):1019-28.
[121] Chenu S, Grégoire A, Malykh Y, et al. Reduction of CMP-N-acetylneuraminic acid hydroxylase activity in engineered Chinese hamster ovary cells using an antisense-RNA strategy. Biochim Biophys Acta. 2003;1622(2):133-44.
[122] Ghaderi D, Taylor RE, Padler-Karavani V, et al. Implications of the presence of N-glycolylneuraminic acid in recombinant therapeutic glycoproteins. Nat Biotechnol. 2010;28(8):863-7.
[123] Ghaderi D, Zhang M, Hurtado-Ziola N, et al. Production platforms for biotherapeutic glycoproteins. Occurrence, impact, and challenges of non-human sialylation. Biotechnol Genet Eng Rev. 2012;28:147-75.
[124] Wong NS, Yap MG, Wang DI. Enhancing recombinant glycoprotein sialylation through CMP-sialic acid transporter over expression in Chinese hamster ovary cells. Biotechnol Bioeng. 2006;93(5):1005-16.
[125] Jeong YT, Choi O, Son YD, et al. Enhanced sialylation of recombinant erythropoietin in genetically engineered Chinese-hamster ovary cells. Biotechnol Appl Biochem. 2009;52(4):283-91.
[126] Khatra BS, Herries DG, Brew K. Some kinetic properties of human-milk galactosyl transferase. Eur J Biochem. 1974;44(2):537-60.
[127] Abeijon C, Mandon EC, Hirschberg CB. Transporters of nucleotide sugars, nucleotide sulfate and ATP in the Golgi apparatus. Trends Biochem Sci. 1997;22(6):203-7.
[128] Wang TF, Guidotti G. Golgi localization and functional expression of human uridine diphosphatase. J Biol Chem. 1998;273(18):11392-9.
[129] Uccelletti D, Pascoli A, Farina F, et al. APY-1, a novel Caenorhabditis elegans apyrase involved in unfolded protein response signalling and stress responses. Mol Biol Cell. 2008;19(4):1337-45.

[130] Collins B, Guo T, Thiruneelakantapillai L, et al. Production of glycoproteins. Patent Application WO 2010138502, filed 05/2009.
[131] Ferrara C, Grau S, Jäger C, et al. Unique carbohydrate-carbohydrate interactions are required for high affinity binding between FcgammaRIII and antibodies lacking core fucose. Proc Natl Acad Sci USA. 2011;108(31):12669-74.
[132] Subramaniam JM, Whiteside G, McKeage K, et al. Mogamulizumab: first global approval. Drugs. 2012;72(9):1293-8.
[133] Beck A, Reichert JM. Marketing approval of mogamulizumab: a triumph for glyco-engineering. MAbs. 2012;4(4):419-25.
[134] North SJ, Huang HH, Sundaram S, et al. Glycomics profiling of Chinese hamster ovary cell glycosylation mutants reveals N-glycans of a novel size and complexity. J Biol Chem. 2010;285(8):5759-75.
[135] Bosques CJ, Liu C, Murphy J, et al. Antennary fucosylation in glycoproteins from CHO cells. Patent Application US 20110136682, filed 12/2009.
[136] Sakagami M, Horie K, Nakamoto K, et al. Sialyl Lewis X-polysaccharide conjugates: targeting inflammatory lesions. Bioorg Med Chem Lett. 1998;8(19):2783-6.
[137] Horie K, Sakagami M, Kuramochi K, et al. Enhanced accumulation of sialyl Lewis X-carboxymethylpullulan conjugate in acute inflammatory lesion. Pharm Res. 1999;16(2):314-20.
[138] Mulligan MS, Warner RL, Rittershaus CW, et al. Endothelial targeting and enhanced antiinflammatory effects of complement inhibitors possessing sialyl Lewisx moieties. J Immunol. 1999;162(8):4952-9.
[139] von Horsten HH, Ogorek C, Blanchard V, et al. Production of non-fucosylated antibodies by co-expression of heterologous GDP-6-deoxy-D-lyxo-4-hexulose reductase. Glycobiology. 2010;20(12):1607-18.
[140] Ogorek C, Jordan I, Sandig V, et al. Fucose-targeted glycoengineering of pharmaceutical cell lines. Methods Mol Biol. 2012;907:507-17.
[141] Becker DJ, Lowe JB. Fucose: biosynthesis and biological function in mammals. Glycobiology. 2003;13(7):41R-53R.
[142] Yurchenco PD, Atkinson PH. Fucosyl-glycoprotein and precursor pools in HeLa cells. Biochemistry. 1975;14:3107-14.
[143] Yurchenco PD, Atkinson PH. Equilibration of fucosyl glycoprotein pools in HeLa cells. Biochemistry. 1977;16:944-53.
[144] King JD, Poon KK, Webb NA, et al. The structural basis for catalytic function of GMD and RMD, two closely related enzymes from the GDP-d-rhamnose biosynthesis pathway. FEBS J. 2009;276(10):2686-2700.
[145] Bisso A, Sturla L, Zanardi D, et al. Structural and enzymatic characterization of human recombinant GDP-D-mannose-4,6-dehydratase. FEBS Lett. 1999;456(3):370-4.
[146] Umaña P, Jean-Mairet J, Moudry R, et al. Engineered glycoforms of an antineuroblastoma IgG1 with optimized antibody-dependent cellular cytotoxic activity. Nat Biotechnol. 1999;17(2):176-80.
[147] Shinkawa T, Nakamura K, Yamane N, et al. The absence of fucose but not the presence of galactose or bisecting N-acetylglucosamine of human IgG1 complex-type oligosaccharides shows the critical role of enhancing antibody-dependent cellular cytotoxicity. J Biol Chem. 2003;278(5):3466-73.
[148] Brockhausen I, Schachter H. Glycosyltransferases involved in N- and O-glycan biosynthesis. In: Gabius H-J, Gabius S (eds). Glycosciences: Status and Perspectives. Chapman & Hall, UK, 1997.
[149] Peuscher AV. High level recombinant antibody production in Chinese hamster ovary (CHO) cells and characterisation of the carcinoembryonic antigen (CEA) specific human full-size IgG1

H10. Dissertation RWTH Aachen, Germany, 2011 (http://darwin.bth.rwth-aachen.de/opus3/volltexte/2011/3644).
[150] Yamane-Ohnuki N, Kinoshita S, Inoue-Urakubo M, et al. Establishment of FUT8 knockout Chinese hamster ovary cells: an ideal host cell line for producing completely defucosylated antibodies with enhanced antibody-dependent cellular cytotoxicity. Biotechnol Bioeng. 2004;87(5):614-22.
[151] Torella C, Cepeljnik T, Toll H, et al. Acceptable changes in quality attributes of glycosylated biopharmaceuticals. Nat Biotechnol. 2011;29(4):310-2.
[152] McCamish M, Woollett G. The State of the Art in the Development of Biosimilars. Clin Pharmacol Ther. 2012;91(3):405-17.
[153] Jäger C, Ferrara C, Umaña P, et al. Determination of antibody glycosylation by mass spectrometry. Methods Mol Biol. 2012;901:195-208.
[154] Raju TS, Jordan RE. Galactosylation variations in marketed therapeutic antibodies. MAbs. 2012;4(3):385-91.
[155] Bengea CT, Rives LM. Methods for controlling the galactosylation profile of recombinantly-expressed proteins. Patent Application US 20120276631, filed 04/2011.
[156] Fukuta K, Abe R, Yokomatsu T, et al. The widespread effect of beta 1,4-galactosyltransferase on N-glycan processing. Arch Biochem Biophys. 2001;392(1):79-86.
[157] Weikert S, Papac D, Briggs J, et al. Engineering Chinese hamster ovary cells to maximize sialic acid content of recombinant glycoproteins. Nat Biotechnol. 1999;17(11):1116-21.
[158] Anthony RM, Ravetch JV. A novel role for the IgG Fc glycan: the anti-inflammatory activity of sialylated IgG Fcs. J Clin Immunol. 2010;30(1):S9-14.
[159] Anthony RM, Wermeling F, Ravetch JV. Novel roles for the IgG Fc glycan. Ann N Y Acad Sci. 2012;1253:170-80.
[160] Munzert E, Müthing J, Büntemeyer H, et al. Sialidase activity in culture fluid of Chinese hamster ovary cells during batch culture and its effects on recombinant human antithrombin III integrity. Biotechnol Prog. 1996;12(4):559-63.
[161] Ferrari J, Gunson J, Lofgren J, et al. Chinese hamster ovary cells with constitutively expressed sialidase antisense RNA produce recombinant DNase in batch culture with increased sialic acid. Biotechnol Bioeng. 1998;60(5):589-95.
[162] Wang FS, Wang J. Methods of Recombinant Production of Glycoproteins. Patent Application WO2009126564, filed 04/2009.
[163] Zhang P, Tan DL, Heng D, et al. A functional analysis of N-glycosylation-related genes on sialylation of recombinant erythropoietin in six commonly used mammalian cell lines. Metab Eng. 2010;12(6):526-36.
[164] Maï NE, Donadio-Andréi S, Iss C, et al. Engineering a Human-Like Glycosylation to Produce Therapeutic Glycoproteins Based on 6-Linked Sialylation in CHO Cells. Methods Mol Biol. 2013;988:19-29.
[165] Sathyamurthy M, Lee JS, Park JH, et al. Overexpression of PACEsol improves BMP-7 processing in recombinant CHO cells. J Biotechnol. 2012;164(2):336-9.
[166] Bolt G, Kristensen C, Steenstrup TD. More than one intracellular processing bottleneck delays the secretion of coagulation factor VII. Thromb Haemost. 2008;100(2):204-10.
[167] Sutkeviciute I, Mistiniene E, Sereikaite J, et al. The influence of different glycosylation patterns on factor VII biological activity. Biochimie. 2009;91(9):1123-30.
[168] Halabian R, Roudkenar MH, Esmaeili NS, et al. Establishment of a cell line expressing recombinant factor VII and its subsequent conversion to active form FVIIa through hepsin by genetic engineering method. Vox Sang. 2009;96(4):309-15.
[169] Wasley LC, Rehemtulla A, Bristol JA, et al. PACE/furin can process the vitamin K-dependent pro-factor IX precursor within the secretory pathway. J Biol Chem. 1993;268(12):8458-65.

[170] Himmelspach M, Pfleiderer M, Fischer BE, et al. Recombinant human factor X: high yield expression and the role of furin in proteolytic maturation in vivo and in vitro. Thromb Res. 2000;97(2):51-67.
[171] Wajih N, Owen J, Wallin R. Enhanced functional recombinant factor VII production by HEK 293 cells stably transfected with VKORC1 where the gamma-carboxylase inhibitor calumenin is stably suppressed by shRNA transfection. Thromb Res. 2008;122(3):405-10.
[172] Gillis S, Furie BC, Furie B, et al. Gamma-Carboxyglutamic acids 36 and 40 do not contribute to human factor IX function. Protein Sci. 1997;6(1):185-96.
[173] Wajih N, Sane DC, Hutson SM, et al. Engineering of a recombinant vitamin K-dependent gamma-carboxylation system with enhanced gamma-carboxyglutamic acid forming capacity: evidence for a functional CXXC redox center in the system. J Biol Chem. 200518;280(11):10540-7.
[174] Wajih N, Sane DC, Hutson SM, et al. Engineering of a recombinant vitamin K-dependent gamma-carboxylation system with enhanced gamma-carboxyglutamic acid forming capacity: evidence for a functional CXXC redox center in the system. J Biol Chem. 200518;280(11):10540-7.
[175] Sun YM, Jin DY, Camire RM, et al. Vitamin K epoxide reductase significantly improves carboxylation in a cell line overexpressing factor X. Blood. 2005;106(12):3811-5.
[176] Wajih N, Hutson SM, Wallin R. siRNA silencing of calumenin enhances functional factor IX production. Blood. 2006;108(12):3757-60.

4 Engineering of the Product

4.1 Control of Biotheraputics Glycosylation

Sandra V. Bennun, Kelley M. Heffner, Emily Blake, Andrew Chung, and Michael J. Betenbaugh

4.1.1 Introduction

Many commercial proteins that are critical for treating diseases contain oligosaccharides that modify their structure and properties. Among these we find monoclonal antibodies, enzymes, enzyme inhibitors, hormones, cytokines, growth factors, and Fc-fusion proteins [1]. These glycosylated proteins are produced inside the cell through complex glycosylation pathways that involve oligosaccharide-lipid assembly, precursor glycan transfer to protein, and glycan modification as glycoproteins migrate to the cell surface [2, 3].

For biotherapeutics of commercial relevance, N-glycosylation pathways that assemble proteins are particularly important. O-glycosylation posttranslational modifications are not as frequently studied and characterized but can also be important. Because the glycosylation patterns of mammalian cells are the most similar to the human versions, mammalian cells are widely used expression systems. Indeed, mammalian cells (including CHO, murine myeloma, and HEK cell lines) are major expression platforms for commercial production of glycosylated biotherapeutics. Nevertheless, many alternative expression platforms such as yeast and plants [4, 5] are also actively being investigated and have shown promising results including modifications of the glycosylation patterns [6-8].

Glycosylation control is critical to the biotherapeutics industry given that the most subtle glycan modifications of the oligosaccharides attached to proteins can have significant impact on their key therapeutic properties, including folding, solubility, immunogenicity, biological activity, and clearance rate [9]. However, in mammalian expression systems it is difficult to achieve complete control over glycosylation due mainly to the fact that glycan structures are not directly encoded by the genome. Instead, glycosylation is the product of a complex sequence of enzyme-catalyzed reactions that add, trim, and modify the carbohydrate portion of the protein. Anticipating the result of the action of those glycosylation enzymes working in a cooperative mode is not always intuitive and straightforward, making it difficult to predict a priori the effects of genetic and cell culture manipulations on glycan structure profiles of therapeutic proteins.

In this chapter we focus on glycosylation control of recombinant biotherapeutic glycoproteins. An integrative description of the main N- and O-glycosylation pathways is presented to establish a context for understanding their impact on glycoform profiles. Additionally, we discuss the effects of cell culture media and glycoengineering manipulations on the final glycosylation pattern to improve glycoproteins for superior therapeutic value (see Chapter 3.5). Given the complexity of glycosylation processing, a simultaneous consideration of molecular biology, analytical chemistry, biochemical engineering, and mathematical modeling concepts are necessary for optimal control and manipulation of glycosylation properties of recombinant biotherapeutics produced by mammalian cells.

4.1.2 Processing Pathways for N- and O-glycosylation in Mammalian Expression Systems

In biotechnology, much effort is focused on producing correctly glycosylated proteins for bioactivity with glycoforms without an allergenic immune response. Two major types of glycosylation are predominant on glycoproteins: N-glycosylation and O-glycosylation, with N- glycosylation being the most frequently studied for industrial glycosylated biotherapeutics. These glycosylation types are defined by the type of linkage ("N" and "O") between the protein backbone and the glycan structure (Figure 4.1.1). N-linked glycans are attached to proteins at the nitrogen atom ("N") of the amide group of an asparagine amino acid residue [2, 10]. O-linked glycans are attached to the oxygen atom ("O") of serine (mainly) or threonine [3, 11].

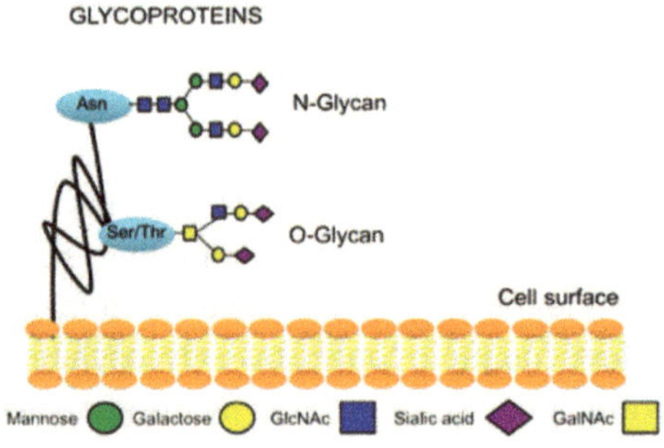

Figure 4.1.1: N-glycosylation is typically more complex than O-glycosylation. In N-glycosylation a N-acetylglucosamine (GlcNac) residue of the glycan structure is β-linked to the amide (N) group of asparagine of a Asn-Xxx-(Ser, Thr) motif, where Xxx is any amino acid except proline. In mucin O-glycosylation a N-acetylgalactosamine (GalNac) residue of the glycan structure is α-linked to the hydroxyl (O) group of serine or threonine.

4.1.2.1 N-linked Glycan Biosynthesis

The process of protein N-glycosylation starts in the endoplasmic reticulum (ER) with the transfer of an assembled glycan from a lipid onto a glycosylation site of a nascent protein in the ER lumen [2] (Figure 4.1.2).

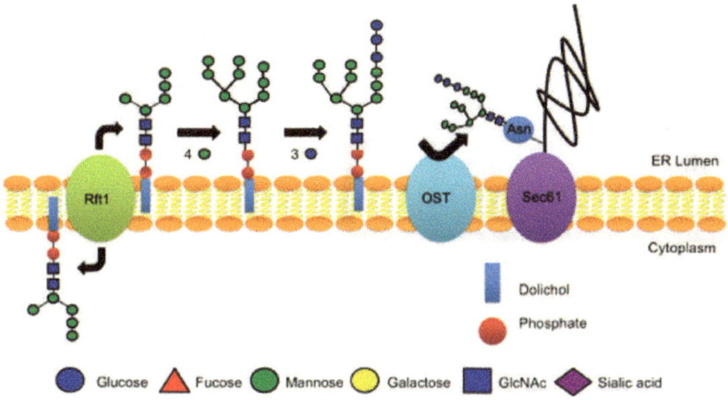

Figure 4.1.2: An oligosacharide structure of GlcNAc (N-acetylglucosamine) and mannose composition forms the polyisoprenoid chain phospholipid-linked oligosaccharide Man5-P-P-dolichol in the cytosol. Next, the oligosaccharide moiety Man5-P-P-dolichol flips from the cytosol to the lumen of the endoplasmic reticulum (ER) where the synthesis of the lipid-glycan moiety (Glc3Man9GlcNAc2-P-P-dolichol) is completed. This step is followed by the *en bloc* transfer of the complex oligosacharide structure (Glc3Man9GlcNAc2) to the asparagine (Asn) residue of the Asn-Xxx-Ser/Thr sequence of the polypeptide chain of the nascent protein by the action of oligosaccharyltransferase (OST) for N-glycosylation.

Some glycan modifications, such as trimming of terminal glucose structures (Figure 4.1.3), take place in the ER facilitating the proper folding of the nascent protein. The protein-bound glycan is trimmed to high mannose glycan structures with 8 and 9 mannose residues (Man8/9-GlcNAc2-protein) in the ER by the enzyme ER α-mannosidase I (ER Man I) and is trimmed further in the Golgi to glycan structures of 5 mannose residues (Man5GlcNAc2) [12, 13].

Subsequent posttranslational glycan modifications occur while the proteins travel through the Golgi apparatus (Figure 4.1.3) generating a diverse set of glycan structures of the subtypes: high-mannose (with 5 or fewer mannose residues), hybrid and complex N-glycans [14, 15] (Figure 4.1.3).

Hybrid glycan structures are formed in the Golgi when a N-acetylglucosamine (GlcNAc) residue is added to the glycan structure, $Man_5GlcNAc_2$, resulting in the formation of $GlcNAcMan_5GlcNAc_2$ (a hybrid structure with 1 antenna). Hybrid structures can be further processed by addition of galactose, sialic acid, and/or core fucose residues, or they can be converted into complex glycans.

Complex glycans are formed by the addition of GlcNAc residues to form bi-, tri- and tetraantennary glycans. These complex glycan structures can undergo further modifications such as elongation of the antennae by lactosamine repeats (Galb1-4GlcNAc) or the addition of galactose and/or sialic acid residues.

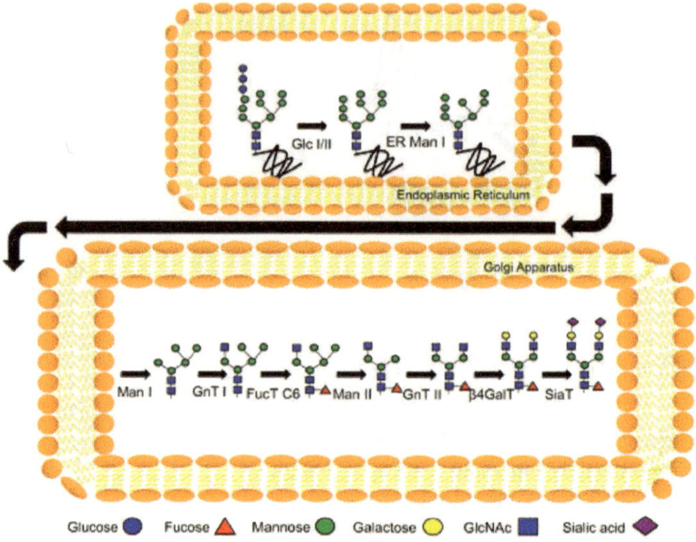

Figure 4.1.3: Schematics of N-glycan biosynthesis pathway in mammalian expression systems. Simplistic representations of the endoplasmic reticulum (ER; upper compartment) and the Golgi apparatus (bottom compartment) are given. The N-glycosylation enzymatic process consist of trimming glucose residues in the ER by α-glucosidase I (Glc I) and α-glucosidase II (Glc II) and subsequent elimination of a specific terminal mannose residue by ER α-mannosidase I (ER Man I). Further, action of N-glycosylation enzymes proceed in the Golgi apparatus by Golgi α(1,2)-mannosidase I (Man I), β-N-acetylglucosaminyltrasferase I (GnTI), core α(1,6) fucosyltransferase (FucTC6), Golgi α(1,3) and α(1,6)-mannosidase II (Man II), β-N-acetylglucosaminyltransferase II (GnTII), β(1,4)-galactosyltransferase (β4GalT), and sialyltransferase (SiaT).

4.1.2.2 O- linked Glycan Biosynthesis

Mucin type O-glycosylation, which begins with the addition of GalNac to serine or threonine (Figure 4.1.1), is the most abundant type of O-glycosylation. Another type of linkage, also abundant in nature, is the GlcNAc(β1-O)Ser/Thr linkage; however, it does not involve further glycosylation processing. In this work we will focus on mucin type O-glycosylation. We will also briefly describe O-mannosylation Man(α1-O)Ser/Thr as later research suggests it is important in CHO cell glycosylation. There are also other types of O-glycosylation defined by other carbohydrate peptide linkages, such as Fuc(α1-O)Ser/Thr, Xyl(β1-O)Ser, Glc(β1-O)Ser, Gal(α1-O)Ser, that can be found in other publications [16].

O-GalNac-linked Glycan (Mucin O-glycosylation) Biosynthesis

If the monosaccharide linked to the (OH) group of Ser/Thr is the N-acetylgalactosamine (GalNAc) sugar residue, then this type of glycosylation is denominated mucin type O-glycosylation because it is found in mucin glycoproteins. Mucin O-glycosylation occurs in the Golgi compartments (Figure 4.1.4).

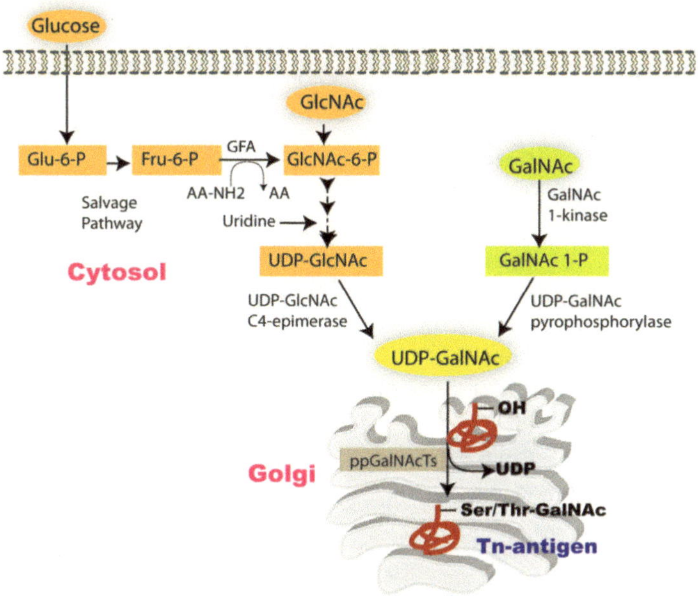

Figure 4.1.4: Schematic of mucin O-glycosylation initiation. UDP-GalNAc is produced endogenously from UDP-GlcNAc, which can be generated from GlcNAc. Alternatively, UDP-GalNAc can be generated from GalNAc by GalNAc 1-kinase and UDP-GalNAc pyrophosphorylase enzymes. UDP-GalNAc is transported into the Golgi providing the sugar residue for the initialization of mucin O-glycosylation. The mucin O-glycosylation initiation is controlled by the catalytic action of the enzyme polypeptide N-acetylgalactosaminyl transferase (pp-GalNAc-T) family. This results in the formation of the core structure Tn antigen GalNAcα1-Ser/Thr in the Golgi apparatus.

The family of polypeptide GalNAc transferases that initiate mucin O-glycosylation by transferring the N-acetylgalactosamine sugar residue to Ser/Thr (forming the Tn antigen- GalNAcα1-Ser/Thr) are localized throughout the Golgi [17]. Additional processing of the Tn antigen glycan structure is show in Figure 4.1.5. Sialylation of the Tn antigen through the α2-6 linkage results in the sTn antigen (NeuAcα2-6GalNAc-Ser/Thr). Further enzymatic glycosylation to add Gal, GlcNAc or GalNAc to the Tn core structure defines at least 8 main mucin O-glycan structures – cores 1 to 8 – found on mammalian glycoproteins [18] (Figure 4.1.5).

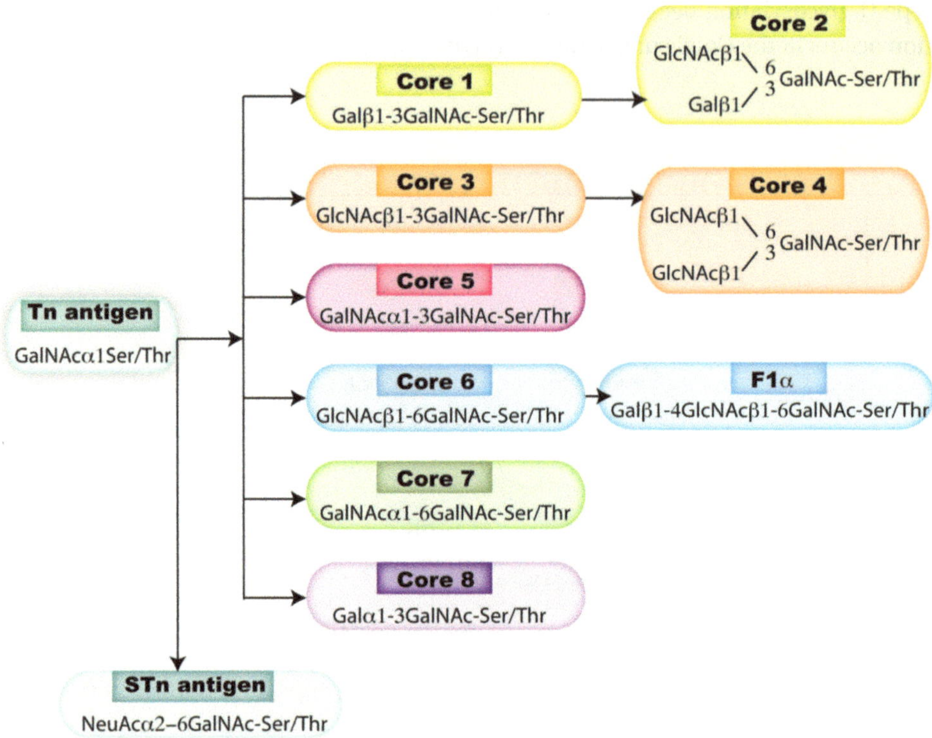

Figure 4.1.5: Further biosynthesis of mucin-type O-linked glycoproteins. Elaboration of the "Tn-antigen" (α-GalNAc-Thr/Ser) by downstream glycosyltransferases generates more complex mucin-type O-linked glycans (core 1 to core 8).

The biosynthesis of the core 1 glycan structure involves the addition of Gal to the GalNAc residue of the Tn antigen through a β1-3 glycosidic linkage by the action of core 1 UDP-Gal:GalNAc-Ser/Th β1,3-galactosyltransferase (core 1 3-Gal-T, EC 2.4.1.122) to form the core 1 disaccharide structure Galβ1-3GalNAcα1-Ser/Thr (Figure 4.1.5). Interestingly, if the same β1-3 linkage to GalNAc residue proceeds with GlcNAc, then core 3 structure is synthesized by the action of (β3Gn-T6). This implies that both, core 1 and core 3 structures, exist simultaneously in a competitive manner. Core 2 and core 4 are formed by addition of GlcNAcβ1-6 to core 1 and core 3 respectively. Core 6 results from GalNAcβ1-6 linkage to the Tn antigen, further modification of core 6 with addition of Gal residue gives the F1α glycan moiety. Similarly, addition of GalNAcα1-3 to the Tn antigen forms core 5, and addition of GalNAcα1-6 to Tn antigen forms core 7. Core 8 is produced by addition of Galα1-3 linkage to the Tn antigen.

O-mannose-linked Glycan Biosynthesis

Whereas all of the O-GalNAc (O-linked N-acetylgalactosamine) pathway steps occur in the Golgi, the first step of O-mannose-linked glycan assembly occurs in the ER and then is completed in the Golgi (Figure 4.1.6). O-mannosylation shown schematically in Figure 4.1.6, consists of activation of the POMT1-POMT2 mannosyltransferase complex, which uses dolichol-P-mannose to add α-mannose to serine (Ser) or threonine (Thr) residues. POMGnT1 then uses UDP-GlcNAc (N-acetylglucosamine) to add a β1,2GlcNAc residue to the mannose. The O-mannose structure is completed by the addition of β1,4-galactose and 2,3-sialic acid residues. The specific transferases have not been identified for these steps. Glucuronic acid, sulfate, and other branches can also be added.

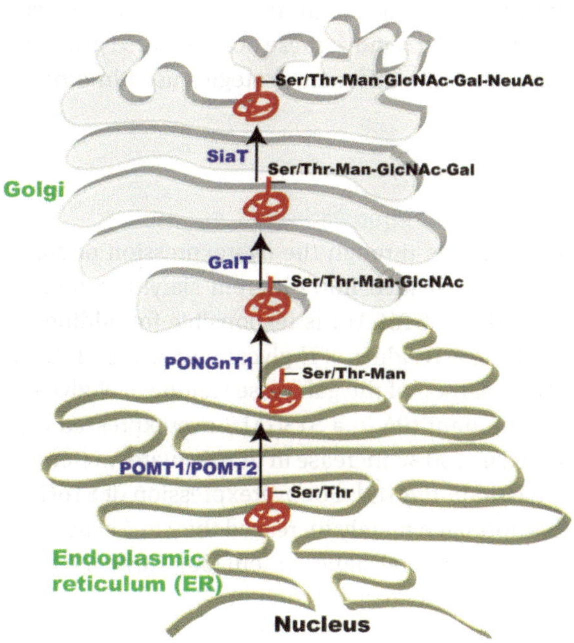

Figure 4.1.6: Schematic of O-Mannosylation synthesis.

4.1.3 Cell Line Glycoengineering for Production of Therapeutics

Protein-based therapeutics have been widely employed as treatments of diseases. In this section, we will discuss various studies in which glycoengineering approaches have been used to obtain desired glycosylation patterns on erythropoietin (EPO), acetylcholinesterase (AChE), and tissue plasminogen activator (tPA). Strategies to increase sialylation will be outlined including expression of 2,3-sialyltransferase (ST3GAL3 and ST3GAL4), 2,6-sialyltransferase (ST6GAL1), CMP sialic acid transporter, and UDP-GlcNAc epimerase-ManNAc kinase (GNE/MNK). In addition, the increase of antenna structures, the knockdown of sialidases that cleave out the terminal sialic acid residues, and the increase of glycan chains by side direct mutagenesis will be also discussed. Furthermore, the elimination of the core fucose residue by siRNA, shRNA, homologous recombination and ZFN to increase the antibody cell dependent cytotoxicity (ADCC), as well as the expression of bisecting glycan structures to enhance ADCC will be addressed.

4.1.3.1 Strategy of Sialylation

Sialylation is critical in controlling pharmacokinetic properties, including the circulatory retention time in the blood. A lack of terminal sialic residues on glycan chains shortens the circulatory retention time due to the asialoglycoprotein receptor, a lectin in hepatocytes that can recognize, capture, and remove these nonsialylated glycoproteins from blood. Following, we summarize the different strategies used in order to alter the sialylation pattern.

Increase Sialylation by Incorporating the Sialylation Pathway

This main approach to improve sialylation is through the overexpression of sialylatransferases. This method can be further divided into different sialyltransferases, including ST6GAL1, ST3GAL3, and ST3GAL4. ST6GAL1 is responsible for adding the sialic acid through a α2-6 linkage to galactose residues of N-glycans, whereas ST3GAL4 and ST3GAL3 add the sialic on the third carbon of the galactose residue of N-glycans.

In 1999, Weikert et al. produced a mutant tPA in a ST3GAL3-overexpressed CHO cell line. The oversialylated tPA resulted in a 30 % increase in the circulatory retention time as compared with the wild type mutant tPA [19]. For overexpression of ST6GAL1 in CHO cells, it was reported that this approach merely increased the 2 to 6 sialylation percentage without a concomitant increase in total sialic content [20]. However, other research suggests ST6GAL1 does increase the total sialic content [21]. For ST3GAL4, it was reported this enzyme has a weaker ability to facilitate N-linked sialylation [22].

In addition to the sialyltransferase gene, Wong et al. overexpressed the CMP sialic acid transporter (CMP-SAT), resulting in a 4 to 16 % increase on sialic content [23]. However, a single gene expression may not substantially enhance the total sialylation content. For this reason, researchers have overexpressed multiple genes from the sialic acid biosynthesis pathway. Jeong et al. coexpressed both, ST3GAL3 and β1-4 galactose transferase (β1-4 GT), and increased the sialylation of erythropoietin (EPO), a heavily glycosylated hormone involved in regulating red blood cell production [24]. The total sialic content was increased from 6.7 to 7.5 mole of sialic acid per mol of EPO. They also compared the sialylation level of single transfection (ST3GAL3), double expression (ST3GAL3 and CMP-sialic acid synthetase (CMP-SAS)), and triple expression (ST3GAL3, CMP-SAS and CMP-sialic acid transporter (CMP-SAT)) [25]. From the above choices, the triple gene expression exhibited the highest sialylation content. Son et al. further increased the sialylation of EPO by triple expression of mutated rat GNE/MNK, CHO CMP-sialic acid transporter and human α2,3-sialyltransferase (α2,3-ST), resulting a 43 % increase in the sialic acid content [26].

Increase Antenna Structures

In 2000, Fukuta et al. attempted to alter glycan branching of human interferon-γ (IFN-γ) by expressing N-acetylglucosaminyltransferases, GnT-IV, and/or GnTV in CHO cells [27]. Wild type CHO cells producing IFN-γ (IM4 cells) typically synthesize biantennary glycan chains and rarely multiantennary glycans [27]. GnT-IV and GnT-V are part of a family of N-acetylglucosaminyltransferases (GnTs) that are involved in forming multi-antennary sugar chains on glycoproteins; therefore, the group introduced these genes to CHO cells in order to produce IFN-γ displaying multi-antennary glycan structures.

Following transfection of GnT-IV and/or GnT-V, they measured the activities of these glycosyltransferases compared to the parental IM4 cell line. The control IM4 cells did not exhibit any detectable GnT-IV activity but did have slight GnT-V activity (0.46 nmol/h/10^6 cells). When transfected with GnT-IV (IM4/IV), these cells had very high GnT-IV activity (17.8 nmol/h/10^6 cells). Similarly, transfecting the cells with GnT-V resulted in high GnT-V activity (IM4/Vh, 7.74 nmol/h/10^6 cells). To create a cell line coexpressing GnT-IV and GnT-V, they transfected the IM4/Vh cells with GnT-IV to generate the IM4/V/IV cell line that had high activities of both GnT-IV (10.7 nmol/h/10^6 cells) and GnT-V (8.52 nmol/h/10^6 cells).

As anticipated by reversed phase HPLC, the control IM4 cell line produced 63.8 % biantennary sugar chains. The glycan chain composition of the IM4/IV cell line consisted of 66.9 % triantennary structures. Similarly, the high expression IM4/Vh cells produced 55.7 % triantennary sugar chains. Finally, tetraantennary structures comprised 56.2 % of the total glycan chain composition of the IM4/V/IV cell line. The addition of GnT-IV and/or GnT-V ultimately succeeded in producing multiantennary sugar chains in the respective cell lines. The extent of sialylation of the glycan chains produced in the various IM4 cell lines was also examined using anion exchange HPLC. Fukuta et al. found no clear correlation between an increase in degree of branching and sialylation. However, the percentage of trisialo and tetrasialo glycans increased in the IM4/Vm, IM4/Vh, and IM4/V/VI cell lines thus indicating a possible relationship between branching degree and extent of sialylation.

Lastly, they analyzed the effect of GnT-IV and/or GnT-V on the glycan branching of cellular proteins in each transfected IM4 line using a DSA lectin blot that detects multiantennary glycoproteins. As expected, the parental IM4 cells did not show activity for DSA as mostly biantennary sugars are produced in this line. However, cell lysates from all other IM4 lines transfected with GnT-IV and/or GnT-V reacted with DSA lectin further confirming the finding that introducing these genes to CHO cells results in the production of predominantly multiantennary sugar chains [27].

Knockdown of Sialidases

It has been found that there are 4 mammalian sialidases, including Neu1, Neu2, Neu3, and Neu4. From these 4 sialidases, Neu2 has been shown to exist in CHO [28]. Ngantung et al. performed RNA interference by using a short hairpin RNA vector containing a siRNA sequence, which effectively knocked down the Neu2 gene and maintained the interferon-γ's sialic acid content even during the death phase [29]. However, this knockdown of Neu2 did not increase the total sialylation. To resolve this problem, Zhang et al. conducted a similar experiment and knocked down both Neu2 and Neu3 gene using shRNA [30]. The result indicated the simultaneous knockdown of both, Neu2 and Neu3, could moderately increase the sialylation.

Increase Glycan Chain by Side Directed Mutagenesis

In 2003, Egrie et al. developed an EPO mutant, darbepoetin alfa, that included 5 N-linked glycans, 2 more than recombinant human EPO (rHuEPO), which contains 3 N-glycans and 1 O-glycan. The overall goal was to examine the effect of increased glycosylation and sialylation on the clearance rate, *in vivo* biological activity, and receptor binding [31]. Previously, Egrie et al. showed that increases in sialic acid content extends the half-life of rHuEPO. Pharmacokinetic studies comparing darbepoetin alfa with rHuEPO indicated that darbepoetin alfa exhibits a slower clearance and longer half-life in both, rats and dogs [32]. For rats, the half-life for darbepoetin alfa was 6.9 h compared with 2.5 h for rHuEPO. Similarly, for dogs, the half-life of darbepoetin alfa was 25 h while that of rHuEPO was only 7.2 h. Egrie et al. did found that the additional glycosylation and sialylation of darbepoetin alfa decreased its affinity for its receptor fourfold compared with rHuEPO. However, darbepoetin alfa has a much higher *in vivo* potency than does rHuEPO. Depending on the dosing frequency of either 1 or 3 times a week, darbepoetin alfa was 3.6 or 13 times more potent than rHuEPO, respectively. They ultimately concluded that despite the decrease in receptor binding, the increase in circulatory half-life resulted in a more potent and more highly active molecule compared to the wild-type rHuEPO [31-33].

Kronman et al. (1995) also studied the effect of protein glycosylation on the circulatory half-life of recombinant human acetylcholinesterase (rHuAChE) [34]. Removal of the sialic acid residues resulted in a significant decrease in half-life. Human acetylcholinesterase (HuAChE) includes 3 glycosylation sites at residues 265, 350, and 464, while fetal bovine serum AChE (FBS-AChE) contains 5 possible sites – the 3 found in HuAChE and 2 additional sites at residues 61 and 541. HuAChE mutants containing 1 glycosylation site (N350Q/N464Q), 4 glycosylation sites (S541N and D61N), and 5 glycosylation sites (S541N/D61N) were created. Removing 2 glycosylation sites from the primary structure resulted in a decreased half-life from 80 min for the wild type enzyme to 45 min for the N350Q/N464Q mutant enzyme. Unexpectedly, the additional glycosylation sites of the other rHuAChE variants did not extend the circulatory half-life of the enzymes. Both tetraglycosylated mutants (S541N and D61N) displayed shorter half-lives of 71 min and 41 min, respectively. Finally, the pentaglycosylated mutant exhibited the shortest half-life of 25 min. They concluded that merely increasing the number of N-glycosylation sites would not always result in a decreased clearance rate [34].

In 2002, Chitlaru et al. produced 16 recombinant AChE structures differing in the number of N-glycosylation sites, their sialylation, and subunit assembly [35]. They used MALDI-TOF mass spectrometry to determine the glycan structures of the different AChE glycoforms in order to examine the effect of N-glycosylation and sialylation on circulatory half-life. These glycoforms, containing between 2 to 5 N-glycosylation sites, were either partially or fully sialylated, and were either dimeric or tetrameric. Because HEK293 cells inefficiently sialylate AChE, they also expressed these AChE glycoforms in genetically modified HEK293 cells with high α-2,6-sialyltransferase activity (293ST-2D6 cells) in order to mimic efficient and total sialylation of the glycoprotein. The AChE variants expressed in this cell line were nearly fully sialylated. They found similar results as in their previous experiment with the various AChE mutants. For undersialylated AChE, increasing the number of glycosylation sites adversely affected the molecules' mean residence times (MRTs). When the rHuAchE mutants were expressed in the 293ST-2D6 cells mimicking total sialylation, the mean residence times (MRTs) were significantly longer than those of the undersialylated forms. Ultimately, they concluded that increasing the number of glycosylation sites positively affects circulatory retention when the AChE molecules are efficiently sialylated.

4.1.3.2 Fucosylation

In addition to sialylation, eliminating fucosylation of the therapeutic antibody is another useful approach, especially for cancer therapy [36]. The antibody can bind to the immune effector cell and induce antibody-dependent cell-mediated cytotoxicity (ADCC). The antibody attached to a cancer cell can be bound by an Fc receptor from the natural killer cell and followed by releasing cytokines that trigger cell apoptosis. Eliminating fucosylation and increasing bisecting GlcNAc levels has been shown to increase antibody ADCC.

The fucosylation pathway is described as follows: GDP-mannose 4,6-dehydratase (GMP) catalyzes GDP-mannose into GDP-4-dehydro-6-deoxy-D-mannose and the GDP-4-dehydro-6-deoxy-D-mannose can further be catalyzed into GDP-fucose and transported into the Golgi apparatus to add on the glycan chain. In mammalian cells, FUT8 encodes α-1,6-fucosyltransferase [37, 38] which produces α-(1,6)-fucosylation on the core (Figure 4.1.7) linked to the innermost N-acetylglucosamine residue of an N-glycan [39]. Antibodies with reduced fucosylation on the constant region (Fc) render stronger ADCC than the fucosylated ones [40-43]. In order to generate and increase bisecting GlcNAC levels of making non-fucosylated antibody, the most widely used approach is through cell glycoengineering to knockdown or knockout the fucosyltransferase gene from the production vehicle. Several approaches have been applied to achieve this goal.

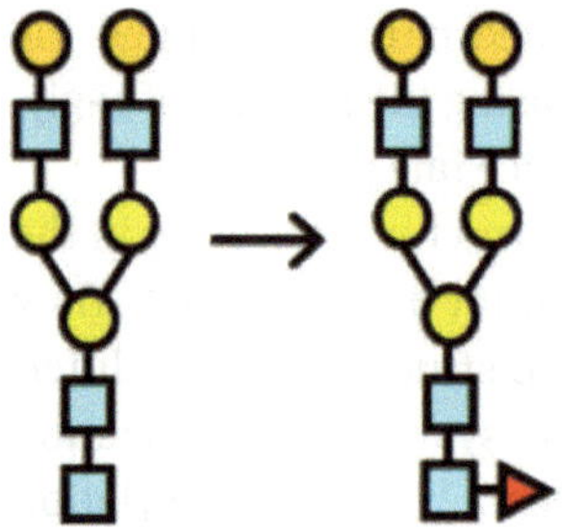

Figure 4.1.7: The fucosylation reaction of FUT8.

One approach has been to utilize the lectin resistance cell line Lec13, a cell line derived from Pro-CHO 5 cell which has a loss of function of GDP-mannose 4,6-dehydratase activity [44] resulting in a 90 % reduction in the number of fucose residues attached to the glycan chain as compared with the parental cell line. Shields et al. created a Lec13 cell line producing anti-Hu4D5 [42]. This humanized anti-IgE IgG1 showed a more prominent ADCC effect on the human breast cancer cell, SK-BR-3.

Nishiya et al. successfully knocked down the FUT8 and GDP-mannose 4,6-dehydratase (GMP) genes from CHO/DG44 cells which produce antibody IgG1 using short hairpin RNA to silence and almost eliminate the fucosylation on the IgG1 antibody [45]. GMP can convert GDP-mannose into GDP-4-dehydro-6-deoxy-D-mannose [46]. The antibodies from the knockdown cell lines exhibited 100 higher ADCC efficacies than the antibodies from the parental CHO cell.

However, the Lec13 cell does not completely eliminate fucosylation and shRNA may not be expressed sufficiently in the host cell without a presence of selection reagent. The application of homologous recombination can circumvent the above bottlenecks. Ohnuki et al. established a FUT8-knockout CHO cell line by the Cre-loxp homologous recombination system [40]. The Anti-HER2 IgG1 produced from the FUT8 complete knockout cell exhibited extensively higher ADCC with respect to the fucosylated one.

The zinc finger is a protein structural motif, which binds to the repetitive sequences of DNA and has been commonly utilized as a genomic editing tool when fused with different functioning domains such as folk nucleases, activators, and suppressors (see Chapter 3.4). Binding to the target sequence and the folk nucleases create a double-strand break (DSB). The gene sequence will be altered through non-homologous end joining of the double-stranded DNA [47]. Unlike other recombination, gene modification, or disruption approaches, which utilize chemical methods, antibiotic selection, or viral vector integration, the user simply performs transient transfection of ZFN-mediated vector and sorts out the target cell. As a result, Cristea et al. created a CHO cell line which has complete knockout of the FUT8 gene [48] using zinc finger nucleases.

Another approach to increase ADCC is overexpression of β1-4-N-acetylglucosaminyltransferase III (GnT III) to produce the bisecting glycan structure (Figure 4.1.8). Davies et al. (2001) established a GnT III overexpressed CHO cell line producing an anti CD-20 antibody [49]. The increase in bisecting glycan chains on the antibody resulted in a twentyfold lower antibody dosage to achieve the same ADCC effect induced by the antibody produced from the parental CHO cell line.

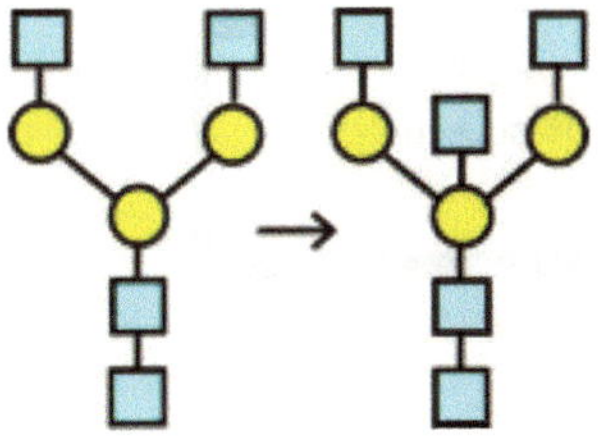

Figure 4.1.8: Overexpression of GnTIII to produce bisecting glycan chain.

Ferrara et al. (2006) combined the cotransfection of GnTIII and the ManII gene to increase both, ADCC and complement-dependent cytotoxicity (CDC) [41]. Overexpression of ManII encoding the Golgi α-mannosidase II enzyme deletes the high mannose type glycan structure (Figure 4.1.9). Coexpressing GnTIII and ManII together with the antibody expression vector generates antibody with mainly hybrid type bisecting nonfucosylated glycans structures. The antibody showed a prominent ADCC activity and a comparable CDC effect with respect to the unmodified antibody.

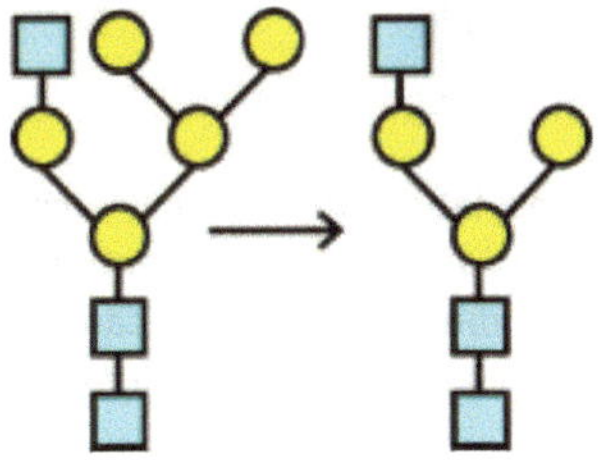

Figure 4.1.9: Overexpression of ManII.

4.1.4 Effect of Cell Culture Media Additives on Protein Glycosylation

4.1.4.1 Effect of Cell Culture Media Additives on Glycosylation

The purpose of this section is to overview the various supplementation strategies and to highlight the effect of these on glycosylation. Many supplements have been shown to affect cell culture glycosylation, and will be presented in Tables 4.1.1 to 4.1.6 and Figures 4.1.10 to 4.1.11. The applications, challenges, and future directions of media additives are also discussed. Often, these strategies are cell line specific.

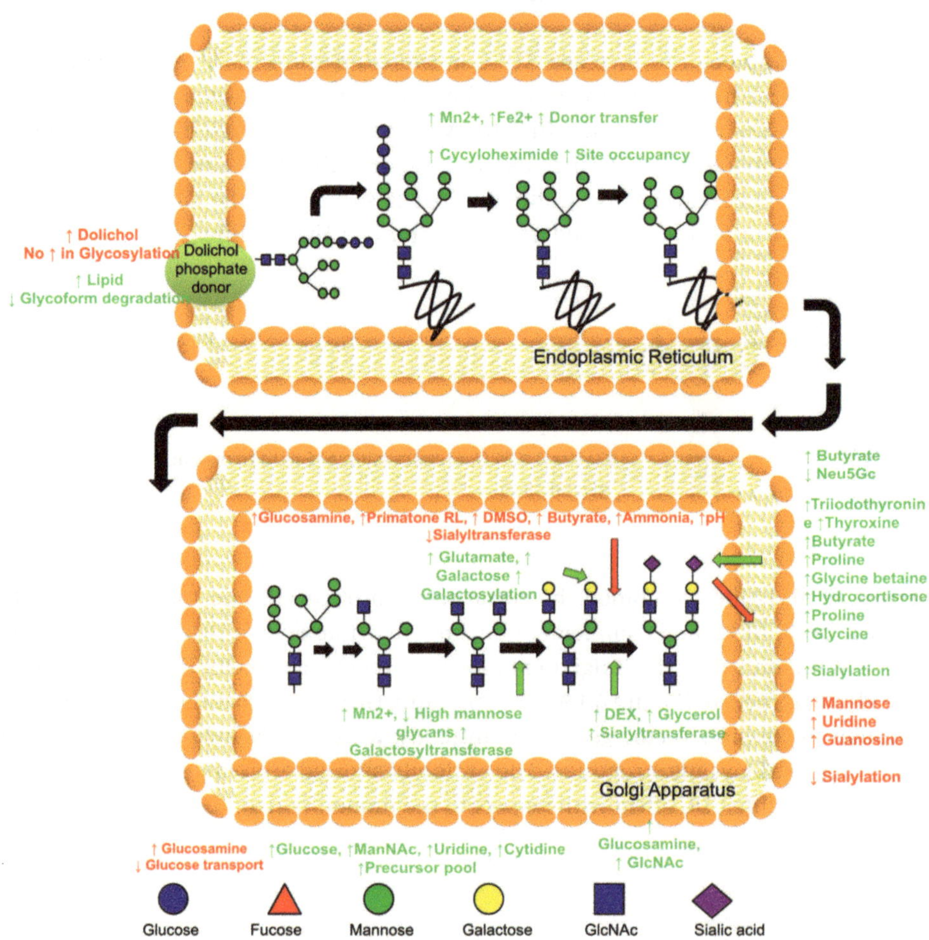

Figure 4.1.10: Overview of how various medium supplements affect glycosylation. The effect of each supplement is shown as green, representing an improvement in glycosylation, or red, representing no improvement or a decline in glycosylation. As described in the text, medium supplements can affect different stages of glycosylation, including initial processing in the endoplasmic reticulum or further processing and sialylation in the Golgi apparatus.

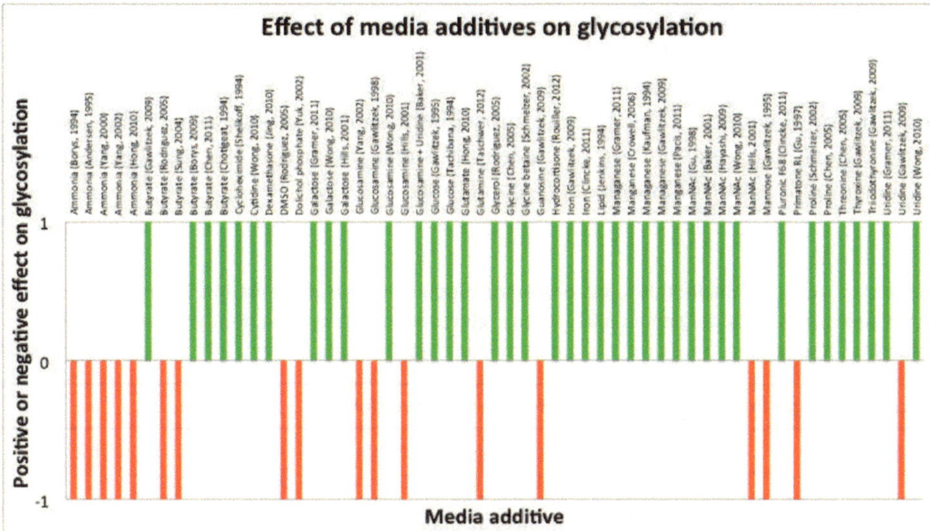

Figure 4.1.11: Summary of the effect of various medium supplements on glycosylation. The effect of each supplement is shown as +1 (green, representing an improvement in glycosylation) or -1 (red, representing no improvement or decrease in glycosylation). The reference is indicated in parentheses.

4.1.4.2 Applications

Medium formulation plays a critical role in cell culture performance. The optimal formulation should achieve high cell growth, high recombinant protein productivity, and maintain consistent glycosylation. Often, different formulations are required for different applications, and the formulation changes for each cell line. For the control of glycosylation, various additives have been tested. As highlighted below and in Figures 4.1.10 and 4.1.11, some additives produce different results in different cell lines while other additives have been used across cell lines to control glycosylation. During industrial process development, cell culture media additives are useful for controlling process robustness while avoiding time delays involved in cell line engineering. Strategies for media supplements require optimization with respect to cell line, recombinant protein produced, and process characteristics.

Nucleotide Sugar Precursors

Nucleotide sugars are activated sugars that are precursors for glycosylation. Generally referred to as UDP-HexNAc, including UDP-N-acetylgalactosamine (UDP-GalNAc) and UDP-N-acetylglucosamine (UDP-GlcNAc) are six-carbon nucleotide sugars with N-acetyl groups. Both UDP-GalNAc and UDP-GlcNAc play roles in transferring GalNAc or GlcNAc to the dolichol phosphate in the cytoplasm and adding GalNAc or GlcNAc to the maturing glycoform in the Golgi. Analysis of nucleotide sugars during cell culture indicated that reducing the UDP-glucose flux also reduces UDP-galactose flux, negatively affecting galactosylation and sialylation [50]. Other examples, including CMP-sialic acid, play critical roles in the subsequent sialylation of galactosylated glycoforms. Thus, supplementation of nucleotide sugar precursors has been investigated as a method for controlling glycosylation (Table 4.1.1).

Table 4.1.1: Effect of nucleotide sugar precursor supplements on glycosylation. Various nucleotide sugar precursors, including ammonia, cytidine, glucosamine, guanosine, N-acetyl mannosamine (ManNAc), and uridine, were investigated as medium supplements for their effect on glycosylation.

Nucleotide Sugar Precursor	Reference	Effect on Glycosylation
Ammonia	[50]	Increasing ammonium chloride concentration inhibits CHO N-linked glycosylation but is dependent on pH
Ammonia	[54]	Increasing ammonium chloride concentration inhibits CHO O-linked sialylation
Ammonia	[51]	Increasing ammonium chloride concentration reduces terminal sialylation and yields more heterogeneous glycoforms
Ammonia	[53]	Increasing ammonium chloride concentration yields more heterogeneous glycoforms
Ammonia	[52]	Increasing ammonium chloride concentration decreases the fraction of galactosylated glycoforms
Cytidine	[58]	Combination of ManNAc and cytidine increase the pool of CMP-sialic acid and sialic acid content of glycoforms
Glucosamine	[53]	Increasing glucosamine concentration decreases sialylation
Glucosamine	[85]	Increasing glucosamine concentration decreases sialylation
Glucosamine	[58]	Glucosamine supplementation with or without uridine increases sialic acid content of glycoforms
Glucosamine	[57]	Glucosamine supplementation is detrimental to glycosylation because the UDP-HexNAc pool increases corresponding to a decrease in galactosylation
Glucosamine + Uridine	[56]	Supplementation of glucosamine in combination with uridine increases the pool of UDP-HexNAc and increases antenarrity of glycoforms in CHO but not NS0 cells. Sialylation is decreased in CHO and NS0 cells
Guanosine	[60]	Guanosine supplementation is detrimental to glycosylation because site occupancy decreases
ManNAc	[86]	ManNAc supplementation improves sialylation by increasing the CMP-sialic acid pool and decreasing the percentage of incompletely sialylated glycoforms

ManNAc	[56]	ManNAc supplementation increases the pool of CMP-sialic acid but does not affect sialylation in CHO or NS0 cells
ManNAc	[62]	ManNAc supplementation improves glycosylation of recombinant protein produced in BHK cells
ManNAc	[58]	ManNAc supplementation with or without cytidine increases sialic acid content of glycoforms
ManNAc	[57]	ManNAc supplementation increases the pool of CMP-sialic acid but does not improve sialylation
Uridine	[59]	Uridine supplementation synergistically increases percentage of G1 relative to G0 glycans with galactose and manganese
Uridine	[60]	Uridine supplementation is detrimental to glycosylation because site occupancy decreases
Uridine	[58]	Uridine supplementation in combination with galactose or glucosamine increases sialic acid content of glycoforms

Ammonia

Ammonia is a toxic byproduct of amino acid metabolism, most often the result of glutamine consumption in mammalian cell culture. However, ammonia is also involved in the synthesis of glucosamine-6-phosphate, a precursor to UDP-GlcNAc. To determine how ammonia concentration affects the degree of N-linked glycosylation, Borys et al. investigated various concentrations of ammonium chloride supplementation at different pH levels and observed that increasing ammonium chloride concentration decreased the percentage of glycosylated protein [51]. The decrease in glycosylated protein was more significant at pH above 7.2 [51]. It was verified that the effect on glycosylation was due to ammonium chloride supplementation and not osmolality [51, 52]. In addition, ammonia inhibits galactosylation, terminal sialylation and increases glycoform heterogeneity [52-54]. As ammonia concentration increases, there is an increase in pH of the trans-Golgi, which can alter the activity of glycosylation enzymes.

Increasing ammonia concentration also affects O-linked glycosylation. Multiple researchers observed a decrease in terminal sialic acid addition as ammonium chloride concentration increased [52, 54, 55]. A decrease in disialylated relative to monosialylated protein was detected [55]. Ammonia is a weak base that minimizes the pH difference between the Golgi and cytoplasm, which leads to the decrease in sialylation that occurs in the trans-Golgi. Other pH dependent effects could affect the activity of sialylation enzymes.

Glucosamine

Glucosamine is the direct precursor to UDP-GlcNAc. The effect of glucosamine supplementation on glycosylation is similar to ammonia. Glucosamine increases glycoform heterogeneity and decreases terminal sialylation [54]. This correlates with an increased pool of UDP-GlcNAc but decreases glucose transport [52]. The decrease in sialylation and increase in UDP-GlcNAc concentration has been confirmed in other experiments [56, 57]. An explanation for the effect of glucosamine may relate to reduced glucose transport, which affects the overall sugar pool and supply of glycosylation precursors to a greater extent than the increased UDP-GlcNAc pool. In one study, glucosamine supplementation in combination with uridine supplementation increased the pool of competing UDP-HexNAc, leading to decreased sialylation [58]. Additionally, conversion of glucosamine to ammonia is detrimental to cell growth.

Not all glucosamine supplementation studies report detrimental effects on glycosylation. Glucosamine supplementation with or without uridine was found to increase sialic acid content and increase the UDP-HexNAc pool by over sixfold [59]. There was a synergistic increase in the UDP-HexNAc pool, although combining glucosamine and uridine supplementation did not increase sialylation any further. [59].

Uridine

Uridine is an important nucleoside precursor involved in transport of sugars to the maturing glycoprotein. Increasing the pool of uridine has been predicted to increase availability of the transporter for glycosylation. Alone and in combination with galactose and manganese chloride, uridine was found to improve galactosylation [60]. In addition, uridine supplementation in combination with galactose or glucosamine supplementation resulted in increased sialic acid content of glycoforms [59]. The uridine nucleotide UTP was linked to glycosylation; a reduction in UTP corresponded to decreased percentage of biantennary glycoforms [50]. In another study, uridine supplementation in combination with glucosamine was shown to increase the UDP-HexNAc pool and improve antennarity of N-glycans [57]. However, this result coincided with a decrease in overall sialylation. Uridine supplementation may help to overcome barriers in glycosylation, such as transport, synthesis, and sugar transfer. However, uridine supplementation has also been found as detrimental to glycosylation as measured by a decrease in N-linked glycoform site occupancy [61].

N-acetyl mannosamine

N-acetyl mannosamine (ManNAc) is involved in sialylation as it is the direct precursor of sialic acid. CMP-sialic acid then transfers sialic acid to the maturing glycoform. ManNAc supplementation aims to overcome membrane permeability barriers that limit the pool of sialic acid and CMP-sialic acid [57, 62]. As a supplement, ManNAc improved sialylation by decreasing the percentage of incompletely sialylated glycoproteins [59, 62]. This effect was related to an over thirtyfold increase in the pool of CMP-sialic acid [59, 62]. In combination with uridine, ManNAc supplementation synergistically increased the CMP-sialic acid pool although there was no further increase in sialic acid [59]. However, a barrier still exists to complete sialylation and may be related to steric hindrance [62]. While ManNAc supplementation can improve glycosylation, there may be a suppression of cell growth [63].

In a study comparing ManNAc supplementation in CHO and NSO cells, ManNAc was similarly shown to increase CMP-sialic acid levels but there was no effect on sialylation [57, 58]. Thus, additional strategies may be required to increase the CMP-sialic acid pool at the direct site of sialylation. As with other supplements, the effects vary among different cell lines and culture conditions.

Guanosine

Guanosine is a precursor to the nucleotide sugar GDP-mannose that adds mannose to maturing glycoforms. Supplementation of guanosine was found to decrease site occupancy of N-linked glycoforms [61]. This decrease limits potential sites for sialylation. Therefore, supplementation of a nucleotide sugar precursor could not effectively overcome glycosylation limitations.

Sugars

During glycosylation, sugars such as glucose, mannose, and fucose, are added to the maturing glycoprotein. Thus, supplementation of simple sugars can increase the availability of these sugars for incorporation onto glycans (Table 4.1.2). The role of nucleotides as well as sugars must be considered.

Glucose

As expected, glucose concentration plays a critical role in glycosylation. In glucose-starved culture, there is an increase in glycoproteins that are truncated [64-66]. Glucose deprivation affects the initial donor formation and subsequently decreases the dolichol phosphate donor-glucose. Thus, glucose concentration affects the heterogeneity of glycoforms and as the concentration is limited, more short chain carbohydrates are produced. When glucose concentration is increased, it is possible to form fully-glycosylated glycoproteins. Thus, maintaining adequate glucose concentration in the extracellular medium is important for consistency of glycoforms.

Table 4.1.2: Effect of sugar supplements on glycosylation. Various sugars, including galactose, glucose, and mannose, were investigated as medium supplements for their effect on glycosylation.

Sugar	Reference	Effect on Glycosylation
Galactose	[59]	Galactose supplementation increases percentage of G1 relative to G0 glycans
Galactose	[58]	Galactose supplementation with or without uridine increases sialic acid content of glycoforms
Galactose	[57]	Galactose supplementation increases the pool of UDP-Gal and has a slight increase on galactosylation
Glucose	[63]	In glucose limited culture, there is increased truncation of glycoforms
Glucose	[64]	Glucose concentration alters the glycosylation pattern
Glucose	[65]	Glucose is required to form fully-glycosylated glycoproteins
Glucose	[87]	In glucose limited culture, lipid-linked oligosaccharide synthesis is altered and glycoforms are truncated
Mannose	[60]	Mannose supplementation is detrimental to glycosylation because site occupancy decreases

Galactose

Galactose adds to glycoproteins processed through the Golgi apparatus. Sialylation requires terminal galactose before sialic acid can be added. Addition of galactose to CHO cell culture alone or in combination with supplements of manganese chloride and uridine resulted in increased galactosylation [58, 60]. A plateau was observed as galactose concentration increased, with the most significant effect on the percentage of G1 (1 galactose) relative to G0 (no galactose) glycoforms [60]. Also, the combination of galactose and uridine synergistically increased sialic acid content [59]. This increase corresponded to a twentyfold increase in the UDP-Hex pool, composed of UDP-galactose and UDP-glucose [59].

Mannose and Fructose

Increasing other sugars has not always improved glycosylation. Substitution of mannose or fructose, or galactose for glucose as the monosaccharide source resulted in similar glycoform patterns [65]. In another experiment, there was no change in site occupancy of N-linked glycoforms upon supplementation of mannose [61]. Increasing the sugar pool may not be sufficient to overcome glycosylation bottlenecks.

Metal Ions

Other supplementation strategies target small molecules that assist in glycosylation. Since metal ions act as cofactors in glycosylation steps, these strategies aim to increase rate-limiting bottlenecks in glycosylation enzyme activity to improve sialylation and to minimize high mannose glycoforms (Table 4.1.3).

Table 4.1.3: Effect of metal ion supplements on glycosylation. Metal ions, including iron and manganese, were investigated as medium supplements for their effect on glycosylation.

Metal	Reference	Effect on Glycosylation
Iron	[60]	Iron supplementation improves glycosylation by increasing site occupancy
Iron	[69]	Iron supplementation improves glycosylation by minimizing the change from fully- to non- glycosylated glycoforms
Managnese	[59]	Manganese supplementation increases percentage of G1 relative to G0 glycans
Manganese	[66]	Manganese supplementation improves galactosylation and decreases percentage of non-sialylated glycoforms
Manganese	[68]	Manganese is important for both N-linked and O-linked glycosylation
Manganese	[60]	Manganese supplementation improves glycosylation by increasing site occupancy
Manganese	[67]	Manganese supplementation decreases the percentage of high mannose, non-sialylated glycoforms

Manganese

Manganese ions are cofactors for galactosyltransferases in glycosylation. In combination with galactose and uridine, manganese chloride was found to improve galactosylation and glycosylation [60, 61]. In addition to galactosylation, manganese supplementation also enhanced sialylation [67, 68]. Specifically, manganese decreases high mannose relative to complex glycoforms [68]. Manganese, a divalent cation, may also assist in the initial transfer of the dolichol phosphate donor to the peptide emerging from the ER [61, 67]. In addition to N-linked glycan synthesis, manganese acts as a cofactor in O-linked glycan synthesis and its depletion is detrimental [69]. The pool of glycoforms becomes more highly galactosylated and homogenous. Manganese supplementation helps to overcome barriers such as transport, synthesis, and sugar transfer.

Iron

Iron is another divalent cation that can act as a cofactor in the initial dolichol phosphate transfer to the emerging peptide. Iron supplementation increased site occupancy of N-linked glycoforms thus improving glycosylation [61, 70]. Indeed, the degree of increased site occupancy was higher for iron than manganese supplementation [70]. Another effect of iron is slowing the conversion of fully- to nonglycosylated glycoforms that occurs over time in cell culture; iron maintains the concentration of fully-glycosylated recombinant proteins [70].

Lipids and Steroids

In addition to sugars and nucleotides, lipids are important components of the glycosylation pathway. Supplementation of lipid and protein blends has been tested as strategies for controlling glycosylation. Also, steroid hormones that bind to the glucocorticoid receptor can affect expression of glycosylation enzymes. Thus, both lipoproteins and steroids have also been investigated as supplements for controlling and increasing glycosylation (Table 4.1.4).

Table 4.1.4: Effect of lipid/steroids supplements on glycosylation. Various lipids and steroids, including cycloheximide, dexamethasone, dolichol phosphate, glycerol, hydrocortisone, thyroxine, and triiodothyronine, were investigated as medium supplements for their effect on glycosylation.

Lipid/Steroids	Reference	Effect on Glycosylation
Cycloheximide	[84]	Cycloheximide supplementation increases percentage of glycosylated glycoprotein
Dexamethasone	[72]	Dexamethasone improves sialic acid addition by sialyltransferases and reduces cleavage by sialidases
Dolichol phosphate	[71]	Dolichol phosphate supplementation increases the donor pool but does not increase glycosylation
Glycerol	[80]	Glycerol supplementation decreases the rate of desialylation and increases sialylation of biantennary glycoforms
Hydrocortisone	[73]	Hydrocortisone supplementation improves glycosylation by increasing sialic acid content of glycoforms
Lipid/Lipoprotein	[70]	A commercial blend of cholesterol and protein decreases the rate of glycoform degradation
Thyroxine	[60]	Thyroxine supplementation improves glycosylation by increasing site occupancy
Triiodothyronine	[60]	Triiodothyronine supplementation improves glycosylation by increasing site occupancy

Lipid and Lipoprotein

The initial step of glycosylation involves the lipid dolichol, which is a phosphate donor. Dolichol phosphate is an important precursor in the initial steps of glycosylation. The lipoprotein ExCyte, a commercial blend of cholesterol and protein, was tested as a supplement for CHO cells producing recombinant protein and was shown to decrease the rate of glycosylation degradation [71]. In the control culture, the percentage of biantennary glycans decreased over time; this trend was slowed down by lipid supplementation [71]. This supplement was compared to another blend containing cholesterol, phospholipids, fatty acids, and protein that also improved glycosylation [71].

In other experiments, supplementation of dolichol phosphate directly was not found to improve glycosylation, as measured by no change in site occupancy of N-linked glycoforms [61, 72]. The increased donor pool was not sufficient for overcoming glycosylation bottlenecks. Although dolichol phosphate inhibition is detrimental to glycosylation, increased dolichol phosphate does not improve glycosylation alone, indicating other pathway bottlenecks.

Dexamethasone

Dexamethasone (DEX) is a glucocorticoid steroid hormone important for increasing expression of glycosyltransferases. DEX supplementation improved sialyltransferase and galactosyltransferase function and reduced cleavage by sialidase [73]. The proportion of asialoglycoforms decreased concomitantly with increases in mono-, di-, tri-, and tetrasialylated glycoforms [73]. While N-linked sialylation was greatly improved by DEX, there was no change in O-linked glycosylation [73]. Thus, DEX may be a useful strategy for producing consistent sialylated glycoprotein.

Hydrocortisone and Predinisolone

In addition to DEX, other glucocorticoids have been investigated as supplements. Both, hydrocortisone and predinisolone, were shown to improve sialylation [73, 74]. Higher concentrations of these glucocorticoids compared to DEX were required to increase sialic acid content to the same extent [73]. Steroid hormones likely affect glycosyltransferases to improve sialylation. As sialylation is often cell line and process dependent, supplementation of glucocorticoids may be beneficial for controlling glycosylation.

Triiodothyronine and Thyroxine

Both, triiodothyronine and thyroxine, are steroid hormones that were evaluated as supplements for CHO cells producing recombinant glycoprotein and found to similarly increase the site occupancy of N-linked glycoforms [61].

Amino Acids

Amino acid supplementation is another strategy for controlling glycosylation. Depleted amino acids are supplemented to improve growth and subsequent productivity and glycosylation. Alternatively, accumulating amino acids are removed or substituted from medium. In some processes, supplementation of depleted amino acids may improve growth but have a negative effect on glycosylation [67]. Thus, amino acid supplementation strategies require optimization (Table 4.1.5).

Table 4.1.5: Effect of amino acid supplements on glycosylation. Various amino acids, including glutamate, glutamine, glycine, proline, and threonine, were investigated as medium supplements for their effect on glycosylation.

Amino Acid	Reference	Effect on Glycosylation
Glutamate	[52]	Replacing glutamine with glutamate improves galactosylation and complement-dependent cytotoxicity
Glutamine	[74]	Adaptation of a cell line to glutamine-free growth improves the antennarity of N-linked glycoforms
Glycine	[76]	Glycine supplementation improves glycosylation by increasing sialic acid content of glycoforms
Proline	[75]	Proline supplementation increases sialylation in hyperosmotic cell culture
Proline	[76]	Proline supplementation improves glycosylation by increasing sialic acid content of glycoforms
Threonine	[76]	Threonine supplementation improves glycosylation by increasing sialic acid content of glycoforms

Glutamine

Glutamine metabolism results in ammonia formation, which negatively affects cell growth. Ammonia may increase the nucleotide sugar precursor pool but glutamine metabolism usually dominates. Adaptation of a cell line to glutamine-free growth resulted in increased antennarity of N-linked glycoforms, suggesting that glutamine is detrimental to glycosylation [75]. Glutamine-free adaptation maintained the pool of nucleotide sugars at a higher concentration that may contribute to increased antennarity [75].

Glutamate

Glutamate can be used to replace glutamine supplementation and reduce ammonia formation. The effect of glutamate on galactosylation was investigated and resulted in increased galactosylation and complement-dependent cytotoxicity activity [53]. The most significant change in antennarity was an increase in G1 relative to G0 glycoforms [53]. Thus, improving galactosylation by glutamate supplementation can enhance antibody effector function.

Proline

Proline, a non-essential amino acid, functions as an osmoprotectant to buffer the osmolality of cultured cells. Proline can also buffer the pH to avoid the negative cellular effects of ammonia. Proline supplementation resulted in increased sialic acid levels [76, 77]. The increased levels of GlcNAc and sialic acid corresponded to decreased ammonia formation and improved cell growth [77]. Thus, the benefit of this amino acid is an osmoprotectant that improves glycosylation.

Glycine and Threonine

In addition to proline, the amino acids glycine and threonine are added to cell culture to reduce ammonia formation. As supplements, both, glycine and threonine, were individually found to improve glycosylation as measured by an increase in galactosylated and sialylated glycoforms [77]. Both, GlcNAc and sialic acid, content increased concomitantly with decreased ammonia formation and improved cell growth [77]. In this case, amino acids, which are usually already formulated in medium for growth and protein productivity, can also affect glycosylation.

Other Supplements

Other supplements primarily added to increase cell growth, viability, or protein productivity have also been investigated for their roles in affecting glycosylation (Table 4.1.6).

Table 4.1.6: Effect of other supplements on glycosylation. Other supplements, including dimethyl sulfoxide (DMSO), glycine betaine, pluronic F68, primatone RL, and sodium butyrate, were investigated for their effect on glycosylation.

Other Supplements	Reference	Effect on Glycosylation
DMSO	[80]	DMSO supplementation reduces sialylation
Glycine betaine	[75]	Glycine betaine supplementation increases sialylation in hyperosmotic cell culture
Pluronic F68	[82]	Pluronic F68 supplementation maintains the percentage of fully-glycosylated glycoforms
Primatone RL	[88]	Primatone RL supplementation improves growth but decreases sialylation
Sodium butyrate	[60]	Sodium butyrate supplementation improves glycosylation by increasing site occupancy
Sodium butyrate	[80]	Sodium butyrate supplementation reduces sialylation

Sodium butyrate	[81]	Sodium butyrate supplementation increases glycoform heterogenity and sialylation decreases
Sodium butyrate	[78]	Sodium butyrate supplementaiton increases the ratio of Neu5Ac to Neu5Gc sialic acid
Sodium butyrate	[79]	Sodium butyrate supplementation improves glycosylation by reducing the Neu5Gc content
Sodium butyrate	[77]	Sodium butyrate supplementation increases protein productivity and sialic acid content

Sodium Butyrate

Supplementation of sodium butyrate is a strategy used for increasing recombinant protein productivity. Butyrate was also investigated for its effect on glycosylation and was found to increase site occupancy of N-linked glycoforms [61, 78]. The increase in site occupancy was linearly correlated with an increase in sialic acid content to suggest that butyrate may increase sialyltransferase activity [61, 78, 79]. Humans produce Neu5Ac sialic acid, whereas CHO cells produce both, Neu5Gc and Neu5Ac. Sodium butyrate supplementation increased the ratio of Neu5Ac to Neu5Gc, which may be implicated in increasing the circulatory half-life of the recombinant protein therapeutic [79, 80]. The decreased synthesis of Neu5Gc may relate to decreased CMP-Neu5Ac hydroxylase activity or NADH supply.

In contrast, in other investigations, butyrate was found to decrease sialylation as measured by decreases in bi- and monosialylated glycoforms [81, 82]. Sodium butyrate supplementation strategies require optimization to balance the protein productivity improvement with effects on product quality.

Glycine Betaine

Glycine betaine supplementation buffers the osmolality of cell culture to protect cell growth and protein productivity. At high osmolality, a beneficial effect of glycine betaine supplementation was observed resulting in increased sialic acid levels [76]. At normal osmolality, glycine betaine actually has a negative effect and reduces sialylation [76]. Thus, this osmoprotectant can also alter glycosylation.

Pluronic F68

Pluronic F68 is a surfactant commonly added to suspension culture to minimize cell damage and subsequently to improve growth, viability, and recombinant protein productivity. This supplement has also been investigated for its role in glycosylation and was found to reverse the decline in fully-glycosylated glycoforms resulting from culture agitation [83]. The positive effect of F68 on glycosylation was recently found to be independent of the cell shear protective effect [83]. Thus, common media additives may play previously unknown roles in controlling and altering glycosylation.

Primatone RL

Most industrial cell culture applications use serum-free medium to produce recombinant glycoproteins. Primatone RL is a serum substitute containing amino acids, vitamins, and trace elements that increases cell growth and recombinant glycoprotein productivity. As a medium supplement, primatone RL was shown to have a negative effect on glycosylation as measured by a decrease in sialylation [84]. Primatone RL did not cause cell lysis, indicating that the decrease in sialylation was not caused by extracellular sialidase cleavage [84] and likely occurs from declines in intracellular sialylation. Thus, although the peptone improved growth and productivity, there was a negative impact on glycosylation.

Glycerol and Dimethyl Sulfoxide (DMSO)

Additional cell culture medium components include glycerol and dimethyl sulfoxide (DMSO), which are added as protein stabilizers to prevent aggregation. Glycerol may also play a role in helping protein synthesis and folding. It was found that glycerol supplementation also affects glycosylation by decreasing the rate of desialylation [81]. In contrast, DMSO supplementation had a negative effect on sialic acid content [81]. Protein stabilizers may have diverse functions, such as growth improvement and reduced aggregation, but the functions must be considered in the context of their effects on glycosylation.

Cycloheximide

Cycloheximide, a protein synthesis inhibitor, was investigated as a medium supplement and found to increase glycoform site occupancy [85]. It was suggested that cycloheximide slows the rate at which protein is processed through the ER and Golgi thus increasing potential time for glycosylation [85]. Overall, glycosylation improved upon cycloheximide supplementation.

4.1.5 Conclusions

4.1.5.1 Summary

The purpose of this chapter is to emphasize the importance of glycosylation control of glycoproteins for the production of biotherapeutics. After reviewing the major types of glycosylation, including N- and O-linked glycosylation, strategies aimed to control glycosylation patterns of therapeutic proteins from the glycoengineering and cell culture control approaches were presented.

Glycoengineering involves the modification of glycosylation by overexpression, knockdown, or knockout of critical genes that affect glycosylation. These strategies aim to improve desired properties of biotherapeutics by most improving galactosylation and sialylation, or eliminating fucosylation. Successful glycoengineering results in improved biotherapeutics circulatory half-life and an increase in ADCC. An alternative to glycoengineering for obtaining desired glycoforms is cell culture process parameter optimization. In mammalian cell culture supplementation of medium components have been shown to control the glycosylation patterns. Various media supplements, including sugar nucleotide precursors, sugars, metal ions, lipids, and amino acids, regulate glycoprotein progression from the ER through the Golgi and affect the final glycoform profile. In spite of contradicting results among cell lines, both, glycoengineering and media supplementation, can be optimized to improve the characteristics of a biopharmaceutical protein. Further understanding of glycosylation through mathematical modeling and experimentation will enable improving biotherapeutics glycosylation.

4.1.5.2 Challenges and Future Direction

Cell line engineering is a promising strategy to control glycosylation. With the availability of genomic resources such as the CHO-K1 genome sequence [86] and the application of systems biology approaches, knowledge driven strategies to design target glycoengineering manipulations will be facilitated. Indeed, the need of these approaches is exemplified by the fact that it is difficult and not intuitive to predict the effects of genetic manipulations on the glycosylation profiles that result from a complex sequence of enzyme-catalyzed reactions.

At the glycosylation level, many variables are in place, such as sugar nucleotide concentrations, glycans concentrations, genes expression levels, enzyme activities, enzyme saturation, and substrate competition. Furthermore, we tend to disregard that many glycosylation types coexist and influence each other. For example, some enzymes influence simultaneously both N- and O-glycosylation. Thus if the concentrations of O-glycans that compete with N-glycans for the same enzymes change significantly, it will affect the concentration of N-glycans because of the less availability of those enzymes for N-glycan glycosylation. Therefore, in understanding glycosylation for improving biotherapeutics performance the use of experiments guided by modeling is critical given the complexity of glycosylation.

At the cell culture level, many supplements, including nucleotide sugar precursors, sugars, metal ions, amino acids, and lipids, have been investigated for their role in achieving desired glycosylation. The sometimes contradictory results indicate that cell culture performance varies, and it is affected by a wide array of factors in addition to the supplements, such as cell line, recombinant protein, and process design. Thus, it can be challenging to decide which supplements may improve a bioprocess. For example, supplementation of glucosamine correlated with an increased pool of UDP-GlcNAc but decreased glucose transport [54] whereas in another example, glucosamine supplementation increased sialic acid content [59]. This presents a challenge in designing experiments. Therefore, evaluating multiple media additives is essential in developing a cell culture performance to control glycosylation. Indeed, there are a vast number of supplements that have already been tested and there are still more novel supplements that could produce beneficial results.

Application of medium formulation and glycoengineering will also require a consideration of process parameters. During mammalian cell culture, the selection and optimization of process parameters can strongly affect cell growth, recombinant protein productivity, and glycosylation. Optimization of parameters such as temperature, pH, dissolved oxygen, feeding strategy, and medium formulation, is important for achieving consistent cell culture performance and to ensure product quality is attained. To further improve cell culture performance, a combination of process parameters, medium formulation, and genetic engineering can be optimized to affect a desirable glycosylation profile.

Thus to achieve glycosylation control requires defining which set of variables are the critical ones for a specific cell line within specific physiological and process conditions. Further understanding of glycosylation through mathematical modeling and experimentation will enable combinations of both supplementation and glycoengineering strategies to increase synergistically the level of glycosylation and maintain a consistent glycoform profile, resulting in superior biotherapeutics.

References

[1] Beck A, et al. Trends in glycosylation, glycoanalysis and glycoengineering of therapeutic antibodies and Fc-fusion proteins. Curr Pharm Biotechnol. 2008;9(6):482-501.

[2] Kornfeld R, Kornfeld S. Assembly of Asparagine-Linked Oligosaccharides. Annu Rev Biochem. 1985;54:631-64.

[3] Jensen PH, et al. Mucin-type O-glycosylation--putting the pieces together. FEBS J. 2010;277(1):81-94.

[4] Lerouge P, et al. N-glycoprotein biosynthesis in plants: recent developments and future trends. Plant Mol Biol. 1998;38(1-2):31-48.

[5] Castilho A, et al. Engineering of sialylated mucin-type O-glycosylation in plants. J Biol Chem. 2012;287(43):36518-26.

[6] Hamilton SR, et al. Production of complex human glycoproteins in yeast. Science. 2003;301(5637):1244-6.

[7] Li H, et al. Optimization of humanized IgGs in glycoengineered Pichia pastoris. Nat Biotechnol. 2006;24(2):210-5.

[8] Hamilton SR, et al. Humanization of yeast to produce complex terminally sialylated glycoproteins. Science. 2006;313(5792):1441-3.

[9] Jenkins N, et al. Getting the glycosylation right: Implications for the biotechnology industry. Nat Biotechnol. 1996;14(8):975-81.

[10] Herscovics A, et al. Structure of the Class IN-glycan processing alpha 1,2-mannosidase from Saccharomyces cerevisiae. Glycobiology. 1999;9(10):1117.
[11] Stalnaker SH, et al. Mammalian O-mannosylation: unsolved questions of structure/function. Curr Opin Struct Biol. 2011;21(5):603-9.
[12] Herscovics A. Structure and function of Class I alpha 1,2-mannosidases involved in glycoprotein synthesis and endoplasmic reticulum quality control. Biochimie. 2001;83(8):757-62.
[13] Krambeck FJ, et al. A mathematical model to derive N-glycan structures and cellular enzyme activities from mass spectrometric data. Glycobiology. 2009;19(11):1163-75.
[14] Bennun SV, et al. Glycoengineering and Modeling of Protein N-glycosylation. In: Al-Rubeai M (ed). Cell Engineering. London: Springer; 2008.
[15] Krambeck FJ, Betenbaugh MJ. A mathematical model of N-linked glycosylation. Biotechnol Bioeng. 2005;92(6):711-28.
[16] Wittmann V. Glycoproteins: Occurrence and Significance. In: Bertram F-R, Tatsuta K, Thiem J (eds). Glycoscience: Springer; 2008.
[17] Rottger S, et al. Localization of three human polypeptide GalNAc-transferases in HeLa cells suggests initiation of O-linked glycosylation throughout the Golgi apparatus. J Cell Sci. 1998;111(1):45-60.
[18] Brockhausen I, et al. Glycoproteins and their relationship to human disease. Acta Anatomica. 1998;161(1-4):36-78.
[19] Weikert S, et al. Engineering Chinese hamster ovary cells to maximize sialic acid content of recombinant glycoproteins. Nat Biotechnol. 1999;17(11):1116-21.
[20] Bragonzi A, et al. A new Chinese hamster ovary cell line expressing alpha2,6-sialyltransferase used as universal host for the production of human-like sialylated recombinant glycoproteins. Biochim Biophys Acta. 2000;1474(3):273-82.
[21] Fukuta K, et al. Genetic engineering of CHO cells producing human interferon-gamma by transfection of sialyltransferases. Glycoconj J. 2000;17(12):895-904.
[22] Zhang P, et al. A functional analysis of N-glycosylation-related genes on sialylation of recombinant erythropoietin in six commonly used mammalian cell lines. Metab Eng. 2010;12(6):526-36.
[23] Wong NS, et al. Enhancing recombinant glycoprotein sialylation through CMP-sialic acid transporter over expression in Chinese hamster ovary cells. Biotechnol Bioeng. 2006;93(5):1005-16.
[24] Jeong YT, et al. Enhanced sialylation of recombinant erythropoietin in CHO cells by human glycosyltransferase expression. J Microbiol Biotechnol. 2008;18(12):1945-52.
[25] Jeong YT, et al. Enhanced sialylation of recombinant erythropoietin in genetically engineered Chinese-hamster ovary cells. Biotechnol Appl Biochem. 2009;52(4):283-91.
[26] Son YD, et al. Enhanced sialylation of recombinant human erythropoietin in Chinese hamster ovary cells by combinatorial engineering of selected genes. Glycobiology. 2011;21(8):1019-28.
[27] Fukuta K, et al. Remodeling of sugar chain structures of human interferon-gamma. Glycobiology. 2000;10(4):421-30.
[28] Muthing J, Burg M. Characterization of cytosolic sialidase from Chinese hamster ovary cells: part II. Substrate specificity for gangliosides. Carbohydr Res. 2001;330(3):347-56.
[29] Ngantung FA, et al. RNA interference of sialidase improves glycoprotein sialic acid content consistency. Biotechnol Bioeng. 2006;95(1):106-19.
[30] Zhang M, et al. Enhancing glycoprotein sialylation by targeted gene silencing in mammalian cells. Biotechnol Bioeng. 2010;105(6):1094-105.
[31] Egrie JC, et al. Darbepoetin alfa has a longer circulating half-life and greater in vivo potency than recombinant human erythropoietin. Exp Hematol. 2003;31(4):290-9.

[32] Egrie JC GJ, et al. The role of carbohydrate on the biological activity of erythropoietin. Glycoconjugate J. 1993;10(4):263.
[33] Egrie JC, Browne JK. Development and characterization of novel erythropoiesis stimulating protein (NESP). Br J Cancer. 2001;84 Suppl 1:3-10.
[34] Kronman C, et al. Involvement of oligomerization, N-glycosylation and sialylation in the clearance of cholinesterases from the circulation. Biochem J. 1995;311 (Pt 3):959-67.
[35] Chitlaru T, et al. Overloading and removal of N-glycosylation targets on human acetylcholinesterase: effects on glycan composition and circulatory residence time. Biochem J. 2002;363(Pt 3):619-31.
[36] Hossler P, et al. Optimal and consistent protein glycosylation in mammalian cell culture. Glycobiology. 2009;19(9):936-49.
[37] Uozumi N, et al. Purification and cDNA cloning of porcine brain GDP-L-Fuc:N-acetyl-beta-D-glucosaminide alpha1-->6fucosyltransferase. J Biol Chem. 1996;271(44):27810-7.
[38] Hayashi H, et al. Molecular cloning of mouse alpha-1,6-fucosyltransferase and expression of its mRNA in the developing cerebrum. DNA Seq. 2000;11(1-2):91-6.
[39] Miyoshi E, et al. Expression of alpha1-6 fucosyltransferase in rat tissues and human cancer cell lines. Int J Cancer. 1997;72(6):1117-21.
[40] Yamane-Ohnuki N, et al. Establishment of FUT8 knockout Chinese hamster ovary cells: an ideal host cell line for producing completely defucosylated antibodies with enhanced antibody-dependent cellular cytotoxicity. Biotechnol Bioeng. 2004;87(5):614-22.
[41] Ferrara C, et al. Modulation of therapeutic antibody effector functions by glycosylation engineering: influence of Golgi enzyme localization domain and co-expression of heterologous beta1, 4-N-acetylglucosaminyltransferase III and Golgi alpha-mannosidase II. Biotechnol Bioeng. 2006;93(5):851-61.
[42] Shields RL, et al. Lack of fucose on human IgG1 N-linked oligosaccharide improves binding to human Fcgamma RIII and antibody-dependent cellular toxicity. J Biol Chem. 2002;277(30):26733-40.
[43] Suzuki E, et al. A nonfucosylated anti-HER2 antibody augments antibody-dependent cellular cytotoxicity in breast cancer patients. Clin Cancer Res. 2007;13(6):1875-82.
[44] Ripka J, et al. Two Chinese hamster ovary glycosylation mutants affected in the conversion of GDP-mannose to GDP-fucose. Arch Biochem Biophys. 1986;249(2):533-45.
[45] Imai-Nishiya H, et al. Double knockdown of alpha1,6-fucosyltransferase (FUT8) and GDP-mannose 4,6-dehydratase (GMD) in antibody-producing cells: a new strategy for generating fully non-fucosylated therapeutic antibodies with enhanced ADCC. BMC Biotechnol. 2007;7:84.
[46] Sullivan FX, et al. Molecular cloning of human GDP-mannose 4,6-dehydratase and reconstitution of GDP-fucose biosynthesis in vitro. J Biol Chem. 1998;273(14):8193-202.
[47] Urnov FD, et al. Genome editing with engineered zinc finger nucleases. Nat Rev Genet. 2010 Sep;11(9):636-46.
[48] Cristea S, et al. In vivo cleavage of transgene donors promotes nuclease-mediated targeted integration. Biotechnol Bioeng. 2013;110(3):871-80.
[49] Davies J, Jet al. Expression of GnTIII in a recombinant anti-CD20 CHO production cell line: Expression of antibodies with altered glycoforms leads to an increase in ADCC through higher affinity for FC gamma RIII. Biotechnol Bioeng. 2001;74(4):288-94.
[50] Borys MC, et al. Ammonia affects the glycosylation patterns of recombinant mouse placental lactogen-I by chinese hamster ovary cells in a pH-dependent manner. Biotechnol Bioeng. 1994;43(6):505-14.
[51] Yang M, Butler M. Effects of ammonia on CHO cell growth, erythropoietin production, and glycosylation. Biotechnol Bioeng. 2000;68(4):370-80.

[52] Hong JK, et al. Substitution of glutamine by glutamate enhances production and galactosylation of recombinant IgG in Chinese hamster ovary cells. Appl Microbiol Biotechnol. 2010;88(4):869-76.
[53] Yang M, Butler M. Effects of ammonia and glucosamine on the heterogeneity of erythropoietin glycoforms. Biotechnol Prog. 2002;18(1):129-38.
[54] Andersen DC, Goochee CF. The effect of ammonia on the O-linked glycosylation of granulocyte colony-stimulating factor produced by chinese hamster ovary cells. Biotechnol Bioeng. 1995;47(1):96-105.
[55] Mason AB, et al. Expression of Glycosylated and Nonglycosylated Human Transferrin in Mammalian-Cells - Characterization of the Recombinant Proteins with Comparison to 3 Commercially Available Transferrins. Biochemistry. 1993;32(20):5472-9.
[56] Baker KN, et al. Metabolic control of recombinant protein N-glycan processing in NS0 and CHO cells. Biotechnol Bioeng. 2001;73(3):188-202.
[57] Hills AE, et al. Metabolic control of recombinant monoclonal antibody N-glycosylation in GS-NS0 cells. Biotechnology and Bioengineering. 2001;75(2):239-51.
[58] Wong NS, et al. An investigation of intracellular glycosylation activities in CHO cells: effects of nucleotide sugar precursor feeding. Biotechnol Bioeng. 2010;107(2):321-36.
[59] Gramer MJ, et al. Modulation of antibody galactosylation through feeding of uridine, manganese chloride, and galactose. Biotechnol Bioeng. 2011;108(7):1591-602.
[60] Gawlitzek M, et al. Identification of cell culture conditions to control N-glycosylation site-occupancy of recombinant glycoproteins expressed in CHO cells. Biotechnol Bioeng. 2009;103(6):1164-75.
[61] Gu X, Wang DI. Improvement of interferon-gamma sialylation in Chinese hamster ovary cell culture by feeding of N-acetylmannosamine. Biotechnol Bioeng. 1998;58(6):642-8.
[62] Hayashi M, et al. Effects of sugar chain precursors on recombinant protein production in BHK cells. Animal Cell Technol : Basic Appl Aspects. 2009;15:81-6.
[63] Gawlitzek M, et al. Effect of different cell culture conditions on the polypeptide integrity and N-glycosylation of a recombinant model glycoprotein. Biotechnol Bioeng. 1995 ;46(6):536544.
[64] Tachibana H, et al. Changes of monosaccharide availability of human hybridoma lead to alteration of biological properties of human monoclonal antibody. Cytotechnology. 1994;16(3):151-7.
[65] Hayter PM, et al. Glucose-limited chemostat culture of Chinese hamster ovary cells producing recombinant human interferon-gamma. Biotechnol Bioeng. 1992;39(3):327-35.
[66] Crowell CK, et al. Amino acid and manganese supplementation modulates the glycosylation state of erythropoietin in a CHO culture system. Biotechnol Bioeng. 2007;96(3):538-49.
[67] Pacis E, et al. Effects of cell culture conditions on antibody N-linked glycosylation-what affects high mannose 5 glycoform. Biotechnol Bioeng. 2011 (Epub ahead of print).
[68] Kaufman RJ, et al. Depletion of manganese within the secretory pathway inhibits O-linked glycosylation in mammalian cells. Biochemistry. 1994;33(33):9813-9.
[69] Clincke MF, et al. Effect of iron sources on the glycosylation macroheterogeneity of human recombinant IFN-gamma produced by CHO cells during batch processes. BMC Proc. 2011;5(8):114.
[70] Jenkins N, et al. Effect of lipid supplements on the production and glycosylation of recombinant interferon-gamma expressed in CHO cells. Cytotechnology. 1994;15(1-3):209-15.
[71] Yuk IH, Wang DI. Glycosylation by Chinese hamster ovary cells in dolichol phosphate-supplemented cultures. Biotechnol Appl Biochem. 2002;36(2):141-7.
[72] Jing Y, et al. Sialylation enhancement of CTLA4-Ig fusion protein in Chinese hamster ovary cells by dexamethasone. Biotechnology and Bioengineering. 2010;107(3):488-96.

[73] Rouiller Y, et al. Effect of hydrocortisone on the production and glycosylation of an Fc-fusion protein in CHO cell cultures. Biotechnol Prog. 2012;28(3):803-13.
[74] Taschwer M, et al. Growth, productivity and protein glycosylation in a CHO EpoFc producer cell line adapted to glutamine-free growth. J Biotechnol. 2012;157(2):295-303.
[75] Schmelzer AE, Miller WM. Effects of osmoprotectant compounds on NCAM polysialylation under hyperosmotic stress and elevated pCO(2). Biotechnol Bioeng. 2002;77(4):359-68.
[76] Chen P, Harcum SW. Effects of amino acid additions on ammonium stressed CHO cells. J Biotechnol. 2005;117(3):277-86.
[77] Chotigeat W, et al. Role of environmental conditions on the expression levels, glycoform pattern and levels of sialyltransferase for hFSH produced by recombinant CHO cells. Cytotec hnology.1994;15(1-3):217-21.
[78] Borys MC, et al. Effects of culture conditions on N-glycolylneuraminic acid (Neu5Gc) content of a recombinant fusion protein produced in CHO cells. Biotechnol Bioeng. 2009;105(6):1048-57.
[79] Chen F, et al. The combined effect of sodium butyrate and low culture temperature on the production, sialylation, and biological activity of an antibody produced in CHO cells. Biotechnol and Bioproc Eng. 2011;16(6):1157-65.
[80] Rodriguez J, et al. Enhanced production of monomeric interferon-beta by CHO cells through the control of culture conditions. Biotechnol Prog. 2005;21(1):22-30.
[81] Sung YH, et al. Effect of sodium butyrate on the production, heterogeneity and biological activity of human thrombopoietin by recombinant Chinese hamster ovary cells. J Biotechnol. 2004;112(3):323-335.
[82] Clincke MF, et al. Effect of surfactant pluronic F-68 on CHO cell growth, metabolism, production, and glycosylation of human recombinant IFN-gamma in mild operating conditions. Biotechnol Prog. 2011;27(1):181-90.
[83] Gu X, et al. Influence of Primatone RL supplementation on sialylation of recombinant human interferon-gamma produced by Chinese hamster ovary cell culture using serum-free media. Biotechnol Bioeng. 1997;56(4):353-60.
[84] Shelikoff M, et al. The effect of protein synthesis inhibitors on the glycosylation site occupancy of recombinant human prolactin. Cytotechnology. 1994;15(1-3):195-208.
[85] Gawlitzek M, et al. Ammonium ion and glucosamine dependent increases of oligosaccharide complexity in recombinant glycoproteins secreted from cultivated BHK-21 cells. Biotechnology and Bioengineering. 1998;57(5):518-28.
[86] Gu X, Wang DI. Improvement of interferon-gamma sialylation in Chinese hamster ovary cell culture by feeding of N-acetylmannosamine. Biotechnology and Bioengineering. 1998;58(6):642-8.
[87] Rearick JI, et al. Glucose starvation alters lipid-linked oligosaccharide biosynthesis in Chinese hamster ovary cells. J Biol Chem. 1981 Jun 25;256(12):6255-61.
[88] Gu X, et al. Influence of Primatone RL supplementation on sialylation of recombinant human interferon-gamma produced by Chinese hamster ovary cell culture using serum-free media. Biotechnology and Bioengineering. 1997 Nov 20;56(4):353-60.
[89] Xu X, et al. The genomic sequence of the Chinese hamster ovary (CHO)-K1 cell line. Nature Biotechnology. 2011;29(8):735-41

4.2 Manufacturing of Complex Biotherapeutic Protein Products: Medical Need and Rational for Monoclonal Antibody Mixtures, Multispecific Formats, and Fc-fusion Proteins

Lore Florin, Sören K. Rasmussen, Torben P. Frandsen, and Anne B. Tolstrup

4.2.1 Introduction

The therapeutic and commercial success of monoclonal antibodies (mAbs) is undisputable with approximately 30 molecules approved for market. Also, the Ig-fusion proteins have proven their value and several have entered the market, with the prominent example of Enbrel (etanercept) as one of the most sold biologics to date [1, 2].

Nevertheless, limitations to the effect of mAb treatment have also been observed, and a confirmed clinical need prevails to develop more potent and broadly applicable therapies, not least for treatment of so-called complex diseases. The scientific background for these limitations may well rely in the inherent specificity of the mAbs for a single epitope, and the next generation of antibody-based therapeutics therefore aims to overcome these limitations by enabling targeting of two or more epitopes [3]. A good example of a complex disease could be an infection caused by a virus or a bacterium. Here, mAb treatment targeting one epitope, e.g., on the surface of a bacterium, has historically shown little effect and with the exception of Synagis (palivizumab) [4] there are no mAbs on the market today for treatment of infections. Cancer is another broad indication where curative or even durable effects of current mAb therapies often are lacking and where broadening the potency of the drugs through targeting several epitopes could be envisioned to enable improved patient responses. Despite the fact that use of plasma-derived immunoglobulins no longer can be considered state-of-the-art, such multivalent drugs have nevertheless inspired researchers to explore new drug development concepts where more than one epitope is targeted as they are known to act in concert to optimally recruit and exert effector functions. This has resulted in a trend in developing new and more efficient biologicals, such as bi- or multispecific formats as well as exploiting the application of recombinant antibody mixtures as a logical next generation step. Other types of optimized molecules, like glyco-engineered formats with enhanced ADCC activity or antibody drug conjugated (ADC) molecules, goes beyond the scope of this chapter and will be dealt with elsewhere.

The bi- or multispecific formats are designed to target two or more rather than just one epitope present on a target molecule or cell (for example a cancer cell) and in addition, some of the smarter multispecific formats will also bind a target on an immune cell, for example CD3 on T cells, enabling this cell to be brought into close contact with the malignant cell. In 2009, the first and so far only multispecific format, catumaxomab, targeting EpCAM (epithelial cell adhesion molecule) and CD3, was approved for treatment of malignant ascites [5]. Complementary to this, antibody mixtures represents an option to freely combine several well-characterized mAbs into a mixture, dependent on the molecular target profile in a given disease. Conceptually, antibody mixtures take advantage of the best properties of two existing drug classes, namely the specificity of the mAbs combined with the diversity of the immunoglobulins targeting multiple epitopes. In 2012, the first mAb combination, trastuzumab and pertuzumab, was approved by FDA. The mixture consists of two of Genentech's previously approved mAbs targeting different epitopes on the HER-2 receptor, and it was approved for treatment of HER-2 positive metastatic breast cancer together with docetaxel [1]. There are more examples of ongoing clinical trials with antibody combinations targeting two different epitopes, e.g., Symphogen is running a Phase II trial with Sym004, a mixture of two mAbs which both binds the EGFR surface receptor [6], and companies like Merrimack and Xoma are also pursuing clinical trials of antibody mixtures comprising 2-3 antibodies. Also larger mixtures have been advanced in the clinic such as a mixture of 25 mAbs for treatment of ITP [7]. Notably, antibody mixtures represent combinations of a naturally occurring molecule format which, as long as they are human or humanized, have a relatively low risk of inducing immunogenicity in patients, whereas bi- and multispecific formats may suffer from increased immunogenicity due to their engineered structures. A third class of molecules that will be covered in this chapter comprise the Fc-fusion proteins typically consisting of the extracellular domain of a naturally occurring receptor fused to the constant domain of an IgG molecule to enhance manufacturability, molecule stabil-

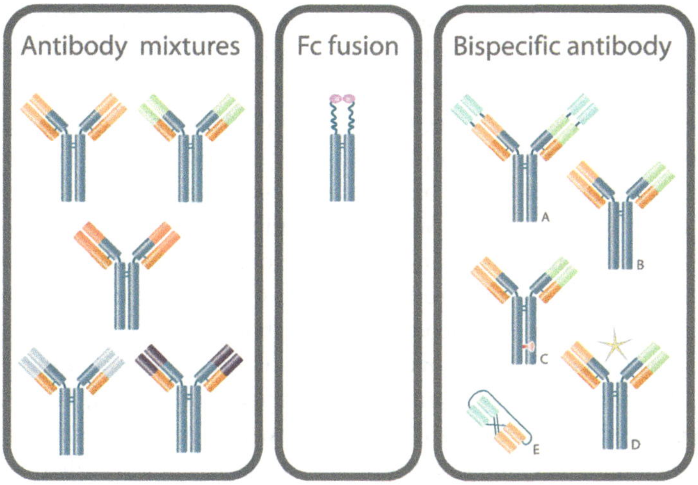

Figure 4.2.1: Illustration of 3 classes of next generation biological therapeutics; antibody mixtures, Fc fusion proteins and multispecific formats. In the latter group, multiple different formats exists, of which some examples are shown. A) dual variable domain, B) bispecific antibody, C) "knob-in-hole" bispecific antibody, D) antibody drug conjugates, E) diabody.

ity, and half-life. A graphic overview of the general structural formats of the next generation therapeutics covered below, namely 1) antibody mixtures, 2) Fc-fusion proteins, and 3) multispecific formats (examples A: dual variable domain, B: bispecific, C: knob-in-hole, D: antibody drug conjugates, E: diabody) is depicted in Figure 4.2.1.

An increased number of both, multispecifics and antibody mixtures, are expected to enter clinical trials in the coming years. A recent review of ongoing trials was provides by May et al. [8], please visit clinicaltrials.gov for the most current information.

Both of the newer classes of drugs – multispecifics and antibody mixtures – represent new challenges to CMC manufacturing, characterization, and release. Here, investigators can take advantage of the technological advances made in animal cell culture-based antibody manufacturing over the past 15 to 20 years. The challenges and currently applied approaches to overcome them will be addressed in the next sections.

4.2.2 Recombinant Antibody Mixtures

4.2.2.1 Manufacturing Approaches for Recombinant Antibody Mixtures

Different manufacturing strategies are currently used to produce antibody mixtures, where the mixing of the individual molecules occurs either during a defined manufacturing process step or where the actual production is performed simultaneously (Figure 4.2.2). Antibody cocktails represent completely individual drug products, manufactured and released as individual drug products that are administered to the patients as a combination (Figure 4.2.2(1)). Recombinant antibody mixtures, typically containing a low number of constituent antibodies, can be produced using a manufacturing approach where the individual drug substances are manufactured and released individually and subsequently mixed at the drug product stage prior to administration of patients (Figure 4.2.2(2)). Recombinant antibody mixtures produced simultaneously can be manufactured using different strategies from a single expression clone (Figure 4.2.2(3)) or using a single-batch manufacturing approach starting from a polyclonal cell bank (Figure 4.2.2(4)). Mixtures of antibodies with three different specificities have thus been obtained using the Oligoclonic™ technology developed by the company Merus where a single producer clone is obtained by transfection using a mixture of expression plasmids encoding one antibody light chain and 2-3 heavy chains. This strategy leads to generation of a mixture of three different mAbs of the types AA, AB, and BB when two heavy chains are transfected, and AA, AB, AC, BB, BC, and CC when three heavy chains are transfected each with a unique specificity (Figure 4.2.2(3)). Published data on clone stability studies suggest that the individual clones are stable over extended generations with respect to growth, production, and ratio of the different molecules [9]. The single-batch manufacturing approach (Figure 4.2.2(4)) also offers the possibility of simultaneous manufacturing of several constituent antibodies, however, with few limitations to the desired composition as compared to the Oligoclonic™ approach. To date, up to as many as 25 antibodies have been produced using this strategy [10]. Very consistent manufacturing has been demonstrated using a polyclonal cell banking concept where the individual stable manufacturing cell lines are mixed at the cell banking stage and the polyclonal cell bank is selected for optimal growth and manufacturing consistency [10-12].

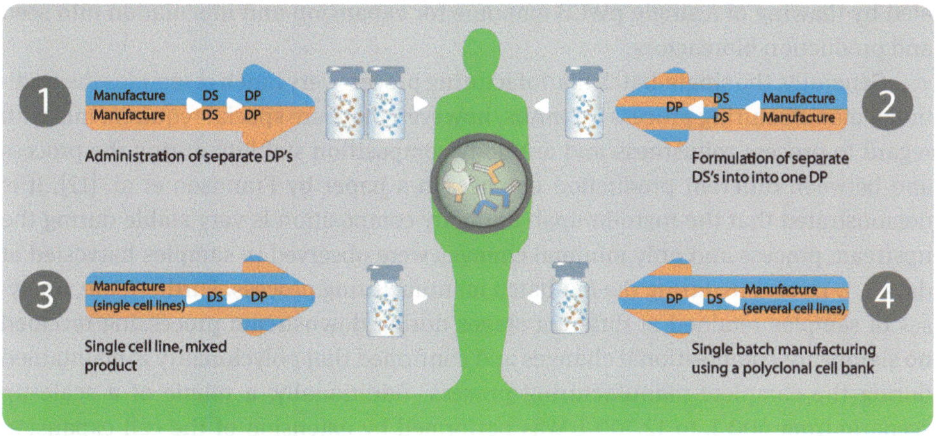

Figure 4.2.2: Manufacturing strategies for antibody mixtures.

4.2.2.2 Single-batch Manufacturing, First Generation

The single-batch production concept of recombinant antibody mixtures or recombinant polyclonal antibodies will be described in more detail below, using a historical perspective. The idea was originally described about 15 years ago by Sharon and colleagues [13] who performed experiments where DNA fragments derived from yeast or phages libraries were transferred in bulk to mammalian expression plasmids before transfection into a given host for expression. Subsequently, the single-batch manufacturing was further developed by the Danish company Symphogen A/S which in 2006 described the development of a complex recombinant antibody mixture, rozrolimupab, composed of 25 different recombinant IgG's showing reactivity towards Rhesus D (RhD) [10]. The recombinant product is intended for replacement of plasma-derived anti-RhD hyperimmune immunoglobulin currently used for treatment of idiopathic thrombocytopenic purpura (ITP).

Rozrolimupab, a first-in-class recombinant polyclonal antibody, was manufactured in CHO-Flp-In cells using a first generation of the so-called Sympress technology for single-batch manufacturing [10] The expression platform was based on a site-specific integration technology where the expression construct is integrated in one copy at a single predefined genomic locus. In theory, the generation of "genetically identical" production cell lines would minimize differences in genomic position effects, production capacities, and growth rates. CHO Flp-In cell lines expressing each of the 25 specific anti-RhD antibodies were generated individually. Subsequently, all cell lines were thawed and expanded simultaneously in compliance with GMP (good manufacturing practice) and mixed with equal cell numbers to generate a polyclonal master cell bank (pMCB) and a polyclonal working cell bank (pWCB). A fed-batch manufacturing process was employed for production of the antibody mixture, initiated by thawing of a single pWCB ampoule for expansion and inoculation into seed and production bioreactors.

Generally, the single-batch manufacturing process very much resembles the traditional manufacturing process for mAbs, however, there are special requirements with regard to process robustness and antibody composition stability during the process and between different production batches. In a paper by Frandsen et al. [12], it is demonstrated that the rozrolimupab antibody composition is very stable during the upstream process and only minimal changes were observed in samples harvested at days 3, 7, 11, and 16 during the fed-batch manufacturing process. Furthermore, analyses of samples retained at different stages during downstream processing revealed no significant compositional changes and confirmed that polyclonality is maintained during the complete manufacturing process. Additionally, a mimic of a scale up scenario from 400 L to 12,500 L was performed by extension of the cell expansion phase with approximately 6 cell generations. Comparison of antibody compositions by use of cation exchange chromatographic fingerprint profiling as analytic method showed highly similar compositions of product derived from "normal" expansion and extended expansion, respectively, with only minor changes in chromatographic peak heights. Finally, analysis of four 400 L batches in a GMP manufacturing campaign showed very similar cation exchange profiles again strongly indicating a very consistent batch-to-batch manufacturing process.

4.2.2.3 Single-batch Manufacturing using an Optimized Second Generation Approach

Overall, the first generation Sympress technology provided a robust and reliable manufacturing platform for rather complex antibody mixtures composed of up the 25 different antibodies. However, there were some drawbacks of the technology with regard to manufacturability. Firstly, the employed parental CHO Flp-In cell line produced relatively low levels of recombinant antibody, and secondly, a large quantitative difference between individual antibodies in the composition was observed by both, cation exchange chromatography, terminal restriction enzyme fragment length polymorphism (T-RFLP) analyses [11], and by light chain LC-MS [14]. This strongly suggested that although site-specific integration technology would ensure similar genetic environments for all expression constructs, the primary sequence of the various antibodies would still result in individual expression levels of each antibody. For diseases with a higher drug demand and a need for a more tightly specified antibody ratio, as required for example for many cancer indications, an expression system enabling a higher antibody production rate and a tighter control over antibody composition was required.

Several different expression systems were evaluated for a second generation Sympress expression platform and surprisingly, it turned out that the most successful results were obtained using random integration in a CHO DG44 derived host cell line termed ECHO [11]. The initial procedures for generation of cell lines are relatively similar to procedures used for mAbs where an expression construct containing the recombinant antibody genes and an appropriate selection marker is introduced into the host cell line by standard transfection methods. Transfection is performed separately with each of the individual antibody expression vectors and methotrexate is used for selection of stable cell lines. The resulting stable pools are single-cell cloned by FACS and high-expressing clones are expanded and frozen as individual research cell banks (RCBs). The RCBs are thoroughly examined with regard to parameters such as growth, production characteristics, and antibody quality attributes. Selected candidates are progressed to the next step where polyclonal cell banks are generated.

For the first generation Sympress technology, a cell bank strategy based on transfer of individual RCBs to a GMP facility for subsequent banking of "polyclonal" cell banks generation was employed. An optimized concept has been introduced for the second generation Sympress technology, namely a three-tiered cell banking schedule. The RCBs are initially expanded and mixed to generate a 'polyclonal' research cell bank (pRCB). This bank then serves to generate the pMCB and pWCB, both to support development under research (non-GMP) conditions, and to facilitate the pMCB and pWCB generation under GMP compliant conditions [15]. The reproducibility and robustness of the three-tiered cell banking system (pRCB to pMCB to pWCB) was evaluated by comparing the antibody composition produced from two different pWCBs generated from the same pMCB producing a mixture of 6 antibodies. Several ampoules from each pWCB were thawed, expanded and subjected to a fed-batch process at 5 L scale. The antibodies were purified and the antibody distribution determined by cation exchange chromatography both within the same pWCB and between different pWCB, was very similar. The batch-to-batch variation has been further examined in small scale bioreactors in several research projects and found to be low and acceptable, both with regard to antibody composition and upstream characteristics such a cell number and viability [11, 15].

Thorough determination of growth characteristic of individual RCBs is of paramount importance for generation of cell cultures robustly expressing defined antibody mixtures. To demonstrate that this is indeed feasible, two RCBs producing two distinct antibodies were selected based on growth and production characteristics to examine how mixing of the two RCBs at different ratios would affect the antibody composition. The two cell lines were thawed and mixed in five different rations and frozen as pMCBs. The pMCBs were subsequently resuscitated, expanded, and examined in a fed-batch process. Antibody distributions were determined and the results clearly showed that the relative amount of each antibody could be controlled by mixing the cell lines in an appropriate ratio [16].

The single-batch manufacturing may suffer from some disadvantages, such as the need for a very thorough selection of individual cell lines with desired characteristics for the polyclonal cell banks and a more restricted manufacturing process with for example a narrow window for the seed train length. Antibody mixtures derived from the second generation Sympress platform have not yet (2013) entered the clinic, however, several products are in the pipeline and the second generation single-batch manufacturing approach may well provide a consistent and cost efficient expression platform for antibody mixtures in the future.

4.2.2.4 CMC (Chemistry, Manufacturing and Control) Challenges

As regulatory guidance for the control, release, and characterization of recombinant antibody mixtures has not yet been established, the strategy for recombinant antibody mixtures has been based on the existing regulatory guidance for mAbs. Thus, to a large extent and whenever technically feasible, the developed analytical methodologies for recombinant antibody mixtures resemble that of mAbs. Some of the developed analytical methods that have been successfully applied for recombinant antibody mixtures are essentially independent of the number of constituent antibodies, performed basically as for mAbs, and the analytical results provides an average characteristics for the recombinant antibody mixture in question. Examples of such analytical methods include, e.g., SDS-PAGE and Western blotting which provide information on the identity and presence of potential product related contaminants. A separation technique like capillary electrophoresis is performed again essentially as for mAbs and provides valuable information on the presence of potential product related contaminants. For this example, analysis of the interim references of the individual antibodies were used to demonstrate that minor peaks present in the electropherograms under reducing conditions were caused by differences in migration between the light chains [12]. Recombinant antibody mixtures are typically made with an identical constant region, with the antigen specificity relying in the different identified, screened, and chosen variable regions which together are selected for improved synergistic effect towards a given target (s). The applied release and characterization strategy for recombinant antibody mixtures developed by Symphogen take advantage of the identical constant regions for the constituent antibodies. While peptide mapping for mAbs are typically used to provide identity of the primary structure and the potential presence and identity of posttranslational modifications, this method for recombinant antibody mixtures will provide the same level of information for the constant regions and potentially parts of the variable regions. For the very complex recombinant antibody mixture of 25 anti-RhD antibodies mentioned above, the 10 most dominating peaks in the peptide maps were identified, thoroughly characterized using mass spectrometry, and used for release as an identity method for the constant region [12].

Various control strategies to identify and quantify the amount of individual constituent antibodies in a given recombinant antibody mixture have been developed. Very different analytical methodologies have been applied for such analysis and the successfully developed methods have been used to, e.g., profile manufacturing consistency of individual constituent antibodies, measure potency of subsets of antibodies, or to monitor the profile of mRNA levels of the expressed antibodies in the mixture. Chromatography based methods like CIEX or RP-HPLC have been widely developed and used for profiling of recombinant antibody mixtures [12], applied as important project specific identity methods and used during process development, for the profiling of manufacturing consistency, release, and characterization. Reference standards of the individual mAbs are typically applied during the development and qualification/validation of such product specific identity methods. Very often, an early version of the product specific identity methods is also applied during the selection process of the recombinant antibody mixture lead candidate to ease the development process for the product specific identity method. Dependent on the nature and complexity of the mixture, chromatography based identity methods may provide the exact quantity of the individual constituent antibodies. For more complex types of recombinant antibody mixtures, mass spectrometry-based methods have been used to identify and provide the relative quantity of individual constituent antibodies for product candidates containing as many as 25 antibodies [14]. For this particular assay, size exclusion chromatography were used to separate the pool of heavy and light chains, for the subsequently LC-MS analysis of the pool of light chain allowing identification and relative quantification.

4.2.2.5 Summary Antibody Mixtures

As described in the previous section there is an increasing amount of interest in antibody mixtures as well as in antibody cocktails, which is based primarily on the medical need for more complex and more effective drugs targeting several different epitopes. This new class of antibody therapeutic has created a need for understanding how to manufacture and characterize antibodies from a CMC perspective.

Manufacturing of antibody mixtures may be based on many different approaches including separate manufacture and release of individual antibodies to more complex approaches such as producing as many as 25 antibodies in a single-batch manufacturing process (Figure 4.2.2). Several manufacturing approaches may be feasible but cost considerations, manufacturing capacities and process robustness and characteristics will of course play an important role when deciding for the most desirable approach [15].

The CMC strategies for quality control of antibody mixtures can be challenging, but with rozrolimupab with a finalized global clinical phase 2 trial as a pioneer a possible path for regulatory acceptance of the single-batch manufacturing approach has been established. The CMC strategy is built on guidance for mAbs supplemented with product specific assays to determine antibody composition and composition variability. In 2010, FDA issued a draft guidance for industry entitled "Co-development of two or more un-marketed investigational drugs for use in combination" with guidance on the preclinical and clinical development of combination products and this may help more drugs to enter clinical trials.

In conclusion, combinations of antibodies constitute a very interesting and efficient new class of therapeutics against serious diseases such as cancer and infections, and different antibody mixtures are likely to enter the market within the next decade.

4.2.3 Multispecifics

Multispecific molecule formats share many of the advantages that polyclonal antibodies and antibody mixtures offer as modern approaches to the treatment of complex diseases with an unmet medical need. Their binding moieties can be designed to recognize several different target molecules or different epitopes on a single target, thereby enhancing the affinity through avidity. They can be used to address several molecules acting in the same or different disease relevant pathways to achieve additive or synergistic effects. Thus, they share the benefits of the aforementioned classes of new molecule formats to increase the affinity, avidity and potency to actively combat the redundant nature of many disease mechanisms. The structural difference relies in the feature that in bi- and higher order multispecific formats, the ability to bind to two or more targets is comprised in a single molecule.

Through the structural linkage of several target-engaging components in one molecule, multispecifics offer additional functionalities that go beyond the modalities that mixtures of individual unlinked molecules can provide. For example, multispecifics can be applied to link several molecules on one or different cells, e.g., to force clustering of cell surface receptors to induce signalling or by bridging immune cells such as natural killer cells to tumor cells to promote efficient cell killing [17, 18]. Another important feature of multispecifics could be a more selective delivery of a therapeutic or toxic payload to a specific cell-type, tissue or to tumour cells. Newer approaches to cross the blood-brain-barrier, which has been a long known challenge for biological therapeutics, rely on multispecific molecules wherein one 'arm' recognizes a receptor on endothelial cells that will mediate the endothelial transcytosis of the molecule into the brain, where the second arm will then function to recognize its target to mediate the actual therapeutic function [19]. A lot of potential is also expected from using bispecifics as vehicles for ADC approaches. By designing bispecifics that would only bind with high affinity to tumour cells that simultaneously express two or more surface markers, researchers hope to be able to more clearly differentiate between a tumor and a healthy cell (that might also express the target even though at lower amounts) and thereby minimize 'on target' cytotoxic effects that in the past have prevented the safe use of these highly potent strategies [20-22]. Furthermore, the target space might not be limited to extracellular targets: Several groups are working on approaches to use multispecific molecules to address intracellular targets, by engaging endocytosis receptors with one binding moiety to induce uptake, build in a mechanism to enable endosomal escape into the cytoplasm or nucleus and use a different binding moiety to selectively engage the target [23]. Those approaches have not entered clinical stages yet, but they illustrate the broad versatility of ideas for applications that can be built into multispecifics to open an entirely new target space to therapy.

At the same time, the fact that all target engaging domains are covalently linked together into one structure also means that the molecule design determines a fixed stoichiometry (e.g. 1:1 in a bispecific) that cannot be titrated or adjusted later on to obtain a potentially more efficient ratio, as well as a fixed distance between the different binding moieties. Therefore, particular care needs to be taken to study or predict the distribution and relative abundances of all targets and the affinities of the corresponding binding domains to ensure that the molecule will be well balanced to effectively bind all desired targets in order to exploit its full potential.

4.2.3.1 Multispecific Formats

Bi- or multispecific molecules can in principle be generated either by fusion of two or several hybridoma cells, by biochemical methods such as reduction/reoxidation or through conjugation or by genetic engineering. The first two approaches usually result in mixtures of paired chains, which – to a defined statistical proportion – will contain a fraction of the desired combination of heavy and light chains which will then have to be purified selectively from the mixture of undesired chain combinations (e.g., AB is desired, while AA, BB, and unpaired chains needs to be removed). In fact, the first marketed bispecific antibody, catumaxomab (recognizing EpCAM and CD3 for the treatment of malignant ascites) was developed by the fusion of mouse IgG2a and rat IgG2b hybridoma cells [24, 25].

The majority of multispecific molecules, however, have been generated by genetic engineering and, in the last 20 years, more than 45 different formats have been developed that are in or approaching clinical trials [26]. Depending on their structure and origin, those formats can roughly be divided into four classes (see also Table 4.2.1):

1. First, there are the IgG-like formats that maintain the structural framework of natural antibodies including the Fc-region. Binding sites with additional specificities can be incorporated by fusing single-chain variable fragments (scFv), diabodies, Fab fragments or derivatives to the N- or C-termini of the IgG to obtain symmetric multispecific molecules. Another approach is based on the heterodimeric assembly of two different heavy chains with either a common or different light chains, leading to asymmetric structures. Several strategies have been developed to facilitate heterodimerization and to prevent the formation of homodimers, including engineering of the amino acids at the dimerization interface in

the CH3 domain ('knobs-into-holes'), introduction of charged residues to provoke repulsion of identical chains or making use of segments from other Ig subtypes, e.g., IgA [27-29].
2. The second group comprises smaller formats built from combinations of antibody-derived binding domains that are connected by flexible linkers. Examples include chains of two or more Fabs or scFvs that are the building blocks of diabodies [30] and BiTEs [31], respectively, or tandem assemblies of even smaller binding units from different species such as the VHH domain antibodies from camelids [32].
3. Thirdly, various scaffold formats have been designed from binding-domains of different intra- or extracellular proteins (for recent review, see [33]). This very diverse third group includes anticalins, ubiquitins, adnectins, DARPins, fynomers, centyrins, and many more that share the common properties of being of small size (usually <20 kDa), compact and usually very stable and easy to express in microbial systems. As a downside, these small molecules are usually rapidly cleared from the circulation, which often makes it necessary to employ half-life extending strategies to maintain effective concentrations and enable patient convenient dosing (reviewed in [34]).
4. The fourth group comprises approaches combining the benefits of the previous groups by using a fully functional antibody and 'decorating' or 'loading' its termini with scaffolds of different specificities. By fusing those small binders to virtually every N- and C-terminus of heavy and light chains, it is possible to create tetra-, penta- and even higher order multispecific molecules that have been shown to maintain the specificity and affinity of the parental antibody and simultaneously engage up to five or more targets.

4.2.3.2 Design Considerations

The variety of options that is represented by this colourful plethora of bi- and multispecific formats on the one hand opens an amazingly variable field of new possibilities to address disease mechanisms. At the same time, it also reflects the various considerations that inspired their design and refinement and those could be as complex as the disease pathologies they are supposed to address. Scientific understanding of the disease condition and its underlying mechanisms will usually be the initial drivers to try a bi- or multispecific approach versus standard monotherapy and define the number of target structures that the molecule should be able to bind. One of the basic decisions will then be whether to choose an antibody-like format, which offers the structural stability of an evolutionary proven molecule or to select a more complex or alternative scaffold-based structure. Scaffolds might be preferred if long circulation half-life is not anticipated to be required or even not desired due to the anticipated high biological potency of the binder combination. The risk of immunogenicity and a strong antidrug antibody (ADA) response could be expected to be higher for formats

Table 4.2.1: Classes of multispecific formats.

	Format	Company/Sponsor	Reference
I)	**IgG-like molecules containing an Fc-portion**		
	Triomabs	Trion Pharma	[24, 35]
	DART	MacroGenics Inc.	[36]
	Mab2	F-star	[37]
	Biclonics	Merus	[38]
	Duobody	Genmab	[39]
	DVD-Ig	Abbot	[40]
	Dopple-Mabs/cross-Mabs	Roche	[41]
II)	**IgG-derived withouth Fc-portion**		
	Small, versatile combination of binding domains, connected by flexible linkers; usually short half-life		
	Nanobodies	Ablynx	[32]
	ScFv	Public domain	
	BiTE	Micromet/Amgen	[31]
	TandAb	Affimed	[42]
III)	**Non-IgG (very diverse group; further, see [17] for review)**		
	Small, usually very stable ‘scaffold’ modules derived from intra- or extracellular protein binding domains		
	Anticalins	Pieris	[43]
	DarPins	MolecularPartners	[44]
	Affibodies	Affibody	[45, 46]
	Fynomers	Covagen	[47]
	Adnectins	Adnexus, Bristol-Myers Squibb R&D Company	[48]
IV)	**Combinations = ‘decorated’ antibodies**		
	Small binding units (e.g., from class III) attached to an IgG (H and/or L chain) or fusions of molecules of the non-Fc class to an Fc portion to increase half-life and/or effector function		
	Zybodies	Zygenia http://www.zyngenia.com/	[49, 50]
	Fyomer/IgGs	Covagen	[51]

that are more far away from naturally occurring structures or if employed building blocks are intracellularly derived. However, there does not seem to be a clear rule as even pentameric antibodies decorated with Fynomers, derived from binding domains of the intracellular Fyn kinases, have been reported to show very good serum half-life and no ADA in initial pharmacokinetic studies in monkeys [47, 51]. Unless short half-life is acceptable, extension technologies will have to be considered for the scaffold-based and smaller antibody-derived formats to maintain effective concentrations and prevent high dosing frequency. Tissue distribution and overall abundance ratios of all target structures will have to be carefully weighted and built into the final design. Different binding affinities and anticipated biologic response further factor into the balance of functional binding units (1:1 or different) to achieve optimal potency. Furthermore, orientation and order of the binding domains within the molecule have been shown to affect both, binding affinity and function, as well as CMC characteristics including expression, folding, and stability. For asymmetric formats consisting of different heavy chains, means to force heterodimerization or alternatively to facilitate enrichment of the desired AB chain combinations from the arising mixture may be built in to support effectiveness in manufacturing. Linkers between domains can be designed to provide more or less flexibility to the structure. Both, their length and sequence, can be optimized to ensure proper assembly and thermodynamic stability. At the same time, accessible linker loops can make a format vulnerable towards proteolytic cleavage, which can be a challenge particularly for complex designs requiring use of several linkers.

Additional considerations related to each individual molecule format goes beyond the scope of this chapter and shall not be further detailed. In general, a successful design will combine desired potency and biologic response through efficient target engagement with minimal immunogenicity and a molecular framework with favorable CMC characteristics that allow efficient manufacturing with acceptable quality attributes.

4.2.3.3 CMC Challenges

Dependent on the structure of the multispecific molecule, expression can be done either in microbial or in mammalian cells. Expression of small, nonglycosylated formats can be performed in bacteria or yeast (preferred for albumin fusion proteins) while bigger, more complex formats, which could present challenges for refolding, and of course glycosylated proteins, should be expressed in mammalian systems. In reality, for some of the new formats, a clear rational for decisions on host cell often becomes fluid and, therefore, many researchers prefer to go for the empirical approach and first try a broader panel of host cells ranging from bacteria to CHO cells before deciding on the most suitable host for a given molecule. As this book concerns animal cell culture, focus will be kept on mammalian expression in the following.

Non-Fc-containing Formats

CMC-challenges for multispecifics lacking an Fc-domain, for example BiTE molecules, do not really differ from the challenges associated with manufacturing of any other novel biological entity (NBE). As many of the formats are built on a naturally occurring protein or protein domain of a relatively small size, decent expression levels can often be achieved without too much effort. An overall challenge is that in contrast to IgG manufacturing, no common platform is available. Thus, several different host cells including microbials may have to be considered, and this is often best done empirically. A specific challenge compared to the IgG formats concerns purification, which obviously cannot take advantage of the protein A-binding properties of the IgG molecule. Chromatographic or filtering steps may take advantage of an inherent specific affinity – which can be costly – and/or rely on more unspecific means such as hydrophobicity or charge properties of the molecule. Addition of tags, for example a His-tag, is another option, which can be considered.

Fc-domain-containing Formats

Multispecific formats relying on incorporation of a Fc-domain in their structure are built from several chains, for example two different HCs combined with one or two LCs. For some formats (Ig-DARTs, DVDs), the HC has been extended with a second antigen binding site added either N-terminally on top of the variable domain, or as a binding domain fused to the C-terminal of the Fc region (Figure 4.2.1c). Such molecules are typically designed using naturally occurring IgG domains as the additional building blocks. While this likely will facilitate manufacturing and folding, the challenge exist that as the molecule becomes relatively large they may still be prone to difficulties with respect to manufacturability, e.g., due to problems with folding and associated formation of aggregates and/or with molecule stability.

Other formats more closely resembling natural IgGs in their length and structure rely on formation of heterodimers between two different HCs (and potentially two different LCs) to generate more than one binding sites. This concept has led to a challenge known as the “chain association issue” which has been reviewed recently [29]. Basically, the challenge is to favor formation of the one desired heterodimeric molecule over the variety of formats that can be generated through random combination of multiple chains. Popularly speaking, the goal is to form the structure AB from the constructs AA and BB when the simplest setup of two HCs and a common LC comprise the molecule building blocks. An early elegant idea was the knob-in-hole structure (reviewed in [27]) where formation of the desired heterodimer from two homodimer constructs was facilitated by designing the molecules with a bulky “knob” (key) structure in the CH3 part of one chain fitting into a “hole” (or lock) in the corresponding site on the other chain. Other approaches with a similar aim rely on introducing charge differences in the same regions of the Fc-domain.

CMC challenges for this approach comprise obtaining a balanced expression of the different HC chains, as well as development of purification techniques, which allow purifying the desired "AB" format with high yield. Molecule stability may also be an issue of concern here, but this can and should be addressed prior to final selection of the lead candidate by doing accelerated stability studies. Manufacturing of most multichain formats rely on formation of the desired functional structure within the host cell, taking advantage of the cellular folding machinery. However, some companies have also pursued methods where formation of the bispecific is done after independent expression of two monoclonals by mixing them during the purification process (see below).

4.2.3.4 Critical Manufacturing Steps

Vector Design

The first step of expression includes building a vector construct(s) encoding the gene of interest. Dependent on the format and number of chains needed to form the molecule this may be a more or less complex exercise. Some constructs like, e.g., the Ig-DARTs or the DVD formats are quite large, exceeding the size of mAbs. Others are engineered to be asymmetric in their form, with special demands for formation of the desired format during manufacturing. Considerations have to be taken on whether one or more vectors are used for the multichain formats. Whereas one vector encoding all chains may be the simplest approach, e.g., limiting necessary selection markers to one, it does give constrains to the size of the molecules, which again may influence transfection efficiency. Further, the one-vector approach does not allow for titration of optimal ratios between the different chains. For formats where the concept of forming the bispecific molecule depends on heterodimer formation within the cell, such an optimization of the ratio may be crucial for obtaining sufficient yields of the purified molecule.

Purification

The purification process represents several challenges dependent on the format, as eluted to above. For non-IgGs, a platform will not be at hand, thus a full purification process has to be developed. To do this, it is usually very helpful to collect as much knowledge as possible of the molecule structure and its physico-chemical properties already during the research phase, so that ideas for the purification process are at hand when CMC-development starts. For IgG-like formats, it is often possible to take advantage of the protein A-binding properties during the first capture step. But as we have seen above, separation techniques to remove undesired chain formats (AA, BB) from the desired one (AB) needs to be established. This may be addressed using for example reverse phase HPLC or cation exchange chromatography to separate the various formats. Notably, dependent on successful selection of a host cell delivering as high amounts as possible of the AB format, the manufacturing yield may be compromised.

A few developers are pursuing a strategy where two independently expressed mAbs differing in charge are expressed independently, purified and subsequently mixed under mildly reducing conditions to allow formation of the heterodimer, an example being Genmab's DuoBody format [52]. The main challenge with this strategy, apart from the need shared with the heterodimerization-dependent formats described above, is to develop a step during the downstream purification where the molecules can be mixed together under appropriate conditions for formation of the heterodimer, followed by development of one or more polishing steps to remove any remaining undesired homodimers.

Analytics

High-throughput method(s) for selectively detecting titer of the desired format needs to be developed prior to start of cell line development. This is true for non-IgG formats, but also for the IgG-like heterodimer approaches, as a simple IgG titer determination typically used for monoclonal antibodies will not be sufficient here to distinguish the desired AB format from AA and AB. The availability of such a specific method is critical for selection of the optimal production clone as well as the best upstream production and downstream purification process. When the molecule has been produced and purified, it is of course essential to analyse the product quality attributes including product and process related impurities. Substantial analytical challenges are associated with the successful development of such new formats since platform methods typically cannot be applied. Thus, analytical methods need to be established and, dependent on the nature of the multispecific, some or even most of these have to be fully developed. Also, it is obviously crucial to perform functional testing both *in vitro* and *in vivo*, including half-life determination of the molecule. Functional assays will often be available from the research phase that led to selection of the lead candidate, but they have to be transferred, qualified and established as release or characterization assays as needed.

Outlook

The past decade has seen an impressive number of multispecific formats being developed and progressed into clinical proof-of-concept studies. From their beginnings as mere alternatives for combination therapy, new strategies have emerged that hold the promise to open new therapeutic concepts for disease treatment. First examples in clinical trials have shown encouraging results and demonstrated that multispecifics can indeed provide improvement over standard therapy. However, there are still open questions around these new modular design drugs, particularly with regards to maintenance of circulation half-life and the risk of inducing undesired immunogenicity in humans. Many of the newer formats are only now starting to approach clinical studies aiming to prove superiority. The field is progressing fast and it will be interesting to see the contribution of multispecific formats to advance science and disease treatment.

4.2.3.5 IgG-fusion Proteins

The IgG-fusion protein is yet another format which has been successfully brought to the clinic for therapeutic use [2]. As of 2012, 7 molecules have been approved for the market and several molecules are in late stage clinical trials [53]. Thus, compared to the above described next generation molecules of which only one has made it into the market so far [5], this class of drugs is already well established. Notably, the commercially most successful therapeutic on the market is Enbrel, which with a record sale of 7,3 billion U. S. dollars in 2010 is the best-selling biologics to date. Enbrel (etanercept) is a fusion protein consisting of the external domain of the TNF-α2 receptor fused to an IgG1 Fc-region. Its target is TNF-α, an important target in rheumatoid arthritis therapy. mAbs such as Remicade (infliximab) target the same molecule, however, Enbrel has since market entry of both Remicade and Enbrel in 1998 outbeated sales of the mAb. Another example of a marketed Fc-fusion molecule is Orencia (abatercept) which was brought to the market in 2005, also for treatment of rheumatoid arthritis. It is comprised of the extracellular domain of the CTLA-4 molecule fused to an IgG1 Fc-molecule. CTLA-4 binds B7-1 and B7-2 on antigen-presenting cells, thereby inhibiting T cell activation through blocking of B7 interaction with CD28 on T cells. Advantages of developing these molecules as efficient therapeutics are that a high-affinity naturally occurring binding domain – e.g., the TNF-α receptor extracellular domain – can be efficiently manufactured by removing its transmembrane and intracellular regions and instead adding the binding domain to an IgG constant region. The advantages of fusing an Fc-region to a smaller protein domain are first of all an enhanced half-life in the body resulting from the Fc-region binding to the neonatal FcRn protein and a slower renal clearance associated with the larger size of the molecule. Furthermore, binding to Fc-receptors on immune cells may be an advantage for the molecule mechanism of action dependent on the specific indication.

CMC Challenges

From the CMC perspective, Fc-fusion molecules are typically well behaved and relatively easy to express, compared to an average protein. As the molecules always are glycosylated, at least in the Fc-region, the expression system of choice is mammalian cells, typically CHO cells. Folding of the Fc-region occur independent of the fusion protein, which may support stabilization and also efficient secretion of the molecules. With respect to downstream processing, the Fc-region enables use of simple and well-established protein A/G chromatography-mediated purification techniques. There are of course large individual differences between molecules, relating to the structure of the target binding part of the fusion protein. Typically, posttranslational modifications such as extensive glycosylation can represent a manufacturing challenge with respect to consistency, as can unpaired cysteines present in the molecule potentially destablilizing the molecule. Both, the TNF-αR and the CTLA-4 parts of the etanercept and abatacept, comprise a number of glycosylation sites, which calls for careful process development of a robust manufacturing process leading to reproducible production of molecules with a consistent glycopattern.

4.2.4 Concluding Remarks

Unmet medical needs are continuously driving the scientific community and the biopharma industry to develop new and innovative ideas for improved treatment modalities. In the present chapter, we have reviewed some of the currently pursued strategies for treatment of complex diseases with alternatives to monoclonal antibodies, such as recombinant antibody mixtures, Fc-fusions, or the fast-growing and very diverse panel of multispecific formats. Whereas the antibody mixtures and the multispecifics concepts represent new, but very promising approaches, the Fc-fusion molecules have already proven their value in the clinic. A number of CMC challenges obviously accompany each of the novel formats, and solutions to these are very actively being pursued. The use of animal cell culture technology as the most prominent platform for manufacturing of the new molecules and the huge efforts invested in optimizing cells and manufacturing processes for efficient and high-quality production may very well drive this technology to new levels of scientific understanding and performance in the years to come.

References

[1] Beck A, Carter PJ, Gerber HP, et al. 8(th) Annual European Antibody Congress 2012: November 27-28, 2012, Geneva, Switzerland. MAbs 5(3) (2013) 339-57.
[2] Beck A, Reichert JM. Therapeutic Fc-fusion proteins and peptides as successful alternatives to antibodies. MAbs 3(5) (2011) 415-6.
[3] Li J, Zhu Z. Research and development of next generation of antibody-based therapeutics. Acta Pharmacol Sin 31 (2010) 1198-207.
[4] Saylor C,Dadachova E, Casadevall A. Monoclonal antibody-based therapies for microbial diseases. Vaccine 27(6) (2009) G38-46.
[5] Seimetz D, Lindhofer H, Bokemeyer C. Development and approval of the trifunctional antibody catumaxomab (anti-EpCAM x anti-CD3) as a targeted cancer immunotherapy. Cancer Treat Rev 36(6) (2010) 458-67.
[6] Pedersen MW, Jacobsen HJ, Koefoed K., et al. Sym004: a novel synergistic anti-epidermal growth factor receptor antibody mixture with superior anticancer efficacy. Cancer Res 70(2) (2010) 588-97.

[7] Robak T, Windyga J, Trelinski J, et al. Rozrolimupab, a mixture of 25 recombinant human monoclonal RhD antibodies, in the treatment of primary immune thrombocytopenia. Blood 120(18) (2012) 3670-6.
[8] May C, Sapra P, Gerber HP, Advances in bispecific biotherapeutics for the treatment of cancer. Biochem Pharmacol 84(9) (2012) 1105-12.
[9] de Kruif J, Kramer A, Nijhuis R, et al. Generation of stable cell clones expressing mixtures of human antibodies. Biotechnol Bioeng 106 (2010) 741-50.
[10] Wiberg FC, Rasmussen SK, Frandsen TP, et al. Production of target-specific recombinant human polyclonal antibodies in mammalian cells. Biotechnol Bioeng 94 (2006) 396-405.
[11] Nielsen LS, Baer A, Muller C, et al. Single-batch production of recombinant human polyclonal antibodies. Mol Biotechnol (2010) 45(3) 257-66.
[12] Frandsen TP, Naested H, Rasmussen SK, et al. Consistent manufacturing and quality control of a highly complex recombinant polyclonal antibody product for human therapeutic use. Biotechnol Bioeng (2011) 108(9) 2171-81.
[13] Sharon J, Sarantopoulos S, Den W, et al. Recombinant polyclonal antibody libraries. Comb Chem High Throughput Screen. 3 (2000) 185-196.
[14] Persson P, Engstrom A, Rasmussen LK, et al. Development of mass spectrometry based techniques for the identification and determination of compositional variability in recombinant polyclonal antibody products. Anal Chem 82 (2010) 7274-82.
[15] Rasmussen SK, Næsted H, Müller C, et al. Recombinant antibody mixtures: production strategies and cost considerations. Arch Biochem Biophys 526(2) (2012) 139-45.
[16] Rasmussen SK, Nielsen LS, Müller C, et al. Recombinant antibody mixtures; optimization of cell line generation and single-batch manufacturing processes. BMC Proc 22(5) (2011) Suppl 8:O2.
[17] Kontermann R. Dual targeting strategies with bispecific antibodies. MAbs 4(2) (2012) 182-97).
[18] Weidle UH, Tiefenthaler G, Weiss EH, et al. The intriguing options of multispecific antibody formats for treatment of cancer. Cancer Genomics Proteomics 10(1) (2013) 1-18.
[19] Yu YJ, Zhang Y, Kenrick M,et al. Boosting brain uptake of a therapeutic antibody by reducing its affinity for a transcytosis target. Sci Transl Med 3(84) (2011) 84ra44.
[20] Robinson MK, Hodge KM, Horak E, et al. Targeting ErbB2 and ErbB3 with a bispecific single-chain Fv enhances targeting selectivity and induces a therapeutic effect in vitro. Br J Cancer 99(9) (2008) 1415-25.
[21] Rossi EA, Rossi DL, Stein R, et al. A bispecific antibody-IFNalpha2b immunocytokine targeting CD20 and HLA-DR is highly toxic to human lymphoma and multiple myeloma cells. Cancer Res 70(19) (2010) 7600-9.
[22] Schubert I, Kellner C, Stein C, et al. A single-chain triplebody with specificity for CD19 and CD33 mediates effective lysis of mixed lineage leukemia cells by dual targeting. MAbs 3(1) (2011) 21-30.
[23] Weisbart RH, Gera JF, Chan G, et al. A cell-penetrating bispecific antibody for therapeutic regulation of intracellular targets. Mol Cancer Ther 11(10) (2012) 2169-73.
[24] Lindhofer H, Mocikat R, Steipe B, et al. Preferential species-restricted heavy/light chain pairing in rat/mouse quadromas. Implications for a single-step purification of bispecific antibodies. J Immunol 155(1) (1995) 219-25.
[25] Bokemeyer C. Catumaxomab--trifunctional anti-EpCAM antibody used to treat malignant ascites. Expert Opin Biol Ther 10(8) (2010) 1259-69.
[26] Müller D, Kontermann RE. Bispecific antibodies for cancer immunotherapy: Current perspectives. BioDrugs 24(2) (2010) 89-98.
[27] Carter P. Bispecific human IgG by design. J Immunol Methods 248(1-2) (2001) 7-15.

[28] Davis JH, Aperlo C, Li Y, et al. SEEDbodies: fusion proteins based on strand-exchange engineered domain (SEED) CH3 heterodimers in an Fc analogue platform for asymmetric binders or immunofusions and bispecific antibodies. Protein Eng Des Sel 23(4) (2010) 195-202.
[29] Klein C, Sustmann C, Thomas M, et al. Progress in overcoming the chain association issue in bispecific heterodimeric IgG antibodies. MAbs. 4(6) (2012) 653-63.
[30] Holliger P, Prospero T, Winter G. “Diabodies”: small bivalent and bispecific antibody fragments. Proc Natl Acad Sci USA 90(14) (1993) 6444-8.
[31] Baeuerle PA, Kufer P, Bargou R. BiTE: Teaching antibodies to engage T-cells for cancer therapy. Curr Opin Mol Ther 11(1) (2009) 22-30.
[32] Van Bockstaele F, Holz JB, Revets H. The development of nanobodies for therapeutic applications. Curr Opin Investig Drugs 10(11) (2009) 1212-24.
[33] Wurch T, Pierré A, Depil S. Novel protein scaffolds as emerging therapeutic proteins: from discovery to clinical proof-of-concept. Trends Biotechnol 30(11) (2012) 575-82.
[34] Kontermann RE. Strategies for extended serum half-life of protein therapeutics. Curr Opin Biotechnol 22(6) (2011) 868-76.
[35] Seimetz D. Novel monoclonal antibodies for cancer treatment: the trifunctional antibody catumaxomab (Removab). J Cancer 2 (2011) 309-16.
[36] Johnson S, Burke S, Huang L, et al. Effector cell recruitment with novel Fv-based dual-affinity re-targeting protein leads to potent tumor cytolysis and in vivo B-cell depletion. J Mol Biol 399(3) (2010) 436-49.
[37] Wozniak-Knopp G, Bartl S, Bauer A, et al. Introducing antigen-binding sites in structural loops of immunoglobulin constant domains: Fc fragments with engineered HER2/neu binding sites and antibody properties. Protein Eng Des Sel. 23(4) (2010) 289-97.
[38] Logtenberg T. Antibody cocktails: next-generation biopharmaceuticals with improved potency. Trends Biotechnol 25(9) (2007) 390-4.
[39] Labrijn AF, Meesters JI, de Goeij BE, et al. Efficient generation of stable bispecific IgG1.by controlled Fab-arm exchange. Proc Natl Acad Sci USA 110(13) (2013) 5145-50.
[40] Wu C, Ying H, Grinnell C, et al. Simultaneous targeting of multiple disease mediators by a dual-variable-domain immunoglobulin. Nat Biotechnol 25(11) (2007) 1290-7.
[41] Schaefer W, Regula JT, Bähner M, et al. Immunoglobulin domain crossover as a generic approach for the production of bispecific IgG antibodies. Proc Natl Acad Sci USA 108 (2011) 11187-92.
[42] Kipriyanov SM, Moldenhauer G, Schuhmacher J, et al. Bispecific tandem diabody for tumor therapy with improved antigen binding and pharmacokinetics. J Mol Biol 293(1) (1999) 41-56.
[43] Skerra A. Alternative binding proteins: anticalins – harnessing the structural plasticity of the lipocalin ligand pocket to engineer novel binding activities. FEBS J 275(11) (2008) 2677-83.
[44] Boersma YL, Pluckthun A. DARPins and other repeat protein scaffolds: advances in engineering and applications. Curr Opin Biotechnol 22(6) (2011) 849-57.
[45] Nygren PA. Alternative binding proteins: affibody binding proteins developed from a small three-helix bundle scaffold. FEBS J 275(11) (2008) 2668-76.
[46] Friedman M, Lindström S, Ekerljung L, et al. Engineering and characterization of a bispecific HER2 x EGFR-binding affibody molecule. Biotechnol Appl Biochem 54(2) 2009 121-31.
[47] Grabulovski D, Kaspar M, Neri D. A novel, non-immunogenic Fyn SH3-derived binding protein with tumor vascular targeting properties. J Biol Chem 282(5) (2007) 3196-204.
[48] Lipovsek D. Adnectins: engineered target-binding protein therapeutics. Protein Eng Des Sel 24(1-2) (2011) 3-9.
[49] Kanakaraj P, Puffer BA, Yao XT, et al. Simultaneous targeting of TNF and Ang2 with a novel bispecific antibody enhances efficacy in an in vivo model of arthritis. MAbs 4(5) (2012) 600-13.

[50] Lafleur DW, Abramyan D, Kanakaraj P, et al. Monoclonal antibody therapeutics with up to five specificities: Functional enhancement through fusion of target-specific peptides. MAbs. 5(2) (2013) 208-18.
[51] Bertschinger J. Engineering Bi-Specific Fnomer Antibody Fusions for Therapeutic Applications. PEGs Europe 2012 (Oral presentation; decorated antibodies Covagen).
[52] Strop P, Ho WH, Boustany LM, et al. Generating bispecific human IgG1 and IgG2 antibodies from any antibody pair. J Mol Biol 420(3) (2012) 204-19.
[53] Czajkowsky DM, Hu J, Shao Z, et al. Fc-fusion proteins: new developments and future perspectives. EMBO Mol Med 4(10) (2012) 1015-28.

5 Basic Aspects of Animal Cell Cultivation

5.1 Physiology and Metabolism of Animal Cells for Production

Venkata Tayi and Michael Butler

5.1.1 Introduction

Animal cells require carbohydrate source like glucose and amino acids for biomass synthesis and their metabolic adenosine triphosphate (ATP) and nicotinamide adenine dinucleotide phosphate (NADPH) generation. If a recombinant product is produced from amplified genes in the cells, the net demand for these energy sources would be much higher. The high consumption of energy sources often accompanies with the production of high level of inhibitory metabolites, which alter cellular metabolism and productivity/quality of therapeutic proteins.

5.1.2 The State of Cells during Phases of Culture

Animal cells typically proliferate based on the environmental conditions and availability of nutrients in addition to the availability of physical substrate for anchorage-

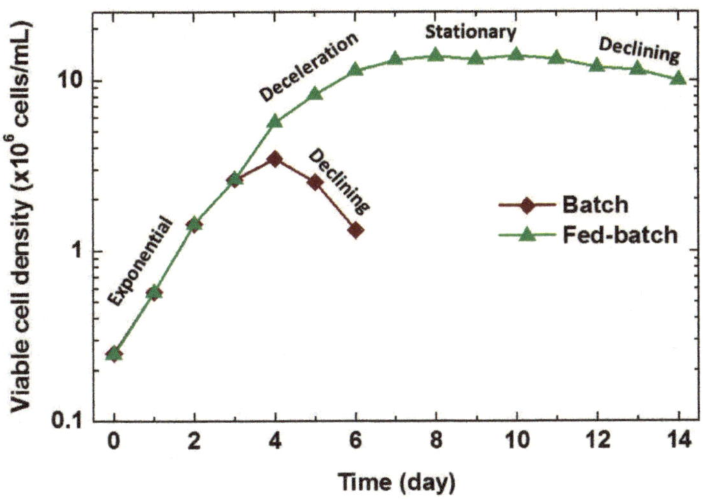

Figure 5.1.1: Growth profiles of a Chinese hamster ovary cell line (CHO-DXB11 derived cell line producing a monoclonal antibody) cultured in batch and fed-batch modes. The feed media supplementation for fed-batch culture was started from day 3.

dependent cells. The typical pattern of growth exhibited by animal cells in batch and fed-batch cultures is presented in Figure 5.1.1.

5.1.2.1 Lag Phase

After seeding, cells exhibit a lag phase during which they adapt to a new environment and do not proliferate. The length of the lag phase depends at least on the seeding density, the state of the cells at which seeding occurs, and the change of medium components. The cells seeded from an exponentially growing culture typically show a shorter or no lag phase compared to the one seeded from the quiescent state. The cells seeded at lower density require a longer time for conditioning the medium and hence exhibit a longer lag phase. Also, change of environmental conditions like media can have great influence on the lag phase. The seeding of cells in the media with different components can exhibit a lag phase until the cells adapt to metabolic changes with respect to new components. Reduced lag phase can result in quicker proliferation and hence production of therapeutic proteins in less time.

5.1.2.2 Exponential Phase

This state is considered to be an active proliferating state in which cells divide exponentially. Almost all the cells enter into cell cycle and at any given time cells are randomly distributed in all phases of cell cycle. All the metabolic pathways are active in this phase and the components of cells are synthesized at a constant rate. The intrinsic doubling time of a particular cell line can be obtained from this phase. In a typical fed-batch culture, the exponential phase is extended to reach maximum cell density by feeding the cells with concentrated nutrient medium. The growth-associated products from cells are produced at the highest rate in this phase.

5.1.2.3 Stationary Phase

In this phase, the cell densities remain fairly constant either due to cell maintenance without dividing or the growth rate is balanced by the death rate. This phase occurs as a result of various conditions like limited nutrients, accumulation of toxic products, or hypoxia. For adherent cell lines, this phase may also occur due to a limitation in the available growth surface. In some cases, the stationary phase is associated with the balance of cell survival mechanism by autophagy and cell death by apoptosis. The cell specific productivity of nongrowth associated products like monoclonal antibodies will be higher in this phase as the cells spend their energy in protein production rather than division, provided all the essential amino acids and nutrients are available for protein synthesis. The stationary phase is often short or negligible when cells are cultured in batch mode. In fed-batch and perfusion processes, this stationary phase is extended by feeding with essential nutrients or removing the toxic metabolites in the latter case. This has the desirable effect of maximizing the cumulative product concentration.

5.1.2.4 Decline Phase

As the nutrients are depleted and/or the inhibitory metabolites reach a critical concentration, the death rate dominates the growth and cells decline at a faster rate. An accelerated death rate is observed in batch as compared to fed-batch culture (Figure 5.1.1). The death mechanisms such as apoptosis or necrosis become very active in this phase. Necrosis is a death response due to exposure to sudden harsh environmental conditions like hypoxia or prolonged stationary phase. On the other hand, apoptosis is a programmed cell death in which cells respond to environmental conditions and proceed with a cascade of reactions, mediated by caspase enzymes, which lead to cell death.

5.1.3 Energy Sources and Their Metabolism

5.1.3.1 Glucose

Glucose is the preferred carbon source for the cultivation of animal cells, though other carbon sources like galactose, fructose, and mannose can be utilized. The hexose transporters GLUT1, GLUT3, and GLUT4 have high affinity for glucose (K_m ~ 2 to 5 mmol L^{-1}), lower affinity for galactose, and no affinity for fructose. Fructose enters cells via the GLUT5 transporter. GLUT1 is expressed in almost all cells [1] and its transcription is regulated by serine/threonine kinase AKT1 [2]. Glucose passes through the glycolytic pathway at a high rate in animal cells [3] when compared to other sugar sources [4, 5].

Glucose enters into glycolysis which generates intermediates for biosynthesis and free energy (Figure 5.1.2).

One glucose molecule converts to 2 pyruvate molecules with a net yield of 2 ATP and 2 nicotinamide adenine dinucleotide (NADH) molecules. The complete oxidation of 2 pyruvate molecules to carbon dioxide in mitochondria can result in the generation of ~36 ATP molecules. However, the conversion of pyruvate to lactate by anaerobic metabolism in animal cells is a waste of available energy and only necessary to regenerate NAD^+. The glucose consumption rate and the corresponding lactate production rate by a Chinese hamster ovary cell line are presented in Figure 5.1.3. As shown in the figure, the glucose consumption and lactate production strictly depends on the amount of glucose available. At reduced glucose levels, cells shift towards energy efficient metabolism by reducing their lactate production. The metabolism with a low lactate yield coefficient (<1) is generally considered energy efficient. The high conversion of pyruvate to lactate occurs predominantly occurs in exponential phase of the cell. Studies with radio isotope tracers revealed that around 60 to 80 % of glucose is converted to lactate in animal cells during exponential phase [3, 6-8]. The demand for

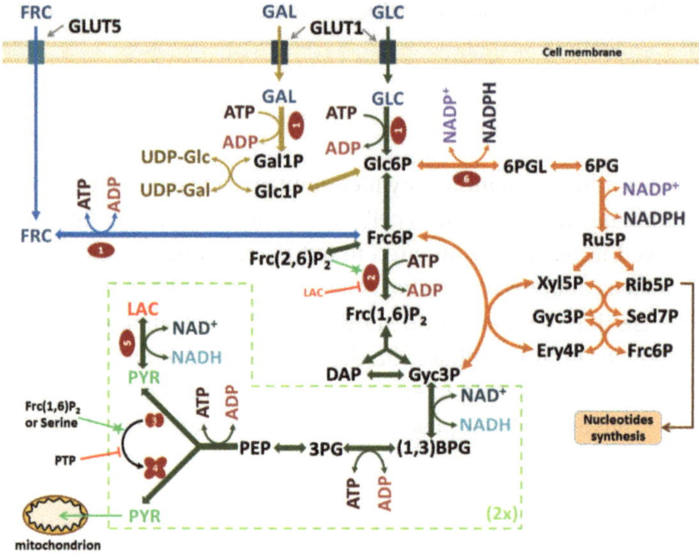

Figure 5.1.2: The pathways of the metabolism of major carbon sources (glucose, galactose, and fructose) in mammalian cells. Glycolysis is represented in dark green and pentose phosphate pathway in orange. The red ovals with numbers representing key enzymes of these pathways are 1: hexokinase, 2: phosphofructo kinase, 3: pyruvate kinase-M2 isoform (dimer), 4: pyruvate kinase-M2 isoform (tetramer), 5: lactate dehydrogenase, and 6: glucose-6-phophate dehydrogenase. ((1,3)BPG = 1,3-biphosphoglycerate; 3PG = 3-phosphoglycerate; 6PGL = 6-phosphogluconolactone; 6PG = 6-phosphogluconate; ADP = adenosine diphosphate; ATP = adenosine triphosphate; DAP = dihydroxyacetone phosphate; Ery4P = erythrose-4-phosphate; FRC = fructose; Frc(1,6)P_2 = Fructose-(1,6)-biphosphate; Frc(2,6)P_2 = fructose-(2,6)-biphosphate; Frc6P = fructose-6-phosphate; GAL = galactose; Gal1P = galactose-1-phosphate; GLC = glucose; Glc1P = glucose-1-phospahte; Glc6P = glucose-6-phosphate; GLN = glutamine; GLUT1 = glucose transporter 1; GLUT5 = glucose transporter 5; Gyc3P = glyceraldehyde-3-phosphate; LAC = lactate; PEP = phophoenolpyruvate; PTP = phosphorylated tyrosine peptide; PYR = pyruvate; Rib5P = ribose-5-phosphate; Ru5P = ribulose-5-phosphate; Sed7P = sedoheptulose-7-phosphate; Xyl5P = xylulose-5-phosphate.)

a cytoplasmic pool of NAD^+ to maintain glycolytic flux and generate ATP might lead to utilization of NADH by converting pyruvate to lactate. The activity of the malate-aspartate NADH shuttle, responsible for regeneration of NADH and proton import into mitochondria, also determines the fate of pyruvate conversion to lactate [9].

Glycolysis is a well regulated process by key enzymes (Figure 5.1.2). Hexokinase (HK) has more affinity towards glucose compared to other sugar sources. Phosphofructokinase 1 (PFK-1) is allosterically inhibited by high levels of ATP thus reducing its affinity towards fructose-6-phosphate (Frc6P). The allosteric inhibition of PFK-1 by

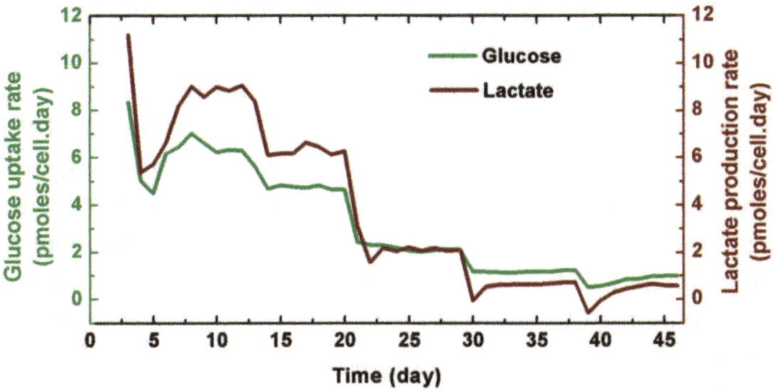

Figure 5.1.3: Lactate production in response to glucose availability by a Chinese hamster ovary cell line (CHO-DXB11 derived cell line producing a monoclonal antibody) cultured in a 3L bioreactor. Bioreactor was operated in continuous mode and the glucose concentration in feed medium was changed at days 2, 12, 20, 29 and 38 to reach 5 different steady states. The profiles show that lactate production is strictly dependent on the amount of glucose available and the glucose consumption rate.

accumulated lactate was found to regulate glycolysis in muscles [10]. Lactate down-regulates the PFK-1 enzyme thus leading to accumulation of fructose-(2,6)-bisphosphate (Frc(2,6)P_2), a byproduct which does not metabolize further but is a potent activator of PFK1. Evidence suggests that pyruvate kinase (PK), the enzyme which catalyzes the irreversible conversion of phosphoenolpyruvate (PEP) to pyruvate, plays a major role in regulation of glycolysis. The M2 isoform of PK (PK-M2), which is expressed exclusively in proliferating cells, is responsible for aerobic glycolysis and conversion to lactate [11]. PK-M2 is less active but allosterically activated by upstream fructose-(1,6)-bisphosphate (Frc(1,6)P_2) to form a more active tetramer [12]. Recently, serine was also found to be a natural ligand for PK-M2 and promotes the formation of active tetrameric form by binding at an allosteric pocket [13]. It was speculated that under serine limiting conditions, the less active PK-M2 results in the diversion of glucose towards biosynthesis of serine. The formation of active tetrameric form is inhibited by either binding of a tyrosine-phosphorylated peptide to it [14] or direct phosphorylation of a tyrosine residue on the enzyme itself [15]. The active tetrameric form of PK-M2 facilitates the transport of pyruvate into mitochondria while the inactive form favors its conversion to lactate. Recently, it was shown that the increase in PK-M2 activity in CHO cells at late stages of cell culture correlated with the decrease in growth rate, decrease in lactate production, and increase in glucose-derived TCA cycle intermediates indicating that this enzyme is also regulated in cells producing recombinant proteins [16].

Glc-6-P, an intermediate metabolite of the glycolysis pathway, is the key precursor for pentose phosphate pathway (PPP). The purpose of PPP is to produce NADPH for anabolic processes and ribose-5-phosphate units for the synthesis of nucleotides. The flux through ox-PPP was found to be minimal or negligible during early exponential phase accounting for ~1 to 2 % of consumed glucose in CHO cells [17, 18], ~4 % in murine hybridomas [8] and ~8 % in BHK cells [3]. The additional NADPH required for the synthesis of lipids during early exponential phase might be generated from other reactions mediated by malic enzyme 1 (ME-1) or isocitrate dehydrogenase. The flux of glucose through ox-PPP increases significantly in late exponential or stationary phase. An estimated 30 to 100 % of glucose taken up passes through the ox-PPP pathway before entering glycolysis as Frc6P [7, 16, 18-20]. There would be minimal or no biosynthesis of cells during the stationary phase indicating that the increase in flux is entirely to generate NADPH required during this phase. In fact, it was observed that G6PDH activity was higher than hexokinase activity resulting in the cyclic mode of ox-PPP activity by conversion of fructose-6-phosphate to glucose-6-phosphate [18]. It was hypothesized that the function of NADPH generation is to combat oxidative stress through regeneration of reduced glutathione from the oxidized form. Glutathione is an antioxidant that plays a key role in scavenging of reactive oxygen radicals generated from oxidative phosphorylation. Alternatively, glutathione also plays a major role in disulfide bond formation of proteins. It was also noted that an increase in PPP flux correlated with the increase in productivity during stationary phase in antibody-producing CHO cells. NADPH is necessary for lipid synthesis for the secretary process. This indicates that the generation of NADPH through PPP is necessary to regenerate reduced glutathione and lipid synthesis for secretion, especially in high producer cell lines.

5.1.3.2 Glutamine

Glutamine provides both, anaplerotic and ATP/NADPH demands for proliferating cells. Glutamine is consumed at much higher rates than other amino acids. Glutamine alone can provide 35 to 70 % of cellular energy indicating that it can be the main energy source for animal cells [21-24]. The process of glutamine conversion to products like CO_2, alanine, aspartate, and lactate by various catabolic pathways is called glutaminolysis (Figure 5.1.4). Glutamine enters into the tricarboxylic acid (TCA) cycle by conversion to α-ketoglutarate in 2 steps. First, glutamine is converted to glutamate in cytoplasm by a glutaminase-mediated reaction in which an ammonium ion is released from glutamine. Glutamate is then converted to α-ketoglutarate by glutamate dehydrogenase in which conversion occurs by releasing another ammonium ion. This results in generation of 2 ammonium molecules from one glutamine molecule to form α-ketoglutarate. This glutaminolysis is an energy efficient process, which involves complete oxidation of glutamine to CO_2 and generates 24 ATP and 3 NADPH molecules. This pathway is especially preferred when the glucose does not provide the necessary energy demands.

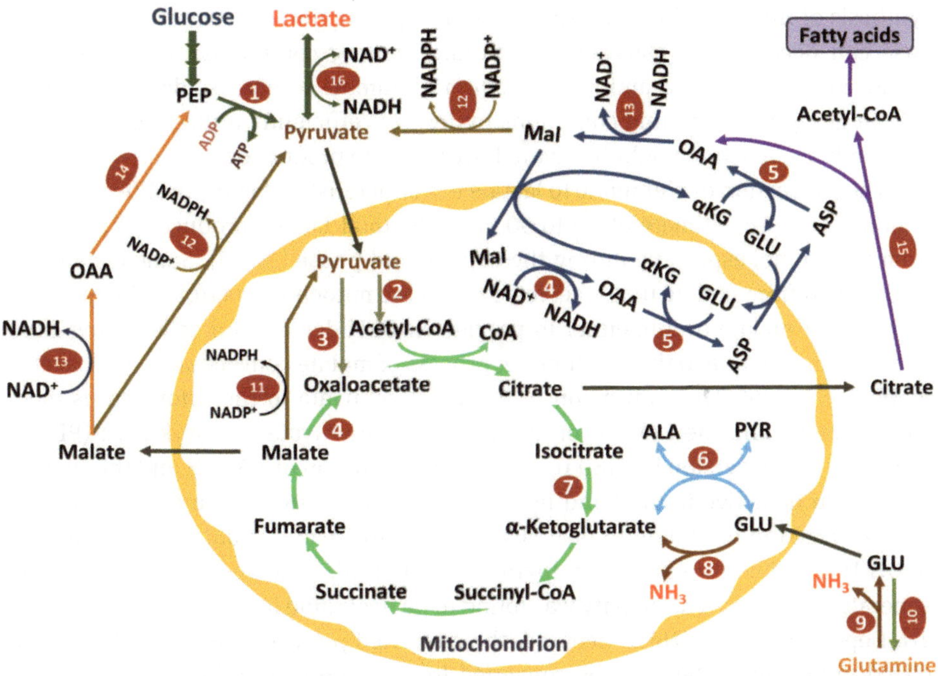

Figure 5.1.4: Metabolic networks of pyruvate and amino acids (especially glutamine) oxidation by TCA cycle in mammalian cells. TCA cycle is represented in green color, malate-aspartate shuttle is represented in dark blue and lipid synthesis is represented in purple. The red ovals with numbers representing key enzymes of these pathways are 1: pyruvate kinase (M2 isoform), 2: pyruvate dehydrogenase, 3: pyruvate carboxylase, 4: malate dehydrogenase II, 5: aspartate aminotransaminase, 6: alanine aminotransaminase, 7: isocitrate dehydrogenase II, 8: glutamate dehydrogenase, 9: glutaminase, 10: glutamine synthetase, 11: malic enzyme II, 12: malic enzyme I, 13: malate dehydrogenase I, 14: phosphoenolpyruvate carboxykinase, 15: citrate lyase, 16: lactate dehydrogenase. (αKG = α-ketoglutarate; ALA = alanine; ASP = aspartate; GLU = glutamate; Mal = malate; OAA = oxaloacetate; PEP = phosphoenolpyruvate; PYR = pyruvate.)

Alternatively, glutamate can be converted to α-ketoglutarate by amino-transaminase reactions by transferring its ammonium ion to pyruvate or oxaloacetate. This results in conversion of pyruvate to alanine and oxaloacetate to aspartate. Since pyruvate and oxaloacetate are part of TCA cycle, this partial oxidation of glutamine can lead to generation of only 6 to 9 ATP molecules. However, these pathways result in generation of only one ammonium ion from glutamine to form α-ketoglutarate. The production of alanine and aspartate may be a cellular response to sequester ammonium ions generated due to excess glutamine consumption. Studies with labeling experiments showed that glutamate participates more actively in pyruvate transaminase reaction than aspartate transaminase [16]. On the other hand, aspartate is a precursor for the synthesis of nucleic acids and asparagine. The production of aspartate by transaminase can occur under asparagine/aspartate limiting conditions. Since alanine is produced by transfer of amine group from glutamate to glucose-derived pyruvate [25], under glucose limiting conditions pyruvate is efficiently utilized and aspartate transaminase reaction might be favored over alanine production.

The conversion of glutamine to lactate is characteristic of many proliferating cells (Figure 5.1.4). An estimate of up to 30 % of the total lactate produced is generated from glutamine, especially during the exponential phase [16, 24]. Some lactate may be obtained from malate that is secreted out of the mitochondria from TCA cycle. The secretion of malate occurs either to provide NADPH demand for cells during exponential phase or due to the rate limiting process of malate conversion to oxaloacetate in TCA cycle [26]. The conversion of malate to pyruvate in the cytoplasm is mediated by malic enzyme I (ME-I) or by phosphoenolpyruvate carboxykinase (PEPCK) enzyme. The conversion by ME-I results in the generation of NADPH and this enzyme is found to be active in CHO and hybridoma cell lines [16, 27]. An alternative route is by conversion of malate to oxaloacetate by cytoplasmic malate dehydrogenase I (MDH-I) and then to phosphoenolpyruvate by PEPCK. However, this route does not generate NADPH and its activity was found to be negligible in many cell lines [3, 7, 16, 22]. Depending on the NAD^+ demand, the converted pyruvate can be further metabolized to lactate. Malate can be converted to pyruvate within the mitochondria by malic enzyme II (ME-II). This provides NADPH in mitochondria and might facilitate complete oxidation of glutamine. This isoform of malic enzyme is found to be very active in cancer cells compared to normal cell lines [28]. The increase in activity of this enzyme correlated with an increase in glutaminase activity suggesting that this enzyme participates actively in oxidation of glutamine and provides energy to tumor cells. This high activity of ME-II enzyme is more beneficial when the activity of malate dehydrogenase 2 (MDH-2) enzyme (catalyzes the conversion of malate to oxaloacetate in mitochondria) is rate limiting for complete oxidation of glutamine.

5.1.3.3 Other Amino Acids

In addition to glutamine, other amino acids also serve as energy sources for animal cells. These are consumed at lesser extent compared to glutamine. The uptake rates of other amino acids by a Chinese hamster ovary cell line were found to be at least eightfold less than the uptake of glutamine (Figure 5.1.5). The availability of sufficient levels of all the amino acids is essential for uninterrupted protein synthesis in production cell lines. The amino acids that cannot be synthesized by all animal cells, generally referred as essential amino acids, are histidine, isoleucine, leucine, lysine, methionine, phenylalanine, threonine, tryptophan, and valine. In addition to these, cysteine, glutamine, and tyrosine are also considered as essential for many animal cell types since these can only be synthesized by specialized cells. The schematic view of the synthesis of nonessential amino acids from essential ones and the contribution

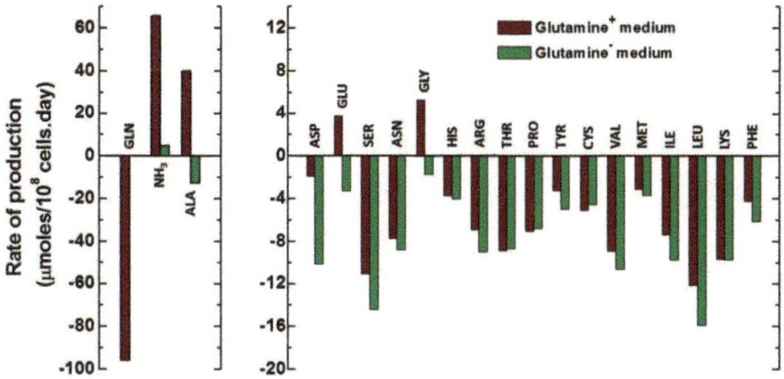

Figure 5.1.5: Shift of amino acid metabolism by a Chinese hamster ovary cell line (CHO-DXB11 derived cell line producing a monoclonal antibody) when adapted to glutamine free medium in a continuously cultured 3 L bioreactor. The values represent consumption/production rates at steady state growth either in glutamine-containing medium or when switched to glutamine-free medium. Negative values represent consumption rates. (ALA = alanine, ARG = arginine, ASN = asparagine, ASP = aspartate, CYS = cysteine, GLN = glutamine, GLU = glutamate, GLY = glycine, HIS = histidine, ILE = isoleucine, LEU = leucine, LYS = lysine; MET = methionine, PHE = phenylalanine, PRO = proline, SER = serine, THR = threonine, TYR = tyrosine, VAL = valine).

of amino acids to the TCA cycle is presented in Figure 5.1.6. These catabolic pathways replenish the TCA cycle intermediates for energy generation especially in the absence of glutamine. As shown in Figure 5.1.5, the consumption rates of other amino acids were increased significantly when CHO cells were adapted to glutamine free medium.

5.1.3.4 Lipids

Lipids serve as energy storage molecules, structural components of cellular membrane, and for secretion machinery for animal cells. The lipids in the context of cell culture represent fatty acids, sterols, phospholipids, triacylglycerols, and sphingolipids. The increased importance of serum-free medium led to the investigation of lipid supplements for certain cell lines. Certain fatty acids can substitute the role of lipids in serum and enhance the growth of cells in serum-free media [29, 30]. Supplementation of serum-free medium with cis-unsaturated fatty acids oleic/linoleic acid provided 58 % enhancement of cell yield for hybridomas [31]. However, the enhancement of either growth or productivity by lipids is dependent on type of hybridoma cell line [32]. The action of phospholipids and fatty acids on enhancement of growth was found to be membrane associated, while antibody enhancement is independent of

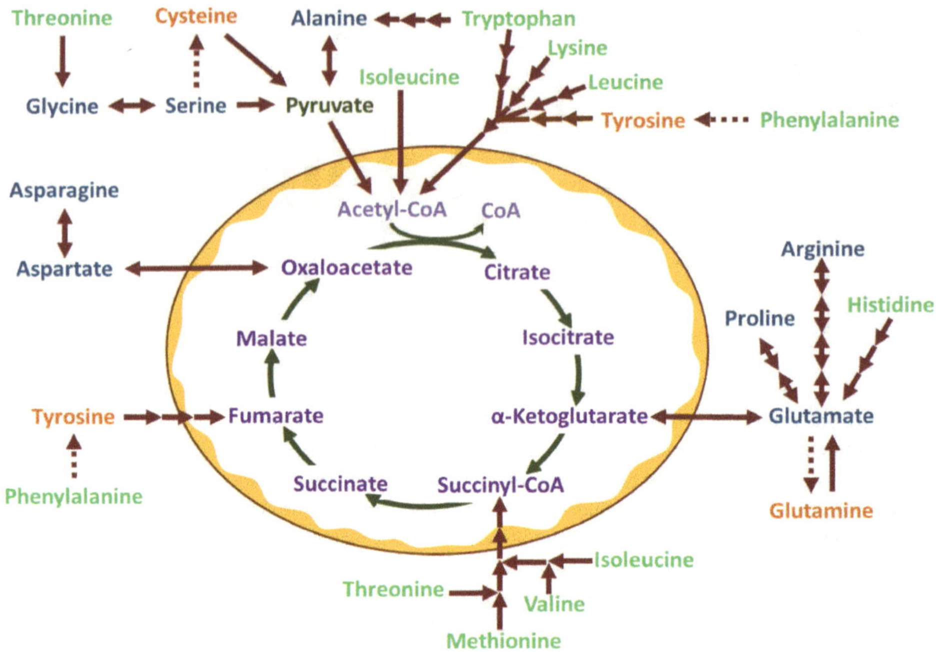

Figure 5.1.6: Schematic view of the pathways of nonessential amino acids synthesis from essential ones and the contribution of amino acids for the generation of TCA cycle intermediates. Nonessential amino acids (represented in blue) can be synthesized from essential amino acids (represented in green) by animal cells. Certain amino acids (represented in orange) are generally considered essential for most animal cells but can be synthesized from other amino acids (shown in dashed arrows) by some animal cells.

changes in membrane dynamics [32]. Lipid supplements also improved growth and productivity of recombinant proteins in CHO and NS0 cell lines [33, 34].

Fatty acids can be metabolized in the TCA cycle following conversion to acetyl-CoA. The oxidation of fatty acids to CO_2 was found to be minimal or negligible indicating that lipids do not serve as a major energy source for the cells [23, 35]. Normal mammalian cells express elongation and desaturation enzymes to synthesize certain fatty acids [36]. Lipid synthesis occurs through from citrate released out of mitochondria from TCA cycle (Figure 5.1.3). Recently, the contribution of glucose and glutamine towards lipid synthesis was investigated by using radiolabelled substrates [7, 25]. By detecting the biosynthesis of palmitate, the most abundant fatty acid in lipids, it was observed that glucose serves as major contributor for fatty acid synthesis up to 80 % [16, 25], while glutamine contributes up to 30 % [7, 25]. The fatty acid synthesis was found to remain high even in the stationary phase, where the growth rate is minimal or zero [7, 16]. This is consistent with the notion of increased NADPH production and citrate lyase (CL) activity during the stationary phase [18]. The fatty acid synthesis may be associated with the requirement of the synthesis of secretion vesicles especially in high production cell lines.

5.1.4 Metabolic Byproducts and Strategies to Reduce Their Production

5.1.4.1 Lactate

Lactate is one of the major byproduct that accumulates in cell culture due to high consumption of glucose. High lactate concentration of above 20 mmol L^{-1} inhibits cell growth and productivity [37-41]. Accumulation of lactate can cause significant reduction in the pH of the culture medium, which inhibits cell growth. However, the control of pH in bioreactors is often associated with the addition of large amounts of alkali to the culture [42]. The combination of lactate accumulation and addition of a high amount of alkali results in increased osmolality of culture media, which can have detrimental effect on cell growth and productivity [38, 43].

Lactate production is influenced by various factors such as the metabolism of sugars, NAD^+/NADH levels, activities of various enzymes, and even components in cell culture medium. Different strategies have been employed to improve the energy metabolism of cells by reducing lactate production. Replacing glucose with alternative sugars like galactose and fructose allowed the cells to survive in culture [4, 5, 21]. Galactose was found to be consumed at lower rates compared to other sugars because of its low affinity to hexokinase. In particular, lactate yield in the presence of galactose is much lower compared to other sugar sources. The consumption of lactate by CHO cells in the presence of galactose was also observed [44, 45]. This is associated with a slow metabolic flux of galactose to pyruvate by an alternative pathway compared to glucose. Also, galactose metabolism yields only 1 ATP molecule compared to 2 by glucose in glycolysis. Other genetic engineering strategies to alter carbohydrate catabolism include manipulation of sugar transporters. The knockdown of glucose transporter GLUT1 in hybridoma cells resulted in significant reduction in glucose uptake and lactate yield [46]. In another work, GLUT5 transporter was stably expressed in CHO cells and fructose was utilized as a sugar source [47]. This resulted in a significant reduction in lactate generation and improved final cell densities.

Downregulation of the lactate dehydrogenase (LDH) enzyme can result in significant reduction in lactate production and improved cell yield and productivity [48-50]. The enhancement of pyruvate diversion into the TCA cycle is a promising approach to improve energy metabolism. Pyruvate dehydrogenase (PDH) catalyzes its conversion to acetyl-CoA to enter into TCA cycle. PDH enzyme is inhibited when it is phosphorylated by pyruvate dehydrogenase kinases (PDHKs). Downregulation of PDHKs improves the energy metabolism of cells and reduces lactate production [50]. Alternatively, pyruvate can enter into TCA cycle by converting to oxaloacetate by pyruvate carboxylase (PYC) enzyme. The PYC enzyme is naturally present in animal cells but studies show that the activity of this enzyme is lower compared to the PDH enzyme [7, 19, 27]. Overexpression of PYC in BHK-21 cells [51, 52], CHO cells [53, 54], and HEK-293 cells [55, 56] has shown profound improvement in energy metabolism by reducing lactate production and enhancement in productivity. The consumption of lactate at later phases of cell culture, especially in prolonged fed-batch cultures, is observed in CHO and NS0 cell lines [57, 58]. One of the factors linked to this behavior is regeneration of NADH to NAD^+ by the malate-aspartate shuttle. A correlation between reduced expression of Aralar1/Trimm8a1 genes and lactate accumulation was observed [9]. Enhancement of malate-aspartate shuttle by overexpressing Aralar1 and Trimm8a1 genes was shown to diminish the lactate accumulation in CHO cells [59]. The lactate consumption maintains a NAD^+/NADH equilibrium due to generation of NAD^+ from oxaloacetate conversion to malate in the malate-aspartate shuttle. The mitochondrial oxidative capacity of the cells also plays a key role on lactate production [60, 61]. A correlation between reduced oxidative capacity of mitochondria and high lactate production was observed in CHO cells [61]. Clones of CHO cells sorted for a low mitochondrial membrane potential showed improved energy metabolism and lower lactate production rates [60]. The presence of copper in the medium also plays a significant role in the transcriptional regulation of lactate production in CHO cells [62, 63]. An optimum concentration of 5 µmol L^{-1} Cu^{2+} in the medium resulted in significant reduction of lactate production by down-regulating lactate dehydrogenase enzyme and improved cell density and product titer [63].

5.1.4.2 Ammonia

Ammonia has a stronger effect than lactate as it inhibits cell growth and viability at concentrations as low as 3 mmol L^{-1} [38-40]. The effect of ammonia was enhanced in the presence of lactate >12 mmol L^{-1} indicating the synergetic effect of ammonia and lactate in the culture [39]. The ammonia generation occurs through amino acid metabolism, especially from the first 2 steps of glutaminolysis. Alternatively, thermal degradation of glutamine to pyrrolidone carboxylic acid also contributes to generation of ammonia in cell culture, since half-life of glutamine at 37 °Celsius and pH 7.2 is ~7 days. The overall mechanism underlying the inhibition of growth by ammonia is not completely known. However, the effect of ammonia and pH is closely related [64, 65]. The pH dependency of ammonium toxicity may reduce or reverse the conversion of glutamate to α-ketoglutarate thus reducing TCA cycle intermediates and ATP generation [66]. Another group argued that increasing ammonia levels reduces intracellular pH and inactivates key enzyme PFK-1 in glycolysis [64]. It has also been suggested that ammonium ions compete with the transport of potassium ions into the cytoplasm via transporter proteins like Na^+/K^+-ATPase and Na^+/K^+ $2Cl^-$ cotransporter [67]. This results in increased demand of maintenance energy to maintain the ion gradient over cytoplasmic membrane and reduces the growth of cells. It has been demonstrated that elevated ammonia levels increases intracellular concentration of

UDP-GlcNAc and UDP-GalNAc [68]. Thus, the growth inhibition could be associated with either decreased intracellular UTP levels or the effect of increased UDP-GlcNAc or UDP-GalNAc levels. This effect can be observed only when glucose is used as the carbon source since Frc6P is needed as an NH_3 acceptor to form UDP-GlcNAc and UDP-GalNAc.

Glutamine is consumed at a high rate in cell culture and its conversion to glutamate results in high ammonia generation. Animal cells can synthesize glutamine from glutamate through glutamine synthetase (GS) enzyme, though its expression varies among different cell types. Cells may be subject to posttranscriptional control by GS upregulation in response to the absence of glutamine. This strategy was used to adapt different cell lines to glutamine-free medium [4, 69-71]. These studies show that culturing of cells in glutamine-free medium resulted in up to 70 % reduction of ammonia accumulation and significant enhancement in viability of cells. As observed from the author's studies, when CHO cells were adapted to glutamine-free medium in continuous culture mode, ammonia generation was reduced drastically and glutamate, glycine and alanine fluxes were changed from secretion to consumption (Figure 5.1.5). The uptake rates of some of the other amino acids were increased to substitute for missing energy source from glutamine. The adaptation of different cells to glutamate might depend upon their glutamate uptake capability rather glutamine synthesis [72]. Alternatively, cells can be adapted to grow in thermally stable medium containing glutamine-based dipeptides. The glutamine consumption from dipeptides depends on the activity of extra- and intracellular peptidases. This provides slower consumption and metabolism of glutamine. Culturing cells in alanine-glutamine or glycine-glutamine dipeptides resulted in >60 % reduction in ammonia production without compromising growth and viability [70, 73]. The activity of glutamine synthetase in hybridomas and myeloma cell lines was found to be negligible and the glutamine-free medium does not support growth of these cell lines [74]. Transfection and expression of GS in these cell lines allowed cells to grow in glutamate-based medium with reduced ammonium formation [46, 74, 75]. Substitution of pyruvate for glutamine [76] or enhancement of pyruvate flux into TCA cycle by overexpression of PYC enzyme [56] also reduced ammonia production demonstrating the link between glycolysis and TCA cycle.

5.1.5 Factors affecting Cell Metabolism

5.1.5.1 Oxygen

The low solubility of oxygen in water (~7 mg L^{-1}) requires a constant feed of the gas into a culture to ensure that it does not become limiting for cell growth. The dissolved oxygen concentration is normally measured as the percentage with reference to air saturation and is routinely controlled at a set-point of 40 to 60 % in a bioreactor process. Cell growth and metabolism is normally unaffected within this range. However, there can be a direct effect on metabolism at high (hyperoxia) or low (hypoxia) oxygen levels. At low dissolved oxygen (<1 % DO) the activity of the electron transport chain is reduced resulting in accumulated NADH and enhanced anaerobic metabolism [77]. Hypoxia-inducing factor 1 (HIF-1), one of the transcriptional regulators of glycolysis and glutaminolysis, becomes active under hypoxic conditions. HIF-1 enhances the expression of PDHK, which inactivates PDH enzyme, and promotes a metabolic switch to anaerobic glycolysis in tumor cells [15, 78]. HIF-1 was also found to upregulate the apoptotic factors BCL2/adenovirus E1B 19 kd-interacting protein (BINIP3) and NIX and induce cell death in tumor cells [79]. However, limited studies have been performed to investigate the role of HIF-1 in CHO cells [80, 81]. A high oxygen level may produce reactive oxygen species that can induce cell death [82]. However, a high oxygen level (~100 % DO) may also be associated with the induction of increased activity of antioxidant enzymes in the cell such as glutathione peroxidase, glutathione S-transferase, and superoxide dismutase, presumably as a defense against reactive oxygen species [83]. Furthermore, it was found in a murine hybridoma at a high oxygen level that there was an increased utilization of glucose, but surprisingly only anaerobically to lactate, suggesting that the TCA cycle may be sensitive to oxygen toxicity [84].

The oxygen uptake rate (OUR) is a useful measure of the extent of oxidative metabolism and can be determined on-line through use of an oxygen probe and used for the control of nutrient feeding [85]. Analysis of OUR with the carbon dioxide evolution rate (CER) can lead to a measurement of the respiratory quotient (RQ) which is characteristic of the metabolic profile of the cell.

$$RQ = CER/OUR$$

Data for this type of determination can be acquired by on-line gas analysis [27]. Complete oxidation of glutamine or glucose would result in an RQ of 0.83 or 1.0 respectively [86]. Assuming that energy requirements are provided by glutamine at 55 % and glucose by 45 % [22], this leads to a theoretical RQ of 0.9 for complete oxidation. Higher RQ values would be characteristic of incomplete oxidation [87].

5.1.5.2 Carbon dioxide

Carbon dioxide (CO_2) is the end product of the complete oxidation of energy substrates such as glucose and glutamine. The physiological range of pCO_2 for cell culture is 31 to 54 mm Hg. In cell culture CO_2 may accumulate unless it is allowed to dissipate through the liquid surface or through gas sparging. This process of CO_2 stripping is of particular concern in large-scale bioreactors in which there may be a low surface to volume ratio [88]. Significant accumulation of CO_2 up to a pCO_2 level of ~220 mm Hg can occur, especially in large-scale, high-density cell culture processes [89-91]. High levels of pCO_2 measured >105 mm Hg in media have been shown to inhibit the cell growth rate up to 60 % and to cause low productivity [89, 90]. This was confirmed in another study in which a dose-dependent inhibition of growth was shown at CO_2 levels within the range of 36 to 250 mm Hg [92]. Under glucose limiting conditions, the increased pCO_2 level to ~140 mm Hg enhanced growth rate of CHO cells by 1.7-fold, but this beneficial effect was lost and the growth rate was reduced significantly as the pCO_2 was increased further to 220 mm Hg [90]. Recent evidence suggests that elevated pCO_2 impairs cell growth of human cells by causing mitochondrial dysfunction [93]. This may be due to upregulation of miRNA-183 that decreases isocitrate dehydrogenase 2 (IDH-2) enzyme levels responsible for conversion of isocitrate to α-ketoglutarate in TCA cycle.

The effect of pCO_2 is associated with the pH and the osmolality of culture. The presence of bicarbonate buffer balances the pH in the culture medium by equilibrating with CO_2 either from respiration or externally controlled.

$$CO_2 + H_2O \leftrightarrow HCO_3^- + H^+$$

Because of high CO_2 production from respiration and lactate accumulation at high cell densities, bicarbonate buffering may fail and result in pH reduction. But in controlled bioreactors, the pH is maintained by addition of base which results in increased osmolality. The decrease in pH (uncontrolled culture) or increase in osmolality (controlled culture) can have a detrimental effect on the cell growth and viability. Studies show that the growth rate of hybridoma cells was reduced significantly up to 50 % when osmolality was increased from the physiological range of ~320 to 435 mOsm kg^{-1} [94, 95]. A similar effect of reduction in growth rate and increase in death rate was observed in CHO cells when cultured at hyper osmolality conditions [96, 97]. The acceleration of the death rate by hyper osmolality was due to an enhanced apoptotic response [97]. The combined effect of pCO_2 and osmolality on growth rate reduction of CHO cells was also found to be significant compared to the individual effects [98].

5.1.5.3 Temperature

Culturing cells below physiological temperature of 37 °Celsius is becoming an attractive option to extend culture longevity. The temperature reduction is associated with advantages like reduced growth rate with increased protein production, extension of stationary phase, delay of apoptosis, reduced production of inhibitory byproducts and resistance to shear. CHO cells cultured at 34 °Celsius maintained high viability for 7 days by cell-cycle arrest at G0/G1 phase and enhanced protein production by 35 % [99]. This advantage has been widely used to extend the viable culture time and enhance the protein production [100-105]. A study concluded that increased phosphorylation of protein kinases in CHO cells cultured at 30 °Celsius altered cell cycle regulation [106]. This resulted in decreased growth rate and increased productivity of alkaline phosphatase. CHO cells producing erythropoietin also showed a decrease in growth rate and an increase in cell specific productivity when cultured at reduced temperatures [107]. Enhancement in the activity of glutathione peroxidase antioxidant enzyme and Bcl-2 expression were observed in CHO cells cultured at 30 °Celsius, indicating a defensive response against reactive oxygen species and a delay in apoptosis at low temperature [108].

5.1.6 Factors that Affect Protein Productivity and Quality in Culture

5.1.6.1 Productivity

The cell-specific uptake of a nutrient in culture is dependent upon the media concentration. So, that in a batch culture of cells provided with 25 mmol L^{-1} of glucose, the rate of substrate utilization will be high and lead to excessive accumulation of lactate though glycolysis. This phenomenon is of greater concern for the other major substrate of energy metabolism, glutamine. Typical concentrations of up to 6 mmol L^{-1} in batch cultures will lead to accumulated ammonia that reduces and eventually inhibits cell growth. Fed-batch has become the favored mode of cell culture strategy in industry because of the realization that low concentrations of either of these substrates in the media will lead to lower rates of uptake and consequently a more efficient metabolism [109, 110]. A slow feed of either substrate will ensure a low metabolic yield coefficient ($Y_{P/S}$) defined as

$$Y_{P/S} = \Delta P / \Delta S$$

where ΔP is the change of product concentration and ΔS the change of substrate concentration.

Furthermore, the lower rate of accumulated byproducts ensures that high cell viabilities will be maintained for extended periods of time. Of particular value is the fact that this enhances the overall integrated viable cell density (IVCD) which is directly proportional to the total production of a secreted protein. In fact, the productivity of recombinant protein in a cell culture system can be measured by the product of the cell-specific productivity and the IVCD.

$$P = Q_p \cdot \int_0^t N_\upsilon \cdot \mathrm{d}t$$

where P is the product formed, Q_p is the cell specific productivity, and N_υ is the concentration of viable cells over the time period, t.

The lower rate of accumulated metabolic byproducts (lactate and ammonia) will ensure a high overall IVCD with recombinant protein production continuing for an extended period of time. Figure 5.1.1 shows how a typical profile of a fed-batch process can extend the high viability of cells up to 14 days during which time the cells could secrete a recombinant protein. The current fed-batch strategies in mammalian cell bioprocesses allow the production of recombinant proteins up to 5 g L^{-1}. Cell growth kinetics show a Monod constant (K_m) of 0.2 mmol L^{-1} for glutamine and 0.5 mmol L^{-1} for glucose. The design of fed-batch media with 2 × of these concentrations ensures low rates of lactate and ammonia production with growth rates around 70 to 90 % of the maximum [111].

5.1.6.2 Glycosylation

The variability of the components in the media over the time course of a bioprocess may affect the glycosylation profile of a synthesized recombinant protein. The initial step of N-glycosylation occurs in the endoplasmic reticulum (ER) through the addition of a 14-oligomer oligosaccharide from a dolichol lipid structure. This structure is then trimmed and modified with a series of enzymes found in the ER and Golgi. The integrity of this process is dependent upon the intracellular availability of carbohydrate and nucleotide precursors [112]. It has been shown that as the substrates are utilized in a batch culture so the extent of glycan site occupancy (macroheterogeneity) may vary [113]. This may be due to the availability of substrates or to the energy status of the cell [114].

A critical question that arises in fed-batch cultures is whether the low concentration of substrates affects the glycosylation of the secreted protein. A potential problem of fed-batch strategies based upon low substrate levels is that fluctuations of concentration may occur below a predefined set point between feeding times. This may cause limitations to the intracellular glycosylation process that is dependent upon a supply of sugar and nucleotide-sugar precursors. Limiting substrate availability of glutamine (<0.1 mmol L^{-1}) and glucose (<0.7 mmol L^{-1}) has been shown to result in decreased sialylation and enhanced formation of high mannose structures in the production of γ-interferon [115]. Exposure of cells to media with depleted glucose may reduce site occupancy (macroheterogeneity) by decreasing the synthesis of the consensus 14-oligomer lipid-linked oligosaccharide in the ER [116]. Microheterogeneity including variable galactosylation of synthesized monoclonal antibodies has also been shown at reduced glucose concentrations in the medium [116]. Culture supplementation with specific precursors such as N-acetyl mannosamine may enhance sialylation by boosting the intracellular nucleotide sugar pools [117]. A similar strategy for enhanced galactosylation involves the addition of a mixed cocktail of supplements (uridine, manganese chloride, and galactose) [118], which can be strategically manipulated to obtain a predetermined level of galactosylation [119].

Metabolic byproducts especially ammonia have a profound effect on glycosylation of proteins. Elevated ammonium levels up to 15 mmol L^{-1} increased branching of N-glycans on interleukin-2 [120] or reduced galactosylation and sialylation of tumor necrosis factor (TNF)α-IgG fusion protein by 40 % [121]. Reduction in sialylation of erythropoietin was also observed in the presence of 10 mmol L^{-1} ammonium chloride [122]. The decrease in galactosylation might be due to decreased levels of uridine diphosphate-activated galactose (UDP-Gal) because of limited uridine triphosphate (UTP), since UTP has been utilized for increased synthesis of UDP-GlcNAc/UDP-GalNAc in the presence of NH_3 [120]. Also, it has been reported that increased intracellular UDP-activated N-acetylglucosamine (UDP-GlcNAc) levels impair the transport of cytidine monophosphate-activated N-acetylneuraminic acid (CMP-Neu5Ac) to the Golgi [123] and this can impair sialylation of proteins. However, an ammonia-mediated pH effect on glycosylation has not been excluded [124]. The decrease in galactosylation and sialylation of proteins was also observed at reduced culture pH [125, 126]. The elevated pH of Golgi (because of elevated ammonia or reduced culture pH) lowers the activities of glycosyltransferases [121] or causes mislocalization of glycosyltransferases along Golgi [127].

The DO of a culture may affect the intracellular redox potential that may in turn affect enzymatic reactions such as the glycosylation process. A study of the effect of DO on the galactosylation of the glycan found on a monoclonal antibody indicated a decreased index of galactosylation with reduced DO particularly over the lower range down to 10 % DO. The importance of this finding is that it is essential to control the DO value during cell culture if a consistent glycosylation profile is required for a cell secreted glycoprotein [128]. Accumulating CO_2 in the culture could also affect the glycosylation of a secreted protein through a changed pattern of sialylation [129]. An elevated pCO_2 level of 195 mm Hg in combination with high osmotic stress (435 mOsm kg^{-1}) reduced the galactosylation by 25 % in an IgG2-type mAb [130]. Reduced branching and sialic acid content of EPO was observed when the CHO cells were cultured at 32 °Celsius compared to 37 °Celsius [131]. This might correlate with findings that at lower temperatures the intracellular pool of UDP-GlcNAc and UDP-GalNAc decreases [132].

References

[1] Wood IS, Trayhurn P. Glucose transporters (GLUT and SGLT): expanded families of sugar transport proteins. B J Nutr 2003, 89, 3-9.

[2] Barthel A, Okino ST, Liao JF, et al. Regulation of GLUT1 gene transcription by the serine threonine kinase Akt1. J Biol Chem 1999, 274, 20281-6.

[3] Neermann J, Wagner R. Comparative analysis of glucose and glutamine metabolism in transformed mammalian cell lines, insect and primary liver cells. J Cell Physiol 1996, 166, 152-69.

[4] Altamirano C, Paredes C, Cairo JJ, et al. Improvement of CHO cell culture medium formulation: Simultaneous substitution of glucose and glutamine. Biotechnol Prog 2000, 16, 69-75.

[5] Petch D, Butler M. The effect of alternative carbohydrates on the growth and antibody production of a murine hybridoma. Appl Biochem Biotechnol 1996, 59, 93-104.

[6] Henry O, Jolicoeur M, Kamen A. Unraveling the metabolism of HEK-293 cells using lactate isotopomer analysis. Bioprocess Biosyst Eng 2011, 34, 263-73.

[7] Ahn WS, Antoniewicz MR. Parallel labeling experiments with [1,2-(13)C]glucose and [U-(13)C] glutamine provide new insights into CHO cell metabolism. Metab Eng 2013, 15, 34-47.
[8] Petch D, Butler M. Profile of energy-metabolism in a murine hybridoma: glucose and glutamine utilization. J Cell Physiol 1994, 161, 71-6.
[9] Rubi B, del Arco A, Bartley C, et al. The malate-aspartate NADH shuttle member Aralar1 determines glucose metabolic fate, mitochondrial activity, and insulin secretion in beta cells. J Biol Chem 2004, 279, 55659-66.
[10] Leite TC, Da Silva D, Coelho RG, et al. Lactate favours the dissociation of skeletal muscle 6-phosphofructo-1-kinase tetramers down-regulating the enzyme and muscle glycolysis. Biochem J 2007, 408, 123-30.
[11] Christofk HR, Vander Heiden MG, Harris MH, et al. The M2 splice isoform of pyruvate kinase is important for cancer metabolism and tumour growth. Nature 2008, 452, 230-U74.
[12] Ashizawa K, Willingham MC, Liang CM, et al. In vivo regulation of monomer-tetramer conversion of pyruvate kinase subtype M2 by glucose is mediated via fructose-1,6-bisphosphate. J Biol Chem 1991, 266, 16842-6.
[13] Chaneton B, Hillmann P, Zheng L, et al. Serine is a natural ligand and allosteric activator of pyruvate kinase M2. Nature 2012, 491, 458-62.
[14] Christofk HR, Vander Heiden MG, Wu N, et al. Pyruvate kinase M2 is a phosphotyrosine-binding protein. Nature 2008, 452, 181-U27.
[15] Hitosugi T, Kang S, Heiden MGV, et al. Tyrosine Phosphorylation Inhibits PKM2 to Promote the Warburg Effect and Tumor Growth. Sci Signal 2009, 2, RA73.
[16] Dean J, Reddy P. Metabolic analysis of antibody producing CHO cells in fed-batch production. Biotechnol Bioeng 2013, 110, 1735-47.
[17] Ahn WS, Antoniewicz MR. Metabolic flux analysis of CHO cells at growth and non-growth phases using isotopic tracers and mass spectrometry. Metab Eng 2011, 13, 598-609.
[18] Templeton N, Dean J, Reddy P, et al. Peak antibody production is associated with increased oxidative metabolism in an industrially relevant fed-batch CHO cell culture. Biotechnol Bioeng 2013, 110, 2013-24.
[19] Goudar C, Biener R, Boisart C, et al. Metabolic flux analysis of CHO cells in perfusion culture by metabolite balancing and 2D C-13, H-1 COSY NMR spectroscopy. Metab Eng 2010, 12, 138-49.
[20] Sengupta N, Rose ST, Morgan JA. Metabolic flux analysis of CHO cell metabolism in the late non-growth phase. Biotechnol Bioeng 2011, 108, 82-92.
[21] Reitzer LJ, Wice BM, Kennell D. Evidence that glutamine, not sugar, is the major energy source for cultured HeLa cells. J Biol Chem1979, 254, 2669-76.
[22] Fitzpatrick L, Jenkins HA, Butler M. Glucose and glutamine-metabolism of a murine B-lymphocyte hybridoma grown in batch culture. Appl Biochem Biotechnol 1993, 43, 93-116.
[23] Zielke HR, Zielke CL, Ozand PT. Glutamine: a major energy source for cultured mammalian cells. Fed Proc 1984, 43, 121-5.
[24] Zielke HR, Sumbilla CM, Sevdalian DA, et al. Lactate: a major product of glutamine metabolism by human diploid fibroblasts. J Cell Physiol 1980, 104, 433-41.
[25] DeBerardinis RJ, Mancuso A, Daikhin E, et al. Beyond aerobic glycolysis: Transformed cells can engage in glutamine metabolism that exceeds the requirement for protein and nucleotide synthesis. Proc Natl Acad Sci USA 2007, 104, 19345-50.
[26] Chong WPK, Reddy SG, Yusufi FNK, et al. Metabolomics-driven approach for the improvement of Chinese hamster ovary cell growth: Overexpression of malate dehydrogenase II. J Biotechnol 2010, 147, 116-21.
[27] Bonarius HP, Ozemre A, Timmerarends B, et al. Metabolic-flux analysis of continuously cultured hybridoma cells using (13)CO(2) mass spectrometry in combination with (13)C-lactate nuclear

magnetic resonance spectroscopy and metabolite balancing. Biotechnol Bioeng 2001, 74, 528-38.
[28] Sauer LA, Dauchy RT, Nagel WO, et al. Mitochondrial malic enzymes. Mitochondrial NAD(P)+-dependent malic enzyme activity and malate-dependent pyruvate formation are progression-linked in Morris hepatomas. J Biol Chem 1980, 255, 3844-8.
[29] Barnes D, Sato G. Methods for growth of cultured-cells in serum-free medium. Anal Biochem 1980, 102, 255-70.
[30] Spiekerpolet H, Polet H. Requirement of a combination of a saturated and an unsaturated free fatty-acid and a fatty-acid carrier protein for invitro growth of lymphocytes. J Immunol 1981, 126, 949-54.
[31] Butler M, Huzel N. The Effect of Fatty-acids on Hybridoma Cell-growth and Antibody Productivity in Serum-free Cultures. J Biotechnol 1995, 39, 165-73.
[32] Savonniere S, Zeghari N, Miccoli L,et al. Effects of lipid supplementation of culture media on cell growth, antibody production, membrane structure and dynamics in two hybridomas. J Biotechnol 1996, 48, 161-73.
[33] Jenkins N, Castro P, Menon S, et al. Effect of lipid supplements on the production and glycosylation of recombinant interferon-γ expressed in CHO cells. Cytotechnol 1994, 15, 209-15.
[34] Gorfien S, Paul B, Walowitz J,et al. Growth of NS0 cells in protein-free, chemically defined medium. Biotechnol Prog 2000, 16, 682-7.
[35] Butler M, Huzel N, Barnabe N. Unsaturated fatty acids enhance cell yields and perturb the energy metabolism of an antibody-secreting hybridoma. Biochem J 1997, 322, 615-23.
[36] Rosenthal MD. Fatty-acid metabolism of isolated mammalian-cells. Prog in Lipid Res 1987, 26, 87-124.
[37] Lao MS, Toth D. Effects of ammonium and lactate on growth and metabolism of a recombinant Chinese hamster ovary cell culture. Biotechnol Prog 1997, 13, 688-91.
[38] Cruz HJ, Freitas CM, Alves PM, et al. Effects of ammonia and lactate on growth, metabolism, and productivity of BHK cells. Enzyme Microb Technol 2000, 27, 43-52.
[39] Hassell T, Gleave S, Butler M. Growth-inhibition in animal-cell culture: The effect of lactate and ammonia. Appl Biochem Biotechnol 1991, 30, 29-41.
[40] Ozturk SS, Riley MR, Palsson BO. Effects of ammonia and lactate on hybridoma growth, metabolism, and antibody-production. Biotechnol Bioeng 1992, 39, 418-31.
[41] Omasa T, Higashiyama K, Shioya S, et al. Effects of lactate concentration on hybridoma culture in lactate-controlled fed-batch operation. Biotechnol Bioeng 1992, 39, 556-64.
[42] Langheinrich C, Nienow AW. Control of pH in large-scale, free suspension animal cell bioreactors: Alkali addition and pH excursions. Biotechnol Bioeng 1999, 66, 171-9.
[43] Li F, Vijayasankaran N, Shen A, et al. Cell culture processes for monoclonal antibody production. Mabs 2010, 2, 466-79.
[44] Vergara M, Becerra S, Berrios J, et al. Metabolic behavior of CHO cells cultured in perfusion with handling simultaneous of low temperature and source of carbon. N Biotechnol 2009, 25, S222-S.
[45] Altamirano C, Illanes A, Becerra S, et al. Considerations on the lactate consumption by CHO cells in the presence of galactose. J of Biotechnol 2006, 125, 547-56.
[46] Paredes C, Prats E, Cairo JJ, et al. Modification of glucose and glutamine metabolism in hybridoma cells through metabolic engineering. Cytotechnol 1999, 30, 85-93.
[47] Wlaschin KF, Hu W-S. Engineering cell metabolism for high-density cell culture via manipulation of sugar transport. J of Biotechnol 2007, 131, 168-76.
[48] Chen KQ, Liu Q, Xie LZ, et al. Engineering of a mammalian cell line for reduction of lactate formation and high monoclonal antibody production. Biotechnol Bioeng 2001, 72, 55-61.

[49] Kim SH, Lee GM. Down-regulation of lactate dehydrogenase-A by siRNAs for reduced lactic acid formation of Chinese hamster ovary cells producing thrombopoietin. Appl Microbiol Biotechnol 2007, 74, 152-9.
[50] Zhou M, Crawford Y, Ng D, et al. Decreasing lactate level and increasing antibody production in Chinese Hamster Ovary cells (CHO) by reducing the expression of lactate dehydrogenase and pyruvate dehydrogenase kinases. J Biotechnol 2011, 153, 27-34.
[51] Irani N, Wirth M, van den Heuvel J, et al. Improvement of the primary metabolism of cell cultures by introducing a new cytoplasmic pyruvate carboxylase reaction. Biotechnol Bioeng 1999, 66, 238-46.
[52] Irani N, Beccaria AJ, Wagner R. Expression of recombinant cytoplasmic yeast pyruvate carboxylase for the improvement of the production of human erythropoietin by recombinant BHK-21 cells. J Biotechnol 2002, 93, 269-82.
[53] Kim SH, Lee GM. Functional expression of human pyruvate carboxylase for reduced lactic acid formation of Chinese hamster ovary cells (DG44). Appl Microbiol Biotechnol 2007, 76, 659-65.
[54] Fogolin MB, Wagner R, Etcheverrigaray M, et al. Impact of temperature reduction and expression of yeast pyruvate carboxylase on hGM-CSF-producing CHO cells. J Biotechnol 2004, 109, 179-91.
[55] Elias CB, Carpentier E, Durocher Y, et al. Improving glucose and gluamiue metabolism of human HEK 293 and Trchoplusia ni insect cells engineered to express a cytosolic pyruvate carboxylase enzyme. Biotechnol Prog 2003, 19, 90-7.
[56] Henry O, Durocher Y. Enhanced glycoprotein production in HEK-293 cells expressing pyruvate carboxylase. Metab Eng 2011, 13, 499-507.
[57] Ma N, Ellet J, Okediadi C, et al. A Single Nutrient Feed Supports Both Chemically Defined NS0 and CHO Fed-Batch Processes: Improved Productivity and Lactate Metabolism. Biotechnol Prog 2009, 25, 1353-63.
[58] Pascoe DE, Arnott D, Papoutsakis ET, et al. Proteome analysis of anti body-producing CHO cell lines with different metabolic profiles. Biotechnol Bioeng 2007, 98, 391-410.
[59] Zagari F, Stettler M, Baldi L, et al. High expression of the aspartate–glutamate carrier Aralar1 favors lactate consumption in CHO cell culture. Pharma Bioprocess 2013, 1, 19-27.
[60] Hinterkörner G, Brugger G, Müller D, et al. Improvement of the energy metabolism of recombinant CHO cells by cell sorting for reduced mitochondrial membrane potential. J Biotechnol 2007, 129, 651-7.
[61] Zagari F, Jordan M, Stettler M, et al. Lactate metabolism shift in CHO cell culture: the role of mitochondrial oxidative activity. N Biotechnol 2013, 30, 238-45.
[62] Luo J, Vijayasankaran N, Autsen J, et al. Comparative metabolite analysis to understand lactate metabolism shift in Chinese hamster ovary cell culture process. Biotechnol and Bioeng 2012, 109, 146-56.
[63] Qian Y, Khattak SF, Xing Z, et al. Cell Culture and Gene Transcription Effects of Copper Sulfate on Chinese Hamster Ovary Cells. Biotechnol Prog 2011, 27, 1190-4.
[64] McQueen A, Bailey JE. Effect of ammonium ion and extracellular pH on hybridoma cell metabolism and antibody production. Biotechnol Bioeng 1990, 35, 1067-77.
[65] Schneider M, Marison IW, von Stockar U. The importance of ammonia in mammalian cell culture. J Biotechnol 1996, 46, 161-85.
[66] Doyle C, Butler M. The effect of pH on the toxicity of ammonia to a murine hybridoma. J Biotechnol 1990, 15, 91-100.
[67] Martinelle K, Haggstrom L. Mechanisms of ammonia and ammonium ion toxicity in animal cells: transport across cell membranes. J Biotechnol 1993, 30, 339-50.
[68] Ryll T, Valley U, Wagner R. Biochemistry of growth inhibition by ammonium ions in mammalian cells. Biotechnol Bioeng 1994, 44, 184-93.

[69] Hassell T, Butler M. Adaptation to nonammoniagenic medium and selective substrate feeding lead to enhanced yields in animal cell cultures. J of Cell Science 1990, 96, 501-8.
[70] Christie A, Butler M. The adaptation of BHK cells to a non-ammoniagenic glutamate-based culture medium. Biotechnol Bioeng 1999, 64, 298-309.
[71] Taschwer M, Hackl M, Hernandez Bort JA, et al. Growth, productivity and protein glycosylation in a CHO EpoFc producer cell line adapted to glutamine-free growth. J Biotechnol 2012, 157, 295-303.
[72] McDermott RH, Butler M. Uptake of glutamate, not glutamine-synthetase, regulates adaptation of mammalian-cells to glutamine-free medium. J Cell Sci 1993, 104, 51-8.
[73] Christie A, Butler M. Glutamine-based dipeptides are utilized in mammalian-cell culture by extracellular hydrolysis catalyzed by a specific peptidase. J Biotechnol 1994, 37, 277-90.
[74] Bebbington CR, Renner G, Thomson S, et al. High-level expression of a recombinant antibody from myeloma cells using a glutamine synthetase gene as an amplifiable selectable marker. Biotechnol 1992, 10, 169-75.
[75] Dempsey J, Ruddock S, Osborne M, et al. Improved Fermentation Processes for NSO Cell Lines Expressing Human Antibodies and Glutamine Synthetase. Biotechnol Prog 2003, 19, 175-8.
[76] Genzel Y, Ritter JB, Konig S, et al. Substitution of glutamine by pyruvate to reduce ammonia formation and growth inhibition of mammalian cells. Biotechnol Prog 2005, 21, 58-69.
[77] Zupke C, Sinskey AJ, Stephanopoulos G. Intracellular flux analysis applied to the effect of dissolved oxygen on hybridomas. Appl Microbiol Biotechnol 1995, 44, 27-36.
[78] Kim JW, Tchernyshyov I, Semenza GL, et al. HIF-1-mediated expression of pyruvate dehydrogenase kinase: a metabolic switch required for cellular adaptation to hypoxia. Cell Metab 2006, 3, 177-85.
[79] Sowter HM, Ratcliffe PJ, Watson P, et al. HIF-1-dependent regulation of hypoxic induction of the cell death factors BNIP3 and NIX in human tumors. Cancer Res 2001, 61, 6669-73.
[80] Wood SM, Wiesener MS, Yeates KM, et al. Selection and analysis of a mutant cell line defective in the hypoxia-inducible factor-1 alpha-subunit (HIF-1alpha). Characterization of hif-1alpha-dependent and -independent hypoxia-inducible gene expression. J Biol Chem 1998, 273, 8360-8.
[81] Williams KJ, Telfer BA, Airley RE, et al. A protective role for HIF-1 in response to redox manipulation and glucose deprivation: implications for tumorigenesis. Oncogene 2002, 21, 282-90.
[82] Cacciuttolo MA, Trinh L, Lumpkin JA, et al. Hyperoxia induces DNA damage in mammalian cells. Free Radic Biol Med 1993, 14, 267-76.
[83] Jan DC, Petch DA, Huzel N, et al. The effect of dissolved oxygen on the metabolic profile of a murine hybridoma grown in serum-free medium in continuous culture. Biotechnol Bioeng 1997, 54, 153-64.
[84] Jan DCH, Petch DA, Huzel N, et al. The metabolic profile of a murine hybridoma grown in serum free medium at various oxygen concentrations. In: Carrondo MJT, Griffiths B, Moreira JLP (eds). Animal Cell Technology. Kluwer, Dordrecht, 1997, 791-6.
[85] Zhou W, Rehm J, Hu WS. High viable cell concentration fed-batch cultures of hybridoma cells through on-line nutrient feeding. Biotechnol Bioeng 1995, 46, 579-87.
[86] Bonarius HP, Hatzimanikatis V, Meesters KP, et al. Metabolic flux analysis of hybridoma cells in different culture media using mass balances. Biotechnol Bioeng 1996, 50, 299-318.
[87] Xiu ZL, Deckwer WD, Zeng AP. Estimation of rates of oxygen uptake and carbon dioxide evolution of animal cell culture using material and energy balances. Cytotechnol 1999, 29, 159-66.
[88] Matsunaga N, Kano K, Maki Y, et al. Estimation of dissolved carbon dioxide stripping in a large bioreactor using model medium. J Biosci Bioeng 2009, 107, 419-24.

[89] Gray DR, Chen S, Howarth W, et al. CO(2) in large-scale and high-density CHO cell perfusion culture. Cytotechnol 1996, 22, 65-78.
[90] Takuma S, Hirashima C, Piret JM. Dependence on glucose limitation of the pCO(2) influences on CHO cell growth, metabolism and IgG production. Biotechnol Bioeng 2007, 97, 1479-88.
[91] Mostafa SS, Gu XJ. Strategies for improved dCO(2) removal in large-scale fed-batch cultures. Biotechnol Prog 2003, 19, 45-51.
[92] Kimura R, Miller WM. Effects of elevated pCO(2) and/or osmolality on the growth and recombinant tPA production of CHO cells. Biotechnol Bioeng 1996, 52, 152-60.
[93] Vohwinkel CU, Lecuona E, Sun H, et al. Elevated CO(2) levels cause mitochondrial dysfunction and impair cell proliferation. J Biol Chem 2011, 286, 37067-76.
[94] Ozturk SS, Palsson BO. Effect of medium osmolarity on hybridoma growth, metabolism, and antibody-production. Biotechnol Bioeng 1991, 37, 989-93.
[95] deZengotita VM, Schmelzer AE, Miller WM. Characterization of hybridoma cell responses to elevated pCO(2) and osmolality: Intracellular pH, cell size, apoptosis, and metabolism. Biotechnol Bioeng 2002, 77, 369-80.
[96] Kimura R, Miller WM. Effects of elevated pCO2 and/or osmolality on the growth and recombinant tPA production of CHO cells. Biotechnol Bioeng 1996, 52, 152-60.
[97] Kim NS, Lee GM. Response of recombinant Chinese hamster ovary cells to hyperosmotic pressure: effect of Bcl-2 overexpression. J Biotechnol 2002, 95, 237-48.
[98] Zhu MM, Goyal A, Rank DL, et al. Effects of elevated pCO2 and osmolality on growth of CHO cells and production of antibody-fusion protein B1: A case study. Biotechnol Prog 2005, 21, 70-7.
[99] Jenkins N, Hovey A. Temperature control of growth and productivity in mutant chinese-hamster ovary cells synthesizing a recombinant protein. Biotechnol Bioeng 1993, 42, 1029-36.
[100] Kou T-C, Fan L, Zhou Y, et al. Increasing the productivity of TNFR-Fc in GS-CHO cells at reduced culture temperatures. Biotechnol Bioprocess Eng 2011, 16, 136-43.
[101] Rodriguez J, Spearman M, Tharmalingam T, et al. High productivity of human recombinant beta-interferon from a low-temperature perfusion culture. J Biotechnol 2010, 150, 509-18.
[102] Tharmalingam T, Sunley K, Butler M. High yields of monomeric recombinant beta-interferon from macroporous microcarrier cultures under hypothermic conditions. Biotechnol Prog 2008, 24, 832-8.
[103] Sunley K, Tharmalingam T, Butler M. CHO cells adapted to hypothermic growth produce high yields of recombinant beta-interferon. Biotechnol Prog 2008, 24, 898-906.
[104] Yoon SK, Hwang SO, Lee GM. Enhancing effect of low culture temperature on specific antibody productivity of recombinant Chinese hamster ovary cells: Clonal variation. Biotechnol Prog 2004, 20, 1683-8.
[105] Yoon SK, Song JY, Lee GM. Effect of low culture temperature on specific productivity, transcription level, and heterogeneity of erythropoietin in chinese hamster ovary cells. Biotechnol Bioeng 2003, 82, 289-98.
[106] Kaufmann H, Mazur X, Fussenegger M, et al. Influence of low temperature on productivity, proteome and protein phosphorylation of CHO cells. Biotechnol Bioeng 1999, 63, 573-82.
[107] Yoon SK, Song JY, Lee GM. Effect of low culture temperature on specific productivity, transcription level, and heterogeneity of erythropoietin in chinese hamster ovary cells. Biotechnol Bioeng 2003, 82, 289-98.
[108] Slikker W, Desai VG, Duhart H, et al. Hypothermia enhances bcl-2 expression and protects against oxidative stress-induced cell death in Chinese hamster ovary cells. Free Radical Biol Med 2001, 31, 405-11.
[109] Wurm FM. Production of recombinant protein therapeutics in cultivated mammalian cells. Nat Biotechnol 2004, 22, 1393-8.

[110] Butler M. Animal cell cultures: recent achievements and perspectives in the production of biopharmaceuticals. Appl Microbiol Biotechnol 2005, 68, 283-91.
[111] Xie L, Wang DI. Fed-batch cultivation of animal cells using different medium design concepts and feeding strategies. 1994. Biotechnol Bioeng 2006, 95, 270-84.
[112] Butler M. Post translational modification of recombinant proteins. In: Castilho Lea (eds) Animal Cell Technology: from Biopharmaceuticals to Gene Therapy. Taylor & Francis, New York, 2008, 129-46.
[113] Curling EM, Hayter PM, Baines AJ, et al. Recombinant human interferon-gamma. Differences in glycosylation and proteolytic processing lead to heterogeneity in batch culture. Biochem J 1990, 272, 333-7.
[114] Kochanowski N, Blanchard F, Cacan R, et al. Influence of intracellular nucleotide and nucleotide sugar contents on recombinant interferon-gamma glycosylation during batch and fed-batch cultures of CHO cells. Biotechnol Bioeng 2008, 100, 721-33.
[115] Chee Furng Wong D, Tin Kam Wong K, Tang Goh L, et al. Impact of dynamic online fed-batch strategies on metabolism, productivity and N-glycosylation quality in CHO cell cultures. Biotechnol Bioeng 2005, 89, 164-77.
[116] Liu B, Spearman M, Doering J, et al. The availability of glucose to CHO cells affects the intracellular lipid-linked oligosaccharide distribution, site occupancy and the N-glycosylation profile of a monoclonal antibody. J Biotechnol 2013, 170, 17-27.
[117] Gu X, Wang DI. Improvement of interferon-gamma sialylation in Chinese hamster ovary cell culture by feeding of N-acetylmannosamine. Biotechnol Bioeng 1998, 58, 642-8.
[118] Gramer MJ, Eckblad JJ, Donahue R, et al. Modulation of antibody galactosylation through feeding of uridine, manganese chloride, and galactose. Biotechnol Bioeng 2011, 108, 1591-602.
[119] Grainger RK, James DC. CHO cell line specific prediction and control of recombinant monoclonal antibody N-glycosylation. Biotechnol Bioeng 2013, 110, 2970-83.
[120] Gawlitzek M, Valley U, Wagner R. Ammonium ion and glucosamine dependent increases of oligosaccharide complexity in recombinant glycoproteins secreted from cultivated BHK-21 cells. Biotechnol Bioeng 1998, 57, 518-28.
[121] Gawlitzek M, Ryll T, Lofgren J, et al. Ammonium alters N-glycan structures of recombinant TNFR-IgG: degradative versus biosynthetic mechanisms. Biotechnol Bioeng 2000, 68, 637-46.
[122] Yang M, Butler M. Effects of ammonia on CHO cell growth, erythropoietin production, and glycosylation. Biotechnol Bioeng 2000, 68, 370-80.
[123] Pels Rijcken WR, Overdijk B, Van den Eijnden DH, et al. The effect of increasing nucleotide-sugar concentrations on the incorporation of sugars into glycoconjugates in rat hepatocytes. Biochem J 1995, 305, 865-70.
[124] Grammatikos SI, Valley U, Nimtz M, et al. Intracellular UDP-N-acetylhexosamine pool affects N-glycan complexity: A mechanism of ammonium action on protein glycosylation. Biotechnol Prog 1998, 14, 410-9.
[125] Yoon SK, Choi SL, Song JY, et al. Effect of culture pH on erythropoietin production by Chinese hamster ovary cells grown in suspension at 32.5 and 37.0 degrees C. Biotechnol Bioeng 2005, 89, 345-56.
[126] Muthing J, Kemminer SE, Conradt HS, et al. Effects of buffering conditions and culture pH on production rates and glycosylation of clinical phase I anti-melanoma mouse IgG3 monoclonal antibody R24. Biotechnol Bioeng 2003, 83, 321-34.
[127] Rivinoja A, Hassinen A, Kokkonen N, et al. Elevated Golgi pH impairs terminal N-glycosylation by inducing mislocalization of Golgi glycosyltransferases. J Cell Physiol 2009, 220, 144-54.
[128] Kunkel JP, Jan DC, Jamieson JC, et al. Dissolved oxygen concentration in serum-free continuous culture affects N-linked glycosylation of a monoclonal antibody. J Biotechnol 1998, 62, 55-71.

[129] Kimura R, Miller WM. Glycosylation of CHO-derived recombinant tPA produced under elevated pCO2. Biotechnol Prog 1997, 13, 311-7.
[130] Schmelzer AE, Miller WM. Hyperosmotic stress and elevated pCO(2) alter monoclonal antibody charge distribution and monosaccharide content. Biotechnol Prog 2002, 18, 346-53.
[131] Ahn WS, Jeon J-J, Jeong Y-R, et al. Effect of Culture Temperature on Erythropoietin Production and Glycosylation in a Perfusion Culture of Recombinant CHO Cells. Biotechnol Bioeng 2008, 101, 1234-44.
[132] Barnabe N, Butler M. Effect of temperature on nucleotide pools and monoclonal-antibody production in a mouse hybridoma. Biotechnol Bioeng 1994, 44, 1235-45.

5.2 Functional -Omics for Cell Lines and Processes: The -Omics Technologies on the Example of CHO Cells

Karina Brinkrolf, Raimund Hoffrogge, Matthias Hackl, Andreas Tauch, Martina Baumann, Thomas Noll, Alexander Goesmann, Alfred Pühler, and Nicole Borth

5.2.1 Introduction

Since the first market introduction of recombinant therapeutic proteins in 1987 [1] the yield of proteins produced by mammalian cell lines has dramatically increased [2]. Mostly this was due to improvements in bioprocess equipment, bioreactor design, new media formulations, and feed strategies, while the cell specific productivity has not changed to a similar extent over the past 25 years [3]. While initially the focus was mostly on these technical and process related parameters, during the last years the cell, its molecular functioning and regulation, has gained new attention, as it became clear that despite enhanced process control, cultures are still unpredictable and subject to variation, the basis of which is still not well understood [4, 5]. The development of new tools and methods that allow an in-depth understanding of the molecular components of the cells, their function, interaction, and regulation, now enables a new approach to enhancing process yields by a better understanding and control of the cell as the active agent in protein production. The so-called -omics techniques follow the genetic information present in a cell (genomics and epigenetic markers) to the activity of genes in terms of being actively transcribed (transcriptome), the regulation of transcript translation (microRNome) to the actual concentration of proteins (proteome). Finally, as the outcome of the activity of proteins and enzymes, the metabolome, complemented by metabolic flux analysis and metabolic models, is being investigated. Taken together, these different layers of information and regulation and their interaction are studied in a systems biotechnology approach, which differs from systems biology by its focus on the use of these techniques for the optimization of cell lines and processes. In the present chapter the authors will describe the methods and tools available for genome, transcriptome and proteome research and present their application for research on CHO cells as the most frequently used host for industrial production of therapeutic proteins.

5.2.2 Genomics and Sequencing

5.2.2.1 Introduction to CHO Cell Line Genomics and Sequencing

Knowledge of the precise DNA sequence of an organism is of fundamental importance for many areas of basic biological research and in several fields of applied sciences.

Numerous sequencing projects yield the determination of the DNA sequence, in order to reconstruct the complete genome sequence and the gene repertoire of an organism of interest. Several methods have been developed for DNA sequencing that can in principle all be described as biochemical processes to determine the exact order of the 4 nucleobases A, C, G, and T within a DNA molecule. DNA sequencing may be used to determine the nucleobase sequence of individual genes, larger chromosomal regions, entire chromosomes, or complete genomes. In this way, the term "genome" defines the entirety of the genetic information of an organism that is encoded in the DNA [6]. DNA sequencing can be specifically applied to describe the complete information content of nuclear DNA, but it can also be used for the genetic information stored within cell organelles that contain their own DNA molecule, such as the mitochondrial genome. Cells of complex and sexually reproducing organisms contain 2 copies of each chromosome. When it is announced that the genome of such a species has been sequenced, this typically refers to the determination of the DNA sequences of one set of autosomes and one of each type of the sex chromosomes.

The diploid genome of the Chinese hamster, *Cricetulus griseus*, consists of the genetic information of 22 chromosomes: 10 pairs of autosomes and 1 pair of sex chromosomes, the X and Y chromosomes [7]. A basic experimental strategy to sequence such a complex eukaryotic genome is referred to as shotgun method [8]. During shotgun sequencing, the genomic DNA is fragmented randomly into numerous small segments, which are sequenced separately to obtain "reads". Due to the random character of this approach, overlapping reads are obtained for the sequenced DNA molecule(s), when the number of individual sequencing reactions is large enough. The set of reads of a sequencing project can then be assembled by computer programs into a continuous nucleotide sequence.

The shotgun method is the basic principle of different sequencing techniques [9]. One of the first and most common methods is Sanger sequencing. It was developed by Sanger and colleagues already in 1977 (Nobel Prize in 1980) and is based on the enzymatic incorporation of labelled chain-terminating dideoxynucleotides by a DNA polymerase during *in vitro* replication of a template DNA molecule [10]. For Sanger sequencing, genomic DNA is usually fragmented into random pieces, cloned into a suitable vector plasmid and thereby amplified in a bacterial host, mostly genetically optimized *Escherichia coli* strains. Plasmid DNA from individual bacterial clones is then purified and the cloned insert is sequenced using vector-encoded priming sites (Figure 5.2.1a). Sanger sequencing has been the most widely used DNA sequencing method for more than 25 years and is still in use for small-scale genome projects. It is able to generate sequence reads of about 800 nucleotides in length [11]. A disadvantage of this technique, however, is that it is time consuming due to the cloning steps required and cost intensive.

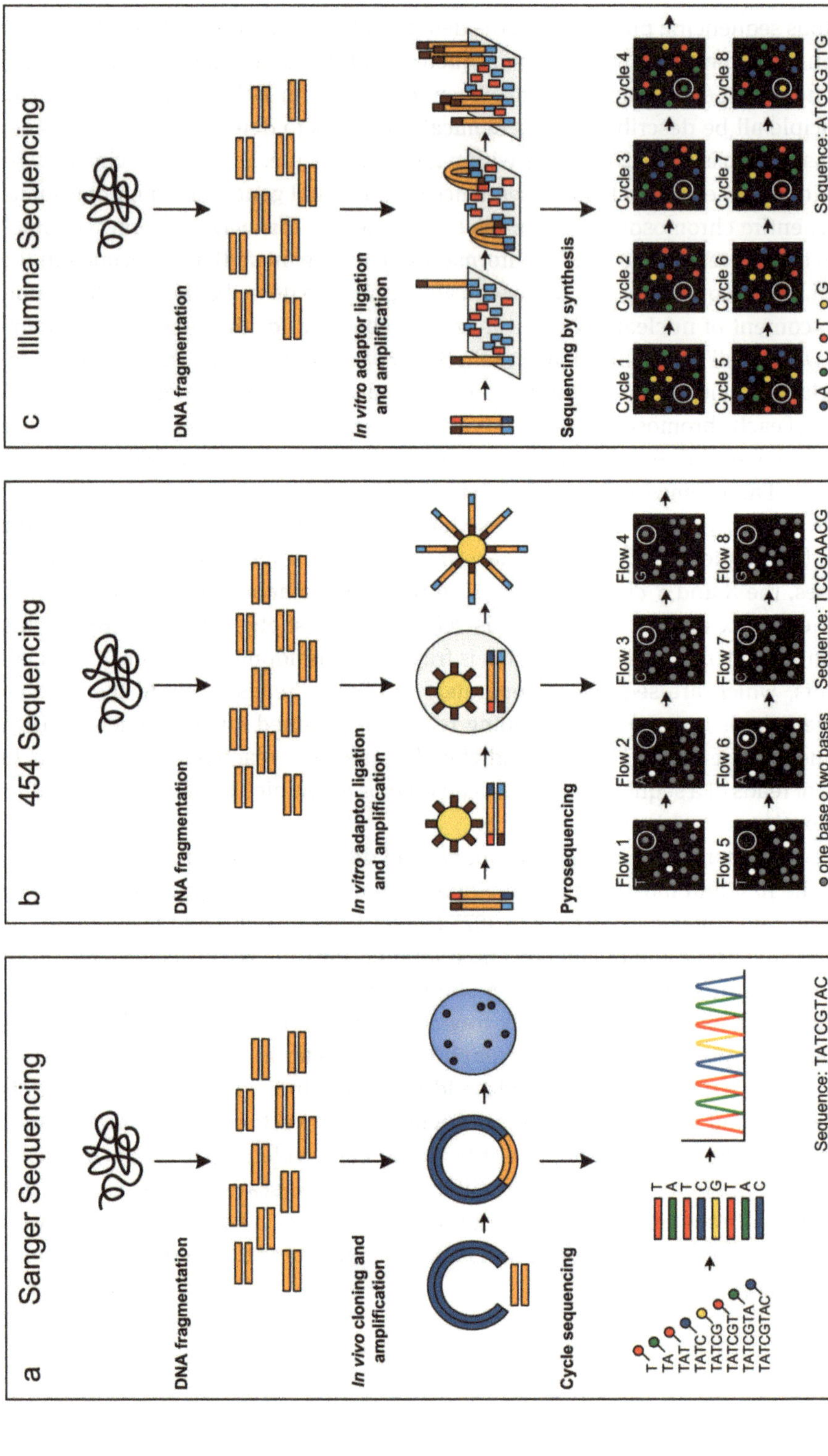
a
Sanger Sequencing
DNA fragmentation
In vivo cloning and amplification
Cycle sequencing
T
TA
TAT
TATC
TATCG
TATCGT
TATCGTA
TATCGTAC
Sequence: TATCGTAC
b
454 Sequencing
DNA fragmentation
In vitro adaptor ligation and amplification
Pyrosequencing
Flow 1
Flow 2
Flow 3
Flow 4
Flow 5
Flow 6
Flow 7
Flow 8
one base
two bases
Sequence: TCCGAACG
c
Illumina Sequencing
DNA fragmentation
In vitro adaptor ligation and amplification
Sequencing by synthesis
Cycle 1
Cycle 2
Cycle 3
Cycle 4
Cycle 5
Cycle 6
Cycle 7
Cycle 8
A C T G
Sequence: ATGCGTTG

5.2.2.2 Next-generation Sequencing Technologies Used in CHO Cell Line Research

Sanger sequencing has recently been replaced for most applications by high-throughput and ultrafast sequencing methods called next-generation sequencing (NGS) technologies. These technologies have greatly accelerated biological and biotechnological research and discovery [9]. Their high speed is instrumental in decoding the large genomes of numerous eukaryotic organisms, including the genomes of primates [12] and the genomes of CHO cell lines [13, 14]. NGS approaches from 454 Life Sciences and Illumina, for example, are faster than traditional Sanger sequencing, because they eliminate time-consuming cloning steps and separation of DNA molecules [11]. They produce shorter reads than Sanger sequencing, but in a highly parallel manner, resulting in hundreds of thousands or millions of reads in a relatively short time [15]. Due to the extensive data volume generated by NGS technologies, modern sequencing leads to a high coverage of the sequenced DNA, which is the average number of reads representing a given nucleobase in the resulting continuous nucleotide sequence. A high coverage of the sequenced DNA fragment is generally desired in shotgun sequencing projects, because it can overcome errors in the detection of individual nucleobases and therefore affects the specificity of the base calling process [16].

The 454 Genome Sequencer was developed by 454 Life Sciences and was the first NGS platform available as a commercial product [9]. This NGS technology specifically amplifies adaptor-flanked DNA fragments inside water droplets in an oil solution (emulsion PCR). Each droplet contains a single DNA template that is attached to a single primer-coated bead and thus forms a clonal DNA fragment for sequencing after amplification (Figure 5.2.1b). The 454 Genome Sequencer is equipped with a PicoTiterPlate system containing tiny wells that can harbour a single amplicon-bearing bead and sequencing enzymes. The process of 454 Pyrosequencing further uses the

Figure 5.2.1: Workflow of common DNA sequencing techniques. **a)** Sanger sequencing. Genomic DNA is fragmented, ligated into a vector plasmid and transferred to bacterial cells. Single bacterial colonies are selected for vector plasmid DNA isolation. Sequencing reactions generate a ladder of dideoxynucleotides-terminated and fluorophore-labeled products. During electrophoresis the fluorophore-labeled products pass a detector. The detector identifies the dideoxynucleotide incorporated in each band by the four-channel emission spectrum, which is used to generate a sequencing trace. **b)** 454 sequencing. Isolated DNA is fragmented and adaptors are ligated to both ends. Adaptor-ligated fragments are immobilized to beads. Bead-bound fragments are emulsified for fragment amplification together with PCR reagents. Sequencing of the libraries takes place in PicoTiterPlates. The 4 nucleobases are floated over the plate separately, 1 nucleobase per flow. Incorporation of a nucleotide is detected by light emission. The intensity of light emission is used to define the number of bases incorporated in each flow. Nucleobases are washed away after each flow, followed by the next base. **c)** Illumina sequencing. The DNA of interest is fragmented and ligated to adaptors. Adaptor-ligated fragments are attached to the surface of a flow cell. DNA fragments are amplified by bridge-amplification leading to the formation of clusters of identical DNA molecules. Four types of flourophore-labeled nucleotides are used in each sequencing cycle. Incorporated nucleotides are detected by laser. Nonincorporated nucleotides are removed by washing after each cycle. Deblocking allows the incorporation of a new nucleotide in the sequencing cycle.

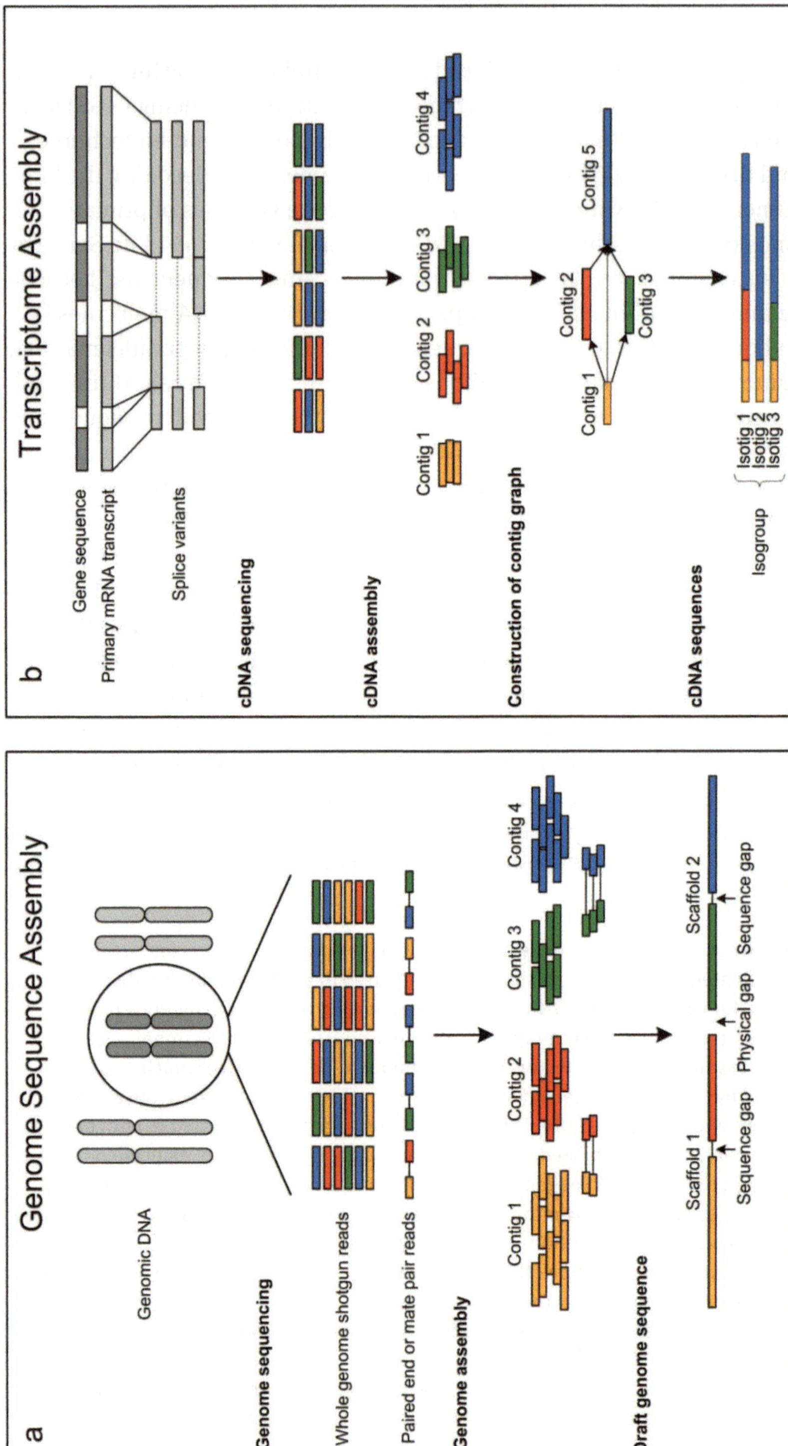
a
Genome Sequence Assembly
Genomic DNA
Genome sequencing
Whole genome shotgun reads
Paired end or mate pair reads
Genome assembly
Contig 1
Contig 2
Contig 3
Contig 4
Draft genome sequence
Scaffold 1
Scaffold 2
Sequence gap
Physical gap
Sequence gap
b
Transcriptome Assembly
Gene sequence
Primary mRNA transcript
Splice variants
cDNA sequencing
cDNA assembly
Contig 1
Contig 2
Contig 3
Contig 4
Construction of contig graph
Contig 1
Contig 2
Contig 3
Contig 5
cDNA sequences
Isogroup
Isotig 1
Isotig 2
Isotig 3

enzyme luciferase to generate light for the detection of the individual nucleotides added to the nascent DNA strand.

Another common NGS platform is based on Illumina's sequencing technology. It uses reversible dye-terminators and cluster technology, which involves the clonal amplification of DNA on a solid surface, the flow cell [9]. In this method, adaptor-ligated DNA fragments and primers are attached to the flow cell and then amplified by DNA polymerase to form local clonal colonies, the so-called clusters (Figure 5.2.1c). To determine the nucleotide sequence of the clonal DNA fragments, 4 types of labelled terminator bases are added and nonincorporated nucleotides are removed by washing. The fluorescently labelled nucleotides are detected by optical imaging and the reversible dye-terminators are then removed from the DNA, thereby facilitating the next sequencing cycle.

5.2.2.3 Sequence Data Assembly and Annotation

The common output of the DNA sequencing process is a set of reads, which differ in length depending on the technology and parameters used for sequencing [17]. To reconstruct the order of nucleobases of the original genomic sequence, these small pieces of DNA are generally aligned and merged to longer DNA fragments in a bioinformatic approach called "genome sequence assembly". Assembly of genomic sequence data results in a set of continuous consensus sequences, the so-called "contigs", which are each built by overlapping reads representing the respective genomic region (Figure 5.2.2a). Besides the generation of regular shotgun sequencing reads, Roche's NGS approaches are complemented by the paired end sequencing technology [18]. Here, both ends of a longer DNA fragment of several kilobases are sequenced and aid to determine the orientation and the relative positions of contigs produced during assembly of shotgun reads. A similar approach is also supported by Illumina's NGS technology, called mate pair sequencing. In both cases, the mean size of the long DNA

Figure 5.2.2: Workflow of sequence data assembly. **a)** Genome sequence assembly. A combination of shotgun sequencing and sequencing of long distance terminal ends is usually applied for sequencing of large genomes. During genome assembly, overlaps between the single reads are detected by bioinformatic tools to deduce continuous consensus sequences, called contigs. Paired end or mate pair reads matching to different contigs help to order contigs into scaffolds. Length of sequence gaps within scaffolds are known due to the fragment size used for library construction. Single scaffolds are divided by physical gaps. There is no further possibility to order scaffolds without additional experimental attempts. **b)** Transcriptome assembly. The starting material for transcriptome sequencing is mRNA that is transcribed *in vitro* into cDNA and used for sequencing library construction. Due to the existence of more than one splice variant of a gene, the resulting reads may include sequence information of more than one exon. This information is used to construct a contig graph. Each possible path through a contig graph is defined as an isotig representing a possible transcript. The total number of all possible paths in a contig graph is defined as an isogroup representing the collection of alternative splice products of a possible gene.

fragments used is experimentally definable and the distance between the 2 terminal reads from each DNA fragment is therefore known. When the 2 terminal reads from one fragment are located on 2 different contigs, this information is used to determine the size of the sequence gap between the contigs. A number of contigs ordered by including paired end or mate pair reads and the respective sequence gaps of known length represent a "scaffold" (Figure 5.2.2a). If this approach is not able to organize all the contigs assembled into a single scaffold, individual scaffolds are generated separated by physical gaps. The paired end or mate pair sequencing technology can also allow for the precise placement of repetitive DNA sequences in the genome. The location of a repetitive DNA sequence is known, if one end read matches the repetitive sequence and its cognate end read is uniquely located in a contig.

Apart from a rapid development in NGS technologies and assembly strategies within the last decade, assembly of large genomes still is a challenging task. *De novo* assembly generates scaffolds of 1 to 2 Mb in average, if the sequencing coverage is high enough. Remaining gaps can be sequenced by a variety of molecular methods. One possibility for rather small gaps is to use contig ends to design PCR primers and to amplify DNA fragments spanning the gap followed by Sanger sequencing [8]. However, current methods are not suitable to gain the complete genome sequence of eukaryotic organisms without sequence gaps, due to the presence of large or highly repetitive DNA regions, for instance. The resulting genome sequence is therefore routinely provided to the public in a high-quality draft status.

A slightly different scenario generating additional genetic information is attributed to the assembly of cDNA sequence data (Figure 5.2.2b). For this application, a sequencing library is constructed from purified mRNA and sequenced by an NGS approach. The application of the GS *De Novo* Assembler software (Newbler assembler) in the cDNA assembly mode generates contigs that are automatically arranged into contig graphs by means of overlapping reads at contig ends [19]. Branches in the contig graphs are associated with alternative splice products of the primary mRNA transcript. Each possible path through a contig graph is defined as an "isotig" that is a synonym for a distinct transcript, i.e., an alternative splice variant of a gene. The total number of all possible paths in a contig graph is defined as an "isogroup" that represents the collection of alternative splice products of a possible gene [20].

Once a genome is sequenced and the DNA sequence data is assembled, the deduced consensus sequence needs to be annotated to provide detailed information about the genes and their gene products. DNA sequence annotation is the two-stage bioinformatic process of identifying the locations of genes in a genome sequence including the positions of intron-exon boundaries (regional annotation), and of providing functional clues of the encoded gene products (functional annotation). Genome annotation is usually started with an automated prediction of genes and gene products by means of computational sequence analysis tools and is complemented by manual curation of the database. In this way, the genome sequence is extensively enriched with additional information at the gene or protein level that is generally stored in genomic databases.

5.2.2.4 cDNA Sequence Data of CHO Cell Lines and the Chinese Hamster

Due to the increased use of CHO cells in industrial production processes, knowledge of the genome and transcriptome sequence of CHO cell lines or the Chinese hamster is an important need for future cell line developments. Only recently information on the transcriptome and genome sequence of CHO cells is starting to be available in public databases (Table 5.2.1). Until 2009, CHO cell transcriptome sequencing attempts have applied Sanger-based sequencing. As one of the first studies, Wlaschin et al. reported 2,602 unique genes sequenced from cDNA libraries of three different CHO cell lines [21]. Of these sequences, 76 % have been annotated as orthologs of sequences stored in the GenBank database at that point of time. Most of these sequences are homologous to mouse, rat, or human transcripts. While long sequence reads are clearly an advantage of Sanger's approach for transcriptome sequencing, a disadvantage of this technique is an underrepresentation of the central part of transcripts, because sequencing from the 5' and 3' end of the library does not always lead to overlapping sequence fragments [22]. With the availability of NGS techniques, the initial set of cDNA sequences has therefore later on been extended to ~28,000 unique sequences combining Sanger-based sequencing with Roche's NGS approach. Sequences derived in both studies have been used to design custom CHO cell Affymetrix arrays for transcriptome analysis [22]. Unfortunately, these sequences have not been released to public databases (Table 5.2.1).

In the mean time, NGS technologies have opened the way for large-scale transcriptome analysis. In 2010, Illumina's sequencing approach has been used by Birzele and coworkers for high throughput sequencing of CHO cell mRNAs. The researchers identified 13,375 CHO cell genes as mouse orthologs, of which ~5,000 genes were novel to the CHO cell model [23]. In addition to the identification of novel genes, this technique also allowed for expression profiling with CHO cells and proved that expression profiling with NGS technologies is feasible, even when a reference genome and public transcriptome data is missing. With this study, large-scale transcriptome sequence data has been deposited to a public database for the first time, even though an assembly of transcriptome data was not presented to the public (Table 5.2.1). A general problem in assembling transcriptome data from NGS technologies is that assembly tools need to address the special features of eukaryotic transcripts like exon and intron structures and alternative splice sites. New tools, such as Roche's GS *De Novo* Assembler, incorporate the needs of transcriptome data to assemble different splice variants of genes. This advantage has been used for the analysis of CHO cell transcriptome data by Becker and coworkers in 2011 [20]. The authors have sequenced and assembled mRNA from different CHO cell lines using Roche's NGS approach and the GS *De Novo* Assembler. By this means, 29,184 isotigs, representing possible transcripts, have been generated that correspond to 24,576 isogroups, representing possible genes. Taxonomic classification has shown that more than 70 % of the transcriptome data is most similar to the transcriptome of mouse, with the rest being most homologous the rat. Reconstruction of the metabolic pathways of the central carbohy-

Table 5.2.1: Sequence data generated for Chinese hamster and CHO cells by Sanger sequencing and next-generation sequencing technologies. (n. a. = no further information available; BAC = bacterial artificial chromosome)

Source Organism	Sequencing Technology	Assembly Tool	Number of Sequences Reported	Description	Data Public	Reference
cDNA						
CHO DXB11 CHO DXB11-IgG CHO DXB11-IFN-γ	Sanger	Phred Phrap Consed	2,602	Unigenes	No	[21]
CHO DG44-IgG Chinese hamster brain Chinese hamster spleen	Sanger, 454	n. a.	~ 28,000	Unigenes	No	[22]
CHO (n. a.)	Illumina	Velvet	158,236,011 13,375	Short read data Mouse orthologs	Yes No	[23]
CHO K1 CHO DUKXB11	454	Newbler, Assembler	29,184 24,576 28,995	Isotigs Isogroups Contigs	No No Yes	[20]
Genome						
CHO-SEAP	Illumina	MAQ	59,083,153 17,883 19.481	Short read data Mouse ortholog genes Rat ortholog genes	Yes No No	[14]
CHO K1	Illumina	SOAP denovo	965,204,627 44,987,709 265,785 14,122 24,383	Short read data genome Short read data transcriptome Contigs Scaffolds Predicted genes	Yes Yes Yes No Yes	[13]
CHO DG44	Sanger	Unassembled data	1,119	Genomic BAC ends	Yes	[25]
Chinese hamster	n. a.	n. a.	1	Mitochondrial genome	Yes	[28]

drate metabolism and biosynthesis routes of sugars used for protein N-glycosylation were shown to be almost completely present within the transcriptome data [24]. From this study, assembled contig data were deposited to the NCBI, the National Center for

Biotechnology Information, making assembled transcriptome data accessible for the first time (Table 5.2.1).

5.2.2.5 Genome Sequence Data of CHO Cell Lines and the Chinese Hamster

With a constant decrease in sequencing costs and a tremendous increase in data output with NGS technologies, it also became feasible to sequence large eukaryotic genomes. Using Illumina's NGS approach, the genomes of the cell lines CHO-SEAP and CHO K1 were both sequenced and assembled in 2011. The CHO-SEAP genome was covered onefold. Assembly of the data therefore had to rely on a reference-guided alignment method using rat and mouse genomes as references [14] (Table 5.2.1). Consequently, the resulting draft genome sequence consists of a relatively high number of 3.57 million contigs summing up to 2.72 Gb of sequence information. The sequencing raw data from this study was additionally mapped against mouse and rat genes. By this means, 17,883 mouse homologs and 19,500 rat homologs were identified for the CHO-SEAP data set. These genes are mainly related to metabolic processes, cellular signalling, and transport. From this study, short read sequencing data is available [14]. In contrast to the CHO-SEAP genome, the CHO K1 genome was sequenced with coverage of more than hundredfold enabling *de novo* assembly of the genome into 265,786 contigs and 14,122 scaffolds. From the given data, the size of the CHO K1 genome was estimated to 2.45 Gb and 24,383 coding regions were predicted. Furthermore, most of the assembled scaffolds were associated with 21 chromosomes isolated by microfluidics in order to identify chromosomal locations of genes. Sequencing raw data as well as genomic sequences were deposited at GenBank, but without the correlation to chromosomes [13] (Table 5.2.1). Despite the high sequence coverage generated and the recent developments in NGS, the thousands and hundred thousands of scaffolds and contigs of the CHO K1 draft genome demonstrate the difficulties of assembling the short read data and scaffolding these into a fully structured eukaryotic genome.

A detailed physical chromosomal map for CHO DG44 was published by Cao and coworkers in 2011 [25] (Table 5.2.1) (see also Chapter 3.2). These authors prepared fluorescence *in situ* hybridization (FISH) imaging probes from selected bacterial artificial chromosomes (BACs) and used Sanger-based BAC-end sequencing in order to construct a physical genome map for CHO DG44. The authors further on used the results from CHO DG44 to analyse chromosomal rearrangements in comparison to CHO K1 cells and Chinese hamster cells, stating that 8 out of 20 chromosomes are relatively stable while the other chromosomes have undergone major rearrangements. Availability of 2 recently accepted publications on the genome sequence of the Chinese hamster will be a valuable source of information to characterize and to study chromosomal rearrangements and stability in CHO cell lines [26, 27].

Besides recent progress in whole genome sequencing of CHO cell lines, the mitochondrial genome of the Chinese hamster was already published in 2007 [28]. The circular genome of 16,284 nucleotides has been established from 6 overlapping PCR

Table 5.2.2: Databases for Chinese hamster and CHO cell sequence information.

Database	Content	Link	Reference
General databases			
NCBI Nucleotide	Core subset of nucleotide sequence records	www.ncbi.nlm.nih.gov/nuccore/?term=cricetulus griseus	[215]
NCBI EST	Expressed sequence Tags records	www.ncbi.nlm.nih.gov/nucest/?term=cricetulus griseus	[217]
NCBI Protein	Protein sequence database	www.ncbi.nlm.nih.gov/protein/?term=cricetulus griseus	[217]
NCBI Genome	Whole genome database	www.ncbi.nlm.nih.gov/genome/?term=cricetulus griseus	[217]
NCBI Gene	Gene-centered information	www.ncbi.nlm.nih.gov/gene/?term=cricetulus griseus	[217]
NCBI SRA	Short read archive	www.ncbi.nlm.nih.gov/sra/?term=cricetulus griseus	[217]
NCBI UniGene	Gene-oriented clusters of transcript sequences	www.ncbi.nlm.nih.gov/unigene/?term=cricetulus griseus	[217]
Chinese Hamster Genome Database	Genome, proteome, transcriptome, and non-coding RNA data, and genome viewer	www.chogenome.org	[29]
Specialized databases			
Genomic tRNA database (GtRNAdb)	tRNAscan-SE analysis	gtrnadb.ucsc.edu	[216]
CHO Proteome Database	Proteomic data from 2D PAGE and MS analysis	http://gofant.com/CHOProteomeMainImage.html	–
CHO Gene Coexpression Database (CGCDB)	Coexpression data of genes and genes correlated to growth rate or cell specific productivity	www.cgcdb.org	–
The Organelle Genome Database (GOBASE)	Data related to mitochondria	gobase.bcm.umontreal.ca	[217]
miRBase	miRNA sequences and annotation	www.mirbase.org	[218]
Long non-coding RNA database (lncrna db)	Annotated lncRNAs	www.lncrnadb.org	[219]
Silva	Comprehensive ribosomal RNA data	www.arb-silver.de	[220]

fragments. In analogy to other mammalian mtDNA, it comprises 13 protein-coding sequences, rRNA genes, and 22 tRNAs. The G-C poor sequence has only few noncoding regions, and predicted genes harbour no introns (Table 5.2.1).

Available sequence data for CHO cell lines and the Chinese hamster are stored in public databases. These databases either present general sequence data or are specialized on one type of data such as miRNAs, tRNAs, or mitochondrial DNA sequences (Table 5.2.2). A database focused on CHO cell specific information is the Chinese Hamster Genome Database (www.CHOgenome.org), representing a comprehensive collection of links to original sequence data and related publications [29].

5.2.2.6 Open Questions

The recently released sequences of multiple CHO cell lines and of 2 Chinese hamster genomes [26, 27] have opened the way to a detailed analysis of the impact of genetic modifications, including chromosome rearrangements, SNPs, insertion, deletions, and amplifications on the process relevant properties of CHO cells, most prominently growth, productivity, product quality, and stability. To enable this type of study, a reliable parental reference genome for the Chinese hamster needs to be developed from the presently available draft genomes. Specifically in view of the frequent chromosomal rearrangements that occur within CHO cells, the ability to assign sequencing reads to chromosomal locations is of prime importance.

5.2.3 Transciptomics

5.2.3.1 Definition

The transcriptome of cells in culture comprises the complete set of RNA molecules that are being transcribed from the genomic DNA present in a cell, including mRNAs, rRNAs, tRNAs, and other noncoding RNAs, such as miRNAs. Although the sequence information for these is laid down in the genome and thus is identical for each cell derived from a given species, the precise pattern of genes and sequences actually transcribed is dependent on the individual cell type, its differentiation status and the culture conditions. The methods used for the analysis of the transcriptome selectively enrich for a certain category of these RNAs, or, by method principle, have a bias for one of them. The most frequently used methods in this field are microarrays and RNA-Seq. All of them measure the expression of specific RNAs present in a given cell pool, thus cannot be used to directly measure transcription itself, as the amount of RNA detected will also depend on the mRNA degradation speed and rate which can differ highly between specific mRNAs, cell types and states. Finally, the expression profile of a cell does not necessarily reflect the protein content in a cell (the proteome), as the actual translation into proteins depends on the activity of translation initiation for each individual mRNA type which is subject to additional regulatory control (for instance by miRNAs) and the stability of the protein in the cell. What transcriptome analyses do reflect is a global picture of the overall activity of genes within a cell. The most pertinent difference between a genome and a transcriptome is that the genome tells us what a cell could in theory do, while the transcriptome reflects what is actually happening within a cell in any given context or at any given time-point.

5.2.3.2 Methods of Transcriptome Analysis

Microarrays

Analysis of the transcriptome by microarrays is currently the most frequently used technique accounting for the largest number of publications in the field. Simply defined, a DNA-microarray (also commonly known as DNA chip) is a collection of microscopic DNA spots (features), attached to a solid surface, which can be hybridized with labelled target molecules. The probe refers to the DNA sequence arrayed on a solid surface by covalent binding to chemically suitable matrices. Each spot contains millions of identical probes. The target is a fluorescently labelled cDNA (copy DNA) or cRNA (complementary RNA) sample, which is hybridized to the microarray surface. Hybridisation between complementary sequences of the target and the immobilized probe will lead to an increase in fluorescence intensity over the background level, which can be measured using a fluorescent scanner to determine the relative abundance of respective nucleic acid sequences in the target sample [30].

In principle, DNA-microarrays can be distinguished based upon properties such as the solid-surface support, the specific method used for probe attachment, the nature of the probe, or target detection.

Amongst other synthetic material, the solid-surface support can be a glass slide, a silicon chip, or a nylon membrane, to which the probes are immobilized as discrete spots. Probe attachment is achieved by 2 predominant methods [31]. The first method is often referred to as the Affymetrix method [32]. In this process, oligonucleotide probes are synthesized directly onto the surface of the microarray (*in situ*-synthesised microarray). As *in situ*-synthesised probes are typically short (20 to 25 mers), multiple distinct probes for one target gene are included, collectively termed as a probe set, to improve sensitivity, specificity, and statistical accuracy. For Affymetrix GeneChips, oligonucleotide probes are synthesized *in situ* by photolithographic fabrication on quartz wafers, using UV light and light-sensitive masking agents to add specific nucleotides to the growing probe [32, 33]. In contrast, the Agilent SurePrint platform employs ink-jet technology for *in situ* manufacturing of probes on glass slides, enabling much longer probes of 60 mers. While longer probes are more specific for individual targets, the Affymetrix platform uses several probes for each individual gene (typically 11 to 20 probes), to increase the reliability of results. Thus the different technologies each have their individual benefits and limitations of which the user should be aware [34-36].

Besides *in situ* synthesis, the second common method for producing microarrays is "printing" or spotting of presynthesised DNA fragments onto a glass slide using high-precision arraying robots. Through the use of chemical linkers that provide functional groups, the probes are covalently coupled to solid slides via modified 3' or 5' ends. The probe spots can be applied by either noncontact (i.e., bubble-jet or ink-jet printer) or contact printing using fine-pointed pins. Due to the relatively large size of the spots, printed microarrays are of lower density (~ 10,000 to 30,000 features) than *in situ*-synthesized microarrays, which can contain over 300,000 individual features (Affymetrix GeneChips: > 10^6 features per microarray) [30, 33].

Agilent arrays are typically used in two-colour hybridisation. The target probes are single stranded cDNA generated through reverse transcription of mRNA extracted from 2 cell populations to be compared. These cDNA samples are labelled using 2 different fluorescent dyes (commonly Cyanine 3 and Cyanine 5) and competitively hybridized onto the same microarray. Relative intensities and ratios of each spot are then analysed using 2 different wavelengths corresponding to the dyes used in order to identify upregulated and downregulated genes [37]. In contrast, Affymetrix GeneChips are limited to one colour. A one-colour design requires 2 separate single-dye hybridizations for comparison of 2 different samples, where the relative mRNA abundance of each target gene must be determined by calculating the relative difference of particular spots on each array [38]. Each technique is associated with advantages and disadvantages that must be considered, like errors involved with the variability in microarray manufacturing when using one-color hybridization or dye specific biases negatively affecting microarray results when using two-colour hybridization [39]. The common mode of analysis is therefore to include duplicate samples for each state and to hybridise each sample twice with a dye swap, thus stained once with each colour. After normalisation, for each sample, the average of both dyes is calculated and used for analysis.

Despite the power to analyse thousands of genes simultaneously and the high degree of flexibility regarding arrayed elements (particularly for custom-tailored DNA-chips), the microarray technology has some limitations. Since microarrays are indirect methods, transcript abundance is deduced from hybridization levels where background noise and cross-hybridization interferes with the accuracy of expression measurements, especially for targets present in low abundance [40, 41]. Other drawbacks are the limited dynamic range for detection due to both background and saturation of signals, as well as dependency on prior knowledge of the genome sequence for probe design. However, advances in surface technology, robotics, and bioinformatics as well as development of new labelling protocols and dyes will continue to further improve the technology in terms of complexity and quality, such as reproducibility, sensitivity, and specificity [38]. An important aspect in this regard is the design of experiment if multiple samples are to be compared (Figure 5.2.3). Correct choice of the design strategy will depend on the specific question to be answered in the experiment. Each design requires elaborate normalisation and statistical evaluation [42].

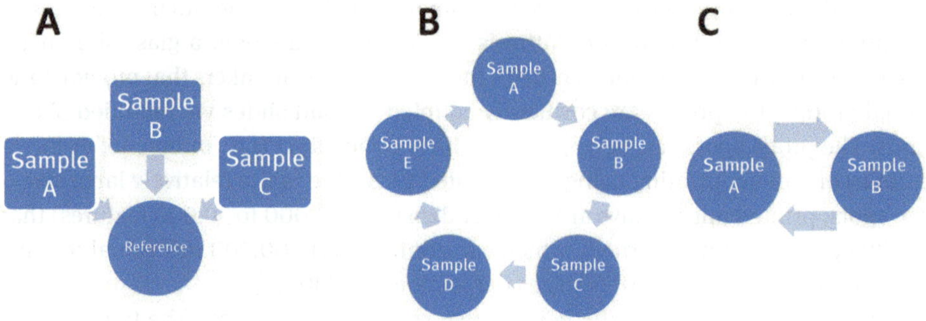

Figure 5.2.3: Design of experiment for microarray studies. **A)** Reference design: this is the prefered design if multiple experiments are to be compared **B)** Ring design: this design reduces the number of arrays to be run and enables comparison of multiple samples
C) Direct comparison: prefered design for evaluation of differences between two samples or states.

RNA-sequencing

NGS technologies have created a new tool with the potential to revolutionize the study of the transcriptome. RNA-seq (RNA sequencing) uses high-throughput sequencing methods to provide an opportunity for characterizing and quantifying the transcription profile of a sample. RNA-seq relies on the generation of short reads of sequence information, which are either aligned to a reference genome or reference transcript or are assembled *de novo* without *a priori* knowledge of the genome. This makes RNA-seq particularly interesting for nonmodel organisms and/or the detection of novel transcripts, bypassing a limitation of microarray analysis [43, 44]. Compared to microarrays, RNA-seq methods have a much larger dynamic range of expression with very low background signal as each sequence obtained can be mapped to a unique site of the genome [45]. Another advantage is the high data resolution. While in hybridization-based approaches resolution is limited to the probe size and allows for a small number of mismatches, RNA-seq offers single-base resolution. Thus, RNA-seq is also useful for distinguishing different isoforms or allele-specific expression and for the identification of novel exons and splice sites, which cannot be detected by DNA-microarrays [46-48]. Besides, RNA-seq data is extremely rich, enabling the study of sequence variations like SNPs (single nucleotide polymorphisms) or RNA editing events. Depending on the experimental goal (e.g., mRNA comparison of 2 samples, identification of novel transcribed elements), the depth of sequencing must be considered. The sequence coverage, i.e., the average number of reads for a given nucleotide, is one of the most important parameters for reliable measurement in RNA-seq studies. Regarding the purpose of analysis, the genome size and transcriptome complexity, the requirement of sequencing depth varies. In principle, the more complex the transcriptome, the more sequencing depth is required for appropriate coverage, which is particularly necessary for detecting low abundant transcripts [49, 50].

Another issue to consider is that most current RNA-seq methods rely on cDNA synthesis, which however was shown to be prone to introducing biases and artefacts, such as template switching or random cDNA synthesis, which may be caused by self-priming due to RNA secondary structures [51, 52]. Furthermore, reverse transcriptases are error-prone due to their lack of a proofreading mechanism [53, 54]. In general, an isolated RNA population is converted to a library of cDNA fragments and sheared in order to reach a size compatible with the sequencing method used (typically <500 bp) [44]. Common fragmentation methods of either RNA or cDNA molecules include RNA hydrolysis, DNAse I treatment, nebulisation, or sonication. When performing fragmentation, it must be considered that each of these methods results in a different bias in the outcome [45, 47].

More recently developed methods involve direct sequencing of RNAs (DRS) to evade this type of artefacts [55]. This technology relies on hybridization of polyadenylated RNA templates to a poly(dT) coated surface for subsequent sequencing by synthesis. Direct RNA sequencing is limited to poly(A)+ transcripts, with sequence data being derived from regions immediately upstream of the 3'polyadenylation site. RNA species that lack natural poly(A) tails (small or noncoding RNAs) can be polyadenylated *in vitro* before analysis. Direct RNA sequencing requires only femtomoles or attomoles of RNA, which could in the long term make DRS useful for identifying the transcriptome of small quantities of cells.

In summary, RNA-seq is a promising technology for the analysis of the transcriptome and is comparable, for some applications even superior, to existing DNA-microarray techniques [56]. Currently, array-based approaches are the primary choice for studying gene expression, but considering the rapidly falling costs of sequencing, RNA-seq has potential to be widely adopted for this purpose. Most importantly, it offers the ability to also analyse the expression of noncoding RNA transcripts and is also suitable for the analysis of transcriptome of cells of species where no known genomic sequence is available.

5.2.3.3 Transcriptomics Applications

As gene expression within a cell is highly dynamic, transcriptomic tools like DNA microarrays or RNAseq can solely give a snapshot of the current cellular state. On basis of differential mRNA expression between a test and reference sample, a number of studies have provided valuable insights into a range of areas including gene discovery, disease characterization and cell culture engineering. The majority of these studies made use of the relatively mature DNA-microarray technology with the objective to identify genes and key pathways that affect special cellular properties. By comparing different cell lines or phenotypes microarray experiments have extensively been used to define transcriptomic signatures, which are sets of genes that can be regarded as markers of respective conditions [57]. Such signatures may be used as markers of diseases or prioritised targets for genetic engineering in order to improve industrial cell culture processes. Such optimisation of processes can have significant impact on productivity and overall yield while reducing cost of production. In this context, the critical properties investigated include, e.g., antiapoptosis, proliferation, and productivity [58-60].

An early application of DNA microarray was used to understand the transcriptomic change upon shifting the metabolism of a mouse hybridoma cell line (MAK) from a high-lactate producing to a low-lactate producing state by controlled nutrient feeding [61]. This study showed a general decrease of transcript levels of many genes involved in glycolysis.

During recent years, great efforts have been made to determine changes at gene levels involved in productivity. Khoo et al. [62] used a mouse array representing 6,400 genes to assess the expression profile of producer and nonproducer NSO myeloma cell lines. Unexpectedly, genes involved in the protein-synthesis pathway were higher expressed in the wild-type cell line. On the other hand, the producer cells had differentially expressed genes for cell cycle-related events and zinc finger proteins, which could be important in coping with the metabolic burden of antibody production. The results suggest that the producer cells could have shifted their synthesis machinery from expressing antigen-presenting proteins to recombinant proteins without an increase in protein translation. Similar studies were accomplished to investigate culture conditions that enhance productivity, including hypothermic cultivation [63-65], hyperosmotic cultivation [66], and exposure to butyric acid [67, 68].

Another approach to increase product yields involves extending the viability of the cells by inhibition of apoptosis. In 2006, Wong et al. [69] reported the identification of key antiapoptotic genes and pathways in batch and fed-batch CHO cell cultures and reported that during periods of high viability most proapoptotic genes were downregulated, while upon decrease in viability apoptotic genes were sequentially upregulated. Four highly active effector genes were identified, including 2 antiapoptotic genes, Faim and Fadd, and 2 proapoptotic genes, Alg-2 and Requiem. Overexpression or knockdown of these genes enabled generation of 4 cell lines, which showed higher apoptosis resistance than the parental cell line [70]. Another study, combining microarray and proteomic expression profiling analysis, identified a list of 21 potential candidate genes associated with a high growth rate phenotype in CHO cells [71]. In a study comparing anchorage-dependent and suspension grown HeLa cells, 2 genes were identified, cdkl3 and cox15, which were upregulated in the faster growing suspension HeLa cells [72]. Enhanced expression of either gene in the anchorage-dependent HeLa cells, as well as in HEK-293 and CHO cell lines, resulted in elevated growth rates.

A present focus in optimising biotechnological processes is media development, including the formulation of serum-free and animal component-free media, where transcriptomic tools may also play a vital role. Allison et al. [73], for example, identified receptors for growth factors, hormones, and cytokines, cell adhesion molecules, and other cell-signalling components. Corresponding ligands or small molecules can be added to the cell culture medium and rationally tested for desired biological effects. Jaluria et al. [74] identified differentially regulated genes associated with serum withdrawal. A HEK-293 cell line was gradually adapted to serum-free medium and analysed at specific serum levels. Based on the results 2 genes with correlated expression patterns, egr1 and gas6, were transfected into different HEK-293 cell lines and enabled faster adaption to serum-free media while improving viabilities and growth rates. In another study, zinc was identified to be an effective replacement for insulin in murine hybridoma cultures [75]. Gene expression profiling indicated no major change in the expression profiles of the insulin-supplemented and zinc-supplemented cultures.

Generally, gene expression profiling can help to identify candidate genes for cell line development, employing an indirect cell engineering strategy [76, 77]. Today's typical fed-batch bioprocess runs over 14 days, covering a variety of cellular states and environmental conditions, with obvious changes in growth rate, productivity, and, most importantly, product quality. The real power of -omics analysis is the ability to monitor these changes [78, 79] and to generate an understanding of the needs of cells that enable them to generate high quality and value products [80].

5.2.4 MiRNAs

5.2.4.1 Background

The gap between the great complexity of cellular functions and limited number of primary RNA transcripts may be explained by sophisticated mechanisms that regulate RNA. Among these mechanisms the posttranscriptional regulation by miRNAs (miRNAs), which are highly conserved small noncoding RNAs, constitutes an important layer in the control of gene expression. By regulating a broad range of targets – similar to transcription factors [81] – miRNAs control the adoption of new phenotypes during differentiation [82] as well as the cellular response to many external and internal stimuli such as hypoxic environment or metabolite availability [83]. This versatility of miRNAs holds promise for their application in biopharmaceutical industry as biomarkers for improved cell line development and as cell engineering targets to enhance the bioreactor performance of mammalian-derived industrial cell factories [84, 85].

5.2.4.2 MiRNA Biogenesis

With a size between 18 and 24 nucleotides mature miRNAs are among the smallest types of noncoding RNAs [86]. They are produced as primary miRNA transcripts from RNA-Pol II promoters or derived from intronic regions, which become cleaved in the nucleus by an RNase enzyme complex consisting of Drosha and Dgcr8 to yield a 50 to 70 nucleotides RNA stem-loop structure termed precursor miRNA (pre-miR). Following the export into the cytoplasm via Exportin-5, pre-miRs are cleaved by the RNase-

III enzyme Dicer, so that a ~22 nucleotides mature miRNA duplex is produced, consisting of a -5p and -3p mature miRNA that are hybridized and exhibit characteristic 3' overhangs.

In some cases primary miRNA structures can contain more than one stem-loop sequence and consequently give rise to several distinct mature miRNA sequences. These miRNAs are commonly referred to as miRNA clusters, which often share similar seed sequences (see below) and therefore biological function. Currently 25,141 mature miRNA sequences from 193 organisms have been annotated including 2,042 human, 1,281 murine and 307 CHO pre-miRNAs. Genomic location, organization as well as the sequence of these miRNAs can be accessed via the miRBase miRNA repository (www.mirbase.org) [87].

5.2.4.3 How miRNAs Control Cellular Processes

Following the dissociation of the miRNA duplex, the guide miRNA directs RISC to the target mRNA sequence through base-pairing with complementary sequences in the 3' UTR [88]. In mammalian cells this miRNA : mRNA interaction process generally relies on imperfect base-pairing, and therefore mRNA slicing by RISC (as observed in plants) is rare. Most mammalian miRNA binding sites consist of a perfect match between nucleotides 2-8 of the mature miRNA, which is also known as the "seed region", followed by mismatches in the central region. Thus, association of RISC with the 3'UTR of a target mRNA can either result in translational repression by disturbing protein interactions between components of the translation initiation complex, or in mRNA destabilization due to deadenylation and decapping of the message. In some cases miRNA binding can occur outside the 3'UTR region [89] or lead to an activation of mRNA translation by masking of elements within the mRNA sequence that promote its decay, such as AU-rich elements [90].

Instead of simple one miRNA : one mRNA binding situations, miRNAs can interact with up to hundred different mRNA species and one mRNA can harbour binding sites for several different miRNAs [91]. Based on this multiplicity of interactions, it has been suggested that up to 60 % of human genes are targeted by one or more miRNAs. While this makes miRNAs powerful regulators of gene expression, it also means that target repression by miRNAs (and consequently their function) is complex to predict, since target interaction not only depends on the thermodynamics of base-pairing, but also on the miRNA : target ratio as well as the presence of other miRNAs targeting the same messages [92]. In addition miRNA activity can be tissue specific [93], with certain miRNAs being highly active only in one distinct cell type such as miR-1 in heart tissue or miR-122 in liver tissue [94]. This complexity explains why miRNA function can change between different species or even cell types. miRNA-21 for instance is known to be overexpressed in many tumour cells resulting in enhanced growth, apoptosis resistance and metastatic potential [95]. However, in endothelial cells it was found that miR-21 is highly overexpressed in senescent cells compared to young dividing cells, and that ectopic expression in normal cells results in growth arrest [96].

For the purpose of improving upstream processes, the identification of miRNAs that control cell growth, stress resistance, protein production, and energy metabolism in CHO cells are of highest interest. While the relevance of miRNAs for these pathways has been extensively studied (oftentimes in relation to cancer and other diseases), the above mentioned complexity of target repression makes it challenging to directly transfer these data to CHO cells. Consequently an independent characterization of miRNA function in CHO cells is necessary starting with the analysis of miRNA expression and correlation to cell phenotype.

5.2.4.4 Quantification, Target Identification, and Manipulation of miRNAs

Compared to measuring the expression of mRNAs that stretch hundreds to thousands of nucleotides, the quantification of miRNAs involves the challenge of detecting short (<24 nucleotides) and sometimes highly similar sequences that differ by only 1 or 2 bases (so-called miRNA families). Three methods are predominately used to quantify miRNA levels: real-time quantitative PCR (qPCR), microarray hybridization, and "digital gene expression" by next-generation sequencing. Each of these platforms faces certain challenges. Hybridization based methods (qPCR and microarray) suffer from the short target length and high variation in GC-content, which limits probe design usually to the full-length complementary miRNA sequence with high variation in melting temperatures. Massive parallel sequencing technologies rely on multistep sample preparation protocols that involve adapter ligation and PCR amplification steps, which can be biased towards certain sequence motifs.

The Gold Standard in miRNA Quantitation: Real-time Quantitative PCR

Real-time quantitative PCR is the method of choice for analysis of miRNA expression as well as for validation of miRNA expression data from other platforms due to its long history of use, high sensitivity and dynamic range, and manageable computational effort for data analysis [97]. Most available miRNA qPCR protocols utilize a 5' tag in the primer for first-strand synthesis reaction, which serves as binding site for the reverse primer during the PCR reaction and results in extended amplicon length. Forward PCR primers bind the mature miRNA sequence and may contain chemical modifications such as locked nucleic acids [98], 2'-O-(2-methoxyethyl)-derivatives for optimized hybridization characteristics [99], or stem-loop primers that allow discrimination between mature and precursor miRNA sequences [100]. The use of miR-specific reverse-transcription reaction (RT) generally yields high specificity and sensitivity but also limits throughput. Mixing of specific RT-primers is a possibility but may affect the performance of these protocols due to interference between primers. Therefore, universal RT reactions have been established by adding a polyA tail to the miRNA sequences prior to RT. This polyA tail is used during RT reaction as template for an oligo dT primer, which is linked to a 5' tag serving as reverse primer binding site during qPCR. Such universal miRNA cDNA pools were the prerequisite for scaling miRNA qPCR experiments to genome level by the development of "qPCR-array" platforms that consist of premanufactured plates (96-well or 384-well) or high density array cards, which were reported to yield highly reproducible results [101].

As with every qPCR experiment, the quality of miRNA qPCR data strongly depends on the correct choice of reference genes for normalization [102]. The use of presumably invariant reference RNAs for normalization such as small nuclear or nucleolar RNAs, or 5S ribosomal RNA is common; however, due to different physicochemical properties invariant miRNA reference genes are better suited for normalization of qPCR data. In general it is advised to use more than one reference gene for normalization, which should be selected using algorithms designed for reference gene evaluation.

High Throughput Analysis of miRNA Expression

The utilization of microarray hybridization technology to profile miRNA expression in 2004 marks the beginning of high-throughput analysis of miRNA transcription [103, 104]. Advances in miRNA microarray technology are characterized by sophisticated designs of capture probes, which must specifically hybridize to short target miRNA sequences – sometimes differing only at a single nucleotide position – at uniform temperature. Similar to qPCR primers, chemical modifications such as locked nucleic acids (LNA) have been introduced [103] as well as hairpin extension and other proprietary designs. As for qPCR, a critical issue for miRNA microarrays is data analysis and especially data normalization. The assumption that compared to the overall number of genes only few genes are differentially expressed and that up- and downregulation of genes are balanced, might not hold true for miRNA experiments. Hence it is important to carefully choose the method of normalization and to control the impact of normalization on the final results [102].

The development of NGS technologies has significantly boosted the discovery of novel miRNA sequences while allowing simultaneous quantification of both known and novel sequences. The digital nature of NGS expression data (i.e., one read equals one count of a specific miRNA) accounts for a higher dynamic range similar to that of real-time qPCR (6 to 7 log scales) and superior to microarray data (up to 4 logs). In contrast to qPCR and even microarrays, however, analysis of small RNA next-generation sequencing data is time consuming due to elaborate preprocessing of raw sequencing data, such as quality filtering and adapter trimming, while adapter ligation and preamplification steps can introduce biases to the quantitative content of NGS data.

A recent study set out to compare the performance of Taqman RT-qPCR, 6 microarray platforms and Illumina NGS for the identification of differentially expressed miRNAs [105]. Analysis of intraarray and interarray signal variation of microarray platforms showed that the robustness varied significantly, with some platforms giving highly reproducible signals while others did not. High concordance between qPCR and NGS data was also obtained by Pradervand et al. who compared three microarray platforms qPCR arrays and NGS [106]. Overall it seems that qPCR and NGS exhibit superior sensitivity to most microarray platforms, which is mirrored in lower compression of fold changes. This might be due to the presence of microarray probes with low T_m values for miRNAs with low GC content, thus reducing hybridization and signal intensity.

High-throughput Analysis of miRNA : mRNA Interactions

MiRNAs exert their biological function by posttranscriptional interaction with target genes, which in most cases results in the repression of mRNA translation or mRNA destabilization, but in rare cases may result in mRNA activation [87]. Consequently, the identification of differentially regulated miRNAs based on any of the above described technologies raises 2 questions: (i) which genes are directly affected by the de-regulation of a miRNA and (ii) does the perturbation of this miRNA in a biological system cause any phenotypic changes? While the second question can be answered by ectopic overexpression or knockdown of miRNAs in a cell, the first question is more difficult to solve. Initially, the identification of miRNA targets involved bioinformatic predictions followed by sequential (gene by gene) validation of miRNA : target interaction using qPCR, Western blot, and luciferase reporter assays. However, since target interaction can depend on as few as 6 nucleotides, which on average can be found every ~4 kb in the genome, confident target prediction is a major challenge. Therefore, 2 methods have recently emerged that allow high-throughput analysis of direct miRNA : target interactions via the isolation of miRNA : mRNA complexes and analysis of the content of these isolates using PCR, microarrays or NGS.

Tandem purification of miRNA targets (TAP-Tar) is a method for isolation of miRISC complexes, which contain argonaute (Ago) proteins, a mature miRNA species and a bound mRNA target [107]. For this purpose a host cell line expressing a tagged version of argonaut proteins (Ago-1) is required, as well as the transfection of a biotinylated miRNA species, for which mRNA targets shall be identified. The isolation of RISC complexes containing the miRNA of interest plus its targets is achieved by sequential immunoprecipitation of Ago-1 proteins and streptavidin-based isolation of the miRNA. For global analysis of miRNA : mRNA complexes isolated by TAP-Tar this method can be combined with high-throughput technologies such as NGS or microarray hybridization. A disadvantage of this method is, however, the need for a host cell line expressing a tagged Ago variant. In addition, the measured interactions are based on the artificial introduction of excess amounts of biotinylated miRNAs that may increase off-target effects and therefore "false-positive" interactions.

High-throughput sequencing combined with crosslinking immunoprecipitation (HITS-CLIP) describes a method suited to directly identify protein-RNA interactions *in vivo*, without perturbing the biological system [108]. Ultraviolet radiation is used to covalently crosslink RNA-protein complexes, which was shown to work for Ago-miRNA-mRNA complexes, thus allowing the isolation of these complexes by immunoprecipitation. NGS analysis of partially digested RNA sequences contained in the purified complexes can give important insights into which mRNA target sequences are actually bound and regulated by miRNAs. Chi et al. performed Ago HITS-CLIP on mouse brain tissue and were able to confirm all previously validated interactions of miR-124, a brain specific miRNA [94]. To this point other studies have applied HITS-CLIP to study the interactions of viral miRNAs in host cells [109] and to get a better insight into the determinants of miRNA : mRNA interactions, which are step-by-step integrated into novel target prediction algorithms [110].

Manipulating miRNA Expression in (CHO) Cells

In the beginnings of miRNA research, targeted overexpression or knockdown studies of miRNAs were performed to explore their biological function in specific tissues or cell types. In the meantime, these powerful molecules have evolved as therapeutic targets for local or systemic substitution or inhibition and have been proposed as target molecules for engineering the phenotype of cell lines used for the production of recombinant therapeutic proteins (see also Chapter 3.5). Here, we will disregard delivery of miRNAs for therapeutic purposes and focus instead on methods for transient and stable knockdown or overexpression of miRNAs in cell lines such as CHO.

The first studies on miRNA expression in CHO cells were done based on the assumption that most miRNA sequences are highly conserved between Chinese hamster and rodents such as mouse or rat [111]. Later, in 2011, 2 groups set out to sequence the CHO miRNome by NGS analysis of the small RNA transcriptome of several different CHO cell lines [112, 113] and, based on miRBase version 14 as reference, reported the transcription of 350 and 365 mature conserved miRNAs in CHO, respectively. After the publication of the CHO-K1 genome reference sequence in 2011, all previously identified CHO miRNAs were aligned against the genome to identify their location as well as precursor structure [114]. This information can now be accessed for 350 CHO miRNAs in miRBase v20 at www.mirbase.org.

For transient overexpression or knockdown, synthetic RNA molecules are transfected into a cell that mimic an endogenous miRNA or sequester it by stable hybridization resulting in reduced biological activity. Synthetic miRNA mimics are double-stranded RNA molecules (dsRNA) with 3' overhangs on either one or both ends that improve stability. In addition chemical modifications have been reported that are designed to increase the efficacy of delivery, stability and guide strand incorporation into the RISC. Synthetic miRNA inhibitors are single-stranded RNA molecules partially or fully complementary to their target sequence. Improved hybridization of these inhibitors to their targets can be achieved by introducing chemically modified nucleotides with 2'-O-methyl or locked nucleic acids. Several commercially available mimic and inhibitor platforms exist, which also offer libraries for high-throughput and genome-scale screening of miRNA function.

Stable engineering of miRNA expression can only be achieved by plasmid based delivery and genomic integration of miRNAs, since synthetic oligonucleotides will be rapidly diluted in fast proliferating CHO cells. To express miRNAs from plasmids, artificial constructs comprising the mature miR-5p and miR-3p sequence and universal loop sequence from a highly expressed miRNA can be synthesized and cloned into specific miRNA expression constructs that contain miRNA flanking regions, RNA-Polymerase II (Pol II) promoter, and a 3'polyadenylation site. This system was also used to transiently assess the influence of 4 different miRNAs on growth and productivity in CHO cells [115]. Alternatively, miRNAs can be PCR-amplified from genomic locations including 100 to 200 nucleotides up- and downstream of the miRNA stem-loop and cloned into similar expression constructs. Recent results indicate that endogenous constructs are superior to artificial constructs in terms of correctly processed mature miRNAs [116]. Knockdown of miRNA activity can be achieved by the expression of RNA transcripts harbouring multiple miRNA binding sites. These so-called sponge constructs sequester mature miRNAs from the cell resulting in lower biological activity [117]. Untranslated (3'UTR) regions of reporter proteins such as green fluorescence protein are frequently used for cloning of sponge constructs, since loss of signal from these reporters indicates sponge function. Similar to sponges are "tough decoys" where miRNA binding sites are sandwiched between an RNA secondary structure consisting of a stem and stem-loop [118] and transcription is controlled by an RNA Polymerase III (Pol III) promoter. Recently, also endogenous forms of miRNA sponges have been identified to be expressed in human as well as rodent brain, as circular RNA transcripts containing multiple perfect binding sites for miR-7 [119].

5.2.4.5 Outlook on Future Developments in the Application of miRNAs in Cell Culture Technology

The understanding and mastering of the described miRNA tools in academic and private research groups has led to interesting insights in miRNA mediated control of growth rate, stress resistance and protein secretion of CHO cells. In addition to the currently published miRNAs (miR-7, miR-17, miR-466) high-throughput profiling experiments and literature research have unveiled a long list of miRNAs worth functional characterization in CHO cells. These results have been extensively reviewed elsewhere and were therefore not discussed in great detail in this section [85, 86, 120, 121]. It will be exciting to see when the first miRNA-based tools will enter cell culture technology and protein production in an industrial process. At this point, cell engineering strategies seem most promising: the short sequence of miRNAs would allow effective incorporation into 3'UTRs or introns of mammalian recombinant protein expression vectors, which would allow simple coexpression of growth or secretion enhancing miRNA species. The great predictive value of miRNA transcript patterns for diagnosis or treatability of human disease raises the question whether cell line selection and development could also benefit from the association of distinct miRNA patterns (biomarkers) with favorable behavior in large-scale processes. Finally several other noncoding RNAs classes are on the verge of being well understood and might show a potential for application in cell culture technology similar to that of miRNAs [122].

5.2.5 Proteomics

5.2.5.1 Background

The term "proteome", introduced by Wilkins [123], as "protein complement of the genome" describes the protein inventory of an organism, organ, tissue, or of specific cells. Even more than the transcriptome, the constitution of the proteome is highly dynamic and is displaying the state of a cell under current internal and environmental impacts. Hence, both intrinsic regulations such as cell cycle, differentiation or apoptosis, as well as the outer milieu with temperature, O_2/CO_2, nutrients, cell signals, drugs, or toxic agents have a direct influence on protein abundances, modifications, and activities. The way from coding genes to the equivalents of biologically active proteins consists of several levels of processing, transport, and regulations: epigenetic DNA modification, transcription, posttranscriptional processing (RNA-splicing etc.), RNA-silencing, nuclear export, translation, posttranslational processing, protein targeting and transport, and finally protein degradation. If for each gene a number of 5 to 6 mRNA splicing forms and 8 to 10 protein modifications is taken into account [124], one can assume that a genome of 25,000 genes represents up to one million of protein variants. Today it is possible to detect a high percentage of the protein quantities of homogenous cells or tissues by up-to-date detection and mass spectrometry (MS) identification techniques. Such profiling of differentially regulated proteins proved to be a valuable tool for the elucidation of cellular events and for target protein screenings. The (quantitative) analysis of the more than 200 known eukaryotic protein modifications (often simply called posttranslational modifications, PTMs) [125] still is a challenge, however. This section deals with classical and novel approaches in the field of cultured animal cell lines, especially producer lines such as CHO or hybridoma cell lines.

5.2.5.2 Methods

For the elucidation of quantitative changes on the proteome level of cells or tissues, techniques are necessary that cover large sets of proteins and that are highly reproducible. Two techniques stand out: gel-based methods, where a quantification or specific detection with predominantly polyacrylamic gel systems is combined with mass spectrometric protein identification, and MS-based techniques, combining MS quantification and identification of proteins. Additionally protein and peptide array techniques have to be mentioned as a growing sector of proteomics methods, however, these will not be treated here.

Sampling and Reduction of Sample Complexity

The reproducibility of sampling has a strong influence on results of proteomics experiments. Not (only) the protein quantity of a set of samples is crucial, but also the avoidance of heterogeneity between biological and technical replicates. In addition, the components of extraction buffers (buffer substances, salts, detergents, reducing agents) have to be chosen well, depending on subsequent proteomics techniques like 2-DE, labeling techniques, and mass spectrometry. Also the possibility of disturbing media components (polymers, antifoaming substances) should be kept in mind and the addition of phosphatase inhibitors, nucleases, and protease inhibitors is recommended. Of course there are diverse strategies to get rid of problematic buffer compounds later on by liquid chromatography or manual methods such as precipitation or the utilization of desalting and size-exclusion columns. However, these require additional steps, which may introduce artefacts.

In most cases the proteomic analysis of crude protein extracts from eukaryotic cells is limited to the more abundant proteins. Hence, interesting cellular compounds such as signaling proteins and transcription factors often are barely detectable for quantification or MS identification. Therefore sample fractionations, based on biochemical and physical characteristics of cellular components or proteins, are crucial for a closer view on cellular events [126]. This can be done for example by differential detergent fractionation [127] or by targeted isolation of organelles such as mitochondria [128]. Nevertheless, the risk of introducing a high technical variance with the increasing number of protein extraction steps remains. This is true even more for enrichment procedures for MS analysis of specific PTMs such as protein phosphorylations [129, 130] or glycosylation [131-133], and the fishing of protein complexes by affinity purifications [134, 135]. Depletion steps, as for instance necessary for blood or other body fluids, usually are not necessary in proteomic approaches with animal cell cultures, except if there is a disturbing effect by a high level of not secreted recombinant protein product, or in case of host cell proteins in growth media that may contain protein supplements or recombinant product.

Gel-based Proteomics Methods (2D-gel Electrophoresis)

Two-dimensional gel electrophoresis (2-DE) [136-138] on polyacrylamide gels, based on the 2 orthogonal parameters “isoelectric point” and “molecular mass”, is a main proteomics technique and is combined with MS protein identification of proteins from the gel [139]. The first dimension, isoelectric focusing, is done under denaturating conditions on strips with immobilized pH gradient (IPG). The second dimension, separation according to the apparent masses, is realized by sodium dodecyl sulphate polyacrylamide gel electrophoresis (SDS PAGE). Usually protein amounts of 50 μg up to 500 μg are used for the 2-DE gel separation. Pre- and postseparation staining is applicable for the display and quantification of protein spots (reviewed in [140]). Traditional postseparation detections are performed with silver nitrate [141], Coomassie Brilliant Blue, or different fluorescence labels [142] which allow a detection limit of a few nanograms. While staining with Coomassie Brilliant Blue or fluorescence labels is highly compatible with mass spectrometry, a specific detaining procedure is necessary for silver gels [143]. With these staining methods a protein spot detection limit of a few nanograms can be achieved. After documentation of gel images by scanners or CCD-cameras, the pictures are loaded to gel analysis programs for gel-matching, spot detection, labeling, and finally quantitative evaluation of protein expression [144].

Quantitative evaluations with reliable statistics became possible with the development of the DIGE method (differential in-gel electrophoresis) [145]. Here 2 samples are labeled with different fluorescence markers before 2-DE separation. A mixture of the samples is applied to the same 2-DE gel and separately detected. Hence, technical differences between independently processed gels are reduced and multiplexing is possible by including a mix of all samples labeled with a third dye. This internal standard for each individual protein drastically improves the reliability [146]. Remaining challenges are the identification of the same protein in multiple spots [147], due to isoforms, posttranslational modifications and degradation of proteins, as well as the fact that frequently a distinct set of proteins is found as regulated regardless of the experiment [148].

Nevertheless, even in times of increasing direct MS quantification techniques, the 2-DE approach keeps its role as a reliable and affordable method for the quantification of up to 1,500 proteins with a high dynamic range. The protein spots stay intact for a long time in the gel matrix, may be stained consecutively with specific procedures, and can be prepared for MS identification even after months. Isoforms of proteins or some PTMs are visible as pI- or mass shift, and only proteins with extreme pI (<3 and >11) or mass (>250 kDa, <10 kDa) and very hydrophobic proteins fall out of the range of the physiochemical parameters of the 2-DE procedure. Therefore 2-D- and the direct MS approaches for protein quantifications are complementary techniques.

Mass Spectrometric Protein Identification, Quantification, and Characterization

The traditional way of identification of single protein spots from 2-DE gels is performed after in-gel digestion by trypsin (or other proteases) generating a sequence dependent mix of peptides of characteristic masses [149]. In the MS measurement the masses of the tryptic fragments are determined with high accuracy. This mass pattern can be used for a so called peptide mass fingerprinting (PMF) for protein identification, i.e., the matching of the experimentally determined mass spectrum to theoretical protein digests deposited in protein databases such as Swiss-Prot/UniProt [150],

NCBI or individual in house sequence databases [151, 152]. Database searches are performed via commercial search engines such as MASCOT [153] and SEQUEST [154] or open source applications like OMSSA [155] and X!Tandem [156].

For storage, management and further evaluation of data and the convenient performance of database searches, specific proteomics data managements systems such as the commercial software ProteinScape (Bruker Daltonik), or open source packages such as ms-LIMS [157] are available. Though identifications by PMF can be simply performed with MALDI-ToF (matrix-assisted laser desorption/ionisation time-of-flight) MS instruments, a higher level of confidence can be obtained by applying MS/MS techniques [158, 159]. Here, patterns of the (tryptic) peptides are generated by fragmentation (MS/MS) of selected (MS) peptides in the mass spectrometer, predominantly at their peptide bonds. The generated spectra provide at least partial information about the amino acid sequence of a peptide and exhibit (with a sufficient peptide number and coverage of the protein) a more reliable basis for protein identifications. Today, not only MS instruments based on ESI principle (electrospray ionization) but also based on MALDI-technique, equipped with different mass analysers and detectors, are used as workhorses for MS and MS/MS measurements of (complex) proteomic samples. Detailed descriptions of different MS instrument types can be found in several reviews (e.g., [160, 161]).

Typically, ESI-instruments are used in combination with nanoHPLCs (with a flow of 200 to 300 nL/min) for so-called tandem MS measurements. In the C18 reverse phase systems, peptides elute with a gradient of increasing organic solution (predominantly acetonitrile) and are directly led to the MS instrument via a "source" that generates a drying spray in an electric field. Alternatively, MALDI variants of this principle collect the eluting peptides in time intervals on consecutive target positions so that a MALDI measurement can be performed later, independent of the nanoLC separation. Usually the generated multiple spectra of ESI- or MALDI-MS/MS are processed as one entire dataset and used for a combined protein database search to identify proteins from complex samples. Thus up to several thousand proteins can be identified in one run.

MS/MS techniques and the development of more precise, sensitive, and fast instruments improved not only the protein identification but also the detection of posttranslational modifications [162] such as phosphorylation [163, 164], acetylation [165], and glycosylation [166-169]. For many of the more than 200 enzymatic modifications of protein structures, the identification via MS is relatively straightforward as according mass-differences of a peptide can be visualized in MS or MS/MS spectra. This can be complicated by low dynamic ranges and liability of the modifications as well as by suppression of ionization, thus requiring a specific preparation. Particularly the direct MS characterization of glycan structures from intact proteins or protein digests is still a challenge [170]. With databases of glycan structures (such as glycoworkbench [171]), and novel MS strategies suitable and faster methods for the analysis of such microheterogeneities are currently being developed.

Both, MS and MS/MS spectra, are utilized for relative mass spectrometric protein quantifications, based on chemical protein- or peptide labeling or label-free approaches. ICAT (isotope-coded affinity tags [172]), ICPL (isotope-coded protein labels [173]), and iTRAQ (isobaric tag for relative and absolute quantitation [174, 175]) are well established *in vitro* labeling strategies. Further developments of labels offer the possibility to perform 4- or 8-plex approaches, allowing a simultaneous identification and quantification of proteins from up to 8 samples. For cell cultures growing in chemically defined media the metabolic/*in vivo* labeling approach via SILAC (stable isotope labeling by amino acids in cell culture) [176] offers a reliable strategy for the relative quantification of 2 or more samples. The main advantage of this method is that there is no additional variance introduced due to sample handling because the pooling is done right after cell sampling before protein extraction.

With the latest generations of mass spectrometers and according bioinformatics tools, convenient and fast label-free protein quantification procedures were evolved. Samples are measured directly after digest (without labeling) in replicate nanoLC-ESI-MS runs for relative quantification using for instance the free software "maxquant" (www.maxquant.org) or commercial packages such as the label-free MS application from Bruker Daltoniks. More complex is the determination of absolute protein quantities, requiring internal peptide or protein standards. Examples are the AQUA method (absolute quantification of proteins and phosphoproteins [177]), QconCAT [178] or absolute SILAC [179]. Finally, the targeted characterization or even quantification of protein modifications, such as phosphorylation and glycosylation, requires specific strategies and is highly labor-intensive [168, 180-183], but in combination with the detection of protein abundances completes the picture of the current state of investigated cells.

5.2.5.3 Proteomic Investigations of Animal Cell Cultures

In the last decade, increasing efforts were made to detect changes on proteome level for elucidation of the cellular background of producer cell lines (especially CHO-cells). Traditionally, many 2-DE proteomics experiments were applied for differential protein expression analysis – with more or less high variation in the number of significantly regulated proteins. This, on the one hand, is depending on the compared cell state or treatment of the culture but, on the other hand, can be traced back to technical differences such as the protein extraction procedure, pH- and mass range of gels, labeling, software evaluation, statistical calculations, and of course applied MS technique. In the following, we will discuss studies performed on protein producing cell lines, mainly hybridoma and CHO cells.

The Butyrate Effect in the Focus of Proteomic Investigations

One of the most frequently studied experimental conditions for cells used in protein production is the effect of butyrate treatment on the productivity of both, mouse hybridoma and CHO cells. The addition of butyrate is frequently used to temporarily increase the specific productivity of such industrial production cell lines [184-186]. One conclusion drawn from these studies is that changes arising in the protein processing machinery have a major impact on productivity, which is in part comparable to result observed with glucose limitation in CHO cells [187]. In [184] most regulated proteins are derived from the functional group of cytoskeleton (6) and nucleotide biosynthetic enzymes (5), followed by three proteins derived from the groups folding or translation. In [185] also GRPs and HSPs were detected in the 2DE approach (in multiple spots) next to cytoskeleton or proteins involved in translation. Notably, there is evidence for a regulation of GRP78 by HDAC and hence HDAC inhibitors [188].

With the assumption that butyrate might have an (indirect) effect on histone modification, it will be interesting to trace epigenetic shifts on DNA methylation level [189] as well as on the histone modification level [190]. Such a detailed MS characterization of protein modifications needs multiple complementary approaches to obtain the entire picture of cellular regulation, however, adequate MS techniques could provide interesting information on this topic in the near future.

Thus, in some parts the butyrate effects could be elucidated by -omics data, but the exact interplay of metabolic and regulatory elements that leads to increased protein production is little understood. Further proteome data, combined with other –omics data, from butyrate treatment, glucose limitation, and media compounds leading to higher productivity should help to find putative intersections between the different effects on CHO cells that lead to increased productivity.

Comparison of Industrially Relevant Producer Cell Lines with Respect to Growth and Productivity

In 2007, Meleady published an overview of large-scale mammalian producer cell line experiments, varying from comparison of low and high producers to experiments with media supplementation, temperature changes, or also metabolic shifts [191]. Herein reviewed works and recently published studies range from the metabolic shift of cells, observed when cells are cultivated at different steady state ratios of glucose consumption to lactate production [192] or under glucose limitation [193], the effect of the overexpression of antiapoptotic proteins [194, 195], changes associated with long-term cultivation [196], to comparisons of different levels of productivity [197]. The later study was extended by further functional analysis to find correlations between productivity and relative protein abundances in functional protein classes [198]. In a comprehensive approach, using DNA microarray, 2DE and iTRAQ data for the comparison of 11 antibody producing NSO cell lines, a more holistic view on cellular events was obtained [199]. Here, a network analysis was performed showing that protein synthesis pathways were altered in high producers at both, transcriptome and proteome levels, including genes that exhibited inconsistent trends between transcript and protein levels. Similar approaches that combine transcriptome and proteome profiling were performed to understand the biology of high productivity or high growth rates in CHO cells [71, 200]. Repeatedly, a basic lack of correlation between mRNA and protein regulations is reported. Detailed data evaluation, e.g., on the level of functional groups, finally leads to interesting insight in cellular events, and for instance a role of biosynthesis elements but also chromatin modifying genes or proteins were reported [200]. Doolan and coworkers could reduce their list of genes and proteins after comparative transcriptomic and proteomic evaluation to an overlap of 21 candidates. A detailed analysis identified, amongst others, valosin-containing protein (VCP) as an interesting key player in the difference of fast to slower growing CHO cultures [71]. A similar approach by the same group led to the identification of productivity associated genes and proteins [201]. Additional studies investigated the effect of temperature shift during production processes [202, 203], on osmoprotective effects of substances (glycine betaine) [204], the response of CHO cells to medium supplementation with hydrolysates [205], apoptosis in CHO cells during prolonged cultivation [206], host cell protein impurities [207], the effect of miRNA engineering on cells [208], as well as specific techniques for the identification of the CHO cell secretome [209].

Recent Developments

Until 2012, the major bottleneck in work with CHO proteomics was the lack of Chinese hamster or CHO sequence data for protein identification and characterization. Hence, most protein identifications until then were based on homologous proteins from other rodentia or mammals. In 2012, 3 groups working with CHO cells demonstrated that the recently available CHO specific sequence information remarkably improved MS based protein identifications of CHO cells by up to 50 %, depending on the sample source and methods used [210]. Additionally, the confidence for a lot of identifications was increased. This gives reason to expect that for instance putative intersections between butyrate and glucose limitation effects on CHO cells or other culture treatments leading to high productivity, may be clarified in near future with further –omics approaches based on a novel generation of MS instruments and bioinformatics tools on the basis of a CHO sequence database [211, 212]. Recently, a comprehensive CHO protein MS identification work was published, based on both, 2DE spot identifications and tandem MS measurements of complex protein samples after fractionation [213]. Currently, data of more than 6,000 protein identifications is being implemented at www.CHOgenome.org.

5.2.5.4 Perspectives of Proteomics of Producer Cell Lines

During the last years, various gel- and MS-based proteomics approaches yielded valuable insights in CHO cell regulation. Future techniques with improved MS instruments and evaluation software will provide tools that allow moving forward from hitherto predominantly relative protein quantifications of only a set of proteins toward comprehensive views on cellular activities. To achieve this aim, reliable gel-based approaches have to be complemented by direct MS quantification methods via labeled and unlabeled approaches with nanoLC-MS and by improved cell or protein fractionation techniques. Together optimized methods could bring future proteomics technologies closer to a level that today already is realized in genome and transcriptome analysis. Additional efforts will be necessary to elucidate protein modifications, activities, interactions, transport, and subcellular localizations, together with the development of adequate bioinformatics tools for data evaluation, integration of the complex cellular events, and modelling. Moreover, for fast quality controls of protein products, methods are required that allow direct ESI-MS analysis of intact product or fast procedures for glyco-peptide characterization during cultivation processes.

5.2.6. Integration of Different -omics Technologies

In the previous sections and chapters, different layers of regulation that are in action in a cell were presented. The challenge for the upcoming years will be the integration of these different layers into a single, comprehensive, and integrated model of the cell to enable a holistic understanding of cellular processes under culture conditions and during protein production. Such a comprehensive view of the cell and its complex regulatory network, also called systems biotechnology, should enable better control over those process relevant properties that determine the economic fate of a production process, and, maybe even more importantly, the product quality of the resulting therapeutic protein. Despite the advances achieved so far, the complexity of regulation of the mammalian genome is huge, with new layers of regulation starting to emerge only recently, such as other types of noncoding RNAs [214]. The release of genomic information for CHO cell lines has increased the possibilities enormously, making this an exciting time to be working in the area.

References

[1] Finkle B. New Medicines from Industry. J Chem Technol Biotechnol 1988;43:313-27.
[2] Aggarwal S. What's fueling the biotech engine 2010 to 2011. Nat Biotechnol 2011;29:1083-9.
[3] Wurm FM. Production of recombinant protein therapeutics in cultivated mammalian cells. Nat Biotechnol 2004;22:1393-8.
[4] Porter AJ, Racher AJ, Preziosi R, et al. Strategies for selecting recombinant CHO cell lines for cGMP manufacturing: improving the efficiency of cell line generation. Biotechnol Progr 2010;26:1455-64.

[5] Pilbrough W, Munro TP, Gray P. Intraclonal protein expression heterogeneity in recombinant CHO cells. PLoS One 2009;4:e8432.
[6] Lederberg J, McCray AT. 'Ome Sweet 'Omics--a genealogical treasury of words. Scientist 2001;15(7):8.
[7] Yang F, O'Brien PC, Ferguson-Smith MA. Comparative chromosome map of the laboratory mouse and Chinese hamster defined by reciprocal chromosome painting. Chromosome Res 2000;8(3):219-27.
[8] Bankier AT. Shotgun DNA sequencing. Methods Mol Biol 2001;167:89-100.
[9] Shendure J, Ji H. Next-generation DNA sequencing. Nat Biotechnol 2008;26(10):1135-45.
[10] Sanger F, Nicklen S, Coulson AR. DNA sequencing with chain-terminating inhibitors. Proc Natl Acad Sci USA 1977;74(12):5463-7.
[11] Simon SA, Zhai J, Nandety RS, et al. Short-read sequencing technologies for transcriptional analyses. Annu Rev Plant Biol 2009;60:305-333.
[12] Marques-Bonet T, Ryder OA, Eichler EE. Sequencing primate genomes: what have we learned? Annu Rev Genomics Hum Genet 2009;10:355-86.
[13] Xu X, Nagarajan H, Lewis NE, et al. The genomic sequence of the Chinese hamster ovary (CHO)-K1 cell line. Nat Biotechnol 2011;29(8):735-41.
[14] Hammond S, Swanberg JC, Kaplarevic M, Lee KH. Genetic sequencing and analysis of a Chinese hamster ovary cell line using Illumina sequencing technology. BMC Genomics 2011;12:67.
[15] Metzker ML. Sequencing technologies - the next generation. Nat Rev Genet 2010;11(1):31-46.
[16] Ledergerber C, Dessimoz C. Base-calling for next-generation sequencing platforms. Brief Bioinform 2011;12(5):489-97.
[17] Kumar S, Blaxter ML. Comparing de novo assemblers for 454 transcriptome data. BMC Genomics 2010,11:571.
[18] Jarvie T, Harkins T. *De novo* assembly and genomic structural variation analysis with genome sequencer FLX 3K long-tag paired end reads. Biotechniques 2008;44(6):829-31.
[19] Droege M, Hill B. The Genome Sequencer FLX System--longer reads, more applications, straight forward bioinformatics and more complete data sets. J Biotechnol 2008;136(1-2):3-10.
[20] Becker J, Hackl M, Rupp O et al. Unraveling the Chinese hamster ovary cell line transcriptome by next-generation sequencing. J Biotechnol 2011;156(3):227-35.
[21] Wlaschin KF, Nissom PM, et al. EST sequencing for gene discovery in Chinese hamster ovary cells. Biotechnol Bioen. 2005;91(5):592-606.
[22] Kantardjieff A, Nissom PM, Chuah SH, et al. Developing genomic platforms for Chinese hamster ovary cells. Biotechnol Adv 2009;27(6):1028-35.
[23] Birzele F, Schaub J, Rust W, et al. Into the unknown: expression profiling without genome sequence information in CHP by next generation sequencing. Nucleic Acids Res 2010;38(12):3999-4010.
[24] Jenkins N, Parekh RB, James DC. Getting the glycosylation right: implications for the biotechnology industry. Nat Biotechnol 1996;14(8):975-81.
[25] Cao Y, Kimura S, Itoi T, et al. Construction of BAC-based physical map and analysis of chromosome rearrangement in Chinese hamster ovary cell lines. Biotechnol Bioeng 2012;109(6):1357-67.
[26] Lewis NE, Liu X, Li Y, et al. Genomic landscapes of Chinese hamster ovary cell lines as revealed by the Cricetulus griseus draft genome. Nat Biotechnol 2013;31(8):759-65.
[27] Brinkrolf K, Rupp O, Laux H, et al. Chinese Hamster genome sequenced from sorted chromosomes. Nat Biotechnol 2013:in press.
[28] Partridge MA, Davidson MM, Hei TK. The complete nucleotide sequence of Chinese hamster (*Cricetulus griseus*) mitochondrial DNA. DNA Seq 2007;18(5):341-6.

[29] Hammond S, Kaplarevic M, Borth N, et al. Chinese Hamster Genome Database: An Online Resource for the CHO Community at www.CHOgenomics.org. Biotechnol Bioeng 2012;109:1353-6.
[30] Miller MB, Tang YW. Basic concepts of microarrays and potential applications in clinical microbiology. Clin Microbiol Rev 2009,22:611-33.
[31] Dolan PL, Wu Y, Ista LK, et al. Robust and efficient synthetic method for forming DNA microarrays. Nucleic Acids Res 2001,29:E107-107.
[32] Fodor SP, Read JL, Pirrung MC, et al. Light-directed, spatially addressable parallel chemical synthesis. Science 1991,251:767-73.
[33] Dalma-Weiszhausz DD, Warrington J, Tanimoto EY, et al. The affymetrix GeneChip platform: an overview. Methods Enzymol 2006,410:3-28.
[34] Sasik R, Woelk CH, Corbeil J. Microarray truths and consequences. J Mol Endocrinol 2004,33:1-9.
[35] Tomiuk S, Hofmann K. Microarray probe selection strategies. Brief Bioinform 2001,2:329-40.
[36] Schena M, Shalon D, Davis RW, et al. Quantitative monitoring of gene expression patterns with a complementary DNA microarray. Science 1995,270:467-70.
[37] Tang T, Francois N, Glatigny A, et al. Expression ratio evaluation in two-colour microarray experiments is significantly improved by correcting image misalignment. Bioinformatics 2007,23:2686-91.
[38] Schulze A, Downward J. Navigating gene expression using microarrays--a technology review. Nat Cell Biol 2001,3:E190-5.
[39] Muyal JP, Singh SK, Fehrenbach H. DNA-microarray technology: comparison of methodological factors of recent technique towards gene expression profiling. Crit Rev Biotechnol 2008,28:239-51.
[40] Allison DB, Cui X, Page GP, et al. Microarray data analysis: from disarray to consolidation and consensus. Nat Rev Genet 2006,7:55-65.
[41] Okoniewski MJ, Miller CJ. Hybridization interactions between probesets in short oligo microarrays lead to spurious correlations. BMC Bioinformatics 2006,7:276.
[42] Tseng GC, Oh MK, Rohlin L, et al. Issues in cDNA microarray analysis: quality filtering, channel normalization, models of variations and assessment of gene effects. Nucleic Acids Res 2001;29:2549-57.
[43] Marioni JC, Mason CE, Mane SM, et al. RNA-seq: an assessment of technical reproducibility and comparison with gene expression arrays. Genome Res 2008,18:1509-17.
[44] Vera JC, Wheat CW, Fescemyer HW, et al. Rapid transcriptome characterization for a nonmodel organism using 454 pyrosequencing. Mol Ecol 2008,17:1636-47.
[45] Wang Z, Gerstein M, Snyder M. RNA-Seq: a revolutionary tool for transcriptomics. Nat Rev Genet 2009,10:57-63.
[46] Pickrell JK, Marioni JC, Pai AA, et al. Understanding mechanisms underlying human gene expression variation with RNA sequencing. Nature 2010,464:768-72.
[47] Mortazavi A, Williams BA, McCue K, et al. Mapping and quantifying mammalian transcriptomes by RNA-Seq. Nat Meth 2008,5:621-8.
[48] Sultan M, Schulz MH, Richard H, et al. A global view of gene activity and alternative splicing by deep sequencing of the human transcriptome. Science 2008,321:956-60.
[49] Wang Y, Ghaffari N, Johnson CD, et al. Evaluation of the coverage and depth of transcriptome by RNA-Seq in chickens. BMC Bioinformatics 2011,12(10):S5.
[50] Cloonan N, Forrest AR, Kolle G, et al. Stem cell transcriptome profiling via massive-scale mRNA sequencing. Nat Meth 2008,5:613-9.
[51] Ozsolak F, Milos PM. RNA sequencing: advances, challenges and opportunities. Nat Rev Genet 2011,12:87-98.

[52] Ozsolak F, Milos PM. Single-molecule direct RNA sequencing without cDNA synthesis. Wiley Interdiscip Rev RNA 2011,2:565-70.
[53] Chen D, Patton JT. Reverse transcriptase adds nontemplated nucleotides to cDNAs during 5'-RACE and primer extension. Biotechniques 2001,30:574-580-2.
[54] Roberts JD, Preston BD, Johnston LA, et al. Fidelity of two retroviral reverse transcriptases during DNA-dependent DNA synthesis in vitro. Mol Cell Biol 1989,9:469-76.
[55] Ozsolak F, Platt AR, Jones DR, et al. Direct RNA sequencing. Nature 2009,461:814-8.
[56] Kogenaru S, Qing Y, Wang N. RNA-seq and microarray complement each other in transcriptome profiling. Bmc Genomics 2012;13,629.
[57] Minguez P, Dopazo J. Assessing the biological significance of gene expression signatures and co-expression modules by studying their network properties. PLoS One 2011,6:e17474.
[58] Jaluria P, Chu C, Betenbaugh M, et al. Cells by design: a mini-review of targeting cell engineering using DNA microarrays. Mol Biotechnol 2008,39:105-11.
[59] Gupta P, Lee KH. Genomics and proteomics in process development: opportunities and challenges. Trends Biotechnol 2007,25:324-30.
[60] Griffin TJ, Seth G, Xie H, et al. Advancing mammalian cell culture engineering using genome-scale technologies. Trends Biotechnol 2007,25:401-8.
[61] Korke R, Gatti Mde L, Lau AL, et al. Large scale gene expression profiling of metabolic shift of mammalian cells in culture. J Biotechnol 2004,107:1-17.
[62] Khoo SH, Falciani F, Al-Rubeai M. A genome-wide transcriptional analysis of producer and non-producer NS0 myeloma cell lines. Biotechnol Appl Biochem 2007,47:85-95.
[63] Swiderek H, Al-Rubeai M. Functional genome-wide analysis of antibody producing NS0 cell line cultivated at different temperatures. Biotechnol Bioeng 2007,98:616-30.
[64] Baik JY, Lee MS, An SR, et al. Initial transcriptome and proteome analyses of low culture temperature-induced expression in CHO cells producing erythropoietin. Biotechnol Bioeng 2006,93:361-71.
[65] Yee JC, Gerdtzen ZP, Hu WS. Comparative transcriptome analysis to unveil genes affecting recombinant protein productivity in mammalian cells. Biotechnol Bioeng 2009,102:246-63.
[66] Shen D, Sharfstein ST. Genome-wide analysis of the transcriptional response of murine hybridomas to osmotic shock. Biotechnol Bioeng 2006,93:132-45.
[67] De Leon Gatti M, Wlaschin KF, Nissom PM, et al. Comparative transcriptional analysis of mouse hybridoma and recombinant Chinese hamster ovary cells undergoing butyrate treatment. J Biosci Bioeng 2007,103:82-91.
[68] Yee JC, de Leon Gatti M, Philp RJ, et al. Genomic and proteomic exploration of CHO and hybridoma cells under sodium butyrate treatment. Biotechnol Bioeng 2008,99:1186-204.
[69] Wong DC, Wong KT, Lee YY, et al. Transcriptional profiling of apoptotic pathways in batch and fed-batch CHO cell cultures. Biotechnol Bioeng 2006,94:373-82.
[70] Wong DC, Wong KT, Nissom PM, et al. Targeting early apoptotic genes in batch and fed-batch CHO cell cultures. Biotechnol Bioeng 2006,95:50-361.
[71] Doolan P, Meleady P, Barron N, et al. Microarray and proteomics expression profiling identifies several candidates, including the valosin-containing protein (VCP), involved in regulating high cellular growth rate in production CHO cell lines. Biotechnol Bioeng 2010,106:42-56.
[72] Jaluria P, Betenbaugh M, Konstantopoulos K, et al. Enhancement of cell proliferation in various mammalian cell lines by gene insertion of a cyclin-dependent kinase homolog. BMC Biotechnol 2007,7:71.
[73] Allison DW, Aboytes KA, Fong DK, et al. Development and optimization of cell culture media-genomic and proteomic approaches. BioProcess Int 2005,3:2-7.

[74] Jaluria P, Konstantopoulos K, Betenbaugh M, et al. Egr1 and Gas6 facilitate the adaptation of HEK-293 cells to serum-free media by conferring enhanced viability and higher growth rates. Biotechnol Bioeng 2008,99:1443-52.
[75] Wong VV, Nissom PM, Sim SL, et al. Zinc as an insulin replacement in hybridoma cultures. Biotechnol Bioeng 2006,93:553-63.
[76] Bailey JE, Sburlati A, Hatzimanikatis V, et al. Inverse metabolic engineering: a strategy for directed genetic engineering of useful phenotypes. Biotechnol Bioeng 2002,79:568-79.
[77] Kuystermans D, Krampe B, Swiderek H, et al. Using cell engineering and omic tools for the improvement of cell culture processes. Cytotechnology 2007,53:3-22.
[78] Hernandez Bort JA, Hackl M, Höflmayer H, et al. Dynamic mRNA and miRNA profiling of CHO-K1 suspension cell cultures. Biotechnol J 2012,7:500-15.
[79] Carlage T, Kshirsagar R, Zang L, et al. Analysis of dynamic changes in the proteome of a Bcl-XL overexpressing Chinese Hamster Ovary cell culture during exponential and stationary phases. Biotechnol Progr 2012;28(3):814-23.
[80] Wong DCF, Wong NSC, Goh JSY, et al. Profiling of N-glycosylation gene expression in CHO cell fed-batch cultures. Biotechnol Bioeng 2010;107(3):516-28.
[81] Hobert O. Gene regulation by transcription factors and miRNAs. Science 2008;319(5871):1785-16.
[82] Mallanna SK, Rizzino A. Emerging roles of miRNAs in the control of embryonic stem cells and the generation of pluripotent stem cells. Dev Biol 2010;344(1):16-25.
[83] Madanecki P, Kapoor N, Bebok Z, et al. Regulation of angiogenesis by hypoxia: the role of miRNA. Cell Mol Biol Lett 2013;18(1):47-57.
[84] Müller D, Katinger H, Grillari J. MiRNAs as targets for engineering of CHO cell factories. Trends Biotechnol 2008;26(7):359-65.
[85] Hackl M, Borth N, Grillari J. miRNAs – pathway engineering of CHO cell factories that avoids translational burdening. Trends Biotechnol 2012;30(8):405-6.
[86] Shukla GC, Singh J, Barik S. miRNAs: Processing, maturation, target recognition and regulatory functions. Mol Cell Pharmacol 2011;3(3):83-92.
[87] Griffiths-Jones S. mirBase: miRNA sequences and annotation. Curr Protoc Bioinformatics 2010;12:12.9.1-10.
[88] Flynt AS, Lai EC. Biological principles of miRNA-mediated regulation: shared themes amid diversity. Nat Rev Genet 2008;9(11):31-842.
[89] Chi SW, Zang JB, Mele A, et al. Argonaute HITS-CLIP decodes miRNA-mRNA interaction maps. Nature 2009;460(7254):479-86.
[90] Jing Q, Huang S, Guth S, et al. Involvement of miRNA in AU-rich element-mediated mRNA in AU-rich element-mediated mRNA instability. Cell 2005;12(5):623-34.
[91] Friedman RC, Farh KK, Burge CB, et al. Most mammalian mRNAs are conserved targets of miRNAs. Genome Res 2009;19(1):92-105.
[92] Arvey A, Larsson E, Sander C, et al. Target mRNA abundance dilutes miRNA and siRNA activity. Mol Syst Biol 2010;6:363.
[93] Mullokandov G, Baccarini A, Ruzo A, et al. High-throughput assessment of miRNA activity and function using miRNA sensor and decoy libraries. Nat Methods 2012;9(8):840-6.
[94] Lagos-Quintana M, Rauhut R, Yalcin A, et al. Identification of tissue-specific miRNAs from mouse. Curr Biol 2002;12(9):735-9.
[95] Medina PP, Nolde M, Slack FJ. OncomiR addiction in an in vivo model of miRNA-21-induced pre-B-cell lymphoma. Nature 2010;467(7311):86-90.
[96] Dellago H, Preschitz-Kammerhofer B, Terlecki-Zaniewicz L, et al. High levels of oncomiR-21 contribute to the senescence-induced growth arrest in normal human cells and its knock-down increases the replicative lifespan. Aging Cell 2013(12):3:446-58.

[97] Meyer SU, Pfaffl MW, Ulbrich SE. Normalization strategies for miRNA profiling experiments: a 'normal' way to a hidden layer of complexity? Biotechnol Lett 2010;32(12):1777-88.
[98] Castoldi M, Schmidt S, Benes V, et al. miChip: an array-based method for miRNA expression profiling using locked nucleic acid capture probes. Nat Protoc 2008;3(2):321-29.
[99] Beuvink I, Kolb FA, Budach, W et al. A novel microarray approach reveals new tissue-specific signatures of known and predicted mammalian miRNAs. Nucleic Acids Res 2007;35(7):e52.
[100] Schmittgen TD, Lee EJ, Jiang J, et al. Real-time PCR quantification of precursor and mature miRNA. Methods 2008;44(1):31-8.
[101] Chen Y, Gelfond JA, McManus LM, et al. Reproducibility of quantitative RT-PCR array in miRNA expression profiling and comparison with microarray analysis. BMC Genomics 2009;10:407.
[102] Bustin SA, Benes V, Garson JA, et al. The MIQE guidelines: minimum information for publication of quantitative real-time PCR experiments. Clin Chem 2009;55(4):611-22.
[103] Nelson PT, Baldwin DA, Scearce LM, et al. Microarray-based, high-throughput gene expression profiling of miRNAs. Nat Methods 2004;1(2):155-61.
[104] Thomson JM, Parker J, Perou CM, et al. A custom microarray platform for analysis of miRNA gene expression. Nat Methods 2004;1(1):47-53.
[105] Git A, Dvinge H, Salmon-Divon M, et al. Systematic comparison of microarray profiling, real-time PCR, and next-generation sequencing technologies for measuring differential miRNA expression. RNA 2010;16(5):991-1006.
[106] Pradervand S, Weber J, Lemoine F, et al. Concordance among digital gene expression, microarrays, and qPCR when measuring differential expression of miRNAs. Biotechniques 2010;48(3):219-22.
[107] Nonne N, Ameyar-Zazoua M, Souidi M, et al. Tandem affinity purification of miRNA target mRNAs (TAP-Tar). Nucleic Acid Res 2010;38(4):e20.
[108] Licatolosi DD, Mele A, Fak JJ, et al. HITS-CLIP yields genome-wide insights into brain alternative RNA processing. Nature 2008;456(7221):464-9.
[109] Haecker I, Gay LA, Yang Y, et al. Ago HITS-CLIP expands understanding of Kaposi's sarcoma-associated herpesvirus miRNA function in primary effusion lymphomas. PLOS Pathol 2012;8(8):e1002884.
[110] Hsieh WJ, Wang H. Human miRNA target identification by RRSM. J Theor Biol 2011;286(1):79-84.
[111] Gammell P, Barron N, Kumar N, et al. Initial identification of low temperature and culture stage induction of miRNA expression in suspension CHO-K1 cells. J Biotechnol 2007;130(3):213-8.
[112] Hackl M, Jakobi T, Blom J, et al. Next-generation sequencing of the Chinese hamster ovary miRNA transcriptome: Identification, annotation and profiling of miRNAs as targets for cellular engineering. J Biotechnol 2011;153(1-2):62-75.
[113] Johnson KC, Jacob NM, Nissom PM, et al. Conserved miRNAs in Chinese hamster ovary cell lines. Biotechnol Bioeng 2011;108(2):475-80.
[114] Hackl M, Jadhav V, Jakobi T, et al. Computational identification of miRNA gene loci and precursor miRNA sequences in CHO cell lines. J Biotechnol 2012;158(3):151-5.
[115] Jadhav V, Hackl M, Hernandez Bort JA, et al. A screening method to assess biological effects of miRNA overexpression in Chinese hamster ovary cells. Biotechnol Bioeng 2012;10(6):1376-85.
[116] Klanert G, Jadhav V, Chanoumidou K, et al. Endogenous miRNA clusters outperform chimeric sequence Clusters in Chinese Hamster Ovary cells. Biotechnol J 2013;9(4):538-44.
[117] Ebert MS, Sharp PA. MiRNA sponges: progress and possibilities. RNA 2010;16(11):2043-50.
[118] Haraguchi T, Ozaki O, Iba H. Vectors expressing efficient RNA decoys achieve the long-term suppression of specific miRNA activity in mammalian cells. Nucleic Acids Res 2009;37(6):e43.
[119] Hansen TB, Jensen TI, Clausen BH, et al. Circular RNAs are a large class of animal RNAs with regulatory potency. Nature 2013;495(7441):333-8.

[120] Barron N, Sanchez N, Kelly P, et al. miRNAs: tiny targets for engineering CHO cell phenotypes? Biotechnol Lett 2011;33:11-21.

[121] Jadhav V, Hackl M, Druz A, et al. CHO miRNA engineering is growing up: Recent successes and future challenges. 2013;31(8):1501-13.

[122] Gerstl MP, Hackl M, Graf AB, et al. Prediction of transcribed PIWI-interacting RNAs from CHO RNAseq data. J Biotechnol 2013;166(1-2):51-7.

[123] Wilkins MR, Sanchez JC, Gooley AA, et al. Progress with proteome projects: why all proteins expressed by a genome should be identified and how to do it. Biotechnol. Genet Eng Rev 1996;13:19-50.

[124] Jensen ON. Modification-specific proteomics: characterization of post-translational modifications by mass spectrometry. Curr Opin Chem Biol 2004,8:33-41.

[125] Walsh CT, Garneau-Tsodikova S, Gatto GJ. Protein posttranslational modifications: the chemistry of proteome diversifications. Angew Chem Int Ed Engl 2005;44:7342-72.

[126] Dreger M. Proteome analysis at the level of subcellular structures. Eur J Biochem 2003;270:589-99.

[127] Simpson RJ. Proteins and proteomics - A laboratory manual. Cold Spring Harbor Laboratory Press; 2003.

[128] Schräder E, Scholz S, Sandig V, et al. Analysis of the mitochondrial subproteome of the human cell line AGE1.HN – a contribution to a systems biology approach. BMC Proc 2011;5(8):P86.

[129] Boersema PJ, Mohammed S, Heck AJR. Phosphopeptide fragmentation and analysis by mass spectrometry. J Mass Spectrom 2009;44:861-78.

[130] Schulze WX. Proteomics approaches to understand protein phosphorylation in pathway modulation. Curr Opin Plant Biol 2010;13:280-7.

[131] Ongay S, Boichenko A, Govorukhina N, et al. Glycopeptide enrichment and separation for protein glycosylation analysis. J Sep Sci 2012;35:2341-72.

[132] Wuhrer M. Glycomics using mass spectrometry. Glycoconj J 2013;30:11-22.

[133] Thaysen-Andersen M, Mysling S, Højrup P. Site-specific glycoprofiling of N-linked glycopeptides using MALDI-TOF MS. strong correlation between signal strength and glycoform quantities. Anal Chem 2009;81:3933-43.

[134] Jankowski A, Zhu P, Marshall JG. Capture of an activated receptor complex from the surface of live cells by affinity receptor chromatography. Anal Biochem 2008;380:235-48.

[135] Bantscheff M, Hopf C, Savitski MM, et al. Chemoproteomics profiling of HDAC inhibitors reveals selective targeting of HDAC complexes. Nat Biotechnol 2011;29:255-65.

[136] O'Farrell PH. High resolution two-dimensional electrophoresis of proteins. J Biol Chem 1975;250:4007-21.

[137] Klose J. Protein mapping by combined isoelectric focusing and electrophoresis of mouse tissues. A novel approach to testing for induced point mutations in mammals. Humangenetik 1975;26:231-43.

[138] Bjellqvist B, Ek K, Righetti PG, et al. Isoelectric focusing in immobilized pH gradients: principle, methodology and some applications. J Biochem Biophys Methods 1982;6:317-39.

[139] Patterson SD, Aebersold RH. Proteomics: the first decade and beyond. Nat Genet 2003;33:311-23.

[140] Berth M, Moser FM, Kolbe M, et al. The state of the art in the analysis of two-dimensional gel electrophoresis images. Appl Microbiol Biotechnol 2007;76:1223-43.

[141] Blum H, Beier H, Gross HJ. Improved silver staining of plant proteins, RNA and DNA in polyacrylamide gels. Electrophoresis 1987;8:93-9.

[142] Berth M, Moser FM, Kolbe M, et al. The state of the art in the analysis of two-dimensional gel electrophoresis images. Appl Microbiol Biotechnol 2007;76:1223-43.

[143] Gharahdaghi F, Weinberg CR, Meagher DA, et al. Mass spectrometric identification of proteins from silver-stained polyacrylamide gel: a method for the removal of silver ions to enhance sensitivity. Electrophoresis 1999;20:601-5.
[144] Berth M, Moser FM, Kolbe M, et al. The state of the art in the analysis of two-dimensional gel electrophoresis images. Appl Microbiol Biotechnol 2007;76:1223-43.
[145] Unlü M, Morgan ME, Minden JS. Difference gel electrophoresis: a single gel method for detecting changes in protein extracts. Electrophoresis 1997;18:2071-7.
[146] Freeman WM, Hemby SE. Proteomics for protein expression profiling in neuroscience. Neurochem Res 2004;29:1065-81.
[147] Baik JY, Joo EJ, Kim YH, et al. Limitations to the comparative proteomic analysis of thrombopoietin producing Chinese hamster ovary cells treated with sodium butyrate. J Biotechnol 2008;133:461-8.
[148] Petrak J, Ivanek R, Toman O, et al.. Déjà vu in proteomics. A hit parade of repeatedly identified differentially expressed proteins. Proteomics 2008;8:1744-9.
[149] Granvogl B, Plöscher M, Eichacker LA. Sample preparation by in-gel digestion for mass spectrometry-based proteomics. Anal Bioanal Chem 2007;389:991-1002.
[150] Wu CH, Apweiler R, Bairoch A, et al. The Universal Protein Resource (UniProt): an expanding universe of protein information. Nucleic Acids Res. 2006;34:D187-91.
[151] Glish GL, Vachet RW. The basics of mass spectrometry in the twenty-first century. Nat Rev Drug Discov 2003;2:140-50.
[152] Steen H, Mann M. The ABC's (and XYZ's) of peptide sequencing Nat Rev Mol Cell Biol 2004;5:699-711.
[153] Perkins DN, Pappin DJ, Creasy DM, et al. Probability-based protein identification by searching sequence databases using mass spectrometry data. Electrophoresis 1999;20:3551-67.
[154] MacCoss MJ, Wu CC, Yates JR. Probability-based validation of protein identifications using a modified SEQUEST algorithm. Anal Chem 2002;74:5593-9.
[155] Geer LY, Markey SP, Kowalak JA, et al. Open mass spectrometry search algorithm. J Proteome Res 2004;3:958-64.
[156] Fenyö D, Beavis RC. A method for assessing the statistical significance of mass spectrometry-based protein identifications using general scoring schemes. Anal Chem 2003;75:768-74.
[157] Helsens K, Colaert N, Barsnes H, et al. ms_lims, a simple yet powerful open source laboratory information management system for MS-driven proteomics. Proteomics 2010;10:1261-4.
[158] Glish GL, Vachet RW. The basics of mass spectrometry in the twenty-first century. Nat Rev Drug Discov 2003;2:140-50.
[159] Steen H, Mann M. The ABC's (and XYZ's) of peptide sequencing. Nat Rev Mol Cell Biol 2004;5:699-711.
[160] Domon B, Aebersold R. Mass spectrometry and protein analysis. Science 2006;312:212-7.
[161] Yates JR, Ruse CI, Nakorchevsky A. Proteomics by mass spectrometry: approaches, advances, and applications. Annu Rev Biomed Eng 2009;11:49-79.
[162] Walsh CT, Garneau-Tsodikova S, Gatto GJ. Protein posttranslational modifications. the chemistry of proteome diversifications. Angew. Chem Int Ed Engl 2005;44:7342-72.
[163] Boersema PJ, Mohammed S, Heck AJR. Phosphopeptide fragmentation and analysis by mass spectrometry. J Mass Spectrom 2009;44:861-78.
[164] Schulze WX. Proteomics approaches to understand protein phosphorylation in pathway modulation. Curr Opin Plant Biol 2010;13:280-7.
[165] Cuomo A, Moretti S, Minucci S, et al. SILAC-based proteomic analysis to dissect the "histone modification signature" of human breast cancer cells. Amino Acids 2011;41:387-99.
[166] Ongay S, Boichenko A, Govorukhina N, et al. Glycopeptide enrichment and separation for protein glycosylation analysis. J Sep Sci 2012;35:2341-72.

[168] Wuhrer M. Glycomics using mass spectrometry. Glycoconj J 2013;30:11-22.
[169] Thaysen-Andersen M, Mysling S, Højrup P. Site-specific glycoprofiling of N-linked glycopeptides using MALDI-TOF MS: strong correlation between signal strength and glycoform quantities. Anal Chem 2009;81:3933-43.
[170] Wuhrer M. Glycomics using mass spectrometry. Glycoconj J 2013;30:11-22.
[171] Damerell D, Ceroni A, Maass K, et al. The GlycanBuilder and GlycoWorkbench glycoinformatics tools: updates and new developments. Biol Chem 2012;393:1357-62.
[172] Gygi SP, Rist B, Gerber SA, etr al. Quantitative analysis of complex protein mixtures using isotope-coded affinity tags. Nat Biotechnol 1999;17:994-9.
[173] Schmidt A, Kellermann J, Lottspeich F. A novel strategy for quantitative proteomics using isotope-coded protein labels. Proteomics 2005;5:4-15.
[174] Ross PL, Huang YN, Marchese JN, et al. Multiplexed protein quantitation in Saccharomyces cerevisiae using amine-reactive isobaric tagging reagents. Mol. Cell Proteomics 2004;3:1154-69.
[175] Burkhart JM, Vaudel M, Zahedi RP, et al. iTRAQ protein quantification: a quality-controlled workflow. Proteomics 2011;11:1125-34.
[176] Ong S, Mann M. Stable isotope labeling by amino acids in cell culture for quantitative proteomics. Meth Mol Biol 2007;359:37-52.
[177] Gerber SA, Rush J, Stemman O, et al. Absolute quantification of proteins and phosphoproteins from cell lysates by tandem MS. Proc Natl Acad Sci USA 2003;100:6940-5.
[178] Pratt JM, Simpson DM, Doherty MK, et al. Multiplexed absolute quantification for proteomics using concatenated signature peptides encoded by QconCAT genes. Nat Protoc 2006;1:1029-43.
[179] Hanke S, Besir H, Oesterhelt D, et al. Absolute SILAC for accurate quantitation of proteins in complex mixtures down to the attomole level. J Proteome Res 2008;7:1118-30.
[180] Boersema PJ, Mohammed S, Heck AJR. Phosphopeptide fragmentation and analysis by mass spectrometry. J Mass Spectrom 2009;44:861-78.
[181] Schulze WX. Proteomics approaches to understand protein phosphorylation in pathway modulation Curr Opin Plant Biol 2010;13:280-7.
[182] Thaysen-Andersen M, Mysling S, Højrup P. Site-specific glycoprofiling of N-linked glycopeptides using MALDI-TOF MS: strong correlation between signal strength and glycoform quantities. Anal Chem 2009;81:3933-43.
[183] Cuomo A, Moretti S, Minucci S, et al. SILAC-based proteomic analysis to dissect the "histone modification signature" of human breast cancer cells. Amino Acids 2011;41:387-99.
[184] van Dyk DD, Misztal DR, Wilkins MR, et al. Identification of cellular changes associated with increased production of human growth hormone in a recombinant Chinese hamster ovary cell line. Proteomics 2003;3:147-56.
[185] Yee JC, Gerdtzen ZP, Hu W. Comparative transcriptome analysis to unveil genes affecting recombinant protein productivity in mammalian cells. Biotechnol Bioeng 2009;102:246-63.
[186] Baik JY, Lee GM. A DIGE approach for the assessment of differential expression of the CHO proteome under sodium butyrate addition: Effect of Bcl-x(L) overexpression. Biotechnol Bioeng 2010;105:358-67.
[187] Wingens M, Gätgens J, Hoffrogge R, et al. Proteomic characterization of a glucose-limited CHO perfusion process-analysis of metabolic changes and increase in productivity. Springer; 2010:265-9.
[188] Baumeister P, Dong D, Fu Y, et al. Transcriptional induction of GRP78/BiP by histone deacetylase inhibitors and resistance to histone deacetylase inhibitor-induced apoptosis. Mol Cancer Ther 2009;8:1086-94.

[189] Wippermann A, Klausing S, Rupp O, et al. Establishment of a CpG island microarray for analyses of genome-wide DNA methylation in Chinese hamster ovary cells. Appl Microbiol Biotechnol 2014;98(2):579-89.

[190] Bartke T, Vermeulen M, Xhemalce B, et al. Nucleosome-interacting proteins regulated by DNA and histone methylation. Cell 2010;143:470-84.

[191] Meleady P. Proteomic profiling of recombinant cells from large-scale mammalian cell culture processes. Cytotechnology 2007;53:23-31.

[192] Korke R, Gatti MdL, Lau ALY, et al. Large scale gene expression profiling of metabolic shift of mammalian cells in culture. J Biotechnol 2004;107:1-17.

[193] Wingens M, Gätgens J, Hoffrogge R, et al. Proteomic characterization of a glucose-limited CHO perfusion process-analysis of metabolic changes and increase in productivity. Springer; 2010:265-9.

[194] Baik JY, Lee GM. A DIGE approach for the assessment of differential expression of the CHO proteome under sodium butyrate addition. Effect of Bcl-x(L) overexpression. Biotechnol Bioeng 2010;105:358-67.

[195] Carlage T, Hincapie M, Zang L, et al. Proteomic profiling of a high-producing Chinese hamster ovary cell culture. Anal Chem 2009;81:7357-62.

[196] Beckmann TF, Krämer O, Klausing S, et al. Effects of high passage cultivation on CHO cells: a global analysis. Appl. Microbiol Biotechnol 2012;94:659-71.

[197] Smales CM, Dinnis DM, Stansfield SH, et al. Comparative proteomic analysis of GS-NS0 murine myeloma cell lines with varying recombinant monoclonal antibody production rate. Biotechnol Bioeng 2004;88:474-88.

[198] Dinnis DM, Stansfield SH, Schlatter S, et al. Functional proteomic analysis of GS-NS0 murine myeloma cell lines with varying recombinant monoclonal antibody production rate. Biotechnol Bioeng 2006;94:830-41.

[199] Seth G, Philp RJ, Lau A, et al. Molecular portrait of high productivity in recombinant NS0 cells. Biotechnol Bioeng 2007;97:933-51.

[200] Nissom PM, Sanny A, Kok YJ, et al. Transcriptome and proteome profiling to understanding the biology of high productivity CHO cells. Mol Biotechnol 2006;34:125-40.

[201] Clarke C, Doolan P, Barron, et al. Predicting cell-specific productivity from CHO gene expression. J Biotechnol 2011;151:2:159-65.

[202] Baik JY, Lee MS, An SR, et al. Initial transcriptome and proteome analyses of low culture temperature-induced expression in CHO cells producing erythropoietin. Biotechnol Bioeng 2006;93:361-71.

[203] Kantardjieff A, Jacob NM, Yee JC, et al. Transcriptome and proteome analysis of Chinese hamster ovary cells under low temperature and butyrate treatment. J Biotechnol 2010;145:143-59.

[204] Kim JY, Kim Y, Lee GM. Differential in-gel electrophoresis (DIGE) analysis of CHO cells under hyperosmotic pressure: osmoprotective effect of glycine betaine addition. Biotechnol Bioeng 2012;109:1395-403.

[205] Kim JY, Kim Y, Han YK, et al. Proteomic understanding of intracellular responses of recombinant Chinese hamster ovary cells cultivated in serum-free medium supplemented with hydrolysates. Appl Microbiol Biotechnol 2011;89:1917-28.

[206] Wei YC, Naderi S, Meshram M, et al. Proteomics analysis of chinese hamster ovary cells undergoing apoptosis during prolonged cultivation. Cytotechnology 2011;63:663-77.

[207] Krawitz DC, Forrest W, Moreno GT, et al. Proteomic studies support the use of multi-product immunoassays to monitor host cell protein impurities. Proteomics 2006;6:94-110.

[208] Meleady P, Gallagher M, Clarke C, et al. Impact of miR-7 over-expression on the proteome of Chinese hamster ovary cells. J Biotechnol 2012;160:251-62.

[209] Slade PG, Hajivandi M, Bartel CM, et al. Identifying the CHO secretome using mucin-type O-linked glycosylation and click-chemistry. J Proteome Res 2012;11:6175-86.
[210] Meleady P, Hoffrogge R, Henry M, et al. Utilization and evaluation of CHO-specific sequence databases for mass spectrometry based proteomics. Biotechnol Bioeng 2012;109:1386-94.
[211] Gupta P, Lee KH. Genomics and proteomics in process development: opportunities and challenges. Trends Biotechnol 2007;25:324-30.
[212] Kildegaard HF, Baycin-Hizal D, Lewis NE, et al. The emerging CHO systems biology era: harnessing the 'omics revolution for biotechnology. Curr Opin Biotechnol 2013;24(6):1102-7.
[213] Baycin-Hizal D, Tabb DL, Chaerkady R, et al. Proteomic analysis of Chinese Hamster Ovary cells. J Proteome Res 2012;11(11):5265-76.
[214] Kornienko AE, Guenzl PM, Barlow DP, et al. Gene regulation by the act of long non-coding RNA transcription. BMC Biology 2013;11:59.
[215] Benson DA, Cavanaugh M, Clark K, et al. GenBank. Nucleic Acids Res 2013;41(Database issue):D36-42.
[216] Lowe TM, Eddy SR. tRNAscan-SE: A program for improved detection of transfer RNA genes in genomic sequence. Nucl Acids Res 1997;25:955-64.
[217] Ashburner M, Ball CA, Blake JA, et al. Gene ontology: tool for the unification of biology. Nat Genet 2000;25(1):25-9.
[218] Kozomara A, Griffiths-Jones S. miRBase: integrating miRNA annotation and deep-sequencing data. NAR 2011;39(Database Issue):D152-D157
[219] Amaral PP, Clark MB, Gascoigne DK, et al. lncRNAdb: a reference database for long noncoding RNAs. Nuc Acid Res 2011;39(1):D146-D151.
[220] Quast C, Pruesse E, Yilmaz P, et al. The SILVA ribosomal RNA gene database project: improved data processing and web-based tools. Nucleic Acids Res 2013;41(D1):D590-D596.

5.3 Nutrient Media for Cell Culture Technology

Claudia Altamirano, Julio Berrios, and Mauricio Vergara

5.3.1 Components of a Typical Culture Medium

A suitable nutrient medium for mammalian cell cultures contains a wide variety of nutrients and must be able to fulfil the cell requirements. In contrast to many microbes that are able to grow in minimal media consisting of carbon and nitrogen sources and salts, mammalian cells additionally require amino acids, lipids, vitamins, and often proteins and growth factors. Since the depletion of some nutrients may trigger apoptosis [1, 2] and reduce productivity, the appropriate nutrient supply must be assured in order to maintain both, culture viability and productivity.

Although every cell type has some specific nutritional requirements that have led to developing several formulations, there are certain nutrients that are normally required by all mammalian cell cultures and are part of the basal media. Eagle's basal medium (BME) or Dulbecco's modified Eagle's medium (DMEM) are some examples of these minimum essential formulations. Sometimes, more specific media can be developed from the basal media by adding other components such as hormones, growth factors, proteins, and/or serum. The following components are part of most formulations and will be described briefly.

5.3.1.1 Water

The water use for media preparation for animal cell culture must fulfil a minimum purity level in order to prevent detrimental effects on cell growth and productivity, but also to avoid undesirable effects on the quality standard of the recombinant protein. Thus, some potentially detrimental components must be removed before the medium preparation. Purification stages normally consist of up to 4 stages including distillation or reverse osmosis, deionisation and micro-filtration (0.02 mm pore size).

5.3.1.2 Glucose

Glucose is added to most formulations as a carbon and energy source (typically 5 to 20 mmol L^{-1}). It is metabolised through glycolysis into pyruvate and provides the carbon flux to several pathways, such as pentose phosphate pathway (PPP) and tricarboxylic acid (TCA) cycle. However, part of pyruvate is reduced to lactate and excreted to the culture medium. This situation causes a lack of energy supply that must be fulfilled by an additional energy source, normally glutamine (2 to 5 mmol L^{-1}), as mentioned below. The lactate accumulation over 20 mmol L^{-1} also may cause growth inhibition or even undesirable production effects. Some studies have proposed to replace glucose by other sugars such as mannose [3, 4] or galactose [5] in order to reduce lactate production but without detrimental effects on the process productivity. A deeper approach of this point will be shown below.

5.3.1.3 Amino Acids

Media formulation normally include at least all 13 essential amino acids for mammalian cells (Arg, Cys, Gln, His, Ile, Leu, Lys, Met, Phe, Thr, Trp, Tyr, and Val). However, it is common to find media formulation containing all amino acids (e.g., RPMI medium), since the nonessential amino acids cannot always been produced by the cells fast enough to support a maximum growth rate [6]. Glutamine is normally supplemented in higher concentrations (3 to 6 mmol L^{-1}) as it is a major energy source. This amino acid, unstable in aqueous solution, produces ammonium accumulation in the medium, either by spontaneous decomposition or by its metabolization, with a detrimental effect on protein production. Its replacement by glutamate [7, 5] or other organic acids [8] has been proposed as alternative to diminish ammonium accumulation. Recently, Taschwer et al. (2012) have studied a CHO cell line adapted to glutamine-free medium to produce a fusion protein from erythropoietin, obtaining a product similar with similar glycosylation pattern and reducing the ammonium production [9]. Some other studies have found formulations supplementing different amino acids by specific amino acid or dipeptide [10] additions or higher concentrations. Mullen et al. (2012) described the addition of cystine (formed by 2 cysteine molecules covalently linked by a disulphide bond), increasing the dimer : monomer ratio of INF-β in CHO cell production [11].

5.3.1.4 Salts

Media formulation contains essential ions such as Na^+, K^+, Ca^{2+}, Mg^{2+}, Cl^-, PO_3^{3-}, and HCO_3^-, among others. These ions are provided by dissolved salts with a total osmolality around 300 to 320 mOsm kg^{-1} [12]. There are also other essential elements such as Fe, Cu, Ni, Mn, Mo, Sn, and Se, among others, that must be present in lower concentrations. They have different metabolic and physiological roles and are essential to cell viability. Some of these elements have shown to have a direct effect on productivity. In this regard, the role that Cu^{2+} plays in protein production has been gaining more interest, since low concentrations of copper sulphate (0.2-50.0 μmol L^{-1}) has been used to obtain an important decrease in lactate accumulation [13, 14].

5.3.1.5 Vitamins

Essential vitamins are normally provided in the basal medium to allow and promote the cell culture growth and production. When added, serum can also provide additional vitamins to the basal medium. However, serum-free (SF) media are preferred in most systems and thus vitamins should be added [15]. Between 8 and 11 vitamins are normally found in basal media and they should be supplemented if cells have an additional or particular requirement [16, 17].

5.3.1.6 Lipids

Besides the roles that lipids play as structural component of biological membranes, source of energy and cell surface signals, they also may have an effect as growth promoter in several cell lines. Lipids such as phosphatidic acid, phosphatidyl choline, phosphatidyl ethanolamine, and phosphatidyl inositol have shown to stimulate the cell growth in several mammalian cell lines [18, 19]. Although some basal media may contain lipids, they can also be added as water-soluble supplements.

5.3.1.7 Serum

The role of serum in mammalian cell cultures is mainly to provide growth factors and proteins substrate cell viability. Fetal bovine serum (FBS) is the most common source used. Since serum is both expensive and, because of its high protein containing, might interfere with the product purification, SF media are preferred for recombinant protein production [16]. In this regard, several means for adapting cells to SF media have been developed [20], allowing lower-cost and safer cultures for pharmaceutical proposes. On the other side, SF media lack of some essential proteins and growth factors that must be added in these media in order to maintain cell growth and viability. This will be further discussed below.

5.3.1.8 Proteins

Serum contains a wide number of proteins with different and specific functions. In SF media, some of these proteins must be added when their role is essential or promotes the cell growth and production. Although the specific protein requirements may vary between cell lines, insulin is added to most SF media formulations. Transferrin is also used as it is necessary for the iron uptake into the cell. Some cell lines are dependent on some proteins for their cultivation such as specific growth factors (e.g., fibroblast, epithelial, or nerve growth factors), but regular cell lines for recombinant protein production do not normally require them.

5.3.1.9 Buffers

The optimum pH range for mammalian cell cultures is normally within 7.0 and 7.2 [12]. Out of this narrow range cell growth and production dramatically decrease and then must be accurately controlled. The addition of bicarbonate generates an appropriate buffer when cells are cultured in a CO_2 atmosphere environment. However, if CO_2 concentration falls (e.g., when flasks are taken out of the incubator) alkalinisation occurs. HEPES buffer is also added as complement to prevent these pH changes.

5.3.2 Defined Nutrient Media

The scientific study of development of culture media for animal cell began in the middle of the twentieth century due to the need of large-scale production of vaccines, requiring the addition of serum as key supplement for growth. Initially, media formulation was made based on blood plasma and other fluids analysis [21]. Eagle (1955) used an alternative strategy that allowed him to identify a defined number of compounds, minimum concentration, which are essential for animal cell growth. This media created by Eagle was named minimum essential medium (MEM); it consisted in 13 amino acids, 8 vitamins, 6 ionic species, and glucose [22]. A dialysated fetal bovine serum supplement carried the undefined factors necessary for growth. Since then, a number of studies conducted to identify the effect of supplied nutrients concentration on the specific growth rate and cell yield began, noting that such effects and parameters vary considerably between cell line, culture system, and conditions used [23].

The development of new media has been critical, mainly by the need to minimise the number of components, maximising yield, or simplifying the interest product purification process.

Animal cells can be grown using completely natural nutrient media (natural fluids) or synthetic media containing a natural component. Developed synthetic media for animal cell can be classified into serum-containing media, serum-free media, chemical defined media, and protein-free media.

5.3.2.1 Serum-containing Media and Serum-free Media

FBS has been the most common supplement for mammalian cell culture. The use of animal serum allows the application of the same medium for the cultivation of different cell lines. However, the use of serum in cell culture has serious problems, especially when the final product is used for therapeutic purposes [24]. Technical considerations and disadvantages for using serum in cell culture applications have been analysed and widely discussed. The main problems related to serum use are: high protein load which compromises product purification process, variable composition, quality control complexity, risk of contamination with viruses, mycoplasma or prions, and high costs. Moreover, animal serum always consists of an unknown composition, with multiple trace concentration components, some of them with positive effects for growth, but also with the presence of cytotoxic compounds [25]. In 1997, the United States Food and Drug Administration (FDA) published a guidance document with strong recommendations towards the reduction or complete elimination of animal origin components used in the manufacture of products regulated by the FDA. Later, in 2001, the European Council (EC) adopted the Regulation No 999/2001, which purpose was to establish principles and guidelines to minimise the risk of transmissible spongiform encephalopathies (TSE) through human or veterinarian drugs. Such regulation applies to all material used in active substances preparation and excipients and includes all material (including culture media) used in the final product production. This strong regulatory position has been the result of efforts towards serum and other animal-derived factors (e.g., albumin, transferrin, and trypsin) elimination and the development of SF formulations. These regulatory issues among the increasing demand of recombinant proteins within the last decade have triggered joint efforts for SF culture media development, which will allow conditions for high productivity and reduce production costs.

Generally, the SF media is specifically formulated to support the growth of a given cell line. In this case, the serum is replaced by defined amounts of growth factors, lipoproteins and other proteins usually present in serum. For example, SF media used for growth of Chinese hamster ovary (CHO) cells containing insulin and transferrin as a supplement [26]. The growth of surface-dependent cells may require the addition of adhesion factors such as fibronectin [27]. In some cases this type of compounds are provided by the addition of animal, vegetable, or microbial protein hydrolysates.

These hydrolysates are oligopeptides, peptides, and amino acids complexes obtained by chemical or enzymatic hydrolysis of casein, albumin, plants, yeasts, or animal tissue. One of the most common hydrolysates is Primatone RL®, a tryptic meat digestion commercially available. Recently, plant hydrolysates have also been used for culture media formulation. These hydrolysates may have an essential role by providing growth factors, antiapoptotic factors or stimulating protein production [28, 29]. However, this type of product has the disadvantage, like serum, to have an undefined and variable composition. This can impact negatively on the validation of the production process.

5.3.2.2 Chemical Defined Media and Protein-free Media

In case of using pure protein additives in the media formulation, like recombinant growth factors produced in bacteria or yeast by genetic engineering along with inorganic and organic compounds, the media are classified as chemically defined. The identification of defined supplements that improve growth, productivity, and consistency is highly attractive for cell culture optimisation strategies. These types of media and specific defined supplements have been extensively studied and several reviews have been published on the subject [30-32].

The formulated media that do not include any kind of protein are classified as protein-free media. This type of medium has a fully known composition, constant quality and composition. It allows high reproducibility processes, facilitates downstream stages and even, in some cases, reduces production costs. However, the adaptation of cell lines to such type of medium is not always successful and its use often affects negatively the overall productivity of the culturing stage [33].

Previous studies show that SF medium implies a case-to-case design based on the individual supplements required for each strain. Unfortunately, to date there is no single supplement of defined composition to replace the benefits of serum and allow the growth of any cell line.

5.3.2.3 Defined Media Optimisation

Due to its complexity and high cost, defined medium optimisation is a key feature for the development of animal cell-based processes. Culture medium optimisation requires the consideration of each component effect evaluated on aspects related to the product, cell line and the associated production process. The determination of the best components mixture and their concentration for a given cell line in a particular process requires a carefully designed strategy, for which different approaches exist.

Growth modalities (batch, fed-batch, or perfusion) will also affect the optimisation of medium strategies. Different formulations are necessary for the various stages of a manufacturing process. During the propagation phase concentrated formulations are used to enable rapid growth. In batch culture a single medium is used for cell growth and for achieving production goals. In fed-batch cultures and perfusion, usually a basal medium is used which supports the first stage of growth, and a feed medium with higher concentrations of nutrients for the production stage. This is usually under conditions where cells are maintained with little or no growth. Therefore, feeding strategies involved in the prolongation of the growth phase and its maintenance should be carefully planned, based on the understanding of growth and the individual metabolism of each recombinant cell line. Improvements in media optimisation and bioprocess control have undoubtedly contributed to viability extension in high-density cultures.

The following examples describe different approaches that have addressed the optimization of animal cell culture medium.

The development of a culture medium based on defined medium mixtures is a widely used strategy where results can be obtained in a short time using commercially available formulations. An example of this approach is the DMEM/F12 formulation. DMEM/F12 enables the growth of a wide range of cell lines both as serum and SF formulations. In case of SF formulations, 4-(2-hydroxyethyl)-1-piperazineethanesulfonic acid (HEPES) buffer is included to compensate the buffering capacity loss caused by serum elimination.

Analysis by components is another approach usually reported [34, 35]. It consists of performing a number of experiments, which evaluate various components or group of components (e.g., vitamin mix) by individually adding different concentrations to a commercially available medium. Commonly cell growth, viability and productivity are evaluated. To determine the importance of the component on cell growth, several cultures are carried out with and without its addition. In a second stage, concentration varies in order to optimise its effect. This procedure is time-consuming and prevents to detect interactions between the components tested [36].

The multiple components analysis combined with a statistical design of experiments, such as Plackett-Burman experimental design, is an efficient way for a direct and rapid determination of effects and the interaction between large numbers of components analysed. A prominent discussion about this strategy is presented by Mandenius and Brundin (2008) [37]. This procedure results in a minimisation of costs and analysis time. Its use for the analysis of the effect of various compounds and supplements for growing different cell lines has proven to be very suitable [38, 39, 16].

Stoichiometric analysis is a further approach for culture media formulation. This means to analyse the concentration of components that changes in the medium during culture. After comparing the composition of the spent media with fresh media, specific rates of consumption and the consumption percentage of each nutrient can be calculates. Based on this information, the culture medium is reformulated, usually fortified with the most consumed nutrients [40, 5]. Recently, this type of approach has been complemented with metabolic models in order to determine nutrient concentrations based on experimental data [41] or theoretical estimation of the specific consumptions (modelling equations) [42, 43]. This methodical strategy is highly dependent on the available analytical capacity.

The rational design of culture media can be defined as a multiple approach, in which methods described above are integrated, considering their advantages and disadvantages. This design strategy does not prevent the work of performing the cell cultures as proof of principle during the development of the culture media. However, it is an approach of higher efficiency without the systemic limitations of a single approach method.

5.3.2.4 Glucose and Glutamine as Main Nutrients in Defined Media

It has been generally accepted that glucose and glutamine are use in the medium as main carbon and energy sources in defined nutrient medium formulations. However, the rapid metabolism of glucose and glutamine taking place in cultured animal cells leads to a very inefficient use of both components, resulting in their rapid depletion from the culture medium and in general an accumulation of the degradation products lactate and ammonium in the medium.

There are 3 main approaches that have been used to limit glucose and glutamine consumption. One is based on substrates that, because of their nature, are less available or slowly metabolised. The second one is based on genetic modifications (cell engineering) for changing the metabolic behaviour by the overexpression of key proteins. The third approach involves the controlled feeding of glucose and glutamine in fed-batch cultures (see below).

The use of galactose or fructose instead of glucose has been shown to be an alternative to prevent lactate overproduction because the affinity of hexokinase for these sugars is approximately twentyfold lower than for glucose [44]. In particular, the use of galactose has been studied as an alternative for reducing lactate production levels [45, 5, 46]. The galactose metabolism is slower compared to the glucose oxidation, allowing a substantial reduction in lactate production. The galactose metabolism has also been shown to increase the cell viability and the overall time of the culture. However, the use of galactose is accompanied by a significant reduction of the cell growth [45]. Noticeably, lactate produced under galactose-containing conditions is remetabolized by the cells. These results have been used as a basis for studying fed-batch [47, 48] and perfusion cultures [49]. It should be noted that there are commercially available media that use galactose as part of their respective formulations.

Efforts to avoid the effects of an increase in the ammonium concentration have basically focused on the substitution of glutamine by another compound able to enter the TCA. The replacement of glutamine by glutamate is one example of this approach, having the following advantages: glutamate has a higher chemical stability than glutamine, glutamine can be directly synthesised from glutamate in CHO cells because they express the glutamine synthetase (GS), and glutamate is less ammoniagenic because it bears only one amine group. Indeed, the replacement of glutamine by glutamate in culture media has proved to be a valuable strategy for CHO cells, resulting in an improvement of the cell growth by 17 % and a reduction of the accumulated ammonium ion by 70 % [45]. Hong et al. (2010) recently showed that this substitution increased the concentration of an IgG produced in CHO cells by a factor of 1.7 [50]. The genetic engineering of hybridoma cells by overexpressing the GS, which converts glutamate to glutamine, allowed a cell growth in glutamine-free medium [51], resulting in similar effects described above for the nutrient replacement strategies (e.g., glutamine by glutamate).

The substitution of glutamine by dipeptides such as alanyl-glutamine and glycyl-glutamine has also been used to reduce ammonium ion production (see also Chapter 5.1). These dipeptides, those are more stable than the amino acids in the culture medium, are intra- or extracellularly cleaved by peptidases, thereby releasing glutamine and either alanine or glycine. The availability of glutamine is therefore dependent on the peptidase activity. In hybridoma cell cultures, high cell yields were obtained in the presence of 6 mmol L^{-1} Ala-Gln or 20 mmol L^{-1} Gly-Gln [52]. The final cell yield under Gly-Gln conditions was 14 % higher than in glutamine alone. The specific consumption rates of glucose and 6 amino acids were reduced, and, additionally, the accumulation of ammonia and lactate was significantly lower [52]. GlutaMAX™ is a commercially available product for medium formulation-based on the alanyl-glutamine dipeptide.

5.3.3 Feeds for Fed-batch Cultures

The production of recombinant proteins with mammalian cells for the use as human therapeutics requires the development of highly competitive processes to obtain highest levels of product while maintaining the maximum standards of product quality. This is due to the increased demand for new products and the requirement of an optimal cost efficiency in the manufacturing of new biotherapeutics.

To achieve a high production level, a several efforts have focused on improving the productivity in fed-batch cultures, including the following: strategies based on monitoring and controlling various culture parameters (pH, O_2, cell concentration) in order to optimise the frequency and the volume of the feeding media, its formulation and the use of hydrolysates from vegetable sources, the use of genetically modified strains, and the manipulation of the culture environment. All these approaches alone and together resulted in optimised cultures, achieving higher productivities and a better understanding of the metabolism and physiology of the cells.

5.3.3.1 Operational Fundaments

The fed-batch culture is a form of semicontinuous mode, which is very attractive for the application in biopharmaceutical manufacturing due to its operational simplicity and competitive advantages compared to the batch culture. An initial culture medium is loaded into the bioreactor where the cells are inoculated and after a well-defined time nutrients or parts thereof are fed into the system until the end of the operation. In this period, the culture volume increases continuously due to the absence of a discharge flow. Thus, the cells are constantly receiving nutrients and oxygen. At the same time, product of interest and products of the metabolism, such as lactate and ammonium are accumulated. In comparison to the batch system, the end of cell growth primarily occurs due to the inhibitory effect of accumulated end products, rather than to the depletion of nutrients, resulting in a substantial prolongation of the culture period and achieving higher cell and product concentrations (for details, see Chapter 6.2).

This cultivation system is generally initiated by using a reduced volume (e.g., 40 to 50 % of the maximum volume) of the bioreactor, which then increases during the feeding period. The feed consists of a flow of concentrated medium (10 × to 15 ×) or selected components that can operate intermittently or continuously until the end of cultivation. The increase in culture volume is accompanied by a raise in the medium osmolality, which results either from the accumulation of unconsumed nutrients and salts when the complete and concentrated nutrient medium is used for the feed or only from the specific waste metabolites such as the lactate and ammonium and from specific salts when a defined feed mixture is applied.

For example, one strategy to minimise the lactate accumulation and the increased osmolality of the medium is to limit the supply of glucose. Through this, the levels of

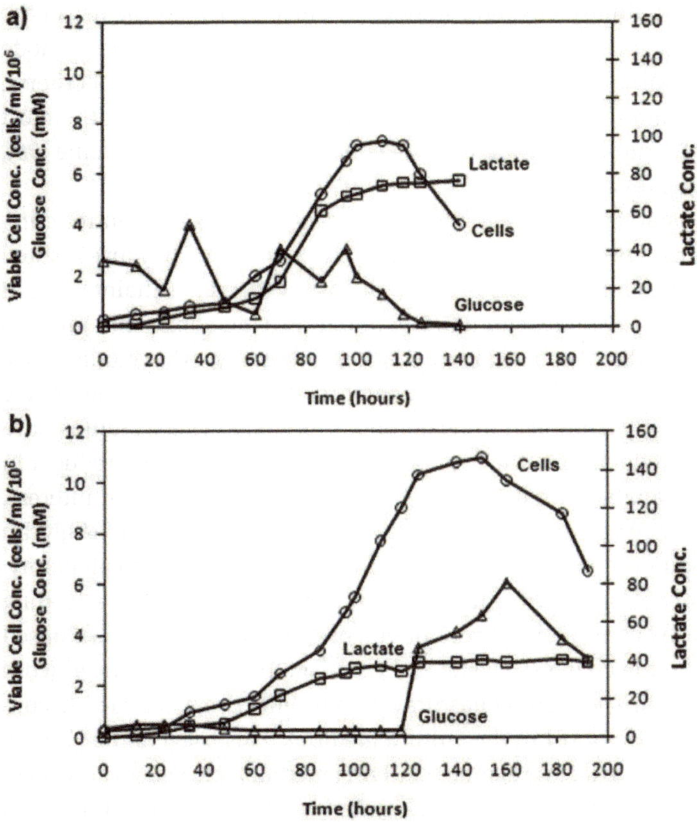

Figure 5.3.1: Time profiles of cell, lactate, and glucose concentration for a hybridoma fed-batch culture with cells growing **a)** high-lactate producing metabolism and **b)** metabolic shift [55].

produced lactate and the base solution to control the pH are lowered, thereby reducing the osmolarity of the medium [53, 54].

5.3.3.2 Feed Strategies

The simplest feeding strategy in the developing of a fed-batch process consists of adding the feed to the culture medium at time intervals ranging from 12 to 24 h. The control criteria, determining the added volume or the flow, are usually triggered by maintaining the concentration of specific nutrients, which are measured off-line, at a constant level. However, this strategy does not allow an operational fine control of the cellular metabolism. To do this, a continuous feeding strategy is required allowing to work in tighter ranges of control criteria and based on a system for online monitoring of the selected control parameters.

The development of optimised feeding strategies, which maximise the use of the fed nutrients, requires a consideration of some basic elements such as feed mode (continuous or intermittent), quantity and frequency of feeding, and control criteria. The latter has attracted more interest and research in recent years. Among the most commonly used control criteria there are glucose concentration [54], glutamine [56], pH [57], and oxygen uptake rate [58].

A large number of studies have used glucose concentration as the control criterion for the development of fed-batch cultures [59, 60, 54, 53, 61, 62]. They investigated the reduction of the lactate production using a glucose feed that is maintained at about 0.3 to 1.5 mmol L^{-1} in the bioreactor [60, 54], initiating a metabolic shift towards a more efficient glucose utilisation (see Figure 5.3.1). Additionally, other studies have investigated how the glucose limitation in fed-batch culture can affect the degree of glycosylation [61] and glycation [62] of recombinant proteins. These studies have shown that, although a reduction in the availability of glucose as a sugar reducer in the culture medium is capable of reducing the glycation of the protein of interest, very low glucose levels may decrease sialylation and affect the extension of N-glycan chains of the recombinant protein, which can be detrimental to its biological quality.

The use of pH as control criterion has allowed the development of a feeding method called HI-end pH-controller delivery of glucose (or shortly HIPDOG), based on the pH change due to the change in lactate metabolism, promoted by the glucose depletion (see Figure 5.3.2). The method has proven to be effective when lactate consumption causes an increase of pH in the culture medium. This triggers the glucose feeding until the cells produce lactate again, acidifying medium and ceasing the feed. This strategy has allowed an increase in the integral of viable cells of 1.28 times and 2.6 times higher relative product titer [57].

Another novel control criteria implemented makes use of an automated system of flow cytometry in order to predict the cessation of cell growth by measuring the population of viable and nonviable cells in culture. This method utilises 2 control systems to trigger the feed: (1) in the cell growth phase, when the concentration of nonviable

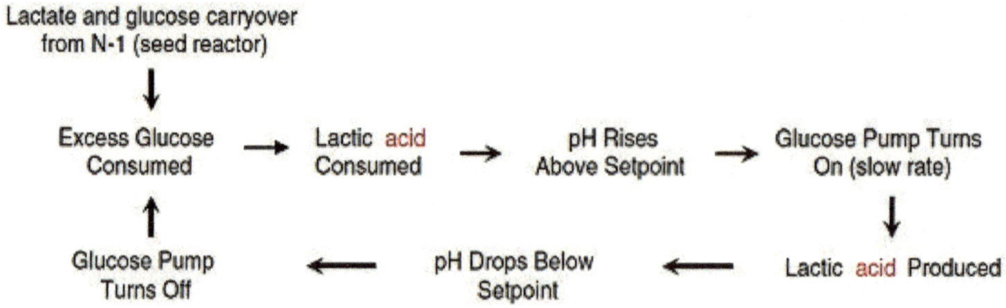

Figure 5.3.2: Pictorial representation of the hypothesized sequence of events in the bulk cell culture fluid during the growth phase of a bioreactor utilizing the HIPDOG control scheme for limiting lactic acid accumulation [57].

cells exceeds 4 times the average concentration of nonviable cells prior to the entry of cell death; (2) in the stationary phase of growth, when the viable cell count stops to increase within 30 h. This strategy was able to predict the entry into the stationary phase in about 40 h, permitting the alimentation of the cells in an appropriate moment and resulting in a significant increase in the total cell count in the reactor [63].

Another variable allowing the adjustment of the feed measured online and often used is the oxygen uptake rate [58]. This parameter is an accurate indicator of the physiological state of the cells, indirectly indicating the specific cellular growth rate at higher or lower oxygen uptake rates, thus allowing the detection of small changes in the metabolism. A recent study has developed a technique for maintaining the concentration of viable cells and the specific growth rate in predefined paths through the indirect control of the oxygen uptake rate, maintaining this at a desired level through the glutamine feed rate [56]. The results showed the excellent reproducibility of various cultures under the action of this control system, allowing the fed-batch cultivation to take less time to reach the predetermined specific growth rate.

5.3.3.3 Feeds and Hydrolysates

Formulation of feeds may be grouped in 2 categories [64]:

1. A concentrate feed, which includes all the nutrients present in the basal medium. The purpose of this feed is to deliver all the nutrients needed to maintain an adequate culture lifetime and productivity without identifying the nutrients that could become limiting components. This strategy has significantly increased cellular concentrations, the cell integral, and the recombinant protein production in CHO cells [65], hybridomas [66], myeloma [67], and 293 human kidney cells [68]. However, despite its effectiveness, the high cost, which may involve feeding all the nutrients of the medium and the high risk of increasing the osmolality of the medium, may be factors against the use of these feeds.
2. Feeding of nutrients that are rapidly consumed by the cells. The purpose of this is to provide only the nutrients of highest demand identified by a specific analysis of the key nutrients affecting cell growth and the production of the protein of interest. This avoids the disadvantages of feeding all the components of the basal medium and also allows directing the cellular metabolism as reported by the controlled feeding of glucose and glutamine in hybridomas [69, 59] and recombinant cells [54, 53]. This kind of feeds usually requires the presence of multiple nutrients, which implies a medium design that considers stoichiometric and energy requirements for the production of biomass and product as well as a "medium proportional design" strategy [67]. This design has been shown by a stoichiometric model based on the composition of the biomass (protein, DNA, RNA, lipids, and carbohydrates), the nutrient composition (glucose, amino acids, and vitamins), the ATP and the product yield, being able to increase the concentration of

viable cells by a factor of 2 and the final concentration of monoclonal antibody by a factor of 10 compared to a batch culture [70]. However, this kind of strategy requires the identification of key nutrients and the determination of optimal composition, especially in the case of compounds such as vitamins, lipids, and trace elements, requiring more effort when working with SF media.

Recent studies have focused on the development of chemically defined media, where feeds containing a nonanimal-derived hydrolysate are replaced by defined formulations. This led to an improved performance of the cultures by an 80 % increase of the product titer and a reduction of the lactate production in the CHO cell line [71] as well as and increasing the mAb titer by 20 % in NS0 cells [72]. Furthermore, Kang et al. (2012) studied the use of chemically synthesized dipeptides as an alternative method for delivering amino acids, particularly those with low solubility such as tyrosine [10]. These tyrosine-containing dipeptides showed a positive effect in fed-batch cultures by increasing the viability of the culture and the titer of product, while reducing the production of lactate and ammonium resulting in an increase of process robustness.

Nevertheless, after the use of animal serum in the past, today plant or yeast hydrolysates are the largest source of nutrients for the supplementation of feeds.

In conclusion, the use of serum and, generally, all animal-derived components is currently prohibited for the manufacturing of biopharmaceuticals (Chapter 4.2). Only biopharmaceuticals exclusively produced under animal-component-free medium conditions will be licensed by the regulatory authorities such as FDA or EMA. In an optimal case, a process rely on a completely chemically-defined formulation for manufacturing but in most cases undefined animal component-free supplements are used such as plant or yeast hydrolysates [73-75].

The main sources for obtaining nonanimal hydrolysates are yeast extracts or parts of plants, such as soybeans, wheat, rice, cotton, pea, cabbage, etc. (Table 5.3.1). These additives of the culture medium contain an undefined mixture of low molecular weight components, such as amino acids, peptides, vitamins, and trace elements [76, 77]. Hydrolysates not only simulate growth factors by improving the cell growth [75, 77], but they also increase the specific productivity of a recombinant protein, partly through a positive effect on the mRNA content and the prolongation of the cellular lifetime at a higher viability [78]. Their role as an inhibitor of some proteases was also demonstrated, thus fulfilling a protective function at the recombinant protein in the culture [73].

Table 5.3.1: Type of protein hydrolysates, used [74].

Vendors	Hydrolysate Type
Becton Dickinson	Phytone peptone
	TC Yestolate
	Yeast extract UF
	Select soytone
	Bacto soytone
Hyclone	Soy hydrolysate
	Wheat gluten hydrolysate
	Yeast extract
	HyQ soy hydrolysate UF
JRH Biosciences	Soy hydrolysate
Sigma	Hy-soy UF
	Wheat-rice hydrolysate
	Wheat hydrolysate
	Yeast hydrolysate
	Rice hydrolysate
	Soy hydrolysate

The positive effect of the supplementation of vegetable hydrolysates on the specific productivity of and the growth of mammalian cells was shown to be dose dependent [28, 74]. In some cases an excessive amount of hydrolysates could also inhibit the cell growth [74], possibly by affecting the balance of nutrients due to the presence of a high concentration of amino acids and oligopeptides in the culture medium [79]. Therefore, it is always necessary to find an optimal concentration of hydrolysates that can maximise the cell concentration, maintaining an adequate viability and an appropriate productive capacity of the cells. Positive or negative effects on growth and productivity were discussed to be related to the constitutions of the hydrolysate, particularly to the fractions obtained after various separation processes [74, 80]. The results obtained suggested that different fractions of a single type of hydrolysate could contain different molecules, which might distinctly contribute to the cell growth or the production of the recombinant protein. In addition, they could exert a detrimental effect on one or another parameter of the culture [80]. More intensive studies on the effect of nonanimal derived hydrolysates and their characterisation have to be done in terms of the overall quantification of the compound families (peptides/amino acids, carbohydrates, salts, phospholipids, polyphenols, etc.) as well as the distribution of their molecular weight and the characterisation of other unknown chemicals.

The cellular response to the use of hydrolysates varied from a metabolic point of view. In some cases, it was observed that the supplementation with hydrolysates was able to either increase or decrease the consumption of the main carbon and energy sources such like glucose and glutamine, directly affecting the production of lactate and ammonia [78, 75]. However, no effect on the consumption and production of these metabolites was also shown [74]. The character of a metabolic response apparently depends on the cell line and the nature and quality of the respective hydrolysate used [77], which has always to be evaluated case by case.

Based on the benefits associated to the supplementation with plant and yeast hydrolysates and the high demand for biopharmaceuticals, many companies have developed product lines based on these animal component-free hydrolysates, including ThermoFischer Scientific, Life Technologies, BD Biosciences, HyClone, Sigma, etc. [74, 81]. All components have been tested primarily in batch culture and in some cases fed-batch culture have been used to evaluate their potential as a supplement for high cell density cultures [82].

In particular, the use of hydrolysates from different sources has been evaluated individually and collectively, on growth, cell viability, and mAb productivity. The combination of some hydrolysates showed a significant increase of mAbs titers, achieving increases by 290 % without impairing the viability of the cultures or the product quality [81]. Other studies, using a feed medium containing glucose and hydrolysate, have shown an increase in the cell concentration, culture lifespan, and yield of the recombinant protein [83]. Furthermore, a study of the production of THIOMABs (antibody with reactive cysteine residues) in fed-batch culture showed that the use of different hydrolysates had no significant effect on the formation of an impurity (3LC, antibody with an extra light chain) associated with these antibodies [84], thus achieving higher cell concentrations (1.25 to 1.5 times) and a higher mAb titer (approximately 1.6 times).

5.3.3.4 Use of Genetic and Environmental Manipulations on Fed-batch System

The fed-batch culture mode is one of the main production systems used in industry. Therefore, different strategies using both, genetic and environmental manipulations, have been evaluated using this mode of cultivation for finding the optimal system for production.

Fed-batch cultivation together with the genetic manipulation of the cell line resulted in a synergistic effect on the process productivity. In particular, overexpression of special proteins inducing a delayed apoptosis, such as Bcl-XL [85], or other proteins related to the assembly and the folding of proteins, such as XBP-1 [86], have been shown to exert a positive effect on the productivity of various recombinant proteins in fed-batch system. In case of a delayed apoptosis by overexpressing the Bcl-XL protein, it was suggested that Bcl-XL exerts an antagonistic interaction with the protein JNK, involved in apoptotic signalling cascade [87], thus delaying the cellular death process and thereby increasing the culture lifetime by 10 %. Overexpression of the XBP-1 protein has a positive effect on the specific productivity of the recombinant protein produced by 40 %, allowing improving the production process without being detrimental to the quality of the protein of interest. XBP-1 is involved in the induction process of the endoplasmic reticulum chaperones and foldases to increase the protein folding capacity of the endoplasmic reticulum [88].

Other approaches resulted in a positive effect on the simplicity and versatility, which enables the manipulation of the environment, by varying culture parameters like temperature, pH, osmolality, and medium design. In this regard, 2 approaches have important implications on the culture performance. Changing the culture temperature to a mild hypothermia (33 to 30 °C) has a positive effect on the productivity of a wide range of recombinant proteins [89, 90, 3] in batch culture. A reduced temperature can positively affect the cellular metabolism, which then has an impact on

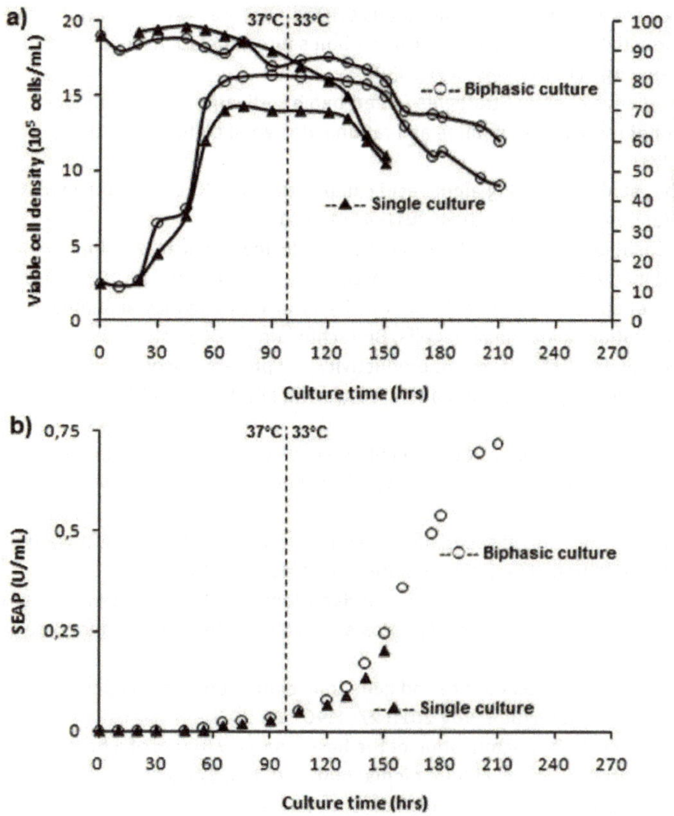

Figure 5.3.3: Comparison of growth, viability and secreted embryonic alkaline phosphatase (SEAP) production between single-temperature and biphasic cultures [91].

the design of the feed medium in a fed-batch cultivation. Applying a biphasic culture strategy which deploys temperature reduction in a fed-batch cultivation achieved significant improvements by extending the lifespan of the culture, increasing the integral of viable cell density (IVCD) also called the integral cumulative cell time by a factor of 1.6 and increasing the specific productivity of the recombinant protein by 2.3 times (see Figure 5.3.3) [91], resulting in a more efficient and productive process [84].

References

[1] Hwang S, Lee G. Nutrient Deprivation Induces Autophagy As Well As Apoptosis in Chinese Hamster Ovary Cell Culture Biotechnol Bioeng 2008; 99: 678-685.

[2] Krampe B, Al-Rubeai M. Cell death in mammalian cell culture: molecular mechanisms and cell line engineering strategies. Cytotechnology 2010; 62: 175-188.

[3] Berrios J, Díaz-Barrera A, Bazán C, Altamirano C. Relationship between tissue plasminogen activator production and specific growth rate in Chinese hamster ovary cells cultured in mannose at low temperature. Biotechnol Lett 2009; 31: 1493-1497.

[4] Berrios J, Altamirano C, Osses N, Gonzalez R. Continuous CHO cell cultures with improved recombinant protein productivity by using mannose as carbon source: Metabolic analysis and scale-up simulation. Chem Eng Sci 2011; 66: 2431-2439.

[5] Altamirano C, Paredes C, Illanes A, et al. Strategies for fed-batch cultivation of t-PA producing CHO cells: substitution of glucose and glutamine and rational design of culture medium. J Biotechnol 2004; 110: 171-179.

[6] Xing Z, Kenty B, Koyrakh I, et al. Optimizing amino acid composition of CHO cell culture media for a fusion protein production. Process Biochem 2011; 46: 1423-1429.

[7] Altamirano C, Illanes A, Casablancas A, et al. Analysis of CHO cells metabolic re-distribution in a glutamate-based defined medium in continuous culture; Biotechnol Prog 2001; 17: 1032-1041.

[8] Genzel Y, Ritter J, Konig S, et al. Substitution of Glutamine by Pyruvate To Reduce Ammonia Formation and Growth Inhibition of Mammalian Cells. Biotechnol Prog 2005; 21: 58-69.

[9] Taschwer M, Hackl M, Hernández J, et al. Growth, productivity and protein glycosylation in a CHO EpoFc producer cell line adapted to glutamine-free growth. J Biotechnol 2012; 157: 295-303.

[10] Kang S, Mullen J, Miranda L, Deshpande R. Utilization of tyrosine- and histidine-containing dipeptides to enhance productivity and culture viability. Biotechnol Bioeng 2012; 109: 2286-2294.

[11] Mullen L, Adams G, Chernajovsky Y. Increased disulphide dimer formation of latent associated peptide fusions of TGF-β by addition of L-cystine. J Biotechnol 2012; 161: 269-277.

[12] deZengotita V, Schmelzer A, Miller W. Characterization of Hybridoma Cell Responses to Elevated pCO2 and Osmolality, Intracellular pH, Cell Size, Apoptosis, and Metabolism. Biotechnol Bioeng 2002; 77: 369-380.

[13] Qian Y, Khattak S, Xing Z, et al. 2011. Cell Culture and gene transciption effects of copper sulfate on Chinese hamster overy cells. Biotechnol Prog 2011; 27: 1190-1194.

[14] Kaschak T, Boyd D, Lu F, et al. 2011 Characterization of the basic charge variants of a human IgG1: effect of copper concentration in cell culture media. mAbs 2011; 3: 577-583.

[15] Van der Valk J, Brunner D, De Smet K, et al. Optimization of chemically defined cell culture media – Replacing fetal bovine serum in mammalian in vitro methods. Toxicol in Vitro 2010; 24: 1053-1063.

[16] Zhang H, Wang H, Liu M, et al. Rational development of a serum-free medium and fed-batch process for a GS-CHO cell line expressing recombinant antibody. Cytotechnology 2013; 65: 363-378.
[17] Kim Do Y, Lee J, Chang H, Oh D. Effects of supplementation of various medium components on Chinese hamster ovary cell cultures producing recombinant antibody. Cytotechnology 2005; 47: 37-49.
[18] Ryan P, Maher V, McCormick J. Modification of MCDB 110 medium to support prolonged growth and consistent high cloning efficiency of diploid human fibroblasts. Exp Cell Res 1987; 172: 318-328.
[19] Chen Z, Ke Y, Chen Y. A serum-free medium for hybridoma cell culture. Cytotechnology 1993; 1: 169-174.
[20] Santos A, Bartolini P, Soares C. Laboratory Production of Human Prolactin from CHO Cells Adapted to Serum-Free Suspension Culture. Appl Biochem Biotechnol 2012; 167: 2212-2224.
[21] Morgan J, Morton H, Parker R. Nutrition of animal cells in tissue culture; initial studies on a synthetic medium. Proc Soc Exp Biol Med 1950; 73: 1-8.
[22] Eagle H. Nutrition needs of mammalian cells in tissue culture. Science 1955; 122: 501-514.
[23] Lambert K, Pirt S. The quantitative requirements of human diploid cells (strain MRC-5) for amino acids, vitamins and serum. J Cell Sci 1975; 17: 397-411.
[24] Werner R. Economic aspect of commercial manufacture of biopharmaceuticals. J Biotechnol 2004; 113: 171-182.
[25] Hesse F, Wagner R. Developments and improvements in the manufacturing of human therapeutics with mammalian cell cultures. Trends Biotechnol 2000; 18: 173-180.
[26] Sunstrom N, Gay R, Wong D, et al. Insulin-like growth factor-I and transferrin mediate growth and survival of Chinese hamster ovary cells. Biotechnol Prog 2000; 16: 698-702.
[27] Butler M, Jenkins H. Nutritional aspects of the growth of animal-cells in culture. J Biotechnol 1989; 12: 97-110.
[28] Burteau C, Verhoeye F, Mols J, et al. Fortification of a protein-free cell culture medium with plant peptones improves cultivation and productivity of an interferon-gamma-producing CHO cell line. In Vitro Cell Dev Biol Anim 2003; 39: 291-296.
[29] Kim S, Lee G. Development of serum-free medium supplemented with hydrolysates for the production of therapeutic antibodies in CHO cell cultures using design of experiments. Appl Microbiol Biotechnol 2009; 83: 639-648.
[30] Kim E, Kim N, Lee G. Development of a serum-free medium for dihydrofolate reductase-deficient Chinese hamster ovary cells (DG44) using a statistical design: beneficial effect of weaning of cells. In Vitro Cell Dev Biol Anim 1999; 35: 178-182.
[31] Keen M, Rapson N. Development of a serum free culture medium for the large scale production of recombinant protein for Chinese hamster ovary cell line. Cytotechnology 1995; 17: 153-163.
[32] Lao M, Shalla C. Development of serum-free medium using computer-assisted factorial design and analysis. Cytotechnology 1996; 22: 25-31.
[33] Schröder M, Matischak K, Friedl P. Serum- and protein-free media formulations for the Chinese hamster ovary cell line DUKXB11. J Biotechnol 2004; 108: 279-292.
[34] Stoll T, Mühlethaler K, von Stockar U, Marison I. Systematic improvement of a chemically defined protein-free medium for hybridoma growth and monoclonal antibody production. J Biotechnol 1996; 45: 166-175.
[35] Kim D, Lee J, Chang H, Oh D. Development of serum-free media for a recombinant CHO cell line producing recombinant antibody. Enzyme Microbiol Technol 2006; 39: 426-433.
[36] Petiot E, Fournier F, Geny C, et al. Rapid screening of serum-free media for the growth of adherent Vero cells by using a small-scale and non-invasive tool. Appl Biochem Biotechnol 2010; 160: 1600-1615.

[37] Mandenius C, Brundin A. Bioprocess optimization using design-of-experiments methodology. Biotechnol Prog 2008; 24: 1191-1203.

[38] Liu C, Chang T. Rational development of serum-free medium for Chinese hamster ovary cells. Process Biochem 2006; 41: 2314-2319.

[39] González-Leal I, Carrillo-Cocom L, Ramírez-Medrano A, et al. Use of a plackett–burman statistical design to determine the effect of selected amino acids on monoclonal antibody production in CHO cells. Biotechnol Prog 2011; 27: 1709–1717.

[40] Xie L, Wang D. Material balance studies on animal cell metabolism using a stoichiometrically based reaction network. Biotechnol Bioeng 1996; 52: 579-590.

[41] Sellick C, Croxford A, Maqsood A. Metabolite profiling of recombinant CHO cells: designing tailored feeding regimes that enhance recombinant antibody production. Biotechnol Bioeng 2011; 108: 3025-3031.

[42] Quek L, Dietmair S, Krömer J, Nielsen L. Metabolic flux analysis in mammalian cell culture. Metab Eng 2010; 12: 161-171.

[43] Sidorenko Y, Wahl A, Dauner M, et al. Comparison of metabolic flux distributions for MDCK cell growth in glutamine- and pyruvate-containing media. Biotechnol Prog 2008; 24: 311-320.

[44] Wagner R. Metabolic Control of Animal Cell Culture Processes. In: Hauser H, Wagner R (ed). Mammalian Cell Biotechnology in Protein Production. De Gruyter, Berlin, 1997: 233-276.

[45] Altamirano C, Paredes C, Cairó J, Gódia F. Improvement of CHO Cell Culture Medium Formulation: Simultaneous Substitution of Glucose and Glutamine. Biotechnol Prog 2000; 16: 69-75.

[46] Wilkens C, Gerdtzen Z. Engineering CHO cells for improved central carbon and energy metabolism. BMC Proceedings. 2011; 5: 120.

[47] Clark K, Griffiths J, Bailey K, Harcum S. Gene-expression profiles for five key glycosylation genes for galactose-fed CHO cells expressing recombinant IL-4/13 cytokine trap. Biotechnol Bioeng 2005; 90: 568-577.

[48] Altamirano C, Illanes A, Becerra S, et al. Considerations on the lactate consumption by CHO cells in the presence of galactose. J Biotechnol 2006; 125: 547-556.

[49] Vergara M, Becerra S, Díaz-Barrera A, et al. Simultaneous environmental manipulations in semi-perfusion cultures of CHO cells producing rh-tPA. Electron J Biotechnol 2012; 15(6). DOI: 10.2225/vol15-issue6-fulltext-2.

[50] Hong J, Yoon S. Substitution of glutamine by glutamate enhances production and galactosylation of recombinant IgG in Chinese hamster ovary cells. Appl Microbiol Biotechnol 2010; 88: 869-876.

[51] Paredes C, Prat E, Cairó J, et al. Modification of glucose and glutamine metabolism in hybridoma cells through metabolic engineering. Cytotechnology 1999; 30: 85-93.

[52] Christie A, Butler M. Glutamine-based dipeptides are utilized in mammalian cell culture by extracellular hydrolysis catalysed by a specific peptidase. J Biotechnol 1994; 37: 277-290.

[53] Sauer P, Burky J, Wesson M, et al. A high-yielding, generic fed-batch cell culture process for production of recombinant antibodies. Biotechnol Bioeng 2000; 67: 587-597.

[54] Cruz H, Moreira J, Carrondo M. Metabolically BHK cell fed-batch cultures. J Biotechnol 2000; 80: 109-118.

[55] Wlaschin K, Hu W. Fedbatch culture and dynamic nutrient feeding. Adv Biochem Eng Biotechnol 2006; 101: 43-74.

[56] Aehle M, Schaepe S, Kuprijanov A, et al. Simple and efficient control of CHO cell cultures. J Biotechnol 2011; 153: 56-61.

[57] Gagnon M, Hiller G, Yen-Tung L, et al. High-end pH-controlled delivery of glucose effectively suppresses lactate accumulation in CHO fed-batch cultures. Biotechnol Bioeng 2011; 108: 1328-1337.

[58] Zhou W, Rehm J, Hu W. High viable cell concentration fed-batch cultures cells through on-line nutrient feeding. Biotechnol Bioeng 1995; 20: 579-587.
[59] Kurokawa H, Park Y, Lidima S, Kobayashi T. Growth characteristics in fed-batch culture of hydridoma cells with control of glucose and glutamine concentrations. Biotechnol Bioeng 1994; 44: 95-103.
[60] Glacken M, Fleischaker R, Sinskey A. Reduction of waste product excretion via nutrient control: Possible strategies for maximizing product and cell yields on serum in cultures of mammalian cells. Bioetchnol Bioeng 1986; 28: 1376-1389.
[61] Wong D, Wong K, Goh L, et al. Impact of dynamic online fed-batch strategies on metabolism, productivity and N-glycosylation quality in CHO cell cultures. Biotechnol Bioeng 2004; 89: 164-177.
[62] Yuk I, Zhang B, Tang Y, et al. Controlling glycation of recombinant antibody in fed-batch cell cultures. Bioetchnol Bioeng 2011; 108: 2600-2610.
[63] Sitton G, Srienc F. Mammalian cell culture scale-up and fed-batch control using automated flow cytometry. J Biotechnol 2008; 135: 174-180.
[64] Bibila T, Robinson D. In pursuit of the optimal fed-batch process for monoclonal antibody production. Biotechnol Prog 1995; 11: 1-13.
[65] Miller D, Frapeau D, Luan Y, et al. Elevated nutrient levels in CHO cell culture: effect on cellular productivity and amino acid uptake rates. Presented at: 205th National Meeting of the American Chemical Society, Denver, 1993.
[66] Ryu J, Lee G. Application of hypoosmolar medium to fed-batch culture of hybridoma cells for improvement of culture longevity. Biotechnol Bioeng 1999; 62: 120-123.
[67] Bibila T, Ranucci C, Glazomitsky K, et al. Monoclonal antibody process development using medium concentrates. Biotechnol Prog 1994; 10: 87-96.
[68] Kyung Y, Hu W. Enhanced productivity of protein C by recombinant human cells in automated fed-batch cultures. Cytotechnology 1995; 17: 109-115.
[69] Glacken M, Huang C, Sinskey A. Mathematical descriptions of hybridoma culture kinetics III. Simulation of fed-batch bioreactor. J. Biotechnol 1989; 10: 39-66.
[70] Xie L, Wang D. Fed-batch cultivation of animal cells using different medium design concepts and feeding strategies. Biotechnol Bioeng 1994; 43: 1175-1189.
[71] Ma N, Ellet J, Okediadi C, Hermes P, et al. A single nutrient feed supports both chemically defined NS0 and CHO fed-batch processes: Improved productivity and lactate metabolism. Bioetchnol Prog 2009; 25: 1353-1363.
[72] Hermes P, Castro C. A fully defined, fed-batch, recombinant NS0 culture process for monoclonal antibody production. Bioetchnol Prog 2010; 26: 1411-1416.
[73] Mols J, Peeters-Joris R, Wattiez S, et al. Recombinant interferon-Y secreted by Chinese hamster ovary-320 cells cultivated in suspension in protein-free media is protected against extracellular proteolysis by the expression of natural protease inhibitors and by the addition of plant protein hydrolysates to the culture medium. In Vitro Cell Dev Biol 2005; 41: 83-91.
[74] Chun B, Kim J, Lee H, Chung N. Usability of size-excluded fractions of soy protein hydrolysates for growth and viability of Chinese hamster ovary cells in protein-free suspension culture. Bioresour Technol 2007; 98: 1000-1005.
[75] Farges B, Chenu S, Marc A, Goergen J. Kinetics of IFN-Y producing CHO cells and other industrially relevant cell lines in rapeseed-supplemented batch cultures. Process Biochem 2008; 43: 945-953.
[76] Nyberg G, Balcarcel R, Follstad B, et al. Metabolism of peptide amino acids by Chinese hamster ovary cells grown in a complex medium. Biotechnol Bioeng 1999; 62: 324-335.

[77] Heidemann R, Zhang C, Qi H, et al. The use of peptones as medium additives for the production of a recombinant therapeutic protein in high density perfusion cultures of mammalian cells. Cytotechnology 2000; 32: 157-167.
[78] Sung Y, Lim S, Chung J, Lee G. Yeast hydrolysate as a low-cost additive to serum-free medium for the production of human thrombopoietin in suspension cultures of Chinese hamster ovary cells. Appl Microbiol Biotechnol 2004; 63: 527-536.
[79] Zhang Y, Zhou Y, Yu J. Effects of peptone on hybridoma growth and monoclonal antibody formation. Cytotechnology 1994; 16: 147-150.
[80] Michiels J, Barbau J, De Boel S, et al. Characterisation of beneficial and detrimental effects of a soy peptone, as an additive for CHO cell cultivation. Process Biochem 2011; 46: 671-681.
[81] Kochanowski N, Siriez G, Roosens S, Malphettes L. Medium and feed optimization for fed-batch production of a monoclonal antibody in CHO cells. Presented at: 22nd European Society for Animal Cell Technology (ESACT) Meeting on Cell Based Technologies, Vienna, 2011.
[82] Lu C, Gonzalez C, Gleason J, et al. A T-flask based screening platform for evaluating and identifying plant hydrolysates for a fed-batch cell culture process. Cytotechnology 2007; 55: 15-29.
[83] Thombre S, Gadgil M. Increase in efficiency of media utilization for recombinant protein production in Chinese hamster ovary culture through dilution. Biotechnol Appl Biochem 2011; 58: 25-31.
[84] Gomez N, Ouyang J, Nguyen M, et al. Effect of temperature, pH, dissolved oxygen, and hydrolysate on the formation of triple light chain antibodies in cell culture. Biotechnol Prog 2010; 26: 1438-1445.
[85] Han Y, Ha T, Kim Y, Lee G. Bcl-xl overexpression delays the onset of autophagy and apoptosis in hyperosmotic recombinant Chinese hamster ovary cultures. J Biotechnol 2011; 156: 52-55.
[86] Becker E, Florin L, Pfizenmaier K, Kaufmann H. An XBP-1 dependent bottle-neck in production of IgG subtype antibodies in chemically defined serum-free Chinese hamster ovary (CHO) fed-batch processes. J Biotechnol 2008; 135: 217-223.
[87] Kharbanda S, Saxena S, Yoshida K, et al. Translocation of SAPK/JNK to mitochondria and interaction with Bcl-xl in response to DNA damage. J Biol Chem 2000; 275: 322-327.
[88] Shaffer A, Shapiro- Shelef M, Iwakoshi N, et al. XBP1, down-stream of Blimp-1, expands the secretory apparatus and other organelles, and increase protein synthesis in plasma cell differentiation. Immunity 2004; 21: 81-93.
[89] Yoon S, Song J, Lee G. Effect of low culture temperature on specific productivity, transcription level, and heterogeneity of erythropoietin in Chinese hamster ovary cells. Biotechnol Bioeng 2003; 82: 289-298.
[90] Trummer E, Fauland K, Seidinger S, et al. Process parameter shifting: Part I. Effect of DOT, pH, and temperature on the performance of Epo-Fc expressing CHO cells cultivated in controlled batch bioreactors. Biotechnol Bioeng 2006; 94: 1034-1044.
[91] Nam J, Ermonval M, Sharfstein S. The effects of microcarrier culture on recombinant CHO cells under biphasic hypothermic culture conditions. Cytotechnology 2009; 59: 81-91.

6 Upstream and Downstream Process Technology

6.1 Bioreactors for Animal Cell Culture

Detlef Eisenkraetzer

6.1.1 Introduction

In this chapter, bioreactors are discussed, that are used for the transformation of substrates to biopharmaceuticals using mammalian cells and that have the capability to control relevant cultivation conditions like pH, temperature and dissolved oxygen content. Systems without control functions, which have to be placed in a controlled environment like an incubator, are not regarded.

Bioreactor technology for the production of therapeutic proteins is based on classic bioreactor systems used in life science industry to produce fine chemicals and proteins using microbial systems. In principle, cell culture bioreactors are modifications of bioreactor systems used for cultivation of microbial cells, which are mainly stirred-tank bioreactors. The stainless steel bioreactor is a perfect system to fulfill the requirements of microbial fermentations:

- Excellent homogenization achieved by a high volumetric power input of >10 kW m^{-3}.
- High mass transfer rates, e.g., an oxygen transfer rate of >5 g L^{-1} h^{-1}.
- Efficient heat transfer with a coefficient >3 kW m^{-2} K^{-1}.

In contrast, animal cells have different requirements for their environment. They are very sensitive to shear forces and gradients in their environment such like the pH-value, the concentration of nutrients and toxic metabolites, as well as the dissolved carbon dioxide level. The classic stirred-tank bioreactor eliminates these gradients by the introduction of high turbulence via the stirrer. Unfortunately, animal cells are not protected by a cell wall and show a higher sensibility to shear stress. Therefore, the power input in cell cultivation systems is usually in a range of 5 to 300 W m^{-3} and much lower as in microbial cultivation systems. Fortunately, the slow growth rate and metabolism of animal cells do not result in an exothermic behavior of the cultivation. The cooling power is less important and therefore plastic is a suitable material for the construction of bioreactors. In the last decade this flexible material enabled the realization of new concepts to solve the engineering conflict of industrial animal cell culture, which is a combination of low shear stress and sufficient homogenization to minimize gradients in bioreactor systems.

Nowadays, 2 major trends are obvious:
- For blockbuster biotherapeutics, the optimization of conventional large-scale stirred-tank stainless steel bioreactors (STR) using state of the art technologies like computational fluid dynamics is still a promising approach. For bioreactors scales of 5,000 to 20,000 L and the related scale-down models, classic bioreactor systems seem to be the preferred technology even in the near future.
- With the increased productivity and the growing importance of tailored medicines, more production processes in a scale below 5,000 L are feasible. Single-use bioreactors (SUB) are here often the system of choice.

This chapter will compare the progress achieved in the characterization and optimization of classic bioreactors with the new approaches in bioreactor design for single-use bioreactors. Moreover, the use of cell retention devices in seed train or production culture to improve productivity is discussed. This review summarizes the available literature and focuses on the engineering aspects of devices for cell cultivations. For basic explanations and comparison, e.g., cultivation strategies (batch, fed-batch, perfusion) please refer to the literature mentioned in the respective Chapters 6.2 and 6.3.

6.1.2 Major Requirements for Animal Cells in Bioreactors

1. Provide a barrier for the environment to ensure a monoseptic process.
2. Access to nutrients in the optimal concentrations. The task of the cultivation system is to minimize gradients, such that nutrients that are added in fed-batch or continuous cultivation are not exceeding inhibitive concentrations. In current high cell density cultivations, often concentrated feeding solutions are used, in which pH, osmolality, and concentrations of components exceed toxic concentrations. To avoid local toxic concentrations an effective macromixing is required.
3. Control dissolved oxygen, temperature, and pH and pCO_2 to optimal values. Animal cells are very sensitive to pH gradients [1] and gradients of 0.1 pH units can have a significant influence on culture performance. The range for acceptable dissolved oxygen (DO) levels is rather large. 40 to 60 % are acceptable for most cell lines [2]. The operation range for temperature is easy to control, as a range of 36 to 38 °C is optimal for most cells and small differences have no significant influence [2]. Most cell lines have a broad range of acceptable pCO_2 which is reported at 4 to 18 % [3, 4] or 5 to 15 % [5]. Moreover, the pCO_2 level influences the glycosylation of secreted proteins via the change of internal pH [6].
4. Homogenization of supplied additives (e.g., nutrients) and cells with limited shear stress. Shear stress results of a force F applied parallel to a surface ($\tau = F/A$ in N m^{-2}). Many data exist on the response of animal cells to laminar shear stress. Ludwig et. al. measured an increase in LDH level already at values from 0.75 to 1.00 N m^{-2} for adherent BHK cells [7]; Kretzmer and Schügerl detected reduced

viability at values >2.7 N m^{-2} [8]. In suspension culture cells are damaged via eddies. As smaller the eddies, as higher is the damage. McQueen measured a critical Kolmogorov eddy length of around 3 μm for a myeloma cell line [9]. Bu-Resh measured critical values of 50 dynes cm^{-2} (500 N m^{-2}) in a turbulent situation [10]. Unfortunately the shear stress itself is not easy to calculate in turbulent systems and these numbers are difficult to apply on bioreactor engineering.

Another useful number to estimate the shear stress resulting from moving components in the bioreactor, is the energy dissipation or the ratio of maximal to average energy dissipation rate in a bioreactor [11]. Mollet and colleagues from Genentech reported in 2008, that in cell culture bioreactors local energy dissipation rates of up to 1,000 W m^{-3} are present and tolerated by the cell culture [12]. Unfortunately, large differences exist between different cell lines. For most cell lines the energy dissipation rate is a suitable criteria to estimate the shear stress, but for some cell lines the tip speed of the stirrer is the preferred criterion. Already in 1984, Präve et. al. recommended a critical stirrer tip speed of 1 m s^{-1} [13] – a value that is still a suitable threshold for critical shear stress caused by the stirrer.

Mechanical damage to animal cells in sparged bioreactors is mainly a result of sparging gases [14, 15] and not of the turbulence introduced by the stirrer. Despite the energy dissipation rates, the velocities, that characterize the sparging process, are rather high. During bubble formation at sparger velocities of 2.5 m s^{-1} can easily be reached [16] and bubble bursting is associated with velocities of over 5 m s^{-1} [17]. These values are even higher as the typical stirrer tip speeds of 1.0 to 1.5 m s^{-1}.

For all these considerations the progress of cell line selection and engineering has to be considered. The energy dissipation values tolerated by industrial cell lines have increased tremendously over the last years. In the late 1990s, typical production cell lines at Roche Diagnostics tolerated up to 150 W m^{-3}. Nowadays, most cell lines tolerate values up to 300 W m^{-3}.

6.1.2.1 Specifics for Adherent Cells

If cell lines cannot be adapted to cultivation in suspension, the cultivation under adherent conditions is still the best choice for industrial production. Different requirements of these systems are:

- Maximal surface area for cultivation per bioreactor volume.
- Higher shear sensitivity, as not only cell damage, but cell removal from the surface is important [7].
- Absence of limitations of cell proliferation and production caused by aggregate formations, contact inhibition, and mass transfer limitations in these aggregates [18, 19].

6.1.3 Classification of Bioreactors by Application and Process Type

6.1.3.1 Bioreactors for Cultivation of Suspended Cells

Most approved pharmaceutical proteins are produced in stirred-tank bioreactors [20].

Classic Stirred-tank Bioreactors

In contrast to the stirred tanks used for microbial cultivations, bioreactors for animal cell cultures differ in design and operating parameters.

In an animal cell culture process, there are significant limitations to fulfill the tasks listed in Table 6.1.1.

Table 6.1.1: Comparison of typical operation parameters for bioreactors.

Parameter	Microbial Process (e.g., *Escherichia coli*)	Animal Cell Culture (e.g., CHO)
Bioreactor geometry	"tall" => height : diameter-ratio >3 to maximize wall surface for heat transfer, gas hold up time and k_La	"flat" => height : diameter-ratio <2 to limit hydrostatic pressure on bottom and to improve axial mixing as most additives are added on top and power input is limited
Aeration (superficial gas velocity)	"high" => superficial gas velocity of 24 to 294 m h^{-1} are typical [21]	"low" => superficial gas velocity at cell culture conditions is only at 4 to 6 m h^{-1} due to the limitations from shear force and foam generation [22, 23, 15]
Power input from stirrer	"high" => up to 10,000 W m^{-3} are usual values [24]	"low" => average values of 5 to 80 W m^{-3} are often reported [12, 15]

To minimize the limitations shown in Figure 6.1.1, several specific devices and modes of operations are used in animal cell culture bioreactors.

- The dispergation of gas in the bioreactor and the
- reduction of gas bubble size cannot be achieved via an effective stirrer and higher stirring rates as it is useful for microbial fermentations because it leads to an increasing shear stress which is induced mainly via small eddies. Therefore, cell culture impellers are usually working in axial flow and are poor gas dispensers. [25]. Langheinrich et al. could show, that oxygen transfer in an 8 m^3 bioreactor with cell culture media follows a correlation of van't Riet and Tramper [26] initially developed to describe the dependency of oxygen transfer rate on superficial gas velocity for bubble columns [27].
- At the given low k_La values in animal cell culture [27] the mode of aeration is the major tool to support the oxygen supply. Several ways to improve oxygen transfer are possible.
  - The use of microspargers is associated with the disadvantage of poor CO_2 removal and the formation of a stable foam via the small bubbles [28].

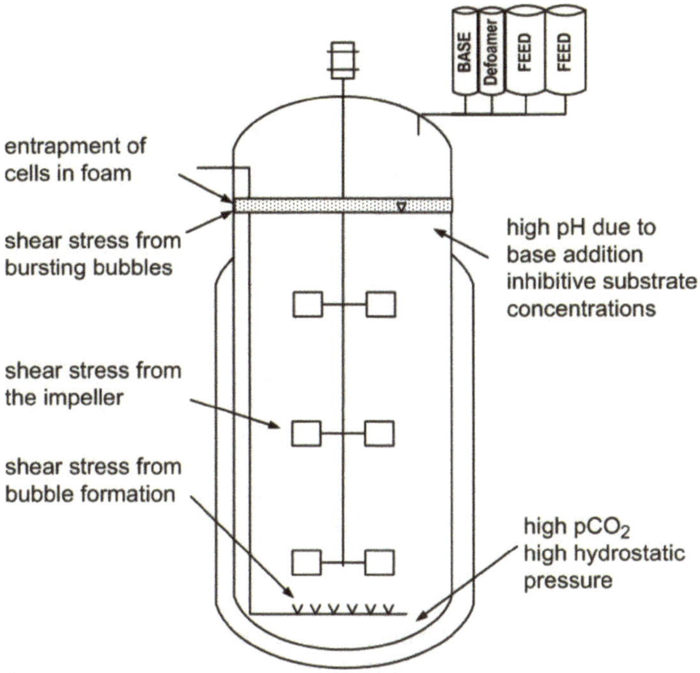

Figure 6.1.1: Limitations for optimal cell culture conditions in stirred-tank bioreactors.

  - The use of increased backpressure results mainly in an increased CO_2 level, because of the higher Henry coefficient of CO_2 compared to O_2 [29].
  - Therefore, the preferred method is the addition of pure O_2 in the aeration stream [30].
- The prevention of toxic dCO_2 levels and the efficient removal of CO_2 are important issues to be solved in large scale bioreactors working in fed-batch mode at high cell density [3, 4]. A very successful method to remove CO_2, supply O_2 and to prevent the formation of a stable foam is the aeration with O_2 via a microsparger and the aeration with nitrogen or air via an open pipe sparger for CO_2 removal. Foam resulting from this operation mode is less stable than foam generated with ring-spargers or microspargers [20]. A controlled pCO_2 level can be managed via the application of an optimized sparging strategy using open pipe, microsparger and adapted impellers, even at a cultivation scale of 15 m^3 [31]. A reliable on line measurement for dCO_2 in combination with specific controllers enables the control of the dCO_2 level via the additional nitrogen flow for CO_2 removal [5]. For CO_2 sensitive processes the consequences in productivity can be dramatic (Figure 6.1.2). Moreover, the dCO_2 effects the glycosylation of the produced proteins via its influence on the internal pH [4-6].

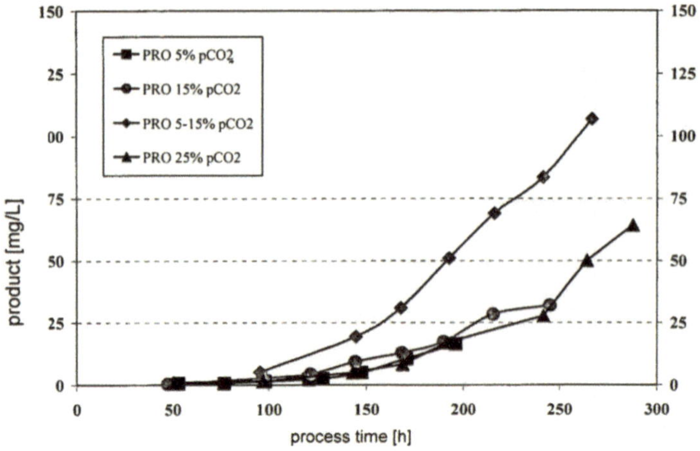

Figure 6.1.2: Product concentration of a process that operates following an optimal dCO_2 range in comparison to processes at low and high dCO2 [5].

- Sieblist from Roche Diagnostics wrote: “Some mass transfer problems need a more structured approach. The removal of CO_2 from large-scale reactors is an important example. In large reactors, sufficient removal rates require high gas throughputs, which however, may influence the transport behavior of the impellers.” [22]. Even under cell culture conditions the overflow of a stirrer is a phenomenon present in large scale cultivations (Figure 6.1.3).

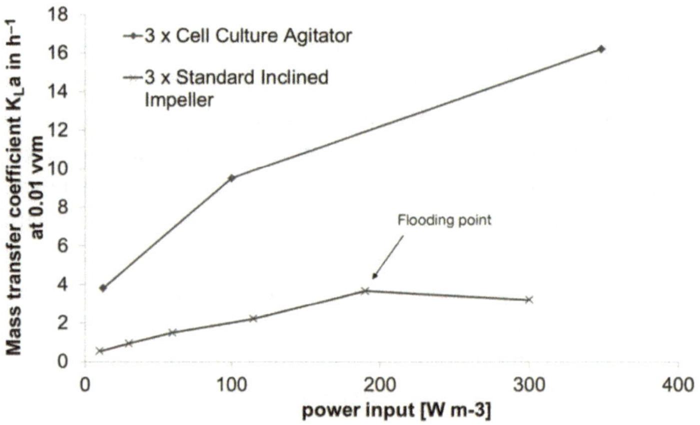

Figure 6.1.3: Mass transfer of impellers under cell culture conditions measured in a 1,500 L bioreactor (adapted from [113]).

– Last but not least the limited power input at animal cell processes influences the behavior of stirrers in animal cell cultivations. To prevent the mentioned negative influence on the impeller behavior, most pharmaceutical processes use optimized inclined impellers like hydrofoil impellers. Examples of used impellers are the Lightnin A320™ (Lightnin Mixers), Maxflo™, HE-3™ (Chemineer) or other impellers specially developed for optimal axial mixing in animal cell culture (Figure 6.1.4 and Figure 6.1.5) [25, 32, 14, 31]. Otherwise very poor transportation of additives from the top of the bioreactor to the bottom is the consequence [22]. An example is shown in Figure 6.1.6: Even for 1,000 L scale, high pH gradients are present if Rushton turbines with a radial flow regime are used. If specialized impellers with an axial flow regime are used, one has to take into account that these impellers achieve the same power input as impellers with radial flow regime at higher shear stress [33]!

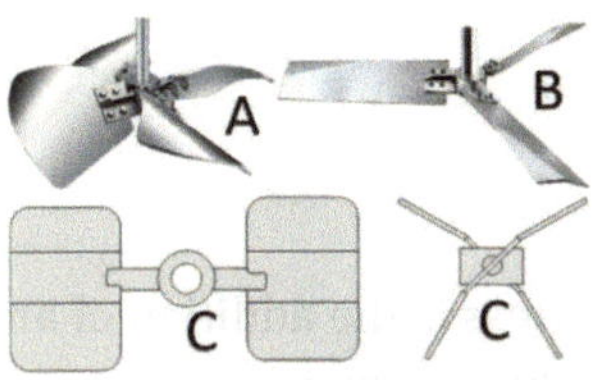

Figure 6.1.4: Maxflow **(a)**, HE-3 **(b)** and cell culture agitator EP1588758 **(c)** (reprint with permission).

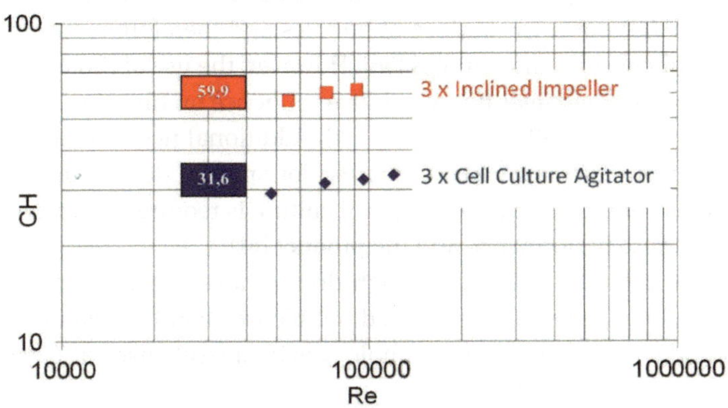

Figure 6.1.5: Homogenization number (mixing time × stirrer speed) of different impellers under cell culture conditions (adapted from [113]).

Summarizing these limitations, the operation of a stirred-tank bioreactor as a "mixed bubble column" is a useful approach: The application of an impeller just for mixing at a constant low power input and the control of oxygen supply and CO_2 removal via the aeration mode using pure oxygen and other gases decouple the shear stress in the bioreactor from the increasing oxygen demand of a growing cell culture [28, 5].

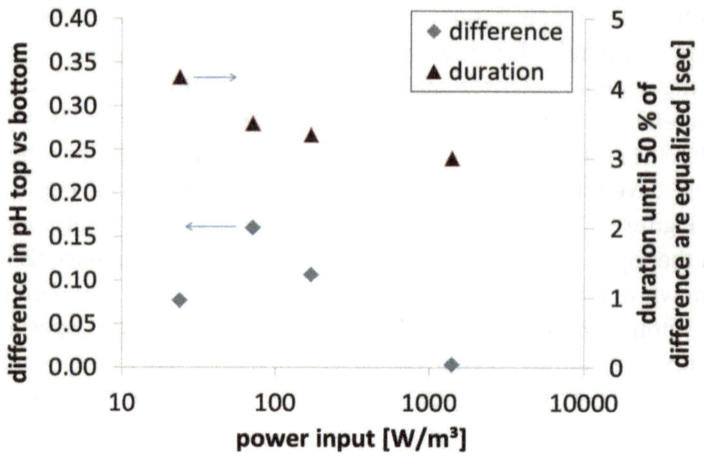

Figure 6.1.6: pH gradients in a 1,500 L stirred-tank bioreactor equipped with 3 Rushton turbines during a pH-shift from 7.2 to 7.0 (data: Detlef Eisenkraetzer, Roche Diagnostics).

Another effective way to meet limitations from shear stress is the addition of amphiphilic block copolymers such like Pluronic F-68® (polyoxyethylene-polyoxypropylene block copolymer, poloxamer) protection agents to the culture medium [34, 35]. Surfactants like Pluronic F-68® cover the surface of bubbles and therefore reduce or eliminate interaction of animal cells and bubbles [36]. However, the use of protecting additives like Pluronic F-68® or defoamer has a negative influence on filtration steps during downstream processing [37, 38] (see Chapter 6.4). Additional negative effects on the cell cultivation performance have been reported for specific defoamers [39, 40]. Therefore, a careful use of protecting agents in cell culture is required to prevent a negative influence on culture performance and the product [41].

If very shear sensitive cells are cultivated which do not have a higher oxygen demand and shear protecting agents cannot be used bubble-free aeration via membranes is another appropriate approach. This technology is intensively used at Bayer Healthcare and well characterized [14, 42].

Airlift Bioreactors

Airlift bioreactors benefit from the density reduction due to sparging to force circulation of a liquid. The circulation can be achieved via a guided internal loop (main application) or via an externally aerated loop. The limitations are characterized in Figure 6.1.7. The operation principles and engineering characteristics have been recently described by Merchuk [43]. The application of airlift bioreactors for the production of biologics is not very common. The only company having a long experience in the use of airlift bioreactors for animal cell culture up to 5 m³ scale is Lonza Biologics [44]. A comparison of the performance and main design parameters between a stirred-tank bioreactor and an airlift bioreactor in m^3-scale is published by Abraham et al. [45]. As the airlift bioreactors are working with high superficial gas velocities to ensure sufficient mixing and mass transfer, the shear stress in airlift bioreactors is comparable to stirred-tank bioreactors [46, 47].

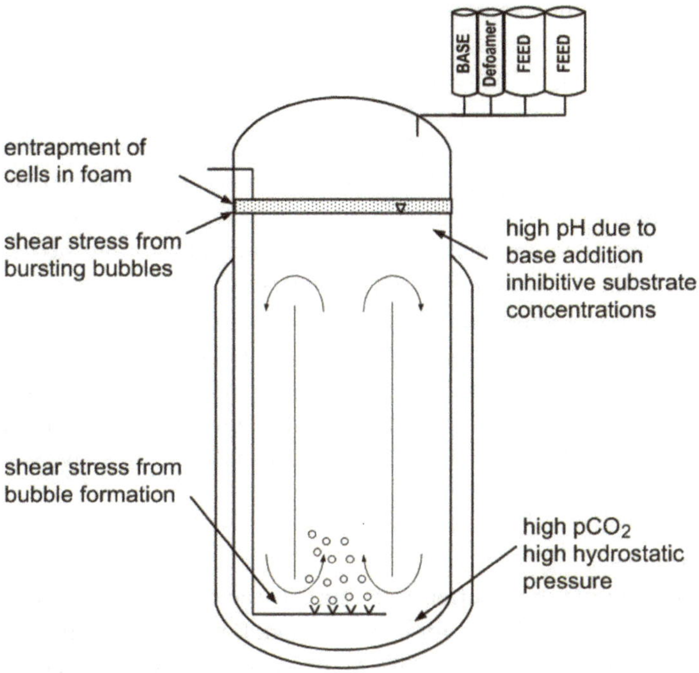

Figure 6.1.7: Limitations for optimal cell culture conditions in an airlift bioreactor.

Devices for Improved Processes in Stirred-tank Bioreactors

The standard process for production of monoclonal antibodies is a fed-batch procedure using the addition of concentrated feed solutions to achieve maximal productivity. Despite this procedure, several engineering strategies are applied to increase the productivity in the product concentration or to shorten the process time. Additionally, several sensible block buster molecules like factor VIII [48] and erythropoietin [49, 50] are produced in continuously perfused cultivation processes using cell retention systems. In industrial applications only differences in size or density serve as retention criteria for animal cells.

A review on the advantages and disadvantages of different cell retention devices for stirred-tank bioreactors was published by Woodside in 1998 [51]. An excellent evaluation and review of useful technologies from an industrial perspective has been published by colleagues from Serono® in 2003 [52]. Very few publications report new data since 2003. Some interesting new aspects are listed below.

Dialysis of Media

The concept of the dialysis of media through a stirred-tank bioreactor is more or less the application of the advantages of a hollow fiber reactor for adherent cells (see p. 403) to a stirred-tank bioreactor system [53]. Modules of membranes permeable for the nutrients and low weight byproducts of the cells, like lactate, are placed in a bioreactor or a liquid stream is pumped through an external loop containing these membranes. As the membrane is impermeable for high molecular weight substances like proteins, very high cell and product concentrations can be reached via the dialysis of the medium. To place the dialysis module outside the bioreactor has some disadvantages: uncontrolled environment (DO, pH, temperature) and higher shear stress for the cells in the circulation pump. The advantage is the access to the module and ability to replace a module, which has blocked due to fouling processes. An industrial dialysis process using modules in the bioreactor was developed by the former Boehringer Mannheim (Roche Diagnostics) up to m^3 scale in the 1980s [54, 55]. The technology has been intensively used and further developed and today a significant bioreactor capacity at Roche Diagnostics is based on this technology [56].

Cell Retention via Centrifuges

A very common way to uncouple the dilution rate in a bioreactor from cell growth of suspended cells is the use of centrifuges for cell retention. Compared to filtration devices, there is no risk of clogging of a filter [57]. Many systems are on the market to enable cell retention on small or large scale. Continuous centrifuges for cell retention are working at low g-forces (<300 × g). After optimization of the applied g-force, it is possible to achieve a separation of viable from dead cells. Viable cells can be retained in the bioreactor and the viability of the culture can be maintained at high values over a long period [58, 18]. Other systems like the CARR® ViaFuge® (Pneumatic Scale Corporation) work at higher g-forces and gently separate all cells. This highly concentrated cell suspension can support a split batch process with a harvest several times a week. Compared to a classic perfusion process with a continuous flow, such a cultivation regime has some advantages for the batch definition and scheduling of downstream operations (see Chapter 6.4).

Cell Retention via Ultrasound

Acoustic filters apply a fixe ultrasonic field that forces cells to aggregate [59]. These aggregates settle at the bottom of the field and either they are continuously washed back to the bioreactor or the field is turned off frequently and the settler is harvested in a "back flush mode" [59, 61]. Even if all publications state the scale-up potential, there are several problems connected with a scale-up. With increasing scale, a power input of at least 60 W in a 400 mL system is needed and a significant loss of efficiency is related of the interaction of the high cell density slurry with the ultrasonic field [62]. The removal of the heat is as well a significant issue. Therefore, no industrial application in large scale has been published [63]. Nevertheless, Evonic Industries has patented perfusion process using an acoustic settler for the production of erythropoietin, but use at large scale by the former Siegfried Biologics (Avesta Biotherapeutics) is not known [50].

Hydrocyclones

Another compact device using density differences as separation criterion are hydrocyclones. Hydrocyclones operate as a solid bowl centrifuge without rotation of the bowl [65]. Instead, the suspension is brought in rotation by being tangentially fed into the cyclone under pressure. The suspension is supplied tangentially into the cyclone and forms a primary vortex along the conical wall. This spinning vortex accelerates as it moves down the tapered cyclone to partly separate the cells. Thus, the underflow is enriched with cells; particularly, larger cells while smaller cells and cell debris are forced to leave the cyclone through its overflow nozzle by forming an upward-spinning secondary vortex in the cyclone core. Hence, hydrocyclones use the same separation principle as centrifuges. However, they have no movable parts as the vortex motion is performed by the energy of the fluid itself, and stable eddies are generated by the high velocity of the tangential flow of the suspension. Moreover, they do not require high precision components, which are expensive to manufacture and operate. These properties suggest hydrocyclones for separating animal cells in continuous perfusion bioreactors.

Whereas they have a long history in microbial fermentations, their use in animal cell culture is still limited. A limiting factor is the significant pressure drop of at least 0.5 bar in the system and the high shear forces. The effect of the shear forces can be reduced by the addition of protection substances and optimization of pressure conditions [64]. In later publications, good results in cell retention with low cell damage combined with the separation of dead (overflow) and living cells (underflow) has been achieved [65]. The scale-up potential is currently limited to 500 L bioreactors, as the separation efficiency decreases with cyclone diameter [64]. A solution would be the operation of parallel systems [64]. Until today, significant applications in industrial scale have not been published.

Cell Retention via Inclined Settler

The state of the art of the settler technology has been fully described by Woodside [51]. In general, all types of inclined plate settlers consist of multiple thin chambers formed between the plates. Suspension from the bioreactor enters the settler at the bottom and in the chambers an overflow of media is flowing towards the harvest tank. In the chambers, cells sediment to the bottom plates and form an underflow downwards to the bottom of the settler. Here, they enter a circulation flow, which carries them back to the bioreactor. An inclined settler is using density difference as separation force. The settling path of the cells is shortened via inclined placement of plates in short distances. In current industrial use, adherence of cells to the plates and fouling in the device is prevented via special surface treatment and the application of vibrations on the plates. Influence on the cell culture in the loop for cell retention is usually reduced via a reduction of the temperature to below 8 °C. Detailed descriptions on the functionality and parameters of inclined settlers are available from manufacturers like Biotechnology Solutions [66]. Whereas in 1998 most data have been based on small scale experiments at universities, meanwhile several publications on industrial use appeared [67, 48]. Systems with a throughput of up to 1,000 L/d are in use [67].

The settler technology is mainly in use for the production of products that are sensitive to degradation in the bioreactor and need a short residence time in it. Publications on proteins that are not stable in culture list alpha-L-iduronidase and N-acetylgalactosamine 4-sulfatase from BioMarin Pharmaceutical and Kogenate® from Bayer HealthCare Pharmaceuticals [48, 42, 68].

A different application of the inclined settler has been recently published by Roche Diagnostics: The retention of the cells in semicontinuously operated seed train bioreactor to achieve higher inoculum cell densities and to shorten the cultivation time in a 14 m^3 production bioreactor [69].

Cell Retention via Filtration

Filtration is the retention of cells via size exclusion – usually achieved via the retention on membranes. A major limitation of these systems is the clogging of the filtration device. This tendency can be reduced by application of different strategies.

Spin filters are rotating cylindrical filters made of stainless steel or disposable plastic material – usually with a pore size of 20 to 50 µm. Via steric hindering of the rotating filter, cell retention is achieved and clogging of the filter is reduced. Nevertheless, clogging of the filter is one of the major drawbacks of the system. A very common version is a simple design of a filtration cylinder directly mounted on the stirrer. Here, the reduction of clogging via an increased rotation speed of the filter is very limited as the increased stirrer speed leads to improved mass transfer and increases cell damage. An alternative is an externally driven spin filter mounted on the stirrer shaft inside the bioreactor – but here, shear stress is as a function of the rotation speed of the cylinder. Nevertheless, the ability to prevent clogging via a back stream of media and the alteration of the rotation speed offer a significant scale-up potential of spin filter units [70]. Industrial applications have been reported from several companies [20].

Cross flow filtration is another alternative technology for cell retention. Like any filtration technology, cross flow filtration devices face the issue of clogged membranes as a scale-up and performance limitation. The XD® technology of DSM Pharmaceuticals uses cross flow filtration devices to achieve extreme cell concentrations up to 3×10^8 mL even at a scale of 500 L [71]. This technology is used by different pharmaceutical companies [72, 73]. In alternating tangential flow (ATF) devices from Refine Technology, media is pushed back frequently though the membranes from the permeate side of the membrane. The prevention of fouling can be achieved and the industrial use and scale up is feasible. Examples of the application of ATF devices in m^3 scale have been reported by CMC Biologics [74].

6.1.3.2 Bioreactors for Cultivation of Immobilized Cells

For the production of vaccines and modern virus-based biologics as well as for medium scale transient transfections, usually adherent cell lines are preferred [75, 76, 34]. Simple conventional cultivation systems are cell factories and roller bottles. These systems can be used for industrial production [77]. Better controlled cultivation systems use cells on microcarriers in stirred-tank bioreactors, fluidized bed bioreactors, fixed bed bioreactors, or the cells are cultivated in hollow fiber systems (Figure 6.1.8). Single-use systems of different complexity are shown in Figure 6.1.9.

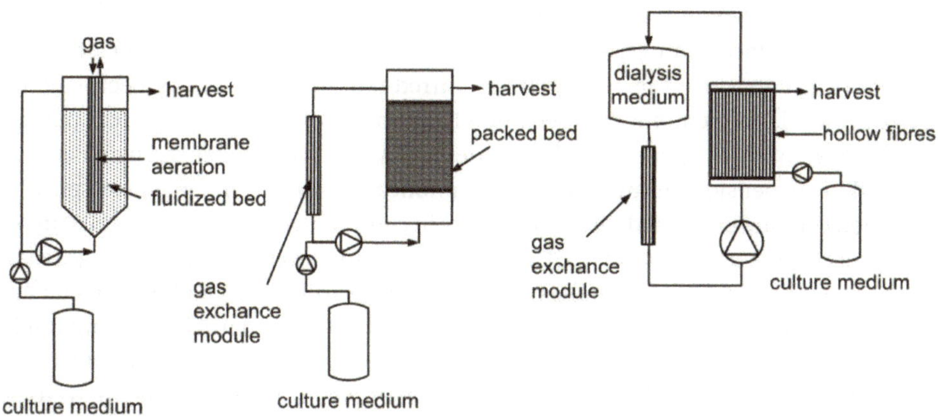

Figure 6.1.8: Principle of perfusion bioreactors for adherent cell culture (fluidized bed, packed bed, and hollow fiber bioreactors).

Especially, for the production of vaccines, single-use bioreactors become more and more the preferred system. Many examples of industrial applications have been reviewed by Eibl and Eibl [78]. Anyhow, industrial production of therapeutic proteins usually relies on the cultivation of suspended cells as the preferred method. Only few

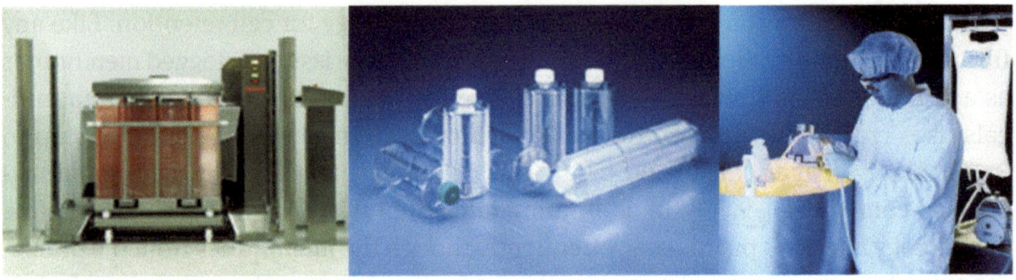

Figure 6.1.9: Disposable systems for the cultivation of animal cells: cell factory, roller bottle, single-use bioreactor (example from Thermo Fisher Scientific).

therapeutic proteins have been produced with adherent cells in the past. The most prominent example is the production of erythropoietin at Amgen [79, 80].

Stirred-tank Bioreactors

Most industrial processes for cultivation of cells on carriers use stirred-tank bioreactors. The cells can be attached to polymer based carriers like Cytodex® (GE Healthcare Life Sciences [81] or macroporous microcarriers. Cell attachment is often supported via charges throughout the immobilization matrix or a collagen layer coupled to the micocarrier surface. An overview of current materials is given by Warnock and Al-Rubeai [82]. Behavior of microcarriers in stirred-tank bioreactors has been studied by Nienow [2]. As mentioned in Section 6.1.2.1, shear sensitivity is a critical parameter. The energy to suspend the microcarriers ($<3\ \mathrm{W\ m^{-3}}$) is lower as the usual energy applied in suspension culture. But several publications report higher shear sensitivity, when cells are cultivated on microcarriers in comparison to suspended cells under identical cultivation conditions [83]. A possible explanation is, that eddies larger than the particles do not damage animal cells [82, 84, 19]. Particles of a higher diameter are more sensitive to shear stress, but the retention of these large particles is comparably easy to achieve using internal or external spin filters [85] or hollow fiber modules [86]. For scale-up strategies it should be considered, that seeded microcarriers can be used to inoculate the next larger bioreactor scale [87].

Fluidized and Packed Bed Bioreactors

In fluidized bed bioreactors, the microcarriers are fluidized via an upward flow generated via an external loop. In contrast, packed bed bioreactors use solid porous material. The circulation principles are usually the same as for fixed bed bioreactors. Both systems operate exclusively in perfusion mode. As gas sparging would influence the homogeneity of the fluidized bed, gas exchange is realized via internal [18] or external

[88] gas exchange membranes. Sampling devices and sensors for DO, pH, and others can be placed in the circulation loop or the bioreactor itself.

Nevertheless, the use of adherent Chinese hamster ovary (CHO) and baby hamster kidney (BHK) cells is often difficult due to the formation of large aggregates in the fluidized bed bioreactors [18, 88]. Compared to fluidized bed bioreactors, packed bed bioreactors have several scale-up limitations. Gradients of DO, pH, and substrates in the packed bed are an inherent feature of the system, as mixing is limited in the packed bed. On the other hand, shear stress is very low. If this is a critical issue for process design, a packed bed bioreactor can be superior to fluidized bed or stirred-tank cultivation [89]. The cell density can be monitored via sampling of carriers and releasing of the cells, e.g., enzymatically via trypsin treatment or by fluorescence based offline assays [90]. Since the use of dielectric spectroscopy has a long history in biomass monitoring [91], meanwhile good results for the online estimation of biomass with dielectric spectroscopy have been reported [90, 92].

Beside possible mass transfer limitations due to large agglomerates, animal cells can stop to proliferate on carriers due to contact limitations. If product formation is coupled to cell proliferation, a dramatic reduction of specific productivity is the consequence. This can inhibit the development of an efficient process [93-95]. The disadvantage of the formation of tissue-like aggregates in these bioreactors offers a new field of application of the technology: tissue culture [96, 97]. So far, no biopharmaceutical proteins from production processes using fluidized or fixed bed bioreactors have been approved, even if the production of licensed products in these systems seems promising [98]. A review on the use of fluidized or fixed bed bioreactors for the production of viruses and viral vectors is available from Généthon [99].

Hollow Fiber Bioreactors

Hollow fiber bioreactors are another system for the cultivation of adherent cells in perfusion mode. Here, the cells are here attached to the outer surface of semipermeable tubes. Fresh media is circulated through the tubes and nutrients and metabolite diffuse through the membranes, whereas proteins are retained in the lumen between the membranes. Therefore high cell densities and product concentrations are reached in the extra capillary volume. Commercial available hollow fiber reactors usually consist of cartridges containing thousands of these tubes. Compared to carrier based culture techniques, the use of a hollow fiber system has the advantage, that the "molecular weight cut off" of the membranes can be adapted to process requirements and that diffusion of media through the membranes and cell layers are better under control. Within the fibers gradients are low, but especially DO gradients in the cell layers are possible [100, 101].

Large scale systems claim to reach a productivity of 10 g L^{-1} antibody per day for the largest available 1 L system [102]. In contrast, a 1,000 L stirred-tank bioreactor with a final product concentration of 1 g L^{-1} after 14 days, characterized by an average daily productivity of only 71 mg L^{-1} d^{-1}, yields in a total productivity of 71 g d^{-1}, which is much higher as in the 10 L hollow fiber system. Typical applications for hollow fiber bioreactors work with less than 5 L/d [63] and therefore these systems are only interesting for sensitive products at low volume and so far only one commercial product is produced in hollow fiber bioreactors by the Cytogen [20].

6.1.4 Single-use Bioreactors

The productivity of industrial cell culture processes has improved dramatically in the last 15 years. Lonza published a fortyfold increase in product concentration and an 240-fold increase of the volumetric productivity for its GS-CHO processes based on metabolically engineered CHO cells expressing a glutamine synthetase (GS) for using glutamine-free medium conditions [44]. Therefore, only the production of antibody blockbuster biopharmaceuticals with a demand in the range of tons per year requires large bioreactor capacity in a scale of 5 to 20 m^3. Many products and the majority of material for clinical studies can be produced in a scale-up to 2,000 L. Moreover, the number of approved products has increased and the trend to tailored medicines requires increased flexibility [20]. This is the reason for the growing market of single-use bioreactors.

Single-use bioreactors consist of a pre-sterilized (usually γ-irradiated) plastic container made of polycarbonate, polyethylene, polystyrene, polytetrafluorethylene, polypropylene, or ethylene vinyl acetate [78]. These single-use containers are placed in holders/frames that support the shape of the reactor and serve often as heat-exchange surface via double jackets or electrical heating. Single-use sensors can be integrated or classic sensors can be placed into the bioreactor via sterile connectors. Not regarded in this chapter are "traditional" single-use systems like the Tecnomouse® from Integra Biosciences [103] or the miniPerm® (Heraeus Sepatech), because these systems are laboratory equipment, that has to be placed in a controlled environment like an incubator. Nevertheless, the differentiation between laboratory equipment and a bioreactor is getting more difficult for single-use systems. Shake flask system equipped with a control for DO, pH, and multiple feed lines integrated in CO_2 incubators exist and have successfully been implemented in industry [104]. These systems are a valid alternative to small-scale bioreactor systems like the Mobius® (EMD Millipore) or the UniVessel® SU (Sartorius).

Single-use bioreactors offer several advantages compared to conventional bioreactors [105-107]:

- Reduced footprint for installation of the systems,
- reduced investment and operational costs,
- reduced CO_2 emissions, far less use of electricity, compressed air and clean steam,
- reduced project realization time for installation of new systems,

- reduced requirements to infrastructure,
- increased flexibility to changing capacity demands.

These advantages are the major driving force for the triumphal progression of single-use bioreactors in the last decade. Since the first successful launch of a single-use bioreactor [108], the variety and quantity of these bioreactors has increased. In general, all single-use bioreactor systems are used for the cultivation of suspended cells or the cultivation of cells using suspended microcarriers. Their characterization is comparable to the classic systems described in Section 6.1.3.1. At the Zurich University of Applied Science several groups are working on the characterization of single-use systems. Several publications from these groups give an overview to the field of single-use technology [109, 110]. Therefore this chapter will focus on some special aspects and systems that are of interest for an industrial application. A classification of existing systems via the mode of mixing is given by the DECHEMA (Figure 6.1.10).

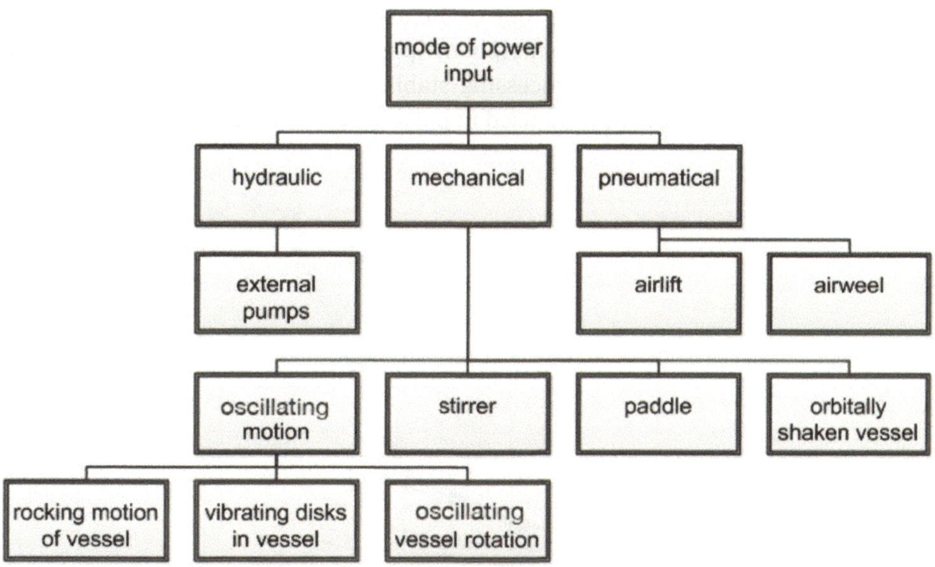

Figure 6.1.10: Classification of single-use bioreactors for cultivation of suspended cells (adapted from [150]).

6.1.4.1 General Specifics of Single-use Bioreactors

Based on their use and the materials they are made of, all single-use bioreactors have common characteristics.

Additional Testing Procedures

Single-use bioreactors are made of plastics. The use of plastic material for the manufacturing of therapeutic proteins requires intense testing of components leached into the product containing culture broth. As bioreactors are used in upstream processing, the requirements with regard to product safety are lower compared to the downstream processes that are closer to the final drug product. Nevertheless, several additional test procedures like integrity tests, bioburden level, transportation safety, sterilization procedures, etc. are required. Guidance on the recommended tests is given by several organizations [111-113].

Leachable Components Influence Process Performance

Negative effects from leachable components can influence the cultivation process dramatically (Figure 6.1.11) [114, 115]. Most leachable tests focus on patient safety and cannot detect the influence on process performance. Few interlaboratory tests have been performed to evaluate the influence on process performance [116]. Recently, colleagues from AMGEN could identify one of the toxic components leaching from bioprocess containers. Bis(2,4-di-tert-butylphenyl)phosphate (bDtBPP) is a degradation product of the trisarylphoshite processing stabilizer IRGAFOS 168® (Ciba Specialty Chemicals), an antioxidant additive used in polyethylene production [117].

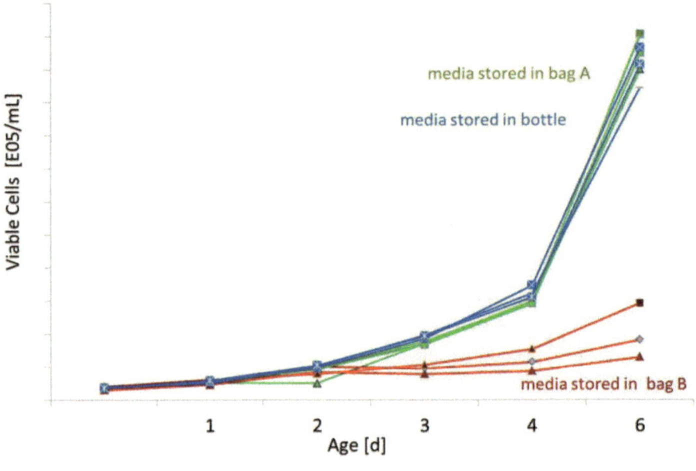

Figure 6.1.11: Influence of media storage in bags on media performance in shaker experiments (data: Christina Fritz, Roche Diagnostics).

Behavior of Media in Plastic Containers

The inner contact layer is often made of polyethylene. Therefore, the adsorption of cholesterol, fatty acids, and other media components can have a critical influence on the process [118]. On the other hand, the storage of feed media in disposable bags instead of tanks equipped with air filter units may have a positive influence on media performance, as oxidation of media components is reduced (Figure 6.1.12).

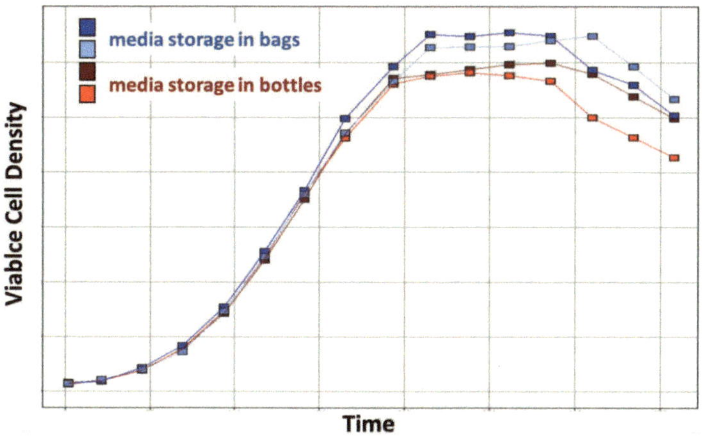

Figure 6.1.12: Influence of feed media storage in bags on media performance in bioreactor experiments (data: Britta Schmidt, Roche Diagnostics).

6.1.4.2 Bioreactors with a Rotating Stirrer

The latest overview on single-use bioreactor models is presented in Table 4.1. in a review from Eibl and Eibl [109]. The most common mode of mixing of single-use bioreactors in this overview is wave induced motion or a rotation stirrer. Regarding stirred single-use bioreactors, some differences with respect to the power input and shear stress are obvious. At Rentschler Biotechnologie a study for a specific mAb process was conducted. Typical power input in the compared single-use bioreactors (HyClone, Xcellerex, Sartorius) was approximately 15 W m^{-3} [119]. Roche Diagnostics published a comparison of engineering numbers from a comparative study [24]. For some parameters like the Newton number, different values are published. For the impeller of the 250 L Hyclone SUB®, Hyclone (Thermo Fisher Scientific) has calculated a Newton value of 2.1. This value is used in many publications, but is rather high compared to other results for inclined impeller that report values between 0.7 and 1.4 for different types of 45° pitched blade impellers [120, 121].

At Roche Diagnostics, impellers in stainless steel bioreactors and different single-use bioreactors have been characterized using identical methods and equipment to measure the power input. The results show that the compared single-use bioreactors are operating exactly in the transition zone from laminar to turbulent flow (Figure 6.1.13). The exact calculation of the power input under process conditions is very difficult because the reactor operates under transient flow conditions at low cell densities and low power input (3 W m^{-3}) and under turbulent conditions at maximal cell concentration and higher power input of 15 W m^{-3}.

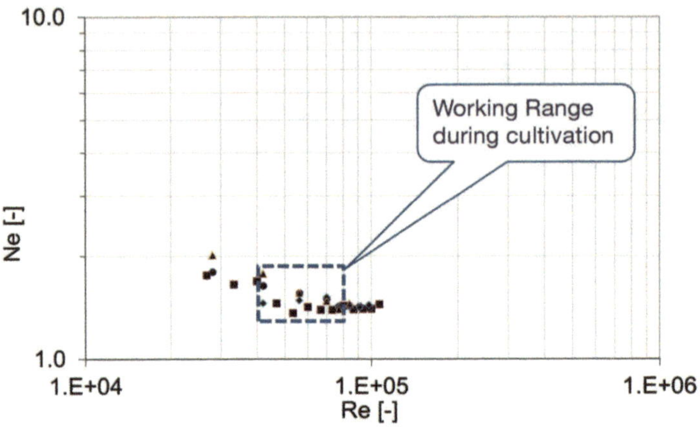

Figure 6.1.13: Newton number of different single-use bioreactors as a function of turbulence in the bioreactor (data: Britta Schmidt and Marion Traverse, Roche Diagnostics).

Most stirred single-use bioreactors are operated at around 15 W m^{-3}. This is related to several constraints:

1. For the shipment of the bioreactor bag from the vendor to the customer it is necessary to fold the empty bag to reduce the volume. Therefore, the bags are designed without baffles. Depending on the location of the stirrer, this tank design leads to a reduction of the power input. At Reynolds numbers typical for single-use bioreactors (>10^4) a reduction by a factor of 4 is characteristic [122]. Experiments at Roche Diagnostics showed only a reduction by a factor of 2 for bioreactors with a centered stirrer system (compare Sartorius single-use bioreactor and stainless steel bioreactor in Table 6.1.2).
2. In combination with the low power input, single-use bioreactors are operated following the regime of bubble columns based on the correlations of van't Riet [123]. To ensure a good mixing time, impellers with axial flow offer advantages. The positive effect is demonstrated for the CultiBag STR™ (Sartorius), which is available with Rushton turbines or inclined impellers (Figure 6.1.14).
3. Even at low power input, the shear stress in unbaffled bioreactors is at least 30 % higher as in baffled bioreactors [124].

Table 6.1.2: Comparison of Newton number measured via identical torque meter method (adapted from [24]).

Bioreactor	Newton number
2 L stainless steel bioreactor (1 inclined impeller; 4 blades d/D = 0.33, 4 baffles)	1.5
250 L Hyclone S.U.B.™ (1 inclined impeller; 3 blades d/D = 0.33, no baffles, stirrer of center placed)	1.4
200 L Sartorius CultiBag™ STR (2 inclined impeller; 3 blades d/D = 0.33, no baffles)	1.3

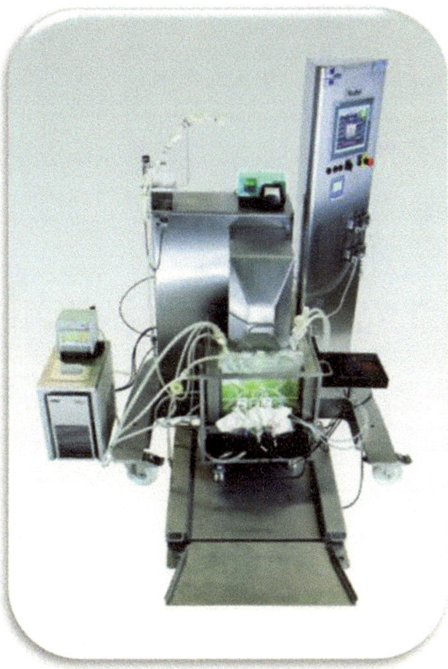

Figure 6.1.14: Mixing time measured in CultiBag® STR using different stirrer combinations (adapted with permission from [23]).

Based on different publications [119, 24, 125], k_La values are lower (maximal 15 to 20 h^{-1}) due to the low power input but sufficient for cell culture applications if pure oxygen is used. Mixing time is usually comparable to classic bioreactors and in most cases sufficient for animal cell culture applications. Few bioreactor systems exist, where the mixing time is significantly higher compared to conventional systems [119, 24] – but these systems differ significantly from a classic design with a stirrer mounted on a stirrer shaft.

6.1.4.3 Bioreactors with a Paddle Mixing System

The absence of baffles results in rotation and formation of a vortex in a cylindrical vessel. A solution for this problem is the use of cubic vessels. The most prominent system is the bioreactor Nucleo™ (ATMI Life Sciences) – a cubic bioreactor with a paddle system to mix and aerate the liquid (Figures 6.1.15 and 6.1.16). Despite or better due to the unusual design, good homogenization numbers of about 40 and k_La values of over 10 h^{-1} at low shear levels can be achieved [24]. Several pharmaceutical companies have successfully evaluated the cubicle as a system for the cultivation of animal cells. [126-130].

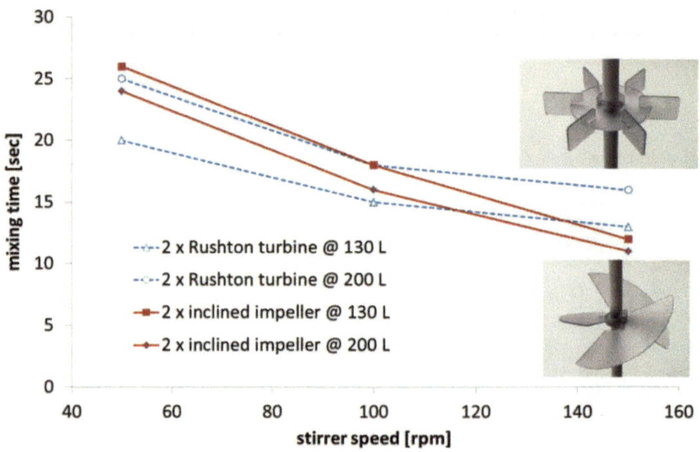

Figure 6.1.15: Nucleo™ 50 L by ATMI LifeSciences.

Figure 6.1.16: Single-use bioreactor CultiBag® STR used for characterization of mixing behaviour using different stirrers [23].

6.1.4.4 Orbitally shaken Bioreactors

The standard cultivation system for animal cells in a CO_2 incubator is the orbitally shaken flask. An obvious scale-up approach is the construction of larger orbitally shaken single-use bioreactors. These systems generally have the advantage of a foamless cultivation due to the use of headspace aeration and the absence of moving parts in the bioreactor. On the other hand the movement of the bioreactor itself requires more flexible connections to the supply systems and a higher footprint compared to stirred bioreactors.

Several systems are used and characterized. Beside their normal application using wave induced motion, orbitally shaken rocker bags are used in universities and companies. Orbitally shaken CultiBags™ have been characterized by the Zurich University of Applied Science Wädenswil by Löffelholz et al. [131]. One of the drawbacks of rocking motion and orbital shaking is obvious: if the movement of the container exceeds a threshold value, the liquid cannot follow the motion moves slower as the container itself. If the liquid gets "out of phase" the power transfer from the vessel to the liquid is reduced (Figure 6.1.17). Moreover, the prediction of the out-of-phase behavior seems difficult at different scales (differences between Figure 6.1.17 and Figure 6.1.18). Data from Merck-Serono showed, that the use of rocking motion for the pillow-design results in a better cell culture performance compared to orbitally shaken bioreactor of the same design [128].

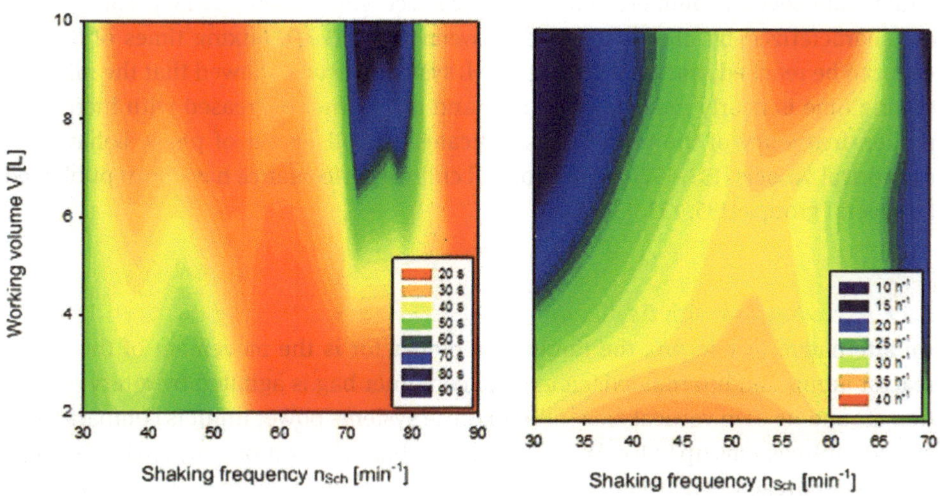

Figure 6.1.17: Mixing time (left picture) and oxygen transfer coefficient k_La (right picture) of an orbitally shaken 20 L CultiBag® using an amplitude of 50 mm. At standard filling level, the liquid motion gets "off phase" at values higher as 70 rpm – mixing time and k_La are not increased, as the power of the fast moving bag cannot be transferred to the slower moving liquid (adapted by Christian Löffelholz).

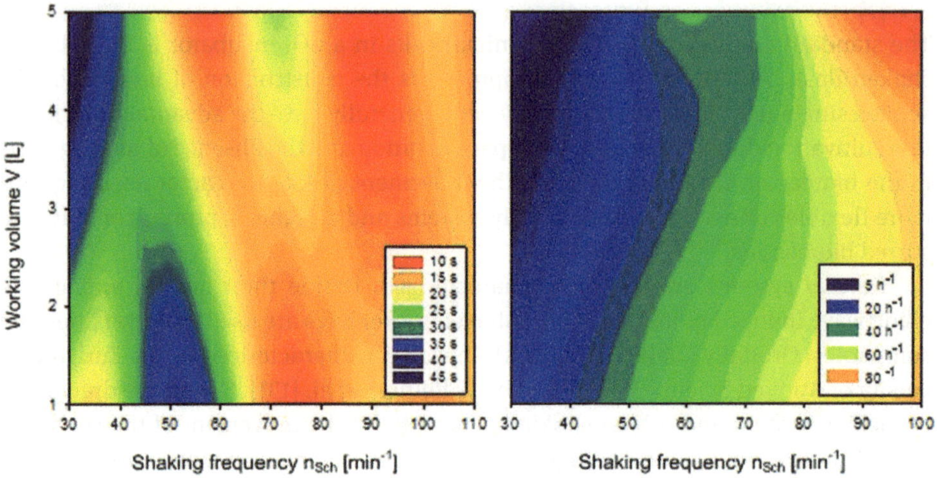

Figure 6.1.18: Mixing time (left picture) and oxygen transfer coefficient k_La (right picture) of an orbitally shaken 10 L CultiBag® using an amplitude of 50 mm. The out-of-phase effect appears at low volume and low shaking frequency (unpublished data by Jennifer Mahlig, Tobias Backoff and Christian Löffelholz, ZHAW).

In contrast to a simple change of the movement for the same bioreactor, the successful application of bioreactor systems designed for orbitally movement has been demonstrated and several commercial systems are successfully used [132-137]. The systems are characterized by high k_La-values between 15 to 45 h^{-1}. Mixing times of around 40 s can be reached, but for different 200 L systems results showed that the minimal mixing time is nearly reached at 60 rpm and not further decreased with increasing motion frequency of the vessel. This indicates as well an out of phase behavior of cylindrical vessels [133, 132]. Results of cell culture performance have been published by several groups [135, 133, 132, 137].

6.1.4.5 Bioreactors with Oscillation

An alternative way to mix the liquid in a bioreactor is the movement of the whole reactor around an internal axis. A cubical container bag is agitated by rotary oscillation around its own vertical axis [138]. In these systems power input is comparable to stirred-tank systems up to 100 W m^{-3} and with submerse aeration only k_La in the range of stirred bioreactors between 30 to 40 h^{-1} can be reached (Figure 6.1.19). Shear stress for these bioreactors was evaluated using the floc destruction method described by Henzler and Biedermann [11] and, therefore, a direct comparison to the publications for classic bioreactor systems from this group is possible: at low power input and surface aeration the shear stress in the single-use bioreactor is very low (floc diameter >150 μm). If high k_La values are required (>30 h^{-1}) and therefore power input of

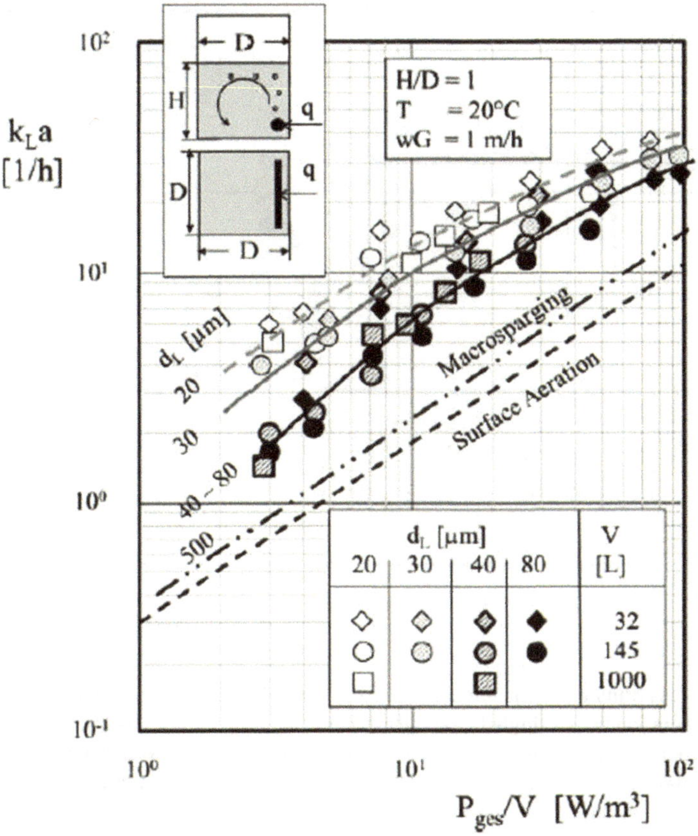

Figure 6.1.19: Volumetric mass transfer coefficient within BaySHAKE® single-use bioreactors (Bayer Healthcare) using bubble aeration with linear microspargers mounted at one side of the container. Variation of pore size d_L from 20 to 80 µm and bioreactor volume from 32 to 1,000 L. The plot contains a comparison with macrosparging (d_L = 500 µm, dashed-dotted line) and surface aeration (dotted line) (reproduced with permission from [138]).

~100 W m^{-3} and microsparging has to be applied, the shear level is in the range reported for sparged stirred tanks (Figure 6.1.20). This is not surprising if one takes into account that the sparging has the major effect on shear stress in the typical cell culture bioreactor (the “stirred bubble column”).

Due to the low power input, good mixing time (<90 s), low shear stress, and high k_La value, the cell culture performance is comparable to conventional systems [138]. As the system is relatively new, few publications on the industrial use exist.

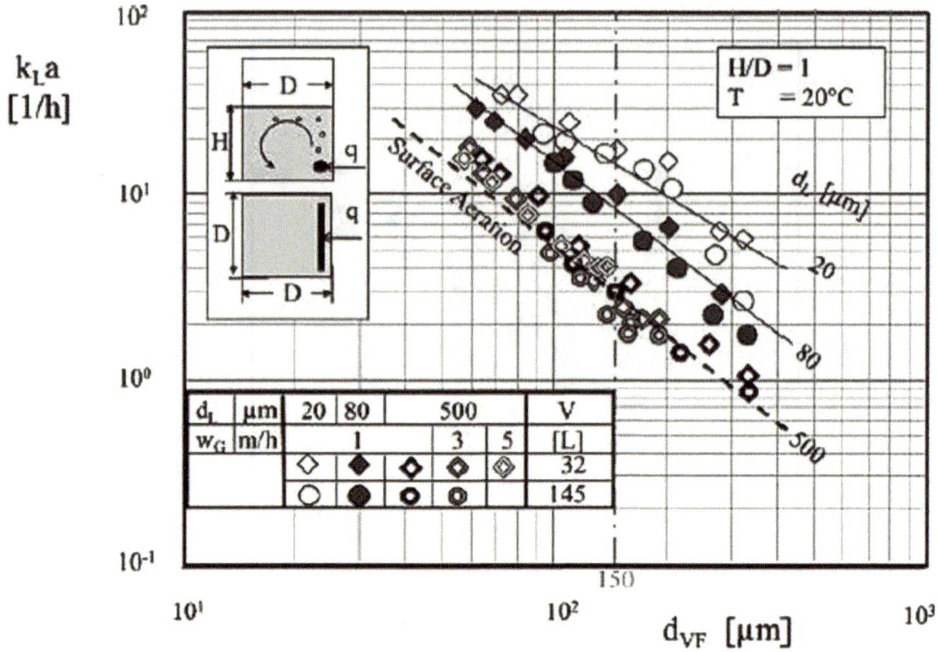

Figure 6.1.20: Volumetric mass transfer coefficient of BaySHAKE® single-use bioreactors (Bayer Healthcare) under bubble aeration with a single linear micro- or macrosparger (pore size between 20 ≤ dL [lm] ≤ 500) mounted in the corner between bioreactor bottom and wall. Data is plotted against the reference floc diameter dVF. Floc diameter indicates shear stress inversely (high diameter = low shear stress; reproduced with permission from [138]).

6.1.4.6 Rocking Motion Type Bioreactors

The standard rocking motion type bioreactor (RTMB) consists of a half filled "pillow" type bag, that undergoes a rocking motion along its length axis. This leads to the formation of a wave, which is responsible for mixing and mass transfer from the gas phase. The first single-use bioreactor, that was used in industrial cell culture was a rocking motion type bioreactor developed by Singh [108]. Even if the k_La values of the first systems have been only 4 h^{-1}, these systems paved the way for single-use bioreactor by their robustness and ease of use. Today, 15 years after introduction, RMTBs are the most common and best characterized single-use bioreactors [139].

Several engineering characterizations have been published in the past. The results can be summarized:

- Power input in RMTBs is comparable to conventional bioreactors. The usual operating range is 20 to 40 W m^{-3} and can be increased up to 80 W m^{-3} [135].
- Mass transfer rates k_La for oxygen reported from 4 h^{-1} [108] up to 8 h^{-1} [140] depending on the used bioreactor type.

- Mixing time for 95 % homogeneity for the complete bioreactor is relative long (100 to 200 s) under normal operation conditions. These poor values are achieved if the tracer is added on the short site of the reactor, orthogonal to the movement of the wave. At the point of addition the local dispersion of added substances is taking place within less than 10 s.
- Just like for all shaken systems at very high rocking speed an “out of phase” behavior is possible, if the bags move faster than the liquid can move inside.

These statements are not valid for the CELL-tainer® (ATMI Life Sciences). Here the rocking motion happens along the short axis of the system and moreover it is combined with a longitudinal motion of the reactor. This leads to significant differences in the engineering numbers:
- Power input ranges from 50 to 400 W m^{-3} [141].
- Mass transfer rates k_La for oxygen are between 300 h^{-1} [108] and 700 h^{-1} [140], depending on the filling level of the bag.
- Mixing time for 95 % homogeneity for the complete bioreactor is less than 10 s.

These high engineering numbers enable cultivation of animal cells at very high cell concentration [142] and microbial organisms [143].

The rocker systems have proven their ability for the cultivation of cells as part of the seed train in several studies [140]. After a transfer from a classic system a short adaptation phase can be observed sometimes (Figure 6.1.21). Schöpke performed an

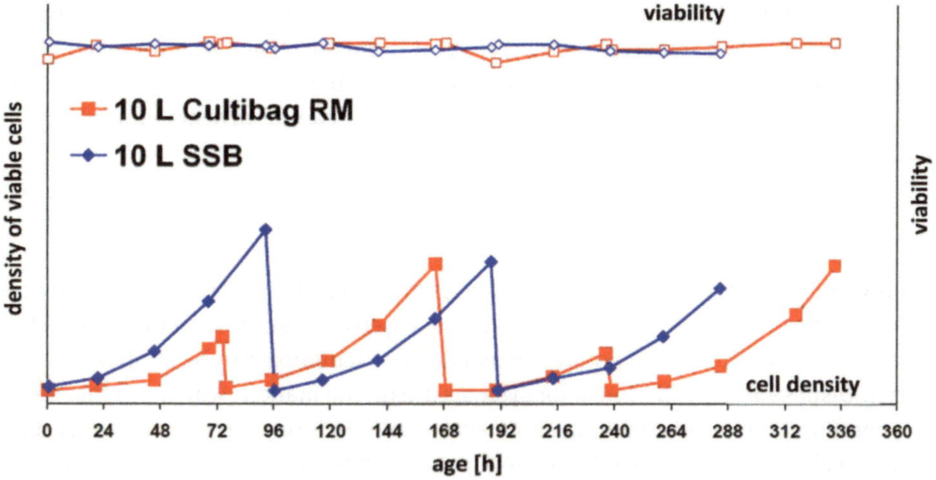

Figure 6.1.21: Comparison of cell density in a solera process performed in a 20 L CultiBag RM® (red) and a conventional 15 L stirred-tank bioreactor (blue) at Roche Diagnostics. Seed train cultivation was performed in shaker flasks at 30 W m^{-3}. The adaptation effect for the first split after is not a regular phenomenon (reproduced with permission from [140]).

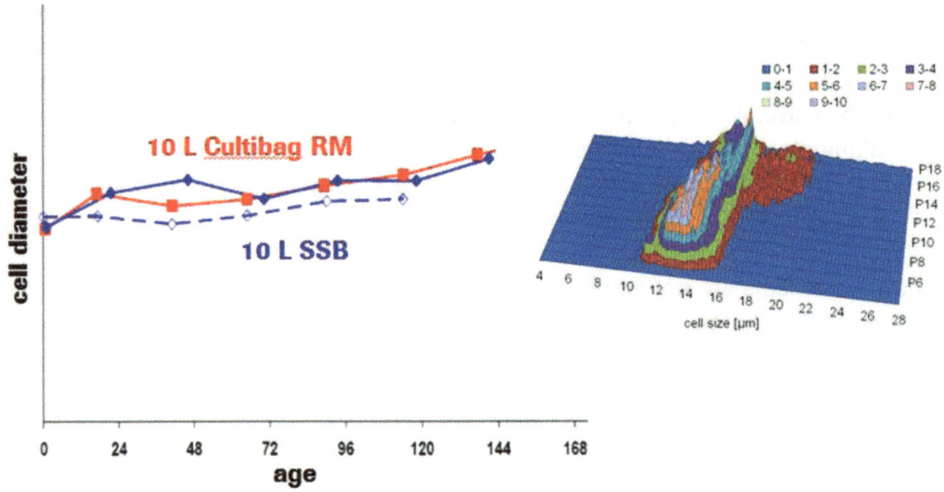

Figure 6.1.22: Comparison of cell size during a cultivation performed in a 20 L CultiBag RM® (red) and a classic 15 L stirred-tank bioreactor (blue) at Roche Diagnostics (adapted with permission from [140]).

intensive comparative characterization engineering parameters and cell culture performance in a 20 L rocking motion type bioreactor using several CHO production cell lines. He could not detect any significant differences in cell growth, viability, cell size (Figure 6.1.22), metabolite and product formation, and substrate consumption. Identical Annexin V levels on the cell membranes during the cultivation indicate even no differences in cell apoptosis [140]. Therefore, it is not surprising that for many pharmaceutical companies the rocking motion single-use bioreactor is the preferred seed train bioreactor as it offers significant advantages in capital investment and operational costs [31].

6.1.4.7 Modifications of Single-use Airlift Bioreactors

Mainly 2 different single-use airlift bioreactors are currently on the market. Airlift bioreactors up to 50 L are available from Cellexus. When a lower aeration rate is used but sufficient circulation of the liquid is needed, the circulation can be supported via a stirrer. Disposable hybrid bioreactor systems are on the market, which use this mode of operation. The published data show acceptable performance. Nevertheless, data on high cell density culture or larger applications have not been published [144].

Another mode to support the mixing of the bioreactor is realized by PBS Biotech. The family of "Air-Wheel" bioreactors is available in scales from 3 to 2,500 L. In these bioreactors an internal wheel is driven by the rising gas bubbles to support mixing of the liquid. The shear stress measured at the wheel is below 0.5 Pa [145], a value that is below the critical value for BHK cells at 0.75 Pa [7]. On the other hand, the average energy dissipation was calculated to be 0.01 $m^2 s^{-3}$ [146] – which is a typical value for shear intense bubble columns. Comparable shear stress levels in stirred tanks are reached at 1 $m^2 s^{-3}$. Successful production of vaccines using shear sensitive anchorage dependent cells indicate that the shear stress from the aeration is lower as in comparable bubble columns without the rotating wheel [146].

6.1.4.8 Single-use Bioreactors for the Cultivation of Adherent Cells

As mentioned above, the production of viral particles or viral vaccines often uses adherent cell lines such like VERO (from Esperanto: verda reno, African green monkey kidney epithelial cells) or MDCK (Madin-Darby canine kidney) and others. Therefore, the use of microcarriers for the cultivation is necessary to provide the surface for cell attachment. GSK Biologicals published an intensive evaluation of single-use bioreactor systems for the above mentioned processes [129]. The evaluation showed, that there exist substantial variances between the bioreactor systems. Large differences in the homogenization number were found and all systems induce higher shear forces due to the low power number of the impellers. Finally, some systems could not be used for production.

For the use of stirred single-use bioreactors for the cultivation of shear sensitive cells on microcarriers, the low power input is a challenge because higher stirrer speed and higher tip speeds are required to bring the microcarriers into suspension. An alternative is the use of packed bed single-use bioreactors for the commercial production of vaccines. A conventional and well-established system for small-scale cultivation is the CellCube® from Corning. Several publications mention the reliable and good performance of these simple systems – but only a few control parameters can be recorded in these bioreactors [89]. The next level of process control can be reached with complete disposable packed bed bioreactors like the CelliGen® BLU by Eppendorf or the iCELLis™ from ATMI LifeSciences. Several groups use these systems for the small scale production of vaccines [147, 148], and they proved that these systems show a comparable performance as classic packed bed systems [149].

6.1.5 Criteria for the Selection of a Bioreactor Type

Several criteria determine the decision of the optimal bioreactor type for a process. Only a few are mentioned here.

1. Lifecycle of the process and the plant
   - When the bioreactor and the facility is dedicated for the production of a single product at least for the next 15 years in a scale over 2 m^3 bioreactor volume, the use of conventional equipment (STR) is usually the optimal choice. Change over procedures does not play an important rule, flexibility

can kept at a low value and robustness of the qualified equipment is high. In this case, a single-use solution is not the preferred choice.
   - In all other cases, most process can be realized in single-use systems and other criteria like compatibility to plastic material determine the decision.
2. Requirements of the cell line
   - For adherent cell lines the best space-time yield seems to exist for fluidized bed bioreactors, fixed bed, and hollow fiber systems. But due to the complexity, these systems are not established in larger scales. Most pharmaceutical companies prefer the cultivation of cells on microcarriers in stirred tanks. These bioreactors are better characterized, easy to operate, and the changeover to a process with suspended cells is possible. Many host cell lines can be characterized by their operation zone with respect to their sensitivity to shear stress, gradients (as a function of homogenization) and mass transfer, e.g., CO_2 removal and O_2 supply. This characterization has a strong influence on the selection of the optimal bioreactor system (Figure 6.1.23).

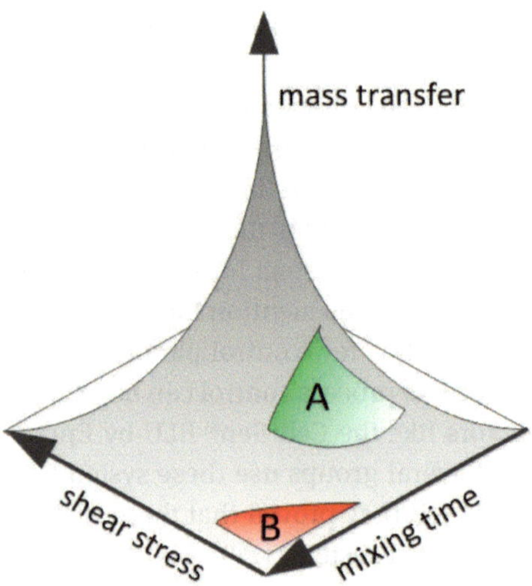

Figure 6.1.23: Examples for operating zones in different processes (A = high sensitivity to gradients and significant need for O_2 supply and CO_2 removal, but tolerance to shear stress; B = low sensitivity to gradients and low oxygen uptake rate but very sensitive to shear stress).

3. Capital expenditure
   - If single-use systems are an alternative with respect to lifecycle and scale, they are often the best choice due to their advantages in economy (investment costs), foot print, and infrastructure requirements. But single-use systems require significant engineering support and surveillance by the customer [24]. In case of process changes requalification of the equipment may be needed, e.g., if the new process conditions are not covered in the initial leachable study. As a consequence, pharmaceutical companies that intensi-

vely use single-use systems are searching for “plastic experts” [114]. Moreover, the dependency from the supplier of the single-use equipment, its supply chain management, and quality assurance (QA) system is significant.

6.1.6 Conclusion

In the last decade, understanding of processes and knowledge about bioengineering for animal cell culture has increased significantly. Especially with the introduction of single-use bioreactor systems a large variety of possible cultivation systems exists. Nevertheless, the best-selling system at production scale is still a stirred-tank bioreactor – either as a complete stainless steel systems or as hybrid system of a stainless steel tank combined with single-use bags or as a completely disposable system (plastic facility). All other systems such as bubble columns, orbitally shaken bags, etc. are still systems with low market penetration. However, some of these cultivation systems have the potential to displace the stainless steel-stirred tank as shown for the rocking motion type bioreactors used as seed train culture vessels. Beside their differences, all bioreactor systems have an optimal operating zone: high k_La and mixing time are usually connected to high shear stress and all kinds of mixing devices have their limitations. Stirrers do not contribute to k_La at low power input and at high stirrer speed they develop gas pillows behind the stirrer blades. On the other hand, shaken bioreactor systems show similar problems such like the out of phase behavior. Nowadays most industrial bioreactor systems are used as multipurpose installations for different processes. A careful characterization of the used equipment can support the determination of the appropriate operating range for the different processes using the same bioreactor type.

References

[1] Wayte J. 2013. Development of mammalian cell culture bioreactor process – Increasing productivity and process robustness (http://bio.lonza.com/uploads/tx_mwaxmarketing-material/Lonza_Posters_GS-NS0_process_improvements.pdf, accessed 15th March 2013).

[2] Nienow A. 2006. Reactor engineering in large scale animal cell culture. Cytotechnology 50: 9-33.

[3] deZengotita V, Kimura R, Miller W. 1998. Effects of CO_2 and osmolality on hybridoma cells: Growth, metabolism and monoclonal antibody production. Cytotechnology 28: 213-227.

[4] deZengotita VM, Schmelzer AE, Miller WM. 2002. Characterization of hybridoma cell responses to elevated pCO_2 and osmolality: Intracellular pH, cell size, apoptosis, and metabolism. Biotechnol Bioeng 77: 369-380.

[5] Eisenkraetzer D, Gaetgens J, Jockwer A, et al. 2009. Method for producing recombinant proteins with a constant content of pCO_2 in the medium. WO/2009/112250.

[6] Kimura R, Miller WM. 1997. Glycosylation of CHO-Derived Recombinant tPA produced under elevated pCO2. Biotechnol Prog 13: 311-317.

[7] Ludwig A, Kretzmer G, Schügerl K. 1992. Determination of a critical shear stress level applied to adherent mammalian cells. Enz Microb Technol 14: 209-213.
[8] Kretzmer G, Schügerl K. 1991. Response of mammalian cells to shear stress. Appl Microbiol Biotechnol 34: 613-616.
[9] McQueen A, Meilhoc E, Bailey J. 1987. Flow effects on the viability and lysis of suspended mammalian cells. Biotechnol Lett 9: 831-836.
[10] bu-Reesh I, Kargi F. 1989. Biological responses of hybridoma cells to defined hydrodynamic shear stress. J Biotechnol 9: 167-178.
[11] Henzler HJ, Biedermann A. 1996. Modelluntersuchungen zur Partikelbeanspruchung in Reaktoren. Chem Ing Tech 68: 1546-1561.
[12] Mollet M, Ma N, Zhao Y, et al. 2004. Bioprocess equipment: Characterization of energy dissipation rate and its potential to damage cells. Biotechnol Prog 20: 1437-1448.
[13] Ruttloff H. 1984. Handbuch der Biotechnologie. Präve P, Faust U, Sittig W, Sukatsch DA (eds). 2. Vol, p 197, Fig. 107 Tab. R. Oldenbourg, München and Wien 1984.
[14] Aunins JG, Henzler HJ. 2001. Aeration in cell culture bioreactors. Biotechnology Set. Wiley-VCH: 219-281.
[15] Henzler HJ, Kauling DJ. 1993. Oxygenation of cell cultures. Bioprocess Eng 9: 61-75.
[16] Joebses I, Martens D, Tramper J. 1991. Lethal events during gas sparging in animal cell culture. Biotechnol Bioeng 37: 484-490.
[17] Boulton-Stone JM, Blake JR. 1993. Gas bubbles bursting at a free surface. J Fluid Mech 254: 437-466.
[18] Eisenkraetzer D. 1999. Production of recombinant glycosyltransferases using Chinese hamster ovary cells – genetic and bioengineering aspects of process development". PhD Thesis, RWTH Aachen/FZ Juelich.
[19] Moreira JL, Alves PM, Aunins JG, Carrondo MJT. 1995. Hydrodynamic effects on BHK cells grown as suspended natural aggregates. Biotechnol Bioeng 46: 351-360.
[20] Chu L, Robinson DK. 2001. Industrial choices for protein production by large-scale cell culture. Cur Opin Biotechnol 12: 180-187.
[21] Junker BH. 2004. Scale-up methodologies for Escherichia coli and yeast fermentation processes. J Biosci Bioeng 97: 347-364.
[22] Schaepe S, Luebbert A, Pohlscheidt M, et al. 2013. Bioreactor performance: Insights into the transport properties of aerated stirred tanks (http://www.americanpharmaceuticalreview.com/Featured-Articles/122317-Bioreactor-Performance-Insights-into-the-Transport-Properties-of-Aerated-Stirred-Tanks, accessed 15th March 2013).
[23] De Wilde D, Noack U, Kahlert W, et al. 2009. Bridging the gap from reusable to single-use manufacturing with stirred, single-use bioreactors. Biopr Intl Suppl: 36-41.
[24] Eisenkraetzer D. 2009. Disposable systems as platform technology for R&D and clinical supply. At: BioProduction, Barcelona, Spain. Biopharm Intl 27.08.2010.
[25] Ibrahim S, Nienow AW. 2004. Suspension of microcarriers for cell culture with axial flow impellers. Chem Eng Res Des 82: 1082-1088.
[26] Schügerl K. 1992. Basic Bioreactor Design. Chem Ing Tech 64: 886.
[27] Langheinrich C, Nienow AW, Eddleston T, et al. 2002. Oxygen transfer in stirred bioreactors under animal cell culture conditions. Food Bioprod Proc 80: 39-44.
[28] Mostafa SS, Gu X. 2003. Strategies for improved dCO_2 removal in large-scale fed-batch cultures. Biotechnol Prog 19: 45-51.
[29] Weiss RF. 1974. Carbon dioxide in water and seawater: the solubility of a non-ideal gas. Marine Chem 2: 203-215.
[30] Chu L, Robinson DK. 2001. Industrial choices for protein production by large-scale cell culture. Cur Opin Biotechnol 12: 180-187.

[31] CMC Biotech Working Group. 2009. A-Mab Case Study - ISPE (www.ispe.org/pqli/a-mab-case-study-version-2.1, accessed 15th March 2013).
[32] Nienow AW. 1997. On impeller circulation and mixing effectiveness in the turbulent flow regime. Chem Eng Sci 52: 2557-2565.
[33] Henzler HJ, Biedermann A. 1996. Modelluntersuchungen zur Partikelbeanspruchung in Reaktoren. Chem Ing Tech 68: 1546-1561.
[34] Xie L, Metallo C, Warren J, et al. 2003. Large-scale propagation of a replication-defective adenovirus vector in stirred-tank bioreactor PER.C6™ cell culture under sparging conditions. Biotechnol Bioeng 83: 45-52.
[35] Murhammer DW, Goochee CF. 1990. Sparged animal cell bioreactors: mechanism of cell damage and pluronic F-68 protection. Biotechnol Prog 6: 391-397.
[36] Jordan M, Eppenberger HM, Sucker H, et al 1994. Interactions between animal cells and gas bubbles: The influence of serum and Pluronic F68 on the physical properties of the bubble surface. Biotechnol Bioeng 43: 446-454.
[37] Schulz C, Vogel JH, Scharfenberg K. 1997. Influence of Pluronic F-68 on the Ultrafiltration of Cell Culture Supernatants. In: Carrondo M, Griffiths B, Moreira JL, eds. Animal Cell Technology. Springer Netherlands: 373-378.
[38] McGregor WC, Weaver JF, Tansey SP. 1988. Antifoam effects on ultrafiltration. Biotechnol Bioeng 31: 385-389.
[39] Hiltner J, Linscheid K-P, Eisenkraetzer. 2009. Applicability of different defoamers in animal cell culture. University of Bonn; Roche Diagnostics GmbH.
[40] Junker B. 2007. Foam and its mitigation in fermentation systems. Biotechnol Prog 23: 767-784.
[41] Hesse F, Wagner R. 2000. Developments and improvements in the manufacturing of human therapeutics with mammalian cell cultures. Trends Biotechnol 18: 173-180.
[42] Henzler H-J, Kauling J, Schmitt F, et al. 2002. A unit and a process for carrying out high cell density fermentation. PCT/EP2002/009620.
[43] Merchuk JC. 2003. Airlift bioreactors: Review of recent advances. Can J Chem Eng 81: 324-337.
[44] Birch JR, Racher AJ. 2006. Antibody production. Adv Drug Deliv Rev 58: 671-685.
[45] Abraham S, Bland H, Westlake A, et al. 2003. Successful transfer and scale up of a mammalian cell culture process producing a monoclonal antibody from a stirred tank reactor to an airlift fermenter (http://bio.lonza.com/uploads/tx_mwaxmarketingmaterial/Lonza_Posters_Large_scale_stir_to_air_transfer.pdf, accessed 15th March 2013)
[46] Molina Grima E, Chisti Y, Moo-Young M. 1997. Characterization of shear rates in airlift bioreactors for animal cell culture. J Biotechnol 54: 195-210.
[47] Chisti Y. 2000. Animal-cell damage in sparged bioreactors. Trends Biotechnol 18: 420-432.
[48] Vogel JH, Nguyen H, Giovannini R, let al. 2012. A new large-scale manufacturing platform for complex biopharmaceuticals. Biotechnol Bioeng 109: 3049-3058.
[49] Laustsen M. 2011. A method for producing a biopolymer (e.g. polypeptide) in a continuous fermentation process. EP20080774069.
[50] Wienand W, Kunz F-R, Reichert D, et al. 2009. Process for the fermentative production of erythropoietin. WO/2009/147175.
[51] Woodside S, Bowen B, Piret J. 1998. Mammalian cell retention devices for stirred perfusion bioreactors. Cytotechnology 28: 163-175.
[52] Voisard D, Meuwly F, Ruffieux PA, et al. 2003. Potential of cell retention techniques for large-scale high-density perfusion culture of suspended mammalian cells. Biotechnol Bioeng 82: 751-765.
[53] Poertner R, Maerkl H. 1998. Dialysis cultures. Appl Microbiol Biotechnol 50: 403-414.
[54] Comer M, Kearns M, Wahl J, et al. 1990. Industrial production of monoclonal antibodies and therapeutic proteins by dialysis fermentation. Cytotechnology 3: 295-299.

[55] Kearns M, Comer MJ, Steegmans U, Jungfer H. 1996. Method and apparatus for cell culture. EP0224800.
[56] Hoffmann A, Lechner M, Reese C. 2012. Dialysis fermenter-bioreactor with dialysis device. US20120301923.
[57] Cruz HJ, Conradt HS, Dunker R, et al. 2002. Process development of a recombinant antibody/interleukin-2 fusion protein expressed in protein-free medium by BHK cells. J Biotechnol 96: 169-183.
[58] Johnson M, Lanthier Sp, Massie B, et al. 1996. Use of the Centritech lab centrifuge for perfusion culture of hybridoma cells in protein-free medium. Biotechnol Prog 12: 855-864.
[59] Gaida T, Doblhoff-Dier O, Strutzenberger Katinger H, et al. 1996. Selective retention of viable cells in ultrasonic resonance field devices. Biotechnol Prog 12: 73-76.
[61] Gorenflo VM, Angepat S, Bowen BD, Piret JM. 2003. Optimization of an acoustic cell filter with a novel air-backflush system. Biotechnol Prog 19: 30-36.
[62] Shirgaonkar IZ, Lanthier S, Kamen A. 2004. Acoustic cell filter: a proven cell retention technology for perfusion of animal cell cultures. Biotechnol Adv 22: 433-444.
[63] Voisard D, Meuwly F, Ruffieux PA, et al. 2003. Potential of cell retention techniques for large-scale high-density perfusion culture of suspended mammalian cells. Biotechnol Bioeng 82: 751-765.
[64] Jockwer A, Medronho R, Wagner R, et al. 2002 The Use of Hydrocyclones for Mammalian Cell Retention in Perfusion Bioreactors. In: Lindner-Olsson E, Chatzissavidou N, Lüllau E (eds). Animal Cell Technology: From Target to Market. 1 ed. Springer: 301-306.
[65] Elsayed E, Medronho R, Wagner R, Deckwer WD. 2006. Use of hydrocyclones for mammalian cell retention: Separation efficiency and cell viability (Part 1). Eng Life Sci 6: 347-354.
[66] Brown PC. 2013. Biotechnology Solutions. Cell settlers for the continuous culture of mammalian cells (http://www.cellsettlers.com/settlers.htm, accessed 15th March 2013).
[67] Shen Y, Yanagimachi K. 2011. CFD-aided cell settler design optimization and scale-up: Effect of geometric design and operational variables on separation performance. Biotechnol Prog 27: 1282-1296.
[68] Pavloff N. 2004. From the laboratory to the patients. Mod Drug Disc: 47-49.
[69] Pohlscheidt M, Jacobs M, Wolf S, Tet al. 2013. Optimizing capacity utilization by large scale 3000 L perfusion in seed train bioreactors. Biotechnol Prog 29: 222-229.
[70] Deo YM, Mahadevan MD, Fuchs R. 1996. Practical considerations in operation and scale-up of spin-filter-based bioreactors for monoclonal antibody production. Biotechnol Prog 12: 57-64.
[71] DSM. 2013. XD process – a revolution in process technology (http://www.dsm.com/en_US/html/dpp/xD_Technology.htm, accessed 15th March 2013).
[72] DSM. 2013. DSM announces license agreement with Amgen (http://www.dsm.com/en_US/html/dpp/news_items/2013_01_DSM_Amgen_Licence.htm, accessed 15th March 2013).
[73] Crucell. 2013. Protein production (http://www.crucell.com/Technology_-_Cell_Technology_-_Applications_-_Proteins, accessed 15th March 2013).
[74] Carstens J, Clarke H, Jensen J. 2009. Perfusion! Jeopardy or the ultimate advantage? (http://www.cmcbio.com/Portals/0/CMC/docs/perfusion.pdf, accessed 15th March 2013).
[75] Le Ru A, Jacob D, Transfiguracion J, et al. 2010. Scalable production of influenza virus in HEK-293 cells for efficient vaccine manufacturing. Vaccine 28: 3661-3671.
[76] Bleckwenn NA, Golding H, Bentley WE, Shiloach J. 2005. Production of recombinant proteins by vaccinia virus in a microcarrier based mammalian cell perfusion bioreactor. Biotechnol Bioeng 90: 663-674.
[77] Schwander E. 2013. Manufacturing vaccines in adherent cell lines using disposable multi-tray bioreactors (http://www.biopharminternational.com/biopharm/article/articleDetail.jsp?id=444994&sk=&date=&%250A%2509%2509%2509&pageID=4, accessed 15th March 2013).

[78] Eibl R, Kaiser S, Lombriser R, Eibl D. 2010. Disposable bioreactors: The current state-of-the-art and recommended applications in biotechnology. Appl Microbiol Biotechnol 86: 41-49.

[79] Griffiths B. 2007. The development of animal cell products: History and overview. Medicines from animal cell culture. John Wiley & Sons; 1-14.

[80] Tsao EI, Bohn MA, Omstead DR, Munster MJ. 1992. Optimization of a roller bottle process for the production of recombinant erythropoietin. Ann New York Acad Sci 665: 127-136.

[81] Geserick C, Bonarius HPJ, Kongerslev L, et al. 2000. Enhanced productivity during controlled proliferation of BHK cells in continuously perfused bioreactors. Biotechnol Bioeng 69: 266-274.

[82] Warnock JN, Al-Rubeai M. 2006. Bioreactor systems for the production of biopharmaceuticals from animal cells. Biotechnol Appl Biochem 45: 1-12.

[83] Nam JH, Ermonval M, Sharfstein ST. 2007. Cell attachment to microcarriers affects growth, metabolic activity, and culture productivity in bioreactor Culture. Biotechnol Prog 23: 652-660.

[84] Croughan MS, Hamel JF, Wang DIC. 1987. Hydrodynamic effects on animal cells grown in microcarrier cultures. Biotechnol Bioeng 29: 130-141.

[85] Stiens LR, Büntemeyer H, Lütkemeyer D, et al. 2000. Development of serum-free bioreactor production of recombinant human thyroid stimulating hormone receptor. Biotechnol Prog 16: 703-709.

[86] Yang JD, Angelillo Y, Chaudhry M, et al. 2000. Achievement of high cell density and high antibody productivity by a controlled-fed perfusion bioreactor process. Biotechnol Bioeng 69: 74-82.

[87] Hu X, Xiao C, Huang Z, et al. 2000. Pilot production of u-PA with porous microcarrier cell culture. Cytotechnology 33: 13-19.

[88] Kratje RB, Wagner R. 1992. Evaluation of production of recombinant human interleukin-2 in fluidized bed bioreactor. Biotechnol Bioeng 39: 233-242.

[89] Merten OW, Cruz PE, Rochette C, et al. 2001. Comparison of different bioreactor systems for the production of high titer retroviral vectors. Biotechnol Prog 17: 326-335.

[90] Justice C, Leber J, Freimark D, et al. 2011. Online- and offline- monitoring of stem cell expansion on microcarrier. Cytotechnology 63: 325-335.

[91] Noll T, Biselli M. 1998. Dielectric spectroscopy in the cultivation of suspended and immobilized hybridoma cells. J Biotechnol 63: 187-198.

[92] Justice C, Brix A, Freimark D, et al. 2011. Process control in cell culture technology using dielectric spectroscopy. Biotechnol Adv 29: 391-401.

[93] Zeng S, Dinter A, Eisenkraetzer D, et al. 1997. Pilot scale expression and purification of soluble protein a tagged α1,6-N-acetylglucosaminyltransferase in CHO cells. Biochem Biophys Res Commun 237: 653-658.

[94] Noll T, Eisenkraetzer D, Kiesewetter A, et al. 2002. Improved product formation in high density Chinese hamster ovary cell cultures transfected at confluency. Biotechnol Letters 4: 861-866.

[95] Dürrschmid MP, Landauer K, Simic G, et al. 2003. Comparison of fluidized bed and ultrasonic cell retention systems for high cell density mammalian cell culture. Biotechnol Prog 19: 1045-1048.

[96] Poertner R, Nagel-Heyer S, Goepfert C, et al. 2005. Bioreactor design for tissue engineering. J Biosci Bioeng 100: 235-245.

[97] Meuwly F, Ruffieux PA, Kadouri A, von Stockar U. 2007. Packed-bed bioreactors for mammalian cell culture: Bioprocess and biomedical applications. Biotechnol Adv 25: 45-56.

[98] Wang MD, Yang M, Huzel N, Butler M. 2002. Erythropoietin production from CHO cells grown by continuous culture in a fluidized-bed bioreactor. Biotechnol Bioeng 77: 194-203.

[99] Merten OW. 2004. State-of-the-art of the production of retroviral vectors. J Gene Med 6: 105-124.

[100] Schonberg JA, Belfort G. 1987. Enhanced nutrient transport in hollow fiber perfusion bioreactors: A theoretical analysis. Biotechnol Prog 3: 80-89.
[101] Piret JM, Devens DA, Cooney CL. 1991. Nutrient and metabolite gradients in mammalian cell hollow fiber bioreactors. Can J Chem Eng 69: 421-428.
[102] Whitford WG, Cadwill JJS. 2011. The potential application of hollow fiber bioreactors to large-scale production. Biopharm Intl Suppl 24: 21-26.
[103] Nagel A, Koch S, Valley U, et al. 1999. Membrane based cell culture systems - an alternative to in vivo production of monoclonal antibodies (http://www.integra-biosciences.com/sites/pdf/cell_culture/clreport3.pdf, accessed 15th March 2013).
[104] Klinger C, Baumann S, Puskeiler R, Jockwer A. 2012. Development of a partly controllable system at shake flask scale. PreSens Application Note (http://www.presens.de/uploads/tx_presensapplicationnotes/120612_APP_SF_Adapter_w_02.pdf, accessed 15th March 2013).
[105] Peuker T, Eibl D. 2011. Biopharmaceutical manufacturing facilities integrating single-use systems. Single-use technology in biopharmaceutical manufacture. John Wiley & Sons: 145-158.
[106] Peuker T, Bogner A. 2011. Equipment design and facility layout for flexible biomanufacturing processes. Eng Life Sci 11: 443-451.
[107] Eisenkraetzer D, Tillich D, Eibl D. 2012. Projektierung von Single-Use Systemen (http://vbu.cme24.de/index.php?id=768, accessed 15th March 2013).
[108] Singh V. 1999. Disposable bioreactor for cell culture using wave-induced agitation. Cytotechnology 30: 149-158.
[109] Eibl R, Löffelholz C, Eibl D. 2011. Single-use bioreactors - an overview. Single-use technology in biopharmaceutical manufacture. John Wiley & Sons: 33-51.
[110] Löffelholz C, Husemann U, Greller G, et al. 2013. Bioengineering parameters for single-use bioreactors: Overview and evaluation of suitable methods. Chem Ing Tech 85: 40-56..
[111] Martin J. 2007. Extractables and leachables subcommittee of the bioprocess systems alliance. Recommendations for extractables and leachables testing. BioProc Intl 5: 36-49.
[112] BPSA. 2007. BPSA Guidelines and standards committee. Bioprocess systems alliance component quality test matrices. Bioproc Intl 5: 1-12
[113] Eisenkraetzer D, Haug A. 2005. Agitator. EP1588758 (A1).
[114] Shanley A. 2011. Some straight talk about single-use bioprocess equipment. BPI Pharma (http://www.pharmamanufacturing.com/blogs/onpharma/at-bpi-some-straight-talk-about-single-use-bioprocess-equipment, accessed 12th May 2014).
[115] Eisenkraetzer D, Greller G, Eibl R. 2012. Verhalten von verschiedenen Zellkulturen in Bags. Sitzung Task Force „Single-Use Technologien in der biopharmazeutischen Produktion", DECHEMA, Frankfurt.
[116] Steiger N, Eibl R. 2013. Interlaboratory test for detection of cytotoxic leachables arising from single-use bags. Chem Ing Tech 1-2: 26-28.
[117] Hammond M, Nunn H, Rogers G, et al. 2013. Identification of a leachable compound detrimental to cell growth in single-use bioprocess containers. PDA J Pharm Sci Technol 67: 123-134.
[118] Okonkowski J, Balasubramanian U, Seamans C, Fries S, Zhang J, Salmon P, Robinson D, Chartrain M. 2007. Cholesterol delivery to NS0 cells: Challenges and solutions in disposable linear low-density polyethylene-based bioreactors. J Biosci Bioeng 103: 50-59.
[119] Dudziak G. 2010. Full plastics – generic antibody manufacturing using total disposable technology (http://rentschler.de/fileadmin/Downloads/Vortraege/Full_Plastics_BioProcess_2010.pdf, accessed 15th March 2013).
[120] Zlokarnik M. 1999. Rührleistung in homogener Flüssigkeit. Rührtechnik: Theorie und Praxis. Springer: 77-79.

[121] Chapple D, Kresta SM, Wall A, Afacan A. 2002. The effect of impeller and tank geometry on power number for a pitched blade turbine. Chem Eng Res Des 80: 364-372.

[122] Brauer H, Mewes D. 1973. Einfluß von Strombrechern auf die Rührerleistung. Chem Ing Tech 45: 461-467.

[123] Van't Riet K. 1979. Review of measuring methods and results in nonviscous gas-liquid mass transfer in stirred vessels. Ind Eng Chem Proc Des Dev 18: 357-364.

[124] Henzler HJ, Biedermann A. 1996. Modelluntersuchungen zur Partikelbeanspruchung in Reaktoren. Chem Ing Tech 68: 1546-1561.

[125] Greller G, Noack U, Brandt S, de Wilde D. 2013. Evaluation of the BIOSTAT® CultiBag STR family in terms of scalability (http://microsite.sartorius.com/fileadmin/Image_Archive/microsite/biostat_cultibag/pdf/10-06-24/Poster_Biotech2010_STR-family.pdf, accessed 15th March 2013).

[126] Goedde A, Reiser S, Russ K, et al. 2013. Characterization of two single-use bioreactors for mammalian cell culture processes (http://rentschler.de/fileadmin/Downloads/Poster/Rentschler-Poster-ESACT-2009_8.pdf, accessed 15th March 2013).

[127] Sinclair A, Monge M. 2010. Implementing disposables technology, delivering innovation, and transforming an organization. Biopharm Intl (http://www.biopharminternational.com/biopharm/article/articleDetail.jsp?id=698527&sk=&date=&pageID=4, accessed 15th March 2013).

[128] Polès-Lahille A, Richard C, Fisch S, et al. 2011. Disposable bioreactors: from process development to production. BMC Proc 5 (S8): 2.

[129] Chaubard J-F, Dessoy S, Ghislain Y, et al. 2010. Disposable bioreactors for viral vaccine production: Challenges and opportunities. Biopharm Intl Suppl (http://www.biopharminternational.com/biopharm/article/articleDetail.jsp?id=695153, accessed 15th March 2013).

[130] Rodriguez R, Castillo J, Giraud S. 2010. Demonstrated performance of a disposable bioreactor with an anchorage-dependent cell line. BioProc Intl 8: 74-78.

[131] Loeffelholz C, Kaiser S, Kraume M, et al. 2014. Dynamic single-use bioreactors used in modern L- and m^3- scale biotechnological processes: Engineering characteristics and scaling up. In: Eibl D, Eibl R (eds). Disposable Bioreactors II. 138 ed. Springer: 1-44.

[132] Weyand J, Herrmann J, de Wilde D, Adams T. 2013. The BIOSTAT® ORB large scale shaken single use bioreactor for the biopharmaceutical industry: Use and applications (http://www.sartorius.com/fileadmin/media/global/products/Appl_Use_of_BIOSTAT-ORB_SBT1012-e.pdf, accessed 15th March 2013).

[133] Anderlei T, Cesana C, de Jesus M. 2009. Shaken bioreactors provide culture alternative. Gene Eng Biotechnol News 29(19).

[134] Zhang X, Bürki CA, Stettler M, et al. 2009. Efficient oxygen transfer by surface aeration in shaken cylindrical containers for mammalian cell cultivation at volumetric scales up to 1000 L. Biochemi Eng J 45: 41-47.

[135] Werner S, Eibl R, Lettenbauer C, et al. 2010. Innovative, non-stirred bioreactors in scales from mL up to 1000 L for suspension cultures of cells using disposable bags and containers–a Swiss contribution. Chimia Intl J Chem 64: 819-823.

[136] Stettler M, Zhang X, Hacker DL, et al. 2007. Novel orbital shake bioreactors for transient production of CHO derived IgGs. Biotechnol Prog 23: 1340-1346.

[137] Muller N, Girard P, Hacker DL, et al. 2005. Orbital shaker technology for the cultivation of mammalian cells in suspension. Biotechnol Bioeng 89: 400-406.

[138] Kauling J, Brod H, Jenne M, et al. 2013. Novel, rotary oscillated, scalable single-use bioreactor technology for the cultivation of animal cells. Chem Ing Tech 1: 127-135.

[139] Eibl R, Werner S, Eibl D. 2010. Bag bioreactor based on wave-induced motion: Characteristics and applications. In: Eibl R, Eibl D (eds). Disposable Bioreactors. 115 ed. Springer: 55-87.

[140] Schoepke B. 2008. Charakterisierung von Einweg-Kultivierungssystemen für tierische Zellkulturen als Vorfermenter. Thesis. University Bielefeld, Germany.
[141] Mass transfer in the CELL-tainer® disposble bioreactor. 2013 (http://www.celltainer.com, accessed 15th March 2013).
[142] Zijlstra GM, Oosterhuis N. 2010. Cultivation of PER.C6-Cells in the novel CELL-Tainer-high-performance disposable bioreactor. In: Noll T (ed). Cells and Culture. 4 ed. Springer: 807-808.
[143] Oosterhuis NMG, van den Berg H. 2011. How multipurpose is a disposable bioreactor? Biopharm Intl 24: 51-56.
[144] Auton K , Bick J, Taylor I. 2007. Application Note: Strategies for the culture of CHO-S cells. Gen Eng Biotechnol News 27(16).
[145] Hashimura Y, Giroux D, Lee B, et al. 2013. Comparison of the novel PBS air-wheel bioreactor against the traditional stirred-tank bioreactor in microcarrier applications. Bioresearch online. IC03247 RevA.
[146] Lee B, Fang D, Croughan M, et al. 2011. Characterization of novel pneumatic mixing for single-use bioreactor application. BMC Proc 5(8): O12.
[147] Dubois S, Lipinski K. 2013. Viral vaccine production with VERO cells in the iCELLis™ nano disposable fixed-bed bioreactor (http://www.bioresearchonline.com/doc.mvc/Viral-Vaccine-Production-With-VERO-Cells-In-0001, accessed 15th March 2013).
[148] Havelange N, Marigliano M, Sainte-Marie M, et al. 2012. Poxvirus production on chicken embryo fibroblasts in iCELLis disposable fixed-bed bioreactor. In: Jenkins N, Barron N, Alves P (eds). Proceedings of the 21st Annual Meeting of the European Society for Animal Cell Technology (ESACT), Dublin, Ireland, June 7-10, 2009. 5 ed. Springer: 719-722.
[149] Hatton T, Shaun B, Rashid K. 2013. CHO cell culture with single-use CelliGen BLU packed-bed Fibra-Cel® Basket (http://www.bioon.com/z/eppendorf/download/appnote_10.pdf).
[150] Eibl D, Eibl R, Köhler P. 2011. DECHEMA Task Force Single-Use-Technologien in der biopharmazeutischen Produktion. Status Minutes (http://dechema.de/biotech_media/Downloads/StatPap_SingleUse_2011.pdf, accessed 15th March 2013).

6.2 High Cell Density Cultivation Process

Natalia Ceaglio, Mariela Bollati-Fogolín, Marcos Oggero, Marina Etcheverrigaray, and Ricardo Kratje

6.2.1 Introduction

Since human tissue plasminogen activator (tPA) became the first approved biotherapeutic obtained from mammalian cells using recombinant DNA technology in 1987 [1], the biopharmaceutical market rapidly expanded based on a development pipeline of several mammalian cell culture-derived drugs. Even more, mammalian cells are still the dominant recombinant protein production systems for clinical applications [2].

In this scenario, high-throughput production processes of therapeutics, expressed in mammalian cells, are necessary to fulfil the market demands. The success of a process depends both on cell productivity (defined as the quantity of the interest protein produced by cell) and the viable cell density in the culture. Considering these parameters, one of the applied strategies to obtain a high recombinant protein yield is the development of culture systems that reach high cell densities and sustain cell viability, leading to an increment of the volumetric productivity.

Animal cells can be grown using different modes of operation, and the maximum cell density that can be achieved will depend on that selection. The method of cultivation defines the environmental and nutritional conditions under which cells will proliferate and synthesize the protein of interest, determining the success of attaining high viable cell density and, consequently, high concentration of the desired product. A general classification distinguishes discontinuous modes (batch, repeated batch and fed-batch) and continuous ones (without cell retention or chemostat, or cell retention, also known as perfusion).

Despite the fact that suspension cultures are by far the most common cell culture format for recombinant protein expression in mammalian cells [3], anchorage-dependent cell lines are also used in the industry for the mentioned purpose. Consequently, the scaling up of a process where cells grow in adherence can be carried out by cultivating them as monolayer in static or gently agitated surfaces or by growing cells attached to polymers spheres that are introduced in bioreactors to maintain them in suspension at high density cultures.

The present chapter comprehensively describes theapproaches for high cell density cultivation of animal cells taking into account strategies to grow them in suspension or immobilized into proper supports.

6.2.2 Achieving High Cell Density Culture Processes Through Different Modes of Feeding

This section will deal with cell culture technology from the point of view of cell density/concentration in both growing forms, suspension or anchorage-dependent growing cells. Maximum cell concentration varies according to the type of culture: batch culture (approx. 2×10^6 mL^{-1}), fed-batch cultures (approx. 5×10^6 mL^{-1}) and concentrations similar to the ones found in tissues (higher than 1×10^7 mL^{-1}), which are achieved with several types of perfusion reactors. In 1987, Tyo and Spier [4] coined the term "dense cultures" to refer specifically to cell cultures that approach the limiting density of packed cells. As mentioned, we will describe the different culture modes involved in growing cells at different cell densities.

6.2.2.1 Batch Culture

In a batch process, the culture vessel or bioreactor is filled with culture medium containing the appropriate amount of substrates and nutrients required for cell growth. Cells are inoculated and they start to grow by metabolizing substrates into biomass and metabolites, including the product of interest. When a determined cell density and product concentration is reached, the culture is interrupted and the cell supernatant is collected for downstream processing.

As no medium is added to the system and the spent medium is not harvested during cultivation, the volume is kept constant throughout the whole process. However, unlike other substrates, oxygen must be supplied continuously because of its low solubility in the medium and other additives (such as acid or base for pH control or antifoaming agents for foam control) or gases (such as CO_2 for media containing sodium bicarbonate as buffer) need to be provided once the fermentation has begun [5-8]. Consequently, the definition of batch culture as a "closed system" is not completely covered.

After a lag phase, whose duration depends on the age of the inoculum and the speed of adaptation to the new environment, cells enter a log phase, in which they display the highest growth rate and can last up to 5 to 6 days in a suitable medium. During this phase, nutrient concentration gradually decreases, whereas toxic metabolites accumulate, so that cell growth is eventually inhibited whether by nutrient limitation (glucose and glutamine) [9, 10] or products toxicity (ammonia, lactate) [11, 12] and, for anchorage-dependent cells, by lack of available surface to adhere [13]. The sensitivity of cells to these factors defines the maximum cell concentration which is possible to attain in a batch culture, which is usually about 1 to 2×10^6 mL^{-1} [14, 15]. Maximum cell concentration is sustained during a short stationary phase, which extends for hours or a couple of days, after which the culture enters the dead phase principally by glucose depletion and viability falls to values below 50 %. Figure 6.2.1a shows the operation mode and the typical time-course profiles of cell density, substrates and products in a batch culture. An example of real data for IFN-α-producing recombinant CHO-K1 cells in batch cultivation is presented in Figure 6.2.1b.

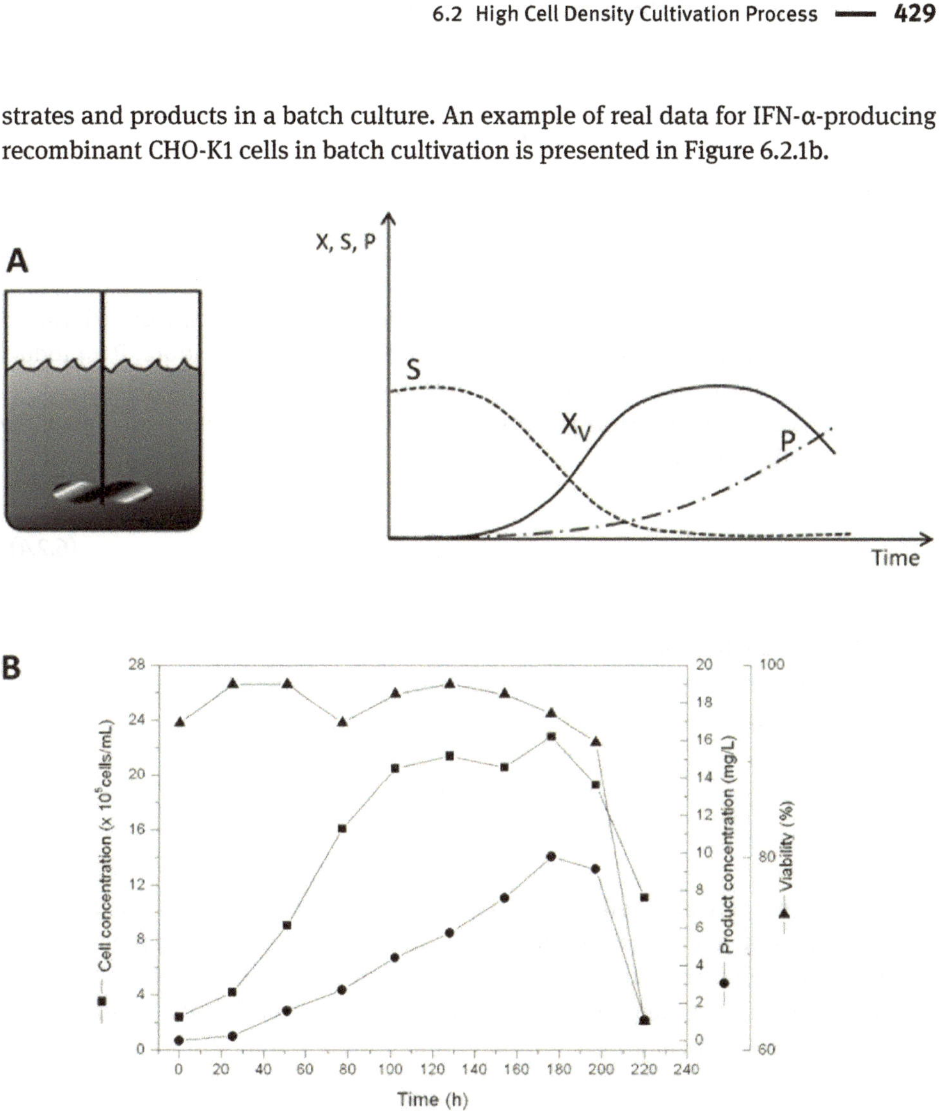

Figure 6.2.1: a) Operation mode and typical time-course profiles of viable cell density (X_v), substrate (S), and product (P) concentrations in batch culture. **b)** Evolution of viable cells, product concentration, and viability of a batch culture of CHO-K1 cells producing an IFN-α-mutein in suspension culture (Cell Culture Laboratory, School of Biochemistry and Biological Sciences, Universidad Nacional del Litoral, Santa Fe, Argentina).

The kinetic behaviour of a batch culture can be described by the following set of equations:

$$\frac{dXv}{dt} = \mu \cdot Xv \qquad (6.2.1)$$

$$\frac{dS}{dt} = -q_S \cdot Xv \tag{6.2.2}$$

$$\frac{dP}{dt} = q_P \cdot Xv \tag{6.2.3}$$

where Xv = viable cell concentration, t = cultivation time, μ = specific growth rate, S = substrate concentration, q_S = specific consumption rate of a substrate, P = product concentration, and q_P = specific formation rate of a product.

Integration of Equation (6.2.1) gives up an expression useful for calculation of the specific growth rate (μ),

$$\mu = \frac{\ln (Xv_2/Xv_1)}{t_2 - t_1} \tag{6.2.4}$$

where t_1 and t_2 are both within the exponential growth phase.

From this expression, it is possible to calculate the population doubling time (PDT) as follows:

$$PDT = \frac{\ln 2}{\mu} \tag{6.2.5}$$

The integral of viable cell density (IVCD) is a useful parameter to calculate specific consumption and/or formation rates (q_S and q_P) of substrates and products [16]. The IVCD represents the cumulative sum of cell concentrations up to certain time or simply the area under the growth curve (concentration of viable cells versus culture time; Equation (6.2.6)).

$$IVCD = \int_0^t Xv \cdot dt \tag{6.2.6}$$

The cell-specific productivity or cell-specific consumption (generally expressed in pg d^{-1}) can be calculated by dividing the product or substrate concentration by the IVCD (Equations 6.2.7 and 6.2.8), either in the exponential stage or for the entire culture period if required [17, 18].

$$q_P = \frac{P}{IVCD} \tag{6.2.7}$$

$$q_S = \frac{S}{IVCD} \tag{6.2.8}$$

Assuming that the specific productivity/consumption is constant throughout the entire culture or at a fixed culture phase, q_P and q_S can be determined as the slope of the P or S against IVCD plot.

Batch culture is the most simple operation mode in terms of system equipment, configuration, and operation and it is considered a reliable process with low risk of contamination or mechanical failure due to minimum disturbance from the outside compared to other methods of cultivation. For these reasons, batch culture is often applied in lab scale in stationary or stirred flasks to characterize kinetics of cell growth, product formation, and substrate consumption and to evaluate optimum culture medium composition and environment at early stages in the development of a recombinant protein production process. Also, batch culture is the method of choice for inoculum preparation in small and intermediate scale bioreactors. Some industries still employ this operation mode at production scales up to 20,000 L.

However, the major limitation of batch culture is the reduced cell yield that can be attained and, consequently, the reduced product concentration at the end of the process. The low productivity of batch operation relies on the fact that the initial concentrations of substrates must be relatively low because of solubility, osmolality, and toxicity issues, resulting in rapid nutrient depletion and inhibition of proliferation [19]. Also, a batch culture runs at a high specific productivity only for a short period of time whether the product is growth-associated or not [20]. In the first case, specific productivity increases along with growth rate and production decreases or finishes when the culture enters the stationary phase, which is the moment when harvest takes place. In the latter case, product formation increases when growth rate decreases and maximum productivity is achieved during stationary phase, which only lasts hours or a few days in a batch culture. Besides, the quality of the desired product should be analyzed in order to define the time of harvest, because cells are subjected to environmental conditions that are continuously changing during the whole process, and that can affect the quality of the synthesized product [3]. Moreover, the yield and quality of the product of interest can vary among different batches, resulting in downstream complications and potential regulatory concerns.

The simplest approach to improve the productivity of a batch process is by carrying out a variant known as repeated batch or batch-reefed. This mode of operation consists in performing a traditional batch culture until the time of harvest of the product, but then only a part of the cell suspension is collected. The remaining part is left in the bioreactor and functions as an inoculum for a new batch cycle, which is started by refilling the culture vessel with fresh medium. This procedure can be repeated several times, but peak cell density is usually lower in the subsequent runs and a decrease in cell growth or product accumulation is observed [21]. The critical point in this operation is to ensure a high viability of the remaining cell suspension at the optimum time of harvest [22]. The whole productivity of the repeated-batch cycles is increased by reducing the time and cost of cleaning and sterilizing the culture system between each batch, shortening the time in which the bioreactor is nonproductive.

6.2.2.2 Fed-batch Culture

The fed-batch mode of operation aims to overcome the major problem of batch culture: depletion of principal nutrients such as glucose and glutamine with the consequent cell growth limitation and death. In fed-batch cultivation, the culture vessel or bioreactor is filled with a volume of culture medium lower than the maximum working volume and started as batch. As substrates are consumed, the culture is fed continuously or intermittently with fresh medium or with a concentrated solution containing key nutrients, with a corresponding increase in the volume of the culture. This controlled feeding aims to fulfil nutritional requirements of cells in order to enable them to grow for a prolonged period of time and, thus, to achieve a higher concentration of

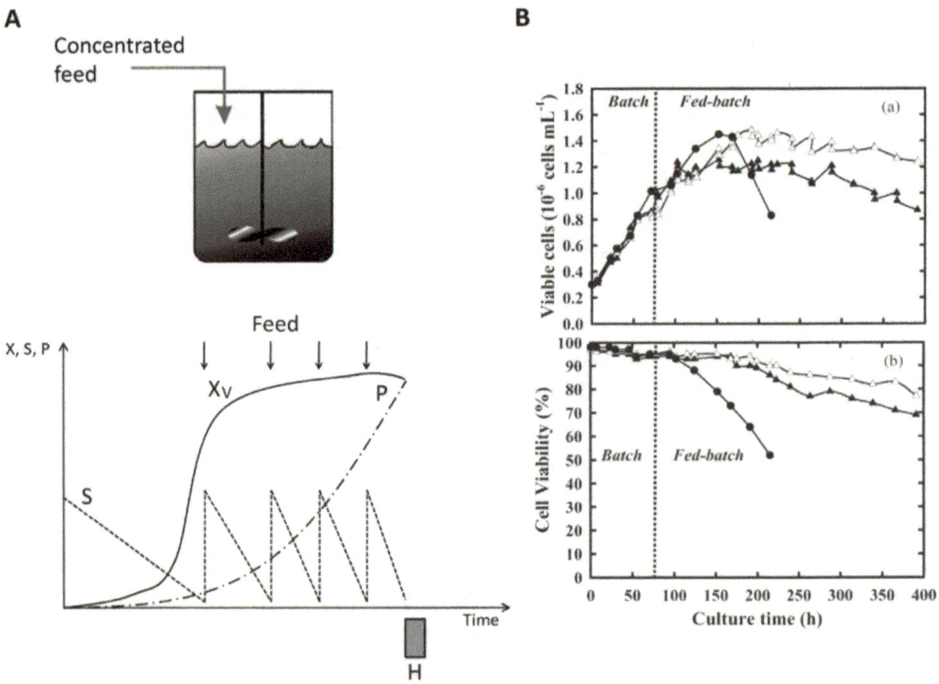

Figure 6.2.2: **a)** Operational configuration and typical fed-batch culture kinetics with viable cell density (X_V), substrate (S), and product (P) concentration profiles. The moment or period of supernatant harvest (H) is also indicated. **b)** Evolution of viable cells (a) and viability (b) of CHO cells producing tPA in batch (•) and fed-batch culture on glutamate using only glucose(▲) or glucose/galactose (Δ) as carbon sources. (Figure 6.2.2b reprinted from [25] with permission from Elsevier.)

the desired product. Since nutrients are not supplied altogether at the beginning of the culture, no osmolality or toxicity issues are encountered, so that feeding allows more substrates to be metabolized by the cells, and biomass and product yields are increased [23].

Compared to batch cultures, fed-batch cultivation is characterized by a longer growth phase (which is particularly important in the production of growth-associated products), a higher viable cell density, a longer stationary phase (which considerably increases productivity of nongrowth associated products) and a higher product concentration. The combination of a higher cell density for a longer period of time results in an increase in IVCD, with an overall higher product yield. However, like in batch culture, the spent medium is not withdrawn from the system, so that toxic waste metabolites (mainly lactate and ammonia) are accumulated and at last the culture environment becomes detrimental for cell growth [24]. Figure 6.2.2a shows the operational configuration and the typical time-course profiles of cell density, substrates and products in fed-batch cultivation. An example of fed-batch cultivation of t-PA producing CHO cells using different substrates as carbon sources [25] is shown in Figure 6.2.2b.

Mass balances for cells, substrates, and products in fed-batch cultures are represented by the following equations:

$$\frac{dXv}{dt} = \left(\mu - \frac{F}{V}\right) \cdot Xv \tag{6.2.9}$$

$$\frac{dS}{dt} = \left(\frac{F}{V}\right) \cdot (S_F - S) - q_S \cdot Xv \tag{6.2.10}$$

$$\frac{dP}{dt} = q_P \cdot Xv - \frac{F}{V} \cdot P \tag{6.2.11}$$

$$\frac{dV}{dt} = F \tag{6.2.12}$$

where F = flow rate of the feed, V = working volume of the bioreactor, and S_F = substrate concentration in the feed.

Designing the appropriate nutrient formulation and feeding strategy represents the major challenge of a fed-batch culture aiming at achieving high titres of the product of interest [26-29]. The composition of the nutrient feed should respond to the culture kinetics and stoichiometry of nutrient utilization in support of cell growth and product formation. Thus, the limiting substrates should be identified in order to be supplied at the correct levels in critical stages of the culture. Usually, the main carbon and energy sources, glucose and glutamine, are included in the feed stream. Also, other components as lipids, phosphate, or amino acids may have to be restored in the bioreactor [30, 31]. If the limiting medium substrates cannot be identified, such as in the case where commercial media including hydrolysates are used, the feed stream can be simply the concentrated basal medium excluding most of the salts in order to avoid osmotic problems. However, it is of major importance to be aware that supplementation for the prevention of nutrient depletion needs to be performed in a way that minimizes the accumulation of the main toxic waste products, lactate and ammonia, which are not only the principal cause of culture death but also affect product quality and quantity [32, 33]. Careful optimization of glucose/glutamine ratio, application of an optimized regimen to maintain low glucose/glutamine levels as well as their total or partial substitution with slowly metabolized substrates are strategies that have been reported to successfully decrease the level of lactate and ammonia in cell cultures [25, 34]. Metabolic flux analysis has been also employed to identify key enzymes in the glucose/glutamine utilization pathway in order to design genetic engineering approaches to produce lower amounts of toxic metabolites [35]. For example, the introduction of the cytosolic yeast pyruvate carboxylase gene into BHK-21, CHO-K1, and HEK293 cells has shown to improve the utilization of glucose and limit the production of lactate and ammonia, although these metabolic changes yielded mixed results with respect to cellular growth and/or productivity [36-40]. On the other hand, Zhou et al. [41], by knocking down the gene expression of lactate dehydrogenase A and pyruvate dehydrogenase kinases in CHO cells expressing a therapeutic monoclonal antibody, showed a reduction in lactate production, an increase of the specific productivity and the volumetric antibody production.

Beyond preventing nutrient depletion and byproduct accumulation, the feed stream can include some specific agents to inhibit apoptosis and thus, extend the culture growth phase. In standard batch conditions, the use of dextran sulfate 5000 – a sulfated polysaccharide with demonstrated antiapoptotic activity – has shown to prolong the life of CHO cells and increase productivity of EPO by 1.8-fold comparing with controls [42]. Other supplements such as suramin [43] or insulin growth factor [44] may provide apoptotic protection. Genetic strategies have also been applied to prevent apoptosis, including the overexpression of antiapoptotic genes such as bcl-2, bcl-xl [45, 46] in the host cell.

In addition, compounds, which enhance the production of the protein of interest, once a high cell density has been reached, can be supplied in the feeding. In biotechnology industry, supplementation of cultures with sodium butyrate, an inhibitor of histone deacetylation, is one of the most successful and widespread strategies to increase glycoprotein synthesis, which is attributed to its ability to arrest the cell cycle progression (G_0/G_1) and thus increase specific productivity of the product of interest [47]. Growth arrest can be also achieved by a decrease in culture temperature; this strategy, known as temperature shift, has also demonstrated an increase in protein specific productivity, although it is strictly dependent on the producer cell line [38, 48].

Nutrient feeding can be performed through different strategies, such as in pulses, steps, at constant flow rate or, more complex but accurate, through exponentially growing feed flow rate. The main purpose of the feeding strategy is to ensure the proper timing and rate of the previously optimized nutrient feeding in order to guarantee the correct provision of essential nutrients while avoiding byproduct accumulation and high osmolality. Two general strategies can be distinguished. In open-loop systems, the feeding strategy is based on mathematical models, which aim to understand kinetics of cell growth and protein production without any feedback control. In closed-loop systems, feeding actions are decided based upon direct or indirect online measurements, which give actual information about cell growth and nutrient consumption [49].

The major advantage of fed-batch culture is its ability to allow product accumulation to a high concentration and its relatively high volumetric productivity. With the careful optimization of the feed formulation and strategy, the culture duration can be extended to approximately 3 weeks from 1 week for simple batch and can result in a 10 times increase in final product concentration. Moreover, when compared to perfusion culture, runs are considerably shorter. This is especially important in cases where the cell line productivity is unstable, as fewer cell doublings occur in fed-batch culture in comparison with perfusion. Also, its short duration allows process validation more easily and in a minimum time. Although fed-batch operation cannot be considered as simple as in batch mode because there is an influx of supplements to the bioreactor, its complexity as well as the risks of contamination is still moderate. Moreover, the reliability and scalability of this mode of culture is comparable to that of batch culture, with applications in industrial scales up to 20,000 L. Less capital investments and technical expertise is required in comparison with perfusion culture. A unique feature of fed-batch mode is its flexibility, which reflects its easiness to adapt for different cell lines and products and it is one of the reasons why established biotech companies have chosen it for large-scale commercial manufacturing. Indeed, generic fed-batch processes with high yield production and low accumulation of toxic byproducts have been developed by large companies for the production of a variety of monoclonal antibodies with important cost and time savings [50, 51].

The main disadvantage of fed-batch process is the long residence time of the product in the bioreactor, which compromises its stability and quality. As molecules synthesized at early times of the culture are not harvested until the end of the process, they are continuously exposed to temporal changes of culture conditions which can adversely affect product consistency, with important concerns from a regulatory viewpoint [52]. For the production of therapeutic glycoproteins it is essential to ensure that a consistent product quality is maintained between batches, because glycosylation can significantly influence clinical safety and efficacy, plasma half-life and immune responses of many enzymes, hormones, and mAbs [53, 54]. During fed-batch culture, the product of interest, which is usually a protein or a glycoprotein, is exposed to proteases and glycosidases that may lead to its degradation or inactivation with a consequent loss of activity. Host cell line and culture parameters such as nutrient content, pH, temperature, oxygen, glucose, ammonia, and osmolality may also have a significant effect on the microheterogeneity of glycan structures found in the recombinant protein, which in some cases directly reflects in its biological activity [55]. Figure 6.2.3 shows how high osmolality has a negative effect on the specific biological activity of recombinant human erythropoietin (rhEPO) produced in CHO-K1 cells. In fed-batch cultures, high specific cell productivity promotes the generation of immature forms of glycosylated proteins, while the extended culture duration may lead to heterogeneous and truncated oligosaccharides resulting from glycosidase released from dead cells [56]. For this reason, fed-batch is considered a forbidden process for the production of proteins, which are unstable at the culture temperature or susceptible to enzymatic cleavage.

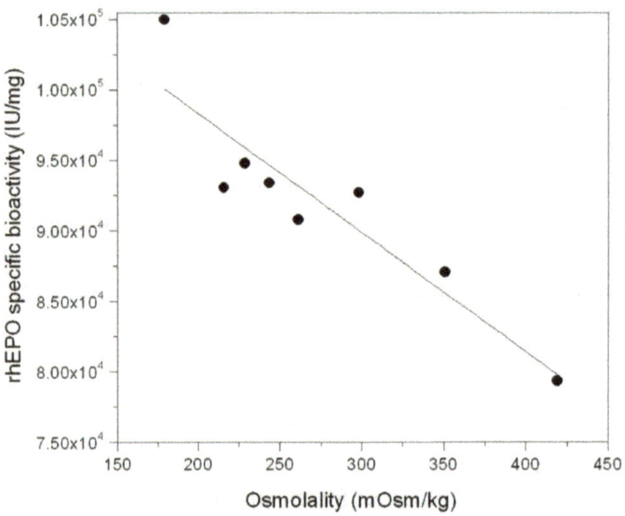

Figure 6.2.3: Effect of osmolality on the specific biological activity of recombinant human erythropoietin (rhEPO). A CHO-K1 producer clone was grown in culture medium supplemented with different concentrations of NaCl and the specific biological activity of rhEPO in the culture supernatant was evaluated (Cell Culture Laboratory, School of Biochemistry and Biological Sciences, Universidad Nacional del Litoral, Santa Fe, Argentina).

6.2.2.3 Continuous Culture without Cell Retention: Chemostat

In a chemostat, fresh nutrient medium is fed at a constant rate to the bioreactor containing the cells, and cells together with spent medium leave the culture at the same rate, so that the steady state is always maintained and the volume of the culture remains constant.

The culture is started as batch and when nutrients are below critical values, feeding of fresh medium and harvest of spent medium and cells are initiated. Growth is controlled by the flow rate of the medium supply and harvest, known as dilution rate (D), which is the medium flow rate divided by the culture volume. When the cell growth rate equals the dilution rate, a steady state of equilibrium is reached and concentrations of cells, nutrients, and products remain constant. The conditions of steady state are always restored if a change occurs, because of the self-regulation of the chemostat system. For example, a temporary decrease in cell concentration will cause a corresponding increase in the growth rate. The operation mode and the typical curves of cell concentration, substrates, and products are shown in Figure 6.2.4a.

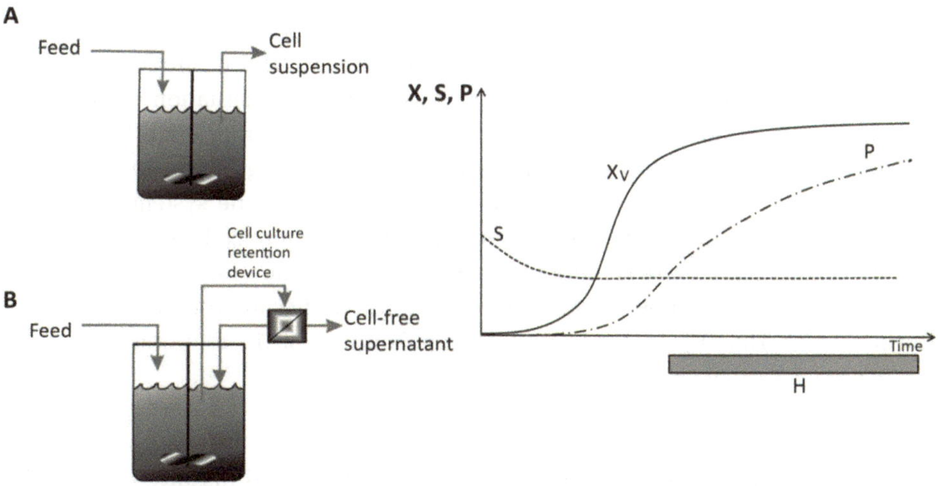

Figure 6.2.4: Mode of operation and typical curves of viable cell concentration (X_v), substrates (S), and products (P) versus time for **a)** continuous culture without cell retention and **b)** continuous culture with cell retention (perfusion).

The behaviour of continuous culture without cell retention is characterized by the following equations:

$$D = \frac{F}{V} \tag{6.2.13}$$

$$\frac{dXv}{dt} = (\mu - D) \cdot Xv \tag{6.2.14}$$

$$\frac{dS}{dt} = D \cdot (S_F - S) - q_S \cdot Xv \tag{6.2.15}$$

$$\frac{dP}{dt} = q_P \cdot Xv - D \cdot P \tag{6.2.16}$$

where D = dilution rate.

In the steady state,

$$\mu = D \tag{6.2.17}$$

So that according to Equation 6.2.14

$$dX/dt = 0$$

Dilution rates can be increased in order to increase the cell concentration but only up to a limit situation, known as bioreactor wash-out. This situation occurs when cells are removed from the bioreactor at an equal or higher rate than the maximal cell growth rate. When dilution rate approaches the maximal growth rate of the cells, duplication of cells cannot compensate their removal from the system and a progressive decrease in cell concentration occurs, until no more cells remain inside in the bioreactor. Thus, there is an upper limit value for dilution rate, called $D_{\text{wash-out}}$ or D_{critical}, which should not be exceeded in continuous cultures to avoid the wash-out of the bioreactor. This $D_{\text{wash-out}}$ can be derived from Equation 6.2.17, adopting μ_{max} for μ:

$$D_{\text{wash-out}} = \mu_{\max} \tag{6.2.18}$$

Chemostat culture is an open system in which nutrients and dissolved oxygen are constantly supplied and waste metabolites and cells are continuously removed, so that there is no inhibition of cell growth and culture can be maintained for a prolonged period of time. However, μ_{max} values for animal cell lines are in the range of 0.02 to 0.04 h^{-1}, so that maximum cell concentration usually cannot exceed 2×10^6 mL^{-1} [57]. Consequently, since D cannot exceed those values, the productivity of the system, which is determined by multiplying the dilution rate by the product concentration (DP, Equation 6.2.16), is low. In other words, proliferating cells abandon the system continuously, making it difficult to attain high cell densities with corresponding low yield of the product of interest.

For this reason, the advantages of the continuous culture without cell retention for industrial applications are much reduced. Indeed, chemostat culture is practically restricted to research and development purposes in small scale, in which they represent a valuable tool, for example, for kinetic studies or metabolic analysis at a fixed μ [58, 59].

6.2.2.4 Continuous Culture with Cell Retention: Perfusion

As chemostat, perfusion is an open continuous system with a fresh medium feeding and a spent medium exit stream that work at the same rate to keep the bioreactor volume constant. However, unlike chemostat, cells do not leave the system with the spent medium but are retained inside the bioreactor by means of a cell retention device. Perfusion tries to overcome the main limitation of chemostat, that is, low product yield because of the low cell densities achieved, due to the loss of proliferating cells in the outlet stream and the limited dilution factor at which it can work to avoid bioreactor wash-out. In perfusion mode, as the cells are kept inside the culture vessel, significantly higher dilution rates can be applied resulting in greater availability of nutrients, so that higher cell concentrations are attained (1×10^7 to 1×10^8 mL^{-1} or even higher) with a consequently higher productivity of the target product [60, 61]. In extreme, the XD® Technology (XD stands for eXtreme Density, DSM Pharmaceutical Products), equipped with an alternating tangential flow (ATF®, Refine Technology; see also Chapter 6.1) cell retention device based on microfiltration, accumulates IgG monoclonal antibodies in the range of 20 to 40 g L^{-1} using Crucells' PER.C6 human cell line at concentrations of 1 to 4×10^8 mL^{-1}. If the culture conditions are maintained optimally, the perfusion process can proceed for a long period of time (weeks or months) with continuous harvests of spent medium containing high concentrations of the protein of interest [62]. Figure 6.2.4b shows the operational configuration of a bioreactor operating in perfusion mode using a cell retention device and the typical profile of cell density, substrates, and products in time. Figure 6.2.5 shows an example of the production of recombinant human BDD-FVIII in CHO cells in perfusion mode using a spin filter as an internal cell retention device.

In perfusion mode, the equations used to describe mass balances of substrates and products are the same as the ones used for chemostat (Equations 6.2.15 and 6.2.16). However, the equation that expresses mass balance for cell density in perfusion is modified to denote cell retention, as follows:

$$\frac{dXv}{dt} = \mu \cdot Xv - \alpha \cdot D \cdot Xv \tag{6.2.19}$$

The coefficient α is defined as the ratio between cell concentration in the perfusate and in the bioreactor:

$$\alpha = \frac{Xv_{\text{perfusate}}}{Xv} \tag{6.2.20}$$

If cell retention is total, that is, no cells leave the bioreactor in the perfusate, then $\alpha = 0$. However, cell retention can be partial in situations where cell bleeding is performed for cell density control, so that α can take values from 0 to 1, this last one indicating no retention of cells at all (chemostat).

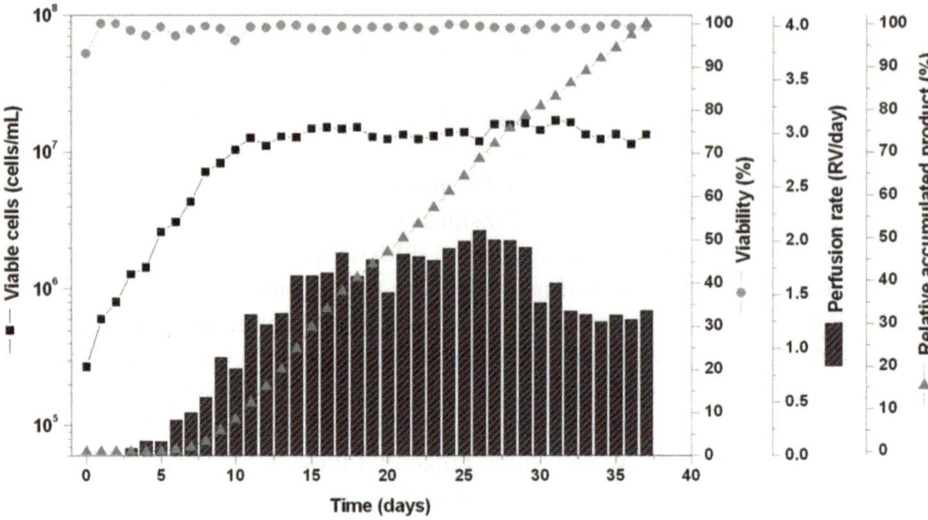

Figure 6.2.5: Evolution of viable cells, viability, perfusion rate, and relative accumulated product of a perfusion culture of suspended CHO-K1 cells producing BDD-FVIII in a 25-L stirred-tank bioreactor using an internal spin filter (Cell Culture Laboratory, School of Biochemistry and Biological Sciences, Universidad Nacional del Litoral, Santa Fe, Argentina).

In perfusion, wash-out of the bioreactor can only occur when α takes high values. If α values are low (0.0 and 0.2), the dilution rate can be increased to values much higher than μ_{max} without falling in a wash-out situation. This results in higher cell densities and productivities between 10 and even 100 higher than those obtained in chemostat processes [50]. In fact, this mode of culture presents the highest volumetric productivities of all systems analysed. The equation for $D_{wash\text{-}out}$ can be derived from equation 6.2.19:

$$\alpha \cdot D_{\text{wash-out}} = \mu_{\text{max}} \tag{6.2.21}$$

A useful variable in perfusion mode is the cell-specific perfusion rate (CSPR):

$$CSPR = \frac{D}{X\upsilon} \tag{6.2.22}$$

The CSPR represents the volume of feed medium per cell per day, and it has a limit value that indicates the minimal medium needed per cell per day for nutritional purposes. CSPR can be used to calculate the titre or product concentration, by replacing Equation 6.2.22 in Equation 6.2.16:

$$P = \frac{q}{CSPR} \tag{6.2.23}$$

In suspension cultures, cell retention can be performed based on differences in particle size, density, or both, by the use of internal or external cell retention devices, such as cross-flow filters, spin filters, settlers, centrifuges, or systems based on tangential flow filtration, among others [63-65]. An ideal cell retention device should operate satisfactorily with a high retention rate during long-term perfusion. It should not affect viability or productivity of cells and, if possible, it should be able to selectively retain viable cells while removing dead cells and debris [66]. The characteristics and applications of various retention devices are discussed in Chapter 6.1. In the case of adherent cells, an immobilization procedure is required for growing cells to high-density using suspension perfusion systems. This can be achieved by adhesion to a solid support through different methodologies, which are described later in this chapter.

Perfusion culture can achieve the highest cell density, the highest volumetric productivity, and the longest culture duration of all modes of culture. As in fed-batch, nutrient limitation is avoided by the continuous supply of medium. Unlike fed-batch, cell debris and inhibitory metabolites are removed from the system, increasing longevity of the culture in pursuit of attaining high cell concentrations. Proteases and glycosidases released to the culture medium by dead cells are also withdrawn, so that inactivation, degradation, and/or deglycosylation of glycoprotein products are minimized. Moreover, as the residence time of the secreted product is very short (in the order of hours), the exposure time to harsh operational conditions (pH, temperature, and released enzymes) is minimal, benefitting the purity and quality of the product. For these reasons, perfusion is the obligatory mode of operation for the production of large quantities of labile or unstable protein products. An example of a successful case of the use of the perfusion technology for commercial applications is the production of antihemophilic factor VIII (Bayer) [67]. Another advantage of the perfusion systems is that, owing its high productivity, the bioreactor can be up to 10 times smaller for the production of the same quantity of product, so that in general it is applied in industrial processes at scales up to 2,000 L.

The main disadvantages of perfusion mode are the higher operational complexity, the prolonged time of process development and validation, and the higher risk of contamination. The most complex aspect compared to fed-batch is the introduction of devices for cell retention, especially at large scales, which demand a more sophisticated bioreactor design so that the complete equipment results more expensive. However, the higher productivities of perfusion cultures allow the use of much more compact bioreactors, with a consequent decrease in cost, which compensates this situation. Also, the risk of equipment failure is high due to potential clogging of the retention device. Nevertheless, perfusion cultures are simpler to develop and operate than fed-batch in at least one aspect: beyond the brief start-up period, only the single steady-state production phase of a perfusion process requires optimization, rather than the continuously changing dynamic culture and feeding of a fed-batch process [63]. As regards process developmental and validation time periods, they are closely associated to the duration of the culture runs: as the perfusion process can last over several weeks or months, the optimization and validation processes are also long. In addition, the long culture runs demand a rigorous evaluation of the genetic stability of the cell line to guarantee constant productivity and quality of the product throughout the whole bioprocess. Finally, the risk of contamination was considered to be high in the past due to the fact that perfusion is an open system, which operates continuously for long periods of time; this disadvantage is nowadays offset by the sophisticated methods for establishment and control of sterility.

6.2.3 Strategies to Grow Immobilized Cells to High Densities

Naturally, most mammalian cell lines require a solid substrate for spreading and further proliferation like it occurs *in vivo*. Although within the last decades several continuous established cell lines have been adapted to grow in suspension (solid substrate free), many of them in large reactors are subjected to physical/mechanical aggressions which can alter their productivity, quality of bioproducts, as well as reproducibility of the process [68, 69]. In an attempt to provide systems that offer large accessible surfaces for cell growth in small-culture volume, a number of techniques have been proposed: the roller bottle system, the stack plate's propagator, the spiral film bottles, the packed bed system, the plate exchanger system. All these systems are nonhomogeneous and suffer from limited potential for scale-up, difficulties to sample, limited potential to measure, and control cultivation parameters [70]. Many attempts have been made in order to overcome these limitations.

Assuming that the ideal situation for *in vitro* cell cultivation process is to try to reproduce what occur in living organisms, where cells are at high densities immobilized in organs and tissues and fed by the blood or lymph flows, in 1923, Carrel introduced for the first time the concept of *in vitro* mammalian cells immobilization [71] and since then, several methods for immobilizing mammalian cells have been described.

Basically, there are 2 techniques for cell immobilization: immurement cultures and entrapment cultures [72-75]. Immurement cultures imply either encapsulation of cells within gels (microencapsulation) or growth of cells within compartments formed by membranes, as it is the case of culturing cells in hollow fibre cartridges. Entrapment cultures are the type of culture in which the cells are retained within an open matrix through which the medium is capable of flowing, as it is the case of entrapping cells on a particle surface or in the interstices of a porous particle (carriers).

Czermak et al. [20] summarized some advantages of the immobilization techniques:

– Attachment of anchorage-dependent cells is the only way these cells can grow.

- Immobilization in macroporous carriers can protect cells against shear stress, promoting the use of serum- or protein-free medium [76].
- Cell densities in immobilized systems are considerably higher compared to those without immobilization (e.g., suspension culture approx. 10^6 to 10^7 mL^{-1}, immobilized culture approx. 10^7 to 10^8 mL^{-1}, tissue approx. 10^9 mL^{-1}) and allow for smaller reactor volumes (cell density of immobilized system is expressed as cell concentration calculated to the total culture volume).
- Immobilization techniques enable preliminary separation of (extracellular) products and cells, easing requirements for downstream.
- Immobilization systems are preferably run in perfusion mode, where the medium feed rate is not dependent on the growth rate of the cells and higher volume-specific productivities can be obtained.

On the other hand, there are still some problems to be solved. Materials have to be selected according to the cell requirements, culture operation mode, and the final intended use of the product. Moreover, the limitations in nutrient and oxygen supply, as well as the removal of toxic metabolites and carbon dioxide are some of the points to work out.

6.2.3.1 Cell Encapsulation

Cell encapsulation or microencapsulation can be used to protect cells against hazardous environmental conditions [74, 77, 78]. The microcapsules can be cultivated in suspension reactors, with direct aeration and in combination with high agitation rates or stirrer speeds without damaging the cells. This approach offers several benefits, including simple handling of the cells, as medium exchange can be easily performed without cell loss, cells are safely protected against shear stress in the capsules, and a favorable microenvironment for the cells within the capsules is developed since self produced growth factors can accumulate and cells switch directly to exponential growth without a significant lag phase. The last one is probably one of most important aspects.

There are 3 basic encapsulation systems available: the bead, the coated bead, and the membrane-coated hollow sphere [77, 79, 80]. Common polymers and appropriate encapsulation devices as well as their application have already been described [78]. Typical size for beads made of polysine-alginate is 300 to 500 µm, and the molecular weight cut-off of these capsule membranes is 60 to 70 kDa [81]. One important issue of concern about is to know the physicochemical characteristics of the immobilization materials and their interaction with media components and products of the cell metabolism. For example, immobilization in Ca-alginate beads often leads to problems when cells release lactate into the medium or when phosphate ions exist as buffering agents in the medium and compete for the binding with the calcium ions. This difficulty led to the development of the coated beads where additional polyelectrolyte layers, which are also stable in the presence of other ions, are applied to the beads.

One disadvantage of this method is the fragile nature of the microcapsule, which is detrimental to scale-up and long-term operation. Another major drawback is low diffusion efficiency resulting in a suboptimal microenvironment for high-density cell growth. Cells grow within capsules up to tissue-like cell density before their growth seems to be stopped by space limitations. Cell densities of 10^8 mL^{-1} are standard. These values can only be reached, though, when the nutrient supply of the cells is adequate and the diameter of the capsules is not too large. In this sense, one of the most critical factors is sufficient supply of dissolved oxygen. If an incomplete oxygen supply occurs, those cells allocated in the centre of the capsules initiate the apoptosis process. Hence, cells with a high consumption rate can only be cultivated in relatively small capsules (<1 mm) whereas cells with a lower consumption rate can be cultivated in larger capsules. Although this approach has been widely used in small-scale cell culture research, especially in the field of tissue engineering, they are not commonly used in cell culture manufacturing process because of the complexity and potentially high cost involved.

6.2.3.2 Carriers for Cell Immobilization

The development of carriers to support cell growth was initially driven by the idea to provide a large-scale production system for adherent cells, e.g., for vaccine production. In the early 1960s facilities for vaccine production used batteries of roller bottles, and the scale-up was performed by increasing the number of roller bottles used. The high labour intensity of this technology is quite obvious and, since it is a nonhomogeneous process, it was difficult to control process parameters and robustness. An important breakthrough was the development of the so called "microcarrier" by van Wezel [82] in the early 1970s, a solid particle with a 100 to 200 μm structure on which cells could grow. For his contribution van Wezel was called "the father of the microcarrier" [83].

In fact, this system combines features of monolayer and suspension cultures. As the microcarriers are suspendable in stirred tanks, it became possible to work with large-scale suspension reactors, nowadays up to 4,000 L or even more. The great success of this technology led to the development of a large number of different microcarriers for suspension culture and furthermore for different reactor systems, especially for fixed-bed and fluidized-bed reactors [73]. The maximum reported size of a bioreactor vessel for cultivating animal cells on microcarriers is 6,000 L. It was used by Baxter Biosciences for influenza production based on Vero cells growing on Cytodex™ microcarriers in stirred bioreactors.

Although initially only positively charged beads were used, it soon became clear that negatively charged or amphoteric materials such as proteins or amino acids polymerized to the surface were equally useful. Eventually numerous different types of microcarrier were developed. The second generation of microcarriers consisted of macroporous beads providing increased surface area for cell attachment and growth by external and interior space. Such microcarriers offer great potential for high cell densities and enhanced productivity for certain production systems, especially recombinant CHO cells. These carriers, which not only provide possibilities for anchorage-dependent cells but also for cells growing in suspension, can be used in homogeneous bioreactors as well as in fluidized or fixed-bed systems [83]. By using microcarriers in simple suspension culture, fluidized or packed bed systems, yields of up to 2×10^8 mL^{-1} are possible.

Several authors summarized the desired features that a carrier should have [73, 83, 84]:
- autoclavable,
- available in large batches for industrial customers,
- high batch-to-batch consistency,
- suitable for a large number of cell lines (adherent and nonadherent cells),
- good long-term stability,
- high surface-to-volume ratio (large multiplication steps),
- available with documents required for approval (drug master or regulatory support files),
- material of nonbiological origin (minimal viral risks),
- macroporous for high cell density and shear force protection,
- efficient diffusion from the medium into the center of the carriers,
- nontoxic, nonimmunogenic matrix,
- size appropriate to reactor system,
- simple immobilization and harvesting of cells,
- mechanical stability,
- uniform size distribution,
- reusable,
- possibility to count cells,
- transferable between vessels (ease of scale-up), and
- transparent.

Even if an "ideal" carrier does not exist, the above list gives some arguments for selecting a carrier or evaluating different suppliers. Although for research purposes aspects such as good growth conditions might be important for production processes, other aspects related to the approval of the process and maintaining the process for the lifespan of the product have to be considered.

Besides the above mentioned advantages of carriers, Lundgren and Blüml [73] listed some disadvantages:
- Some carriers have to be prepared before use.
- Scale-up of a culture using cells harvested from microcarriers is more complex than expansion of a suspension culture.
- Cell harvest from macroporous carriers can be difficult.

- Cell count, especially in the case of macroporous carriers, requires special techniques
- For larger macroporous carriers, mass transfer limitations have to be considered.

Depending on the intended uses, specifically the culture system, the properties of the carriers, in particularly the size, vary significantly (from 10 µm to 5 mm). The term “microcarrier” denotes small beads, either solid or macroporous with a diameter of approximately 100 to 300 µm and a density of 1.02 to 1.04 g mL^{-1}. Larger carriers (0.6 to 1 mm) are used for fluidized and fixed bed reactors due to the higher sedimentation rate. For fixed-bed cultures, larger carriers (3 to 5 mm) are better to prevent blocking of free channels between the carriers. In general, larger carriers have limitations in terms of mass transfer effects, because of the poor solubility of oxygen in the medium and the high uptake rate of the cells.

Many authors described a broad variety of different materials for carriers including dextran, collagen, polymers, glass, and ceramic, among others [73, 85, 86]. In addition, several shapes are also accessible, such as fibres, flat discs, woven discs, cubes, and sphere (Figure 6.2.6). The surface is usually positively or negatively charged to ensure cell attachment. Furthermore, there are noncharged carriers available, which are normally coated with gelatine, collagen, fibronectin, or fibronectin peptides.

The density of the carriers is determined by the intended application. Smooth microcarriers for suspension culture have a density just above the medium (1.02 to 1.04 g mL^{-1}), while materials used for macroporous carriers have densities between 1.04 and 2.50 g/mL.

The surface of most carriers can be either solid or microporous. In the case of microporosity, the pore size is not sufficient to allow cells to grow in the inner parts of the carrier. Micropores are intended to take up larger molecules, to allow the cells to create a microenvironment on the beads and to support cell attachment and growth. For macroporous carriers, the average pore size is between 30 and 400 µm, allowing cell growth within the carriers creating a three-dimensional, almost tissue-like structure at higher densities compared to solid or microporous carriers. These types of carriers are suitable for immobilizing both adherent or nonadherent cell lines, and the three-dimensional architecture stabilizes the cell population, protects against shear stress and decreases the need for external growth factors. Thus, macroporous carriers are preferably used for long-term continuous culture. Due to the high cell density, the different culture technologies applied for macroporous carriers (stirred tanks for macroporous microcarriers, fixed bed, and fluidized bed reactors) are generally run at high perfusion rates. This provides a sufficient nutrient supply as well as removal of toxic metabolites.

In the first experiments, van Wezel used beaded ion exchanger N,N-diethylaminoethylamine (DEAE) Sephadex™ (GE Healthcare) A-50 as a microcarrier. After optimization of the surface charge, Cytodex 1™ was developed for a wide variety of cells. The next major step forward was the development of macroporous microcarriers.

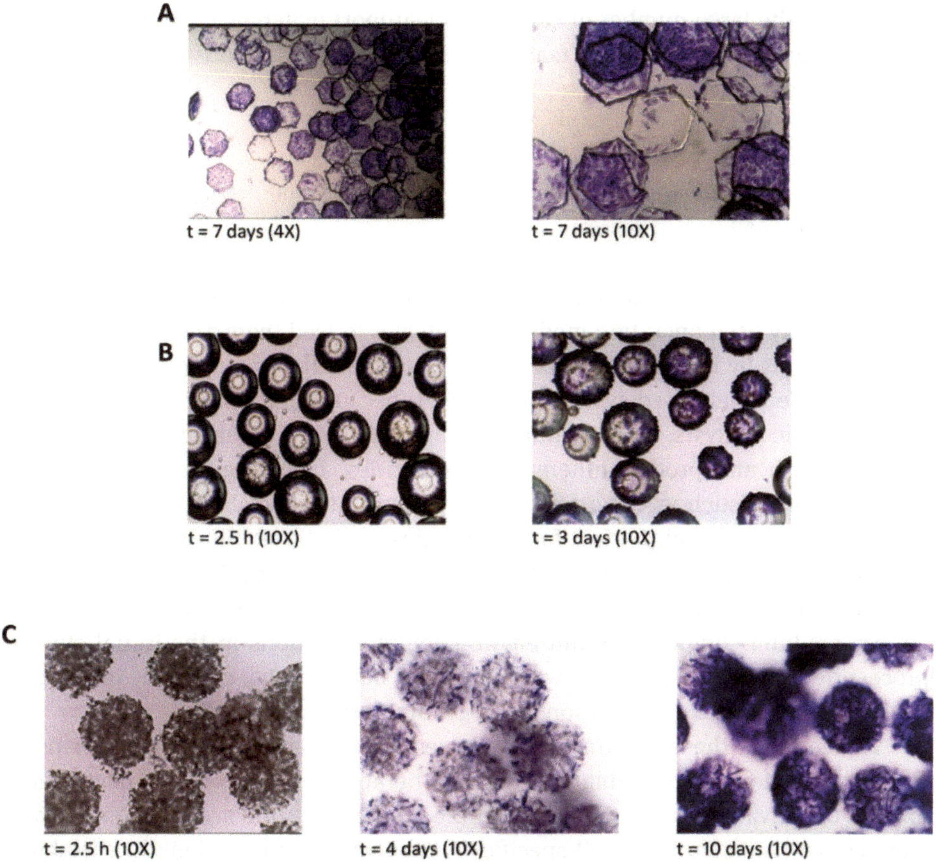

Figure 6.2.6: Growth of recombinant CHO cells producing IFN-β1a in different types of microcarriers. **a)** Seven-day culture of cells on Thermo Scientific™ Nunc™ 2D MicroHex™ microcarriers, which are solid polystyrene particles hexagon shaped, of low density (1.05 g/mL) and accessible area of 0.076 m^2/g, radiation sterilized and ready for use, with no conditioning required.
b) Culture of cells (2.5 h and 3 days) on GE Healthcare Life Sciences Cytodex 1™ carriers, which are dextran spheric beads with positive-charged DEAE groups, of low density (1.03 g/mL), particle size of 60 to 87 µm and prior to use, beads need to be swelled in buffer and autoclave at 120 °C. **c)** Culture of cells (2.5 h, 4 days, and 10 days) on GE Healthcare Life Sciences Cytopore 2™ carriers, which are dextran spheric macroporous beads with positive-charged DEAE groups, matrix cross-linked cotton cellulose, with an average particle size of 230 µm (Cell Culture Laboratory, School of Biochemistry and Biological Sciences, Universidad Nacional del Litoral, Santa Fe, Argentina).

Perfusion technology with macroporous microcarrier is used to produce recombinant proteins [87], monoclonal antibodies [88], or nonlytic virus [89].

Despite the large number of carriers and support materials suggested in the literature, only very few are on the market and even fewer fulfil industrial standards for large-scale manufacturing [73].

6.2.4 Comparison between Culture Modes for Recombinant Protein Production at High Cell Densities

In the biotech industry, fed-batch and perfusion of mammalian cell cultures are the most currently used operation modes as strategies to feed them and produce recombinant proteins by large-scale processes [90], in which the suspension-growing cells constitute the widely used system. In contrast, because of its short duration and its simple kinetic analysis, batch culture is mainly employed at lab scale for the study of cell growth and product-inherent particularities, including characterization of growth kinetics (such as determination of growth rate), metabolic behaviour (substrate consumption and product formation rates), genetic stability, and cell specific productivity and quality of the product of interest. In this way, based on these characteristics, batch culture can be used for clone selection and evaluation of optimum culture medium composition and environmental conditions in the initial steps of process development. Also, as it was mentioned before, batch culture is the method of choice for inoculum preparations of larger bioreactors.

Therefore, in these circumstances a question to answer is fed-batch or perfusion for large demand of animal cell-derived therapeutics? Many works have tried to explain the advantages of one mode over the other taking into account some parameters as benchmarks: from cell-specific productivity, perfusion length, scalability, automation, up to personnel requirements or very subjective parameters according to product, for example [3, 15, 91-97]. The most relevant parameters are summarized in Table 6.2.1. The experience of each working group when operating with fed-batch or perfusion cultures, as well as the importance attributed to each parameter of Table 6.2.1 constitute the critical factors in the selection of the operation mode for the production of recombinant proteins in large-scale processes.

Besides, taking into consideration these parameters, it is feasible to explain the change from an operation mode to another. For example, Meuwly et al. [98] and Lee et al. [99] showed how they changed from perfusion to fed-batch technology mainly based on the higher market demand of a product that largely exceeds the production capacity of the first generation process. They also compared the process productivity and product quality and demonstrated that the process can be changed, particularly if an increase of productivity, scalability, and quality is attained. Nevertheless, the most valuable strategy is to make an exhaustive study to decide on the most proper culture mode during the development procedure in order to avoid a further conversion to a more sustainable procedure when the market demand is the dominant stage of the process.

Table 6.2.1: Comparison of fed-batch and perfusion operation modes taking into account parameters depending on cell culture and/or reactor technology (adapted from [93]).

Depending on	Parameters	Fed-batch	Perfusion
Cell culture/Reactor technology	**Cell density (× 10^6 cells per mL)**	5 - 10	10 - >100
	Cell generations through run	Medium	High
	Product concentration (g/L)	0,5 - 5	0,1 - 1
	Volumetric productivity	High	Low
	Accumulated production	Medium	High
	Throughput product (per reactor unit)	Medium	High
	Scale-up (L)	20,000	2,000
	Culture time (days)	10 - 20	20 - >60
	Product reactor residence time	High	Low
	Product stability/homogeneity	Low	High
	Toxic waste metabolites accumulation	Yes	No
	Cell environment	Change	Constant
	Control of pH	Yes	Yes
	Harvest	Single	Multiple
	Volume harvest	Small	Large
	Lot consistency	Medium	High
	Lot size	Fixed	Pooled
Reactor technology	Technical complexity	Less	More
	Process control/development complexity	Medium	High
	Process development (time per cost)	Medium	High
	Automation requirement	Medium	High
	Technical failure risk	Medium	High
	Total space consumption	Medium	High
	Personnel requirements	Low	Medium
	Regulatory constrictions	Medium	High
	Technology transferability	Medium	Low

Pollok et al. [97] evaluated in detail the performance of fed-batch and perfusion cell culture technologies for the production of monoclonal antibodies in a wide range of titres and scales of work in order to encompass a broad spectrum of possibilities that can truly appear at industrial level. As regards continuous cultures, they studied 2 perfusion technologies: spin filter perfusion and a relative new perfusion technology utilizing alternating tangential flow (ATF). Results showed that economic, environmental, operational, and robustness features at most titres and scales of spin filter perfusion strategy shed light on its restricted application in industrial processes. On the other hand, the ATF perfusion strategy offers economic benefits that make up for its lower robustness, compared to fed-batch strategies. To sum up, ATF perfusion and fed-batch strategies seem to have similar performances.

In general and as concluding remark, the selection of a high cell density system for mammalian cell culture, intended for production of biotherapeutics at industrial scale, is a matter of experimental and modelling studies, rational considerations, a fine knowledge of the selected cell line and the product, and many economically or environmental factors that surround the decision.

References

[1] Walsh G. Biopharmaceutical benchmarks 2006. Nat Biotechnol 2006;24:769-76.
[2] Zhu J. Mammalian cell protein expression for biopharmaceutical production. Biotechnol Adv 2012;30:1158-70.
[3] Wurm FM. Production of recombinant protein therapeutics in cultivated mammalian cells. Nat Biotechnol 2004;22:1393-8.
[4] Tyo MA, Spier RE. Dense cultures of animal cells at the industrial scale. Enzyme and Microbial Technology 1987;9:514-20.
[5] Aehle M, Kuprijanov A, Schaepe S, et al. Simplified off-gas analyses in animal cell cultures for process monitoring and control purposes. Biotechnol Lett 2011;33:2103-10.
[6] Papoutsakis ET. Media additives for protecting freely suspended animal cells against agitation and aeration damage. Trends Biotechnol 1991;9:316-24.
[7] Sieblist C, Jenzsch M, Pohlscheidt M. Influence of pluronic F68 on oxygen mass transfer. Biotechnol Prog 2013;29(5):1278-88.
[8] Zhou H, Purdie J, Wang T, Ouyang A. pH measurement and a rational and practical pH control strategy for high throughput cell culture system. Biotechnol Prog 2010;26:872-80.
[9] Ozturk SS, Palsson BO. Growth, metabolic, and antibody production kinetics of hybridoma cell culture: 1. Analysis of data from controlled batch reactors. Biotechnol Prog 1991;7:471-80.
[10] Takagi M, Hia HC, Jang JH, Yoshida T. Effects of high concentrations of energy sources and metabolites on suspension culture of Chinese hamster ovary cells producing tissue plasminogen activator. J Biosci Bioeng 2001;91:515-21.
[11] Lao MS, Toth D. Effects of ammonium and lactate on growth and metabolism of a recombinant Chinese hamster ovary cell culture. Biotechnol Prog 1997;13:688-91.
[12] Ozturk SS, Riley MR, Palsson BO. Effects of ammonia and lactate on hybridoma growth, metabolism, and antibody production. Biotechnol Bioeng 1992;39:418-31.
[13] Fiore M, Degrassi F. Dimethyl sulfoxide restores contact inhibition-induced growth arrest and inhibits cell density-dependent apoptosis in hamster cells. Exp Cell Res 1999;251:102-10.
[14] Butler M. Animal cell cultures: recent achievements and perspectives in the production of biopharmaceuticals. Appl Microbiol Biotechnol 2005;68:283-91.
[15] Rodrigues ME, Costa AR, Henriques M, et al. Technological progresses in monoclonal antibody production systems. Biotechnol Prog 2010;26:332-51.

[16] Adams D, Korke R, Hu W-S. Application of Stoichiometric and Kinetic Analyses to Characterize Cell Growth and Product Formation. In: Pörtner R (ed). Methods in Biotechnology, Vol 2: Animal Cell Biotechnology: Methods and Protocols. 2nd ed; Springer 2007:269-84.
[17] Renard JM, Spagnoli R, Mazier C, et al. Evidence that monoclonal antibody production kinetics is related to the integral of the viable cells curve in batch systems. Biotechnol Lett 1988;10:91-6.
[18] Li F, Vijayasankaran N, Shen AY, et al. Cell culture processes for monoclonal antibody production. MAbs 2010;2:466-79.
[19] Gerson DF, Kole MM, Ozum B, Oguztoreli MN. Substrate Concentration Control in Bioreactors. Biotechnol Genet Eng Rev 1988;6(1):67-150.
[20] Czermak P, Pörtner R, Brix A. Special Engineering Aspects. In: Eibl R, Eibl D, Pörtner R, et al (eds). Cell and Tissue Reaction Engineering: Principles and Practice. Springer 2009:83-172.
[21] Kaneko Y, Sato R, Aoyagi H. Evaluation of Chinese hamster ovary cell stability during repeated batch culture for large-scale antibody production. J Biosci Bioeng 2010;109:274-80.
[22] Seewoster T, Lehmann J. Cell size distribution as a parameter for the predetermination of exponential growth during repeated batch cultivation of CHO cells. Biotechnol Bioeng 1997;55:793-7.
[23] Xie L, Wang DI. Integrated approaches to the design of media and feeding strategies for fed-batch cultures of animal cells. Trends Biotechnol 1997;15:109-13.
[24] Hassell T, Gleave S, Butler M. Growth inhibition in animal cell culture. The effect of lactate and ammonia. Appl Biochem Biotechnol 1991;30:29-41.
[25] Altamirano C, Paredes C, Illanes A, et al F. Strategies for fed-batch cultivation of t-PA producing CHO cells: substitution of glucose and glutamine and rational design of culture medium. J Biotechnol 2004;110:171-9.
[26] Gorfien SF, Paul W, Judd D, et al. Optimized nutrient additives for fed-batch cultures. Biopharm International 2003;16:34-40.
[27] Huang YM, Hu W, Rustandi E, et al. Maximizing productivity of CHO cell-based fed-batch culture using chemically defined media conditions and typical manufacturing equipment. Biotechnol Prog 2010;26:1400-10.
[28] Fike R. Nutrient Supplementation Strategies for Biopharmaceutical Production: Part 1 – Identifying a Formulation. BioProcess International 2009;7:44-51.
[29] Wlaschin KF, Hu WS. Fedbatch culture and dynamic nutrient feeding. Adv Biochem Eng Biotechnol 2006;101:43-74.
[30] deZengotita VM, Miller WM, Aunins JG, Zhou W. Phosphate feeding improves high-cell-concentration NS0 myeloma culture performance for monoclonal antibody production. Biotechnol Bioeng 2000;69:566-76.
[31] Xing Z, Kenty B, Koyrakh I, et al. Optimizing amino acid composition of CHO cell culture media for a fusion protein production. Process Biochem 2011;46:1423-9.
[32] Schneider M, Marison IW, von Stockar U. The importance of ammonia in mammalian cell culture. J Biotechnol 1996;46:161-85.
[33] Yang M, Butler M. Effects of ammonia on CHO cell growth, erythropoietin production, and glycosylation. Biotechnol Bioeng 2000;68:370-80.
[34] Hegde S, Pant T, Pradhan K, et al. Controlled release of nutrients to mammalian cells cultured in shake flasks. Biotechnol Prog 2012;28:188-95.
[35] Dean J, Reddy P. Metabolic analysis of antibody producing CHO cells in fed-batch production. Biotechnol Bioeng 2012;110:1735-47.
[36] Irani N, Wirth M, van Den Heuvel J, Wagner R. Improvement of the primary metabolism of cell cultures by introducing a new cytoplasmic pyruvate carboxylase reaction. Biotechnol Bioeng 1999;66:238-46.

[37] Irani N, Beccaria AJ, Wagner R. Expression of recombinant cytoplasmic yeast pyruvate carboxylase for the improvement of the production of human erythropoietin by recombinant BHK-21 cells. J Biotechnol 2002;93:269-82.

[38] Fogolín MB, Wagner R, Etcheverrigaray M, Kratje R. Impact of temperature reduction and expression of yeast pyruvate carboxylase on hGM-CSF-producing CHO cells. J Biotechnol 2004;109:179-91.

[39] Elias CB, Carpentier E, Durocher Y, et al. Improving glucose and glutamine metabolism of human HEK 293 and Trichoplusia ni insect cells engineered to express a cytosolic pyruvate carboxylase enzyme. Biotechnol Prog 2003;19:90-7.

[40] Henry O, Durocher Y. Enhanced glycoprotein production in HEK-293 cells expressing pyruvate carboxylase. Metab Eng 2011;13:499-507.

[41] Zhou M, Crawford Y, Ng D, et al. Decreasing lactate level and increasing antibody production in Chinese Hamster Ovary cells (CHO) by reducing the expression of lactate dehydrogenase and pyruvate dehydrogenase kinases. J Biotechnol 2011;153:27-34.

[42] Menvielle JP, Safini N, Tisminetzky SG, Skoko N. Dual role of dextran sulfate 5000 Da as anti-apoptotic and pro-autophagy agent. Mol Biotechnol 2013;54:711-20.

[43] Zanghi JA, Renner WA, Bailey JE, Fussenegger M. The growth factor inhibitor suramin reduces apoptosis and cell aggregation in protein-free CHO cell batch cultures. Biotechnol Prog 2000;16:319-25.

[44] Adamson L, Walum E. Insulin and IGF-1 mediated inhibition of apoptosis in CHO cells grown in suspension in a protein-free medium. Altern Lab Anim 2007;35:349-52.

[45] Lee JS, Ha TK, Park JH, Lee GM. Anti-cell death engineering of CHO cells: co-overexpression of Bcl-2 for apoptosis inhibition, Beclin-1 for autophagy induction. Biotechnol Bioeng 2013;110:2195-207.

[46] Kim YG, Kim JY, Mohan C, Lee GM. Effect of Bcl-xL overexpression on apoptosis and autophagy in recombinant Chinese hamster ovary cells under nutrient-deprived condition. Biotechnol Bioeng 2009;103:757-66.

[47] Damiani R, Almeida BE, Oliveira JE, et al. Enhancement of Human Thyrotropin Synthesis by Sodium Butyrate Addition to Serum-Free CHO Cell Culture. Appl Biochem Biotechnol 2013 2013;171(17):1658-72.

[48] Yoon SK, Kim SH, Lee GM. Effect of low culture temperature on specific productivity and transcription level of anti-4-1BB antibody in recombinant Chinese hamster ovary cells. Biotechnol Prog 2003;19:1383-6.

[49] Pörtner R, Schwabe JO, Frahm B. Evaluation of selected control strategies for fed-batch cultures of a hybridoma cell line. Biotechnol Appl Biochem 2004;40:47-55.

[50] Bibila TA, Robinson DK. In pursuit of the optimal fed-batch process for monoclonal antibody production. Biotechnol Prog 1995;11:1-13.

[51] Sauer PW, Burky JE, Wesson MC, et al. A high-yielding, generic fed-batch cell culture process for production of recombinant antibodies. Biotechnol Bioeng 2000;67:585-97.

[52] Senger RS, Karim MN. Optimization of fed-batch parameters and harvest time of CHO cell cultures for a glycosylated product with multiple mechanisms of inactivation. Biotechnol Bioeng 2007;98:378-90.

[53] Chamorey AL, Magne N, Pivot X, Milano G. Impact of glycosylation on the effect of cytokines. A special focus on oncology. Eur Cytokine Netw 2002;13:154-60.

[54] Li H, d'Anjou M. Pharmacological significance of glycosylation in therapeutic proteins. Curr Opin Biotechnol 2009;20:678-84.

[55] Restelli V, Butler M. The effect of cell culture parameters on protein glycosylation. In: Al-Rubeai M, ed. Cell engineering Kluwer Academic Publishers; 2002:61-92.

[56] Robinson DK, Chan CP, Yu Lp C, et al. Characterization of a recombinant antibody produced in the course of a high yield fed-batch process. Biotechnol Bioeng 1994;44:727-35.
[57] Aehle M, Kuprijanov A, Schaepe S, et al. Increasing batch-to-batch reproducibility of CHO cultures by robust open-loop control. Cytotechnology 2011;63:41-7.
[58] Altamirano C, Illanes A, Casablancas A, et al. Analysis of CHO cells metabolic redistribution in a glutamate-based defined medium in continuous culture. Biotechnol Prog 2001;17:1032-41.
[59] Berríos J, Altamirano C, Osses N, González R. Continuous CHO cell cultures with improved recombinant protein productivity by using mannose as carbon source: Metabolic analysis and scale-up simulation. Chem Eng Sci 2011;66:2431-9.
[60] Ozturk SS. Engineering challenges in high density cell culture systems. Cytotechnology 1996;22:3-16.
[61] Yang JD, Angelillo Y, Chaudhry M, et al. Achievement of high cell density and high antibody productivity by a controlled-fed perfusion bioreactor process. Biotechnol Bioeng 2000;69:74-82.
[62] Mercille S, Johnson M, Lanthier S, et al. Understanding factors that limit the productivity of suspension-based perfusion cultures operated at high medium renewal rates. Biotechnol Bioeng 2000;67:435-50.
[63] Woodside SM, Bowen BD, Piret JM. Mammalian cell retention devices for stirred perfusion bioreactors. Cytotechnology 1998;28:163-75.
[64] Voisard D, Meuwly F, Ruffieux PA, et al. Potential of cell retention techniques for large-scale high-density perfusion culture of suspended mammalian cells. Biotechnol Bioeng 2003;82:751-65.
[65] Castilho LR, Medronho RA. Cell retention devices for suspended-cell perfusion cultures. Adv Biochem Eng Biotechnol 2002;74:129-69.
[66] Zhang X, Wen Y, Yang ST. Modes of Culture/Animal Cells. In: Murray M-Y (ed). Comprehensive Biotechnology, Volume 1: Scientific Fundamentals of Biotechnology. 2nd ed; Academic Press 2011:285-302.
[67] Boedeker B. Recombinant Factor VIII (Kogenate®) for the Treatment of Hemophilia A: The First and Only World-Wide Licensed Recombinant Protein Produced in High-Throughput Perfusion Culture. In: Knäblein J (ed). Modern Biopharmaceuticals: Recent Success Stories. Wiley-VCH 2013:429-43.
[68] Hu WS, Aunins JG. Large-scale mammalian cell culture. Curr Opin Biotechnol 1997;8:148-53.
[69] Papoutsakis ET. Fluid-mechanical damage of animal cells in bioreactors. Trends Biotechnol 1991;9:427-37.
[70] Markvicheva E, Granfils C. Microcarrier for animal cell culture. In: Nedovic V, Willaert R (eds). Fundamentals of Cell Immobilisation Biotechnology, Series: Focus on Biotechnology. Kluwer Academic Publisher 2004:141-61.
[71] Carrel A. A Method for the Physiological Study of Tissues in Vitro. J Exp Med 1923;38:407-18.
[72] Davis JM, Hanak JA. Hollow-fiber cell culture. Methods Mol Biol 1997;75:77-89.
[73] Lundgren B, Blüml G. Microcarriers in cell culture production. In: Subramanian G, Lundgren B, Blüml G (eds). Bioseparation and Bioprocessing – A Handbook. Wiley-VCH 1998:165-222.
[74] Butler M. Animal Cell Culture and Technology. BIOS Scientific Publishers; 2003.
[75] Davis J. Hollow Fiber Cell Culture. In: Pörtner R (ed). Methods in Biotechnology, Vol 2: Animal Cell Biotechnology: Methods and Protocols. 2nd ed; Humana Press 2007:337-52.
[76] Lüdemann I, Pörtner R, Schaefer C, et al. Improvement of the culture stability of non-anchorage-dependent animal cells grown in serum-free media through immobilization. Cytotechnology 1995;19:111-24.
[77] Hübner H. Cell Encapsulation. In: Pörtner R (ed). Methods in Biotechnology, Vol 2: Animal Cell Biotechnology: Methods and Protocols. 2nd ed; Humana Press 2007:179-92.

[78] Hübner H, Buchholz R. Microencapsulation. In: Flickinger M, Drew S (eds). Encyclopedia of Bioprocess Technology: Fermentation, Biocatalysis and Bioseparation. Wiley 1999:1785-98.
[79] Murtas S, Capuani G, Dentini M, et al. Alginate beads as immobilization matrix for hepatocytes perfused in a bioreactor: a physico-chemical characterization. J Biomater Sci Polym Ed 2005;16:829-46.
[80] Sielaff TD, Hu MY, Amiot B, et al. Gel-entrapment bioartificial liver therapy in galactosamine hepatitis. J Surg Res 1995;59:179-84.
[81] Shuler ML, Kargi F. Bioprocess engineering : basic concepts. Prentice Hall; 2002.
[82] van Wezel AL. Growth of cell-strains and primary cells on micro-carriers in homogeneous culture. Nature 1967;216:64-5.
[83] van der Velden-de Groot CA. Microcarrier technology, present status and perspective. Cytotechnology 1995;18:51-6.
[84] Pörtner R, Märkl H. Festbettreaktoren für die Kultur tierischer Zellen. BIOforum 1995;18:449-52
[85] Pörtner R, Platas Barradas OBJ. Cultivation of Mammalian Cells in Fixed-Bed Reactors. In: Pörtner R (ed). Methods in Biotechnology, Vol 2: Animal Cell Biotechnology: Methods and Protocols. 2nd ed.; Humana Press 2007:353-70.
[86] Pörtner R, Platas Barradas OBJ, Fassnacht D, Net al. Fixed Bed Reactors for the Cultivation of Mammalian Cells: Design, Performance and Scale-Up. Open Biotechnol J 2007;1:41-6.
[87] Swamy J. The development of long term, microcarrier-based perfusion process for the production of recombinant proteins. In: IBC Conference, Cell Culture and Upstream Processing. San Diego, USA; 2003.
[88] Müller D, Unterluggauer F, Kreismayr G, et. a. Continuous perfused fluidized bed technology, increased productivities and product concentration. In: Griffiths B, Wurm N (eds). 16th ESACT Meeting: Animal Cell Technology: Products from Cells, Cells as Products. Lugano, Switzerland; 1999.
[89] Reiter M. Vaccine production with a contiuous perfusion system. In: Cell culture and upstream processing conference. Berlin, Germany; 2004.
[90] Kretzmer G. Industrial processes with animal cells. Appl Microbiol Biotechnol 2002;59:135-42.
[91] Birch JR, Racher AJ. Antibody production. Adv Drug Deliv Rev 2006;58:671-85.
[92] Warnock JN, Al-Rubeai M. Bioreactor systems for the production of biopharmaceuticals from animal cells. Biotechnol Appl Biochem 2006;45:1-12.
[93] Whitford WG. Fed-Batch mammalian cell culture in bioproduction. BioProcess Int 2006:30-40.
[94] Meuwly F, Ruffieux PA, Kadouri A, von Stockar U. Packed-bed bioreactors for mammalian cell culture: bioprocess and biomedical applications. Biotechnol Adv 2007;25:45-56.
[95] Jain E, Kumar A. Upstream processes in antibody production: evaluation of critical parameters. Biotechnol Adv 2008;26:46-72.
[96] Hacker DL, De Jesus M, Wurm FM. 25 years of recombinant proteins from reactor-grown cells - where do we go from here? Biotechnol Adv 2009;27:1023-7.
[97] Pollock J, Ho SV, Farid SS. Fed-batch and perfusion culture processes: economic, environmental, and operational feasibility under uncertainty. Biotechnol Bioeng 2012;110:206-19.
[98] Meuwly F, Weber U, Ziegler T, et al. Conversion of a CHO cell culture process from perfusion to fed-batch technology without altering product quality. J Biotechnol 2006;123:106-16.
[99] Lee S-Y, Kwona Y-B, Choa J-M, et al. Effects of process change from perfusion to fed-batch on product comparability for biosimilar monoclonal antibody. Process Biochem 2012;47:1411-8.

6.3 Industrial Cell Culture Process Scale-up Strategies and Considerations

Weiwei Hu and Kelly Wiltberger

6.3.1 Introduction

The industrial large-scale cell culture process has played an important role to support the boom of the biopharmaceutical industry in terms of meeting dynamic commercial demands as well as providing necessary clinical material to advance candidates in the pipeline. Cell culture has undergone an evolution from primarily supporting vaccine production starting in the 1950s to producing therapeutic proteins in the 1970s and 1980s. With the emergence of recombinant protein and monoclonal antibody therapies, it has since transformed into a highly successful industry with annual revenues exceeding 100 billion U. S. dollars [1, 2]. Over the years, many bioreactor systems of varying designs were created. However, the stirred and sparged bioreactor became, and has maintained its standing as, the traditional design for cell culture [3, 4].

The required size of commercial bioreactors is determined by product demand and volumetric productivity. Both of them have changed significantly in last two decades [5]. Bioreactors in the 1,000 to 2,000 L range commonly existed until the mid-1990s and then a movement towards ten-fold larger bioreactors was observed in reaction to the forecasted demand of human monoclonal antibodies in the pipeline assuming batch titers in the milligram per liter range. Multiple large-scale 10,000 to 20,000 L facilities were built with the future product demand in mind (Table 6.3.1). In parallel to the usage of these large-scale facilities, cell culture processes have advanced significantly due to cell line improvement (see Chapter 2) and media optimization. Since the large-scale facility design and construction boom, the average harvest batch titer has increased ten-fold (250 mg/L or 500 mg/L to 2.5 g/L, 5 g/L or at times nearing 10 g/L), thus creating the question of whether large-scale facilities will be needed in the future [5]. Yearly product demands can be achieved with various combinations of process batch output, facility scale, and batch number produced. Figure 6.3.1 illustrates how the increase of titer or demand affects manufacturing strategy. The number of batches required annually can change significantly. Thus, effective facility planning and mitigation strategy development is required and should be prepared in advance.

Recently, a shift back to smaller (i.e., ~ 1,000 L) but much more flexible facilities has gained momentum. Single-use systems and facilities have been highly evaluated and in some cases implemented into GMP manufacturing. The single-use facility concept provides companies with greater flexibility for globalization at reduced capital investment [6, 7]. In addition, most of the current existing biomanufacturing capacity is located in western countries (i.e., in North America and Europe), which has had minimal growth in recent years with the exception of Bristol-Myers Squibb build-

Table 6.3.1: Large-scale cell culture manufacturing facilities [112] (* expected completion date, ** originally built by Biogen Idec sold to Gentech in 2005).

Manufacturing facility	Date facility completed	Capital investment (US $M)	Area (sq ft)	Production bioreactor capacity		
				Number	Size (L)	Total (L)
Genentech – Vacaville, CA, USA	2000	250	310,000	8	12,000	96,000
Imclone – Branchburg BB36, NJ, USA	2001	53	80,000	3	10,000	30,000
Biogen – LSM, RTP, NC, USA	2001	175	245,000	6	15,000	90,000
Boehringer ingelheim expansion – Biberach, Germany	2003	315	–	6	15,000	90,000
Lonza biologics expansion – Portsmouth, NH, USA	2004	207	270,000	3	20,000	60,000
Amgen – BioNext, West Greenwich, RI, USA	2005	500	500,000	9	20,000	180,000
Genentech NIMO** – Oceanside, CA, USA	2005	380	470,000	6	15,000	90,000
Imclone – Branchburg BB50, NJ, USA	2005	260	250,000	9	11,000	99,000
Biogen Idec – Hillerød, Denmark	2007*	350	366,000	6	15,000	90,000
Lonza biologics – Tuas, Singapore	2009*	250	–	4	20,000	80,000
Genentech expansion – Vacaville, CA, USA	2009*	600	380,000	8	25,000	200,000

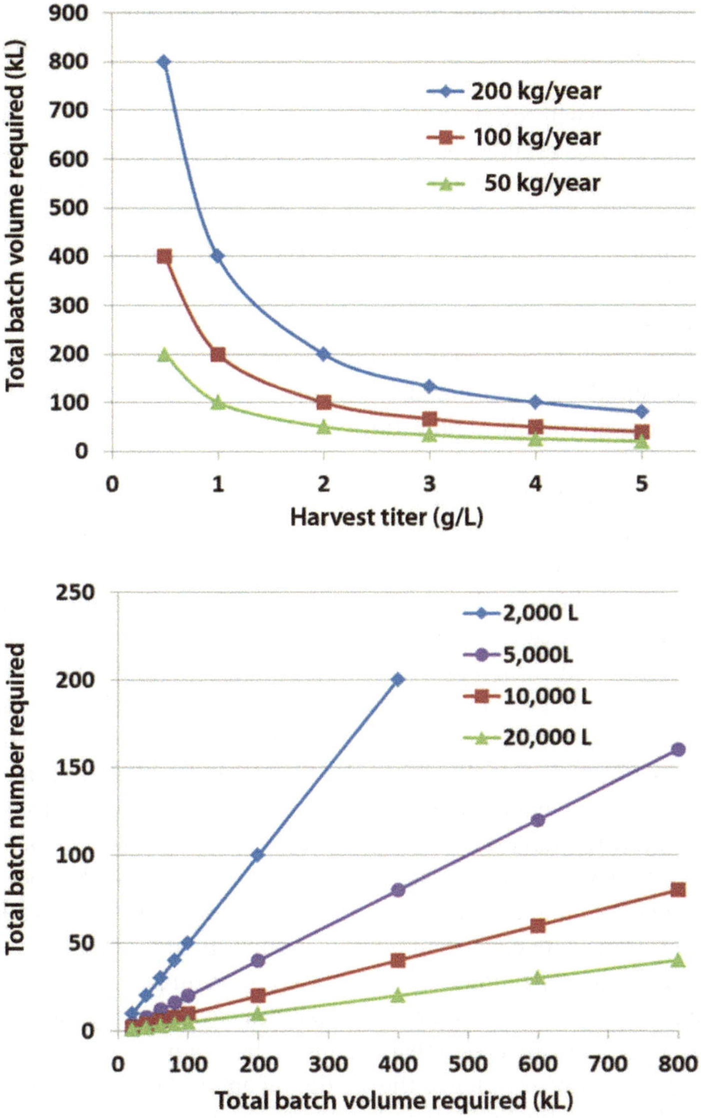

Figure 6.3.1: Impact of annual product demand and harvest titers on bioreactor size and batch numbers (assuming downstream yield at 50 %).

ing a large-scale biologics manufacturing facility in Devens, MA, USA at a cost of 750 million U. S. dollars (2009) and another 250 million U. S. dollars for additional expansion (to be completed in 2015). On the other hand, the emergence of strong interests in biosimilar development as well as the improvement of technologies and process understanding by local professionals has resulted in quick growth of this industry in

Asian countries such as South Korea, China, and Singapore. The trend will likely continue for the area of contract manufacturing business and until there is more clarity on biosimilars in general.

The question of facility size will continue to be debated. Nevertheless, a variety of bioreactor geometries exist across established bioprocessing companies and will continue to be utilized. The scales range from 500 to 2,000 L, 5,000 L, 12,000 L, 16,000 L, and up to 25,000 L. Smaller bioreactors are also used to make clinical or toxicology material. A variety of physical features exist across these systems including:

- vessel aspect ratio (height/diameter),
- impeller type, number, and position,
- impeller shaft options (angled versus vertical, top-mount versus bottom mount),
- sparger types for gas-liquid mass transfer (sintered, open-pipe, and drilled hole),
- baffle capability and design (number and width).

Due to increasing activity of contract manufacturing, strategic collaborations, and acquisitions in the biopharmaceutical industry, the scale-up and technology transfer of cell culture processes to bioreactors with different geometric features at various scales are quite common. When large-scale systems are selected for use, the financial investment per batch is also significantly large. High batch success rate at large scale is critical in these scenarios. However, both dynamic business needs and various process formats inevitably present great challenges. Because of the expensive nature of large-scale batches along with tight schedules, studies are generally not performed at large scale but rather performed at smaller lab scales to prepare for the large scale manufacturing campaign. Thus, having predictive small-scale models where processes can be tested is essential. The largest number of process experiments will be performed at bench-scale (<10 L) with again a lesser number of studies and batches performed at pilot scale (generally ranging from 50 to 1,000 L). The correlation across all scales must be extremely strong with consistent outputs to support large scale manufacturing operations. Thus, the utilization of proper scale-up techniques is critical, which will be explained in detail later. Manufacturing facilities have high capital cost investments associated; once a facility is established, it is in the company's best interest to utilize the facility efficiently and enable the maximum amount of throughput. As previously mentioned, facility flexibility (i.e., be able to run various cell culture processes) is a key piece.

Prior to any cell culture process fit into a large scale facility, it is recommended to understand the bioreactor operating range through characterization studies such as standard mixing and mass transfer tests [8, 9]. The data is used to determine appropriate operational parameters. In addition, large-scale equipment must be robust mechanically to handle the stresses. Assessment of the hydrodynamic forces on the bioreactor is important in order to avoid mechanical breakage. Weak points are assessed and then mitigated. Examples of areas for possible reinforcement include the impeller shaft, the impeller blades, baffles and their point of attachment to the tank wall, as well as the sparger. Large bioreactor vessels with high working volumes running at high agitation rates create a significant hydrodynamic force on these areas within the vessel. A vibrational assessment should also be performed at process conditions to ensure sufficient strength in external areas of the vessel such as reactor legs and supports, impeller motor and gearbox attachment to the reactor, etc.

While understanding of the final production bioreactor is a key aspect, the culture passages leading to that final culture stage is also important to understand and model across scales, which sometimes is overlooked. A standard cell culture expansion train most often will replicate that shown in Figure 6.3.2. The seed train should be performed as consistently as possible across scales. It is recommended to follow and incorporate equivalent seed train bioreactor stages even though extra work is needed to prepare and run these bioreactors. Compared to shake flasks, a bioreactor has better pH and DO control as well as a different hydrodynamic field that the cells experience. Cell culture performance (i.e., cell growth and metabolism profile) might be different if cells from shake flasks are used to inoculate small/lab scale production bioreactors versus a seed bioreactor, which can cause unpredicted large-scale performance and ultimately failure at large scale, subsequently delaying the program significantly in its clinical to commercial lifecycle. Having characterized vessels at pilot scale similar to that recommended for the manufacturing stage is very useful to identify any performance difference as well as sensitivity issues.

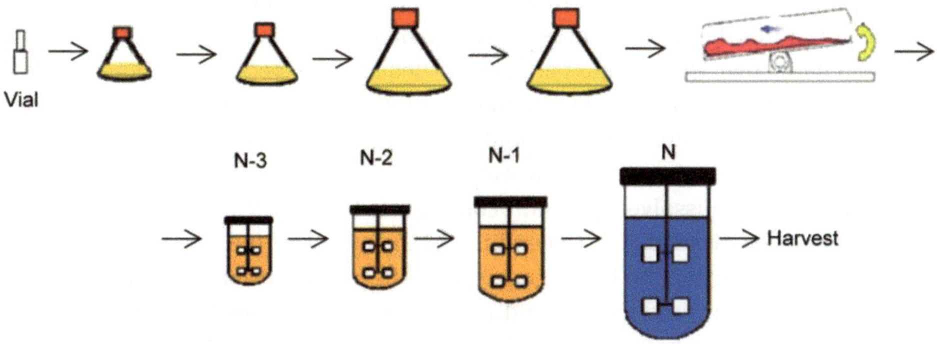

Figure 6.3.2: Typical inoculum expansion in manufacturing scale.

Chinese hamster ovary (CHO) cell lines are commonly used in large scale cell culture processes [10]. Different CHO cell lineage includes DXUB11, DG44, CHO-K1, and CHO-S (see Chapter 2). Besides that, NSO, SP2/0, baby hamster kidney (BHK), human embryonic kidney (HEK), and PER.C6® cell lines have also been applied by the industry to generate commercial and clinical materials at large scale [11]. Cell line differences should be taken into consideration during facility fit and scale-up practices. Extra work might be required and should be incorporated into process development timelines for sensitive cell lines. Risk assessment (such as process robustness) including mitigation strategies should be performed before manufacturing campaigns.

6.3.2 Differences Across Scales

Predictive cross scale models are essential; however, simply applying like for like process settings across scales will not achieve this goal. Some process parameters do lend themselves to direct one for one application. These scale-independent parameters include the setting of temperature, pH, dissolved oxygen (DO), seed density, culture duration, basal/feed media, and fed-batch feeding strategies. However, nutrient consumption rates may vary across scales and thus feed robustness should be performed. Additionally, alignment of feed strategies require taking sample volume into consideration, especially at small scale, when calculating feed percentage to ensure cells are seeing the same amount of fresh nutrients on a per cell or per culture volume basis. Depending on the difference of reactor size, liquid height, and headspace pressure, the oxygen saturation level in the liquid can vary noticeably. If a process or cell line is known to be oxygen sensitive then further understanding of aspects causing the oxygen concentration misalignment is worth pursuing.

Scale-dependent parameters are more challenging to accommodate when establishing cross-scale models. The fundamental change is bioreactor geometry, which will result in inevitable differences in fluid dynamic and bubble behavior. Previous research has shown that certain parameters such as the aspect ratio (liquid height to tank diameter) and impeller to tank diameter ratio have an impact on fluid pattern and gas-liquid interaction, which should be kept at constant values during the design of new bioreactors if possible. However, heterogeneity still exists across scales as fluctuation of pH, dissolved gas, and concentrated substrates in large scale bioreactors cannot be avoided [12]. Depending on the extent of the fluctuation gradient and cell line sensitivity, the heterogeneity can affect cell metabolism such as lactate production, growth, productivity, and product quality. The goal of agitation speed selection is to create a better mixed solution while preventing hydrodynamic damage [13]. Thus, as cross-scale model agitation selection occurs, mixing characteristics and energy dissipation rate (EDR) should be factored in.

The goal of aeration strategy selection is to deliver sufficient oxygen to the cells while removing carbon dioxide to prevent accumulation, which may have negative effects on cell health and productivity. The mass transfer rate depends on the k_La and driving force. The varying liquid height across scales causes bubbles introduced from the sparger at the bottom of the bioreactor to have varying residence times, which eventually changes gas hold-up and composition inside the bubbles [14, 15]. Furthermore, the variation in bubble residence times leads to a varied bubble size profile as the bubble travels up through the bioreactor. Bubble size additionally has an impact on the terminal velocity as shown in Figure 6.3.3 Given the difference in bubble characteristics, matching mass transfer phenomenon across scales is a complicated task, which will be addressed in detail later. On the other hand, the prevention of physical cell damage via bubble rupture should also be addressed [16]. The damage is a function of flow rate and cell-to-bubble attachment depending on bubble residence time. In fact, some cell damage can only be observed at certain scales, which makes it difficult to predict in advance. It is definitely considered to be a process risk.

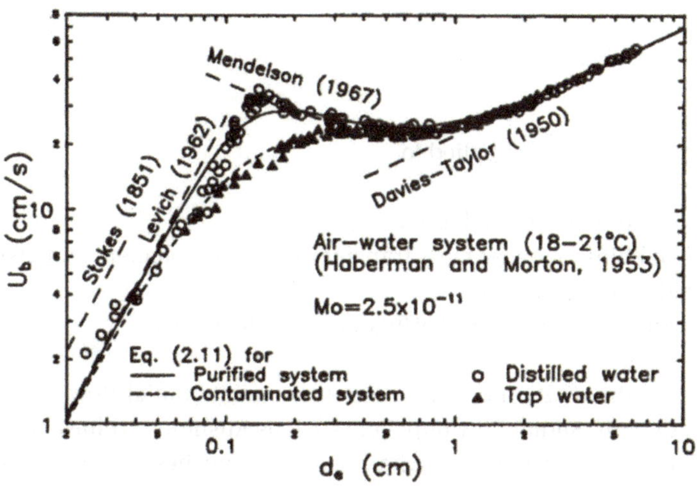

Figure 6.3.3: Effect of bubble size on terminal rising velocities [113].

There are various methods of understanding and aligning mixing, shear, mass transfer and bubble related cell damage across scales, which will be further detailed in the next sections. However, it is impossible to keep every parameter the same during scale up. Therefore, the goal is to identify and focus on the critical parameters for a given process while maintaining other parameters in a range that does not impact cell culture performance.

6.3.3 Bioreactor Agitation and Aeration Strategies

As mentioned above, stirred and sparged tank is the most commonly used type of bioreactor by industry for mammalian cell culture from bench scale to bioreactors greater than 10,000 L. With this increase in size, associated fluid dynamics and bubble behavior change along the way. Additionally, different bioreactor geometries are commonly used for the same process as programs need to be transferred or scaled-up to other facilities. Many times this cannot be avoided due to regulatory drivers or economic and timeline infeasibility. It creates additional hurdles that require a good understanding of critical process parameters for specific cell culture processes. The operation conditions should be adjusted accordingly in order to achieve comparable performance across scales.

Given the importance of agitation and aeration to the macro- and microenvironment cells experience, they are always considered as critical aspects. How to choose suitable strategies under aggressive timelines with limited information is a question frequently asked.

6.3.3.1 Impeller and Agitation

Impeller Type

Mechanical agitation by impeller(s) is applied to
1) keep cells in suspension,
2) provide good liquid mixing (nutrient, base, etc.),
3) disperse and break bubbles for better mass transfer (oxygen supplement and dissolved carbon dioxide (dCO_2) removal),
4) ensure good heat transfer (from heat jacket/blanket to bulk liquid).

Depending on the design, the rotation of the impeller can create radial flow, axial flow, or a combination of both. Radial flow impellers such as Rushton turbines have flat blades, which are set vertically along an agitation shaft. It pushes fluid towards the vessel wall. Axial flow impellers such as high efficiency HE-3 or SC-3 (Chemineer) and A-310/A-510 (Lightnin) have a low solidity ratio, which is defined by a projected area of impeller blades divided by the impeller horizontal cross-sectional area.

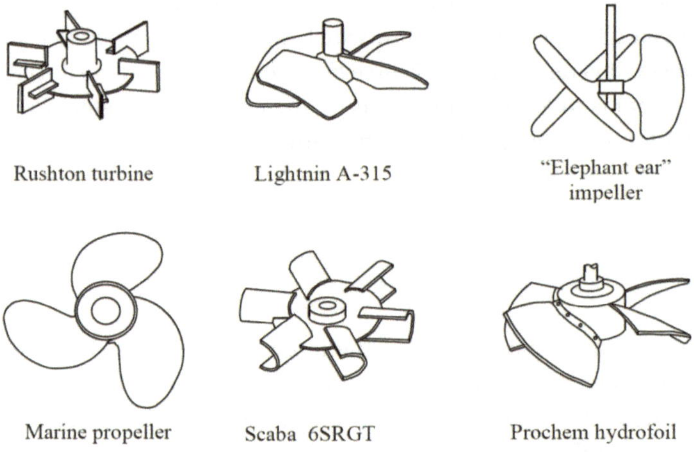

Figure 6.3.4: Impeller configurations commonly used in cell culture processes [114].

They push fluid up or down along the shaft depending on the orientation of blades and rotational direction. Instead of true radial or axial flows, a combination of both has typically been seen in cell culture bioreactors equipped with angle axial flow impellers. The impeller configurations commonly used in cell culture bioreactors are shown in Figure 6.3.4.

Selection of Impeller

Mammalian cells do not have cell walls. They have mechanical support from vascularized solid tissue in native status. So it is logical to consider them as "shear sensitive" when cultured in suspension in a stirred bioreactor. This perception had a big impact on the selection of an impeller and the speed of agitation. It is quite understandable that people prefer to use "low shear" impellers and operate them at a low speed. For example, the "elephant ear impeller" (EE) as shown in Figure 6.3.4 became quite popular in industry for large scale cell culture. Its "low shear performance" claim was due to the features of 1) high solidity ratio, 2) deep blade resulting in high pump capability, 3) the same distance from blades' edge to the drive shaft [17]. Given the popularity of this type of impeller in the cell culture area, the information regarding its fluid dynamic characteristics as well as cell culture performance comparison with other impellers is very limited. It is partially due company confidentiality policies that prevent the sharing of information about equipment and processes used for producing clinical and commercial material. Recently, there was a series of studies [18, 19] performed using particle imaging velcocimetry (PIV), which showed that measured turbulence parameters of an EE impeller, such as turbulent kinetic energy (TKE) and local specific EDRs, are comparable to other types of impellers such as a standard pitched blade turbine (PBT). These impellers cannot be differentiated by "shear" properties. The studies have provided a clear and useful guidance for transferring cell culture processes between bioreactors with different impellers.

On the other hand, tremendous amounts of research have been conducted since the 1980s to understand cell sensitivity to shear [20-23]. All of these studies have demonstrated that mammalian cells can survive high agitation speed (i.e., high shear) when bubbles are not introduced from the surface. Nowadays, it is widely accepted that mammalian cells, especially industrial cell lines, are not as sensitive as previously perceived, which helps to widen the operating range for agitation speed. However, one research area still being debated is the sublethal impact of shear [24]. It has been reported that shear might change the glycosylation pattern of protein [25], while others showed there is no impact using the same method to investigate [26]. The sublethal response seems to be cell line specific.

One thing should be pointed out: industrial cell lines normally have gone through very strict adaptation and selection processes. Prentice et al. has reported on adapting CHO cell lines through "bioreactor evolution" [27]. Even though it was not discussed in the paper, it is believed that cell resistance to shear was significantly enhanced after this evolution. The adaptation process is critical to ensure the cell lines can survive rigorous bioreactor environments at large scale. In addition, cells typically experience a few seed train bioreactor stages before entering the final large-scale production bioreactor. Alternatively, a continuous rolling seed culture in a stirred-tank bioreactor is used by some companies to increase manufacturing batches per each vial thaw. All of these steps can be considered part of adaptation that potentially increases cell resistance to shear.

Besides shear or hydrodynamic characteristics, another important property of impellers is mixing capability, which has been studied extensively by various industries. Mixing time, Θm, is a commonly used parameter to characterize it. It is recommended to check experimental details such as the definition of homogeneity (i.e., 95 or 99 % of final steady state after a step change) and the method used (i.e., pH, conductivity, or color) before comparing mixing time study results since each practice could result in different values. One efficient way to improve mixing (i.e., lower mixing time) is via the installation of baffles. Without a baffle, the flow generated by the impeller is mainly horizontal in its rotation, which has low power input under the same agitation speed and creates a central vortex in most cases [28]. Since the vertical flow is limited, it takes a longer time to reach homogeneity. The installation of a baffle(s) can change the flow pattern by creating more vertical flow along the baffle and bioreactor wall, which can increase power consumption and improve mixing.

Under turbulent flow regime, the mixing time of one impeller can be empirically correlated to power input and tank geometric parameters as shown in Equation 6.3.1 [29]:

$$\theta_m = 5.9(\overline{\varepsilon_T})^{-1/3}\,(D/T)^{-1/3}\,T^{2/3} \tag{6.3.1}$$

This equation is applicable to various impeller types, such as a Rushton Turbine and a Lightnin A310, as long as the tank has an aspect ratio of 1.0. It shows that power consumption, not impeller design, has a great impact on mixing time. Mixing time increases with the increase of tank size, which inevitably changes during scale-up. In order to lower mixing time, high power input and large impeller diameter can be applied. For bioreactor aspect ratios larger than 1.0, mixing time increases significantly with the increase of liquid height as shown in Equation 6.3.2 [29]:

$$\theta_m = \propto (H/D)^{2.43} \tag{6.3.2}$$

However, mixing time does not provide detailed information regarding spatial and temporal transients after another substrate is added. It is determined by flow and circulation rates, which are affected by impeller type [29].

In addition, depending on impeller position and the feed percentage applied to a process, a partially covered impeller scenario may occur during fed-batch cell culture processes. This scenario will cause a liquid splashing issue, especially with angle-mounted impellers. It is not an ideal situation but it is hard to avoid at large scale when multiple impellers are required. Unfortunately, investigation in this area is very rare in terms of how it affects cell culture performance and how to mimic the impact across scales. Therefore, the impeller with less vertical depth has an advantage to minimize the duration of a partially exposed impeller during a batch.

Overall, a variety of impeller designs have been successfully used in large-scale cell culture processes. Even though similarity is preferred, the use of different impellers should not be a roadblock for scale-up and technology transfer. Sandadi et al. have demonstrated there was no significant difference in cell culture performance between Rushton turbine and marine impellers [30]. A Rushton turbine was successfully used for the production of interferon at large scale [31]. Corrales et al. (2006) reported a successful transfer of a cell culture process from pitched blade to Lightnin A510 axial flow impeller at large scale [32]. Meier has reported successful use of Rushton turbines, Lightnin A320s, and "elephant ear impeller" at large scale [33]. Junker et al. reported a successful culture following a fermentor modification [34]. To the best of our knowledge, the negative impact of impeller type on cell culture performance has not been reported in literature, at least for industrially-relevant cell lines.

Agitation Speed

Depending on what scale-up criterion is chosen, agitation speed at large scale can be determined [35]. Constant power consumption per unit volume of fluid (P/V) is widely used for scale-up in industrial mixing applications including cell culture since it has great impact on fluid mixing and other hydrodynamic parameters. Yang et al. showed that constant P/V also resulted in a more reasonable agitation speed from an operating point of view [35]. The typical P/V range for cell culture process is 10 to 150 W/m^3 [13]. It was also reported that 250 W/m^3 did not cause negative impact in cell culture performance [36]. The choice of the exact P/V value for scale-up is quite empirical and generally does not need to change across processes.

In order to check whether the selected P/V is suitable, the size of the smallest turbulent fluid eddies, Kolmogoroff eddy size (λ_k, Equation. 6.3.3), can be compared with cell diameter (i.e., 12 to 25 µm). The energy introduced through the impeller is ultimately converted into heat dissipation in the smallest eddy size. When the size is lower than 2/3 to 1/2 of the cell diameter, detrimental effects might occur.

$$\lambda_k = \left(\frac{(\varepsilon_T)_I}{\upsilon^3}\right)^{1/4} \tag{6.3.3}$$

This theory has successfully explained the damage of cells on the surface for microcarriers since they have large size in the order of 100 µm in diameter [37].

As mentioned above, mixing time is determined by power consumption, which will increase with the increase of scale under the constant P/V. For example, mixing time at small scale or pilot scale is usually around a range of 10 to 60 s, while mixing time at large scale (>10,000 L) can be a few minutes. Long mixing time can cause a high degree of heterogeneity [38, 39], which can potentially affect cell culture performance. Therefore, establishing a minimal mixing requirement at large scale should be considered. Unfortunately, it is not an easy task. There are several studies in literature using specifically designed small scale models with dual bioreactors, which shows pH perturbations have significant negative impact on cell viability [40, 41]. However, the study is more qualitative than quantitative. The response also depends on the cell lines and process used. It is recommended to choose a mid-to-high P/V value in the acceptable range to start with for large-scale runs to reduce the fluid heterogeneity as much as possible. The agitation speed can be adjusted after a few batches at large scale if necessary.

Once P/V value is determined, agitation speed can be easily calculated using Equation 6.3.4 when other parameters are known:

$$\frac{P}{V} = \frac{N_p \times \rho \times N^3 \times D^5}{V} \tag{6.3.4}$$

Power number, Np, is an important dimensionless parameter, which is widely used in mixing operations. It was defined by Rushton et al. in 1950 [42]. Depending on impeller design, Np can vary significantly. In general, radial flow impellers such as a Rushton turbine have high Np (i.e., ~ 5.0) values. True axial flow impellers such as a high efficiency impeller have very low Np (i.e., ~ 0.3) values. The other types are in the middle of the range. Np of "elephant ear" and Lightnin' A315 hydrofoil impeller is around 1.7 and 0.84, respectively [43].

Even though some researchers have tried to estimate Np based on physical fundamentals such as angular momentum balance or drag force analysis [44], it is still recommended to obtain accurate Np value through experimental measurement due to the limitation of theoretical approaches. Experimentally, power consumption can be calculated from measured torque and agitation speed as shown in Equation 6.3.5:

$$P = 2\pi N T_q \tag{6.3.5}$$

The torque on the shaft, which can be measured by a torque transducer installed on the shaft, is proportional to the torque caused by fluid forces. However, extra attention and patience are needed to calibrate and measure the very small torque differences in order to get accurate and repeatable results. Chapple et al. described an experiment set-up (Figure 6.3.5), which makes use of a pillow bearing and flex coupler along with tight temperature control [45]. The torque due to other factors including from the pillow bearing, the shaft movement, and other baseline torque was measured and deducted. The authors have disclosed extensive experiment details and intended to make it a standard method. Zhu et al. have successfully measured power draw using a Coesfeld Viscomix unit (Coesfeld) with a small scale bioreactor [17]. However, the detailed experiment setting was not shared in the paper. Alternatively, King et al. reported the measurement of electrical power consumed by the motor [46]. The conversion requires application of the motor efficiency curve and the other power losses from the gearboxes and bearings. This method sometimes is applied on large-scale tanks; however, it is hard to get an accurate electrical reading at small scale since the impeller power draws only accounts for a small percentage of total power consumption. Thus, the same method cannot be used accurately across scales.

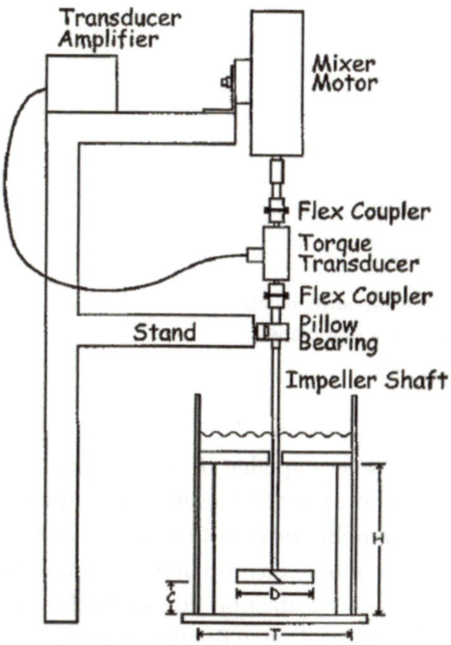

Figure 6.3.5: Schematic diagram of the power number experimental setup [45].

It is important to understand that there are a number of factors that can change the system power number of an impeller. The power consumption of a Rushton turbine is mainly due to form drag, which is affected by the blade geometry such as blade thickness (Figure 6.3.6). With respect to pitched blade impellers, form drag is less important and rather the interaction between the impeller and the proximity of the tank wall (the impeller to tank diameter ratio) plays the key role [45]. For a Rushton turbine, Np drops significantly with the increase of Re when it is between 1 and ~100. While Re is > 100, Np stays relatively constant or increases slightly with the increase of Re [42]. Re within 10^3 and 10^5 also has no impact on the power number of down-pumping four-bladed 45°-pitched blade impellers [45]. In addition, power number can be significantly affected by the number of baffles (Nb) and their width (B). The conventional power number reported has a default configuration with 4 baffles and width equal to T/10. With the decrease of NbB, the power number decreases as shown in Figure 6.3.7.

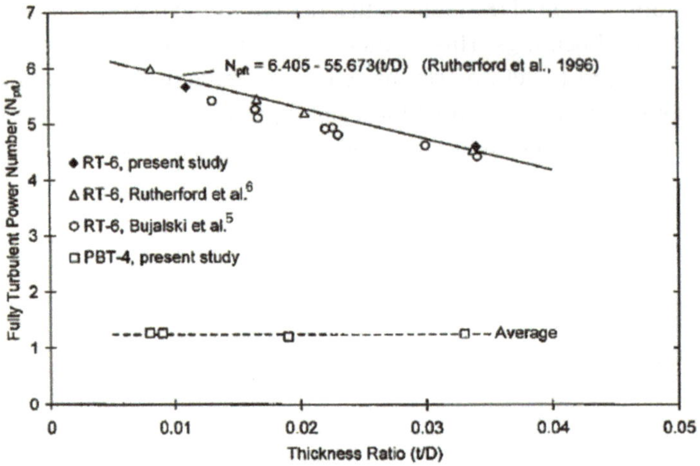

Figure 6.3.6: Effect of blade thickness ratio on fully turbulent power number for a six-bladed Rushton Turbine (RT-6) and a down-pumping four-bladed 45°-pitched blade turbine (PBT-4) [45].

Sparging has a tendency to decrease power draw by impeller since air-filled cavities are formed behind the blades. It is difficult to estimate gassed power number because the decrease is a complex function of impeller design, flow rate, agitation speed, etc. For example, a Rushton turbine has a pronounced drop in power draw and is easily flooded. Impeller flooding means there is no horizontal flow of bubble/gas from impeller to bioreactor wall. The transition from loading to flooding significantly lowers the power draw and gas hold up. However, given the conditions in which a mammalian cell culture bioreactor is normally operated at (i.e., low agitation speed compared to microbial fermentation), the impeller has less impact on the bubble flow pattern and there is no air cavity formed in most cases. Therefore, a gassed power number should be similar to an ungassed power number. Zhu et al. showed that there was no power drop when a small scale bioreactor with either up-pumping or down-pumping impellers was operated at flow rates in the range of 0.01 to 0.05 vvm [17]. However, when the flow rate is increased to 0.5 vvm, which is one magnitude higher than the typical range, the down-pumping EE impeller power draw was decreased by 30 % [17].

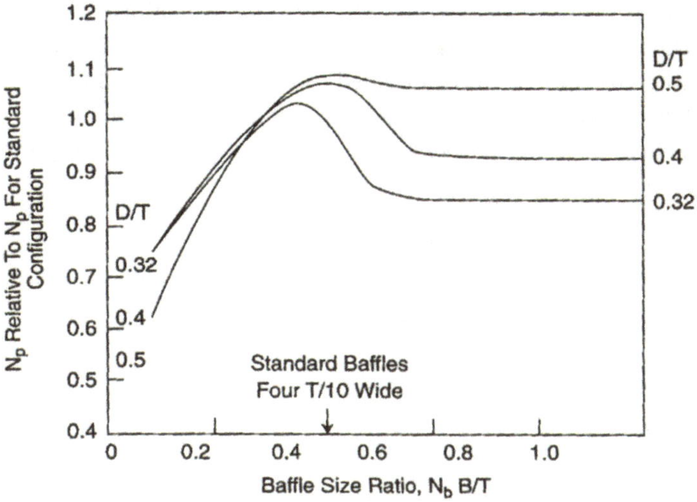

Figure 6.3.7: Effect of baffling and D/T on power number [115].

Overall, it is important to compare the bioreactor of interest with standard configurations and make appropriate adjustments of the power number value if it is not feasible to take a direct measurement. It is important to verify whether the selected agitation speed is within the motor's capability and the vessel will be stable without extensive vibration.

Multiple Impellers

With the increase of bioreactor size (>2,000 L), multiple impellers are recommended and commonly used especially for tanks with high aspect ratios [47]. It is not necessary to keep all of them the same however. The combination of different impeller designs for better gas dispersion and mechanical stability has been reported. In most cases, each impeller creates independent circulation loops. The total mixing time depends on the rate of mass exchange between these loops. The exchange flow rates between impellers are determined by the type of flow generated by each impeller and pumping capability. It is important to check that the impeller distance is suitable so that the circulation of the upper impeller, which normally carries base added from the top of the liquid to control pH, can have a good exchange with the lower loop where the pH probe is located. Vrabel et al. reported that a significant decrease of mixing time was observed when using axial up-pumping style as the top impeller compared to a radial impeller due to the different circulation pattern [48]. Other advantages of up-pumping impellers include [17, 49], 1) effective for gas dispersion and capable to handle high aeration rate without impeller flooding, 2) negligible torque and flow instabilities, 3) efficient foam-breaking capability if placed close to liquid surface.

It has been shown that the distance between 2 impellers has a significant impact on the total power consumption [50]. Basically, the system power number decreases when the distance between is shortened. The detailed profile depends on the impeller type. For unaerated disc turbines, the change of power number is shown in Figure 6.3.8. However, the practical distance between impellers should be less than 2D to avoid dead zones between the impellers.

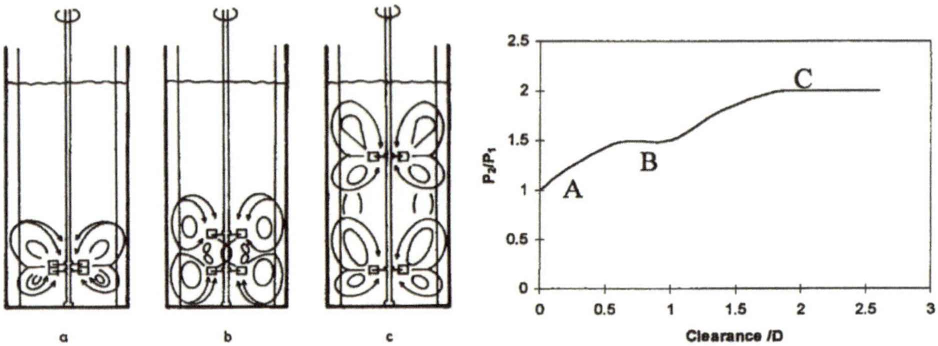

Figure 6.3.8: Impact of impeller spacing on unaearted power draw [47, 50] (P2 = power consumption of dual impeller system, P1 = power consumption of single impeller system).

The active volume or mean circulation zone for axial impellers is only 2/3 of the tank volume. The rest of the volume is governed by macroinstability [51]. The position of the top impeller with respect to the liquid surface is important in terms of base dispersion and dCO_2 stripping. Sigma and Gu have reported that moving the top impeller closer to the surface can significantly improve dCO_2 removal [52], which helps prevent dCO_2 accumulation at large scale where it occurs most.

6.3.3.2 Aeration Strategy

Oxygen is a key nutrient in mammalian cell culture processes. However, the solubility of oxygen in liquid media is very low, around 7 mg/L under typical cell culture conditions. Cell specific oxygen uptake rate (sOUR) depends on the cell line, temperature, pH, and cell metabolism status, and thus may vary significantly between 0.5 to 25 pmol per cell per day [53]. Oxygen is continuously consumed by the cells. Without continuous supplementation of oxygen into the liquid, the dissolved oxygen will be depleted within a few minutes.

In addition, cells produce and release carbon dioxide as a byproduct of cellular metabolism. The typical respiratory quotient (RQ) is close to 1.0 [53]. Compared to oxygen, carbon dioxide solubility is significantly higher. For example, 1.3 g/L CO_2 can be dissolved in water under 1 atm and 30 °C [54]. Furthermore, dCO_2 can react with water to form carbonate acid and subsequently dissociated to hydrogen ion and bicarbonate ion as shown below.

$$CO_{2(g)} + H_2O \xleftrightarrow{K_1} H_2CO_3^* \tag{6.3.6}$$

$$H_2CO_3^* \xleftrightarrow{K_2} HCO_3^- + H^+ \tag{6.3.7}$$

$$HCO_3^- \xleftrightarrow{K_3} CO_3^{2-} + H^+ \tag{6.3.8}$$

K_1, K_2, and K_3 are 1.6, 6.3, and 10.23 respectively at 37 °C and 0.115 molar ionic strength [53]. At cell culture pH around 7.0, HCO_3^- formation is dominating followed by H_2CO_3*. The transfer of bicarbonate to carbonate is negligible. If there is no efficient way to strip dCO_2 out of the liquid, the dCO_2 level can build up easily which will have the following 2 impacts.

1) pH of the extracellular solution will decrease dramatically. Since pH in cell culture processes is controlled tightly, the increase of dCO_2 will eventually result in high base addition and then high osmolality, which will have negative effects [55, 56].
2) Since CO_2 is a nonpolar molecule, it can freely diffuse into cells, which can change the intracellular pH and cause subsequent physiology changes. For example, the protein glycosylation and charge distribution was altered at high dCO_2 [57].

Mass Transfer

Aeration has 2 goals to achieve: oxygen supplementation and dCO_2 removal. According to Fick's first law governing the diffusive transport of the solute across the liquid-side boundary layer, mass transfer rate for oxygen and CO_2 can be expressed as Equation 6.3.9 and Equation 6.3.10, respectively.

$$OTR = k_L a_{(O_2)} \times (C^*_{O_2} - C_{O_2}) \tag{6.3.9}$$

$$CER = k_L a_{(CO_2)} \times (C^*_{CO_2} - C_{CO_2}) \tag{6.3.10}$$

According to van't Riet's correlation [58], $k_L a$ increases with the increase of power input and superficial gas velocity as shown in Equation 6.3.11.

$$k_La = \alpha \times (P/V)^{\beta} \times v_s^{\gamma} \qquad (6.3.11)$$

α, β, and γ are constants. Under typical cell culture operating conditions, P/V has less impact on mass transfer compared to superficial gas velocity determined by the sparging rate for a given bioreactor diameter. For example, β is 0.47 and γ is 0.8 in a study conducted by Xing et al. (2009) using an industrial 5,000 L bioreactor for oxygen transfer [39]. Under the same aeration rate (vvm), defined as volume of gas per volume of liquid per minute, high k_La is normally obtained due to high v_s. Fundamentally, it is due to high gas residence time and high gas hold-up. The use of different impellers or combinations of impellers can also change mass transfer performance. Arjunwadkar et al. demonstrated that the disc turbines (DT) – pitched blade turbine downstream (PTD) combination – has the best performance at a relatively low power input [59]. Other researchers have shown up-pumping impellers to be good in terms of mass transfer capability [60, 61].

Among various methods reported for k_La measurement including the hydrogen peroxide steady-state method [62], the dynamic gassing-out gassing-in method is commonly used by industry due to its experimental simplicity and low cost. However, in order to obtain reliable data, special focus is needed on execution details including the data analysis. The same basic method can be performed, but if details such as vessel conditions, buffer expiration, etc. are not applied consistently, the data generated will have significant experimental error across studies. For example, the response time of the DO probe should be assessed before the experiment. Jorjani and Ozturk have shown its impact on measured k_La [63]. A long DO probe response time is a more significant problem when mass transfer is high such as in microbial fermentation. For cell culture processes, where high k_La is around 10 h^{-1}, DO response time up to 30 to 40 s is acceptable. A step change in bioreactor pressure is recommended to check DO response time *in situ* [64]. The data analysis technique such as how to determine the start and end point of data to include also have an impact on results. Mass transfer rate is also determined by the composition of cell culture media, which has different surface tension, ionic strength, anti-foam, etc. It is better to run the test using liquid with similar properties to media and at process temperature.

Driving force is another factor determining how fast the mass transfer rate is. A common and easy way to increase it is supplying pure oxygen into inlet gas. Driving force can be heterogeneously distributed in a large-scale bioreactor since the DO level differs if the bulk liquid is not well mixed. It is not a large concern since mixing time is usually significantly smaller than that of oxygen transfer. Therefore, oxygen transfer is not a bottle neck from a time constant analysis point of view [65]. In addition, the driving force decreases as liquid height increases because the gas composition changes (oxygen concentration decreases and carbon dioxide concentration increases) as bubbles rise through the bioreactor. Despite this, it should not be a problem for mammalian cell culture processes since the bubble oxygen percentage difference between the inlet and exhaust is small.

However, CO_2 percentage in a gas bubble changes significantly when the bubble rises from the bottom to the top of the reactor vessel. For small bubbles such as microbubbles with diameters around a few hundred micron generated by sintered sparger, it only takes a few seconds to reach an equilibrium status between the liquid media and the bubble. Then, those bubbles lose the ability to remove additional dCO_2 while oxygen transfer still occurs. In this case (i.e., CO_2 saturated in exhaust gas), the total gas sparge flow rate allowed is critical, which can determine how much dCO_2 can be removed, while other parameters such as agitation speed will not have an impact [66]. On the other hand, the outlet gas flow rate can be significantly different than the inlet if the bubble size is small and the oxygen concentration in the bubble is high. This will lead to dCO_2 accumulation in the bulk liquid. In an extreme case, microbubbles with pure oxygen will be totally dissolved before they reach liquid surface so that the outlet gas flow rate is zero. When the bubble size increases, it takes longer than bubble residence time allows to fully saturate a bubble with CO_2. In this case, mass transfer rate can have an impact on the dCO_2 level. Because it is hard to obtain an accurate driving force for CO_2, the measurement of k_La (CO_2) is seldom reported. The value can be converted from k_La (O_2) based on the mass transfer theories shown below for a mobile or rigid interface [67, 68].

$$k_L^{\text{mobile}} = 1.13\sqrt{\frac{u_s}{d}}\,D_g^{1/2} \tag{6.3.12}$$

$$k_L^{\text{rigid}} = 0.6\sqrt{\frac{u_s}{d}}\,D_g^{2/3}\,\nu^{-1/6} \tag{6.3.13}$$

The difference of k_L between O_2 and CO_2 results from the mass diffusivity (D), which is reversely correlated to molecular weight.

There are 2 aeration strategies to control DO level at a desired range, total flow control or cascade. A minimal flow is typically applied at the beginning of cell culture process even when DO level is still high. This is to prevent liquid backflow into the sparger and gas piping which can cause cell damage and nozzle clogging. When DO drops below the set point in the total flow control model, air sparge rate increases until it reaches a preset value (total flow cap). Then oxygen sparge turns on to further supply the oxygen demanded while air sparge rate drops so that the total flow rate is maintained as a constant level but the concentration of oxygen in the gas delivered has increased. With the cascade model, when the air sparge rate reaches a preset value (air cap), it will stay at this level and oxygen sparge will be added on top, causing the total flow of sparge gas to continue to increase. An illustration of the control methods is shown in Fig. 6.3.9. Both methods have been successfully used by industry. The total flow control model is relatively popular since it is easy to scale-up/scale-down.

However, maintaining tight DO control can be challenging since 2 flow rates of varied oxygen concentrations have to be adjusted simultaneously.

In both cases, the dCO_2 level is not directly controlled compared to DO. With the increase of bioreactor size (liquid column height), the mass transfer capability of oxygen increases which will result in a lower gas sparge rate. Also the driving force for CO_2 removal decreases at large scale. Unfortunately, both of these cause high dCO_2 accumulation. As a result, the air cap or total flow rate should be carefully chosen to ensure sufficient dCO_2 removal, which becomes more important as the bioreactor size increases.

A dual sparger system has been reported that can potentially decouple O_2 supplementation and dCO_2 removal through a combination of microbubbles with oxygen

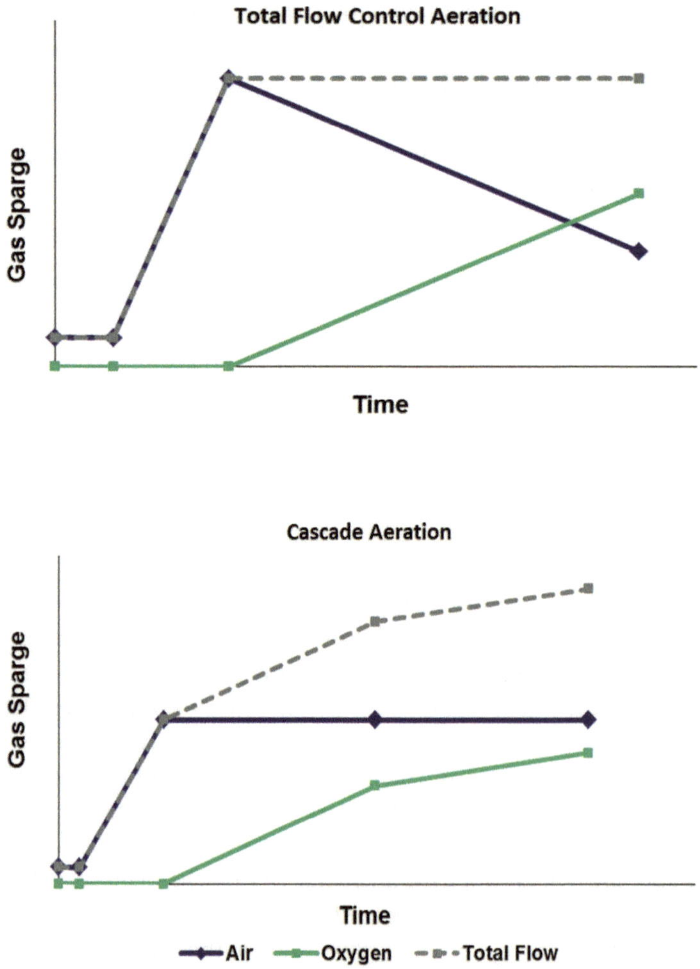

Figure 6.3.9: Illustration of bioreactor aeration control strategies.

enriched air or pure oxygen and large bubbles with air [69]. It makes use of the strength of each format to support higher mass transfer demand at low flow rates. However, it increases the complexity of the sparger system which might require additional facility work (i.e., piping) as well as process complexity in terms of how the automation determines the flow rate of pure oxygen and air required to maintain a DO setpoint. Scale up of a dual sparger system is even more complicated and has not been reported.

Surface aeration at large scale generally has less impact on overall mass transfer rate. At the beginning of the cell culture process, cell mass and oxygen demand is low. Therefore, the DO and dCO_2 profiles are mainly determined by surface aeration. It is typical to see a quick DO drop and relatively high dCO_2 level at beginning. However, if the upper impeller is close to the liquid surface or partially submerged as discussed above, surface aeration will increase significantly causing a much slower drop in DO over time. However, these observations are quite empirical without good mathematical modeling or correlations due to the instability of the surface. The impact also highly depends on the impeller type, size, and location, which generate some challenges in developing a valid scale-down model.

The sparger location in relationship to the impeller(s) also plays a significant role. The gas distribution can be maximized by locating the gas entrance under the impeller zone. It is also important to consider the reactor skid piping diameter and design lines large enough as to not be restricted to support higher cell mass. As the total gas entering the reactor increases, the exhaust capacity also must increase. Considering the exhaust piping size as well as the final exhaust location is valuable to prevent pressure build-up in the tank and prevent high oxygen concentration exposure near electrical or other flammable items from a safety perspective.

Bubble Related Cell Damage

Another important aspect related to aeration strategy is bubble-associated cell damage. Sparge aeration is commonly used in large-scale bioreactors because of simplicity. Numerous studies have shown that cells have a tendency of attaching to the rising bubble surface and can be severely damaged by the rupture of these bubbles at the gas disengagement zone [16]. The hydrodynamic forces generated during bubble rupture are substantially higher than that in the impeller region [70, 71]. Figure 6.3.10 shows that the EDR generated during bubble rupture decreases with the increase of bubble size. It is easy for large bubbles to break at the liquid surface while small bubbles have to be significantly higher than the liquid level to trigger it. Pressure inside a small bubble is also larger than that of large one. As shown previously, a large bubble has a higher terminal velocity than some small bubbles as shown in Figure 6.3.3. Therefore, the residence time of small bubbles in a bioreactor is much longer, which also creates more chance for cell-to-bubble attachment.

In order to alleviate this damage, protective additives such as pluronic F-68 and polyvinyl alcohol (PVA) have to be used in cell culture media. There is still a debate

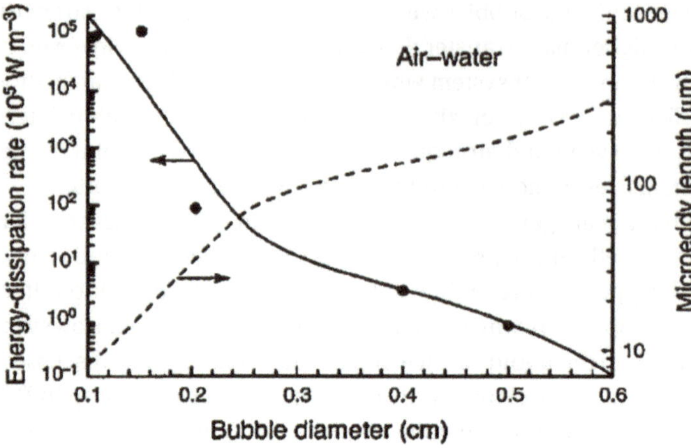

Figure 6.3.10: Effect of bubble diameter on maximum EDR during bubble rupture and corresponding microeddy length (same as Kolmogoroff eddy size) [16, 70].

about the fundamental mechanisms of these additives [72]. While it is known that pluronic F-68 can penetrate the cell membrane and change its properties, the most commonly accepted theory is that these polymer surfactants can alleviate cell-to-bubble attachment [73, 74]. So, when a bubble ruptures at the liquid surface the number of cells impacted is lowered. However, it cannot totally prevent cells from being damaged. Its effectiveness also depends on bubble size. Meier et al. showed that pluronic F-68 is less protective for cell attachment to small bubbles [75]. Therefore, the impact of bubble size should be taken into consideration during sparger selection, which again increases the complexity of determining an appropriate aeration strategy for a given process. A recent study showed that pluronic F-68 can be consumed by cells. However, the amount consumed is too little to have any significant impact [76]. Lot-to-lot variation of pluronic F-68 was observed and documented more than 25 years ago [77]; it was found that 2 specific lots could not protect cells from bubble damage. Given the importance of pluronic F-68, there is a strong need to identify the root cause and improve the consistency, which requires interdisciplinary collaborations such as polymer synthesis, colloid chemistry, cell physiology, and analytical method development.

Despite the significant impact to bubble dynamics, sparger design has not been investigated sufficiently in the past. The types of spargers typically used in a cell culture bioreactor, sometimes in combination, are a sintered sparger (10 to 100 µm), a drilled hole sparger (0.2 to 2.0 mm), and an open pipe (>5.0 mm) [9, 52]. As discussed before, each one has advantages and disadvantages. The sintered sparger has significant oxygen transfer capability because of high surface to volume ratio, but its ability to remove dCO_2 is limited due to the quick saturation of CO_2 in bubbles and low flow rate.

Small bubbles also cause more problems in cell damage and foam. Due to the concern of cleaning, industry practice includes the replacement of the sintered material on a sintered sparger after each run. Open pipe has advantages such as low cell damage and foam. But it also has low mass transfer rate, which might not be able to support the high cell density cell cultures. The drilled-hole sparger with bubble size in the middle has a potential of balancing all the requirements and challenges, which is getting more popular with the increase of cell culture intensity. However, it has its own unique problem. One of them is high gas entrance velocity (GEV), which has shown negative impact on cell growth and viability [78-80].

GEV is a parameter, describing how quick the bubbles go through the sparger nozzles, which can significantly change bubble formation and breakup in sparger zone. Thus, the sparger has to be properly designed during scale-up including an increase of nozzle number or size. For sensitive cell lines, utilization of the total flow rate DO control strategy will create a known maximum GEV which can be selected in accordance with the cell line of interest. The mechanism of how the cells are damaged by high GEV in the sparger zone is not well understood. Mathematical modeling (a second order moment (SOM) bubble-liquid two-phase turbulent model based on the two-fluid continuum approach) has shown that traditional EDR cannot explain the damage [81]. Subsequently, a newly proposed correlation expression, stress-induced turbulent energy production (STEP), was found to correlate well with the unusually high cell death rate at the gas inlet region [81]. Furthermore, hole clogging can be an issue, which can dramatically change GEV and bubble formation. Therefore, the selected aeration strategy and sparger design are important process parameters and should be optimized to prevent cell damage from happening in the sparger zone.

Foam Issue

Cell culture solutions have the tendency to foam when gas sparging is occurring because they contain a surfactant(s) such as pluronic F-68 or PVA. Cells also produce and release recombinant and host cell proteins into the culture solution, which can also act as surfactants due to their amphiphilic structure. At the end of cell culture process, dead cells can release intracellular contents such as DNA into the liquid, which make the foam stable and hard to break. The formation of foam in a bioreactor should be restricted to prevent a bioreactor foam-out and potential contamination. The impact of foam on the cell culture process is complicated. A small level of stable foam on the liquid surface can allow cells to drain back to the liquid phase and prevent severe cell damage by bubble rupture. However, excessive foam can be detrimental to cells, which get trapped in the foam layer since it is not a well-controlled environment. Different cell culture performance might be observed across the scales if the foam behavior and treatment are different. It is often observed that yellow cell debris or dead cells accumulate at the top of the foam layer around bioreactor wall.

The culture viability may drop noticeably when antifoam is added by bringing those dead cells back to bulk liquid.

The foam level in the bioreactor is related to bubble size, gas sparge flow rate, and liquid properties such as surface tension etc. Generally, foam formed by small bubbles has high liquid content. The surface of small bubbles is more rigid, which makes liquid drainage and bubble coalescence difficult. Therefore, the bubbles can last for a longer period of time. Mostafa and Gu have showed that even at significantly lower flow rates smaller bubbles still caused more foam problems compared to large bubbles [52]. Additionally, it is challenging to suppress the foam generated from small bubbles via antifoam.

As a mitigation strategy, antifoam is commonly used in the cell culture process. Its composition normally contains oil droplets (i.e., silicone oils, mineral oils), hydrophobic solid particles (silica, Al_2O_3, TiO_2), or a mixture of both. Typically antifoam globules have a size of ~5 μm in radius, which contains silica particles with a radius around 5 nm. The mixture of oil and particle was found to be more efficient than each of the individual components. The solid particle has properties that can lower the entry barrier to an asymmetric oil-water-air film ("pin effect") as well as facilitate bridge formation in the film thinning process, especially in the early stages of the defoaming process [82]. The oil droplet ensures the antifoam globules are deformable, which is critical for the bridging-mediated mechanism. Antifoam can lose its function with time as demonstrated in Figure 6.3.11. Therefore, multiple antifoam additions are typically required during a batch, especially toward the end of cell culture process when viability is low. However, the optimization of antifoam addition in terms of fre-

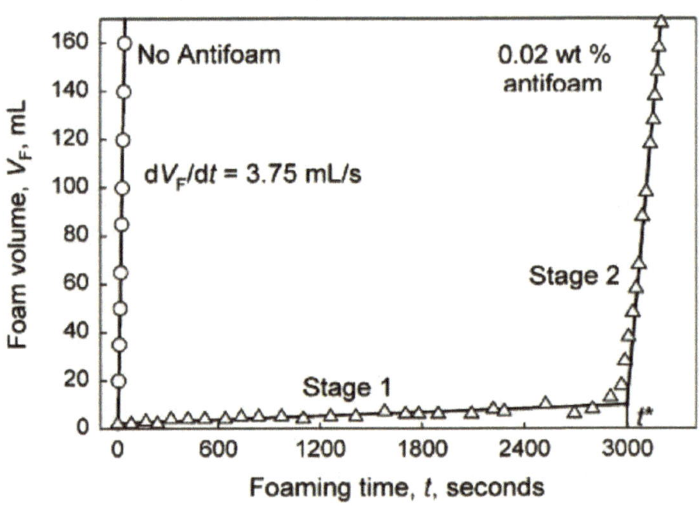

Figure 6.3.11: An example of foam evolution in the presence of silicone oil-silica compound [82].

quency and amount has not been reported. The maximum antifoam allowance and downstream clearance capability should be assessed before scale-up.

However, the addition of antifoam during the cell culture process has a significant negative impact on the mass transfer rate [83]. It has been reported that k_La decreases by more than 50 % with the use of antifoam [84]. The antifoam can contaminate the bubble surface, which can dramatically increase the resistance for gas-liquid mass transfer [85, 86]. A sharp spike in the gas sparge rate spike is always observed after antifoam addition in order to maintain the target DO level, which can cause more foam and thus perpetuate a vicious circle.

Overall, bioreactor aeration is a complicated task. In which, oxygen supplementation, dCO_2 removal, cell damage, and foaming should be balanced and controlled through-sparger selection or design as well as sparge control strategy selection.

6.3.4 Computational Fluid Dynamics (CFD) Simulations

In recent years, CFD has been applied to bioreactor characterization. A few good review papers about CFD applications in this area have been published [87, 88]. In general, CFD simulation consists of mesh constructions, boundary definition, solving various governing equations for momentum, mass, and energy balance, and data analysis. CFD is a useful tool, which provides us another way to understand bioreactor behavior and study the difference across scales. However it is recommended to understand modeling capability, assumptions, and limitations, which can help interpret and make use of CFD results. There are a number of software options available, though all require a skilled user to enable the delivery of useful data. The model should be validated with experimental data points if possible.

The application of CFD work in the area of industrial cell culture mainly focuses on where experimental measurements cannot be performed due to technical challenges, resources constrains, and GMP restrictions. For example, heterogeneity is a significant concern in a large-scale tank. The addition of baffles as well as the usage of an angled shaft impeller can improve mixing and increase uniformity across the tank. But, there will always be pockets, zones, or gradients across the liquid as gases and solutions are introduced. Experimental studies can be performed to characterize mixing time, but these only provide a holistic view of the vessel in a given condition. In this scenario, single-phase turbulence CFD modeling is very useful. Modeling pH spikes observed in the base addition zone and hydrodynamic force distribution near the impeller tip or battles can greatly improve overall understanding in a zonal fashion across the vessel. Compared to average P/V, CFD can provide more detailed information of EDR distribution. Vakili and Esfahany (2009) showed that EDR in a zone around the baffles is less than that in the impeller zone [89], which helped eliminate the concern of cell damage by baffles. In addition, the impact of dual impeller positions and the presence of gas on power number were also investigated with CFD modeling [90]. In other cases, varying impeller designs and their positions in a large scale bioreactor can be studied with CFD simulation to obtain relative information, which can significantly save resources and work compared to experimental testing which may not yield the same level of information as CFD.

Simulating mass transfer simulation is a more complicated task. Even though the selection of single average bubble size across the vessel, via a couple discrete phase model (DPM), can easily create matching values versus experimental data, it is preferred to use population balance equations to reflect bubble breakage and coalescence, which is closer to a real case scenario with a broad bubble size distribution [91]. Rathore et al. showed that without the application of the population balance method, the simulation result can be misleading, yielding very high gas volume fractions at the top of tank [92]. In addition, their work showed that mass transfer rate is more than tenfold higher in the impeller zone versus other zones in the bulk liquid. Therefore, it is reasonable to speculate that the DO is heterogeneously distributed, which is much higher in the zone close to the impellers, especially the bottom impeller closest to the sparger. Another benefit of utilizing CFD is adapting process parameters to align with sensitive cell lines. A scale-down model can be developed accordingly for a DO sensitive cell line based on the information gathered through CFD simulation.

6.3.5 Challenges and Perspectives

While the focus has traditionally been on engineering principles to build scale-up/scale-down models, there are other very important aspects of the process requiring attention to allow for success. Successful scale-up requires the integration of established engineering principles as well as raw material and cell line robustness. The respective cell line and raw materials are critical inputs into a process that traditionally has caused numerous scale-up surprises. A predictive scale-up model would not allow for such surprises. Thus, these inputs require attention upfront. Cell age is known to play a role in cellular metabolism and protein production, and in some cases depends on the level of adaptation prior to cell bank creation [27]. As a result, cell line stability understanding is recommended [93]. In the absence of the full understanding of the impact of cell age, performing experiments of equivalent cell age (tracked via cell population doublings) is recommended.

Media component raw materials are known for having lot-to-lot variability even with current chemically defined medium. This variability consequently results in inconsistent cell culture performance [77, 94-98]. Variation may be caused by a change in vendor's milling process, lot-to-lot variability of raw materials from the vendor's respective supply chain, impurities introduced via carryover across vessels used to make other chemical components, etc. Small differences in raw materials can lead to metabolic shifts and these shifts may be up or down depending on the cell line and process, causing additional complexity in finding root causes of process variability [99]. GMP lots of critical raw materials are thus often use-tested at lab scales to prevent undesirable performance in manufacturing. Though this activity is valuable, it does not always have the perceived predictive power. It has also been proven that the impact of raw materials may be scale dependent even when using a qualified scale-down model for the use. Thus, it is critical to characterize raw materials as much as possible and understand the possible variation each may generate by utilizing and testing a large array of lots during process development and over the program's lifecycle.

The difference in media prep and storage across scales can also result in unexpected cell culture performance. For example, the degree of CO_2 degassing from media solution containing sodium bicarbonate can vary significantly due to a difference in mixing intensity, which can cause pH drift and buffer capacity loss if a pH adjustment step is included during media prep. Often, media are highly sensitive to temperature and oxidation [100]. In order to ensure expected process performance, the ability to store filtered solutions unmixed at a specified temperature and with an inert gas in the headspace is valuable. It has been also reported that media chemistry such as cysteine oxidation can occur differently across scales, which caused media precipitation only at certain scales [101]. Since some media components are sensitive to light, the media prep and storage at small scale should be carefully defined in order to mimic the large scale scenario [102]. The container and its cleaning procedure can also have an impact on media quality. It is recommended to run satellite small-scale bioreactors with large-scale media if possible to decouple the impact of media and process, which can provide critical information during troubleshooting.

Robust automation is valuable to obtain consistency and efficiency in today's manufacturing facility. Automation of numerous activities provides a repeatable method of execution and also minimizes the manual labor required to execute a batch. Vessel cleaning, steaming, pressure testing, as well as uploading of reactor process parameters are all activities generally automated. Electronic documentation and on-line trending can also improve facility efficiency. Rapid review and data analysis can be performed as well as accurately tracking operations. Having raw materials and consumables also embedded into the electronic tracking method is highly valuable for inventory management efficiency as well as batch performance troubleshooting. As previous discussed, raw materials can play a significant role in determining performance and batch-to-batch variability. Thus, this data needs to be mined and analyzed. Having information at your fingertips will save time and allow fast responses and resolutions.

Manufacturing facilities ultimately must maintain flexibility. Processes will continue to intensify and facilities must be able to adapt and flex accordingly to accommodate novel process requirements as well as maximize facility capacity. For example, many of today's fed-batch cultures have high feed volume. A facility today must have large feed tanks to match the requirements of the large-scale bioreactors (up to 60 % total volume of the bioreactor). Going forward, a greater need in media/feed tank volume will be necessary as continuous processing becomes more prevalent.

Continuous cell culture processing has been historically used to produce products, which were not stable under batch culture conditions. However, this platform and concept has gained more attention in recent years for achieving high volumetric productivity in conjunction with stable products such as monoclonal antibodies (MAb) [103, 104]. This theory is supported by the fact that other matured industries such as the automobile and chemical industries all use continuous manufacturing to streamline the process flow and lower the cost. Warikoo et al. have successfully demonstrated 2 CHO perfusion cultures achieving 50 to 60 × 10^6 viable cells per mL for 60 days [105]. However, there are still technical hurdles to overcome in order to make it a successful and robust process format for commercial operations. Besides media improvement to minimize perfusion rate, balancing the mass transfer requirement without hurting the cells as discussed above is critical. In this case, high viability is even more important as these cultures must be maintained for long periods of time.

Besides the production stage, perfusion can also be integrated at the N-1 stage in a typical fed-batch process to increase seed density. In most manufacturing facilities, the production tank is the bottle neck for overall output which includes several days of cell growth prior to reaching peak cell density. Application of perfusion in the N-1 stage can reduce the time to peak cell density once within the production stage by starting the production phase at a high initial cell density and overall allowing for a higher facility throughput [106]. N-1 perfusion also has a potential to increase peak cell density and titer achieved in the production state. The use of perfusion technology to make high-density cell bank, which can shorten cell expansion duration from vial thaw to production stage has been also reported [107].

Given the incorporation of unique cell lines, raw material variation, increased automation, and new process formats, creating a cross-scale system that simulates the critical process aspects providing a similar microenvironment can be challenging. Currently, the industry is trying to achieve 500 to 1,000 mg/L/d volumetric productivity, which definitely requires high cell mass. Peak density of 30 and 50 to 100 × 10^6 cells per mL has been reported in fed-batch and perfusion cell culture process respectively in recent years [104, 105, 108]. As cell culture densities have risen and processes continue to intensify, scale-up becomes more challenging and requires adaptive facilities (both lab and manufacturing). High cell masses place significant demands on both, mixing and mass transfer, including high agitation rates and substantial sparge rates. Unfortunately all of these process parameter inputs have possible risks associated with them as discussed in previous sections. Finding a balanced solution of providing a homogeneous liquid solution with sufficient oxygen supplementation and carbon dioxide stripping without causing cell damage and foaming is complicated. Furthermore, this will become even harder to achieve for future further intensified cell culture processes. In the future, mass transfer will likely be the limiting factor determining the maximum volumetric productivity that bioreactor can produce. The following aspects should be considered as potential areas for further improvements.

1. Screening cell lines to make them insensitive to bubble rupture,
2. exploration of highly potent protective surfactants,
3. application of rational and optimized sparger designs,
4. deriving an effective antifoam solution and usage method with minimal impact on mass transfer.

Most scale-up issues encountered are cell line specific. Overall process and scale-up success rates can potentially be improved through the creation of a very robust cell line that is adapted to the bubble shear environments and thus not negatively impacted. These activities must begin during the cell line generation and selection process. Also generating and applying better understanding of historically used raw materials such as serums and hydrolysates to identify a surfactant more effective than pluronic F-68 in preventing bubble rupture damage at high flow rates and cell densities is desirable [109, 110]. The impact of sparger designs on bubble formation and size distribution also should be examined, which can be optimized to achieve the desired bubble size under a broad range of aeration rates. Additionally, developing a novel antifoam solution and addition strategy that would not impact mass transfer aspects would also be extremely valuable in terms of expanding the sparge rate operation range [111].

Notation

θ_m	mixing time (s)
(ε_T)	mean specific energy dissipation rate (m^2/s^3)
$(\varepsilon_T)_l$	local specific energy dissipation rate (m^2/s^3)
D	impeller diameter (m)
T	Tank diameter (m)
H	Liquid height (m)
λ_k	Kolmogoroff eddy size
υ	kinematic viscosity (m^2/s)
N_p	power number
P	power (W)
V	volume (m^3)
ρ	liquid density (kg/m^3)
N	impeller speed (s^{-1})
T_q	torque (Nm)
k_La	volumetric mass transfer coefficient (s^{-1})
D_g	gas diffusivity in the liquid (m^2/s)
v_s	superficial velocity (m/s)
OTR	oxygen transfer rate ($g/(m^3\cdot s)$)
CTR	carbon dioxide transfer rate ($g/(m^3\cdot s)$)
u_s	slip velocity (m/s)
d	bubble diameter (m)
k_L	liquid-side mass transfer coefficient (m/s)
C	bulk liquid concentration (g/m^3)
C*	the equilibrium liquid concentration (g/m^3)

References

[1] Griffiths B. The Development of Animal Cell Products: History and Overview: Medicines from Animal Cell Culture. John Wiley & Sons; 2007:1-14.
[2] Walsh G. Biopharmaceutical benchmarks 2010. Nat Biotechnol 2010;28:917-24.
[3] Warnock JN, Al-Rubeai M. Bioreactor systems for the production of biopharmaceuticals from animal cells. Biotechnol Appl Biochem 2006;45:1-12.

[4] Zhou T-C, Zhou W-W, Hu W, et al. Bioreactors, Cell Culture, Commercial Production. Encyclopedia of Industrial Biotechnology. John Wiley & Sons; 2009.
[5] Wurm FM. Production of recombinant protein therapeutics in cultivated mammalian cells. Nat Biotechnol 2004;22:1393-8.
[6] Krishnan R, Chen H. A comprehensive strategy to evaluate single-use bioreactors for pilot-scale cell culture production. Am Pharm Rev 2012;15:15-6.
[7] Shukla AA, Gottschalk U. Single-use disposable technologies for biopharmaceutical manufacturing. Trends Biotechnol 2013;31:147-54.
[8] Varley J, Birch J. Reactor design for large scale suspension animal cell culture. Cytotechnology 1999;29:177-205.
[9] Marks DM. Equipment design considerations for large scale cell culture. Cytotechnology 2003;42:21-33.
[10] Jayapal K, Wlaschin K, Hu W-S, Yap M. Recombinant Protein Therapeutics from Cho Cells – 20 Years and Counting. CHO Consortium: SBE Special Edition; 2007:40-7.
[11] Chu L, Robinson DK. Industrial choices for protein production by large-scale cell culture. Curr Opin Biotechnol 2001;12:180-7.
[12] Lara AR, Galindo E, Ramirez OT, Palomares LA. Living with heterogeneities in bioreactors. Understanding the effects of environmental gradients on cells. Mol Biotechnol 2006;34:355-81.
[13] Nienow AW. Reactor engineering in large scale animal cell culture. Cytotechnology 2006;50:9-33.
[14] Sieblist C, Jenzsch M, Pohlscheidt M, Lubbert A. Insights into large-scale cell-culture reactors: I. Liquid mixing and oxygen supply. Biotechnol J 2011;6:1532-46.
[15] Sieblist C, Hageholz O, Aehle M, et al. Insights into large-scale cell-culture reactors: II. Gas-phase mixing and CO(2) stripping. Biotechnol J 2011;6:1547-56.
[16] Chisti Y. Animal-cell damage in sparged bioreactors. Trends Biotechnol 2000;18:420-32.
[17] Zhu H, Nienow AW, Bujalski W, Simmons MJH. Mixing studies in a model aerated bioreactor equipped with an up- or a down-pumping 'Elephant Ear' agitator: Power, hold-up and aerated flow field measurements. Chem Eng Res Des 2009;87:307-17.
[18] Simmons MJH, Zhu H, Bujalski W, et al. Mixing in a Model Bioreactor Using Agitators with a High Solidity Ratio and Deep Blades. Chem Eng Res Des 2007;85:551-9.
[19] Nienow AWZH, Bujalski W, Simmons MJH. Are there any 'low Shear' agitators for animal cell culture? Cell Culture Engineering XI, April 13-18, 2008; Queensland, Australia.
[20] Oh SKW, Nienow AW, Al-Rubeai M, Emery AN. The effects of agitation intensity with and without continuous sparging on the growth and antibody production of hybridoma cells. J Biotechnol 1989;12:45-61.
[21] Oh SK, Nienow AW, al-Rubeai M, Emery AN. Further studies of the culture of mouse hybridomas in an agitated bioreactor with and without continuous sparging. J Biotechnol 1992;22:245-70.
[22] Kunas KT, Papoutsakis ET. Damage mechanisms of suspended animal cells in agitated bioreactors with and without bubble entrainment. Biotechnol Bioeng 1990;36:476-83.
[23] Thomas CR, al-Rubeai M, Zhang Z. Prediction of mechanical damage to animal cells in turbulence. Cytotechnology 1994;15:329-35.
[24] Hu W, Berdugo C, Chalmers JJ. The potential of hydrodynamic damage to animal cells of industrial relevance: current understanding. Cytotechnology 2011;63:445-60.
[25] Godoy-Silva R, Chalmers JJ, Casnocha SA, et al. Physiological responses of CHO cells to repetitive hydrodynamic stress. Biotechnol Bioeng 2009;103:1103-17.
[26] Meier S. Scale-down studies of the effect of hydrodynamic forces on CHO cells: Implications for industrial production conditions Cell Culture Engineering XIII, April 22-27 2012; Scottsdale, Arizona, USA.

[27] Prentice HL, Ehrenfels BN, Sisk WP. Improving performance of mammalian cells in fed-batch processes through "bioreactor evolution". Biotechnol Prog 2007;23:458-64.
[28] Assirelli M, Bujalski W, Eaglesham A, Nienow AW. Macro- and micromixing studies in an unbaffled vessel agitated by a Rushton turbine. Chem Eng Sci 2008;63:35-46.
[29] Nienow AW. On impeller circulation and mixing effectiveness in the turbulent flow regime. Chem Eng Sci 1997;52:2557-65.
[30] Sandadi S, Pedersen H, Bowers JS, Rendeiro D. A comprehensive comparison of mixing, mass transfer, Chinese hamster ovary cell growth, and antibody production using Rushton turbine and marine impellers. Bioprocess Biosys Eng 2011;34:819-32.
[31] Pullen KF, Johnson MD, Phillips AW, et al. Very large scale suspension cultures of mammalian cells. Dev Biol Stand 1985;60:175-7.
[32] Corrales M, Fallon E, Yraceburn M, et al. Cell culture scale-up and start-up strategies for NIMO, Genentech's new manufacturing facility. American Chemical Society; 2006:BIOT-269.
[33] Meier S. Cell culture scale-up: mixing, mass transfer, and use of appropriate scale-down models. Biochem engineering XIV, 2005 July 10-14; Harrison Hot Springs, BC, Canada.
[34] Junker BH, Hunt G, Burgess B, et al. Modified microbial fermenter performance in animal cell culture and its implications for flexible fermenter design. Bioprocess Eng 1994;11:57-63.
[35] Yang JD, Lu C, Stasny B, et al. Fed-batch bioreactor process scale-up from 3-L to 2,500-L scale for monoclonal antibody production from cell culture. Biotechnol Bioeng 2007;98:141-54.
[36] Nienow AW, Langheinrich C, Stevenson NC, et al. Homogenisation and oxygen transfer rates in large agitated and sparged animal cell bioreactors: Some implications for growth and production. Cytotechnology 1996;22:87-94.
[37] Croughan MS, Hamel J-F, Wang DIC. Hydrodynamic effects on animal cells grown in microcarrier cultures. Biotechnol Bioeng 1987;29:130-41.
[38] Langheinrich C, Nienow AW, Eddleston T, et al. Liquid Homogenization Studies in Animal Cell Bioreactors of up to 8 m3 in Volume. Food Bioprod Process 1998;76:107-16.
[39] Xing Z, Kenty BM, Li ZJ, Lee SS. Scale-up analysis for a CHO cell culture process in large-scale bioreactors. Biotechnol Bioeng 2009;103:733-46.
[40] Osman JJ, Birch J, Varley J. The response of GS-NS0 myeloma cells to pH shifts and pH perturbations. Biotechnol Bioeng 2001;75:63-73.
[41] Osman JJ, Birch J, Varley J. The response of GS-NS0 myeloma cells to single and multiple pH perturbations. Biotechnol Bioeng 2002;79:398-407.
[42] Rushton JHCEW, Everett HJ. Power characteristics of mixing impellers. Chem Eng Prog 1950;46:395-476.
[43] Nienow AW, Flickinger MC. Impeller Selection for Animal Cell Culture. Encyclopedia of Industrial Biotechnology. John Wiley & Sons; 2009.
[44] Wu J, Pullum L. Performance analysis of axial-flow mixing impellers. AIChE J 2000;46:489-98.
[45] Chapple D, Kresta SM, Wall A, Afacan A. The Effect of Impeller and Tank Geometry on Power Number for a Pitched Blade Turbine. Chem Eng Res Des 2002;80:364-72.
[46] King RL, Hiller RA, Tatterson GB. Power consumption in a mixer. AIChE J 1988;34:506-9.
[47] Gogate PR, Beenackers AA, Pandit AB. Multiple-impeller systems with a special emphasis on bioreactors: a critical review. Biochem Eng J 2000;6:109-44.
[48] Vrábel P, van der Lans RGJM, Luyben KCAM, et al. Mixing in large-scale vessels stirred with multiple radial or radial and axial up-pumping impellers: modelling and measurements. Chem Eng Sci 2000;55:5881-96.
[49] Nienow AW, Bujalski W. The Versatility of Up-Pumping Hydrofoil Agitators. Chem Eng Res Des 2004;82:1073-81.
[50] Hudcova V, Machon V, Nienow AW. Gas-liquid dispersion with dual Rushton turbine impellers. Biotechnol Bioeng 1989;34:617-28.

[51] Bittorf KJ, Kresta SM. Active volume of mean circulation for stirred tanks agitated with axial impellers. Chem Eng Sci 2000;55:1325-35.
[52] Mostafa SS, Gu X. Strategies for Improved dCO_2 Removal in Large-Scale Fed-Batch Cultures. Biotechnol Prog 2003;19:45-51.
[53] Goudar CT, Piret JM, Konstantinov KB. Estimating cell specific oxygen uptake and carbon dioxide production rates for mammalian cells in perfusion culture. Biotechnol Prog 2011;27:1347-57.
[54] Duan Z, Sun R. An improved model calculating CO_2 solubility in pure water and aqueous NaCl solutions from 273 to 533 K and from 0 to 2000 bar. Chem Geol 2003;193:257-71.
[55] Zanghi JA, Schmelzer AE, Mendoza TP, et al. Bicarbonate concentration and osmolality are key determinants in the inhibition of CHO cell polysialylation under elevated pCO(2) or pH. Biotechnol Bioeng 1999;65:182-91.
[56] Zhu MM, Goyal A, Rank DL, et al. Effects of elevated pCO_2 and osmolality on growth of CHO cells and production of antibody-fusion protein B1: a case study. Biotechnol Prog 2005;21:70-7.
[57] Schmelzer AE, Miller WM. Hyperosmotic stress and elevated pCO_2 alter monoclonal antibody charge distribution and monosaccharide content. Biotechnol Prog 2002;18:346-53.
[58] van't Riet K. Review of Measuring Methods and Results in Nonviscous Gas-Liquid Mass Transfer in Stirred Vessels. Ind Eng Chem Process Des Dev 1979;18:357-64.
[59] Arjunwadkar SJ, Saravanan K, Pandit AB, Kulkarni PR. Optimizing the impeller combination for maximum hold-up with minimum power consumption. Biochem Eng J 1998;1:25-30.
[60] Aubin J, Le Sauze N, Bertrand J, et al. PIV measurements of flow in an aerated tank stirred by a down- and an up-pumping axial flow impeller. Exp Therm Fluid Sci 2004;28:447-56.
[61] Sardeing R, Aubin J, Xuereb C. Gas-Liquid Mass Transfer: A Comparison of Down- and Up–pumping Axial Flow Impellers with Radial Impellers. Chem Eng Res Des 2004;82:1589-96.
[62] Vasconcelos JMT, Nienow AW, Martin T, et al. Alternative Ways of Applying the Hydrogen Peroxide Steady State Method of KLa Measurement. Chem Eng Res Des 1997;75:467-72.
[63] Jorjani P, Ozturk SS. Effects of cell density and temperature on oxygen consumption rate for different mammalian cell lines. Biotechnol Bioeng 1999;64:349-56.
[64] Bellucci JJ, Hamaker KH. Evaluation of oxygen transfer rates in stirred-tank bioreactors for clinical manufacturing. Biotechnol Prog 2011;27:368-76.
[65] Doran P. Design of reactors for plant cells and organs. Bioprocess Des Control 1993:115-68.
[66] Royce PNC, Thornhill NF. Estimation of dissolved carbon dioxide concentrations in aerobic fermentations. AIChE J 1991;37:1680-6.
[67] Bird RB, Stewart WE, Lighfoot EN. Transport phenomena. Wiley; 1960.
[68] Griffith RM. Mass transfer from drops and bubbles. Chem Eng Sci 1960;12:198-213.
[69] Hu W, Mollet M, Rathman J, Chalmers J. Two Circuits or One? Should O_2 addition and CO_2 removal be uncoupled? .Cell Culture Engineering IX, March 7-12 2004; Cancun, Mexico.
[70] Boulton-Stone JM, Blake JR. Gas bubbles bursting at a free surface. J Fluid Mech 1993;254:437-66.
[71] Garcia-Briones MA, Brodkey RS, Chalmers JJ. Computer simulations of the rupture of a gas bubble at a gas—liquid interface and its implications in animal cell damage. Chem Eng Sci 1994;49:2301-20.
[72] Wu J. Mechanisms of animal cell damage associated with gas bubbles and cell protection by medium additives. J Biotechnol 1995;43:81-94.
[73] Trinh K, Garcia-Briones M, Chalmers JJ, Hink F. Quantification of damage to suspended insect cells as a result of bubble rupture. Biotechnol Bioeng 1994;43:37-45.
[74] Chattopadhyay D, Rathman JF, Chalmers JJ. Thermodynamic approach to explain cell adhesion to air-medium interfaces. Biotechnol Bioeng 1995;48:649-58.
[75] Meier SJ, Hatton TA, Wang DI. Cell death from bursting bubbles: role of cell attachment to rising bubbles in sparged reactors. Biotechnol Bioeng 1999;62:468-78.

[76] Gigout A, Buschmann MD, Jolicoeur M. The fate of Pluronic F-68 in chondrocytes and CHO cells. Biotechnol Bioeng 2008;100:975-87.
[77] Passini CA, Goochee CF. Response of a Mouse Hybridoma Cell Line to Heat Shock, Agitation, and Sparging. Biotechnol Prog 1989;5:175-88.
[78] Barbosa MJ, Albrecht M, Wijffels RH. Hydrodynamic stress and lethal events in sparged microalgae cultures. Biotechnol Bioeng 2003;83:112-20.
[79] Barbosa MJ, Hadiyanto, Wijffels RH. Overcoming shear stress of microalgae cultures in sparged photobioreactors. Biotechnol Bioeng 2004;85:78-85.
[80] Zhu Y, Cuenca JV, Zhou W, Varma A. NSO cell damage by high gas velocity sparging in protein-free and cholesterol-free cultures. Biotechnol Bioeng 2008;101:751-60.
[81] Liu Y, Hu W, Wiltberger K, et al. Effects of Bubble-Liquid Two-Phase Turbulent Hydrodynamics on Cell Damage in Sparged Bioreactor. Biotechnol Prog 2014;30:48-58.
[82] Denkov ND. Mechanisms of Foam Destruction by Oil-Based Antifoams. Langmuir 2004;20:9463-505.
[83] Morão A, Maia CI, Fonseca MMR, et al. Effect of antifoam addition on gas-liquid mass transfer in stirred fermenters. Bioprocess Eng 1999;20:165-72.
[84] Vasconcelos JMT, Rodrigues JML, Orvalho SCP, et al. Effect of contaminants on mass transfer coefficients in bubble column and airlift contactors. Chem Eng Sci 2003;58:1431-40.
[85] Sardeing R, Painmanakul P, Hébrard G. Effect of surfactants on liquid-side mass transfer coefficients in gas-liquid systems: A first step to modeling. Chem Eng Sci 2006;61:6249-60.
[86] Hebrard G, Zeng J, Loubiere K. Effect of surfactants on liquid side mass transfer coefficients: A new insight. Chem Eng J 2009;148:132-8.
[87] Sharma C, Malhotra D, Rathore AS. Review of computational fluid dynamics applications in biotechnology processes. Biotechnol Prog 2011;27:1497-510.
[88] Kelly WJ. Using computational fluid dynamics to characterize and improve bioreactor performance. Biotechnol Appl Biochem 2008;49:225-38.
[89] Vakili MH, Esfahany MN. CFD analysis of turbulence in a baffled stirred tank, a three-compartment model. Chem Eng Sci 2009;64:351-62.
[90] Khopkar AR, Tanguy PA. CFD simulation of gas-liquid flows in stirred vessel equipped with dual rushton turbines: influence of parallel, merging and diverging flow configurations. Chem Eng Sci 2008;63:3810-20.
[91] Zhang H, Zhang K, Fan S. CFD simulation coupled with population balance equations for aerated stirred bioreactors. Eng Life Sci 2009;9:421-30.
[92] Rathore AS, Sharma C, Persad AA. Use of computational fluid dynamics as a tool for establishing process design space for mixing in a bioreactor. Biotechnol Prog 2012;28:382-91.
[93] Bailey LA, Hatton D, Field R, Dickson AJ. Determination of Chinese hamster ovary cell line stability and recombinant antibody expression during long-term culture. Biotechnol Bioeng 2012;109:2093-103.
[94] Luo Y, Chen G. Combined approach of NMR and chemometrics for screening peptones used in the cell culture medium for the production of a recombinant therapeutic protein. Biotechnol Bioeng 2007;97:1654-9.
[95] Lee HW, Christie A, Liu JJ, Yoon S. Estimation of raw material performance in mammalian cell culture using near infrared spectra combined with chemometrics approaches. Biotechnol Prog 2012;28:824-32.
[96] Kirdar AO, Chen G, Weidner J, Rathore AS. Application of near-infrared (NIR) spectroscopy for screening of raw materials used in the cell culture medium for the production of a recombinant therapeutic protein. Biotechnol Prog 2010;26:527-31.

[97] Jose GE, Folque F, Menezes JC, et al. Predicting mab product yields from cultivation media components, using near-infrared and 2D-fluorescence spectroscopies. Biotechnol Prog 2011;27:1339-46.

[98] Lanan M. QbD for Raw Materials. Quality by Design for Biopharmaceuticals. John Wiley & Sons; 2008:193-209.

[99] Luo J, Vijayasankaran N, Autsen J, et al. Comparative metabolite analysis to understand lactate metabolism shift in Chinese hamster ovary cell culture process. Biotechnol Bioeng 2012;109:146-56.

[100] Halliwell B. Oxidative stress in cell culture: an under-appreciated problem? FEBS Lett 2003;540:3-6.

[101] Nyberg G. Improving our understanding of raw materials and their impact on cell culture processes. Cell Culture Engineering XIII, April 22-27 2012; Scottsdale, Arizona, USA.

[102] Zang L, Frenkel R, Simeone J, et al. Metabolomics profiling of cell culture media leading to the identification of riboflavin photosensitized degradation of tryptophan causing slow growth in cell culture. Anal Chem 2011;83:5422-30.

[103] Konstantinov K, Goudar C, Ng M, et al. The "push-to-low" approach for optimization of high-density perfusion cultures of animal cells. Adv Biochem Eng/Biotechnol 2006;101:75-98.

[104] Clincke MF, Molleryd C, Samani PK, et al. Very high density of Chinese hamster ovary cells in perfusion by alternating tangential flow or tangential flow filtration in WAVE bioreactor-part II: Applications for antibody production and cryopreservation. Biotechnol Prog 2013;29(3):768-77.

[105] Warikoo V, Godawat R, Brower K, et al. Integrated continuous production of recombinant therapeutic proteins. Biotechnol Bioeng 2012;109:3018-29.

[106] Padawer I, Ling WL, Bai Y. Case Study: An accelerated 8-day monoclonal antibody production process based on high seeding densities. Biotechnol Prog 2013.

[107] Tao Y, Shih J, Sinacore M, et al. Development and implementation of a perfusion-based high cell density cell banking process. Biotechnol Prog 2011;27:824-9.

[108] Huang YM, Hu W, Rustandi E, et al. Maximizing productivity of CHO cell-based fed-batch culture using chemically defined media conditions and typical manufacturing equipment. Biotechnol Prog 2010;26:1400-10.

[109] Hu W, Rathman JJ, Chalmers JJ. An investigation of small-molecule surfactants to potentially replace pluronic F-68 for reducing bubble-associated cell damage. Biotechnol Bioeng 2008;101:119-27.

[110] Ma N, Chalmers JJ, Aunins JG, et al. Quantitative studies of cell-bubble interactions and cell damage at different pluronic F-68 and cell concentrations. Biotechnol Prog 2004;20:1183-91.

[111] Calik P, Ileri N, Erdinc BI, et al. Novel antifoam for fermentation processes: fluorocarbon-hydrocarbon hybrid unsymmetrical bolaform surfactant. Langmuir 2005;21:8613-9.

[112] Farid SS. Process economics of industrial monoclonal antibody manufacture. J Chromatog B 2007;848:8-18.

[113] Fan LS, Tsuchiya K. Bubble wake dynamics in liquids and liquid-solid suspensions. Butterworth-Heinemann; 1990.

[114] Godoy-Silva R, Berdugo C, Chalmers JJ, Flickinger MC. Aeration, Mixing, and Hydrodynamics, Animal Cell Bioreactors. Encyclopedia of Industrial Biotechnology. John Wiley & Sons; 2009.

[115] Hemrajani RR, Tatterson GB. Mechanically Stirred Vessels. Handbook of Industrial Mixing. John Wiley & Sons; 2004:345-90.

6.4 Extraction and Purification of Biologics from Cell Culture: Monoclonal Antibody Downstream Processing

Nuno Fontes, Ruoheng Zhang, and Jens H. Vogel

6.4.1 Generic Antibody Purification

6.4.1.1 Early Immunoglobulin and Monoclonal Antibody Purification Processes

Antibody purification processes have changed significantly since the first use of immunoglobulin products from human plasma for the treatment of immune deficiency [1]. Initial processes for the production of immunoglobulins were based on the Cohn process developed in the 1950s. These processes made no use of chromatography and were instead based on a series of fractional precipitation steps using cold ethanol. In plasma fractionation, precipitation is still the preferred method. Nevertheless, some of processes developed more recently for the production of intravenous immunoglobulin (IVIG) from plasma, already include chromatography steps downstream of the initial fractional precipitation steps [2].

The purification processes used for the production of the first monoclonal antibodies (mAbs) approved for human therapeutic use in the 1980s and early 1990s had very little in common with the processes used for production of immunoglobulins from plasma. Initial mAb processes displayed very diverse purification schemes and were generally more conservative in regards to clearance of process related impurities. This reflects the significant increase in process and clinical knowledge gathered over the last 20 years, which gave the industry and the health authorities an improved understanding of critical product quality attributes (CQA). Early mAb purification processes employed various types of microfiltration and depth filtration for harvest of the cell culture fluid. Although Protein A affinity chromatography has become the industry-wide standard for direct capture and purification of mAbs and Fc fusion proteins from the harvested cell culture fluid [3], some early processes used Protein G instead. Protein A is a 42 kDa type I membrane protein from the bacterium *Staphylococcus aureus* with high specificity for the Fc region of antibodies. In addition to this main binding site, some immunoglobulins, particularly those containing heavy chains from the human vH3 gene family, have been shown to have an additional binding site for Protein A on their heavy chain variable domain [3]. Protein G is a 65 kDa or 58 kDa (depending on the subtype) streptococcal membrane protein with strong affinity for immunoglobulins and albumin. It differs from Protein A in its binding specificity to immunoglobulins of different subtypes and origins. For bioprocessing applications, the albumin-binding site is deleted from recombinant protein G. Another key difference in early mAb purification schemes, compared to today's practices, was the placement of the affinity step downstream of a nonaffinity conventional chromatography capture step, with the objective to protect the expensive affinity resins from the relatively crude harvested cell culture fluid [1]. Size-exclusion chromatography (SEC) was also prevalent in early processes as both a purification and formulation step, but its poor scalability led to its replacement with alternative steps, namely with cation exchange (CIEX) for aggregate removal and with ultrafiltration (UF) and diafiltration (DF) for formulation. Implementation of ultrafiltration in mAb processes was initially met with skepticism due to the concern that the antibody molecule could be sheared and damaged by the tangential flow. Routine use of ultrafiltration in mAb processes over the last 20 years, as well as theoretical considerations and shear studies, have dismissed those concerns and demonstrated that mAb molecules are too small to be affected by the relatively mild shearing stress present in common ultrafiltration conditions. Exposure to air-liquid interfaces, mostly created by pump cavitation is instead the common cause for mAb degradation in ultrafiltration processes [4], which can be avoided with the right choice of equipment and processing conditions.

Since most mAb production has been carried out using mammalian cell-based host systems, special consideration has been given to the implementation of viral reduction steps. While traditional processes commonly used organic solvents and/or detergents for viral inactivation, modern viral reduction approaches have converged towards the use of acid inactivation for endogenous virus and retrovirus and of viral filtration and chromatography unit operations for both, endogenous and adventitious viruses. This topic will be addressed in more detail in Section 6.4.2.

Another common feature of early processes that changed over time was that the number of chromatography steps, which decreased from 4 or more to 2 or 3. Finally, given that in the early days upstream titers were low, sizing of purification unit operations has shifted from maximizing the volumetric productivity to maximizing capacity. It was fairly common for chromatography steps to be operated at load densities significantly below the respective dynamic binding capacities, which in turn were considerably lower than those observed with current chromatography resins. Finally, the early large diversity of mAb protein sequence frameworks spanning the natural evolution from murine and chimeric to modern fully humanized mAbs, led to a wider range of mAb purification schemes [1].

6.4.1.2 Industry Standard mAb Purification Platform

Purification platforms for mAb production are generic purification schemes that take advantage of common properties of mAbs, namely the strong binding affinity of the Fc region to Protein A and their relatively basic character, to maximize the efficiency of purification process development, transfer, and implementation. Process platforms enable minimization of the development time, effort, and costs required to achieve target product quality, impurity levels, yield and facility fit. Biopharmaceutical companies with monoclonal antibody pipelines expressed in Chinese hamster ovary (CHO) cell lines have arrived at very similar purification platforms. Evidence of the convergence of downstream platforms comes mostly from information shared at conferences. A number of recent publications further support this claim [1, 2, 5-11] and give extensive information and details on the industry standard process platform. In this section our aim is to provide only a generic overview of the industry purification process platform flow sheet, shown in Figure 6.4.1.

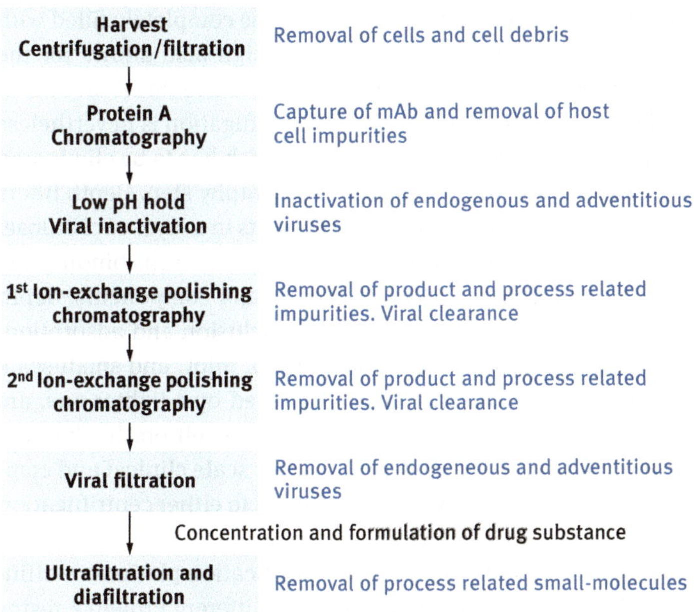

Figure 6.4.1: Industry-standard monoclonal antibody purification platform.

Given that mAbs are expressed into the cell culture supernatant, the first step consists in the removal of the host cells and cell debris and recovery of the supernatant containing the soluble mAb. Current platforms use centrifugation coupled with depth filtration followed by microfiltration, prior to loading of the Protein A column, which is the first chromatography step. Early purification schemes had successfully employed tangential flow filtration (TFF) as a cell culture fluid harvest technique. However, the combined increase in cell densities, specific productivity and cell culture duration, aimed at maximizing cell culture titer, led to a corresponding increase in impurity levels and in the percentage of solids present in cell culture fluid. The increase in the sub-micron debris load made implementation of TFF more challenging and led to a preference for centrifugation coupled with depth filtration as the current industry standard for harvest of the cell culture fluid [11]. An optimized centrifugation process minimizes cell lysis (and hence generation of additional cell debris as well as release of intracellular impurities and proteases) and maximizes sedimentation of submicron particles and product yield. Recent improvements in harvest operations made use of fully hermetically sealed, bottom-fed centrifuges, which completely eliminate the air-liquid interface commonly present in standard nonhermetic centrifuges. This interface is known to be harmful to the integrity of the mammalian cells and to be the cause of additional cell debris when not appropriately managed by the use of backpressure. Fully hermetic centrifuges employ mechanical seals that fully isolate the product fluid from outside air and enable the machine to be completely filled with liquid eliminating all air-liquid interfaces. This modern design also allows for the elimination of valves and pumps from the inlet piping which contributes to an overall reduction in extent of cell lysis by more than 50 % [12]. Centrifugation is nevertheless unable to completely remove the submicron particle load, which has to be eliminated by depth filtration prior to the loading of the first chromatography step. Depth filters come in many different varieties, but a common design consists in a layer of cellulose, a porous filter-aid such as diatomaceous earth, and a polymeric resin binding the two together that carries a positive charge. Based on these major components, depth filters remove impurities and particulate material by size exclusion and adsorption, and provide essential protection to membrane filters. For lab, pilot, and small-scale clinical applications, 2 types of depth filters, a coarse followed by a tighter one, are commonly placed in series and loaded directly with the whole cell broth, skipping the centrifugation or tangential flow filtration step. For larger scale clinical and commercial operations depth filtration is most commonly coupled to either centrifugation or depth filtration.

The second unit operation in the industry standard purification platform is affinity chromatography using a Protein A ligand. There are many different Protein A resins available on the market. The major determinants of resin performance are properties such as the origin and type of Protein A ligand (native, recombinant, or recombinant modified alkali tolerant), the chemical nature of the solid support (controlled porous glass, agarose-derived, or poly styrene-divinylbenzene), the type of ligand-support coupling (single- versus multipoint attachment), the pore and particle size, and the ligand density. Protein A is one of the most expensive raw materials utilized for the manufacturing of mAbs (only matched or surpassed by the cost of parvovirus filters). For that reason, Protein A columns are commonly undersized and operated in multiple cycles per batch. As mentioned later in this chapter, resin selection is a key part of process development. There is no universally best resin, but rather a case-by-case best fit for the specific application and manufacturing capabilities. Load capacity and operating flow rate have to be balanced to provide the optimal number of cycles that enable fit to the existing pool tank volumetric capacity, operation within the stability of the load and pool material, and that the process meets the run rate needs of the production site. Given the very high affinity between the Fc region of mAbs for the Protein A ligand, adsorption is effective within a relatively wide range of loading conditions (pH, ionic strength, temperature). Elution is typically achieved by dropping the pH to less than 4, a condition which causes protonation of key residues in both the

Protein A ligand and mAb binding sites and leads to a significant decrease in binding affinity by electrostatic repulsion of like positive charges. The mAb is eluted from the Protein A resin and then held for a predefined period of time (usually for a minimum of 30 min) under acidic conditions (pH <4) for the inactivation of enveloped endogenous and adventitious viruses. The viral inactivation pool is then titrated to a higher pH for either increased pool stability and/or for setting it to appropriate loading conditions for the subsequent polishing step. This is the point of greatest diversity in current mAb purification platforms. Generically, platforms consist of 2 successive polishing steps downstream of Protein A, usually consisting of a cation exchange resin (CIEX) operated under bind-and-elute conditions and an anion exchange resin (AIEX) operated under flow through conditions. The order of these steps may not be fixed, even within the same company, and the specific resin type may also change for different mAbs. There is a very large number of ion exchange resins on the market. The main resin characteristics that determine step performance (yield, load capacity, purification factor) are the ligand type, ligand density, ligand coupling chemistry, pore size, particle size, and particle rigidity determined by its chemical nature. Most of the industry has adopted a 3 column purification process, but there are also successful platforms based on 2 column processes, one of the most notorious of which is that developed and adopted by Wyeth (now Pfizer), which operates the anion exchange step in a weak-partitioning mode [13] characterized by loading under conditions that promote stronger binding of impurities. Weak partitioning chromatography preserves the isocratic operation typical of common flow-through anion exchange and despite significant binding of the product to the resin, yields exceed 95 % due to the high loadings and short washes. The main function of the chromatographic polishing step(s) downstream of Protein A is reduction of process and product related impurities, of which the most important are cell culture media components, leached Protein A ligand, DNA, host cell proteins (HCP) and product aggregates. When anion exchange is operated in the industry standard flow-through mode, CIEX is the workhorse for removal of leached Protein A (removed in the flow-through fraction) and product aggregates (retained in the column). Viral filtration in normal filtration mode is in most cases operated downstream of the 2 chromatographic polishing steps, but there are companies that place the virus filtration between those 2 steps. Over the last decade, parvovirus filtration has replaced retrovirus filtration as the industry standard. A variety of virus filters and protective prefilters are also available from different vendors. Parvovirus filters, along with Protein A resins, are amongst the most expensive raw materials, together typically accounting for 50 to 70 % of the downstream raw material costs.

The final step in the industry standard platform is an ultrafiltration and diafiltration performed by tangential flow filtration. The purpose of this step is to concentrate the product to the desired final drug substance protein concentration, to remove small molecule process related impurities, and to formulate the drug substance into a buffer of the desired composition. The purified liquid drug substance is then usually stored frozen from −20 to −80 °C before being thawed and filled into the final drug product format which is usually a vial, a pre-filled syringe, or a delivery auto-injector.

6.4.1.3 mAb Purification Process Development Approaches

Drivers

The primary driver for purification process development is achieving the required purity to support safety and efficacy *in vivo*. Secondary drivers include the speed of process development, process yield, process robustness, and scalability. The relative priority of the secondary drivers is influenced by several factors; the most important are the stage of clinical development, the market size, and the competitive landscape. During early-stage development, the focus is on minimizing the period of time from the generation of the cell line to the delivery of drug product to the first-in-human clinical trials. Most biopharmaceutical companies report similar durations for this stage of development, ranging from 16 to 19 months from DNA to product in the clinic. Manufacture of drug product is on the critical path and therefore, speed of process development is usually the main secondary driver. As the product succeeds in the initial stages of clinical development and proceeds to phase III, the focus shifts towards maximizing cell culture titer and process yield and minimizing cost of goods.

Early-stage Purification Development

A common trend in biopharmaceutical development is the establishment of a molecular and manufacturability assessment team on the interface between late-stage research and early-stage process development. The main goal of this cross-functional team is to screen and select the best preclinical molecule candidates generated by the research group, from the perspective of best fit to the company's early-stage upstream, downstream and formulation platforms. There is no process development during this phase. Transient cell culture pools are typically processed through a predetermined process platform and assessed for fit based on critical process and quality attributes such as yield, product related impurities such as aggregate and charge variants, host cell protein levels, product stability, and drug substance viscosity. Coordination between the manufacturability assessment team and the early development organization is important to enable integration of the information gathered during the molecular assessment into the early stage of process development. Most companies use the same manufacturing process to supply the preclinical toxicology and the phases I and II clinical trials.

Early-stage purification process development for mAb processes has become increasingly templated over the last 20 years, and most biopharmaceutical companies make use of well-established purification platforms as previously mentioned. Purification process development usually starts by performing full run-throughs of the first available harvested cell culture pools, using the company's early-stage platform. After evaluation of initial in-process and final product quality attributes from a run-through of the unchanged process platform, priorities are set on the unit operations that require tuning to achieve the target product quality profile. High-throughput screening has been increasingly utilized for process optimization work.

Harvest operations are usually performed under standard centrifugation conditions and oversized depth filtration. In many instances, centrifugation is skipped for the relatively smaller scale (100 to 500 L bioreactor volume) early-stage processes and only implemented for the phase III process. Protein A affinity chromatography, the workhorse of mAb purification, is such a robust unit operation for mAb purification, that commonly does not require any process development for early-stage applications. Minimal tuning may include optimization of elution conditions to achieve the desired pool pH for viral inactivation without the need for pool pH titration, and screening of templated postload wash buffers for reduction of the level of host cell proteins. The flow-through anion exchange step downstream of Protein A also requires minimal process development for well-behaved mAbs with relatively high isoelectric points. This typically includes the determination of the binding capacity, based on host cell protein breakthrough, and minor tuning of the load pH and ionic strength to ensure a reasonable load capacity. The bind-and-elute cation exchange step typically requires the greatest amount of work during early-stage development. The goals of this step is to further reduce process-related impurity levels, namely leached Protein A that is not removed by anion exchange, and to reduce the level of product related impurities, especially aggregates. Gradient elution is frequently a choice for CIEX, given that it maximizes impurity resolution while minimizing process development efforts, which outweighs its somewhat higher complexity compared to step elution. Loading and postloading wash conditions may also be tuned, and most companies have adopted simplified methodologies that rely on the screening of templated buffers.

Viral filtration development for early-stage processes consists in determining filter mass throughput, defined by the upper limit of product mass or load volume per unit membrane area. The amount of effort required to develop the final ultrafiltration tangential flow filtration step depends on the target drug product concentration and on the biophysical properties of the product. High concentration liquid formulations, monoclonal antibodies, with a propensity to form viscous solutions and to aggregate, require the greatest amount of effort.

Late-stage Purification Development

Phase III Process Development Goals

Late-stage purification development is usually defined by the development of the phase III purification process. It is also highly desirable, from the regulatory perspective, to make no changes in the manufacturing process between phase III and the commercial process, to ensure product comparability and to avoid the requirement for additional clinical studies. Therefore, the phase III process development should be conducted with focus on facility fit and commercial manufacturing implementation. In reality, minor process changes and tuning of process parameter ranges are frequently implemented for the commercial process, based on the lessons learned from the extensive process knowledge gathered during the process characterization and validation stages. However, these changes usually represent a fine-tuning of the phase III process, and major changes such as the implementation of new resins and filters and scale changes should be avoided to prevent the requirement for additional comparability work.

One of the greatest challenges during late-stage purification process development, especially for companies with large manufacturing network, is that manufacturing site selection and knowledge of the available manufacturing equipment and capabilities does not occur until quite late in the development cycle. In these instances, process development must take a risk-based approach and decide between development of an optimal process for the most likely manufacturing site, that may fit poorly in other sites, or development of a process that is generally sub-optimal across the manufacturing network but could be fit in every site without requiring major changes to the process and the facility. To support process design efforts, facility fit tools are commonly used throughout late-stage process development.

Phase III purification process development typically leverages the data generated during the development of the phase I/II manufacturing process and further optimize it based on the company process development knowledge and experience.

The major goal for phase III development is developing a robust process that meets the required product quality, impurity levels, viral clearance, process yield, plant fit, and is able to recover the maximum titer and volume of harvested cell culture fluid.

The development of the phase III process should be focused on the following high-level strategic objectives:

- Product quality and process yields
  - Harvested cell culture fluid, intermediate pools, and purified bulk substance are analyzed for product quality and impurity levels (CHO protein, DNA, leached Protein A, other process related impurities) to meet all standard regulatory requirements and target quality attributes.
  - The process must enable purification of harvested cell culture fluid with titers spanning the range observed during development with yields that enable achievement of target cost of goods.
- Process safety and compliance
  - The process should minimize the use of hazardous chemicals and meet all requirements for endotoxin and bioburden.
- Process robustness and scalability
  - The process should focus on use of standard unit operations, which are proven and scalable.

- Buffers are designed to minimize requirement for titrations, and buffer specifications should allow for normal assay variability and minimize the use of corrosive chemicals.
- Process intermediates are evaluated for stability at ambient temperature, at 2 to 8 °C and/or for freeze-thaw stability as appropriate. This enables use of reliable and representative feedstocks for development, sample representativeness for analytical assays and increase in process knowledge, which ultimately leads to process robustness.

Phase III Process Development Strategy

As described above, purification platforms typically include 3 column chromatography steps (Protein A capture followed by 2 ion exchange steps), a viral inactivation step immediately downstream of Protein A, a viral filtration step downstream of the ion exchange steps, and a final UF/DF step. For each step, mass throughput requirements, pool purity targets, expected manufacturing constraints including pool and buffer tank limitations must be considered.

Phase III development usually starts by running a scale-down version of the phase I/II process to establish a benchmark and reference point for phase III development. Based on these initial experiments, a phase III process development plan is established to guide the experimental approach and define process development priorities (ex whether additional resin screening will be required and the extent of development and optimization required for each step).

Below we provide some of the strategic considerations and key aspects for the development of chromatography steps in general and for each step in particular.

General Chromatography Process Development Considerations, Strategy, and Sequence

- Resin Selection: leverage phase I/II data, consider vendor quality, industry and in-house experience, long-term availability, dynamic binding capacity, packing characteristics, cost.
- Column bed height, diameter, and operational flow rates: columns may be sized for capacity (allowing for residence times that enable processing of upper range of cell culture titers), for throughput to enable achievement of the target run rate, and for plant fit, to enable efficient scale-up.
- Column packing should be performed according to vendor recommendations. Height equivalent for a theoretical plate (HETP) and asymmetry should be determined at all scales and phases of development for data gathering to support future scale-down model qualifications (conducted during process characterization and validation).

- Load density, defined as mass of product loaded per volume of resin, is determined based on dynamic binding capacity experiments, on mass throughput and pool purity requirements.
- Buffer systems: one should implement industry standard nonhazardous buffer systems, with robust specifications aimed at meeting process targets.
- Buffer parameters (concentration, pH, and conductivity): a Design of Experiments (DOE) approach is typically used to efficiently test and define buffer parameters for all chromatography steps.
- Phase durations: for each stage of the step, phase durations are optimized to balance buffer consumption with process robustness.
- Elution strategy: for bind-and-elute steps, step elution is preferred over gradient elution for large-scale implementation. Several rounds of a DOE based approach; using high-throughput screening may be used to define the elution strategy.
- Pooling criteria: should be tested and developed in coordination with the elution strategy to determine the conditions required to achieve the optimal pool pH and conductivity (with consideration for pool stability) and minimize pool volume. A preferred strategy will be to define a starting optical density (OD) and an ending OD. Whenever required based on pool volume limitations, volume based pooling criteria may instead be defined.

Step Specific Considerations and Key Experimental Strategy
Protein A Affinity Chromatography Step

- Protein A serves as the capture step providing a significant reduction in impurity levels.
- Resin selection: select and test resins that meet the process and manufacturing requirements. Facility fit models provide very valuable guidance in the determination of the desired resin properties, namely minimum binding capacity and pressure-flow characteristics. There is a wide range of Protein A resins available on the market. The main choice is usually between high-flow rigid particle resins that enable high flow rates and execution of a higher number of Protein A cycles at a lower mass load, or a lower flow-rate compressible particle resin that enables high mass throughput through a lower number of cycles executed at higher mass loads. Companies must evaluate the best balance between the 2 options and choose the one that best fits the target manufacturing plant or that provides the greatest flexibility if the manufacturing site is not known at time of development.
- Dynamic binding capacity (DBC) versus residence time to maximize load density and mass throughput.
- Protein A postload wash buffer development to maximize impurity clearance and pool stability.
- Low pH elution strategy to minimize pool volume and enable downstream viral inactivation without need for titration.

- Maximize volumetric throughput on load and pool filters.
- Track column differential pressure and periodically monitor DBC as an early indicator for column lifetime.

Low pH Viral Inactivation Step

- Assess pool stability (aggregation, fragmentation, charge variants, and any other relevant product quality attributes) under worst-case conditions.

Cation Exchange Step (Bind-and-elute)

- Cation exchange is developed as a polishing step, enabling an overall reduction of process- and product-related impurity levels and is mainly focused on assurance of the required product quality. Cation exchange is commonly the main step for controlling aggregate levels and product variants.
- Alternative resins may be screened (DOE approach for minimization of pool aggregate levels).
- Development of packing procedures: cation exchange resins are often challenging to pack, and careful consideration should be devoted to definition of the packing procedure to ensure a successful scale-up and transfer.
- Dynamic binding capacity (DBC) versus residence time to maximize load density and throughput: high-capacity resins with known capability for aggregate reduction should be considered to enable purification of the upper range of cell culture titers while ensuring target monomer levels.
- Selection and development of buffer systems that maximize the efficiency of column equilibration and provide strong buffering conditions at the operating pH.
- Optimize wash and elution conditions for robust clearance of size and charge variants and impurity clearance (e.g.. ensure clearance of leached Protein A).

Anion Exchange Step (Flow-through)

- Anion-exchange is a key unit operation for viral clearance and for overall reduction of process impurities (CHO proteins, DNA).
- Resins should be screened with particular focus on viral clearance and host-cell protein removal efficacy.
- Maximize virus LRVs by evaluating alternative resins, process parameters, buffer systems, and load densities.
- Dynamic binding capacity experiments should be conducted to determine the maximum load density. Since the product does not bind, DBC is determined based on CHO proteins breakthrough and confirmed to meet the required viral clearance targets.

- Preference is usually given to operating the step under flow-through mode for IP considerations. However, in case there are additional purification challenges, a more powerful weak partitioning mode may be considered to provide additional purification ability. Nevertheless, this strategy is protected with IP and will require a license from a third party for commercial applications.

Parvovirus Filtration Step

- Virus filtration is used as part of the strategy to achieve the overall viral clearance target. (For more details see Section 6.4.2 below.)
- The step is operated in normal flow mode. The objective is to minimize the filter area required to enable filtration of a target load volume within a predefined period of time (typically 3 to 4 h).
- Membrane chromatography-based prefilters may be used to increase step throughput.
- Screen for main filter and prefilter, using mass throughput as main criteria. Select the combination that best suits the selected manufacturing site if known or that provide the greatest flexibility.
- Identify best step placement (between ion exchange steps or downstream of ion exchange steps).
- Maximize filter and prefilter capacities to minimize COGs impact (balance with process time).
- Include screening and optimization of load material, buffer system, pH and conductivity, which must be coordinated with the development of the elution scheme for the step upstream of viral filtration.
- Maximize viral log reduction value (LRV) for selected model viruses.
- Investigate effect of feedstock conditions on throughput (temperature, hold time, number, and conditions of freeze-thaw cycles).

Ultrafiltration and Diafiltration Step (UFDF)

- The UF/DF step concentrates and formulates the drug substance to the desired final concentration and buffer composition. In addition, it also provides clearance of small molecules (process related impurities).
- Flux excursions consisting of permeate flux determinations over a range of transmembrane pressures are conducted at different bulk concentrations to provide the background information necessary to define the process parameters (bulk concentration during diafiltration, control strategy, etc.).
- Selection of membrane vendor, type, and molecular weight cut-off.
- Selection of appropriate membrane area.
- Optimization of process and equipment design to enable high concentration formulations when needed.

- Optimization of product concentration for the dialfiltration step.
- Optimization of the number of diavolumes to achieve sufficient small molecule clearance and maximum recovery yield.
- Setting of final concentration step target concentration and buffer/formulation addition strategy.
- Consideration of Donnan effects to ensure adequate diafiltration buffer composition and final pool additions/titrations to achieve the desired drug substance pH and excipient concentrations.

6.4.2 Viral Safety Control of Monoclonal Antibody Manufacturing

6.4.2.1 Viral Safety Control Strategy

Biotechnology products such as mAbs, fusion proteins, and cytokines are cell-derived products. The commonly used cells are bacterial, yeast, and mammalian. Commonly used mammalian cell lines such as CHO, NSO are susceptible to viral infection. Potential viral contamination is from endogenous viruses and adventitious viruses.

Potential sources of contamination for cell-derived products are from raw materials, cell lines, and cell banks. In addition, adventitious virus may be inadvertently introduced during the manufacturing process through personnel, pest or compressed air.

During the manufacturing of biotechnology products such as mAbs, appropriate control of viral safety is required to ensure product quality and patient safety and to meet regulatory requirements. However, rare events of viral contamination in the past have caused severe impact to the product supply, production facility downtime, and significant business loss for the affected companies. Therefore, multititers of viral safety control strategy are essential to ensure viral safety in biologics products manufacturing. These controls, as described below, include testing of the cell banks and product, raw material controls, viral clearance studies, as well as cleaning and segregation practices in manufacturing processes [14, 15].

Testing of Cell Banks, Preharvest Cell Culture Fluid, and Bulk Drug Substance

The testing is first performed on the master cell bank and the working cell bank. The testing is further performed on the cell culture fluid prior to the harvest.

- Master cell bank (MCB): extensive characterization to demonstrate low risk for presence of contaminants. Evaluation for endogenous agents; testing for in vitro and in vivo adventitious agents.
- Working cell bank (WCB): more limited testing but still includes in vitro adventitious agents and other viral assays associated with the specific cell type and manufacturing process.
- Preharvest cell culture fluid: tests for endogenous virus-like particles.

Raw Material Control

- Source control: vendor inspection, audit, and qualification.
- Raw material viral risk assessment program: avoid the use of raw materials of animal or human origin.
- High temperature short time (HTST) treatment is suitable for heat stable large volume solution. Both UV-C and nanofiltration can be used for heat labile components.
- Raw material testing program which is a risk based approach, as testing of raw material is difficult and probably subject to false negative. Raw materials potential contaminated are not homogeneous.

6.4.2.2 Downstream Process Viral Clearance Capability

Direct testing throughout the purification process would only detect known and tested viruses and low level of infectious viruses may fail to be detected by direct testing. Viral clearance studies provide evidence that the manufacturing process will effectively remove or inactivate viruses to a target number of LRVs.

6.4.2.3 Cleaning and Segregation Practices in Manufacturing Processes

Equipment Cleaning and Product Changeover

To prevent viral contamination, equipment cleaning and product changeover need to be performed. The manufacturer needs to demonstrate that cleaning and disinfection agents are effective against both, virus and bacteria. During cleaning validation studies, parameters such as contact time and temperature should be evaluated. Other controls include positive pressure cell culture suites, high efficiency particulate adsorption filter- (HEPA-)filtered single pass airflow, dedicated equipment, and execution of changeover activities on equipment on a campaign basis.

Material and Personnel Flow

- Expand cell bank in dedicated suite. Closed system from seed bioreactor through harvest. Harvest and purification in separate suites. Segregate trains in multiproduct facility.
- Require gowning in areas based on activities performed.
- Monitor cell culture processes for indicators of infection, e.g., changes in growth patterns, viability, metabolites, monitor expansion growth and viability. Compare in-process measurement trends in real-time through multivariate data acquisition, analysis, and historical trend comparison.

6.4.2.4 Virus Removal and Inactivation Steps

Many techniques have been demonstrated to inactivate or remove virus (see Table 6.4.1).

Table 6.4.1: Viral removal and activation methodologies.

Inactivation	Partitioning
1. Heat treatment	5. Precipitation
2. Solvent and detergent treatment	6. Chromatography
3. Low pH treatment	7. Viral filtration
4. UV-C	

Heat Treatment

Viruses can be inactivated at high temperatures. Generally, enveloped viruses are more sensitive than nonenveloped viruses, and the longer the exposure to high temperature the greater the extent of inactivation.

Solvent and Detergent Treatment

Solvent and detergent inactivation is effective against enveloped viruses. While the viral protein coat and genome may still be functional, the lipid membrane, which is essential for docking of the virus to the host cell, is dissolved, and consequently the virus becomes noninfectious. Solvent and detergent inactivation steps should be placed in the beginning of the purification process, so that the solvent and detergent used can be removed from the product stream by the purification steps downstream. Common detergents are tri(n-butyl) phosphate (TNBP), tween 80, and Triton X-100. The concentrations typically range from 0.3 to 1 % and the exposure times from 30 to 60 min.

Low pH Treatment

Low pH treatment has been demonstrated as a robust step for inactivation of enveloped viruses and it is commonly used in mAb purification processes. The effectiveness of the step depends on the pH, temperature, and duration of the step. Typical mAb process conditions are pH <3.5 for 30 min at ambient temperature.

UV-C

Ultraviolet light inactivates viruses by targeting the nucleic acid in the genome. Viral sensitivity to the UV-C is propositional to genome size. While UV treatment is one of the major methods used for treatment of cell culture media, for viral contamination control of media raw materials, it is typically not used in process streams due to measurable and/or potential negative impact on product quality.

Precipitation

At 20 % ethanol concentration, enveloped viruses are generally inactivated in approximately 5 min. Precipitation can be effective against both enveloped and nonenveloped viruses. However, it is rarely implemented in mAb processes.

Chromatography

Chromatography removes viruses through partitioning. Although it is usually difficult to understand the exact mechanism of viral removal, chromatography is very effective and very commonly used in mAb processes. Viral removal is highly dependent on the resin, protein, and buffer conditions used. Chromatography can be effective against both, enveloped and nonenveloped viruses.

Viral Filtration

Viral filtration removes viruses based on their size, while permitting flow through of the target protein. During the process, filter integrity must be determined by an integrity testing method before and after use if possible. For some filter types, the integrity testing procedure is destructive and it can only be performed postuse. Depending on the pore size of the filter, it can be effective to both enveloped and nonenveloped viruses. Virus filtration is a very robust platform method, typically does not affect product quality, and in particular for smaller proteins and mAbs typically allows for achievement of high product yield.

6.4.2.5 Viral Clearance Studies

The purpose of a viral clearance study is to demonstrate viral clearance capability of selective purification steps in a downstream process. Purification steps, for which viral clearance is to be claimed, are challenged with model viruses to demonstrate the ability of the purification process to remove and inactive virus of a certain type. The total virus reduction is calculated as the sum of the log10 reduction factors (LRF) determined for each step.

Viral clearance studies must be performed under good laboratory practices (GLP) guidelines and conducted in laboratories of at least biosafety level 2. A study plan must be prepared and signed by the test facility and the study sponsor. The study plan describes how spiking, inactivation, sampling, and testing is to be performed. Viral testing laboratories typically issue viral toxicity, interference, and viral clearance test results to the sponsor. Sponsor in turn, documents the entire viral clearance study into summary report. Viral clearance studies are used to support regulatory filings.
A typical procedure for a viral clearance study includes the following activities:

- Choice of appropriate model viruses,
- scale-down model qualification,

- contract the viral testing laboratory (when capability not available in-house),
- the contracted laboratory provides work plan including validation study, virus testing, and cytotoxicity/interference study,
- sponsor generates a virus clearance protocol,
- generation of feedstock materials for each of the relevant steps,
- performing cytoxicity/interference study and execute virus clearance study,
- contracted laboratory provides the data report,
- sponsor write and approve virus clearance report,
- sponsor includes the virus clearance results in CMC section of the regulatory submissions.

Study Design

In most of the mAb downstream processes, steps such as low pH treatment, detergent treatment and nanofiltration are specially designed for removal or inactivating of the potential viruses. In addition, selective chromatography steps such as anion exchange and affinity chromatography steps are oftentimes included in a viral clearance study. Only specific steps in downstream process more likely to provide viral clearance are evaluated in the study. These steps need to have different and independent viral clearance mechanisms. In addition, the step must be properly scaled down.

Choice of Viruses

In a viral clearance study, model viruses are used (see Table 6.4.2). Viruses with a wide range of physico-chemical properties as below are used to evaluate the robustness of the process:
- DNA/RNA genome,
- lipid-enveloped and nonenveloped,
- large, intermediate, and small size,
- from very resistant to easily inactivated.

Table 6.4.2: Virus for mAb phase I study.

Virus	XMuLV (Xenotropic Murine Leukemia Virus)	MMV (Murine Min Virus)
Genome	RNA	DNA
Envelope	Yes	No
Family	Retro	Parvo
Size (nm)	80 to 110	20 to 25
Resistance	Low	High

Model virus selection also requires suitable culture system, availability of high titer stocks, and reliable methods for qualification. Using high-titer virus stocks will help to achieve higher LRVs, but virus stock may form aggregates, which may enhance physical removal and decrease inactivation. Typically, a 5 to 10 % spiking ratio (virus: sample in volume) is used.

Cytotoxicity and Viral Interference

Most common viral testing methods are the cell culture-based infection assay and real time polymerase chain reaction (PCR). The PCR method can be used to evaluate steps where inactivation and removal are working at the same time. Virus quantitation is conducted by exposing healthy cells to dilutions of the testing sample containing viruses. Virus replicates in the cells and produces a visible effect that can be measured. By testing sufficient dilutions, one can accurately determine the total number of infectious virus particles present. In some cases, a large volume testing is used to increase the testing sensitivity to achieve high log reduction. Nevertheless, before the actual viral clearance study is executed, toxicity of the sample (without the virus) must be evaluated to prevent confounding with viral replication effects. The toxicity assessment is performed by exposing the cells to dilutions of the test sample effect prior to the spiking with virus. In addition, sample components may interfere with the infection assay when virus is present at low concentrations. Any cytotoxic effect induced by the test item on the indicator cells must be prevented. If toxicity or interference is observed, the test item must be diluted to the point where no cytotoxicity or interference is observed.

To achieve the highest possible log reduction of virus, typically a high titer of viral stock and high spike ratio of virus stock solution to study loading samples are used. Recently, purified virus stock solutions are increasingly used. This is particularly important for steps such as viral filtration, where the impact of spiked viruses to filter throughput is minimized.

Scale-down Model

Viral clearance studies are not feasible at manufacturing scale; therefore, a scale-down model of the manufacturing process must be used for this purpose. Comparable process performance of the scale-down model versus manufacturing process at scale, of the steps to be validated for viral clearance, needs to be demonstrated. This is achieved by using scale independent parameters such as pH, conductivity, column bed height, and linear velocity. In the study, the scale dependent parameters such as column diameter and viral filter area are evaluated for any potential impact to the process. Qualification of the scale-down model is achieved by comparing process performance indicators such as chromatograms, flux decay, and step yields. In addition, the relevant product quality attributes such as monomer content and residual DNA content are analyzed to demonstrate the validity of the scale-down model.

For a process aimed for commercialization, viral clearance is assessed during the viral clearance study using resins at both, the beginning and at the end of the resin lifetime. In this case, a viral carry over study is also needed.

Viral Clearance Steps

In viral clearance studies, the virus stock solution is spiked onto load samples of either chromatography or filtration steps. An aliquot of the virus-spiked sample is held for the duration of the experiment including the hold of the pool sample post column or filtration steps. This hold sample is used to determine any activity loss over the time of the experiment.

6.4.2.6 Log_{10} Reduction Value (LRV)

Viral clearance factors are commonly expressed in units of log_{10} reduction value. LRV is calculated by the equation:

LRV = log_{10} [(input virus titer × input virus volume) / (output virus titer × output virus volume)].

For example: LRV = log_{10} [(10^8 IU mL^{-1} × 20 mL) / (10^2 IU mL^{-1} × 10 mL] = log_{10} (2 × 10^6) = 6.3.

Total LRV of a purification process is the sum of LRV for each of the purification steps evaluated during the viral clearance study. Table 6.4.3 is an example of virus clearance log reduction factor summary for both, murine leukemia virus (MuLV) and mouse min virus (MMV).

Table 6.4.3: Log_{10} reduction factor summary of a purification process.

Process Step	xMuLV	MMV
Affinity chromatography	3.5	2.9
Low pH viral inactivation	5.4	Not tested
Ion exchange chromatography	≥ 5.1	≥ 4.0
Virus filtration	≥ 5.4	≥ 5.9
Total process viral clearance []	≥ 19.4	≥ 12.8

6.4.2.7 Viral Safety Risk Assessment

The retroviral safety risk assessment is typically done by calculating the available log safety margin between the endogenous retroviral like particle load in the preharvested cell culture fluid and the total LRV determined by the viral clearance study.

As an example, using an electron microscopic evaluation of the production bioreactor preharvest cell culture fluid from multiple full scales runs, the virus like particle (VLP) titer was estimated to be 2.6 × 106 mL^{-1}.

Assuming a titer of product of 1.8 mg mL^{-1} in the preharvest cell culture fluid, a 70 % minimum purification yield and a maximum dose of 400 mg, the volume of preharvest cell culture fluid needed per dose of the product is 176.4 mL (400 mg / (1.80 mg mL^{-1} × 0.70)).

This could potentially result in 4.58 × 108 VLPs per dose (8.7 $\log_{10}$) if no virus clearance occurred in the downstream process. Since the total cumulative process reduction of xMuLV was ≥19.4 $\log_{10}$ (Table 6.4.3), the safety margin is calculated at 19.4 – 8.7 = 10.7 $\log_{10}$, which demonstrates an adequate control of retroviruses, by the manufacturing process (see Table 6.4.4 for a summary of the example above).

6.4.2.8 Viral Clearance Study for a Commercial Process

As products move from early development to commercial production, the patient exposure is significantly increased and the regulatory expectation increases.

For investigational new drug (IND) applications, testing of 2 model viruses is adequate for most cell-derived products such as monoclonal antibody. However, during the process characterization and validation studies to support biological license application (BLA), typically 4 viruses are included in the virus clearance study. In this case, 2 replicate experiments are performed at the minimum and the lowest value is used in calculation of total LRV.

During the process characterization and validation studies, resin reuse studies are performed to demonstrate that the chromatography step has acceptable performance in controlling product variants and process related impurities throughout the resin life time. In addition, viral clearance studies using resin from the end of resin lifetime are performed to determine if the virus clearance capability is maintained or impacted by resin reuse.

In viral clearance studies conducted to support market applications, carry-over analysis of viruses from one batch to the other is typically included in the study and the regulatory filing.

In viral clearance studies it is important to analyze all intermediate fractions so that the viral mass balance can be closed. Observation of viral mass balance improves the chances for understanding the viral removal mechanism, which is of critical importance to address operating deviations in routine commercial manufacturing.

Table 6.4.4: Retrovirus (xMuLV) safety margin calculation.

Description	Value/Calculation
Maximum clinical dose	400 mg
Minimum titer in preharvest cell culture	1.8 mg mL^{-1}
Downstream process yield	70 %
Volume of production bioreactor bulk harvest per maximum clinical dose	400 mg / (1.80 mg mL^{-1} × 0.70) = 176.4 mL
Retrovirus-like particles in production bioreactor preharvest	≤ 2.6×10^6 VLPs mL^{-1}
Virus-like particles per maximum clinical dose	2.6×10^6 VLPs mL^{-1} × 176.4 mL = 4.58×10^8 VLPs
	$4.58 \times 10^8 \equiv 8.7\ \log_{10}$
Total process viral clearance	≥ 19.4 $\log_{10}$
Safety margin	19.4 $\log_{10}$ − 8.7 $\log_{10}$ = 10.7 $\log_{10}$

6.4.2.9 General Regulatory Expectation for Viral Safety Control

The viral contamination of the cell lines or adventitious introduction of virus during production could have serious impact on manufacturing and clinical development of biological products. Therefore, virus testing and assessment of virus removal and inactivation by the manufacturing process are required during the development of biologics [16].

It is a regulatory expectation that the MCB, WCB and cells at the limit of *in vitro* cell age are tested for the presence of retroviruses. The assays include infectivity assays, transmission electron microscopy (TEM) analysis and *in vivo* assays.

In addition to testing of cell banks and bulk drug substance, evaluation of viral removal and inactivation capability by the downstream process is required for biologics product development. For inactivation steps, samples should be taken at different time points so that inactivation kinetics can be shown [14]. Model viruses for viral clearance studies should represent viruses that may contaminate the product and should display of a wide range of physico-chemical properties in order to test the ability of the downstream process to eliminate a large range of viruses [16]. It is recommended that the purification scheme includes at least one virus inactivation step [17].

The validity of the scale-down should be demonstrated. For chromatography steps, column bed-height, linear velocity, contact time, buffer and resin types, pH, temperature, and concentration of protein should all be shown to be representative of production-scale manufacturing.

As a general guidance, it is recommended that the purification process includes at least 2 orthogonal (i.e., based on different mechanisms) robust (greater than 4 LRVs) virus removal steps [18]. It is important to show that not only viruses are removed, but that there is excess capacity for virus removal built into the purification process to assure an appropriate level of safety for the final product.

6.4.3 Alternative Purification Technologies and Emerging New mAb Platforms

6.4.3.1 Alternative Purification Technologies

As discussed above in greater detail (Section 6.4.1), platform approaches decrease the investment per mAb product, improve development efforts, simplify raw material procurement, improve technology transfer, and streamline the regulatory aspects of processing. Kelley argues that the state of the art of mAb development and manufacturing has all hallmarks of a highly industrialized family of manufacturing processes [1]. He further defends that companies have no need to stray away from conventional production technologies because use of current separations media combined with a focus on facility fit has shown that many plants can be debottlenecked to support titers of up to 5 g L^{-1}. Still according to the same author, “attractive returns coupled with potential losses in revenue resulting from delays in product approval have made companies focus on time to market rather than on improving process economics”. The concept that companies have no need to look into novel purification technologies may be argued in the context of mAb development in large biopharmaceutical companies, where COGS represent a small fraction of the overall cost of development. The same conclusion however, is not applicable for Contract Manufacturing Organizations (CMOs), for which development and manufacturing costs represent a significant fraction of the overall cost structure. In this context, development and implementation of novel processing technologies, that either lead to shorter development timelines and/or reduced operating costs, may have a significant impact to the company’s profitability and competitiveness in the marketplace. The purpose of this section is to provide a generic overview of alternative downstream processing technologies that are not commonly part of the industry standard mAb purification platforms. Some of these technologies have been utilized in specific applications with particularly challenging mAb products, others are becoming part of new alternative purification platform approaches and may one day become part of a possible future industry standard.

Alternative Chromatographic Approaches: Chemistries and Operating Modes

Hydrophobic Interaction Chromatography

Hydrophobic interaction chromatography (HIC) is a useful technique to separate proteins based on differences in their surface hydrophobicity and it can be operated both in bind-and-elute and flow-through modes. To induce binding of the protein of interest to the stationary phase, the load material is conditioned with a salt, usually kosmotropic, very commonly with ammonium or sodium sulfate, to a moderate salt concentration, below that used for protein precipitation. Addition of the salt to the load material causes a decrease in the entropy of the solvent water by preferential solvation of the salt ions (and preferential dehydration of the protein surface). This leads to a further unfavorable interaction between solvent water molecules and exposed hydrophobic residues on the surface of proteins, which will become more stable in the vicinity of the hydrophobic ligands of the resin and therefore binding to it. Preferential binding of proteins to the resin relative to protein-protein binding occurs because of the greater hydrophobicity of the resin ligands compared to that of the protein residues. Elution from HIC resins is typically achieved by a gradient of decreasing salt concentration. Companies usually turn to the use of HIC resins, deviating from the standard mAB platforms, when facing challenges with the removal of product variants, commonly product aggregates hard to remove by conventional CIEX. The HIC resins most commonly used in bioprocessing applications contain both, butyl or phenyl ligands, and the step is usually placed downstream of the conventional ion exchange step(s) to ensure a cleaner load material.

Mixed-mode Chromatography

As the name implies, mixed-mode chromatography results from the use of resins with ligands that can interact with proteins simultaneously via multiple types of chemical interactions, namely via electrostatic interaction, hydrophobic interaction, and hydrogen bonding. The combined use of multiple interaction modes leads to an increase in selectivity and in the ability to resolve species otherwise difficult to separate by conventional means. Commercial mixed mode resins exist in both AIEX and CIEX formats, which enable operation at both, high and low pH, and in a wide range of ionic strengths, due to the hydrophobic character of the stationary phase. Elution is usually achieved by a change in pH (decrease in pH in resins with anion exchange character and increase in resins with cation exchange character) to induce repulsion of like charges in the resin and in the protein of interest [19].

Hydrophobic Charge Induction Chromatography

Hydrophobic charge induction chromatography was developed to enable salt-independent adsorption of proteins on hydrophobic resins without the use of the harsh elution conditions (organic solvents) required with reversed phase chromatography. Binding of proteins to the stationary phase does not require use of a kosmotropic salt, as with hydrophobic interaction chromatography, because of the much higher hydrophobicity of the ligands. Elution is achieved by a decrease in pH that induces the creation of a like repulsive charge in an ionizable ligand. This elution mechanism resembles that of Protein A chromatography, but the technique is not specific for antibodies. This chromatography mode has been shown to have greater selectivity than HIC in regards to separation of host cell proteins [20].

Ceramic Hydroxyapatite Chromatography

Ceramic hydroxyapatite (CHP) is an inorganic media containing positively charged calcium (C sites) and negatively charged phosphate groups (P sites). The binding mechanism is complex and cannot be explained simply in terms of ion exchange interactions. Evidence of this, is that the chloride ion is unable to effectively elute acidic proteins off of C sites, nevertheless sodium ions are effective in the elution of basic proteins from P sites. Therefore, sodium chloride is a very effective component for the preferential elution of acidic proteins from CHP resins. Elution of basic proteins from C sites is achieved by use of phosphate ions. CHP using a sodium phosphate buffer has been effectively used as a polishing step for the removal of aggregates and leached Protein A in a mAb purification process [21]. CHP resins are made of an inorganic matrix, which is very unstable and dissolves under low pH conditions, and in the presence of metal chelators such as citrate and EDTA. Ensuring a long resin lifetime is usually a big challenge and companies avoid its use unless there are no other means of achieving a required separation. Even though CHP is excellent from the perspective of biocompatibility, to the author's knowledge it is only employed in a single manufacturing process for an approved mAb product.

Continuous and Semicontinuous Chromatography

Clinical and commercial manufacture of monoclonal antibodies is commonly operated in batch mode. Typically, the cell culture process is carried out in a single bioreactor, usually for a period of 11 to 14 days, and then it is harvested and purified as a single batch. Chromatography columns are commonly sized to the lowest column volume that enables sufficient capacity to process the maximum expected product mass. In other instances, columns are sized to allow a volumetric throughput that enables processing of the maximum expected load volume within the facility run rate target. Due to its high cost and relatively low dynamic binding capacities, Protein A affinity chromatography columns are usually undersized and operated in multiple cycles per batch. Batch mode, while practical and simple from the perspective of automation procedures, is suboptimal in regards to productivity and utilization of raw materials, namely resin capacity and buffer volume. Continuous chromatography offers an opportunity to significantly increase productivity and to decrease cost of goods (COGS). Since true continuous chromatography, with the stationary and mobile phases moving in counter-current, is technically hard to implement, a process denominated as simulated moving bed (SMB) was developed to simulate that countercurrent motion by a concerted switch of the inlet and outlet valves in a multicolumn set-up. One advantage of SMB over conventional batch chromatography is that a partial separation of components is sufficient to attain a high purity with high yields. In batch chromatography, product purity must come at a cost to yield in the absence of baseline separation between the product and the impurities. Another advantage of SMB over batch chromatography is that the capacity and volume of the resin is fully utilized, since the flow-through postbreakthrough phase is diverted for binding into another column. A disadvantage of the SMB process, originally developed for the separation of small molecules, namely of racemic mixtures, is that it only allows for the separation of 2 fractions and that it does not allow for the use of gradient elution. For that reason, to the authors' knowledge there are currently no industrial SMB processes implemented for the commercial manufacture of monoclonal antibodies. Protein A affinity chromatography is in essence a 2 fraction separation application. However, in this case there is baseline resolution between product and impurities and the SMB process offers no advantage from the perspective of the purification factor versus process yield. The disadvantages of batch chromatography and the practical limitations of SMB chromatography led to the development of a multicolumn countercurrent solvent gradient purification (MCSGP), which combines the advantages of batch chromatography and SMB [22]. The MCSGP process is a hybrid between continuous and batch chromatography that enables three-fraction separations, with buffer gradients, high efficiency, and low buffer consumption. This process can be operated with any conventional mode of stationary phase and was originally developed for the separation of peptides by reversed-phase chromatography. Explaining the principle of the MCSGP process is beyond the scope of this section and can be found elsewhere [22]. The process can be operated in a continuous or semi-continuous mode, using 6 or 3 column segments, respectively.

MCSGP has been successfully utilized for the purification of monoclonal antibodies in both, capture and polishing applications, and it has been shown that this process can manufacture mAbs with up to tenfold lower levels of impurities compared to batch processes. [22]. Mahajan et al. have also shown that semicontinuous chromatography is very effective in improving resin utilization [23]. In this study, product yield and quality were evaluated and compared with a batch process. The authors concluded that the continuous operation showed the potential to reduce both resin volume and buffer consumption by approximately 40 %. It was also acknowledged that the system hardware and the process control were more complex. In a different study [24], Pollock et al. explored the potential of semicontinuous chromatography to reduce clinical and commercial manufacturing costs. Their work predicted that semicontinuous chromatography using periodic counter-current (PCC) capture technology, could potentially offer more significant savings in direct costs for early-stage clinical manufacturing than for late-stage or commercial manufacturing.

Non-chromatographic Approaches

Membrane Adsorbers

Synthetic microporous membrane adsorbers have been developed to overcome some of the limitations with conventional resin chromatography [25, 26]. In membrane chromatography, mass transfer is predominantly limited by convection rather than by diffusion as in standard column chromatography; therefore, dynamic binding capacity is essentially independent of the operating flow rate within a wide operational range. In addition, the effective bed height of membrane adsorbers on the market today is typically in the millimeter range only, resulting in very low pressure drops and therefore enabling very high flow rates, but low packing densities in terms of matrix volume/device volume. For that reason, membrane chromatography is better suited for applications requiring fast processing of large volumes of material. Membrane adsorbers are available in some of the conventional ligand chemistries found with packed bed chromatography, particularly ion exchange, but also in mixed mode and affinity. Membrane adsorbers are available in 3 types: flat sheet, hollow fibers, and radial flow. Flat sheet is the most common format and since the volume of the largest commercially available membrane chromatography devices is still relatively small, compared to the range of volumes commonly used for packed bed chromatography, most of the applications are operated in flow-through mode, to capture impurities which are present in much smaller concentrations, compared to that of the product of interest. Feasibility of membrane chromatography for flow-through applications, including purification of monoclonal antibodies, has been widely demonstrated. In many instances membrane adsorbers show comparable performance to that of packed beds, at comparable or lower operating cost, even when the membranes are operated in a single-use mode. Savings come mainly from a simpler operation requiring fewer resources since there is no need for column packing, resin cleaning, regeneration procedures, HETP, and asymmetry testing, and the buffer volumes are considerably smaller. Anion exchange membrane adsorbers were the first to be assessed for purification of monoclonal antibodies, for their superior ability to remove DNA, viruses and host-cell proteins in flow-through mode. Cation exchangers have also been evaluated and implemented, but rather than replacing the conventional packed bed cation exchange steps, they have been mostly used as prefilters in parvovirus filtration, to protect and extend the capacity of the nanofiltration layer.

Bind-and-elute mode membrane chromatography has been demonstrated in commercial scale operation for capturing complex and labile protein drugs from relatively dilute cell culture harvests [27], where the dramatic speed advantage more than outweighs the comparatively low total capacity/adsorber capsule. For monoclonal antibodies, higher capacities per unit would be required for reasonable bind-and-elute process designs. Nevertheless, a recent study [28] has shown that ion exchangers, including membrane devices, have the potential to be successfully operated under product binding conditions in an “overload” mode. In these applications, the dynamic binding capacity is largely exceeded (up to tenfold), but the step yields is still high because the absolute amount of product retained in the relatively small membrane devise is essentially negligible. What was intriguing with this study was that the purification performance of ion exchangers is largely retained under overload conditions, presumably because of a much higher affinity of the impurities (usually product aggregates and host cell proteins) for the ligand compared to that of the mAb, therefore displacing it and ensuring a large capacity for the retention of those impurities. Interestingly, this behavior was not exclusive of membrane adsorbers and it was also observed with conventional packed beds, showing that there is an opportunity to significantly reduce the size of packed columns. Operation of cation exchangers in overload mode was demonstrated to be successful in replacing conventional cation exchange for the purification of monoclonal antibodies [28].

Precipitation

Chromatography is an extremely powerful purification technique for process-scale separations, particularly in regards to its resolution potential. Nevertheless, there are many other alternative techniques that have over the years been successfully applied in the purification of biologics. The main motivation to evaluate alternatives to chromatography include the potential to replace it with simpler and cheaper unit operations, especially considering the high costs of the Protein A resins used to capture mAbs from the cell culture. Additional drivers for studying alternative techniques include the identification of new, scalable, and economic methods for the reduction of impurity levels upstream of the first chromatography step in a purification scheme. Aqueous two-phase extraction, three-phase partitioning, crystallization, high-performance tangential flow filtration and precipitation are some of the most important alternatives to chromatography that have been studies in regards to a possible implementation in the purification of monoclonal antibodies. Reviewing these alternative techniques is out of scope for this chapter, nevertheless, a recent resurgence in the investigation of precipitation methods for the purification of monoclonal antibodies by both, academia and the industry, is worth mentioning here. As mentioned previously, the early days of immunoglobulin purification relied heavily on precipitation methods based on the Cohn fraction process using cold ethanol. It is interesting that 60 years later precipitation is once again getting attention and being considered for implementation at process-scale. Glynn et al. [29] provided a short review of process-scale precipitation of impurities in mammalian cell culture broth. The precipitation agents covered in the review were ammonium sulfate, polymers (including PEI and PEG), ionic liquids, cationic detergents, ethacridine, and caprylic acid. Purification factor results vary with the precipitating agent, and the study reported up to twentyfold reduction in host cell proteins and up to 164,000-fold reduction in DNA, with process yields in the 90 to 95 % range. The authors concluded by stating that coupling of precipitation with centrifugation or depth-filtration still requires additional development and optimization work to become economically competitive with standard chromatographic techniques. In a very recent publication by a leading biotech company, Giese et al. [30] evaluated the purification of mAbs by precipitating impurities with polyethylene glycol (PEG) with the purpose of developing a two-chromatography column purification process. The authors claimed that PEG was effective in the removal of host cell proteins, aggregates, leached Protein A and host cell DNA, and that effective and robust precipitation performance was observed with a range of different mAbs. The authors concluded that a 2 column process in combination with PEG precipitation resulted in acceptable overall process yields and drug substance product quality for both IgG1 and IgG4 mAbs with a broad range of isoelectric points. The authors further claimed successful scale-up of the precipitation-based method. In another recent study, Oelmeier et al. used [31] centrifugal partitioning chromatography as a means of multi-step extraction to remove host cell proteins using aqueous two-phase systems. This was conducted in combination with PEG precipitation of the monoclonal antibody of interest, followed by resolubilization to increase the mAb concentration, remove the phase forming polymer, and further reduce HCP. The authors claimed the overall process removed 99.4 % of the initial HCP levels with a process yield of 93 %.

6.4.3.2 Evolution of the mAb Purification Platform: the Case for Continuous Processing and Disposable Technologies

In the last decade, monoclonal antibody production has been characterized by increasing upstream titers, increasing cell densities, higher concentration formulations, shrinking development timelines, and a focus on cost reduction. These challenges have been met by: the development of resins with higher dynamic capacities and improved flow-pressure properties, the development and implementation of highly efficient downstream platforms, execution of fast process development by use of high-throughput screening, reduction in the number of process steps and overall process simplification to improve facility fit.

Many alternative technologies, both chromatography- and nonchromatography-based, have been evaluated and developed for purification of monoclonal antibodies, and in many cases, proof of principle for alternative process platforms have been demonstrated by both, academia and industry. A major focus of alternative approaches has been the replacement of Protein A and the development of nonaffinity platforms, with the argument that Protein A is a very expensive raw material. Additional efforts for improvements in the mAb purification platform have been focused on the reduction of the number of purification steps, the replacement of packed beds by alternative technologies, such as precipitation and membrane adsorbers, and on a preference for flow-through operating modes, more amenable to implementation in continuous/semicontinuous platforms. From the combination of these trends one would expect the industry mAb purification platform to evolve towards a nonaffinity-based capture, by selective precipitation of the host cell impurities, or of the mAb product, followed by a series of polishing nonaffinity membrane-based flow-through steps, viral filtration, and a final ultrafiltration and diafiltration step. While this conceptual future platform is a possibility within technical reach, mature biopharmaceutical companies, with a large installed network of stainless steel-based manufacturing facilities, are very unlikely to stray away from using the conventional affinity-based mAb platform within the next decade.

From the point of view of large biopharmaceutical companies, the reliability offered by the current affinity based platform, which includes the assurance of fast and predictable process development timeline, a low risk regulatory strategy, and robust manufacturing, will be hard to match by any new alternative platform. In 2 different publications [1, 5] Kelley discussed this point to a great level of detail and offered convincing arguments in its favor. The very common statement that Protein A is a very expensive raw material, which is the main motivation for the investigation of nonaffinity based capture technologies, while true from the strict perspective of the price per volume of resin, is perhaps overstated in the literature and not correctly framed in the context of the total cost of production. Assuming an approximate price of 10,000 U. S. dollars per L of Protein A resin, and that due to compression factor and manipulation losses one may require an additional 20 % of resin, the cost of packed Protein A is currently around 12,000 U. S. dollars per L. Assuming an average typical load density of 35 g L^{-1}, a 200-cycle lifetime and an overall process yield of 85 % for all purification steps downstream of Protein A, one arrives at an approximate cost of 2 U. S. dollars g^{-1} (U. S. dollars of Protein A resin per gram of commercial bulk drug substance manufactured). Assuming an overall downstream raw material cost of 8 U. S. dollars g^{-1} [5], the cost of Protein A amounts to ca. 25 % of the total downstream raw material costs, which in fact constitutes a significant fraction. Nevertheless, if one considers the tremendous selectivity and purification factor offered of Protein A, the cost of the unit operation seems to be more justifiable. Kelley further argued that Protein A costs are a very small fraction of the overall COGS. Assuming a typical COGS range of 200 to 400 U. S. dollars g^{-1} [5], Protein A would represent only up to 1.0 % of the overall drug substance manufacturing costs. The relative significance of the cost of Protein A may be better put into perspective if one considers that the total cost of Protein A (2 U. S. dollars g^{-1}) could be offset and paid off by a mere increase of 1 % in either the average cell culture titer or in the overall process yield. In an hypothetical scenario assuming a typical cell culture titer and overall process yield of 2.0 g L^{-1}, and 80 %, respectively, the total cost of Protein A could be offset by an increase of only 0.02 g L^{-1} in the average titer or by an increase of only 0.8 % in the overall process yield. From this perspective, one must carefully reconsider whether the investment and the effort required to develop and implement a new purification technology, aimed at replacing Protein A, is really worth it from a cost perspective. These considerations suggest that, from the perspective of improving the cost of goods, new technology development efforts should instead focus on improving the most impactful cost factors mentioned above: cell culture productivity and the overall process yield.

The previous discussion assumes a commercial production scenario. For clinical production, the relative cost of Protein A is substantially higher, because the resin is only used for a limited number of cycles and then discarded well before reaching the resin lifetime. If one assumes an average of 5 Protein A cycles per batch and a clinical campaign with 4 batches, the Protein A resin is only cycled 20 times, approximately 10 % of its lifetime of 200 cycles. In this scenario, using the same cost assumptions of the previous example, the cost of Protein A would increase to approximately 20 U. S. dollars g^{-1}, a ten-fold increase relative to its commercial operating cost. Nevertheless, the cost of many other raw materials, namely of the other reusable resins and filters, also increases, even more so proportionally, since they are usually only operated for one cycle per batch, therefore being discarded at a much relatively earlier point of their lifetime. Taking the example of a modern cation exchanger resin or mixed-mode resin, assuming a shipped price of ca. 2,500 U. S. dollars L^{-1}, a 20 % resin overage (due to compression and handling) leading to approximately 3,000 U. S. dollars L^{-1} of packed bed, an average load density of 50 g L^{-1}, an overall process yield of 88 % downstream of that step, and a resin lifetime of 100 cycles, would result in an overall cost of 0.68 U. S. dollars per gram of purified drug bulk substance for commercial applications (compared to approximately 2 U. S. dollars per gram Protein A). However, for clinical applications, this resin would only be utilized for 4 cycles (1 cycle per batch) before being discarded, leading to an overall cost of 17 U. S. dollars g^{-1}, which would correspond to a 25-fold increase in cost, relative to the commercial operation, and more importantly, would bring its operating cost very close to that of Protein A (20 U. S. dollars g^{-1}). This second example, based on a clinical production scenario, further challenges the validity of the argument for replacing Protein A with cheaper alternative technologies.

The discussion above suggests, that from the perspective of drug substance manufacturers, the new bioprocess technology development efforts, namely on new raw materials for improvement of the mAb manufacturing platform, should focus on increasing the denominator side of the COGS fraction (the "grams"), rather than on decreasing the numerator (the money). It can in fact be argued, that the greatest future platform improvements will come from the implementation of technologies that enable maximum process intensification (increase in titer, yield, productivity, run rates), leading to the dilution of the large fixed operating cost by a larger quantity of drug substance manufactured per unit of time, coming from either larger batches or from a higher number of runs per year. In this context, improvements in cell culture productivity are obviously one of the most important cost determining factors, and should therefore always be a focus of improvements. For downstream, the focus should be on improving yield and product mass throughput. Yield may improve by either by the development of new technologies enabling improved resolution of impurities at higher yields or by optimizing manufacturing procedures to improve yields with current technology. Improvements in throughput are more likely to come from the development and implementation of continuous or semicontinuous processing technologies. It is well known that continuous processing enables savings in raw materials, namely on Protein A affinity chromatography, by the use of much smaller columns and significantly higher resin utilization. However, given that savings in raw materials have a very modest impact to the overall COGS, as previously discussed, the most important contribution of continuous processing comes from significantly higher productivities. When the overall plant productivity bottleneck is downstream, this leads to a real increase in the facility drug substance mass output, largely offsetting the relatively small increase in the overall fixed cost (from the capital investment on continuous processing equipment), and therefore improving COGS.

For new manufacturing facilities, or for utilization of available space within existing manufacturing facilities, the focus for platform improvement should be on technologies that enable a lower fixed capital cost. In this context, single-use disposable technologies are the most efficient solution to achieve the result.

Another trend and driver for platform evolution has been to reduce the development timeline and the number of process steps. Reducing the development time is of the most critical importance; especially for early-stage applications where process development is typically on the critical path. To this regard, developments in high-throughput processing systems (HTPS) have been instrumental in enabling shorter process development timelines and correspondingly a shorter time to the clinic. This has in fact been the greatest driver for the development and implementation of platform processes. In some instances, especially for well-behaved mAbs, early-stage process development is almost completely eliminated by direct implementation of process platforms without any optimization. In this context, the early-stage mAb development efforts have been so streamlined that it is difficult to conceive further contraction of the timelines.

The other common driver of reducing the number of process steps makes economic sense for commercial processes only if it translates into a reduction in required manufacturing floor plan and that extra available space can be used for other productive activity. Advantages related to savings in raw materials, from elimination of a process step, are not as significant for commercial processes as previously described. Another common argument is that fewer process steps take less effort and time to develop. This is a fallacy, because in reality it takes more work and more time to develop a process that meets the same critical product quality attributes and impurity levels with one less step. From this perspective, reducing the number of process steps makes more sense for late-stage applications (phase III process development) when process development activities are not on the critical path to the clinic and there is ample time to develop and optimize the downstream process. Furthermore, savings in manufacturing floor plan and improvements in facility fit have the greatest impact in large-scale commercial operations. For early-stage applications, reducing the number of steps should not be a priority driver for new platform improvements. In fact, one could argue that a possible platform evolution may even consider increasing the number of steps, if that would significantly increase the robustness of impurity removal, increase the assurance of final product quality, and decrease process development time by decreasing the extra-effort required to achieve the same product quality and impurity levels with fewer steps.

In conclusion, one can argue that the tremendous efforts and investment by industry and academia on the research and development of new purification unit operations aimed at replacing Protein A and eliminating process steps, are likely misplaced. The reason is that variable costs are a very small fraction of the overall production costs, and decreasing the number of process steps comes at a cost of increasing development timelines and regulatory/quality risks.

Nevertheless, the development of new purification technologies can be attractive in particular with regards to expanding the global purification toolbox for the purification of non-mAb complex proteins, which are increasingly fed into the global biologics pipelines. For mAbs, a much more rational argument can be made to focus innovation efforts and resources on the development and implementation of new technologies and processes that lead to improvements in the most impactful components of COGS, namely on fixed cost, batch size, and productivity. To this point, continuous processing and intensification of disposable operations seem to be the right direction for the near future of mAb purification platforms.

References

[1] Kelley B. Industrialization of mAb production technology. The bioprocessing industry at a crossroads. mAbs 2009;1:443-52.

[2] Kelley B, Blank G, Lee A. Downstream processing of monoclonal antibodies: current practices and future opportunities. In: Gottschalck U (ed). Process Scale Purification of Antibodies. Wiley & Sons 2009:1-23.

[3] Ghose S, McNerney T, Hubbard B. Protein A Affinity Chromatography for Capture and Purification of Monoclonal Antibodies and Fc-Fusion Proteins: Practical Considerations for Process Development. In: Shukla A, Etzel MR, Gadam S (eds). Process Scale Bioseparations for the Biopharmaceutical Industry. Taylor and Francis 2006:463-90

[4] Bee JS, Stevenson JL, Mehta B, et al. Response of a Concentrated Monoclonal Antibody Formulation to High Shear. Biotechnol Bioeng 2009;103:936-43.

[5] Kelley B. Very large scale monoclonal antibody production: The case for conventional unit operations. Biotechnol Prog 2007;23:995-1008.

[6] Shukla AA, Hubbard B, Tressel T, et al. Downstream processing of monoclonal antibodies – Application of platform approaches. J Chromatogr B Analyt Technol Biomed Life Sci 2007;848:28-39.

[7] Shukla AA, Thommes J. Recent advances in large-scale production of monoclonal antibodies and related proteins. Trends Biotechnol 2010;28:253-61.

[8] Rathore AS, Godavarti R, Kumar V, Tugcu N. Evolution of the Monoclonal Antibody Purification Platform. Biopharm Intl. 2013;26:32-7.

[9] Glynn J, Hagerty T, Pabst T, et al. The Development and Application of a Monoclonal Antibody Purification Platform. Biopharm Int Suppl 2009;Mar:16-20.

[10] Trexler-Schmidt M, Sze-Khoo S, Cothran AR, et al. Purification Strategies to Process 5 g/L Titers of Monoclonal Antibodies. Biopharm Int Supp. 2009;Mar:8-15.

[11] Lui, HF, Ma J, Winter C, Bayer R. Recovery and purification process development for monoclonal antibody production. mAbs 2010;2:480-99.

[12] Schmidt M. Antibody degradation (disulfide reduction) in CHO production process. Antibody Development & Production Conference, Carlsbad CA, IBC 2009.

[13] Kelley BD, Tobler SA, Brown P, et al. Weak partitioning chromatography for anion exchange purification of monoclonal antibodies. Biotech Bioeng 2008;101:553-66.

[14] Aranha H. Viral clearance strategy for biopharmaceutical safety. Bio-production, Munich, Germany, 2004.
[15] Zapata G, Zhang R, Leininger E. Strategy for viral clearance and clearance validation for biologics manufactur. Bioprocess International, Berlin, Germany, 2005.
[16] Internatioal Conference on Haromisation of Technical Requirements for Registration of Pharmaceuticals for Human Use (ICH), Viral safety evaluation of biotechnology products derived from cell lines of human or animal origin Q5A1. http://www.ich.org/fileadmin/Public_Web_Site/ICH_Products/Guidelines/Quality/Q5A_R1/Step4/Q5A_R1__Guideline.pdf, accessed 13th May 2014.
[17] Center for Biologics Evaluation and Research (CBER), Points to Consider in the Characterization of Cell Lines Used to Produce Biologicals (1993). http://www.fda.gov/downloads/biologicsbloodvaccines/guidancecomplianceregulatoryinformation/otherrecommendationsformanufacturers/ucm062745.pdf, accessed 13th May 2014.
[18] Center for Biologics Evaluation and Research (CBER), Points to Consider in the Manufacture and Testing of Monoclonal Antibody Products for Human Use (1997). http://www.fda.gov/downloads/biologicsbloodvaccines/guidancecomplianceregulatoryinformation/otherrecommendationsformanufacturers/ucm153182.pdf, accessed 13th May 2014.
[19] Toueille M, Uzel A, Depoisier J-F, Gantier R. Designing new monoclonal antibody purification processes using mixed-mode chromatography sorbents. J Chromatogr B Analyt Technol Biomed Life Sci 2011;879:836-43.
[20] Shukla AA, Yigzaw Y. Modes of preparative chromatography. In: Shukla A, Etzel MR, Gadam S (eds). Process Scale Bioseparations for the Biopharmaceutical Industry. Taylor and Francis 2006:491-505.
[21] Franklin SG. Removal of aggregate from an IgG 4 product using CHT ceramic hydroxyapatite. BioProcess Int 2003;1:50-1.
[22] Muller-Spath T, Morbidelli M. Continuous chromatography for the purification of monoclonal antibodies. In: Gottschalck U (ed). Process Scale Purification of Antibodies. Wiley & Sons 2009:223-38.
[23] Mahajan E, George A, Wolk B. Improving affinity chromatography resin efficiency using semi-continuous chromatography. J Chromatogr A 2012;1227:154-62.
[24] Pollock J, Bolton G, Coffman J, et al. Optimising the design and operation of semi-continuous affinity chromatography for clinical and commercial manufacture. J Chromatogr A 2013;1284:17-27.
[25] Gottschalk U, Fruehholz-Fischer S, Reif O. Membrane adsorbers: a cutting edge process technology at the threshold. Bioprocess Int 2004;May 56-65.
[26] Ghosh R. Protein separation using membrane chromatography: opportunities and challenges. J Chromatogr A 2002;952:13-27.
[27] Vogel JH, Nguyen H, Giovannini R, et al. A New Large-Scale Manufacturing Platform for Complex Biopharmaceuticals. Biotechnol Bioeng 2012; 109:3049-58.
[28] Liu H F, McCooey B, Duarte T, et al. Exploration of overloaded cation exchange chromatography for monoclonal antibody purification. J Chromatogr A 2011;1218:6943-52.
[29] Glynn J. Process-scale precipitation of impurities in mammalian cell culture broth. In: Gottschalck U (ed). Process Scale Purification of Antibodies. Wiley & Sons 2009: 309-24.
[30] Giese G, Myrold A, Gorrell J, Persson J. Purification of antibodies by precipitating impurities using Polyethylene Glycol to enable a two chromatography step process. J Chromatogr B 2013;938:14-21.
[31] Oelmeier SA, Ladd-Effio C, Hubbuch J. Alternative separation steps for monoclonal antibody purification: combination of centrifugal partitioning chromatography and precipitation. J Chromatogr A 2013;1319:118-26.

7 Monitoring and Control of Processes and Products

7.1 Concepts and Technologies for Advanced Process Monitoring and Control

John Paul Smelko, Thomas Ryll, and Dethardt Müller

7.1.1 Introduction

Dating back from the 1980s to present, mammalian cell culture processes continue to evolve in order to maximize protein production, improve product quality, and accelerate development and commercialization. A wide range of techniques have been used from simple batch to fed-batch to the more complex perfusion culture as a way to optimize the process [1]. Fed-batch is most prevalent in cell culture processes today due to the operational ease, flexibility, and robustness it provides. As a direct result of ongoing optimizations over the last 30 years, cell culture process yields have increased 1,000-fold from 10 mg/L to 10 g/L today [2, 3]. The rise in protein titers was made possible by boosting cell mass while maintaining cell viability and productivity via finely tuned fed-batch processing [3, 4]. A new frontier to further improve cell culture process robustness and consistency is advanced process control. In this chapter, we review approaches to monitoring and controlling cell culture variables during culture production processes. Over the last 15 years, more sophisticated monitoring and control approaches have found their way into cell culture process optimization and control making use of automated sampling with metabolite analytics, new sensor technology, and multivariate data analysis [5-8]. Therefore, to improve process robustness and ease of control, integrating the use of advanced *in situ* sensors would eliminate the need for culture sampling and would allow modeling and monitoring parameters that go beyond pH and DO and include cell mass, culture metabolites, product concentration, and product quality.

In the following sections we will review aspects of automated sampling and in-line metabolite analysis, dielectric spectroscopy, fluorescence spectroscopy, near infrared spectroscopy, and Raman spectroscopy in context of monitoring and control of cell culture production processes. In particular, the availability and use of multivariate data analysis systems has enabled the acquisition and use of complex data streams, which in combination with spectroscopic sensors has enabled the building of process models that can depict a variety of process parameters including metabolites and cell mass.

Either direct measurements or modeling using internal or external sensors can be used to further automate and optimize today's culture production processes. As one outcome, we expect that in the near future culture sampling can be reduced significantly while maintaining improved culture control. Benefits will include: reduction of labor, improved awareness and oversight by operational staff, faster more informed investigations, reduced performance variability, improved consistency of product quality, and ultimately automation of process control aspects such as feeding, parameter changes, transfers, harvesting, and the like. In the end, we expect significant improvements to robustness and consistency of process performance and product quality through development and implementation of these advanced monitoring and control strategies.

7.1.1.1 PAT, QBD and Process Monitoring and Control Technologies

In 2004, the FDA created the process analytical technology (PAT) initiative in an effort to guide improvements towards control of specific critical quality attributes in which certain critical process parameters are monitored or even controlled real-time [9]. In doing so, quality will be scrutinized throughout the process resulting in a heightened level of understanding and process control, earlier detection of problems, and an environment for continuous process improvements. The PAT initiative has led to advancements in probe technologies, real time data collection, analysis software, and advances in process control.

Quality by design (QbD) encompasses a wider concept of building quality into the product through understanding of critical quality attributes and process parameters rather than by testing at the end of the process (ICH Q8 [10]). Today's expectation is that products are designed with quality aspects in mind and that manufacturing processes are well understood to enable control of critical quality attributes.

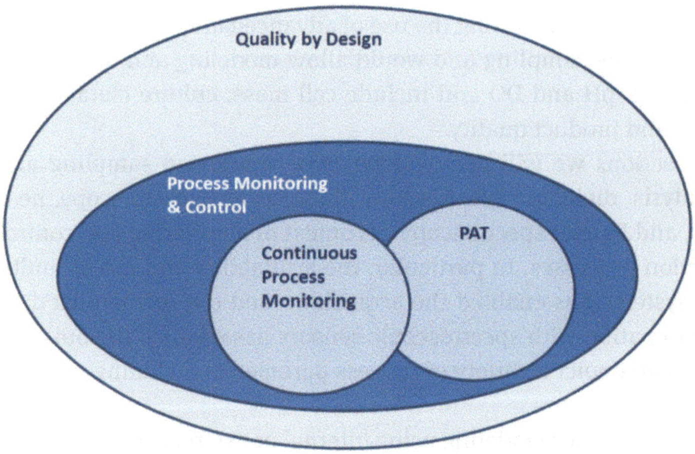

Figure 7.1.1: Depiction of QbD, PAT, and process monitoring and control spaces.

Concepts and applications of QbD, PAT and process monitoring and control are often mentioned and used in similar context. Process monitoring and control is fundamental to PAT and QBD (see Figure 7.1.1 for a depiction of associated relationships). We want to clarify here, that this chapter is focusing on technologies and approaches being developed for the next generation of process monitoring and control. In this respect, such technologies will offer opportunities for improved on-line data acquisition and process guidance and thus can and will be used as PAT tools. However, we are not focusing on PAT and its specific opportunities and challenges such as parametric release. When appropriate, we will mention applicability of monitoring and control technologies discussed towards PAT like process control [11]. Process control and PAT, if applicable, are fundamental concepts within QBD. Aspects of QBD and associated process characterization are detailed in Chapter 7.3.

7.1.2 Process Monitoring and Control

7.1.2.1 History

Process parameters are monitored and controlled during processing to guide the process and maintain control within acceptable parameter levels. For cell culture processes, monitoring is historically done either by use of internal sensors for temperature (T), pH and dissolved oxygen (DO), or through sampling and off-line analysis on a daily basis. Typical off-line acquired parameters include metabolites such as glucose, lactate, ammonia and amino acids, product concentration and cell counts, as well as culture viability. Temperature, pH, and DO are typically controlled using simple feed-back control loops that include the sensor, a controller, and control elements such as valves, pumps, heating or cooling elements (Figure 7.1.2). Such feedback control is achieved using set points and controller limits (often called "dead bands"). For example, if the controlled parameter is within the control limits, no action is taken. Conversely, if a control limit is reached, action is taken by activating a response element resulting in changes to gas sparge mixtures, pumping acid or base, and the like. In most cases, proportional-integral-derivative controller logic is used (PID control) [12].

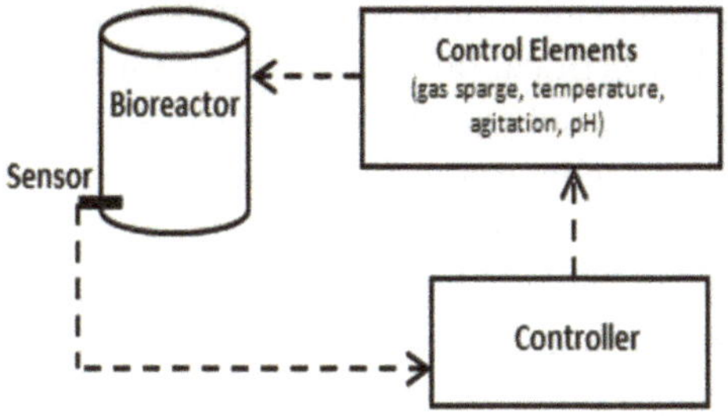

Figure 7.1.2: Simple feedback control loop used for controlling elements such as temperature, DO, and pH.

The use of additional sensors, such as for dielectric spectroscopy or other laser-based spectroscopic techniques, demands further processing of the signal before action can be taken. As a result, such technologies are primarily used for monitoring of production processes. The additional information such sensors can deliver can greatly improve troubleshooting and can help with optimization. Use of multivariate data processing enables the building of process models and the resulting algorithms can ultimately be used for feedback control. This requires implementation of so called "middleware" that enables processing of complex data streams (spectra) and translation into signals that can be fed back into process control systems. This integration may not be trivial since when applied to GMP processes will require CFR Part 11 compliance. A simple depiction of such control loops is shown in Figure 7.1.3.

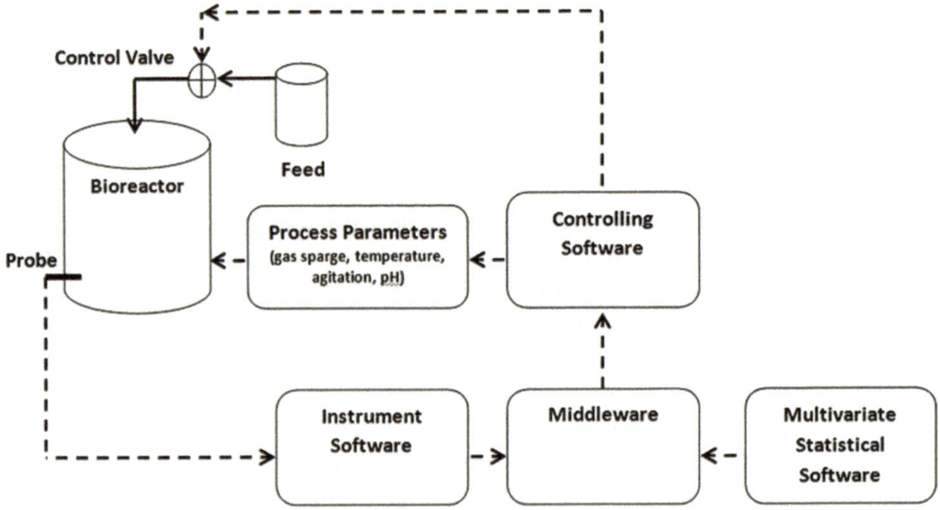

Figure 7.1.3: Advanced feedback control loop using a spectroscopic sensor and complex data streams.

7.1.2.2 Data Analysis and Model Building

Data Analysis

Historically, a large number of process parameters are monitored and analyzed individually (univariate). It's not uncommon to have critical process decisions based off univariate data. For example, cell growth is monitored and process decisions may be made based on cell mass limits or just based on predetermined time limits.

The data generated from a spectroscopic sensor such as NIR or Raman during a mammalian cell culture run can be immense. Organizing, interpreting, and correlating these data can at first seem overwhelming. A cell culture broth contains complex mixtures of nutrients, metabolites, culture additives such as polyols and antifoam, waste products, host cell proteins, and the recombinant product in the presence of live and dead cells. This complexity can be simplified by the use of chemometrics. Chemometrics is the data driven extraction of relationships within complex chemical systems. Chemometrics can further be described as techniques that can be used to correlate spectral changes with specific constituent concentrations through the use of multivariate calibrations. Multivariate analysis can be used to correlate multiple peaks within the spectra to one or multiple constituents at once [13-15].

Principal component analysis (PCA), invented in 1901 by Pearson, is one of the more commonly used chemometric techniques. PCA consists of multiple principal components in which the first component captures the maximum variance in the data and each succeeding component has the highest variance possible under the constraint that it is orthogonal to the preceding component. Analysis of each component via scores or loading plots leads to an improved understanding of the origin of performance variation. This technique is used widely in the industry in multiple areas, one of which is evaluating batch-to-batch performance or raw material screening of multiple lots. However, a different chemometric technique is better equipped to handle *in situ* monitoring and prediction of a mammalian cell culture bioreactor processes [16, 17].

Partial least squares (PLS) was introduced by Hermon Wold a Swedish statistician, and then developed with his son, Svante Wold. PLS finds the linear regression model by projecting the predicted variables versus the observed variables to a new area. PLS regression is most widely used in the field of chemometrics for analysis of complex spectral signatures with a variety of mediums. PLS analysis can be used to predict process outputs and these predictions can then be used to control the process using a feedback algorithm based off a key constituent model derived from a spectroscopic sensor inserted into the bioreactor. Soft independent modeling of class analogy (SIMCA), developed by Wold and Sjostrom in 1977, uses PCA and PLS techniques in combination with today's advanced multivariate statistical software, which creates a powerful tool that is currently being used in the industry [18].

These statistical techniques, along with many others, help simplify the data analysis process by evaluating the information in a multivariate format, which allows for improved correlations and the development of accurate models describing complex environments and interactions within a cell culture process.

Model Building

The process of building multivariate models for mammalian cell culture constituents using spectroscopy requires multiple steps along with an iterative approach. Building a model can vary in many ways depending on the data set, the constituent being modeled, and the application. However, a general approach for modeling cell culture process constituents *in situ* can be described as follows.

1. During cell culture run:
   - Collect spectra using analyzer software.
   - Collect cell culture (reference) samples and analyze off-line for growth, metabolite, and product quality.
     - Multiple samples per day are desired to build stronger models within a shorter time frame.
     - Take samples before and after feed additions (minimum). Auto sampling systems can be used to take upwards of 8 to 12 samples a day.
2. After run is completed:
   - Match spectra collected during run to offline reference samples based on timestamp of sample (using Excel, for example).
3. Model generation:
   - Input data into statistical software (MatLab or SIMCA-P+).
   - Apply preprocessing techniques (2 examples below, others can be applied):
     - 1st derivative Savitsky-Golay smoothing,
     - Standard normal variate (SNV).
   - Apply specific region selection, if applicable.
   - Remove outliers.
   - Perform statistical evaluation using software.
   - Create calibration model(s) per constituent.
4. Model validation:
   - Apply calibration model(s) to independent run.
   - Evaluate model accuracy using root mean square errors of estimation (RMSEE) and root mean square errors of prediction (RMSEP).
5. Repeat steps 3 and 4 for each constituent.

Once the modeling process is complete, application of the model can then take place. It is during this process that model refinement may occur over a period of months or years. Model modification may be required if the process boundaries shift or the current model limits are routinely exceeded. When limits of the current model are exceeded, the statistical probability that the model will be accurate diminishes greatly. Depending on how many runs are used in the calibration model and the variability within those runs, model modifications may also be required if there are unforeseen interactions that were not captured within the original calibration data set. Finally, incorporation of new scientific and chemometric understanding beyond that of the existing model may be desired to improve the model further. A model in general is never set in stone unless it is built upon a very stable process or a large data set that spans a wide range of variability. Therefore, continued model maintenance is always preferred to ensure proper batch-to-batch accuracy, repeatability and reliability.

Offline analytical equipment accuracy is essential for robust models, since the model is built by comparing online spectra to offline analytical results obtained from sampling the bioreactor, i.e., the model can only be as good as the data it is built with. This key principal is paramount and must be considered strongly. In the biotechnology industry, different offline analytical equipment is typically available for determining key constituent concentrations within a bioreactor. Oftentimes, the equipment that provides the highest accuracy is expensive to run and is in high demand, causing competition for this resource with other parts of the business. Reliability of the equipment can vary if not closely monitored and well maintained. Therefore, one must consider a number of variables, including time to analysis, cost per sample, accuracy, precision, linearity, specificity, range, equipment availability, throughput, ease of use, and maintenance (down time) when selecting the right off-line analytical instrument. Once the instrument is chosen, the final step is to incorporate the analytical equipment error into the model, which will help define the models accuracy. For example, if the analytical equipment can measure glucose within +/- 0.2 g/L then the resulting model can only be as accurate. Offline instrument repeatability, reliability, accuracy and precision are often times overlooked when initially exploring the use of spectroscopy for cell culture monitoring. However, this is one of the key pillars (reference data) that must be reliable if modeling is truly intended to be part of a validated, GMP, FDA regulated system.

7.1.2.3 Culture Feeding Control

Nutrient feeding in fed-batch processes typically consist of bolus additions of concentrated feeds, either daily or continuous, or at some predefined feeding schedule. A predefined feeding schedule is unable to consider process immanent variability originating from aspects such as raw material lot-to-lot variation, scale-up effects, variable seeding density, or cell age. Oftentimes some of these factors are not observed during development due to limited number of runs and short timelines but rather during a manufacturing campaign of an approved or late stage drug. Within the last several years, ongoing research has led to a new form of feeding that may bridge the gap and provide the next evolution of feeding strategies for the future. This type of feeding is called dynamic feeding. Dynamic feeding is based on real time nutrient demands of the process. Rather than feeding bolus shots where the nutrient load will spike and come down over the course of a day or days, the exact amount of nutrient is fed to maintain proper cell health, proliferation, and in some cases improvement in product quality attributes. In order to accomplish dynamic feeding, key metabolites and cell mass values need to be ascertained from the bioreactor at hourly intervals (4 to 6 h) rather than daily as the industry standard. To make this possible certain technologies had to evolve. Automated sampling is one such technology that has made advancements over the past decade with the improvement of microcomputers, software algorithms, miniaturization of hardware, and sterile sampling techniques. By using such a technology, metabolites can be monitored near real time allowing for future evolution of fed-batch processes in the form of dynamic feeding and feedback-based process control [5, 6].

Dynamic Feeding

Using a variety of feeding algorithms, dynamic feeding can be used to tailor the nutrient delivery in quantities specific to the cell's needs for optimal growth, productivity, and product quality [19, 20]. Dynamic feeding was publicized as the earliest 2000 by Sauer et al. [21] and later by Gong et al. (2006) [22] and Kuwae et al. (2005) [23]. Their efforts were mainly focused on achieving low levels of glucose and glutamine in an effort to reduce byproduct accumulation during the run, specifically for lactate and ammonia. This was accomplished by investigating cellular growth and nutrient uptake by manually sampling the reactor more frequently during a run, performing a calculation, and then feeding. This process was entirely manual and would not be considered a manufacturing friendly approach given the operational complexity and number of manual steps required.

Automated feeding was studied by Zhou (1994) [24] and Zhou et al. (1995) [25], using a different approach correlating oxygen uptake rate (OUR) with glucose consumption using the assumed stoichiometric ratio between glucose and oxygen. The feeding strategy was used to maintain glucose at a low level. One of the intended benefits of this approach was a reduction in manual sampling, due to the development of the OUR measurement. Nevertheless, periodic manual sampling was still required for glucose in order to update the glucose to oxygen ratio as it changed throughout the run.

Future efforts towards automation were described by Chee Furng Wong et al. (2005) [26] and Lee et al. (2003) [27] where the advancement of an automated aseptic sampling loop was used to monitor and then feed the reactor to maintain low glutamine levels. In 2010, dynamic feeding was used by Huang et al. [3] in the form of integral viable cell concentration to optimize feeding based on cell mass, resulting in titers exceeding 10 g/L. Automated sampling was successfully demonstrated by Bayer Technologies' BaychroMAT® system during a long-term perfusion bioreactor run with mammalian cell culture to monitor cell density, viability, glucose, and lactate concentrations [28]. Recently, Lu et al. (2013) [6] published findings of an advanced dynamic feeding method incorporating an automated sampler integrated with an on-line metabolite and cell density analyzer, as well as a dielectric spectroscopy probe to predict viable cell density (VCD). Two distinct dynamic feeding methods were used, a predictive method and a feedback-based algorithm. A chemically defined feed medium was optimized based on an iterative approach. During this process, ratios of feed components were based off a key metabolite indicator, asparagine. Since asparagine could not be measured via the on-line metabolite instrument, another indicator, glucose, was used because the ratio between asparagine and glucose consumption rates remained constant. Key findings from the study demonstrated a twofold increase in product titer with a similar product quality profile using the dynamic feed method based off glucose. This was accomplished with little manual intervention letting the automated software and hardware perform the sampling, analysis, and feedback control.

Dynamic feeding with feedback control allows for automatic adjustment of the feed rate according to culture performance. This approach allows the culture to always receive the right amount of nutrients at the correct time and avoids typical fed-batch over and under feeding due to standard predetermined feeding practices. Therefore, it can be suggested that dynamic feeding can positively impact culture performance and correct some of the raw material, process and product variability seen during extended manufacturing campaigns. If used correctly, literature has also shown that not only can a more consistent process be achieved but also a process that outperforms typical fed-batch processes in productivity and product quality [3-6, 26].

7.1.3 Culture Monitoring Technologies

7.1.3.1 Automated Sampling and On-line Metabolite Monitoring Hardware

Dynamic feeding strategies can be difficult to implement if performed manually, just recently an automated approach has been demonstrated at lab scale. But can or should this approach be used at manufacturing scale [6]? Before this question can be answered, careful consideration of how automated sampling systems work including the associated on-line metabolite instruments they are integrated with must be reviewed.

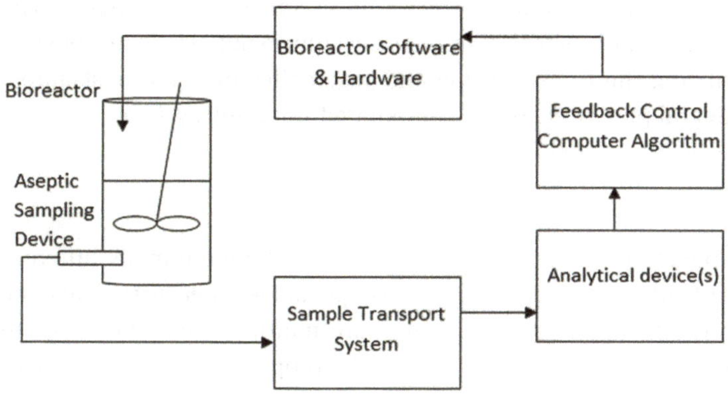

Figure 7.1.4: Automated sampling and on-line metabolite process flow diagram.

Feedback-based control via automated sampling and on-line metabolite analysis can be separated into 5 critical parts: aseptic sampling device, sample transport system, analytical device(s), feedback control computer algorithm, and bioreactor software and hardware to execute the feed(s) (Figure 7.1.4).

Aseptic Sampling Device

The aseptic sampling device has been the most challenging from a design and robustness standpoint, but within the past 5 years multiple vendors have developed robust designs. These valves normally are sterilized in one of 2 ways; either by direct steam injection followed by sterilized air to cool the valve chamber or via alcohol based solution followed by a saline flush. Some sample valves can pull upwards of 6 to 12 samples a day depending on how many devices are connected into one analyzer unit. Typical sample volumes can range from 5 to 50 ml depending on instrument design. The valve is the single most important part to an automated sampling system. If bioreactor sterility cannot be consistently maintained in a robust manner when used in an automated process, the technology has no use in manufacturing or development labs.

Sample Transport System

The transport system's main function is to move the culture from the aseptic sampling device, in a controlled manner, to the on-line metabolite instruments connected to the system. These on-line metabolite systems can either be integrated with the sample transport system or can be separate devices. There are multiple ways to transport a sample and each system accomplishes this in its own unique way. Some examples of sample transport are: air to push, syringe to pull, or combination acting at the same time. When desiring gas composition or pH measurements, careful consideration must be made to properly avoid degassing of the sample during transport. The sample transport system not only acts to transport the culture but also is responsible for cleaning and sterilizing the transport line and assemblies pre- and postsample thus it is a key and integral part of the overall automated sampling system.

Analytical Devices

Depending on the application and constituent of interest, there are many different analytical devices to consider. Some devices are very specialized measuring only one constituent but at very high accuracy, others measure multiple constituents spanning various ranges and levels of accuracy. Choice of the proper analytical equipment for use should depend on the following: accuracy, precision, reliability, and maintenance; all with respect to the process constituent or lever that is to be controlled. An ideal choice satisfies all criterions, but this is not always possible. Alternatively, multiple instruments may be required. For example, high-performance liquid chromatography (HPLC) can be used for titer measurement, and a metabolite analyzer for glucose and lactate concentrations. These analytical instruments and the data generated from them must be robust and fit within the GMP and FDA regulated guidelines for PAT and process control [9]. Also, the values outputted often determine the batch performance so its paramount these values are accurately reproduced and the instruments perform consistently and reliably. Therefore, the analytical device(s) are one of the most scrutinized parts of the automated sampling and on-line metabolite monitoring system.

Feedback Control Computer Algorithm

The algorithm used during feedback control can vary widely depending on the application. Some published applications involve tightly controlling key nutrients that lead to either higher productivity or potentially more controlled product quality attributes [3-6, 26]. In general, nutrient concentrations are analyzed, key nutrient consumption rates are calculated, and a feed volume or rate is determined. Lu et al. (2013) [6] provide a detailed example using such equations.

Bioreactor Software and Hardware

Bioreactor software can consist of a multitude of different platforms. Two of the most popular platforms used in the biotech industry are DeltaV and programmable logic controller (PLC). These programs can take a series of commands or inputs from an outside source (analytical instrument for example). Normally, the feedback control computer algorithm is coded in the bioreactor software. Therefore, the output from the analytical device is sent to the bioreactor software and taken as an input and fed into the feedback control algorithm and an output is generated. This output can be of many different forms depending on the application. Classically, a pump or control valve will be actuated to allow a certain nutrient of choice to enter the bioreactor at a predetermined rate, quantity, and time interval as set by the algorithm. When all software and hardware is working in harmony the system is self-managed and only requires minimal human intervention to execute the specified task.

Case Study

The following case study is based on early feedback control methodology studied by Tsang et al. at Biogen Idec (US 8,318,416 B2) [5]. Evaluated was the use of an automated sampling and feedback control system to monitor surrogate nutrients and optimize productivity of a CHO cell culture process. Some but not all findings are discussed below.

Media and feed optimization studies suggested glutamate or glucose could be used as a surrogate marker for automatic addition of complex feed without changing media formulations. Glutamate was found to be consumed in stoichiometric ratio as other amino acids and glucose was used as an indicator of cell growth and hence nutrient trends. The hypothesis was, by utilizing the automated monitoring and feedback control capabilities of the current technology, glutamate and glucose could be tightly controlled at specific concentrations for the duration of the run. Percentage of complex feed fed was based off this hypothesis in which different combinations of glutamate and glucose concentration set points were studied. Cell culture processes evaluated in this study used varying types of feed material for complex feed. Some processes used hydrolysate and others used chemically defined feeds.

The following equipment was used for automated sampling, monitoring, and feedback control: NOVA Bio-Profile 400 Analyzer equipped with NOVA online autosampler system (NOAS), Applikon 5L glass vessels, and separate pumps for feeding. The NOVA (NOAS) was scheduled to sample every 3 h. This frequency was chosen (via calculation of sample volume) to ensure sufficient nutrient level control and to reduce volume loss from sampling. The NOVA BDM computer offers simple feedback control, and was programmed to maintain a chosen nutrient at desired level by addition of feed when needed. Each nutrient feed addition was calculated to boost the chosen nutrient component by 0.5 unit (0.5 mM glutamate or 0.5 g/L glucose). The software was also programmed to resample 30 min after feed. If the nutrient level was still below the threshold, the software will signal the pump for an additional dose.

Study results showed total feed percentage when maintaining the specified glutamate concentration throughout the run resulted in three times more nutrient feed being consumed (Figures 7.1.5 and 7.1.6). In addition, titer increased by 35 % within the normal culture duration and up to 60 % during extended culture duration (Figure 7.1.7). Extra feed volume, required to support growth and productivity, indicates certain key metabolites were being depleted. By using automated feedback control of glutamate, the culture did not experience the normal fluctuations or temporary depletion in nutrient levels as would be typical with some fed-batch processes. Maintaining glutamate at a predefined level can be accomplished at manufacturing scale by either a daily calculation based off offline glutamate reading (manual process), using automated monitoring and feedback control (automated process), or by using development data generated with automated feedback control to improve fed-batch process (hybrid approach). The hybrid approach was shown to deliver the same amount of antibody titer gains as the automated feedback approach (Figures 7.1.8 and 7.1.9).

Future advances in feed components and media development have resulted in reduction in feed volumes and an increase in protein titers since initial work in 2007 by Tsang et al. (unpublished).

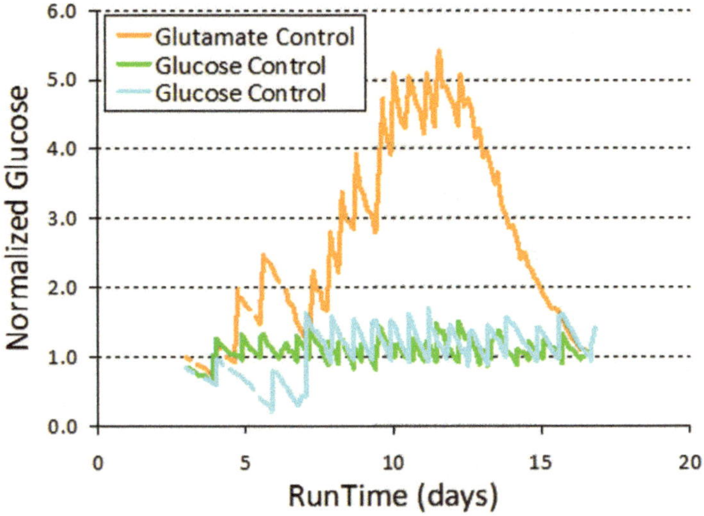

Figure 7.1.5: Normalized glucose feed versus time. Total glucose feed addition is five times higher when glutamate is controlled at constant level throughout process.

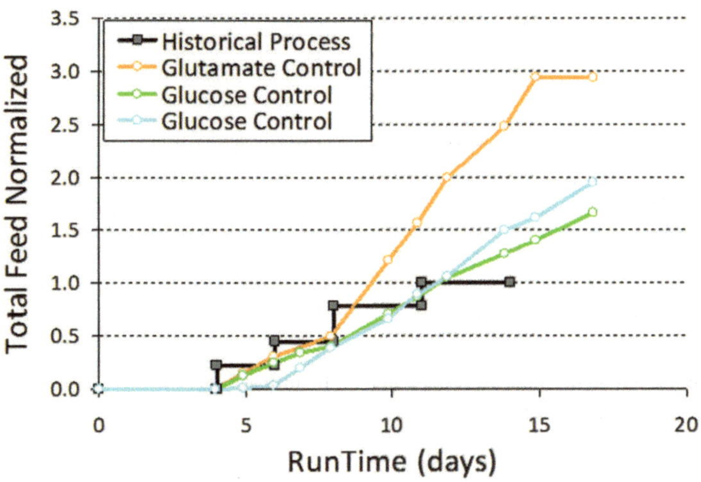

Figure 7.1.6: Total feed comparison versus time for control strategies tested.

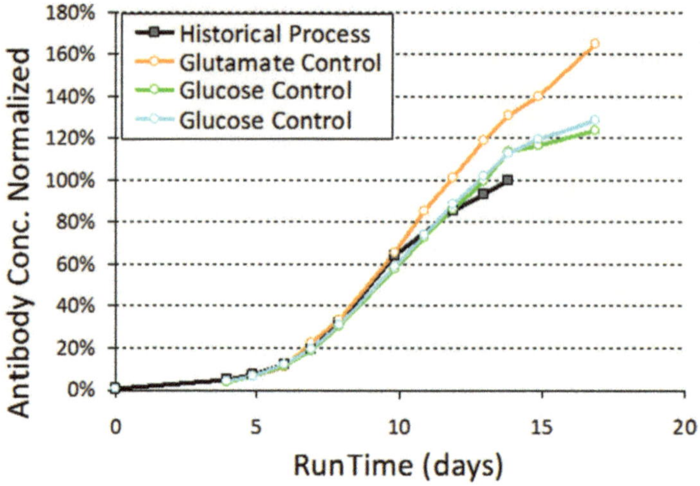

Figure 7.1.7: Antibody concentration normalized versus time.

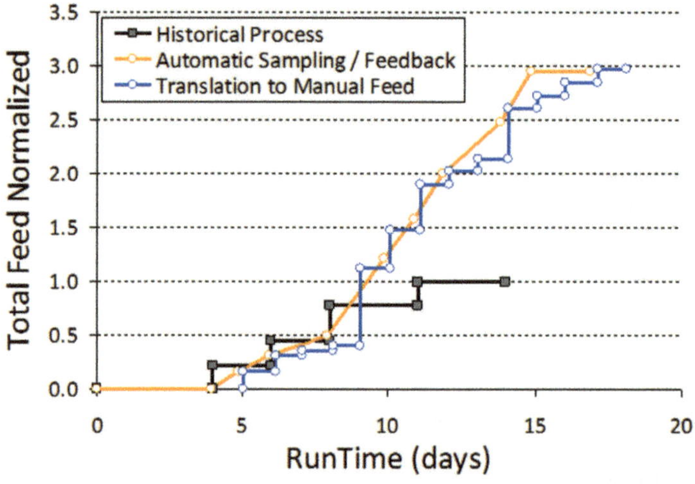

Figure 7.1.8: Comparison of total feed normalized versus time for historical, automated, and translated manual feed processes.

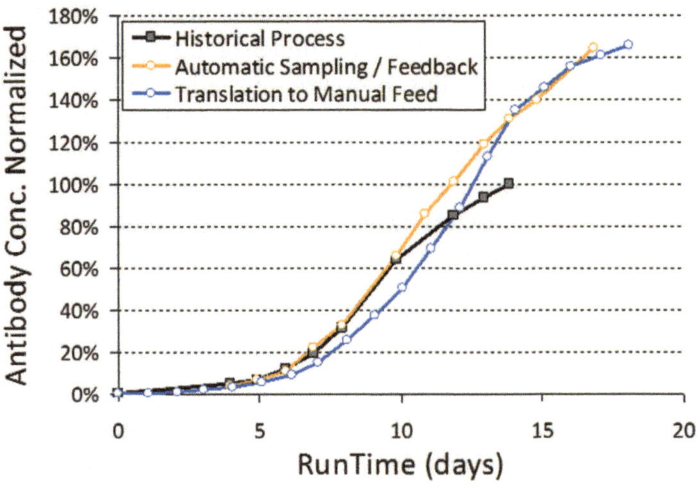

Figure 7.1.9: Comparison of antibody concentration normalized versus time for historical, automated, and translated manual feed processes.

Additional Uses

Simple utilization of just the automated sampling system can result in a reduction or elimination in manual sampling by the scientist or operators during a cell culture run. The benefits of this approach are that it offers a reduction in manpower and can provide more data with fewer operators than the traditional once a day manual sampling protocols. This method also provides the opportunity to increase sampling frequency and analysis during a product run. If used properly, an automated sampling system can generate upwards of 6 to 12 times the amount of data per day, which can lead to further understanding of process parameter impact on performance and product quality and can aid process optimization and robustness. The additional data can also be coupled with other technologies such as spectroscopic analysis via Raman or near-infrared (NIR) to generate more robust models in a shortened amount of time which could potentially lead to cost savings in time and labor during the model development phase [14].

Future of Automated Sampling and On-Line Metabolite Monitoring

Looking into the future, automated sampling definitely has its place in bioprocesses. Given the benefits of gathering data (cell mass, metabolites, titer) at shortened time intervals certainly provides benefits in the form of process understanding and control and with added paybacks of reduced time and labor. Now, whether automated sampling and on-line metabolic monitoring will become mainstream in manufacturing is yet to be determined. There are multiple drivers to use this technology in a development setting to fine tune processes and gain scientific understanding through optimized feeding routines such as dynamic feeding. However, translating the equipment and process flow into a manufacturing and regulated environment may prove more challenging. The simple fact that having multiple moving parts (aseptic sampling device, sample transport system, analytical instruments) to complete the task rather than one (i.e., spectroscopic probe in reactor) increases the risk for failure and thus the largest hurdle for the technology to overcome.

7.1.3.2 Dielectric Spectroscopy (Radio-Frequency Impedance Spectroscopy)

Background

Since the introduction of the 'biomass monitor' by Davey and his colleagues in 1993 [29], dielectric spectroscopy, also called radio-frequency impedance (RFI) spectroscopy, found broad acceptance as a PAT tool for monitoring viable mammalian cells in biotechnological processes from laboratory into industrial scale. Making use of the dielectric properties of viable cells, dielectric spectroscopy measures the impedance of a medium exposed to an alternating electrical field as a function of frequency. Viable cells suspended in a conducting ionic solution such as cell culture medium thereby interact with the alternating electrical field like tiny capacitors. They build up a temporarily separated charge (i.e., polarization) due to ion movement within the cytoplasm, which is bounded by the nonconductive, double-layered cell mem-

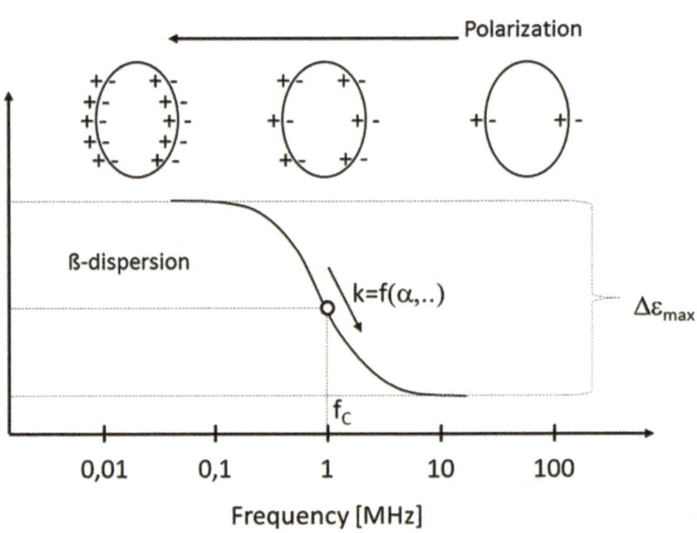

Figure 7.1.10: β-dispersion of dipolar cells in an alternating electrical field.

brane. The degree of polarization thereby strongly depends on the frequency of the alternating electrical field. At frequencies below 0.1 MHz complete cell polarization takes place and therefore the impedance of the cell suspension, the so-called permittivity, is at maximum. In contrast, at frequencies exceeding 10 MHz permittivity is at minimum due to insufficient time available for cell polarization. The characteristic drop of permittivity within the frequency range between 0.1 and 10 MHz, the so-called β-dispersion, is characteristic for dipolar elements such as viable cells. It is defined by tree parameters, the difference in permittivity between high and low frequencies ($\Delta\epsilon_{max}$), the characteristic frequency at $\Delta\epsilon_{max}/2$ (f_C) and the slope of the tangent through the point of inflexion (k) which is a function of the so-called Cole-Cole factor [30] (Figure 7.1.10).

According to the following Equations (7.1.1) to (7.1.3),

$$\Delta\varepsilon_{max} = \frac{9}{4} \cdot r \cdot P \cdot C_M \tag{7.1.1}$$

$$P = \frac{4}{3} \cdot \pi \cdot r^3 \cdot N \tag{7.1.2}$$

$$f_C = \left(2 \cdot \pi \cdot r \cdot C_M \left(\frac{1}{\sigma_i} + \frac{1}{2\sigma_m}\right)\right)^{-1} \tag{7.1.3}$$

the permittivity increment $\Delta\epsilon_{max}$ is a function of the cell radius r (m), the specific membrane capacity of the cell C_M (Fm^{-2}) and the biovolume P (–), which is proportional to the cell concentration N (m^{-3}). The characteristic frequency f_C is related to r and C_M and additionally depends on the intracellular conductivity σ_i (Sm^{-1}) as well as the conductivity of the medium σ_M (Sm^{-1}). Furthermore, it is presumed that the Cole-Cole factor α, is a direct measure for homogeneity of the cell population, as a widening of cell size distribution causes a flattening of the β-dispersion curve. According to Figure 7.1.11, permittivity increases with increasing biomass (cell concentration, cell size), whereas the characteristic frequency f_C may vary due to changes in cell size or conductivity. A more detailed description of the basic principles of dielectric spectroscopy can be found by different authors [29, 31].

Application

Compared to nonviable cells, the permittivity of viable cells with an intact membrane is quite high. Therefore, dielectric spectroscopy, unlike many other optical techniques used for biomass monitoring, is less susceptible to particles (e.g., gas bubbles, dead or leaking cells, cell debris, microcarrier) and fouling effects. According to this, die-

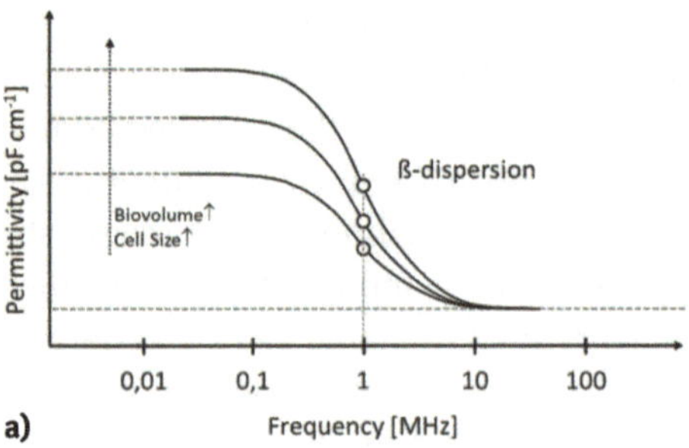

a)

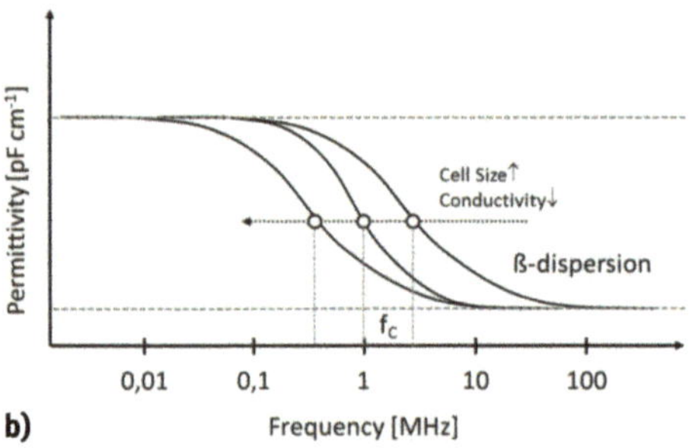

b)

Figure 7.1.11: Impact of **a)** biovolume, **a, b)** cell size, and **b)** intra-/extracellular conductivity on β-dispersion.

lectric spectroscopy has been widely explored and successfully applied for biomass monitoring of both microbial cultures (bacteria, fungi, yeast) and mammalian cell cultures. A summary of selected reports is given in Table 7.1.1.

Industrially relevant biomass monitoring systems based on dielectric spectroscopy are available. In most of the studies either the Biomass Monitor (Aber Instruments) or the Fogale Biomass System (Fogale) are used. Both suppliers offer reusable probes for standard operation in a stainless steel or glass bioreactor as well as disposable probes for use in disposable bioreactor systems [7, 44].

Case Study: Dielectric Spectroscopy-based Closed Loop Control of Biomass in a Perfusion Process

Aim of the study was to use dielectric spectroscopy (Fogale Biomass Monitor) not only as a tool for monitoring viable CHO cell concentrations in a perfusion process, but to further set up an automated loop for controlling the viable cell concentration at a designated level.

Table 7.1.1: Selected reports on successful application of dielectric spectroscopy for biomass monitoring.

Biological System	Cultivation	Target	Reference
Bacteria			
Streptomyces	Mycelial	Biomass	Fehrenbach et al. [32]
Escherichia coli	Suspension	Biomass	Clementschitsch and Bayer [33]
Yeast			
Saccharomyces cerevisiae	Suspension	Biomass	Harris et al.
Saccharomyces cerevisiae, Pichia pastoris	Suspension	Biomass	Fehrenbach et al. [32]
Unspecified	Suspension	Death kinetic	Patel and Markx [35]
Saccharomyces cerevisiae	Immobilized	Biomass	Salter et al. [36]
Saccharomyces cerevisiae	Suspension	Cell cycle	Gheorghiu and Asami [37]
Fungi			
Monascus spec.	Mycelial	Cell concentration	Krairak et al. [38]
Mammalian Cells			
Hybridoma	Suspension, immobilized	Viable cell concentration	Noll and Biselli [39]
CHO	Microcarrier, fixed-bed	Viable cell concentration	Ducommun et al. [40]
CHO	Suspension	Viable cell concentration	Cannizzaro et al. [41], Opel et al. [42]
CHO	Microcarrier	Viable cell concentration	Carvell and Dowd [43]

Prior to application in perfusion, the general suitability of the Fogale Biomass Monitor towards real-time, *in situ* monitoring of viable CHO cells was tested in simple batch processes using a recombinant CHO cell line. Thereby, the permittivity increment within a constant frequency range between 1 and 10 MHz ($\Delta\varepsilon_{fogale}$), which is nearly $\Delta\varepsilon_{max}/2$, was continuously measured and used for correlation with the so-called biovolume. The biovolume is directly proportional to the viable cell concentration, provided that cell size remains constant. In several repeated batch cultivations (exemplarily shown for one batch cultivation in Figure 7.1.12) a good correlation between permittivity and viable cell concentration could be demonstrated during growth and early decline phase. During late decline phase, where progressive cell death was observed, the permittivity signal drifted apart, as it could also be observed for the characteristic frequency f_C (not shown).

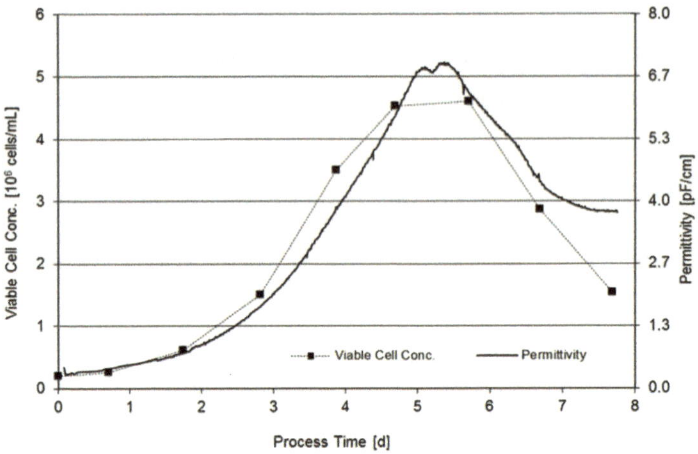

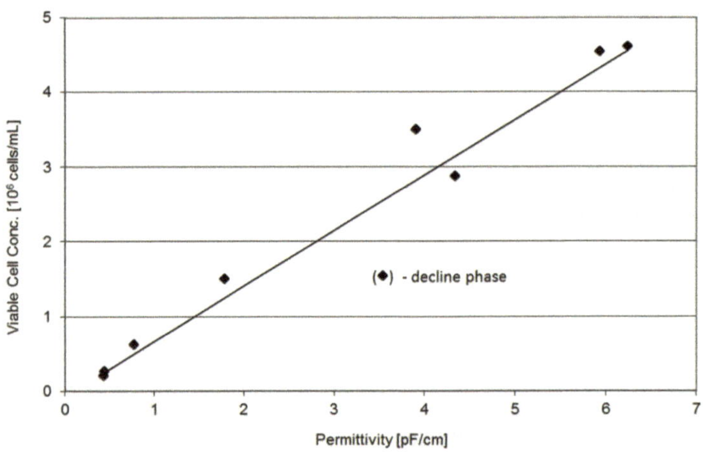

Figure 7.1.12: Progression of viable cell concentration and permittivity in a standard batch cultivation (left). Correlation of viable cell concentration and permittivity (right).

After positive evaluation of the Fogale Biomass Monitor a perfusion process was set up, using the same recombinant CHO cell line and default process parameters. Cell retention was achieved by alternate tangential flow (ATF) filtration (Refine Technologies). As filtration systems like the ATF operate at maximum cell separation efficiency (up to 100 %), biomass bleeding was activated in addition to perfusion to maintain cell concentration on a constant level. In a first step, dielectric spectroscopy was used only for biomass monitoring, while biomass bleeding was performed manually and discontinuously according to off-line biomass measurements. It could be demonstrated, that even at higher viable cell concentrations of about 20 million cells per mL correlation with the real-time permittivity measurements was excellent (Figure 7.1.13). In contrast to the batch process, where signal drifting occurred during the decline phase, both characteristic frequency f_C and tangent slope k remained constant under steady-state conditions in perfusion.

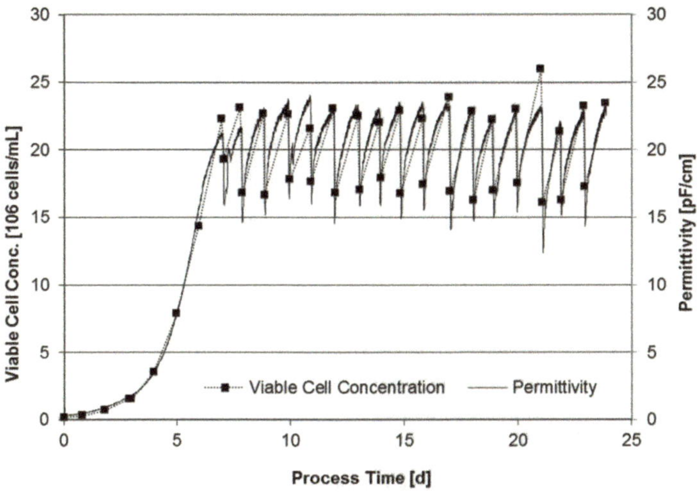

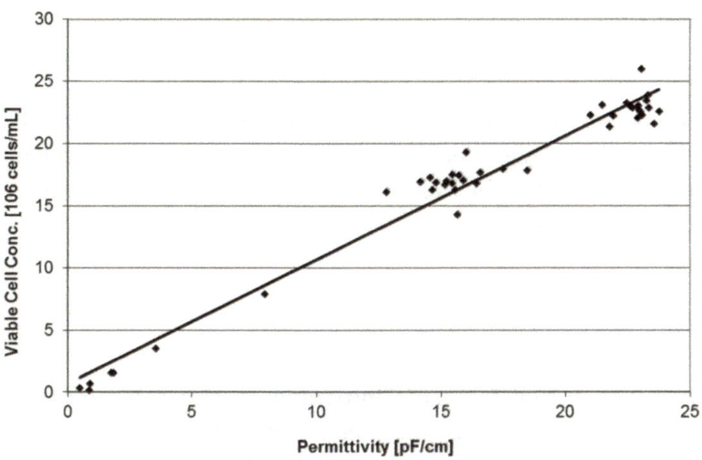

Figure 7.1.13: ATF-based CHO perfusion process applying dielectric spectroscopy for biomass monitoring; manual bleeding regime (left). Correlation of viable cell concentration and permittivity (right).

In a second step, permittivity was integrated as input signal into a closed-loop on/off controller, which automatically activated or inactivated bleeding whenever the permittivity signal exceeded or fell short of the controlled band. Thus, cell concentration could be controlled at set point within a 10 % band (Figure 7.1.14) for most of the time. Recalibration of the biomass monitor, which was easy to handle, was necessary after 15 and 20 days due to changes in the signal ratio between permittivity and cell concentration.

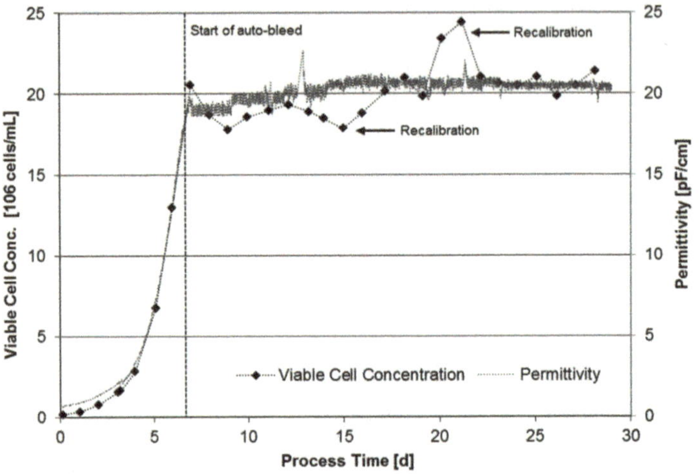

Figure 7.1.14: Control of viable cell concentration in an ATF-based perfusion process using the dielectric spectroscopy signal (permittivity) for automated biomass bleeding. Recalibration after 15 and 20 days due to changes in the permittivity/cell concentration ratio.

In conclusion, dielectric spectroscopy was successfully applied to monitor viable cell concentration in a perfusion-based high-density process. Furthermore, the excellent correlation between permittivity and viable cell concentration enabled to implementing a simple closed-loop biomass control.

7.1.3.3 Fluorescence Spectroscopy

Background

Fluorescence spectroscopy is one of the most intriguing spectroscopic methods gaining more and more importance in mammalian cell-based bioprocess monitoring. The comparably high sensitivity enables to detect significant changes in the emission spectra, even when only slight variations in the monitored process occur. This in turn requires reinforced calibration and chemometrics to extract the relevant information from the process. However, several studies show that it is well worth the effort since, unlike other spectroscopic tools, fluorescence spectroscopy offers a two-dimensional analysis by measuring fluorescent emission spectra at different excitation wavelengths.

Fluorescence is a luminescent phenomenon characterized by a nonthermal dissipation of energy in form of light emission. Upon excitation with UV or visible light, nonbinding electrons of a molecule are raised to a higher energetic level. This quantum mechanical effect can be described in terms of a transition from a singlet ground state to a singlet excited state at different vibrational levels (Figure 7.1.15a), where nonbinding electrons of a molecule remain paired. After nonradiative deactivation to the vibrational ground state, the molecule relaxes to the stable singlet ground stage either by nonradiative or radiative deactivation. In case of nonradiative deactivation (Figure 7.1.15b), electrons release the energy surplus to the environment in form of heat without any luminescent effect. In case of radiative deactivation, energy is released in form of a photon. Luminescence owing to the direct relaxation from the excited to the ground singlet state is called fluorescence (Figure 7.1.15c). It mainly occurs in molecules with rigid structure like aromatic compounds, hydrocarbons, compounds with carbonyl groups, etc. Fluorescence disappears within nanoseconds after excitation stops.

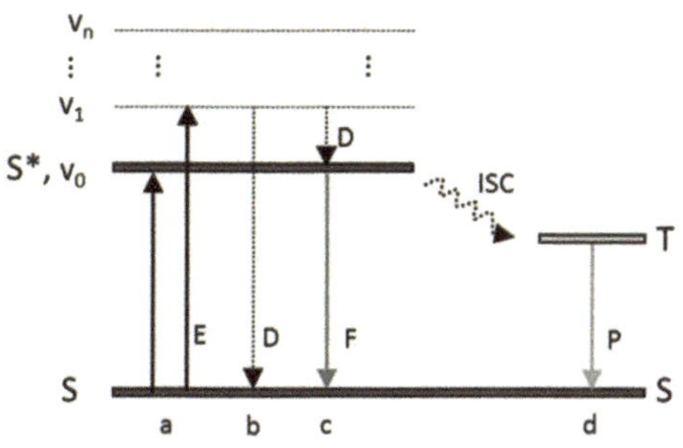

Figure 7.1.15: Simplified Jablonski diagram illustrating the electronic states of a molecule and its transitions. **a)** Excitation of a molecule from the singlet ground state (S) to its singlet excited state (S*) at different vibrational levels (v). **b)** Nonradiative deactivation. **c)** Fluorescence. **d)** Phosphorescence upon intersystem crossing (ISC) to the triplet state (T).

Another form of luminescence occurs when nonbinding electrons of a molecule change from the excited singlet state to a less energetic triplet state in which the electrons are unpaired. This nonradiative electron transition, which in the sense of quantum mechanics is a forbidden mechanism, is called intersystem crossing. Relaxation from the triplet state to the singlet ground state is called phosphorescence (Figure 7.1.15d). It may persist for seconds or even hours after excitation stops.

Application

As already mentioned, fluorescence occurs in molecules with rigid structure (e.g., aromatic compounds, hydrocarbons, or compounds containing carbonyl groups). As regards the metabolism of prokaryotic and eukaryotic cells many of the key metabolites such as vitamins, cofactors and proteins exhibit these properties (Figure 7.1.16). In preliminary studies NAD(P)H became the main target for fluorimetric analysis, which was not surprising because of its ubiquity in living cells and organisms. One of the first reports dates back to 1957, when Duysens and Amesz measured intracellular NAD(P)H levels in starved yeast cells upon addition of ethanol and glucose [45]. In further studies, NADPH dependent culture fluorescence was successfully applied to monitor the biomass of yeast, fungi, and bacteria in culture [46, 47] assuming a constant intracellular concentration of NAD(P)H. Furthermore, fluorimetry was used for bioreactor characterization as regards mixing time, oxygen transfer, and homogeneity, using unmetabolizable fluorescent dyes with similar excitation and emission ranges than NAD(P)H [48, 49].

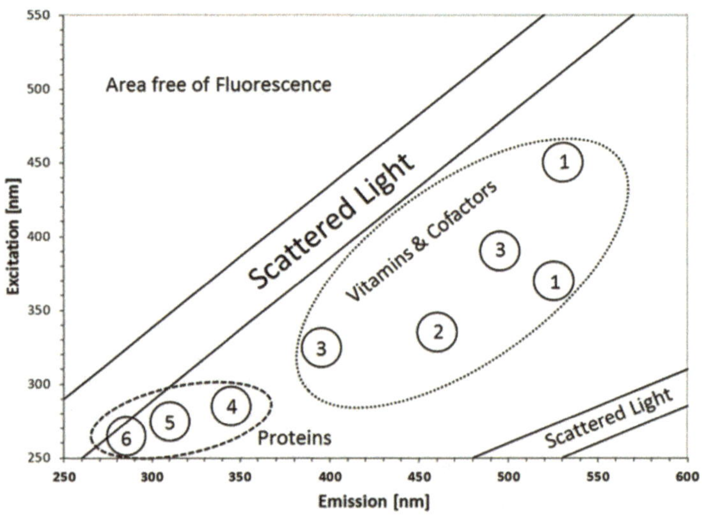

Figure 7.1.16: Biogenic fluorophores in a 2D fluorescence spectrum (according to Schulmann). 1) riboflavin, FAD, FMN; 2) NAD(P)H; 3) pyridoxine, pyridoxamine, pyridoxal-5-phosphate; 4) tryptophan; 5) tyrosine; 6) phenylalanine.

It took until the late 1990s to introduce two-dimensional fluorescence spectroscopy as a noninvasive technology for bioprocess monitoring [50, 51]. In contrast to the fluorimetry systems used in preliminary studies in which fluorescence measurement was limited to the range of NAD(P)H fluorescence (excitation, 340 to 360 nm; emission, 450 to 460 nm), two-dimensional fluorescence spectroscopy covers a wide range of excitation and emission wavelengths [50, 52, 53]. It thus enables to explore the whole fluorescent spectrum of living cells and increases the sensitivity of the method significantly. As well as other spectroscopic methods, two-dimensional fluorescence spectroscopy provides an enormous amount of real-time data that need to be analyzed using multivariate chemometric techniques in order to extract relevant information from the spectra. In particular the application of principal component analysis (PCA) and partial least squares projection to latent structures (PLS), which is the regression extension of PCA [54], helped to pave the way for successful application of two-dimensional fluorescence spectroscopy as a PAT tool in bioprocess monitoring. Multiple reports from the last recent years emphasize the suitability of two-dimensional fluorescence spectroscopy in different fields of biotechnological application comprising waste water treatment [55-57], food technology [58], as well as biopharmaceutical technology using fungi [54, 59, 60], yeast [50, 51 ,61], and bacteria [51, 62-64]. During the last recent years, stronger focus has been put on the *in situ* monitoring of recombinant mammalian cell-based bioprocesses [65-70]. Various two-dimensional fluorescence spectrometers are available on the market that can be connected to a bioreactor via fiber optic. For industrial application the BioView® sensor (Delta Light & Optics) is predominantly used.

Case Study: Model-based Prediction of Fed-batch Processes using Two-dimensional Fluorescence Spectroscopy

The aim of this study was to investigate the capability of two-dimensional fluorescence spectroscopy to describe and predict the progression of a bioprocess in a predefined range. For this purpose, seven fed-batch cultivations were carried out in shake-flasks following a standard protocol with discontinuous feeding on a daily basis starting on day 3 of the cultivation. Six cultivations were used to calibrate a multivariate model (PLS) based on the 2D spectra (X) and on standard in-process controls (Y). One cultivation was performed following the same protocol in order to examine the predictability (test-set validation). An additional cultivation was performed as batch process without feeding to demonstrate that process changes, which have not been calibrated within the model, can be detected. In Figure 7.1.17 part of the in-process control (IPC) data are shown for all of the fed-batch cultivations. Additional run #8, which was cultivated in batch mode without feeding (Figure 7.1.17, dotted line in dark grey), can be clearly distinguished from the seven fed-batch cultivations. In order to build up a multivariate model, the following IPC data were included to the Y-matrix: viable cell concentration, cell viability, product titer and metabolite levels of glucose, lactate, glutamine, glutamate, ammonia, as well as process data as osmolality, pH, pCO_2, and pO_2. Fluorescence measurements were performed offline, using a SpectraMax M5e Reader (Molecular Devices). Samples were measured in a 96 well format using a top reading procedure with a wavelength step width of 20 nm. The excitation wavelength (λ_{Ex}) was in a range between 285 and 485 nm and the emission wavelength (λ_{Em}) in a range between 305 to 545 nm. Due to the occurrence of signal oversaturation in the area of the characteristic first-order Rayleigh scattering ($\lambda_{Em} = \lambda_{Ex}$) and second-order Rayleigh scattering ($\lambda_{Em} = 2\lambda_{Ex}$), emission wavelengths in a range of ±40 nm of the exci-

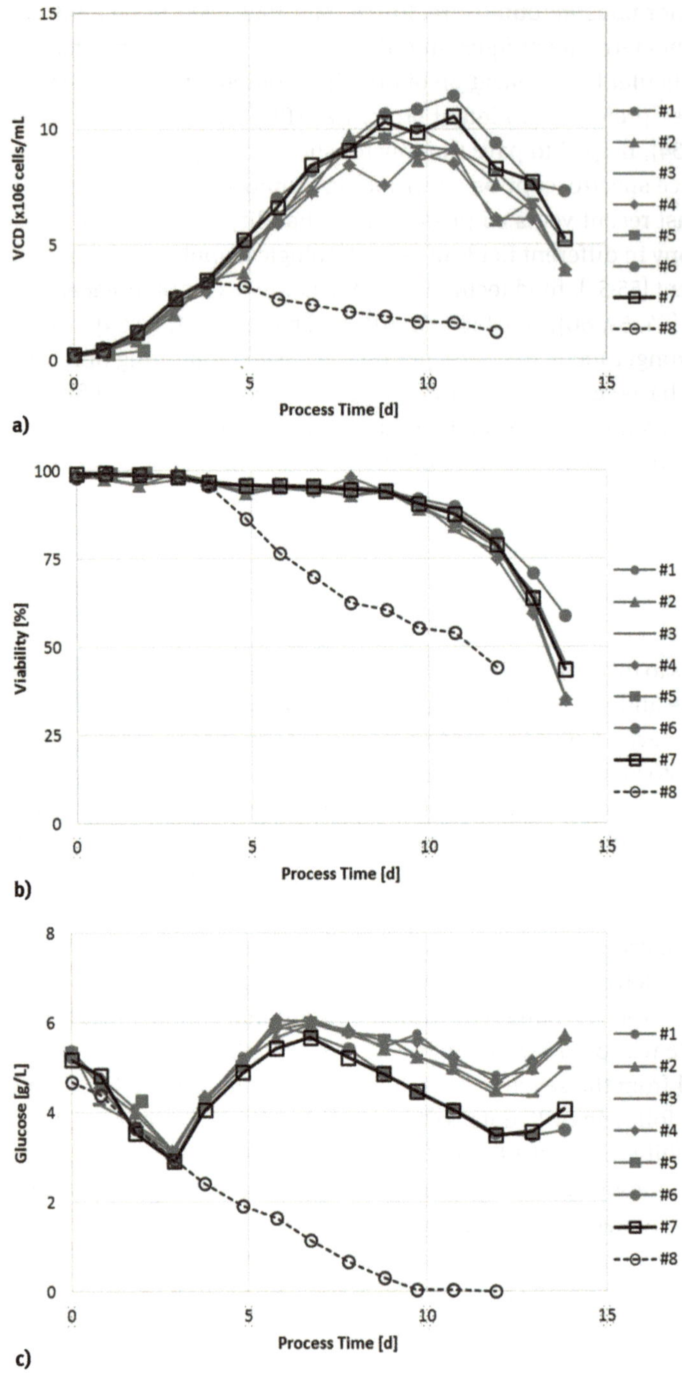

Figure 7.1.17: Cultivation of recombinant CHO cells in fed-batch shaker flasks. **a)** Viable cell concentration, **b)** cell viability, **c)** glucose,

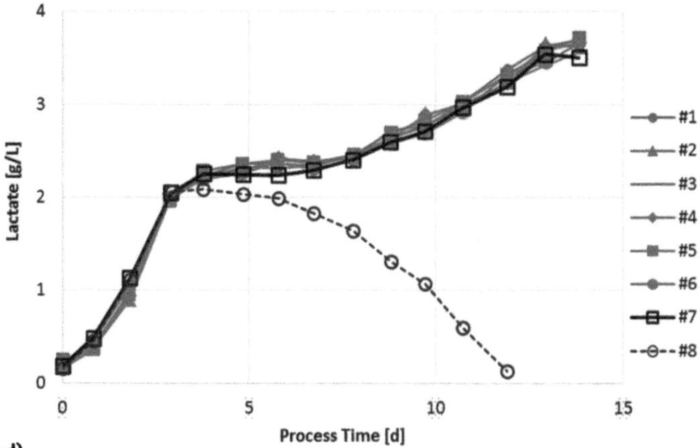

d)

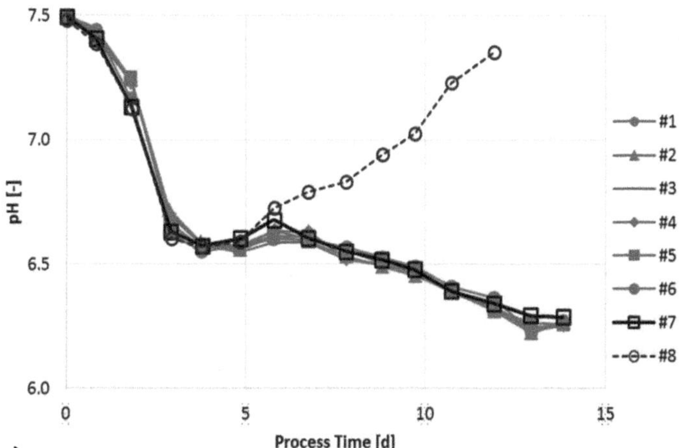

e)

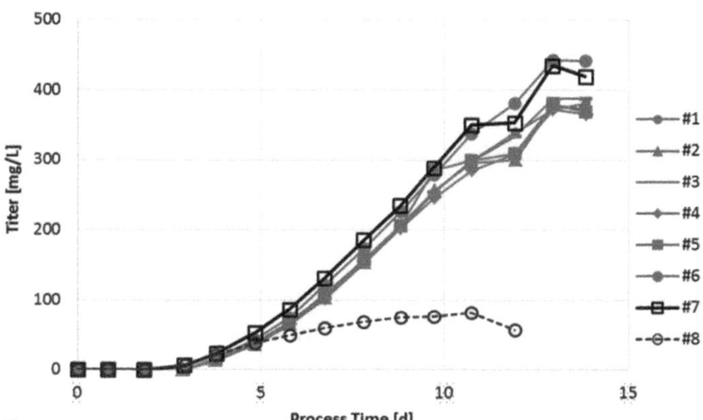

f)

d) lactate, **e)** pH, **f)** titer. Run #1 to #6 (light grey) are used for multivariate model calibration, run #7 (black) and run #8 (dark grey, dotted line) are used for model validation.

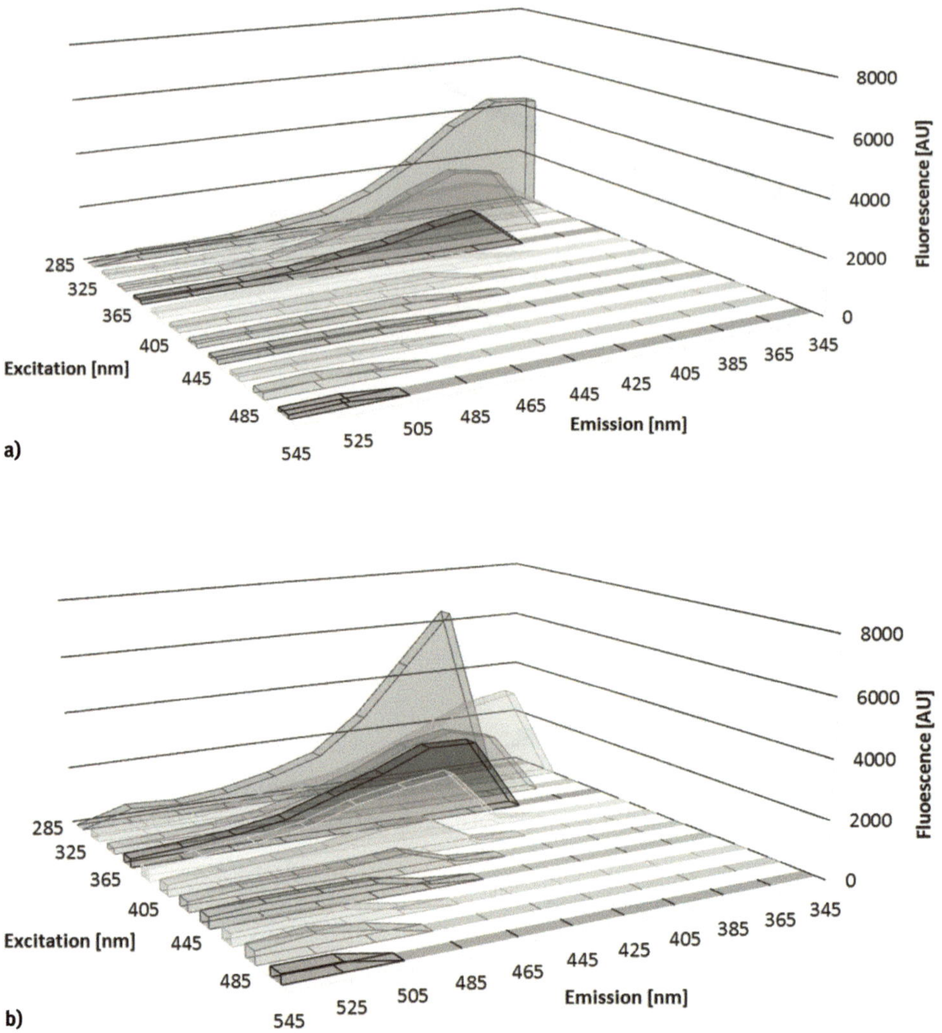

Figure 7.1.18: 2D fluorescence spectra from **a)** day 0 and **b)** day 14 of cultivation.

tation wavelength were not considered for modeling. 2D fluorescence spectra from the start and end of the cultivation process are exemplarily shown for one culture in Figure 7.1.18.

In a first step, both data sets 2D fluorescence spectra (X) and IPCs (Y) of the calibration runs (#1-6) were analyzed with principal component analysis using SIMCA (Umetrics). It could be demonstrated that the variance of the data correlates with the state of the process (Figure 7.1.19). The PCA score plots for both data show a clus-

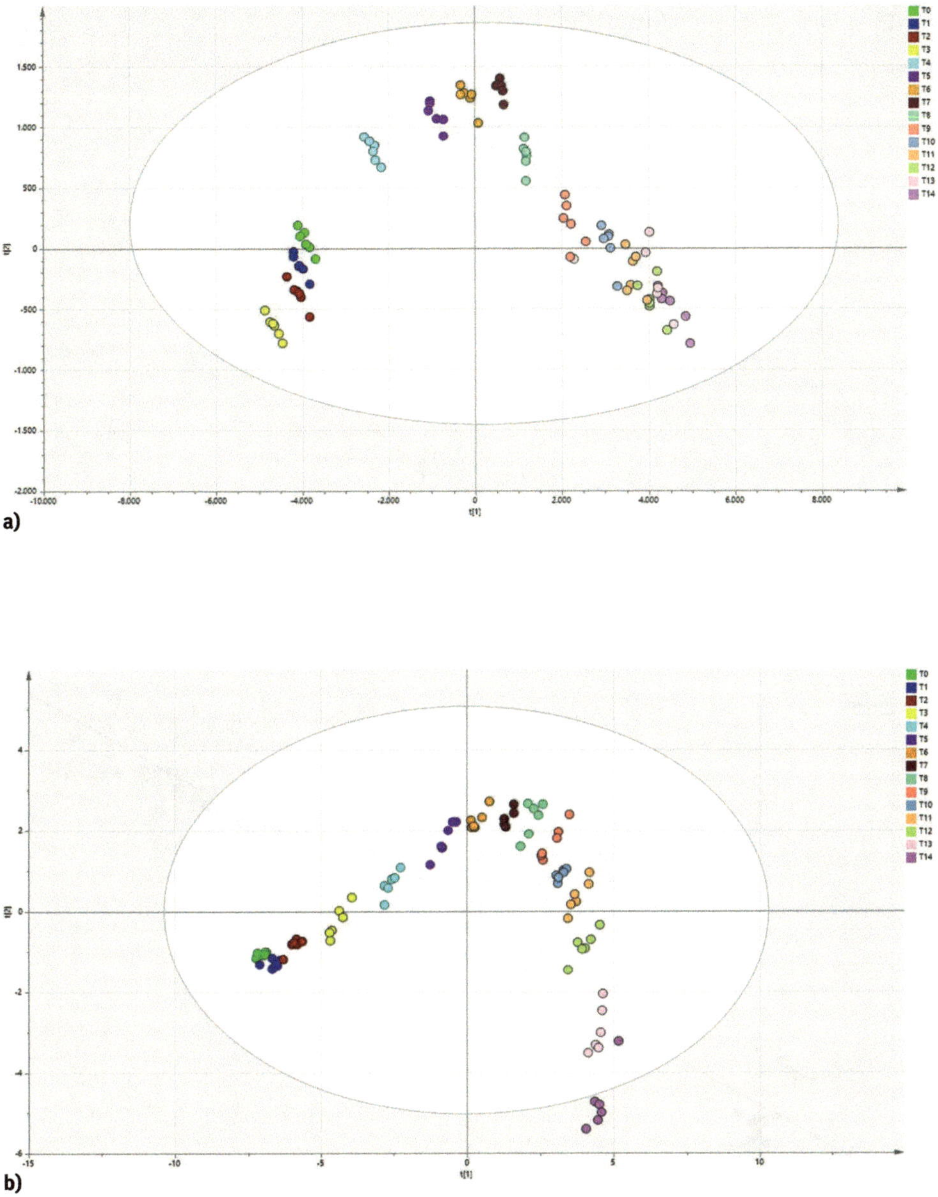

Figure 7.1.19: PCA score plots for **a)** 2D spectra and **b)** IPCs showing the variance as a function of process time points from T0 to T14. ure 7.1.18: 2D fluorescence spectra from a) day 0 and b) day 14 of cultivation.

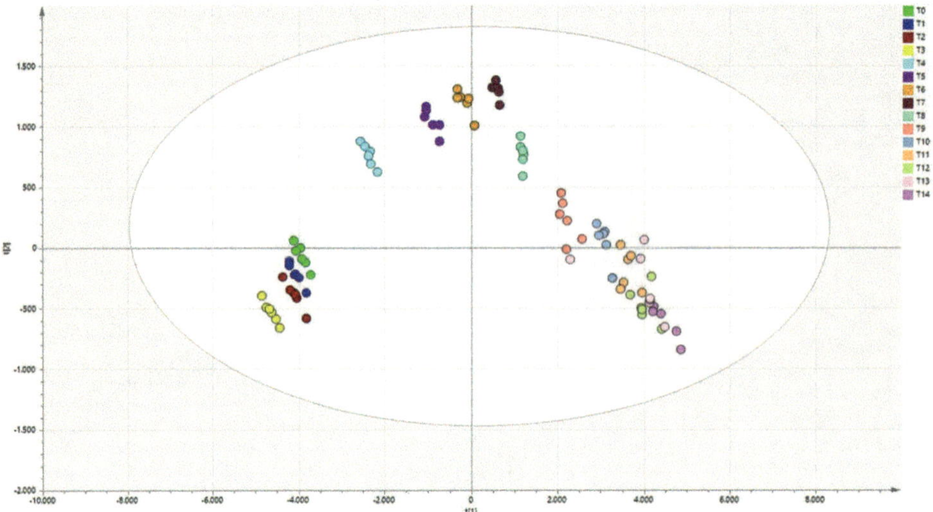

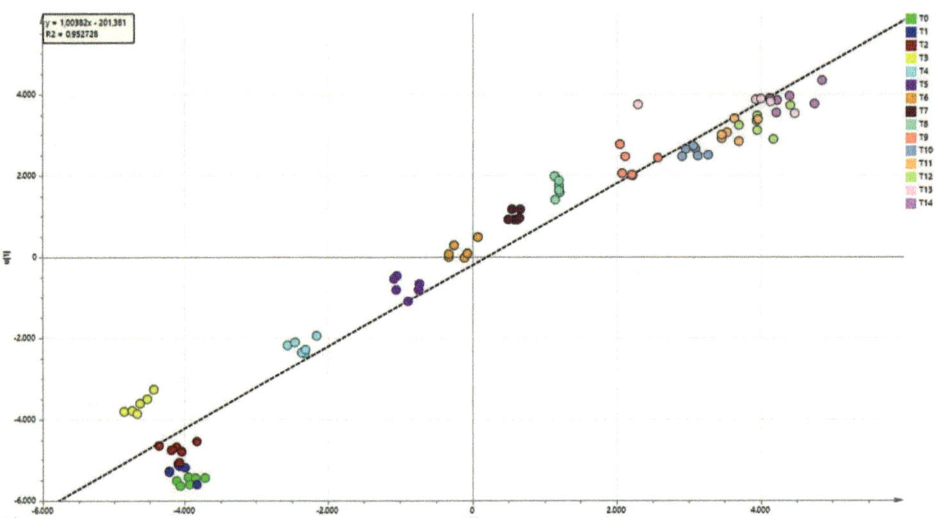

Figure 7.1.20: **a)** PLS score plot including both data sets X and Y. **b)** Relationship between the first principal component t[1] of X and u[1] of Y.

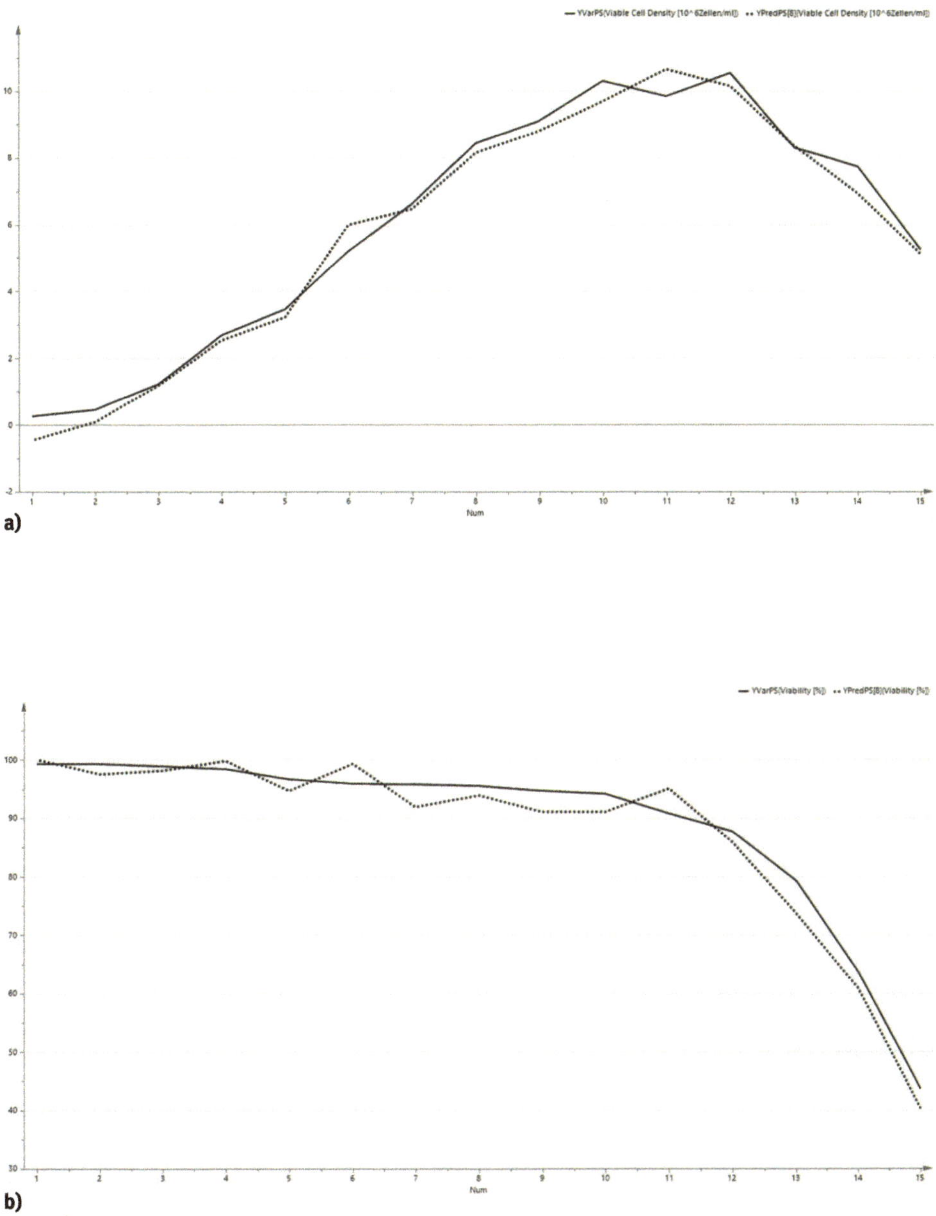

Figure 7.1.21: Prediction plots for IPC data (Y) from the 2D Spectra (X) regarding run #7; true value (straight line), predicted value (dotted line). **a)** Viable cell concentration, **b)** cell viability.

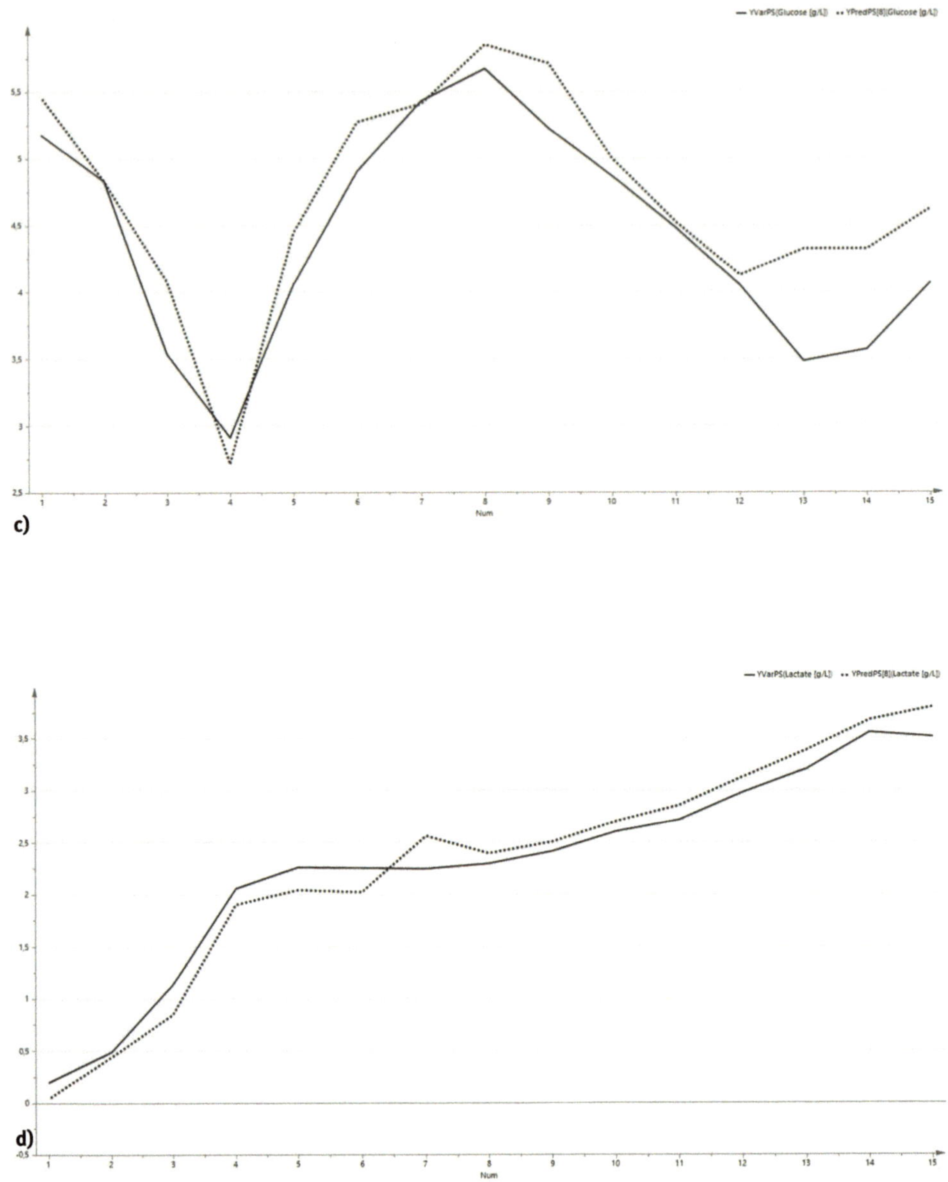

Figure 7.1.21: Prediction plots for IPC data (Y) from the 2D Spectra (X) regarding run #7; true value (straight line), predicted value (dotted line). **c)** glucose, **d)** lactate.

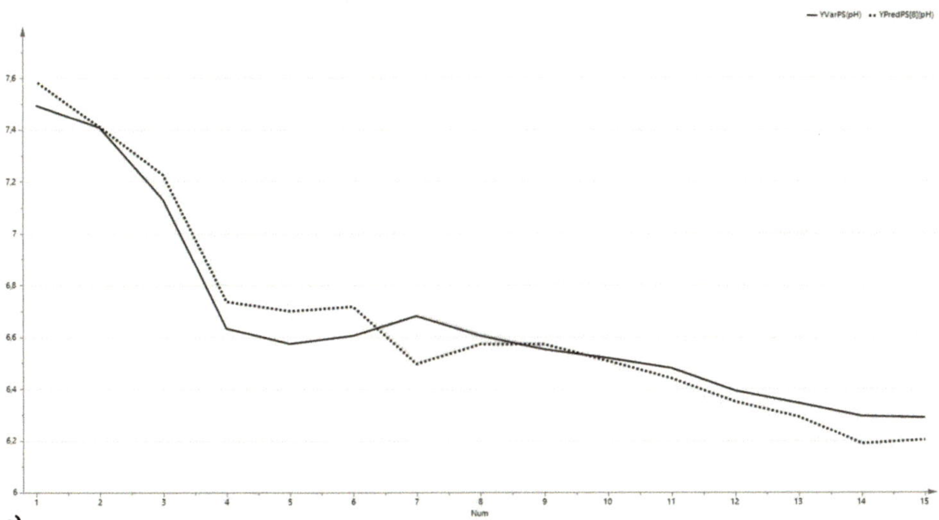

e)

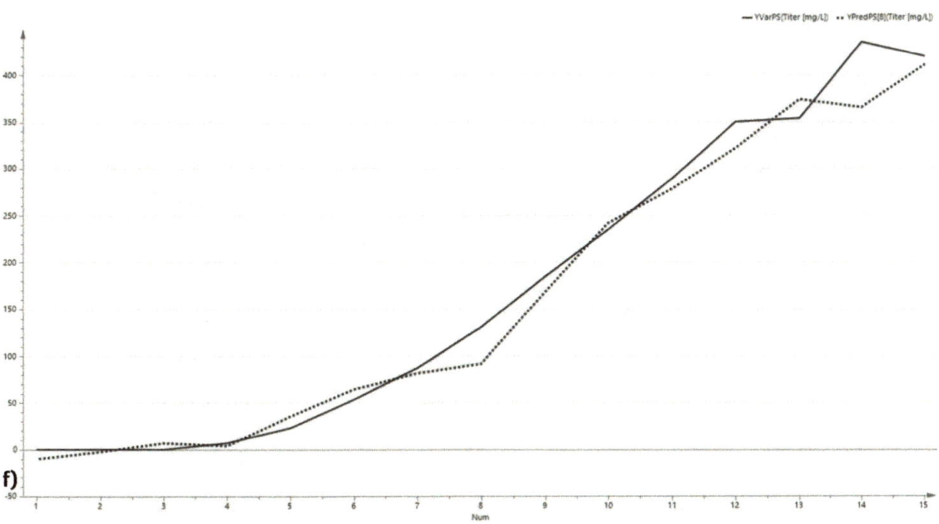

f)

Figure 7.1.21: Prediction plots for IPC data (Y) from the 2D Spectra (X) regarding run #7; true value (straight line), predicted value (dotted line). **e)** pH, **f)** titer.

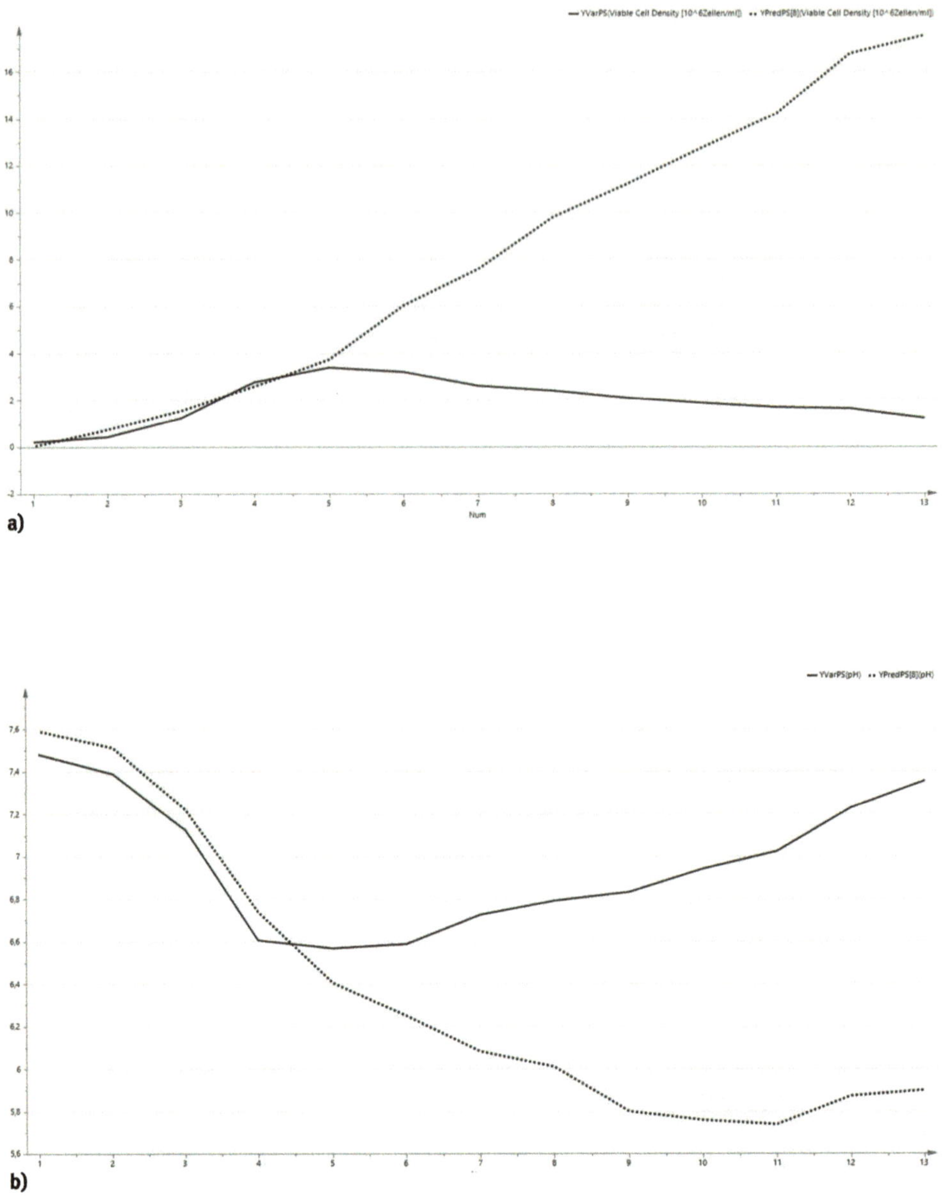

Figure 7.1.22: Prediction plots for IPC data (Y) from the 2D Spectra (X) regarding run #8; true value (straight line), predicted value (dotted line). **a)** Viable cell concentration, **b)** pH.

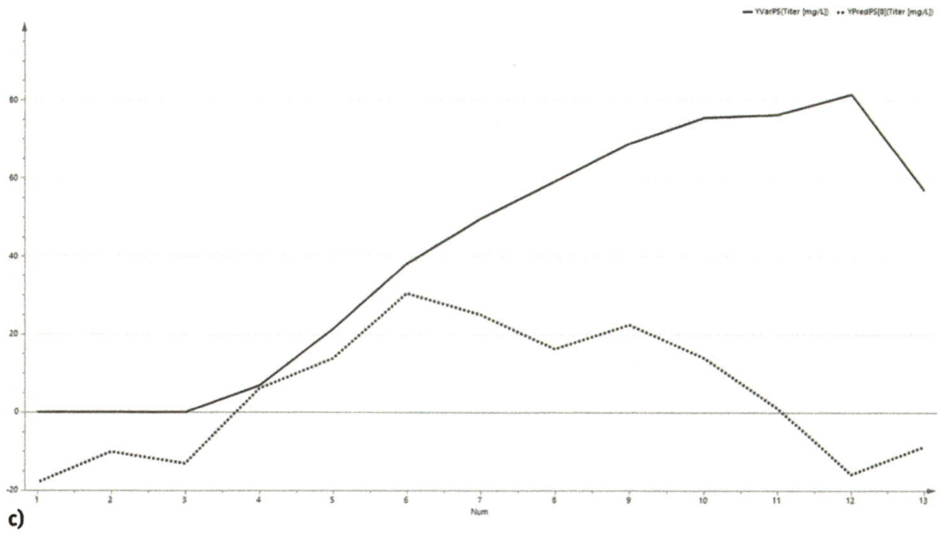

c)

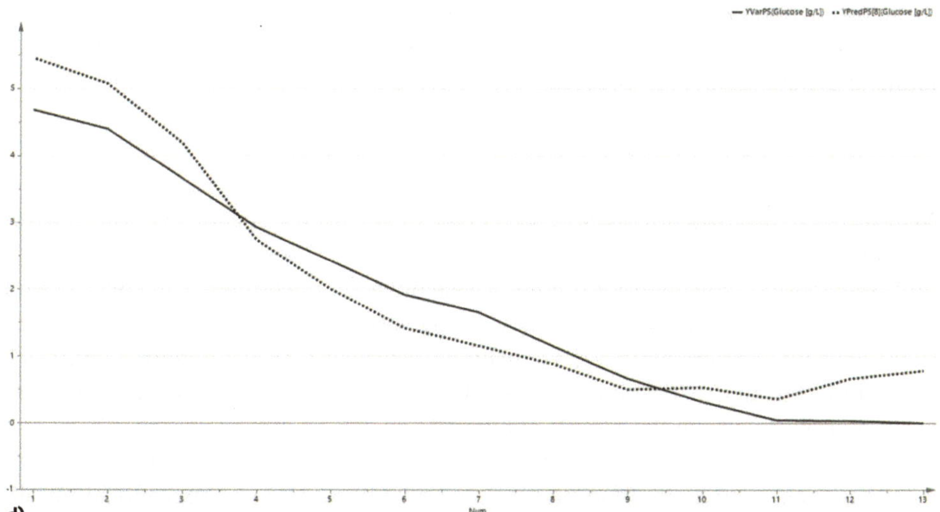

d)

Figure 7.1.22: Prediction plots for IPC data (Y) from the 2D Spectra (X) regarding run #8; true value (straight line), predicted value (dotted line). **c)** titer, **d)** glucose.

tering of the six cultivation replicates according to the process time. As regards the 2D fluorescence spectra data resolution deteriorates from day 9 resulting in a data overlay during the last 2 days of cultivation. Variance of the IPC data at day 14, which is located outside the ellipse (model confidentiality) could no longer be explained by the first and second principle component.

In a second step, both data sets, X and Y, were analyzed using PLS in order to demonstrate that the variance of Y can be explained by the variance of X (Figure 7.1.20). A linear regression coefficient of more than 0.95 was calculated for the linear relationship between the first principal components t[1] (X) and u[1] (Y). Based on this calibration model, a test set validation was performed. It could be demonstrated (Figure 7.1.21) that IPC data (Y) of a similar cultivation (run #7) can be predicted with high accuracy from the model using the 2D spectra (X).

A second test set validation was carried out using run #8, which, unlike the other cultivations, was performed as a batch cultivation. The results in Figure 7.1.22 clearly show that for most of the IPC data a prediction is not possible, which is not surprising, considering the fact that batch process data were not included to the model. Interestingly, a good prediction was found for glucose, which may be due to a direct correlation between the glucose level and a subset of fluorescence data.

According to the results of this case study, two-dimensional fluorescence spectroscopy emerged as a powerful tool for bioprocess monitoring, which is in accordance with other reports. In this study, the sensitivity of this method combined with multivariate chemometrics enabled to monitor a process within its capability space.

7.1.3.4 Near-infrared Spectroscopy

Near-infrared spectroscopy (NIRS), targets the near-infrared spectrum of the electromagnetic spectrum (780 to 2,500 nm) (Figure 7.1.23). NIRS is established from overtone and combination molecular vibrations. These molecular overtones and combination bands can be very broad and complex yielding very complicated mathematical solutions for useful interpretation [71]. For example, see Figure 7.1.24. Therefore, multivariate techniques such as PCA and PLS are often required to deconvolute and interpret spectroscopic data from the sample of interest.

NIRS has been used for many years in other industries such as cosmetics, chemical processing, petroleum, food processing, agriculture, and pharmaceutical. Application to the biotechnology industry has typically focused around raw material identification and characterization [72 ,73] with more recent research focusing on *in situ* applications in microbial and cell culture bioprocesses [74]. In 2006, a comprehensive review of near-infrared (NIR) monitoring was applied to bioprocess by Scarff et al. looking into the evolution of NIR technology with respect to type of fermentation process [75]. Later in 2009, Cervera et al. wrote an excellent review called “Application of Near-Infrared Spectroscopy for Monitoring and Control of Cell Culture and Fermentation” [74]. In this review Cervera et al. investigated important practical aspects that

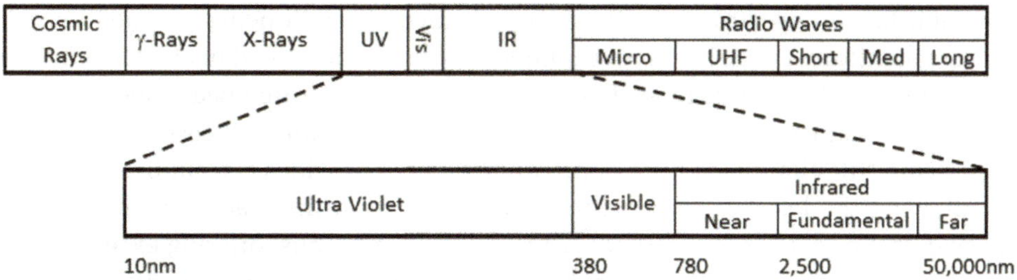

Figure 7.1.23: NIR in the electromagnetic spectrum.

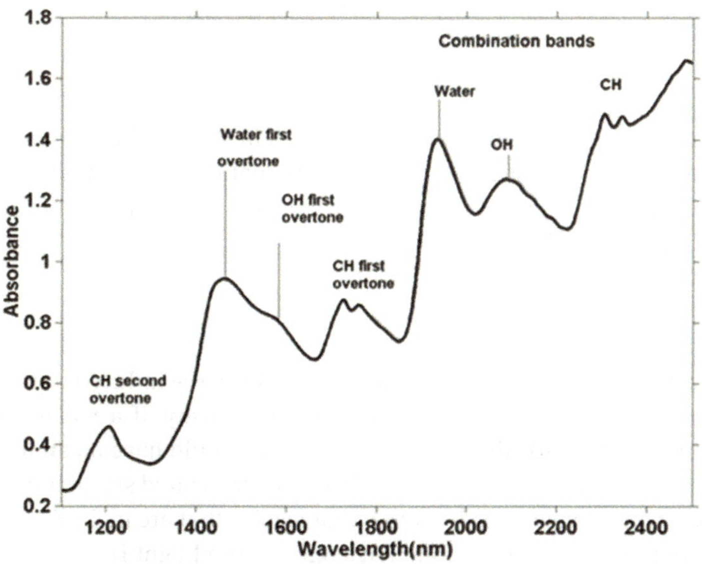

Figure 7.1.24: Example of NIR spectrum showcasing overtones and combination bands.

challenge NIRS implementation as an online advance control solution. Such aspects discussed were sampling, modeling of biomass concentration, influence of microorganism morphology on the spectra, effects of the hydrodynamic conditions in the fermenter, temperature influence, instrument settings, and signal optimization [74].

To date, NIRS still faces many challenges to monitor high intensity fed batch mammalian cell culture broth *in situ* and has not yet been proven to be a robust and accepted on line monitoring tool. Its main challenges are water interference and relative low sensitivity when *in situ*. Many articles demonstrate NIR capability at line or off line but to date there has not been overwhelming evidence to demonstrate the same or higher value to alternative technologies such as fluorescence, dielectric, and Raman spectroscopy. However, advances are still being pursued for cell culture monitoring with NIRS. An example of which is described in an recent article by Sandor et al. [76] where an Ingold-port adaption of a free-beam spectrometer is used for real time monitoring of total cell count, viability, glucose, and lactate. Free-beam NIR eliminates fiber optic cabling which provides certain benefits such as allowing the laser to be generated and collected at the reactor, using a diode array detector instead of a cathode array detector and, thus, offering excellent signal-to-noise ratio over conventional NIRS configurations for improved cell culture correlations. Therefore, as the technology, science, and understanding continue to evolve, NIRS could become an advanced monitoring and control tool for cell culture applications in the future.

7.1.3.5 Raman Spectroscopy

Background

The Raman effect, named after one of its discoverers Raman in 1928, is the exchange of vibrational energy due to inelastic scattering of monochromatic light by a molecule [77]. Raman bands can only be observed if there is a change in polarizability of the molecular vibrations, represented by the equation

$$\mu = \alpha E,$$

where α is the polarizability, E the incident electric field, and μ is the induced dipole moment. Polarizability is directly related to Raman signal intensity. If a molecule exhibited no change in polarizability, then the dipole moment would have no amplitude modulation and no recordable Raman signal [78]. Raman measured signal intensity is very weak, leaving only a small portion 1 in 10^5 molecules that are inelastically scattered when compared to the intense elastic Rayleigh scattered light [79]. Inelastic scattering occurs when the frequency of monochromatic light, normally from a laser source, changes when interacting with a sample. The photons are absorbed by the sample and then reemitted at a frequency either with more (anti-Stokes) or less (Stokes) energy when compared with the original monochromatic frequency (Figure 7.1.25).

Anti-Stokes is the stronger signal and is almost always used. The frequency (wavelength) change depends on many factors such as atom mass, geometric configuration, and strength of the chemical bonds. Ultimately, the resulting shift contains information about the vibrational, rotational, and other low frequency transitions in molecules [78].

The Raman spectrum is the plot of the intensity of Raman scattered radiation as a function of its frequency difference from the incident radiation (usually in units of wavenumbers, cm^{-1}). Raman is complementary to mid-infrared (MIR) but with different intensities and selectivity.

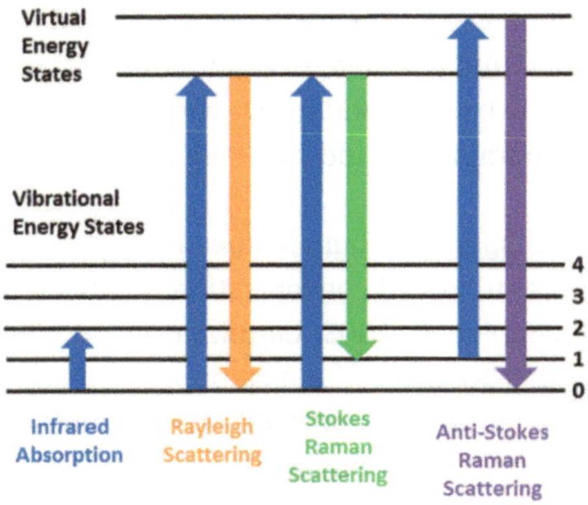

Figure 7.1.25: Virtual energy states diagram.

Raman technology has made enormous strides over the last 4 decades. Early on in the late 1970s and early 1980s, Raman spectroscopy was delegated to the research and development labs only. This was due to large laser sources that were unreliable, hard to maintain, and expensive. Rapid technological advances were seen in the mid 1980s to the mid 1990s such as Fourier transform spectroscopy, charge couple device (CCD) detectors, holographic and dielectric filters, near-infrared (NIR) lasers, and advanced dispersive spectrometers. However, the technology was still costly and large in size. This changed with advancement of semiconductor lasers, optical miniaturization techniques, compact data processing capabilities, and probability-based statistical analysis programs. Additional key developments in the last decade have been with *in vivo* applications, the advent of optical fiber probes and nonlinear spectroscopy. These advancements over the past 4 decades have led to the adoption of Raman in various industries spanning chemical, forensic science, nanotechnology, semiconductors, pharmaceutical, and now biotechnology.

The application of Raman spectroscopy in monitoring aqueous bioprocesses has a unique advantage over other spectroscopic techniques because it can provide highly specific vibrational fingerprints of various chemical compounds and is relatively insensitive to water. MIR and NIR are other common spectroscopic techniques utilized; however, water peaks can dominate the spectral signal. When using NIR and MIR, the signal to noise ratio is reduced dramatically making it more difficult to resolve certain critical constituents present in low concentrations in cell culture media and feed [13, 15]. Water produces a very weak Raman scatter. The impact of water on the Raman signal can only be seen at the end of the spectrum (2,900 to 4,000 cm^{-1}) when used during in-process bioreactor monitoring. Raman laser excitation wavelengths can vary depending on the application but typically for bioreactor monitoring 785 nm excitation laser source is the desired choice. A 785 nm laser can provide the optimal balance between detection sensitivity and fluorescence rejection for mammalian cell culture processes. If *in situ* fluorescence is high, a longer excitation wavelength such as a 1,064 nm laser can be used. The opposite can be said for detection of gases with Raman, the shorter excitation wavelength is normally optimal allowing for heightened sensitivity with very low background florescence. So choosing the correct excitation laser wavelength is paramount and the solution is highly dependent on the application.

In 2002, McGovern et al. [15] utilized dispersive Raman spectroscopy and advanced chemometric data analysis techniques to model concentration of gibberellic acid in offline whole fermentation broth samples. In 2003, Cannizzaro et al. [79] described how a Raman model was able to predict an intracellular carotenoid concentration from a fed-batch yeast process. Lee et al. [80] in 2004 was able to demonstrate models to predict glucose, acetate, formate, lactate, and phenylalanine concentrations at the same time with *Escherichia coli* bioprocess with varied results. In 2011, 2 papers were published (Abu-Absi et al. [81] and Moretto et al. [14]) documenting successful models generated for prediction of glucose, lactate, glutamine, glutamate, ammonium, osmolality, viable cell density (VCD), and total cell density (TCD) using CHO cell lines. Most recently, Whelan et al. (2012) [8] published an article demonstrating Raman PLS models for cell culture growth and metabolites that could successfully be transferred from 3 to 15 L bioreactor scale.

Case Study

The following case study is based on work conducted at over several years using different CHO cell lines and cultures. The data set covers multiple bioreactor scales from 3 L (lab), to 200 L (pilot), to 2,000 L (manufacturing) scale. The following Figures 7.1.26

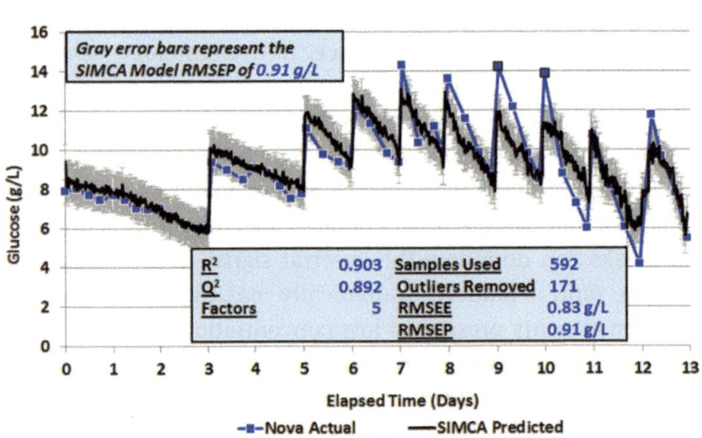

Figure 7.1.26: Glucose prediction using Raman spectroscopy compared to off-line measured values using a NOVA bioanalyzer. (RMSEP = root mean square errors of prediction, RMSEE = root mean square errors of estimation.)

to 7.1.28 depict models for glucose, lactate, and ammonium for a CHO based mammalian cell culture process using a similar method as the general one described in data analysis and model building section (see p. 527)[82].

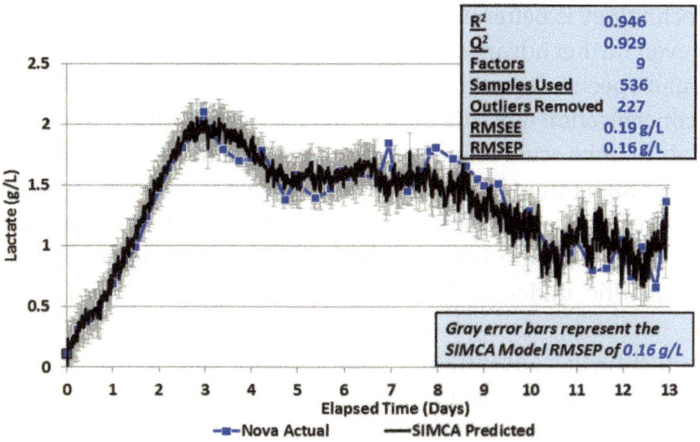

Figure 7.1.27: Lactate prediction using Raman spectroscopy compared to off-line measured values using a NOVA bioanalyzer.

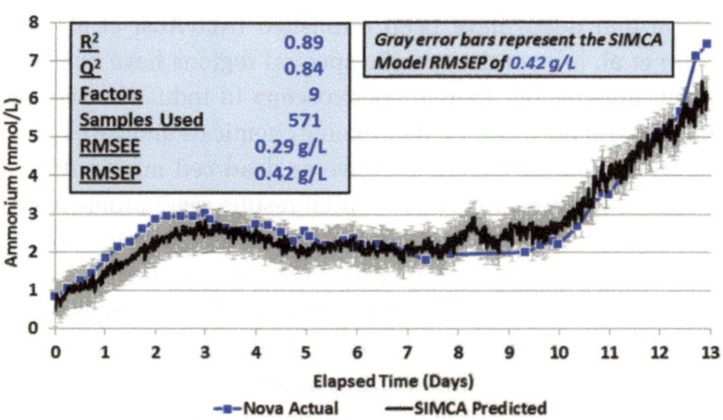

Figure 7.1.28: Ammonium prediction using Raman spectroscopy compared to off-line measured values using a NOVA bioanalyzer.

Cell Culture Challenges using Raman Spectroscopy

The application of Raman spectroscopy in cell culture has enormous promise but there are still challenges that must be understood and overcome. In general, Raman is in the infancy stage of understanding and application in the biotechnology industry. Since the turn of the century, a handful of Raman studies related to the biotechnology industry have been published and of that only a small subset with regards to *in situ* monitoring and control (Abu-Absi et al. [81] and Moretto et al. [14]). As the industry matures and Raman technology is better understood, the adoption threshold should diminish allowing for even further advancement of the technology.

Some specific Raman spectroscopy challenges in industry today include *in situ* low-level constituent concentration quantification, cell mass models (live and dead), scaling models from lab to large scale manufacturing, offline analytical equipment accuracy, an approach for high throughput model building that is cost and resource efficient, and integration of Raman equipment and software for real time feedback control. Some of these challenges are being studied and have a defined solution whereas others lack scientific understanding and further experimentation must occur.

In situ low-level constituent concentration quantification can be a challenge when trying to read micro molar amino acid concentrations during a cell culture run. The Raman peaks can sometimes be discerned if measured without cells present, but *in situ* remains a challenge. This is due to the background noise generated by cell culture broth. Though data to this end has not yet been published, research is ongoing. In the future, if Raman spectroscopy could read amino acid levels real time during a cell culture run, feeds could be optimized to promote higher cell growth and improved product quality profiles.

Cell mass models (live and dead) have been published (Abu-Absi et al. [81], Moretto et al. [14], Whelan et al. [8]) yet the specific spectral regions have not been described and are not well understood. Raman spectroscopy in industrial settings has been applied to synthesis and purification of inorganic chemicals and gases only; applying Raman spectroscopy to accurately model live or dead cell mass within a reactor is an area that requires further exploration. Initial results seem promising as shown in Figure 7.1.29; however, further understanding of what specific region in the spectra is correlating with viable cell density still needs to be explored further.

Scaling models from lab to large-scale manufacturing have been discussed in literature by Whelan et al. [8] going from 3 to 15 L scale. However, when increasing the scale by 1,000 times to large-scale manufacturing other challenges may arise. With a general approach, Raman spectroscopy can be applied to achieve models that are statistically sound, but when tested across various scales (different cell lines) and over multiple manufacturing campaigns Raman spectroscopy-based modeling has not yet been demonstrated to be robust. We have recently conducted experiments at 3 L, 200 L, and 2,000 L scales and investigated cross scale robustness (same program at all scales) [82]. Certain constituent models hold up very well across scales such as glucose and lactate, meaning a small-scale model will have a comparable RMSEP as a model that includes data from all scales (3 L, 200 L, 2,000 L). Conversely, other constituents that are not as well understood, such as viable

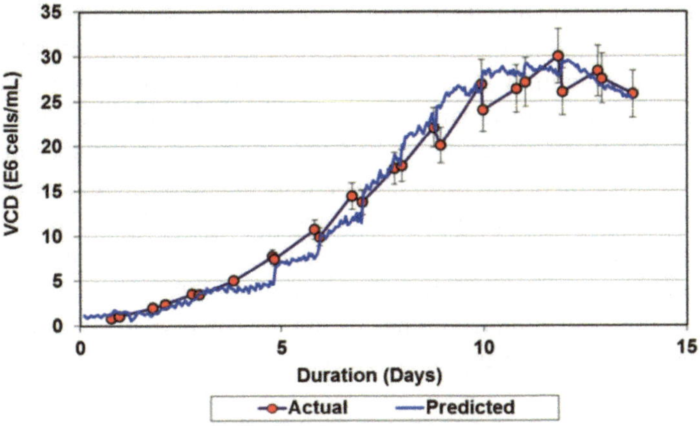

Figure 7.1.29: Raman model predicting VCD for a CHO cell culture process at 200 L scale. Off-line measurements where analyzed using a Cedex cell counter. Error bars represent known Cedex variability of ± 10 %.

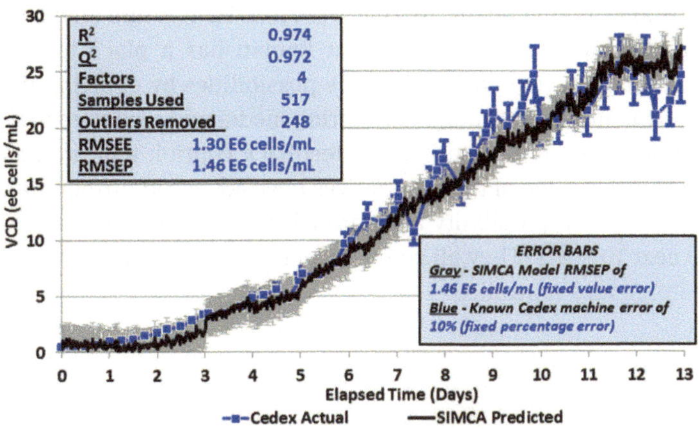

Figure 7.1.30: Cross scale model (SCM) l predicting VCD at 2,000 L scale. Calibration data set contains 3 L, 200 L, and 2,000 L data.

cell density and viability, tend to show slightly larger variability; however, acceptable results can be achieved if the target scale data is included in the calibration set (Figure 7.1.30). Ideally future-modeling techniques will be able to use small-scale data to accurately predict large scale performance for VCD and viability. This gap can possibly be addressed by understanding the underlying spectral linkages between viable and nonviable cell mass. Once this is understood, a more robust Raman model can be built. A VCD Raman model can vary among 3 L, 200 L, and 2,000 L scales especially when using a broad part of the Raman spectrum. In contrast, other constituents, such as glucose and lactate, hold up well across scales with the variability between scales being much less (<10 %) for these constituents when compared to VCD, TCD, and ammonia. The majority of this variability is tied to understanding the specific wavenumbers for a given constituent and the understanding of how scale-up factors such as sparge rate, agitation, and turbidity could impact signal intensity and therefore Raman model accuracy. When a broad range of the spectrum is used, the number of factors affecting Raman-based model accuracy and predictability are amplified. Future scientific exploration and understanding in this area will yield improved mammalian cell culture Raman models at production scale.

Future of Raman Spectroscopy in Biotech

There are many spectroscopic techniques that can be applied to a mammalian cell culture process, all with certain advantages and disadvantages. Raman spectroscopy provides some unique benefits that should propel research and eventual adoption of this technology. There are still many hurdles to overcome with Raman, but the core science and application to the industry has been proven. With future advancements and scientific understanding, we believe that Raman has a place in PAT process control. Looking forward, Raman has multiple possibilities by applying the PAT approach such as raw material screening, monitoring media and feed prep variability, cell culture and purification monitoring and feedback control, and product quality monitoring and control. Some of these initiatives have advanced further than others, but all are in the realm of possibility and through technological and scientific advances over the next 10 years. They all could become real solutions for how a biotech process is monitored and controlled.

7.1.4 Conclusions and Future Vision

The desire to bring better products to patients in need will further foster development of more robust and well-controlled bioprocesses. We believe that advanced process control will be an integral part of future cell culture-based manufacturing processes in support of this vision. Monitoring and control technologies and approaches reviewed in this chapter have the potential to significantly enhance process knowledge and offer opportunities for PAT like process control. Ultimately, control based on on-line or at-line acquired information aims to govern process control parameters to achieve consistent processing within the cell and its environment.

The quality of prediction will strongly depend on the quality and complexity of the multivariate model. Available software solutions in context of *in situ* spectroscopy will without doubt offer powerful PAT tools for control of performance and product quality parameters in real-time.

In particular, spectroscopic methods are also used to screen and characterize raw materials that enter the production process. The combination of enhanced raw material control with advanced process control is expected to reduce process variability through addressing these two major factors leading to variability in outcomes.

Although it is not certain how the use of these process models and advanced feedback control will be implemented into a GMP compliant cell culture process, we do not foresee any fundamental challenges that would prevent the use of spectroscopic sensors and associated process models and control.

References

[1] Bibila TA, Robinson DK. In pursuit of the optimal fed-batch process for monoclonal antibody production. Biotechnol Prog 1995;11:1-13.

[2] Wurm FM. Production of recombinant protein therapeutics in cultivated mammalian cells. Nat Biotech 2004;22:1393-8.

[3] Huang YM, Hu W, Rustandi E, et al. Maximizing productivity of CHO cell-based fed-batch culture using chemically defined media conditions and typical manufacturing equipment. Biotechnol Prog 2010;26:1400-10.

[4] Yu M, Hu Z, Pacis E, et al. Understanding the intracellular effect of enhanced nutrient feeding toward high titer antibody production process. Biotechnol Bioeng 2011;108:1078-88.

[5] Tsang V, Wang AX, Yusuf-Makagiansar H, iet al. Nutrient monitoring and feedback control for increased bioproduct production. United States of America patent US 8,318,416 B2. 2012.

[6] Lu F, Toh PC, Burnett I, et al. Automated dynamic fed-batch process and media optimization for high productivity cell culture process development. Biotechnol Bioeng 2013;110:191-205.

[7] Logan D, Carvell J. A biomass monitor for disposable bioreactors. BioProcess Inter 2011;9:48-54.

[8] Whelan J, Craven S, Glennon B. In situ Raman spectroscopy for simultaneous monitoring of multiple process parameters in mammalian cell culture bioreactors. Biotechnol Progr 2012;28:1355-62.

[9] PAT-A Framework for Innovative Pharmaceutical Development, Manufacturing, and Quality Assurance. Guideance for Industry. 2004 (http://www.fda.gov/downloads/drugs/guidancecomplianceregulatoryinfomation/guidances/UCM070305.pdf, accessed 15th May 2014).

[10] Administration UFaD. International Conference on Harmonization (ICH) Q8(R2): Pharmaceutical Development. INTERNATIONAL CONFERENCE ON HARMONISATION OF TECHNICAL REQUIREMENTS FOR REGISTRATION OF PHARMACEUTICALS FOR HUMAN USE; August 2009.

[11] Boudreau MA, McMillan GK. New Directions in Bioprocess Modeling and Control: Maximizing Process Analytical Technology Benefit. ISA 2007.

[12] Goodwin GC, Graebe SF, Salgado ME. Control System Design. Prentice Hall 2000.

[13] Li B, Ryan PW, Ray BH, et al. Rapid characterization and quality control of complex cell culture media solutions using raman spectroscopy and chemometrics. Biotechnol Bioeng 2010;107:290-301.

[14] Moretto J, Smelko J, Cuellar M, et al. Process Raman Spectroscopy for In-Line CHO Cell Culture Monitoring. Am Pharm Rev 2011;5:18-25.

[15] McGovern AC, Broadhurst D, Taylor J, et al. Monitoring of complex industrial bioprocesses for metabolite concentrations using modern spectroscopies and machine learning: application to gibberellic acid production. Biotechnol Bioeng 2002;78:527-38.

[16] Pearson K. On lines and planes of closest fit to systems of points in space. Philos Mag 1901;13:559-72.
[17] Geladi P. Chemometrics in spectroscopy. Part 1. Classical chemometrics. Spectrochim Acta B 2003;58:767-82.
[18] Wold S, Sjostrom M. SIMCA" A method for analyzing chemical data in terms of similarity and analogy: In: Kowalski BR, Chemometrics: Theory and Application. Amercian Chemical Societey 1977:243-82.
[19] Wlaschin KF, Hu WS. Fedbatch culture and dynamic nutrient feeding. Adv Biochem Eng/ Biotechnol 2006;101:43-74.
[20] Zhang L, Shen H, Zhang Y. Fed-batch culture of hybridoma cells in serum-free medium using an optimized feeding strategy. J Chemi Tech Biotech 2004;79:171-81.
[21] Sauer PW, Burky JE, Wesson MC, et al. A high-yielding, generic fed-batch cell culture process for production of recombinant antibodies. Biotechnol Bioeng 2000;67:585-97.
[22] Gong X, Li D, Li X, et al. Fed-batch culture optimization of a growth-associated hybridoma cell line in chemically defined protein-free media. Cytotechnology 2006;52:25-38.
[23] Kuwae S, Ohda T, Tamashima H, et al. Development of a fed-batch culture process for enhanced production of recombinant human antithrombin by Chinese hamster ovary cells. J Biosc Bioeng 2005;100:502-10.
[24] Zhou W, Hu WS. On-line characterization of a hybridoma cell culture process. Biotechnol Bioeng 1994;44:170-7.
[25] Zhou W, Rehm J, Hu WS. High viable cell concentration fed-batch cultures of hybridoma cells through on-line nutrient feeding. Biotechnol Bioeng 1995;46:579-87.
[26] Chee Furng Wong D, Tin Kam Wong K, Tang Goh L, et al. Impact of dynamic online fed-batch strategies on metabolism, productivity and N-glycosylation quality in CHO cell cultures. Biotechnol Bioeng 2005;89:164-77.
[27] Lee YY, Yap MG, Hu WS, Wong KT. Low-glutamine fed-batch cultures of 293-HEK serum-free suspension cells for adenovirus production. Biotechnol Prog 2003;19:501-9.
[28] Chong L, Saghafi M, Knappe C, et al. Robust on-line sampling and analysis during long-term perfusion cultivation of mammalian cells. J Biotechnol 2013;165:133-7.
[29] Davey CL, Davey HM, Kell DB, Todd RW. Introduction to the dielectric estimation of cellular biomass in real time, with special emphasis on measurements at high volume fractions. Anal Chim Acta 1993;279:155-61.
[30] Cole KS, Cole RH. Dispersion and Absorption in Dielectrics I. Alternating Current Characteristics. J Chem Phys 1941;9:341-51.
[31] Yardley JE, Kell DB, Barrett J, Davey CL. On-line, real-time measurements of cellular biomass using dielectric spectroscopy. Biotechnol Genet Eng Rev 2000;17:3-35.
[32] Fehrenbach R, Comberbach M, Petre JO. On-line biomass monitoring by capacitance measurement. J Biotechnol 1992;23:303-14.
[33] Clementschitsch F, Bayer K. Improvement of bioprocess monitoring: development of novel concepts. Microb Cell Factories 2006;5:19.
[34] Harris CM, Todd RW, Bungard SJ, et al. Dielectric permittivity of microbial suspensions at radio frequencies: a novel method for the real-time estimation of microbial biomass. Enzyme Microb Technol 1987;9:181-6.
[35] Patel P, Markx GH. Dielectric measurement of cell death. Enzyme Microb Technol 2008;43:463-70.
[36] Salter GJ, Kell DB, Ash LA, et al. Hydrodynamic deposition: a novel method of cell immobilization. Enzyme Microb Technol 1990;12:419-30.
[37] Gheorghiu E, Asami K. Monitoring cell cycle by impedance spectroscopy: experimental and theoretical aspects. Bioelectrochem Bioenerg 1998;45:139-43.

[38] Krairak S, Yamamura K, Nakajima M, et al. On-line monitoring of fungal cell concentration by dielectric spectroscopy. J Biotechnol 1999;69:115-23.
[39] Noll T, Biselli M. Dielectric spectroscopy in the cultivation of suspended and immobilized hybridoma cells. J Biotechnol 1998;63:187-98.
[40] Ducommun P, Bolzonella I, Rhiel M, et al. On-line determination of animal cell concentration. Biotechnol Bioeng 2001;72:515-22.
[41] Cannizzaro C, Gugerli R, Marison I, von Stockar U. On-line biomass monitoring of CHO perfusion culture with scanning dielectric spectroscopy. Biotechnol Bioeng 2003;84:597-610.
[42] Opel CF, Li J, Amanullah A. Quantitative modeling of viable cell density, cell size, intracellular conductivity, and membrane capacitance in batch and fed-batch CHO processes using dielectric spectroscopy. Biotechnol Prog 2010;26:1187-99.
[43] Carvell JP, Dowd JE. On-line Measurements and Control of Viable Cell Density in Cell Culture Manufacturing Processes using Radio-frequency Impedance. Cytotechnology 2006;50:35-48.
[44] Nacke T, Barthel A, Frense D, et al. Application of High Frequency Sensors for Contactless Monitoring in Disposable Bioreactors. Chem Ing Tech 2013;85:179-85.
[45] Duysens LN, Amesz J. Fluorescence spectrophotometry of reduced phosphopyridine nucleotide in intact cells in the near-ultraviolet and visible region. Biochim Biophys Acta 1957;24:19-26.
[46] Scheper T, Schügerl K. Culture fluorescence studies on aerobic continuous cultures of Saccharomyces cerevisiae. Appl Microbiol Biotechnol 1986;23:440-4.
[47] Scheper T, Lorenz T, Schmidt W, Schügerl K. Measurement of culture fluorescence during the cultivation of Penicillium chrysogenum and Zymomonas mobilis. J Biotechnol 1986;3:231-8.
[48] Scheper T, Schügerl K. Characterization of bioreactors by in-situ fluorometry. J Biotechnol 1986;3:221-9.
[49] Gschwend K, Beyeler W, Fiechter A. Detection of reactor nonhomogeneities by measuring culture fluorescence. Biotechnol Bioeng 1983;25:2789-93.
[50] Lindemann C, Marose S, Nielsen HO, Scheper T. 2-Dimensional fluorescence spectroscopy for on-line bioprocess monitoring. Sens Actuators B 1998;51:273-7.
[51] Marose S, Lindemann C, Scheper T. Two-dimensional fluorescence spectroscopy: a new tool for on-line bioprocess monitoring. Biotechnol Prog 1998;14:63-74.
[52] Marose S, Lindemann C, Ulber R, Scheper T. Optical sensor systems for bioprocess monitoring. Trends Biotechnol 1999;17:30-4.
[53] Lindemann C, Marose S, Scheper T, et al. Two-dimensional fluorescence spectroscopy for application in biotechnology. Proc SPIE-Int Soc Opt Eng 1999;3534:89-91.
[54] Ganzlin M, Marose S, Lu X, et al. In situ multi-wavelength fluorescence spectroscopy as effective tool to simultaneously monitor spore germination, metabolic activity and quantitative protein production in recombinant Aspergillus niger fed-batch cultures. J Biotechnol 2007;132:461-8.
[55] Henderson RK, Baker A, Murphy KR, et al. Fluorescence as a potential monitoring tool for recycled water systems: a review. Water Res 2009;43:863-81.
[56] Galinha CF, Carvalho G, Portugal CAM, et al. Two-dimensional fluorescence as a fingerprinting tool for monitoring wastewater treatment systems. J Chem Technol Biotechnol 2011;86:985-92.
[57] Galinha CF, Carvalho G, Portugal CA, et al. Real-time monitoring of membrane bioreactors with 2D-fluorescence data and statistically based models. Water Sci Technol 2011;63:1381-8.
[58] Grote B, Zense T, Hitzmann B. 2D-fluorescence and multivariate data analysis for monitoring of sourdough fermentation process. Food Control 2014;38:8-18.
[59] Boehl D, Solle D, Toussaint HJ, et al. Application of fluorescence spectroscopy for on-line bioprocess monitoring and control. Proc SPIE 2001:50-7.
[60] Hantelmann K, Kollecker M, Hull D, et al. Two-dimensional fluorescence spectroscopy: a novel approach for controlling fed-batch cultivations. J Biotechnol 2006;121:410-7.

[61] Bogomolov A, Grasser T, Hessling M. In-line monitoring of Saccharomyces cerevisiae fermentation with a fluorescence probe: new approaches to data collection and analysis. J Chemom 2011;25:389-99.
[62] Clementschitsch F, Jurgen K, Florentina P, Karl B. Sensor combination and chemometric modelling for improved process monitoring in recombinant E. coli fed-batch cultivations. J Biotechnol 2005;120:183-96.
[63] Rhee JI, Kang T-H. On-line process monitoring and chemometric modeling with 2D fluorescence spectra obtained in recombinant *E. coli* fermentations. Process Biochem 2007;42:1124-34.
[64] Jain G, Jayaraman G, Kökpinar Ö, et al. On-line monitoring of recombinant bacterial cultures using multi-wavelength fluorescence spectroscopy. Biochem Eng J 2011;58–59:133-9.
[65] Calvet A, Li B, Ryder AG. Rapid quantification of tryptophan and tyrosine in chemically defined cell culture media using fluorescence spectroscopy. J Pharm Biomed Anal 2012;71:89-98.
[66] Teixeira AP, Duarte TM, Carrondo MJ, Alves PM. Synchronous fluorescence spectroscopy as a novel tool to enable PAT applications in bioprocesses. Biotechnol Bioeng 2011;108:1852-61.
[67] Jose GE, Folque F, Menezes JC, et al. Predicting Mab product yields from cultivation media components, using near-infrared and 2D-fluorescence spectroscopies. Biotechnol Prog 2011;27:1339-46.
[68] Teixeira AP, Oliveira R, Alves PM, Carrondo MJ. Advances in on-line monitoring and control of mammalian cell cultures: Supporting the PAT initiative. Biotechnol Adv 2009;27:726-32.
[69] Teixeira AP, Portugal CA, Carinhas N, et al. In situ 2D fluorometry and chemometric monitoring of mammalian cell cultures. Biotechnol Bioeng 2009;102:1098-106.
[70] Hakemeyer C, Strauss U, Werz S, et al. Near-infrared and two-dimensional fluorescence spectroscopy monitoring of monoclonal antibody fermentation media quality: aged media decreases cell growth. Biotechnol J 2013;8:835-46.
[71] Bokobza L. Near infrared spectroscopy. J Near Infrared Spectrosc 1998;6:3-17.
[72] Lee HW, Christie A, Liu JJ, Yoon S. Estimation of raw material performance in mammalian cell culture using near infrared spectra combined with chemometrics approaches. Biotechnol Prog 2012;28:824-32.
[73] Kemper MS, Luchetta LM. A guide to raw material analysis using near infrared spectroscopy. J Near Infrared Spectrosc 2003;11:155-74.
[74] Cervera AE, Petersen N, Lantz AE, et al. Application of near-infrared spectroscopy for monitoring and control of cell culture and fermentation. Biotechnol Prog 2009;25:1561-81.
[75] Scarff M, Arnold SA, Harvey LM, McNeil B. Near Infrared Spectroscopy for Bioprocess Monitoring and Control: Current Status and Future Trends. Crit Rev Biotechnol 2006;26:17-39.
[76] Sandor M, Rudinger F, Solle D, et al. NIR Spectroscopy for Process Monitoring and Control in Mammalian Cell Cultivation. BioProcess Int 2013;11:40-9.
[77] Raman CV, Krishnan KS. A New Type of Secondary Radation. Nature 1928;121:2.
[78] Larkin PJ. IR and Raman Spectroscopy Principles and Spectral Interpretation. Elsevier 2011.
[79] Cannizzaro C, Rhiel M, Marison I, von Stockar U. On-line monitoring of Phaffia rhodozyma fed-batch process with in situ dispersive Raman spectroscopy. Biotechnol Bioeng 2003;83:668-80.
[80] Lee H, Boccazzi P, Gorrett N, et al. In situ bioprocess monitoring of *Escherichia coli* bioreactions using Raman Spectroscopy. Vib Spectrosc 2004:131-7.
[81] Abu-Absi NR, Kenty BM, Cuellar ME, et al. Real time monitoring of multiple parameters in mammalian cell culture bioreactors using an in-line Raman spectroscopy probe. Biotechnol Bioeng 2011;108:1215-21.
[82] Berry B, Moretto J, Smelko JP, Wiltberger K. Cross-Scale Predictive Modeling of CHO Cell Culture Growth and Metabolites using Raman Spectroscopy and Multivariate Analysis (to be submitted).

7.2 Analytical Techniques and Quality Control of Protein Therapeutics

Amy Hong Que, Yang Wang, and Steven Ziyi Kan

7.2.1 Protein Analytical Techniques

Therapeutic proteins are usually expressed in mammalian cell lines and are produced containing several variants with differences in modifications or isoforms, i.e., with microheterogeneity [1, 2]. Microheterogeneity can result from posttranslational modifications and processing, environment alteration, storage, and incorrect translation of the target protein [1, 3]. Common sources of heterogeneity include glycosylation, partial carboxypeptidase processing of heavy chain (HC) C-terminal lysine residues [4], amino terminal modifications (e.g., to pyroglutamate), asparagine (Asn) deamidation or isomerization [5], oxidation, fragmentation [6], aggregation, sequence variants [7], phosphorylation and sulfation, etc. These variations in protein composition may impact the activity and stability of biotherapeutics. Characterization and monitoring stability of therapeutic proteins are therefore regarded as essential for demonstrating safety and efficacy of biotherapeutics, and are required by the United States Food and Drug Administration (FDA) and other regulatory agencies.

The microheterogeneity can be mainly divided into a few different categories: (1) size variants, (2) charge variants, and (3) glycoforms due to posttranslational modification.

This section describes the characterization of a therapeutic protein with emphasis on the following physicochemical attributes: size heterogeneity, charge heterogeneity, glycoform heterogeneity, primary structure, biological activity, and common process impurities.

7.2.1.1 Size Heterogeneity

Size microheterogeneity is mainly caused by protein aggregation and fragmentation. Aggregation of therapeutic proteins is formed during drug manufacturing and storage, and it is considered to be a critical attribute because of the known risk of enhanced potential of undesired immunogenicity. It needs to be characterized and controlled stringently during the development and production in order to prevent loss of drug efficacy or adverse immunogenic effects. The size exclusion chromatography (SEC), sodium dodecyl sulfate polyacrylamide gel electrophoresis (SDS-PAGE), and capillary electrophoresis with SDS (CE-SDS) are commonly used as purity assays to analyze protein size variants.

Size Exclusion Chromatography (SEC)

SEC provides quantitative information about the molecular size distribution of a native protein, including aggregates, monomer, and fragments.

Size variants are separated commonly using a TSK G3000SW$_{xl}$ column (7.8 × 300 mm, Tosoh) or equivalent, eluted isocratically with a mobile phase consisting of 0.2 mol L^{-1} potassium phosphate (pH 6.2 to 7.0). The separation is conducted at ambient temperature with a typical flow rate of 0.5 mL min^{-1}. Proteins eluted from column are detected with either optical absorbance at 280 nm or intrinsic fluorescence (excitation at 280 nm and emission at 340 nm). Sodium sulfate or potassium chloride may be added to the buffer to increase buffer ionic strength in order to minimize undesired nonspecific interaction with the SEC column. TSK-gel SW type of columns can also be operated under these conditions using denaturing settings, such as with SDS containing eluents, or formulation containing detergents such as Tween or Triton.

A typical SEC profile of a monoclonal antibody (mAb) is shown in Figure 7.2.1. Aggregates, monomer, and fragments are analyzed and summarized as the contents of high molecular weights (HMW) species, monomer, and low molecular weight (LMW) species, respectively.

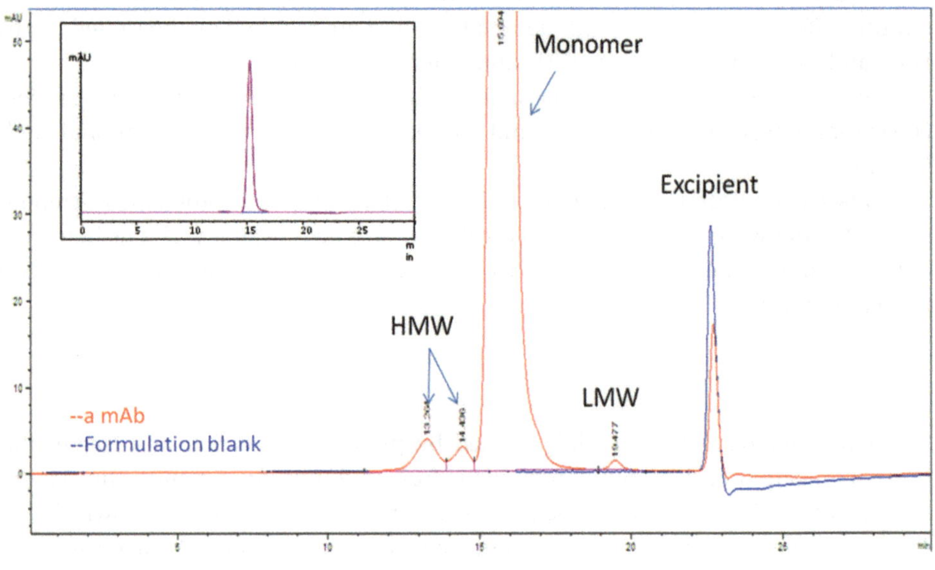

Figure 7.2.1: A typical SEC profile of a mAb.

Sodium Dodecyl Sulfate Polyacrylamide Gel Electrophoresis (SDS-PAGE)

As an essential and classical biochemical technique, SDS-PAGE analysis has been commonly used for many years to assess overall product purity in the development of biopharmaceutical products for lot release, stability testing, batch-to-batch consistency confirmation, and product characterization. Although these methods are often semiquantitative, they are powerful tools for detecting minor protein impurities.

SDS is an anionic detergent applied to a protein sample to denature or linearize proteins and to impart a negative charge to the linearized proteins. Proteins are denatured by heating in the presence of SDS-PAGE sample buffer. In most proteins, the binding of SDS to the polypeptide chain imparts an even distribution of charge per unit mass, thereby resulting in a fractionation by approximate size during electrophoresis. Nonreduced samples can be alkylated with iodoacetamide to prevent artifacts caused by disulfide bond shuffling, while reduced samples are treated with dithiothreitol (DTT) or β-mercaptoethanol (BME), which further denatures the proteins by reducing disulfide linkages, thus overcoming some forms of residual protein folding, and breaking up quaternary protein structural linkage (oligomeric subunits). The prepared samples and molecular weight standards are separated on a polyacrylamide or other gels in electrophoresis. The protein components were then visualized by Coomassie, or silver, or SYPRO Ruby staining, the approximate molecular weights of the samples are assessed based on comparison with those of the molecular weight markers.

CE-SDS UV/Florescence Detection

Capillary electrophoresis (CE)-SDS method has advantages over classical SDS-PAGE in assay precision, linearity, specificity, separation resolution, robustness, automation, etc., which makes CE-SDS assay easier to be validated and to be used in quality control (QC) environment [9-12]. The method has played a critical role in purity analysis and structural characterizations. Many applications of CE-SDS rely on UV detection, which is equivalent to SDS-Coomassie staining sensitivity. However, certain companies, including Genentech, have chosen a laser induced fluorescence (LIF) detection strategy in order to increase detection sensitivity. Figure 7.2.2 shows an example data for a CE-SDS-LIF analysis of a protein in both reduced and nonreduced modes. Labeling of proteins with 5-carboxytetramethylrhodamine succinimidylester (5-TAMRA.SE) can significantly decrease gel interference during online detection while providing up to a hundredfold increase in sensitivity.

7.2.1.2 Charge Heterogeneity

Manufacturing control and stability monitoring of therapeutic proteins are regarded as essential for ensuring safety and efficacy of these drugs and are required by the FDA and other regulatory agencies. Regulatory agencies routinely require assessment of charge heterogeneity of biotechnology products. Charge variants of recombinant proteins often include oxidation, aspartic isomerization, asparagine deamidation, amino terminal modifications (e.g., to pyroglutamate), truncation of the C-terminus,

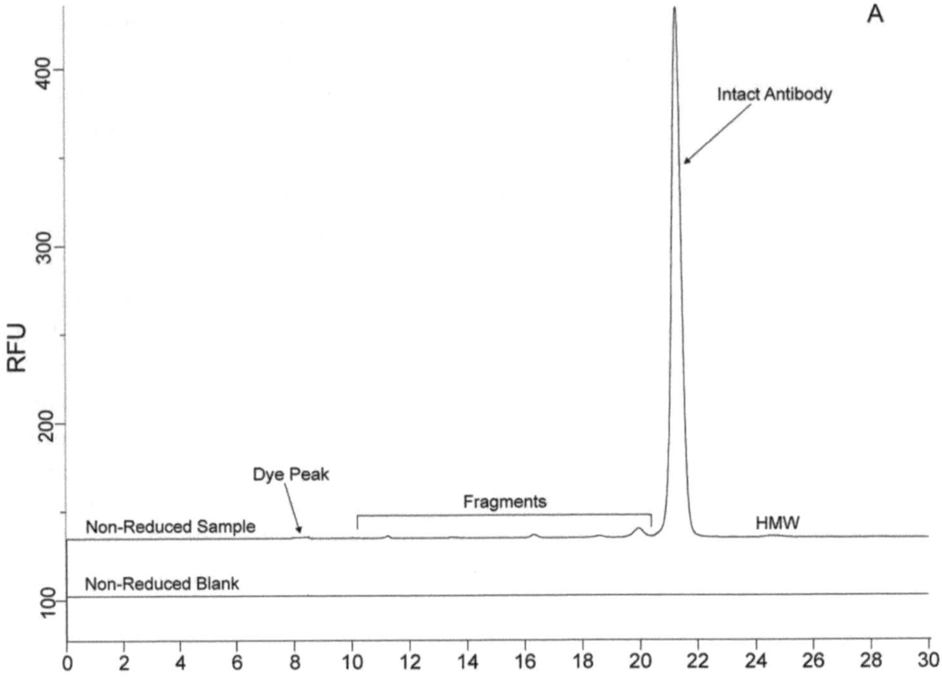
A
Intact Antibody
Dye Peak
Fragments
Non-Reduced Sample
HMW
Non-Reduced Blank
RFU
Migration time (min)
0
2
4
6
8
10
12
14
16
18
20
22
24
26
28
30
100
200
300
400

a)

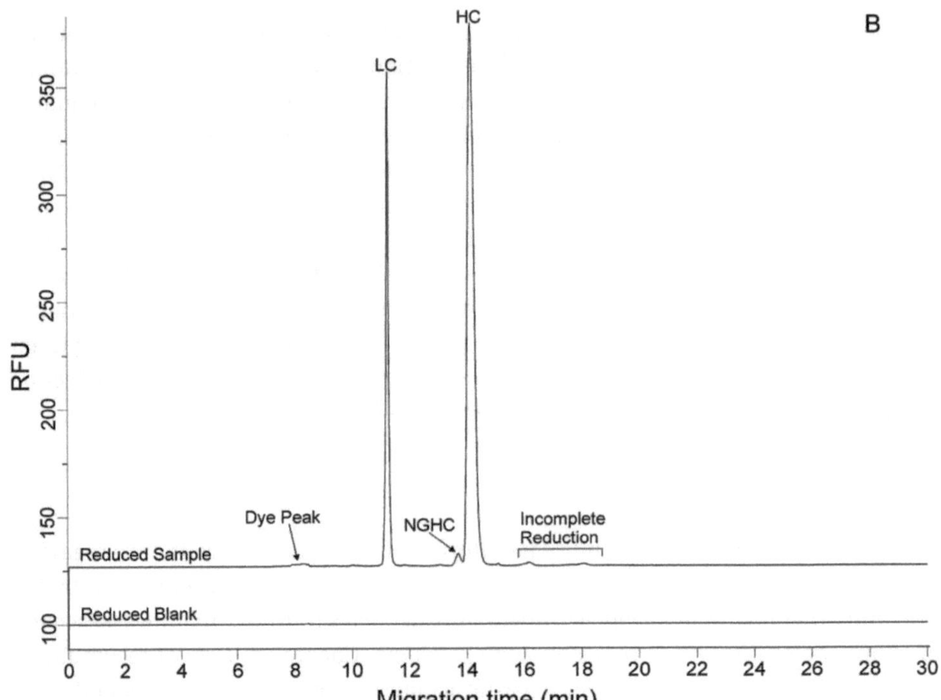
B
HC
LC
Dye Peak
NGHC
Incomplete
Reduction
Reduced Sample
Reduced Blank
RFU
Migration time (min)
0
2
4
6
8
10
12
14
16
18
20
22
24
26
28
30
100
150
200
250
300
350

b)

and posttranslational modifications such as sialylation, sulfation, and phosphorylation. These changes are considered to have the potential to affect critical functionality and need to be characterized and controlled during the development and production of a therapeutic protein. Ion exchange chromatography (IEC) and capillary isoelectric focusing (cIEF) are commonly used to analyze protein charge variants. The 2 methods are complementary to each other and each method has its own advantages and disadvantages. IEC is more robust and easy with fraction collection while cIEF often provides better resolution. The separation mechanisms of 2 methods are different and, therefore, probe somewhat different properties of protein charge variants.

Ion Exchange Chromatography (IEC)

IEC separation relies on surface charge-charge interactions between the exposed charge groups of protein and charged functional groups in column matrix. Consequently, the method is more sensitive to the protein structure, i.e., the extent of exposure and arrangement of the charge groups on proteins. For example, ProPac® WCX-10 column (Dionex) is a weak cation exchange column commonly used for the separation and analysis of mAbs, in which positively charged ions on exposed protein surface bind to a negatively charged functional group on column and are eluted by a salt or pH gradient [13-14].

Figure 7.2.3 shows a typical separation of a monoclonal antibody (mAb X) using ProPac WCX-10 column. This data illustrates that this mAb contains 2 variants in the acidic region (Peak 1 and 2), main peak (Peak 3), and 4 variants in the basic region (Peak 4, 5, 6, and 7). After treatment with carboxypeptidase (CpB) which is an enzyme specifically cleaving C-terminal lysine, Peaks 5, 6, and 7 disappeared in native mAb while main peak (Peak 3) and a basic variant (Peak 4) increased. It indicates that Peaks 5, 6, and 7 are related to the 1 or 2 C-terminal lysines in heavy chains, where Peak 3 and 4 represent the variant contains des-lysine. The formation of pyro-glutamine at the N-terminal makes the protein more acidic, as a result Peak 3 containing 2 pyro-glutamine and Peak 4 containing 1 pyro-glutamine are separated in IEC. Therefore, the main product contains pyro-glutamine at both N-terminus and des-lysine at both C-terminus of 2 heavy chains (Peak 3 in Figure 7.2.3). Four basic variants are: pyro-glutamine at 1 N-terminal and des-lysine at both C-terminus of 2 heavy chains

Figure 7.2.2: CE-SDS separations of **a)** nonreduced an **b)** reduced preparations of a 5-TAMRA SE-labeled rMAb sample. Separation conditions were as follows: ProteomeLabTM PA 800 instrument equipped with LIF detection; 50 µm ID, 375 µm OD uncoated fused-silica capillary with effective length 21.2 cm and total length 31.2 cm; both anode and cathode buffers were Beckman Coulter CE-SDS gel solutions. Samples were injected at a constant electric field of 160 V cm^{-1} for 20 s and electrophoresed at 480 V cm^{-1} (32.5 µA) and 40 °C (reprinted from [12], with permission).

(Peak 4), pyro-glutamine at both N-terminus and lysine at 1 C-terminal of 2 heavy chains (Peak 5), combination components of pyro-glutamine at 1 N-terminal and des-lysine at 1 C-terminal of 2 heavy chains and pyro-glutamine at both N-terminus and lysine at both C-terminus of 2 heavy chains (Peak 6), and pyro-glutamine at 1 N-terminus and lysine at both C-terminus of 2 heavy chains (Peak 7).

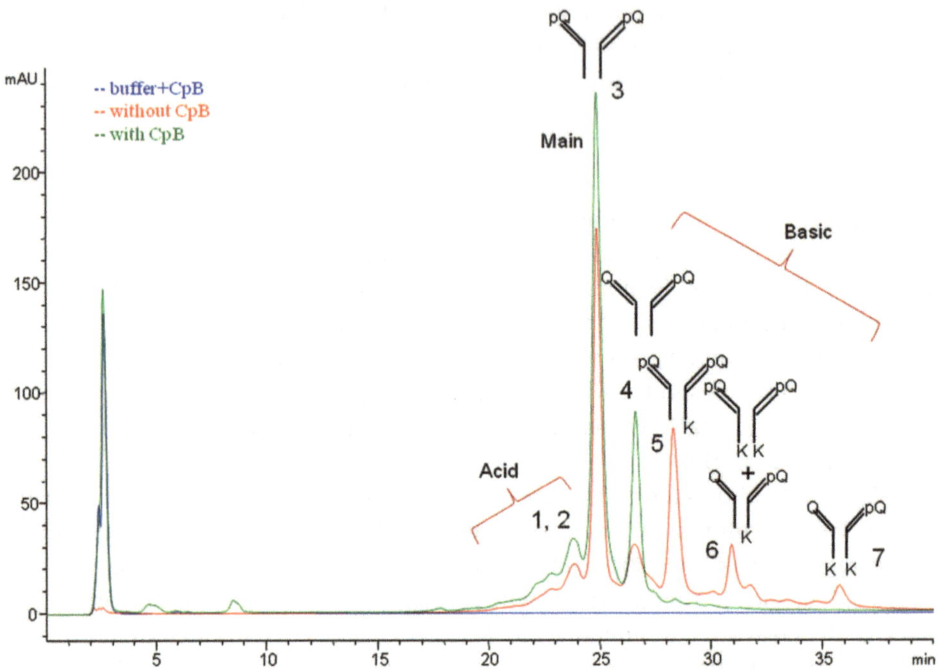

Figure 7.2.3: A cation-exchange chromatogram of a mAb A; native mAb and mAb treated with Carboxypeptidase B.

Capillary Isoelectric Focusing (cIEF)

Due to the mechanism of separation, the development of an IEC method is typically extensive and product-specific. IEC assays require the optimization of complex separation parameters including column type, mobile phase composition, pH, salt, temperature, and gradient. In contrast, cIEF techniques offer the advantages of faster analysis time and development of generic methods for multiple products, which are desirable in today's arena of fast paced therapeutic protein development. Convenient protein and mAb chemistry kits are commercially available from BECKMAN COULTER PA800 and ProteinSimple's iCE3, which facilitate not only method development but also integration into quality control.

cIEF separation is based on net charge heterogeneity of a protein or pI, an intrinsic property instead of a property of interacting with matrix. It combines the separation fidelity of gel-based IEF with the automation and quantitation of capillary column sample injection and detection. ProteinSimple (acquired the formerly Convergent Bioscience) was the first to commercialize the whole-column detection cIEF technique in the iCE IEF analyzer instrument, named as imaged cIEF (icIEF). This unique technology eliminates the need for a mobilization step of on-column detection technique. Figure 7.2.4 shows the principle of performing cIEF using ProteinSimple's iCE3 system [15]. In performing cIEF with the iCE instrument, protein samples are first premixed with carrier ampholytes, additives, and pI markers. The mixture is injected to fill the entire capillary column. A separation voltage is applied to the anolyte and catholyte. Under the voltage, a pH gradient is created within the column. Proteins are separated and focused along the capillary column based on their apparent pI. The whole-column detector monitors the IEF process in a real-time fashion within the separation column, and the focusing time can be optimized in a single sample run. At the end of the focusing process, all the focused protein zones within the column are recorded by a charge-coupled device (CCD) digital camera detector through protein absorbance around 280 nm without disturbing the separation resolution. Finally, the column is washed and ready for the next sample injection. Any sample precipitation and aggregation during focusing can be observed. Different additives are easily selected to improve reproducibility when issues are identified. The biggest advantage of whole-column detection in cIEF is its utilities in fast and easy method development because of the ability to monitor the IEF process within the separation column, though it has less structural information.

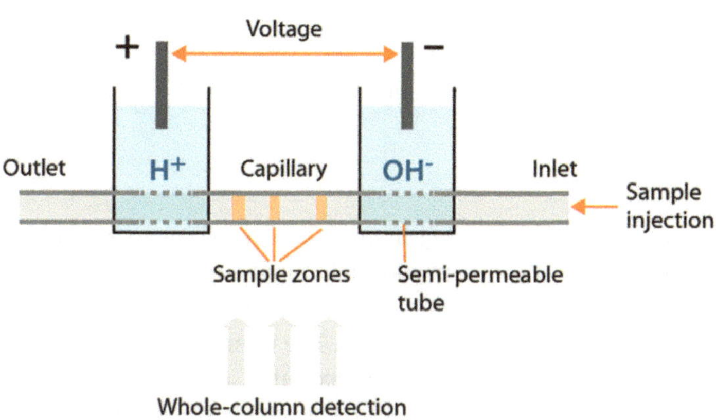

Figure 7.2.4: Principle of performing image cIEF using ProteinSimple's iCE3 system.

Figure 7.2.5 illustrates the powerful separation efficiency of imaged cIEF technology as a tool for the analysis of charge variants of a monoclonal antibody. The ampholyte solution consisted of pharmalyte pH 3 to 10 and 2 pI markers: 5.85 and 8.79. A fluorocarbon-coated capillary cartridge was used for the separation. The anolyte was 80 mmol L^{-1} phosphoric acid, and the catholyte was 100 mmol L^{-1} sodium hydroxide, both in 0.1 % methyl cellulose. Additionally, cIEF can be used as an identity assay on the basis of high resolution and reproducibility. In this study, the pI of the major component was determined to be 7.5.

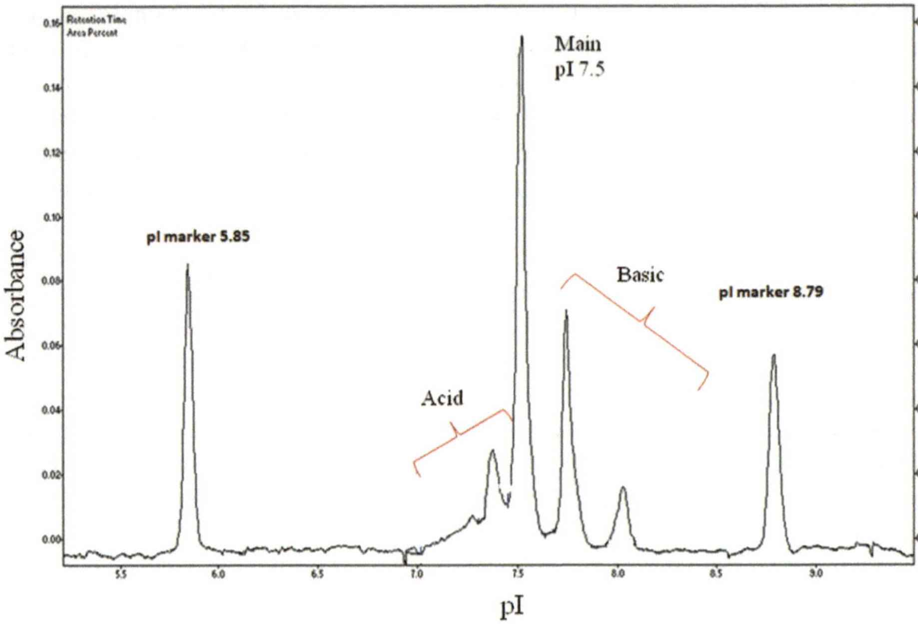

Figure 7.2.5: Determination of charge variants and pI of mAb B by imaged cIEF.

7.2.1.3 Glycosylation Heterogeneity

Protein therapeutics is usually produced using mammalian cell lines in a complex cell culture process. Glycosylation is a common cotranslational and posttranslational modification of a protein primary structure, which results in the covalent attachment of oligosaccharide chains to the polypeptide backbone. The phenomenon adds great variability in proteins in terms of attachement sites, extent of the attachment, oligosaccharide sequence and composition. Even within the same glycosylation site of the same protein, the composition of oligosaccharide chains can vary. Glycoforms are defined as the glycoproteins with identical amino acid sequences, but with different oligosaccharide-chain location, composition, and structure. O-linked glycosylation (sugar side chains attached via serine or threonine residues, no concensus sequence) or N-linked glycosylation (sugar side chains attached via asparagine residues, consensus sequence Asn-Xxx-Ser/Thr) can influence protein stability, ligand binding, bioactivity, immunogenicity, and serum half-life [16].

Immunoglobulin G (IgG) antibodies are important class of protein therapeutics. They are composed of 2 heavy chains (HC) and 2 light chains (LC) that form the Fab (fragment antigen binding) and Fc regions. Naturally occurring IgG is glycosylated in the Fc region at asparagine (Asn) residue 297. The N-glycan has predominantly the biantennary complex structure with the core heptasaccharide of $GlcNAc_2$-Man_3-$GlcNAc_2$ (Figure 7.2.6). Varieties of oligosaccharides of antibodies include the presence or absence of galactose, bisecting N-acetylglucosamine (GlcNAc), or sialic acid at the nonreducing terminal and also the presence of a fucose at the reducing terminal (see Figure 7.2.6).

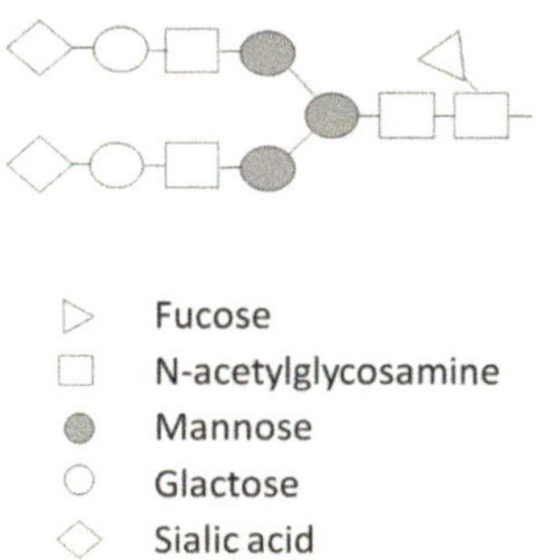

Figure 7.2.6: N-linked oligosaccharides present in human IgG.

Carbohydrate chains in mAb pharmaceuticals play important roles for their biological activities, such as antibody-dependent cellular cytotoxicity, whereas the oligosaccharide profile is easily changed depending on the manufacturing conditions, such as different cell culture parameters. Irrespective of potential biological consequences, regulatory agencies require a demonstration of oligosaccharide consistency across manufacturing campaigns or process changes, in part because that glycosylation is often sensitive to cell culture conditions, thereby serving as a potential indicator for changes in other sources of heterogeneity [17].

Various analytical methods have been developed for the analysis of N-linked and O-linked glycans.

Enzymatic Release and Fluorescent Derivatization of Oligosaccharides

Structures of oligosaccharides linked to the core protein and their distribution should be evaluated, preferably quantitatively. Because of the presence of other heterogeneities, including those of protein backbone such as the processing of a C-terminal lysine residue, oxidation of a methionine residue and pyroglutamination of an N-terminal glutamine residue, direct evaluation of glycoform heterogeneity of protein therapeutics quantitatively is somewhat difficult and often require high-end mass spectrometry technique. Therefore, an indirect, quantitative method is often used.

Oligosaccharides can be released by enzymatic digestion using peptide-N-glycosidase F (PNGase F). PNGase F is capable of releasing almost all oligosaccharides from a protein at the amide linkage between GlcNAc and asparagine residue unless there exists α(1-3) core fucosylated. Figure 7.2.7 is a scheme of deglycosylation reaction of N-glycans with PNGase F. The asparagine residue, from which the glycan is removed, is deaminated to aspartic acid. Reduction and denaturation are often required for complete release of oligosaccharides in some glycoproteins. However, N-linked oligosaccharides of IgG are usually completely released by simple digestion of native IgG samples.

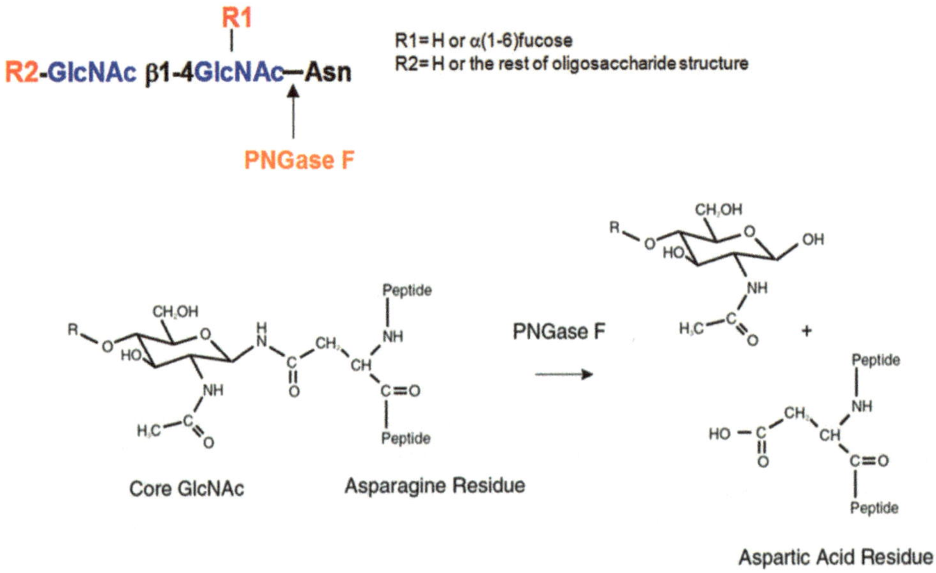

Figure 7.2.7: Deglycosylation reaction of N-glycans with PNGase F.

Released oligosaccharides are usually required to be derivatized with fluorophoric reagents to improve detection sensitivity and resolution by capillary electrophoresis (CE) and/or high performance liquid chromatography (HPLC) method. Because glycans have low spectral signal in both UV and visible regions, it is often necessary to label them in order to enhance detection. N-linked oligosaccharides are usually fluorescently labeled with 2-aminobenzoic acid (2-AA) or 2-aminobenzamide (2-AB) prior to HPLC-fluorescence detection and matrix-assisted laser desorption ionization time of flight mass spectrometry (MALDI-TOF-MS) profiling, or with 8-aminopyrene-1,3,6-trisulfonate (APTS) prior to capillary electrophoresis and laser-induced fluorescence detection (CE-LIF). These labeling reagents provide valuable tools for glycan analysis, due to their sensitivity and stability when covalently bound to glycans [18].

Derivatization of glycans is based on Schiff's base reaction, as shown in Figure 7.2.8. Oligosaccharides with a free reducing sugar exist in equilibrium between the cyclic (closed ring) and acyclic (open ring) structures. A stable Schiff's base is formed when the carbonyl carbon of an acyclic reducing sugar is linked to the amine moiety of the dye in a nucleophilic manner. Following formation of the Schiff's base, the resulting imine group is reduced using sodium cyanoborohydride as a reductant, resulting in a stable labeled glycans that can be excited with light at 330 nm and emit fluorescent light at 420 nm.

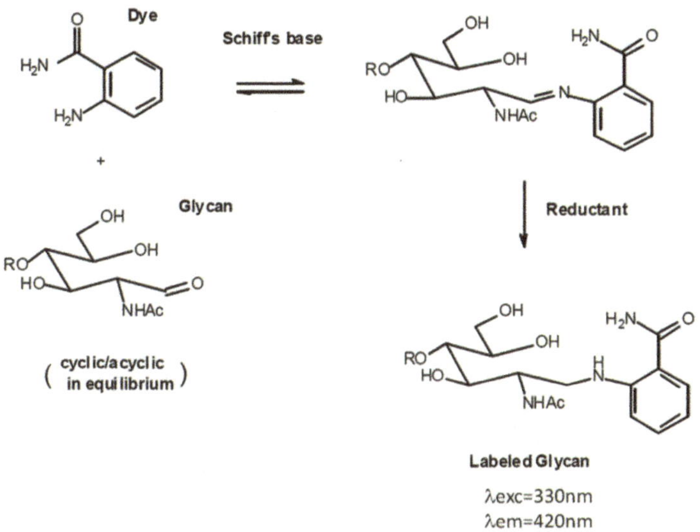

Figure 7.2.8: Reductive amination derivatization approach for oligosaccharide labeling.

Oligosaccharide Profiling by Capillary Electrophoresis: Laser-induced Fluorescent (CE-LIF)

A fluorescent reagent, 1-aminopyrene-3,6,8-trisulfonic acid, trisodium salt (APTS), for CE-LIF of the released oligosaccharides has been used as a simple and convenient method for the analysis of oligosaccharides by CE [19-21], and the analysis kit is commercially available (Beckman Coulter). Since APTS has 3 sulfate groups in a highly fluorescent pyrene ring, APTS derivatives of oligosaccharides can be successfully analyzed fast with high resolution based on their highly negative charges. The method is particularly superior to HPLC in resolving power and speed. Figure 7.2.9 shows the analysis of oligosaccharides in the commercially available antibody pharmaceutical trastuzumab, as an example [20]. Four oligosaccharides typically present in mAbs are clearly separated and the analysis is completed within 6 min. The use of the APTS method has been reported for the quantitative analysis of oligosaccharides derived from various proteins.

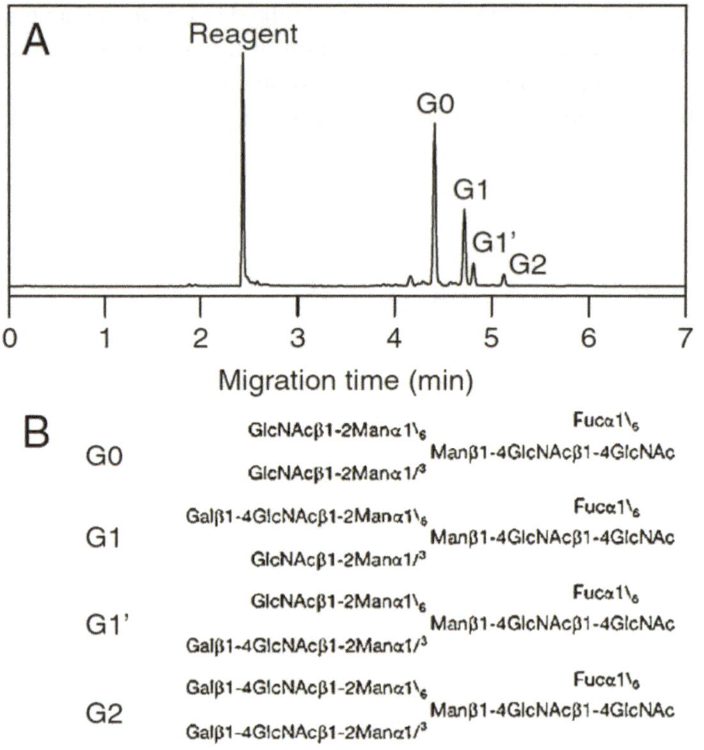

Figure 7.2.9: **a)** Analysis of APTS-labeled oligosaccharides derived from trastuzumab by CE-LIF. **b)** Major oligosaccharides in trastuzumab. Analytical conditions: CE system, ProteomeLab PA800; running buffer, 50 mmol L^{-1} Tris-acetate buffer (pH 7.0) containing 0.5 % polyethylene glycol (M = 70,000); capillary, DB-1 (full length 30 cm, effective length 20 cm, 50 mm id); injection: 0.5 psi, 5 s; separation, –18 kV, 25 °C; detection, Ar laser excitation fluorescence detection (excitation wavelength 488 nm, fluorescence emission wavelength 520 nm) (reprinted from [20], with permission).

Oligosaccharide Profiling by HPLC

Hydrophilic-interaction chromatography (HIC) is a well-known and reliable technique that can effectively separate fluorescently labeled glycans. The separation mechanism is based on the subtle difference in hydrophilicity between individual glycan and polar stationary phase, namely that glycans with stronger hydrophilicity elutes later. The most successful HIC resin or matrix for this purpose is amide-modified silica, which delivers good retention and selectivity for both neutral and (multiple) charged glycans in 1 separation. This type of stationary phase is currently preferred for the separation of 2-AB modified glycans [22].

Figure 7.2.10 shows that 2-AB glycans released from a monoclonal antibody are separated by a TSK-gel amide-80 column and detected by a fluorescence detector at excitation wavelength 320 nm and emission wavelength 420 nm. The antibody contains 1 N-linked oligosaccharide site per heavy chain, which is located in the CH2 domain of the Fc portion of the molecule. The relative distribution of the oligosaccharides found at this site was determined by HPLC after enzymatic release with PNGase F and fluorescent labeling with 2-AB. The N-linked oligosaccharides present in this antibody are typical of those observed on other CHO-produced monoclonal antibodies, and corresponding graphical representations of the oligosaccharide structures are shown in Figure 7.2.9.

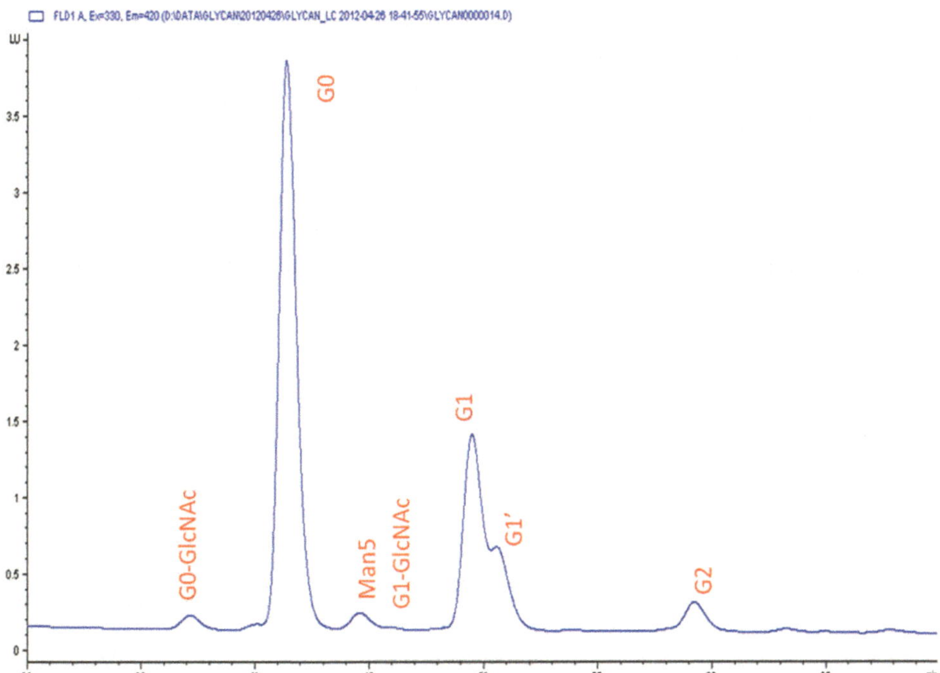

Figure 7.2.10: Analysis of 2-AB-labeled oligosaccharides derived from a monoclonal antibody by HPLC.

Waters' ACQUITY Ultra Performance Liquid Chromatography (UPLC) BEH glycan column is specifically designed for the use on Waters' ACQUITY UPLC system to give improved separation resolution, sensitivity, and speed compared to traditional LC techniques. It can provide users with more certain identification and reliable quantitation [23]. UPLC is based on the use of columns packed with smaller particles and instruments optimized for use with these high-resolution columns.

Oligosaccharide Profiling by Matrix-assisted Laser Desorption Ionization Time of Flight Mass Spectrometry (MALDI-TOF MS)

The advances in modern mass spectrometers (MS) with high resolution and mass accuracy have led to its use in analyzing glycans/glycoproteins for profiling and structural studies. MALDI-TOF MS is ideally suited for carbohydrate analysis, and therefore, it is the most widely used MS technique in carbohydrate analysis. Mass spectra can be obtained from underivatized and derivatized glycans. The information gained from MALDI-MS is the mass of individual molecular components and this can be used to assign putative monosaccharide structures present in a pure oligosaccharide since the mass of a monosaccharide is measured with a high degree of accuracy. Like HPLC, MALDI-MS is widely used in conjunction with exoglycosidase enzymes for structural and linkage analysis of glycans.

Many matrixes have been successfully utilized for the analysis of glycoproteins [24]. 2,5-dihydroxybenzonic acid (DHB) is the most common matrix for neutral glycans. Some useful matrices for acidic glycans include 6-aza-2-thiothymine (ATT) and 2,4,6-trihydroxyacetophenone (THAP). Glycopeptides are more effectively examined by α-cyano-4-hydroxycinnamic acid (4-CHCA), either alone, or mixed with nitrocellulose. Sinapinic acid appears to be the most appropriate matrix for larger proteins and glycoproteins.

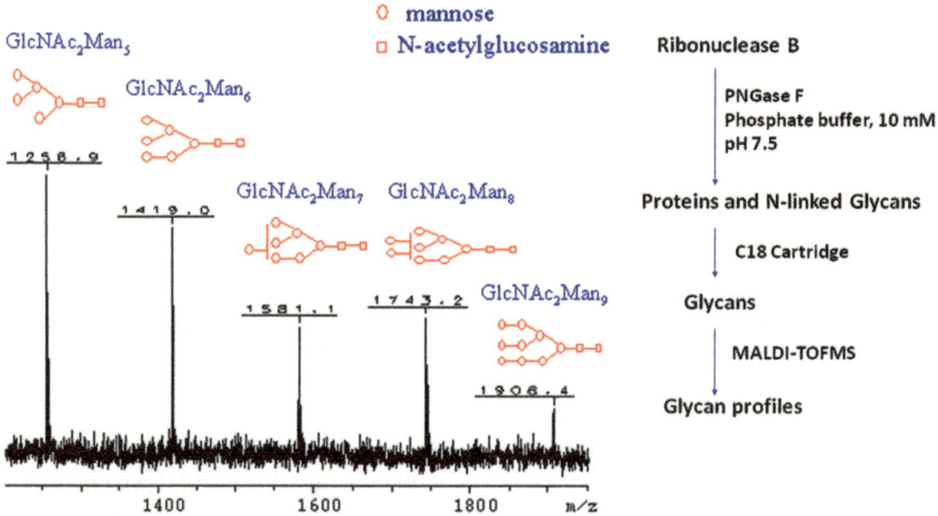

Figure 7.2.11: MALDI-TOF MS of N-glycans cleaved from ribonuclease B.

Figure 7.2.11 shows a glycans profiling by MALDI-MS. Ribonuclease B has 1 N-linked glycosylation site, including 5 different glycans/glycoforms: $GlcNAc_2Man_5$, $GlcNAc_2Man_6$, $GlcNAc_2Man_7$, $GlcNAc_2Man_8$, and $GlcNAc_2Man_9$. $GlcNAc_2Man_7$ and $GlcNAc_2Man_8$ contain isomers considering that the terminal mannose could be located at different branches. After deglycosylation, released glycans from ribonuclease B were purified by C18 cartridge, followed by MALDI-MS analysis.

7.2.1.4 Primary Structure

The expected primary structure of a protein can usually be confirmed by a single or a combination of techniques, such as N-terminal sequence analysis, mass spectrometric analysis of intact or reduced proteins, and liquid chromatography-mass spectrometric (LC-MS) analysis of peptide map post- and endoproteinase treatment.

Terminal Amino Acid Sequence

Terminal amino acid analysis is performed to identify the nature and homogeneity of the amino- and carboxy-terminal amino acids. If the desired product is found by terminal sequencing method to be heterogeneous with respect to the terminal amino acids, the relative amounts of the variant forms should be determined using an appropriate analytical procedure. The sequence of these terminal amino acids should be compared with the terminal amino acid sequence deduced from the gene sequence of the desired product.

N-terminal sequence analysis provides confirmation that the purified recombinant protein has the expected amino-terminal sequence as predicted by the cDNA sequence. Due to the limited sequence coverage, it often serves the purpose of protein identification. Typically, the proteins are subjected to 10 to 15 cycles of Edman degradation using standard methods on protein sequencer.

Mass Measurement of Intact or Reduced Proteins

Mass spectrometric (MS) analysis of the intact or reduced protein with either electrospray ionization time of flight mass spectrometry (ESI-TOF MS) or linear MALDI-TOF MS can confirm the structural integrity of the molecule [25] and characterize the modifications.

Figure 7.1.12 shows the mass spectra of intact and reduced monoclonal antibody analyzed by ESI-MS. After reduction of the disulfide bonds with Tris(2-carboxyethyl) phosphine hydrochloride (TCEP), the reduced samples are desalted by reversed-phase high-performance liquid chromatography (RP-HPLC) for direct online ESI-MS analysis. Molecular masses were obtained for the light chain and the various heavy chain glycoforms, and the observed masses correspond to the predicted masses, confirming expected amino acid sequence of the light and heavy chains. The major masses obtained for the heavy chain correspond to the predicted masses of heavy chain without the carboxy-terminal lysine residue and with either a G0 or G1 oligosaccharide.

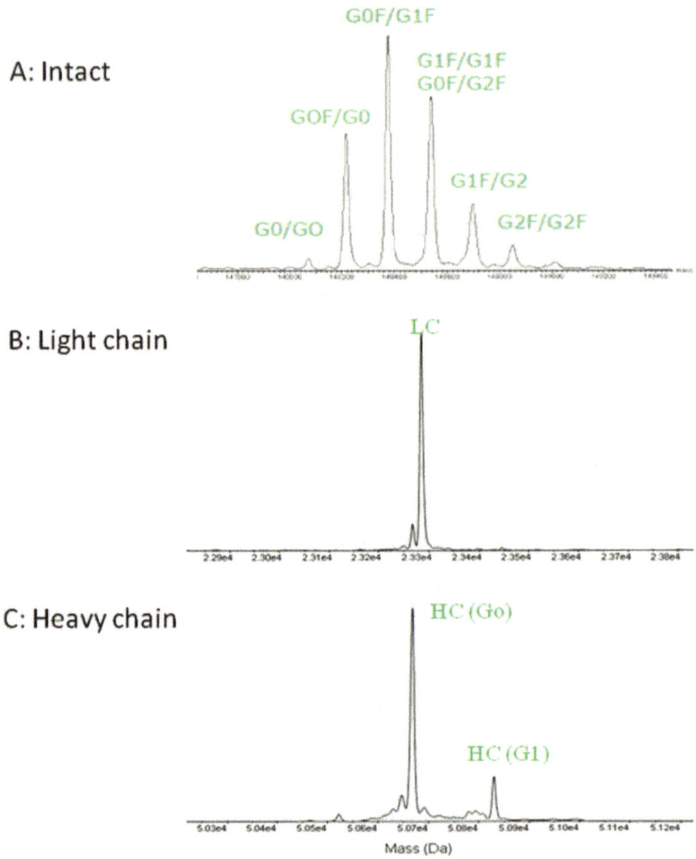

Figure 7.2.12: ESI-MS of intact and reduced mAb.

Peptide Mapping

Peptide map analysis using liquid chromatography tandem mass spectrometry (LC-MS-MS) is used to verify the primary structure of a recombinant protein, to demonstrate lot-to-lot consistency, to monitor protein stability, and to characterizing sequence variants including single or multiple amino acid substitutions [7].

Figure 7.2.13 illustrates an example using peptide mapping to detect one site of amino acid modification (E->H). The peptide mappings of 2 proteins are identical except the last peak (Peak 1 and Peak 2). Based on their masses obtained from mass spectra of the corresponding products (data not shown), Peak 1 (top) corresponds to peptide D_1YFPEPVTVSWNSGALTSGV$\mathbf{E}_{21}$TFPAVLQSSGLYSLSSVVTVPSSSLGTQTYICNVNHKPSNTK, while Peak 2 corresponds peptide D_1YFPEPVTVSWNSGALTSGV$\mathbf{H}_{21}$TFPAVLQSSGLYSLSSVVTVPSSSLGTQTYICNVNHKPSNTK, indicating the existence of one amino acid modification (E_{21}->H). In this case, the modification was resulted from incorrect cDNA sequence. In this study, protein was digested with trypsin after subjecting the protein to reduction and alkylation sulfitolysis, a process that both reduces the disulfide bonds and alters the resultant-free thiols to produce the nonreactive thiosulfonate derivatives. The resulting peptides were separated by RP-HPLC-MS using a 5 μm, 300 Jupiter C18 column (4.6 × 250 mm) at a flow rate of 0.25 mL min^{-1}, with mobile phases containing 0.1 % TFA in water (solvent A) and 0.09 % TFA in 90 % acetonitrile (solvent B).

Peptide mapping by LC-MS (MS/MS) is commonly used to characterize amino acid sequence coverage of a protein, N-terminal and C-terminal modifications, glycosylation sites, disulfide bond linkages, degradation pathways such as deamidation, oxidation, etc.

Peak 1: DYFPEPVTVSWNSGALTSGVETFPAVLQSSGLYSLSSVVTVPSSSLGTQTYICNVNHKPSNTK

Peak 2: DYFPEPVTVSWNSGALTSGVHTFPAVLQSSGLYSLSSVVTVPSSSLGTQTYICNVNHKPSNTK

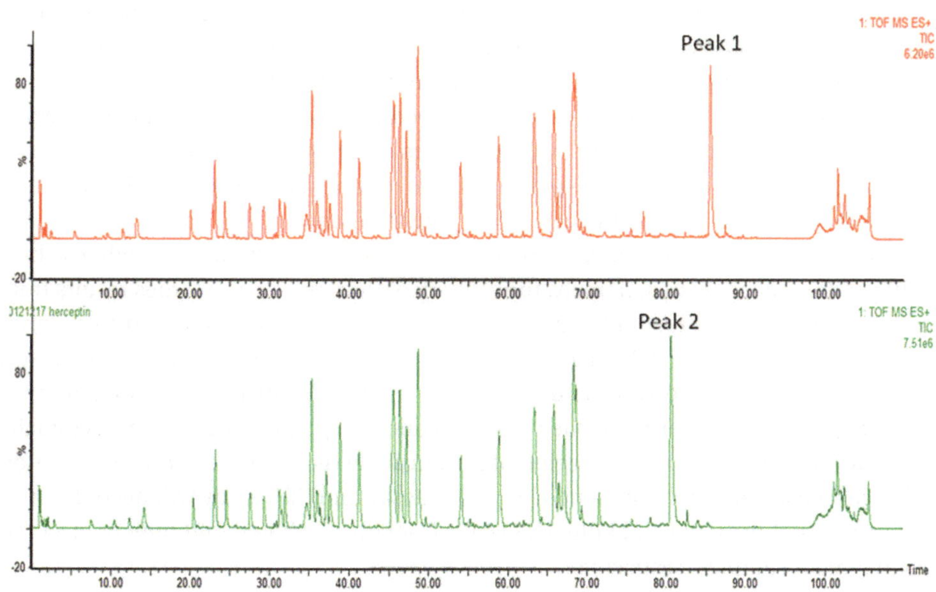

Figure 7.2.13: Peptide mapping (TIC chromatograms) of 2 proteins with one site of amino acid replacement.

7.2.1.5 Functional Assessment

Functional assessment techniques measure the biological activities or potencies of therapeutic proteins; therefore, they are often called bioassays or potency assays. From regulatory perspective, potency is defined as the specific ability or capacity of the product, as indicated by appropriate laboratory tests or by adequately controlled clinical data obtained through the administration of the product in the manner intended, to affect a given result.

For therapeutic proteins, 2 major types of assays have been used for functional assessment or evaluation of biological activity: cell-based biologic assays and non cell-based target-binding assays. Target binding assays, such as those measuring specific and functional relevant antibody-antigen and ligand-receptor association, may be an alternative in early product development. For latter phase, regulatory agencies generally require the use of cell-based activity assays, since they often consider that cell-based assays are more reflective of the *in vivo* situation and are recommended until a clear correlation with the alternatives can be established by extensive evidence and data. Sometimes they both are used in parallel during development, because of the larger variability of the cell-based (bioactivity) assays.

Analytical methods or formats to quantitatively assess specific molecular interactions relevant to mechanism of action (MOA) are numerous, and the major 2 categories are enzyme-linked immunosorbent assay (ELISA) type and biosensor type, as shown in Figure 7.2.14. Both have advantages and disadvantages and can be implemented relatively easily, given good, critical reagents are properly developed.

The relevant cellular responses potentially being measured in cell-based bioassays for potency assessment are also numerous, including readouts such as phosphorylation of intracellular substrates, proliferation, calcium mobilization, adhesion, cell death and so on. In certain cases, one may need to develop cell lines or transfected cell or pseudovirus to express relevant receptors or reporter constructs. Alternatively, for mAb, the ability to block a response originated from a receptor/ligand interaction on the target cells may form the basis for a potency assay.

In addition, since biological activity of therapeutic antibodies is mediated by 2 independent mechanisms, namely, the efficacy of therapeutic antibodies results from a specificity for the target antigen and from biological activities referred to as 'antibody effector functions', which are activated by the formation of immune complexes. Therefore, bioassays need to be developed to assess the effector functions of antibody-dependent cellular cytotoxicity (ADCC) and complement-dependent cytotoxicity (CDC) through the constant region of the antibody (Fc) and of the direct induction of target cell killing. Development of an ADCC assay and CDC assay will need the critical reagents of fresh natural killer (NK) cells from human blood and human or rabbit complement samples, respectively.

Generally, bioassays have significant variability and a limited dynamic range for their activity curves. Such problems can make development and validation of these assays difficult (i.e., ADCC assay using NK cells) and FDA understands such difficulties. Nonetheless, we will recommend such assays because they are critical to understanding the importance of the *in vivo* responses of therapeutic proteins.

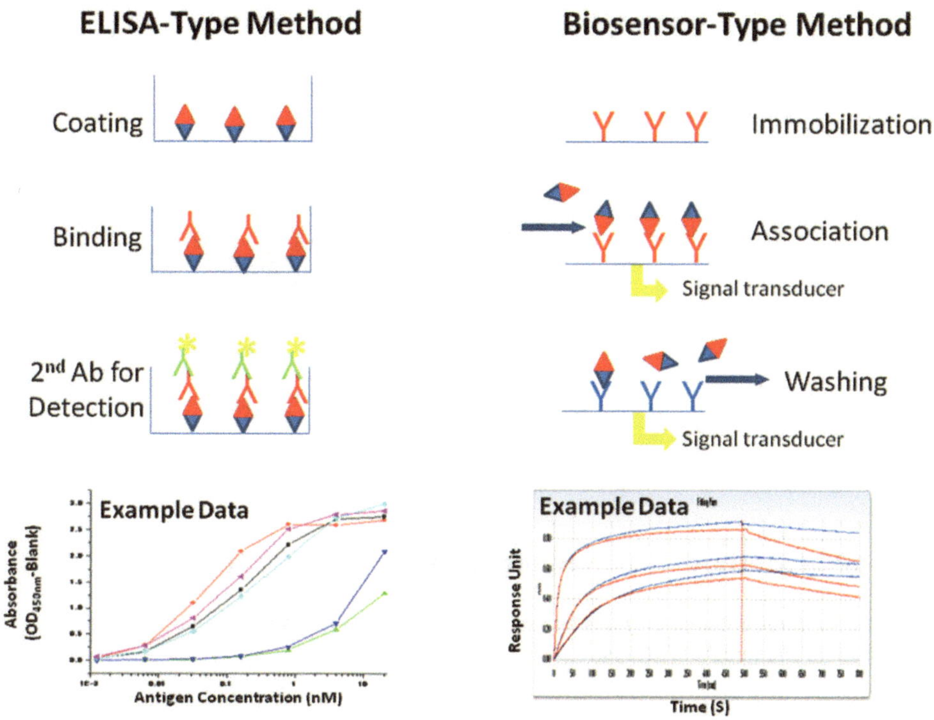

Figure 7.2.14: Two different methods for assessment specific molecular interactions: ELISA and biosensor assay.

7.2.1.6 Impurity Assessment

Ensuring high quality is essential for biopharmaceutical manufacturing, which includes the careful evaluate of the impurities in therapeutic protein products. The impurities can be divided into 2 main groups, those related to the protein product itself as some variants discussed above and those not related to the product. Product-related impurities include those isoforms or protein with modifications that are proved to be nonactive or undesired because of safety or efficacy concerns, for which relevant analytical methods for these types of impurities have been discussed above.

The not product-related impurities are those that are not derived from the expressed therapeutic proteins. They can be various types, depending on the expression system and what could be added or produced during the manufacturing process, two are common for all protein products and they are residual DNAs and proteins

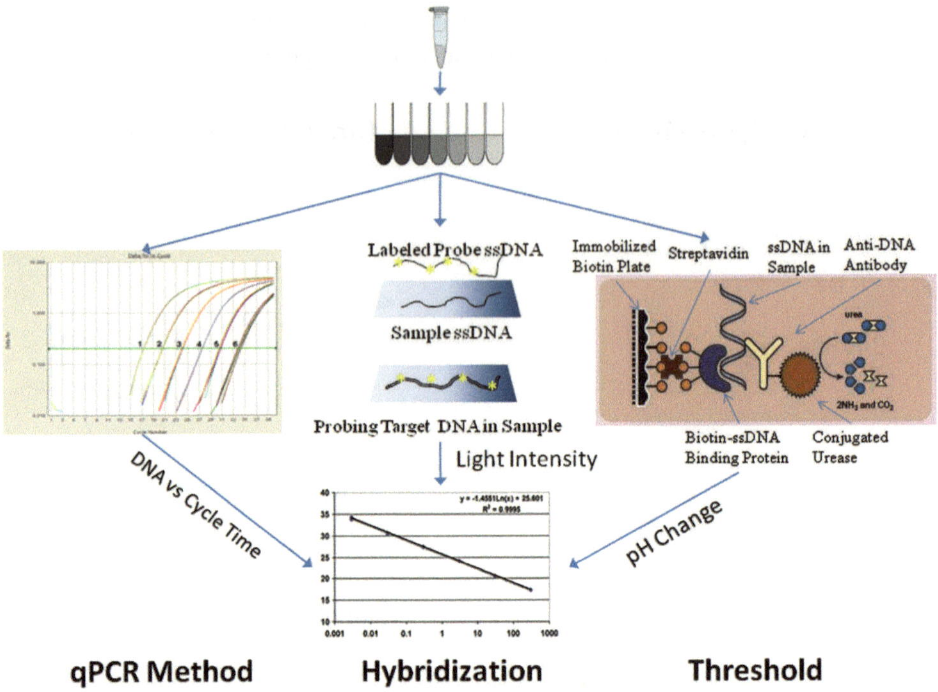

Figure 7.2.15: Three different methods for assessment the host-cell residual DNA content: qPCR, hybridization, and threshold assays.

from the host cells. For antibody based therapeutic proteins, a third common type of impurities is the residual Protein A fall off from the use of affinity chromatography with Protein A as resin component.

Host cell residual DNA can be evaluated with 1 of 3 commonly used methods, quantitative polymerase chain reaction (qPCR), hybridization, and antibody based methods (see Figure 7.2.15), and all of which offer adequate sensitivities. For qPCR and hybridization methods, both will need the development of specific DNA probes, while for the threshold method it is relatively universal for measuring total DNA content.

Both host cell proteins and leached Protein A need to be evaluated through a quantitative ELISA method (as shown in Figure 7.2.14). Though the analytical methods are relatively straight forward for both residues, for host-cell protein evaluation the choice and development of the reagents (i.e., host-cell and process specific polyclonal antibodies and relevant protein impurity standard) used in the protein impurity assays are of challenges in terms of ensuring most, if not all, of the host-cell protein impurities can be detected. The typical approach is to use a commercially available reagent set for the early product development, and at some point, for example when clinical proof of concept is achieved, one could invest time and resources to develop a set of product-specific host-cell protein ELISA reagents, where company- and product-specific antigen is used to generate polyclonal antibody reagent.

7.2.2 Quality Control

7.2.2.1 Product Release

For clinical application, products are released according to a set of predefined specifications to ensure its intended use and consistency. ICH (International Conference on Harmonization) Q6B defines specifications as a list of test, references to analytical procedures, and appropriate acceptance criteria, which are numerical limits, ranges, or other criteria for the tests described [26]. A material specification establishes a set of criteria to which a drug substance, drug product or other materials at different stages of their manufacture are quality-controlled for their intended use. Specifications are one part of a total control strategy designed to ensure product quality and consistency. Other parts of this strategy include thorough product characterization during development, upon which many of the specifications are based, adherence to the current good manufacturing practices (cGMP), a validated manufacturing process, raw materials testing, in-process testing, stability testing, etc.

Generally, the following tests and acceptance criteria are considered applicable to all drug substances and drug products: appearance, identity, purity and impurities, potency, quantity, and several general pharmacopoeial tests. Pharmacopoeial tests (e.g., endotoxin measurement) should also be performed on the drug substance, where appropriate. Additional drug substance specific acceptance criteria may also be necessary, such as limits for those organic solvents, detergents, virus deactivation reagents, antibiotics that are applied during the manufacturing process, to ensure the clearance of these impurities to a safety level in the drug substance. Pharmacopoeial requirements apply to the relevant dosage forms. Typical tests found in the pharmacopoeia include, but are not limited to, sterility, endotoxin, microbial limits, recoverable volume in container, particulate matter, uniformity of dosage units, and moisture content for lyophilized drug products. If appropriate, testing for uniformity of dosage units may be performed as in-process controls with corresponding acceptance criteria. Formulation excipients, such as the added surfactants, sugars, antioxidants, amino acids, are also expected in the drug product specification to ensure the correct concentrations in the finished products.

Table 7.2.1 is a list of exemplary platform specifications for drug substance and drug product of mAbs.

Table 7.2.1: A platform specification of drug substance and drug product for mABs.

Attributes	Drug Substance		Drug Product	
	Test Methods	Acceptance Criteria	Test Methods	Acceptance Criteria
Quantity	Protein concentration	XX -XX mg mL^{-1}	Protein concentration	XX-XX mg per vial
Potency	Cell-based or binding bioassay	XX-XX % (relative potency)	Cell-based or binding bioassay	XX-XX % (relative potency)
Identity	Peptide mapping	Consistent with reference	Identity method	Positive
Purity	SEC (%)	≥ XX	SEC (%)	≥ XX
	CE-SDS (%)	≥ XX	CE-SDS (%)	≥ XX
	IEC (%)	Main peak(s) ≥ XX	IEC	Main peak(s) ≥ XX
General	pH	Target ± XX	pH	Target ± XX
	Osmolality	Target ± XX	Osmolality	Target ± XX
	Content of surfactant (%)	Target ± XX	Content of surfactant (%)	Target ± XX
	NA		Extractable volume	≥ mL
			Visible/subvisible particles	≤ XX
Impurity	Bioburden	≤ XX CFU in 10 mL	Sterility	No growth
	Residual DNA	< XX ng per dose	Bacterial endotoxin	≤ XX EU mL^{-1}
	Residual Protein A	< XX ng per dose	NA	
	HCP	< XX %		

7.2.2.2 Stability Study

The purpose of stability testing is to investigate how the quality of a drug product or drug substance is affected by different environmental factors such as time, temperature, humidity, and light, to establish a retest period for the drug substance, and to determine a shelf life for the drug product and recommended storage conditions. ICH quality guidelines Q1A [27] suggests that drug product batches applied in the clinic should be covered by a stability protocol for duration greater than its actual clinical use. The requirement may be satisfied by data from other relevant lots but the data must meet the representative criteria in terms of production process and storage duration and conditions.

When analyzing stability samples, stability-indicating analytical methods should be applied. These methods must be validated to show their suitability in analyzing both impurities and degradation products in the stability samples. Stress testing in accelerated manner should be performed when validating the analytical procedures. Heat, acid/base, oxidation, agitation, and photo exposure are typical stress conditions for these accelerated studies. These studies are meant to identify the degradation products and degradation pathways of the drug substance, and more importantly, to demonstrate that selected analytical methods are capable of analyzing the degradation products, i.e., stability-indicating. Then, a stability monitoring program must include the stability-indicating methods chosen based on upon the physical and chemical properties of the drug molecule, to address the attributes susceptible to change during storage and likely to affect the quality, safety, and efficacy of the drug in development.

Batch selection to be included in the stability program is often based on the drug development stage. Typically, the first preclinical batch or the first GMP batch is placed in the formal stability study during toxicology and phase I studies. Major changes in production process or formulation could necessitate additional batch of products be put on stability studies in addition to other studies. For final marketing application submission, it is recommended that at least 3 primary batches of drug substance and 3 drug product batches produced at minimum of pilot scale using a procedure that simulates the final manufacturing process are included in the stability program to generate at least 12 months of long-term stability data.

In general for regulatory submission, long-term, real-time stability testing is performed at the proposed storage condition. Accelerated storage condition should

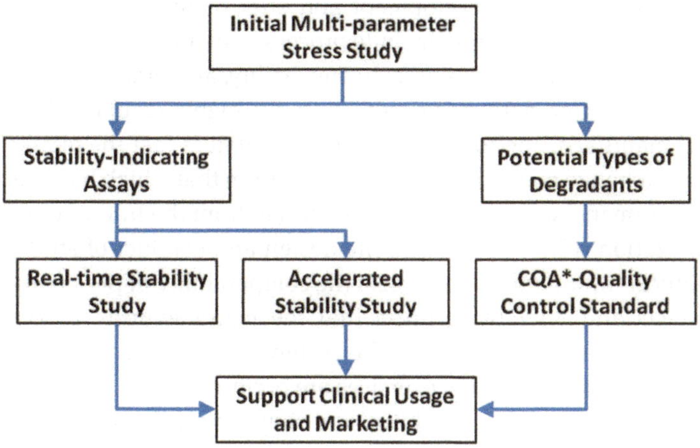

Figure 7.2.16: Strategy of stability studies of biological therapeutics.

also be included to generate around 6 months of data to evaluate the effect of short-term excursions outside of the labeled storage conditions. For example, 12 months of long-term stability results at storage condition of 5 °C should be accompanied by 6 months of results at 25 °C or higher. Long-term stability testing should be performed every 3 months during the first year, every 6 months during the second year, and annually thereafter. Accelerated study should be designed with the consideration of sample's properties, but typically tested at the third and sixth months of storage. The overall concept of the stability study program is summarized in Figure 7.2.16.

7.2.2.3 Bio-comparability and Bio-similarity Assessments

It is worth to clarify the terminology first: A comparability exercise assesses products after a change has been implemented to a given manufacturing process by the same manufacturer and includes an assessment of the differences in manufacturing process controls. Comparability can be defined mathematically as the precondition for 2 products based on known parameters; it is demonstrated when the distribution of the postchange data set is similar to that of the prechange data set. The data sets of interest are those, which speak to clinical safety and efficacy.

A similarity exercise assesses products from 2 different manufacturing processes by 2 different manufacturers. Under the current regulations, similarity is demonstrated with both preclinical and clinical bioequivalence regardless if the *in vitro* data are subset of the innovator's product data set, whereas in comparability demonstration, clinic study can be reduced in scope or avoided. A biological product intended as a biosimilar is compared with a reference biological product that is already marketed for that indication. Similarity of these products cannot be established at present time by only physicochemical and *in vitro* methods.

Manufacturers of biotechnological products must frequently introduce manufacturing changes throughout clinical development and after obtaining marketing authorization approval. Reasons for doing so include process scale-up to meet supply demands, the need to move to a different manufacturing facility, and process optimization for improved quality and/or yield. It is necessary in the highly-regulated business of pharmaceutical manufacturing to demonstrate to regulators that the quality, safety, and efficacy of postchange material does not decline from that which has been previously approved for human use. Guidance document Q5E from the International Conference on Harmonization (ICH) states "... in the design and conduct of studies used to collect the technical information to establish the comparability of pre-change and post-change products and, thereby, confirm that the manufacturing process changes did not have an adverse impact on the quality, safety, and efficacy of the drug product" [28]. For this reason, comparability exercises are performed.

What is required to demonstrate comparability and similarity will vary depending on the level of complexity of the manufacturing change(s), knowledge about MOA, and the phase of development (i.e., before or after clinical proof of concept or post licensure), and for similarity, the extent of *in vitro* characterization and the scope of planned clinic studies. Different from biosimilary demonstration, "if a manufacturer can provide assurance of comparability through analytical studies alone, nonclinical or clinical studies with the post-change product might not be warranted" (ICH Q5E) [28]. The general considerations are depicted in Figure 7.2.17.

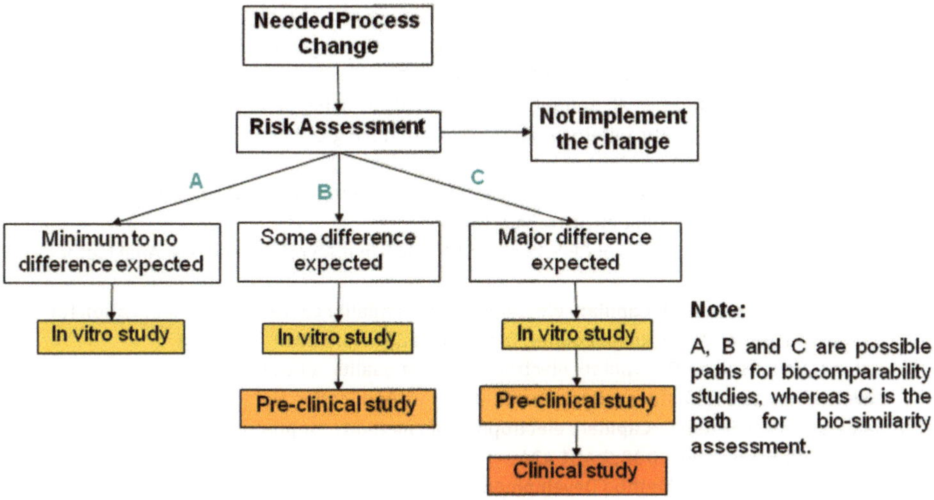

Figure 7.2.17: General approaches for biocomparability and biosimilarity assessments.

7.2.3 Summary Remark

Analytical development for therapeutic protein and quality control are step-wise, development-stage specific and systematic approach, which begins with predefined critical quality attributes to ensure safety and efficacy. Throughout the process, science and risk-management considerations are applied to gain product and process understanding and ultimately quality control. Candidate analytical methods are carefully screened, assessed and developed along with appropriate, ever stringent specifications in hope to establish methods suitable for intended purpose, and further qualified, verified, and validated for quality "gate-keeping" by product release and stability monitoring.

References

[1] Ahrer K, Jungbauer A. Chromatographic and electrophoretic characterization of protein variants. J Chromatogr B 2006;841(1-2):110-122.

[2] Harris RJ, Shire SJ, Winter C. Commercial manufacturing scale formulation and analytical characterization of therapeutic recombinant antibodies. Drug Dev Res 2004;61:137-154.
[3] Harris RJ. Heterogeneity of recombinant antibodies: Linking structure to function. Dev Biol 2005;122:117-127.
[4] Harris RJ. Processing of C-terminal lysine and arginine residues of proteins isolated from mammalian cell culture. J Chromatogr A 1995;705:129-134.
[5] Harris RJ, Kabakoff B, Macchi FD, et al. Identification of multiple sources of charge heterogeneity in a recombinant antibody. J Chromatogr B 2001;752:233-245.
[6] Cordoba AJ, Shyong BJ, Breen D, Harris RJ. Non-enzymatic hinge region fragmentation of anti-bodies in solution. J Chromatogr B 2005;818:115-121.
[7] Que AH, Zhang B, Yang Y, et al. Sequence variant analysis using peptide mapping By LC–MS/MS: Assessing genetic heterogeneity in MAb-producing cell lines. BioProcess Intl 2010;September:52-60.
[8] Thim L, Vandahl B, Karlsson J, et al. Purification and characterization of a new recombinant factor VIII (N8). Haemophilia 2010;16:349-359.
[9] Shi Y, Lia Z, Lin J. Advantages of CE-SDS over SDS-PAGE in mAb purity analysis, Anal Methods, 2012;4:1637-1642.
[10] Felten C, Salas-solano O. Capillary electrophoresis in quality control: Part I: Application for therapeutic proteins. Beckman Coulter, Inc. Application Note 2010, T-14326A.
[11] Felten C,Salas-solano O. Capillary electrophoresis in quality control: ParT II: CE-SDS: Method development and robustness. Beckman Coulter, Inc. Application Note 2011,B-16385A.
[12] Salas-solano O, Felten C. Capillary electrophoresis methods for pharmaceutical analysis. Separation Sci Technol 2008;9:401-424.
[13] Rao S, Pohl C. Optimal Method Development for Separation of Monoclonal Antibodies Using ProPac Weak Cation-Exchange (ProPac WCX) Columns: Effect of Iron on Column Performance. HPLC Posters, 2009, LPN 2173-01 01/09, Dionex Corporation, http://www.dionex.com/en-us/webdocs/69948-LPN-2173-01_MAb_WCBP-note.pdf (accessed 14th May 2014).
[14] Wang F, Peklansky B, Anderson C, et al. Improved ion-exchanged HPLC method in mab using pH gradient and its comparison with cIEF. J Liquid Chrom Rel Technol 2012;35:1259-1269.
[15] iCE3 System User Guide, P/N 045-127, Revision 3, February 2012, ProteinSimple.
[16] Kamoda S, Kakehi K. Evaluation of glycosylation for quality assurance of antibody pharmaceuticals by capillary electrophoresis. Electrophoresis 2008;29:3595-3604.
[17] Harris RJ, Shire SJ. Winter C. Commercial manufacturing scale formulation and analytical characterization of therapeutic recombinant antibodies. Drug Dev Res 2004;61:137-154.
[18] Bigge JC, Patel TP, Bruce JA, et al. Non-selective and efficient fluorescent labeling of glycans using 2-aminobenzamide and anthranilic acid. Anal Biochem 1995;230:229-238.
[19] Mechref Y, Muzikar J. Novotny MV. Comprehensive assessment of *N*-glycans derived from a murine monoclonal antibody: A case for multimethodological approach. Electrophoresis 2005;26:2034-2046.
[20] Kamoda S, Nomura C. Kinoshita M, et al. Profiling analysis of oligosaccharides in antibody pharmaceuticals by capillary electrophoresis, J Chromatogr A 2004;1050:211-216.
[21] Kamoda S, Kakehi K. Evaluation of glycosylation for quality assurance of antibody pharmaceuticals by capillary electrophoresis. Electrophoresis 2008;29:3595-3604.
[22] Dionex Corporation. Analysis of 2-Aminobenzamide (AB) labeled glycans using HPLC with fluorescence detection, Technical Note 109, LPN 2898, 2011, Sunnyvale, CA, USA.
[23] Gillece-Castro B, Tran KV, Turner JE. High resolution UPLC analysis of 2-AB labelled glycans. The Application Notebook, 2009, Sept.1.
[24] Mechref Y, Novotny MV. Structural investigations of glycoconjugates at high sensitivity. Chem Rev 2002;102:321-369.

[25] Liu H, Gaza-Bulseco G, Chumsae C. Analysis of reduced monoclonal antibodies using size exclusion chromatography coupled with mass spectrometry. J Am Soc Mass Spectrom 2009;20:2258-2264.
[26] International Conference of Harmonization of Technical Requirements for Registration of Pharmaceuticals for Human Use (ICH), Guideline Q6B: Specifications: Test Procedures and Acceptance Criteria for Biotechnological/Biological Products. 1999.
[27] International Conference of Harmonization of Technical Requirements for Registration of Pharmaceuticals for Human Use (ICH), Guideline Q1A(R2): Stability Testing of New Drug Substances and Products. 2003.
[28] International Conference of Harmonization of Technical Requirements for Registration of Pharmaceuticals for Human Use (ICH), Guideline Q5E: Comparability of Biotechnological/ Biological Products Subject To Changes in Their Manufacturing Process. 2004.

7.3 Process Characterization for Upstream and Downstream Process Development

Vijay Janakiraman, Matthew Westoby, and Lynn Conley

7.3.1 Introduction

The approach for development and characterization of biopharmaceutical manufacturing processes has been gradually changing since the introduction of Quality by Design (QbD) in 2002 by the US Food and Drug Administration (FDA) [1] in Pharmaceutical cGMP practices for the 21st century. This initiative's purpose was to enhance and modernize the regulation of pharmaceutical manufacturing and product quality [2]. The 21st century initiative embraced quality risk management and statistical concepts that have been evolving since the 1950's in other industries [3, 4]. These concepts in other industries have led to increased process and product knowledge, reduced risks, and provided opportunities for continuous improvement. This has resulted in improved product quality with less product being rejected and improved manufacturing process efficiencies. The application of QbD principles offers similar potential benefits to biopharmaceutical manufacturing processes.

QbD is defined as a "systematic approach to product development that begins with predefined objectives and emphasizes product and process understanding and process control, based on sound science and quality risk management" [5]. QbD principles and other related concepts are outlined in International Conference on Harmonisation of Technical Requirements for Registration of Pharmaceuticals for Human Use (ICH) guidance documents Q8 (R2), Q9, Q10 and Q11 [5-8]. ICH Q8 introduces the concept of design space that links product and process knowledge to product quality and uses quality risk management. It encourages the use of these concepts throughout the product lifecycle. Design space is defined as "multidimensional combination and interaction of input variables (e.g., material attributes) and process parameters that have been demonstrated to provide assurance of quality. Working within the design space is not considered as a change". This implies the use of statistical methodology such as design of experiments (DoE) to identify process parameters that will impact product quality. The design space is a region of acceptable product quality and, if approved by regulatory agencies, could allow more flexibility to manufacturing processes for process improvements and potentially relief from regulatory post approval change process. Previously, the traditional approach used the proven acceptable range (PAR), which is a "characterized range of a process parameter, for which, operation within this range, while keeping other parameters constant, will result in producing a material meeting relevant quality criteria". Using the traditional approach, the manufacturing process was fixed, relied on in-process and drug substance testing and allowed limited opportunities for process improvement. ICH Q9 discusses tools used in quality risk managements systems. ICH Q10 integrates QbD into the quality system. ICH Q11 provides further elucidation of ICH Q8, Q9, and Q10 for the development and manufacture of drug substances. ICH Q11 clarifies that the "degree of regulatory flexibility is generally predicated on the level of relevant scientific knowledge provided in the application for marketing authorization. Traditional and enhanced approaches are not mutually exclusive. A company can use either a traditional approach or an enhanced approach to drug substance development, or a combination of both". This could be interpreted to mean that companies can use design space as applicable for a complete manufacturing process or portions of the manufacturing process based on relevant scientific knowledge. In addition, the FDA guidance for industry on process validation has been revised and updated in January 2011 to integrate QbD principles [5] and with an emphasis placed on continuous verification and use of QbD principles, post approval in lifecycle management. Numerous publications have been written on QbD, in addition to recognized industry/regulatory collaborations for biopharmaceuticals [9-20].

This chapter will discuss the application of QbD principles to the characterization of upstream and downstream manufacturing processes of monoclonal antibodies and fusion proteins secreted by mammalian cells. The concepts covered here are broad and may be applicable to other expression systems. This chapter will show how quality risk management, scale-down modeling, DoE experimentation, and design space definition can be utilized to characterize and define a robust manufacturing control strategy that will provide the assurance of consistent product quality.

7.3.2 Overall Approach for Characterization of a Drug Substance Manufacturing Process

The goal of pharmaceutical development is to develop robust manufacturing processes that will consistently produce a quality product [6]. Product and process knowledge evolves and increases through the course of the development effort. The timing for initiating process characterization studies is dependent on acquiring enough scientific knowledge to make sound risk-based decisions on process parameters and quality attributes to be evaluated [8, 21]. In recent years, process characterization studies are being performed earlier than before because they are being incorporated into a comprehensive lifecycle approach to gain an earlier understanding of functional relationships between the product and process. An enhanced scientific and risk-based approach for characterization of a late stage drug substance manufacturing process is described in Figure 7.3.1. Characterization activities begin by assembling both prior product and process knowledge from a number of sources that is used to systematically evaluate, understand, and refine the manufacturing process [8]. The product knowledge, such as properties of the molecule and mode of action, data from *in vitro*, nonclinical and early phase clinical studies with the product or similar molecules, as

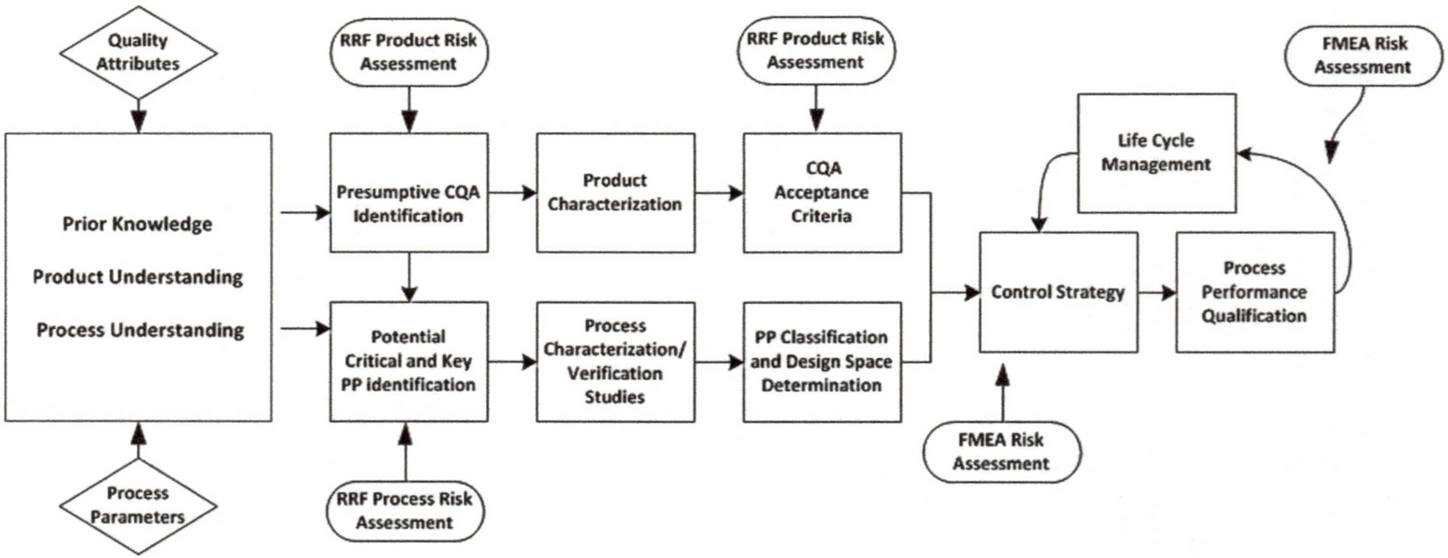

Figure 7.3.1: Description of enhanced scientific and risk based approach for development of a drug substance manufacturing process for a recombinant protein. Critical quality attribute (CQA), process parameter (PP), risk ranking and filtering (RRF) and failure mode and effects analysis (FMEA).

well as information published in the literature, are used in a risk ranking and filtering (RRF) assessment to identify presumptive or preliminary critical quality attributes (CQAs) based on the quality target product profile (QTPP). The outcome of the risk assessment is used to prioritize level of criticality of quality attributes. The CQA criticality analysis aids in determining the level of testing required for in-process intermediates and drug substance and prioritizes the quality attributes that need more in-depth product characterization. Subsequently, CQAs are finalized and acceptance criteria ranges are established that will ensure patient safety and product efficacy.

Process knowledge from early process development and manufacturing experience, platform process formats and scientific literature are used in process risk ranking and filtering assessment to identify potential critical process parameters (CPPs) and/or key process parameters (KPPs) that may have an effect on CQAs for each process step. The identified process parameters are then evaluated for their effect on CQAs through a design of experiments (DoE) approach. Process parameters that impact CQAs are confirmed and predictive statistical models are built. These models are used to define the design space, i.e., process parameter ranges within each individual process step that should meet desired drug substance quality targets. Verification experiments are performed within the characterized region to confirm models are valid and test robustness of each individual process step. Worst case runs are then performed at the proposed action limits of the critical process parameters through the entire process, preferably from cell culture bioreactor to drug substance. This information is used to verify the design space and evaluate robustness between steps for the complete manufacturing process. Input process parameters are classified as critical process parameters (CPP), key process parameters (KPP) or non-key process parameters (N-KPP). CPPs and KPPs are process parameters that impact one or more CQAs and process consistency respectively [21]. Therefore they should be monitored or controlled to ensure the process runs consistently and produces the desired product quality [12, 21]. A failure mode and affects analysis (FMEA) risk assessment is then performed to identify modes of failure associated with controlled input parameters along with prioritization and mitigation of risks. The results of this risk assessment are used as an initial basis of the control strategy for the manufacturing process and the process performance qualification (PPQ) campaign. After PPQ, a follow-up FMEA risk assessment is performed, where the data from PPQ is used to refine the control strategy and finalize it prior to BLA filing. The last step is lifecycle management with continuous process verification and process improvement, but is not in the scope of this chapter. The next few sections will briefly discuss the main steps for process characterization with examples from upstream and downstream processes.

7.3.3 Risk Assessments

Risk assessments are essential tools that are an integral part of quality risk management. Risk assessment is not synonymous with risk management. Quality risk management is a “systematic process for the assessment, control, review and communication of risks to the quality of the drug product” [5]. It is a dynamic process and must be managed throughout the lifecycle of the product, process, or system [22]. Quality risk management includes 4 steps: risk assessment, risk control, risk review and monitoring, and risk communication. Risk assessment is the first step in the quality risk management process and is used to provide a scientific evaluation of risk to product quality and links it to patient safety. ICH Q9 defines risk assessment as “a systematic process of organizing information to support a risk decision to be made within a risk management process. It consists of the identification of hazards, and the analysis and evaluation of risks associated with exposure to those hazards” [5]. It is a means to identify and focus development efforts on the critical elements of a manufacturing process. Three basic questions are asked during the risk assessment process:
- What might go wrong?
- What is the likelihood it will go wrong (probability)?
- What are the consequences (severity)?

The output of a risk assessment is either a quantitative estimate of risk or a qualitative description or narrative of a range of risk. There are many different types of risk assessments and it is important to understand the purpose and desired outcome of a risk assessment in choosing the appropriate tool. Risk assessment tools vary in their

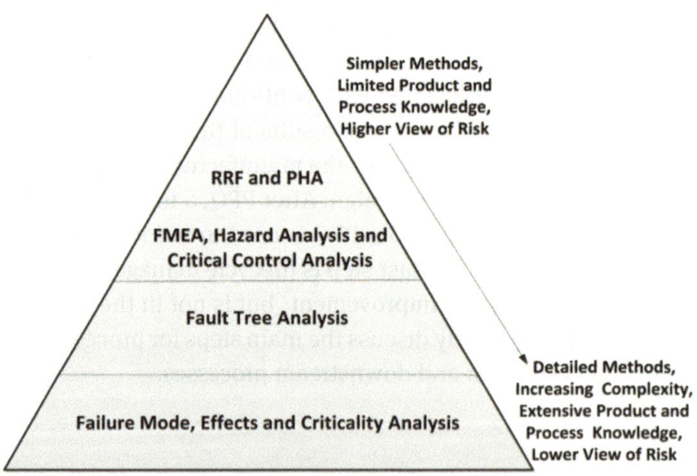

Figure 7.3.2: Summary of risk assessment methods and tools with increasing levels of detail and complexity.

approach and rigor. Figure 7.2.2 shows some of the different types of risk assessments that are listed in ICH Q9 with increasing level of detail and complexity [23].

The 3 types of risk assessments commonly used during process characterization are risk ranking and filtering (RRF), preliminary hazard analysis (PHA) and failure mode and effects analysis (FMEA). All 3 of these risk assessments have been adapted from the original intended use and have been modified for use in the biopharmaceutical industry. Risk ranking and filtering and preliminary hazard analysis are typically used in the early phase of process characterization and are precursors to process and/or product characterization studies. They both use a simpler method of analysis when prior knowledge is limited and take a high level view of risks. FMEA, on the other hand, is a type of risk assessment that uses more detailed analysis and requires more knowledge to evaluate risks. FMEAs have been used before and/or after process characterization studies depending on the level of prior knowledge and its intended use. Application of these 3 common types of risk assessment used for process characterization is discussed.

7.3.3.1 Risk Ranking and Filtering

Risk ranking and filtering is a tool for comparing and ranking risks. Risk ranking of complex systems involves evaluation of multiple diverse quantitative and qualitative factors for each risk. The tool involves breaking down the basic question of risk into as many components as needed to capture factors involved in the risk. “Filters”, in the form of weighting factors or cutoff for risk scores, can be used to scale or fit the risk ranking to project objectives [5]. Several variations of this type of risk assessment is used in early characterization studies to identify process parameters to include in process characterization studies or used to identify potential CQAs. A few of these variations of this type of risk assessment are discussed.

The first application involves evaluation of product quality (PQ) attributes for their potential risk on drug safety and efficacy. PQ attributes are assigned 2 risk scores: the first one is an impact score (which measures the potential impact of the attribute on patient safety and efficacy) and the second one is an uncertainty score (which measures the certainty of the information used to assess the potential impact). An overall risk score, also known as risk priority number (RPN), is then generated for each attribute, by multiplying the impact score and uncertainty score. The set of RPNs form a risk matrix that is used to rank and filter the noncritical attributes from the critical ones using a cutoff RPN. This is used to determine the preliminary level of criticality of quality attributes, which aids in prioritizing the attributes to be evaluated during characterization studies and deciding the level of testing required for in-process intermediates and drug substance during such studies [1, 16].

In the second application, the purpose of the risk assessment is to determine which process parameter poses higher risk to product quality and need to be evaluated during process characterization. Since the initial list of potential process parameters can be quite large, this risk assessment is used to prioritize and reduce the number of process parameters to be studied [19, 24]. The application of this risk assessment is virtually the same as the first variation, except the potential impact of process parameter on CQA or process performance is evaluated instead of the impact of quality attributes on safety and efficacy of drug product. This impact score is similarly multiplied by the uncertainty score to generate an RPN. The subsequent risk matrix generated is applied to determine parameters that can be excluded from process characterization and those that should be included in either a univariate or multivariate study [1, 16].

A variation of the above RRF application is sometimes used to classify the potential impact of process parameters on CQAs, as well as possible interactions with one another. Process parameters are assigned 2 rankings: an impact score (as previously explained), which is multiplied by an interaction score (which measures the potential interaction with other process parameters). The ranking for impact to critical attributes are weighted more severely than impact to lower criticality attributes or process attributes. The overall risk matrix is used to determine if multivariate, univariate, or no further experimental studies are needed [16]. If the knowledge of the impact is unknown or very limited for a certain parameter, the highest risk score is assigned and the parameter is included in a multivariate study [1, 16].

7.3.3.2 Preliminary Hazard Analysis

Preliminary hazard analysis (PHA) is a tool of analysis based on the application of prior knowledge or experience of a hazard or failure to identify future hazardous situations that may cause harm as well as to estimate probability of occurrence. The tool consists of identification of possibility that risk happens, the qualitative evaluation of possible injury, a relative ranking of the hazard using a combination of severity and likelihood of occurrence and identification of possible remedial measures ICH Q9 [5]. In practice, PHA is similar to risk ranking and filtering with severity (similar to impact) being multiplied by likelihood of occurrence, instead of uncertainty of information. PHA has also been used for CQA criticality analysis [16, 25] and for evaluating the potential of process parameter to impact CQA or process performance [22].

7.3.3.3 Failure Modes and Effects Analysis

Failure modes and effects analysis (FMEA) is one of the most widely used risk assessment tools in the biopharmaceutical industry. FMEA is a risk analysis designed to identify potential modes of process failure associated with operational parameters that may affect the product quality (safety and efficacy) and/or process consistency. Risks are assessed based on their possible consequences and their probability of occurrence and detection [26]. Three risk scores, severity of failure, the expected frequency of occurrence, and the likelihood of detecting failure are multiplied to generate an overall risk score. A risk matrix is applied to classify the level of risk as low, medium, or high. If this type of risk assessment is used before process characterization, high-risk parameters are evaluated during DOE studies, medium-risk parameters are considered to be evaluated in a DOE study (with justification if not included), and low-risk parameters are not included in any study. However, there are some who argue that FMEA is not suitable in the early phase of process characterization to identify potential CPPs for subsequent evaluation. The ICH Q8 definition of a CPP is "a process parameter whose variability has an impact on a critical quality attribute". This definition refers only to the severity of harm, but not the probability of occurrence necessary for the definition of a risk. In addition, likelihood of occurrence is difficult to predict in the early stages of development when the availability of data is limited, especially at manufacturing scale, making FMEA less effective. FMEA is more appropriately applied after process characterization, it is used to identify and prioritize process parameters that have higher risks and generate a plan for mitigating and reducing the risks in advance of the process performance qualification (PPQ) runs.

Risk reduction and risk acceptance form the next step in the QRM process. ICH Q9 defines risk acceptance as a decision to accept risk and risk control as actions implementing risk management decisions. Risk reduction and risk acceptance are iterative steps and must result in a documented decision to reduce risk or acknowledge it is as low as reasonably practical and cannot be reduced further [22]. The final step in the QRM process is to communicate risk to key stakeholders and continuously review and monitor during lifecycle of the product.

7.3.3.4 CQA Criticality Analysis

Detailed characterization of a late stage program begins with the patient's safety and efficacy of the drug in mind. The desired quality attributes of the product are identified based on the quality target product profile (QTPP). The QTPP is a prospective summary of the quality characteristics of a drug product that ideally will be achieved to ensure the desired quality, taking into account safety and efficacy of the drug product [6]. From the QTPP, a list of quality attributes is created and prioritized, according to their criticality by using risk ranking and filtering or preliminary hazard analysis types of risk assessment. The evaluation requires a thorough understanding of the properties of the molecule and the mode of action, data from *in vitro*, non-clinical, and early phase clinical studies with the product or similar molecules, as well as information published in the literature. Quality attributes are classified as either Critical Quality Attributes (CQAs) or non-Critical Quality Attributes (non-CQAs) based on this risk assessment. This binary classification is arbitrary because of the complexity of the structure-function relationships of large molecules and in reality is a continuum of criticality [16]. The primary benefit in performing this risk assessment is prioritization of the level of criticality of quality attributes. The CQA criticality analysis is initially used to determine the quality attributes that need more in depth product characterization. It is later used in process characterization to determine the level of testing required for in-process intermediates and drug substance and to define acceptable ranges for safety and efficacy.

The following excerpt describes an example of risk ranking and filtering risk assessment for determining the criticality of quality attributes of a humanized monoclonal antibody. The first step in the process is to list all the product quality attributes of the drug substance that may most likely affect safety and efficacy, which include immunogenicity, potency, and pharmacokinetic/pharmacodynamic (PK/PD). The quality attributes fall into 4 categories: product related substances, product related impurities, contaminants, and process related impurities. The process related impurities could either be derived from the host cell line (such as host cell protein and DNA) or from the raw materials used in the process (such as methotrexate and Protein A leachate). There are also obligatory CQAs in terms of composition and strength that do not need to be included in risk assessment. The risk analysis is performed by assigning each attribute an impact score (on a scale of 4 to 20) and an uncertainty score (on a scale of 2 to 8), which collectively determine the criticality of the attribute. The assumption in performing the risk assessment is that quality attributes are not controlled during the process. The impact score is multiplied by the uncertainty score to give an overall risk score (RPN). Quality attributes are classified as critical if RPN is above 24. Table 7.3.1 shows some of the product risk assessment results for a monoclonal antibody and lists 16 of approximately 50 quality attributes evaluated during risk assessment. In reality, biological products are very complex and the number of potential products variants for a monoclonal antibody that may affect product quality, has been estimated at ~10 [8, 10].

Most quality attributes considered during risk assessment are classified as potential CQAs due to high uncertainty score because no specific data are available for the impact of such attributes. Some quality attributes are always considered CQAs, such as adventitious agents and bioactive process related impurities such as deoxyribonucleid acid (DNA), host-cell protein (HCP), and Protein A, based on potential impact to patient safety [1]. Nonbioactive process related impurities, such as methotrexate and antifoam, are classified as non-CQAs based on the maximum amount of impurity in 1 dose, assuming no clearance through the purification process [16]. Determining if a product related impurity is critical requires clinical studies evaluating the product variant or nonclinical PK studies conducted with purified or enhanced product variant in suitable animal models or toxicity or immunogenicity studies [1, 22]. Reducing the number of CQAs during product characterization studies becomes a major challenge, especially considering potential immunogenicity of a quality attribute. Product risk assessments are often performed early in the development cycle and reevaluated and refined throughout the development lifecycle as more data becomes available [1, 22, 24, 25, 27]. Potential CQAs during development are confirmed as being truly critical based on suitable product data; however, in some cases criticality may never be fully elucidated.

Table 7.3.1: Summary of product risk assessment using risk ranking and filtering tool evaluating potential impact on product safety and efficacy for a monoclonal antibody (PD/PK = pharmacodynamics and pharmacokinetics, SAR = structure-activity relationship, RPN = risk priority number, CQA = critical quality attribute).

Attribute	Impact	Uncertainty	RPN	CQA/ Non-CQA	Potential impact
Aggregates	20	5	160	CQA	Immunogenicity and potency
Low molecular weight	20	5	100	CQA	Potency
Host cell DNA	16	5	80	CQA	Immunogenicity and safety
Host Cell Proteins	16	5	80	CQA	Immunogenicity and safety
Protecn A leachate	16	5	80	CQA	Immunogenicity
Binding potency	20	4	80	CQA	Potency
Adventitious agents (endotoxin, bioburden, viruses)	20	3	60	CQA	Safety
Polysorbate-80	12	5	60	CQA	Potency and immunogenicity
Glycosylation	8	5	40	CQA	Potency and PD/PK
Protein concentration	20	2	40	CQA	Potency
Deamidation of Asn	12	3	36	CQA	Potency
Methionine oxidation	8	2	16	Non-CQA	None, based on SAR experiments
Galactosylation	8	2	16	Non-CQA	None, based on SAR experiments
Methotrexate	8	2	16	Non-CQA	Toxicity
N and C terminal isoforms	4	3	12	Non CQA	None, based on SAR experiments
Antifoam C	4	2	8	Non-CQA	Toxicity

7.3.4 Scale-down Modeling

7.3.4.1 Importance of Scale-down Models for Process Characterization

Scale-down models for manufacturing unit operations are critical tools used for process development and characterization to understand the impact of changes in input parameters on the process and product quality at the manufacturing scale. A sound scale-down model is the first step towards the successful definition of the manufacturing control strategy for input parameters used to keep critical quality attributes (CQAs) within an acceptable range.

Ideally, a manufacturing campaign would run smoothly; however, problems are often encountered as more batches are run for subsequent resupply campaigns. If the root cause analysis requires further experimentation, typically small scale is the preferred choice for such studies due to time and resource limitations of troubleshooting at larger scales. This requires an appropriate scale-down model for the process that is comparable to the manufacturing process in order for the investigation to yield the right conclusions. A strong scale-down model goes beyond achieving comparability at operating setpoints; it also predicts the appropriate response as a function of changes in the process setpoints [28]. This is especially important for process characterization, where experiments are conducted at bench scale in order to understand the impact of changes in input parameters on the process and product quality at the manufacturing scale.

7.3.4.2 Overall Methodology for Scale-down Model Development and Qualification

Figure 7.3.3 outlines the main steps for scale-down model development and qualification, which are explained in the subsequent sections. It may be fair to say the strength of the process characterization package is only as good as the fit of the scale-down model to the manufacturing scale data. This begs the question of what needs to be matched and what are the criteria for a good scale-down model.

Types of Input Parameters for Scale-down Models

Since there are inherent differences between small scale and manufacturing scale, the exercise of developing a scale-down model is not trivial. Not everything can be matched, and not every parameter may affect the process or product quality. There are 2 main types of input parameters for a scale-down model of any unit operation: those that are independent of scale and those that are dependent. Scale independent input parameters must be maintained at the same levels as manufacturing. Examples from an upstream bioreactor include pH setpoint and temperature setpoint. For downstream unit operations such as chromatography, the wash, load, and elution pH and buffer conductivity are independent of scale. Scale-dependent input parameters are maintained at different levels depending on the scale. Some of them, such as feed volumes for bioreactors, scale linearly and are easy to set for a particular scale. However others are nonlinear in nature and require alternate approaches for scaling. Examples include agitation for all culture bioreactor, column diameter for a chromatography step and centrifugation RPM for the clarification step. Typically a complementary parameter that is not routinely measured is used as the scale-down principle to scale such parameters. This will be covered in more detail via examples from upstream and downstream.

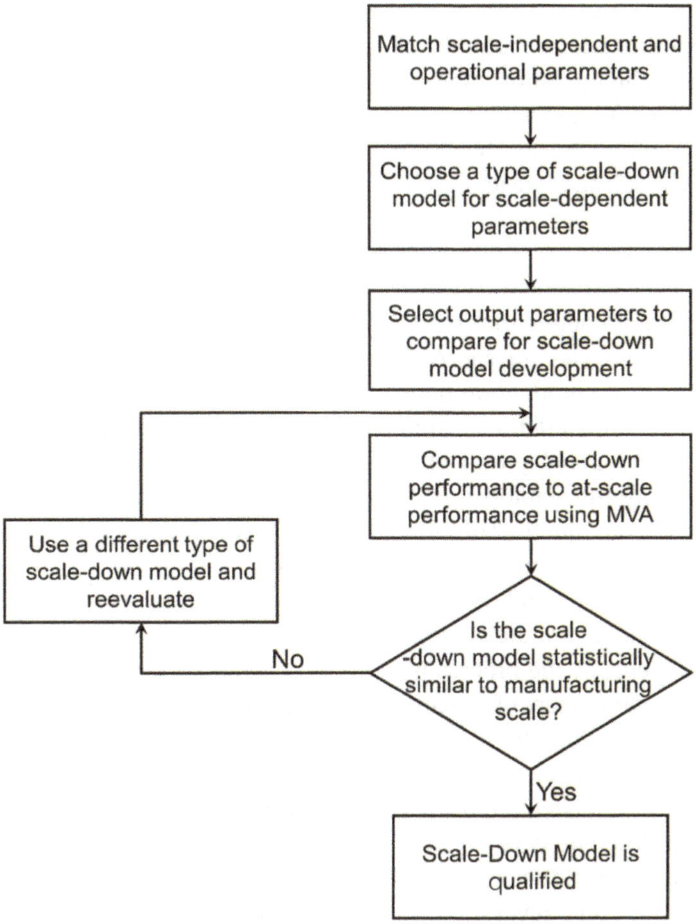

Figure 7.3.3: Main steps for scale-down model development and qualification (MVA = multivariate analysis).

Yet another class of parameters that may be more qualitative in nature are operational parameters and protocols. They are typically scale-independent and must be matched or adopted for the scale-down system. An example is the sequence of steps in a unit operation. If the nutrient feed addition is performed after temperature shift, the same practice must be followed at small scale. The time taken to complete steps, the analytical assay equipment used and dilutions for assays may all have an impact on the final outcome of the operation and must be scrutinized closely. If an operation or step is deemed time sensitive, an effort must be made to mimic manufacturing scale as closely as possible. For example, due to processing volumes, the intermediate hold times between purification steps are usually much longer in manufacturing and should be assessed at small scale for any product quality related changes due to longer hold times. Other operational aspects that cannot be directly matched due to work flow or equipment limitations should be altered for small scale without changing the overall impact of the operation. As an example, for a process that requires cold feed, a temperature controlled feed tank is used in manufacturing. This may be altered at small scale by storing the feed bottles in a temperature controlled cold room and welding them on and off the bioreactor for feed as required.

Selection of Output Parameters for Scale-down Model Development

Typically a set of output parameters that represent the performance of the process have to be matched across scales for successful scale-down model development. For example, for a cell culture production bioreactor process, growth, metabolism, titer and harvest product quality are the most important output parameters. Process outputs are monitored once per day and product quality mainly at harvest. Cell growth relates to cell density and viability on different days as the process progresses. Cell metabolism represents the "quality" of cell growth, and has an impact on process performance. Glucose and oxygen consumed by cells and lactate, ammonia and pCO_2 produced are some examples of metabolic output parameters. The cell growth and health of the cells impact the titer of the process.

In the downstream steps, as the process is more focused on purification of the protein, the performance parameters of interest include the step yield and are more heavily centered on the product quality. The more common attributes that are compared include product high molecular weight (HMW) levels, product purity, product charge isoforms, and process related impurities such as HCP and DNA. The product quality of the protein at different steps all the way from harvest to drug substance impacts the safety and efficacy of the product. Hence, this forms the most important attribute to be matched for scale-down modeling. During process characterization, there is a constant effort to map the relationship of product quality to the process inputs and outputs.

Multivariate Analysis for Qualification of Scale-down Models

After the output parameters to be matched across scales are selected, statistical acceptance criteria are developed to qualify the scale-down model. A common approach used is univariate analysis, where each parameter at small scale is compared to the corresponding value at manufacturing scale using statistical measures or by demonstrating equivalency of the parameters means at the different scales [29]. However, with increasingly complex multidimensional data sets being generated from cell culture and purification processes, it becomes difficult and inefficient to use a univariate approach. For example, cell culture processes are run in continuum, which means for each output parameter there are as many parameters as there are samples taken during the time course of the run. Thus for a 17 day process with

20 output parameters and daily samples, a total of 360 variables have to be compared. The tool of choice to analyze such large datasets with statistical rigor is multivariate analysis (MVA) [30]. This method is able to examine multiple inputs and outputs at the same time and draw correlations or identify the main variables contributing to scale differences.

Briefly, all the variables for the scale-down model are lumped together and converted into orthogonal components using principal component analysis (PCA), which is a form of MVA. Most of the variation in the data is accounted for by the first 2 components of the model, and inclusion of the third component onwards should be considered depending on their significance. Thus the original n-dimensional variable space (360-dimensional in the previous example) is transformed into a two-dimensional component space. Each condition or batch is then represented by a single point in this two-dimensional space, as defined by its component scores, t1 and t2. The training set used for the PCA model mainly consists of data from manufacturing, and a 95 % confidence ellipse is generated. In instances where manufacturing data may not be available, one may use data from the next highest scale (typically pilot scale) that compares well with manufacturing [14]. When the small scale data set is subsequently overlaid on the training model, if the component scores are such that the small scale batches fall within the ellipse, then we could say with 95 % confidence that the scale-down model is statistically similar to the manufacturing scale.

As an example, multivariate analysis was applied to process variables for a cell culture process that was scaled up to 15,000 L scale. The 95 % confidence ellipse scatter plot (Figure 7.3.4a) showed that the small scale runs fell outside the confidence interval. Another way to look at this result is through batch control charts, where the progression of the batch in elapsed time is tracked using component scores. Figure 7.3.4c shows that in the second half of the run, majority of the small-scale batches deviated away from the manufacturing average. Contribution plots were used to identify the main gaps between the 2 scales, which demonstrated that a number of parameters including pH, pCO_2, osmolality, base addition, and titer were different from manufacturing in the second half of the run. After subsequent scale-down modeling experiments, a k_La based scale-down model was developed. Figure 7.3.4b shows the new scatter plot with the scale-down runs within the 95 % confidence ellipse. The batch control charts now show a much better match with the manufacturing average in the second half of the run (Figure 7.3.4d). This exercise qualified the scale-down model to be used for further characterization of this process.

Although it was not included in this example, it is also possible and considered prudent to include product quality parameters in the multivariate analysis models. This will ensure that the scale-down model matches not just process performance compared to manufacturing but also product quality parameters. Finally, multivariate partial least squares (PLS) analysis can be used for correlating 1 or more Y variables with several X variables.

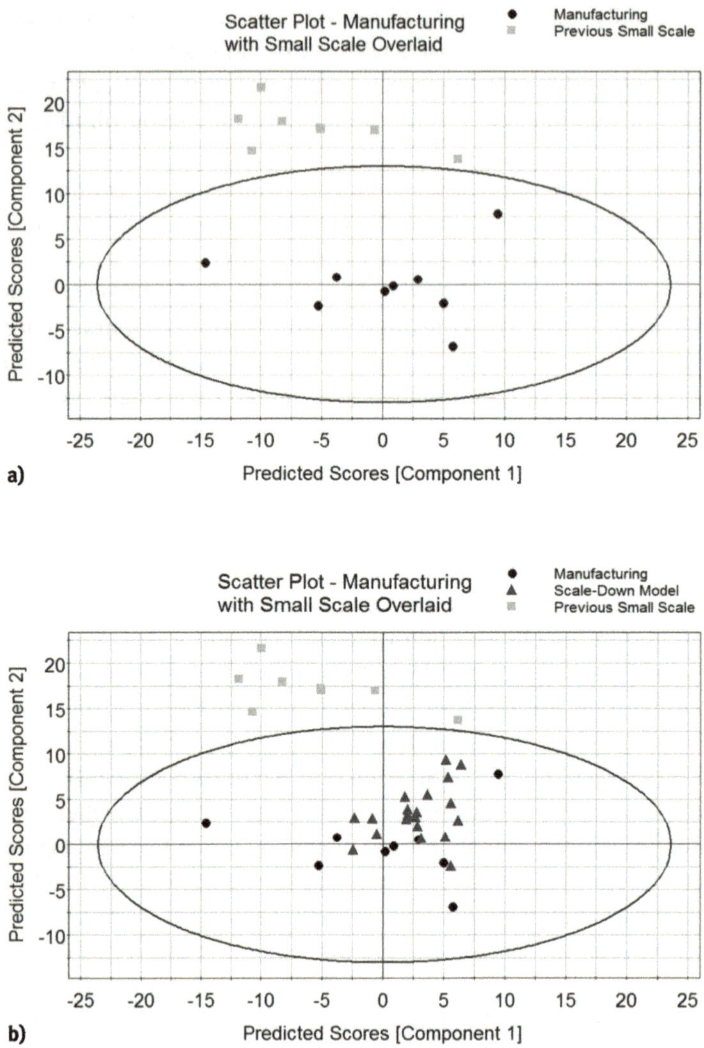

Figure 7.3.4: **a)** 95 % confidence ellipse showing small scale runs (light grey squares) fall outside the confidence interval and hence not statistically similar to manufacturing (black circles). **b)** Scale-down modeling was performed to match process parameters in the second half of the run. Scale-down runs (dark grey triangles) cluster with the manufacturing runs and are within the 95 % confidence ellipse, indicating a qualified scale-down model is generated. Manufacturing data set was used to build the training model over which small scale data was overlaid.

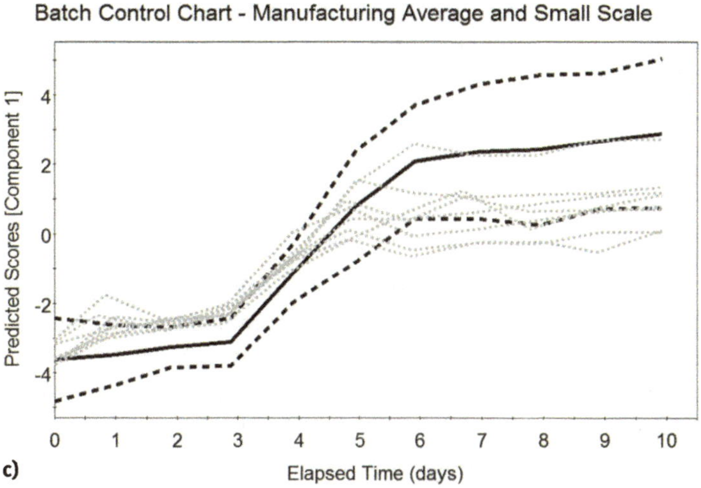

c)

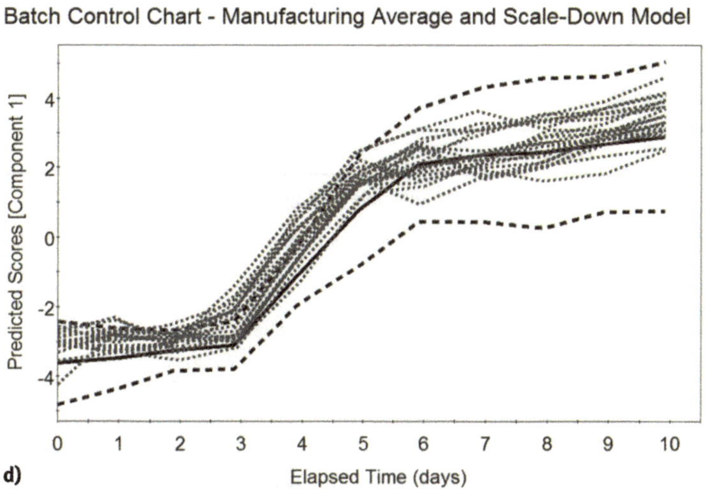

d)

Figure 7.3.4: **c)** Batch control charts showing manufacturing average (black solid line) and +/−3 standard deviations (black dashed lines). Trends for small scale runs (light grey dotted lines) show significant difference from the manufacturing average; runs are closer to the −3 standard deviations. **d)** Batch control charts for scale-down model runs (dark grey dotted lines) match manufacturing average quite well to indicate comparable performance.

7.3.4.3 Examples of Scale-down Models from Upstream and Downstream

Section 7.3.4.2 outlined the general process for development and qualification of scale-down models. This section will focus on scale-down principles used for nonlinear scaling of parameters for specific unit operations from upstream and downstream as examples.

Upstream Process: Cell Culture Bioreactor

Upstream processes for the production of a therapeutic monoclonal antibody or antibody like (Fc) fusion protein are carried out in bioprocess containers, such as shake flasks, wave bags, and bioreactors. A combination of these is typically used to expand the cells from vial thaw, while the final stage is in the production bioreactor, where the product is made. Thus generating a scale-down model for the cell culture production bioreactor forms an important prerequisite for process characterization. Table 7.3.2 summarizes some of the parameters and their types for a typical scale-down model for cell culture bioreactor.

Table 7.3.2: Partial list of parameters for scale-down model of cell culture production bioreactor grouped according to their type (* Power per unit volume or oxygen transfer rates used as scale down criteria for bioreactor agitation).

Type of Parameter	List of Parameters
Scale independent	Temperature, pH, dissolved oxygen set-points, seed density
Scale dependent linear	Sparge caps, overlay, feed volumes
Scale dependent nonlinear	Bioreactor agitation*, ratio of final over initial working volume
Operational	Feed times, sequence of process changes, batch medium storage and equilibration procedures, assay equipment used

A common theme that is true for any scale-down model for the cell culture bioreactor is that when some input parameters are matched some others may be mismatched. Often, this makes scale-down model development an iterative process, until the successful matching of predetermined performance parameters across the scales is achieved. Agitation is a key parameter that scales nonlinearly and 2 principles for scaling agitation are described here.

P/V model: Power per unit volume (P/V) describes the power that is transferred to the cell culture fluid through agitation per volume of fluid in the vessel. This is the classical method of scaling and models bulk fluid flow across scales, which impacts the mixing and shear stress profiles in the bioreactor [31-34]. The power imparted to the cell culture is calculated as

$$\text{Power} = N_p \rho N^3 D^5 ,$$

where N_p is power number, D is impeller diameter, N is agitation speed, and ρ is the density of water = 995 kg m^{-3}.

In this scale-down approach, the agitation for the scale-down model is set by matching P/V across scales. However, this type of scaling may lead to differences in mass transfer coefficients (k_La) for transfer of oxygen from sparged gas to the liquid, which in turn will lead to differences in the total amount of sparge per volume of cell culture per min.

Oxygen transfer rate model (or k_La model): Using the mass transfer coefficient (k_La) as the scale-dependent parameter to match will theoretically lead to the same sparge vvms (volume sparge per volume liquid per min) across scales. This is important because sparge rates affect bubble-related shear and hydrodynamic stress in a bioreactor. It may however lead to higher P/V or higher turbulent shear at small scale, depending on the type of sparger used, hence there is a tradeoff. Practically though, the k_Las differ and so do sparge vvms, but together they give the same oxygen transfer rate (OTR). OTR is a function of k_La and oxygen driving force, as described by the equation

$$OTR = k_L a\,(C^* - C),$$

where C^* is the oxygen concentration in the inlet gas stream and C is the bulk oxygen concentration in the liquid.

k_La can be empirically expressed as a function of agitation and sparge rates as

$$k_L a = cN^{\alpha}Q^{\beta},$$

where c, α and β are empirical constants, N is the agitation (rpm), and Q is the gas flow rate (slpm).

When agitation and sparge rates alone cannot be used to match k_La, another parameter that is sometimes used to modulate the k_La is the sparger type. Typically spargers with smaller hole size lead to higher k_La at the same sparge rates, due to increase in surface area of the bubbles. The graph below shows the impact of sparger hole size on the k_La (Figure 7.3.5). Thus, different sparger types may be used to bring the agitation into an operationally feasible regime for the particular scale-down system.

For shear sensitive cell lines, there may be a threshold agitation, sparge rate or bubble size to avoid in order to minimize shear damage to the cells. In such cases, the sparge composition can be manipulated to increase its enrichment in oxygen to enable the same or higher oxygen driving force to be achieved at lower total sparge rates. The downside of this technique is that the cells will be exposed to different localized concentrations of dissolved oxygen and different normalized sparge rates compared to manufacturing. This strategy was used for an internal project as summarized by Figure 7.3.6. Figure 7.3.6a shows an increase in viable cell density, as the per-

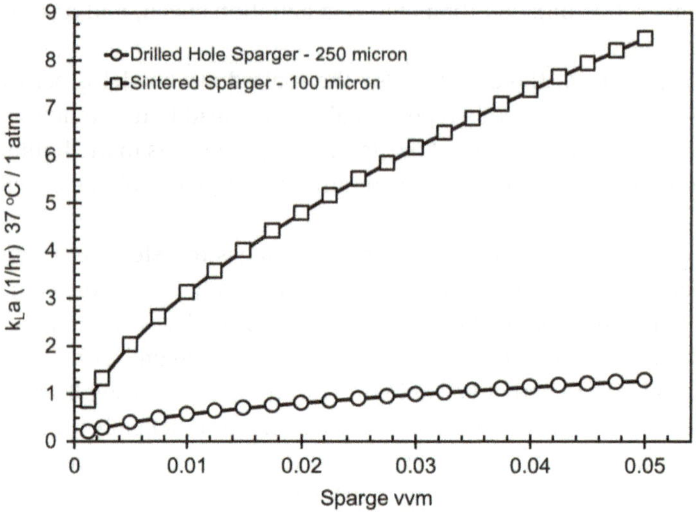

Figure 7.3.5: Calculated mass transfer coefficient (k_La) as a function of sparge vvm for 2 different sparger types. Sparger hole size and type have significant impact on k_La.

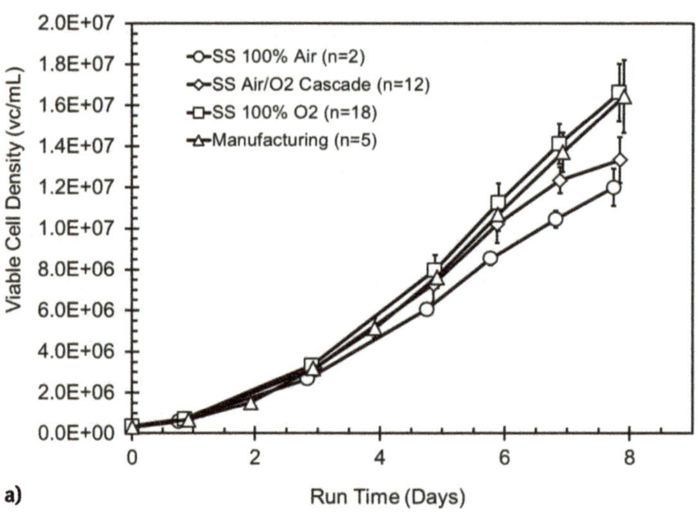

a)

Figure 7.3.6: **a)** Viable cell density trends for different sparge gas compositions at small scale compared to performance in the 2,000 L manufacturing scale. As the percentage of oxygen in the sparge gas increases from 21 % (100 % air) to 100 %, the cells reach higher densities and match manufacturing. The air/O_2 cascade condition corresponds to a P/V model where the percentage of oxygen in the sparge gas varies based on oxygen demand.

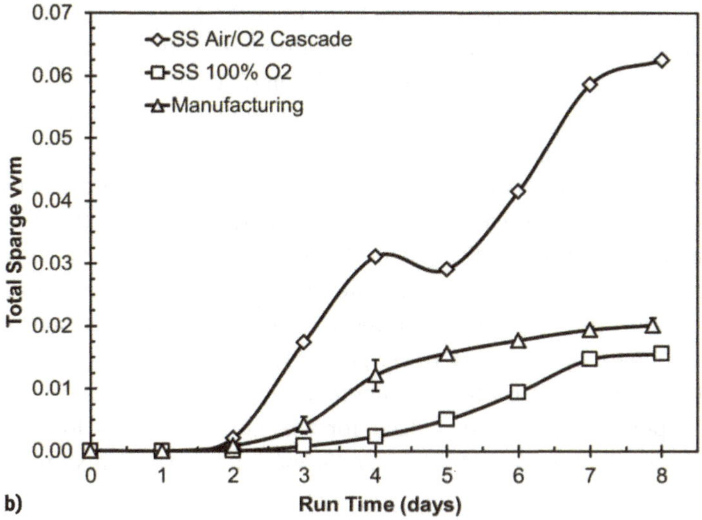

b)

Figure 7.3.6: **b)** Total sparge rates (air + oxygen) on a vvm basis indicate that the air/O_2 cascade condition has more than three times sparge than manufacturing towards the end of the process, while the 100 % O_2 condition is comparable.

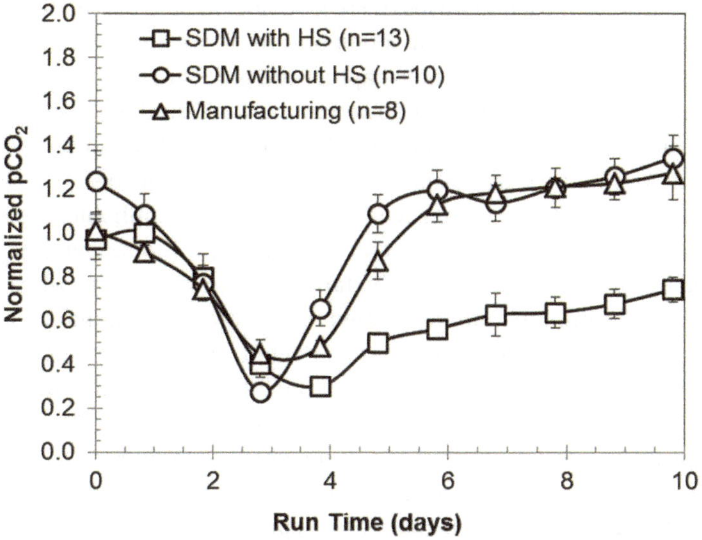

Figure 7.3.7: Example of scale-down modeling, where headspace gas can be manipulated to match pCO_2 stripping rates from manufacturing. Scale-down model without any headspace gas has comparable pCO_2 to manufacturing scale, compared to with headspace gas (SDM = scale-down model, HS = headspace).

centage of oxygen in sparge gas is increased, as this leads to lower total sparge. Figure 7.3.6b shows that the sparge vvms for the 100 % O_2 condition (OTR model) match the manufacturing scale better than the air/O_2 cascade condition (P/V model).

Sparging pure oxygen can sometimes be used for better matching of pCO_2 profile between scales. Li et al. [35] used this approach since the pCO_2 in their small-scale bioreactor was significantly lower than manufacturing scale, due to inherent differences in stripping capabilities. Another approach to match pCO_2 is to turn off headspace gas completely or add a small percentage of CO_2 gas in the headspace. Figure 7.3.7 summarizes the results when this was done for an internal project. The use of no overlay in the scale-down model matched the stripping rates at small scale with those at the manufacturing scale as shown by matching pCO_2. This also resulted in better alignment of bioreactor pH, which is an important parameter affecting product quality.

Additional scale dependent parameters used for scale-down models for cell culture bioreactors include agitator tip speed, energy dissipation rates and mix times.

Downstream Process: Centrifugation and Chromatography

Figure 7.3.8 shows a schematic of a typical downstream purification process for a therapeutic monoclonal antibody or antibody like (Fc) fusion protein. The cell culture bioreactor is harvested using a continuous centrifuge to remove cells. The clarified cell culture is then filtered through depth filtration and polishing filtration trains to further remove smaller particles, which removes the risk of blocking the downstream chromatography columns due to deposition of solids on top of the adsorbent bed. Protein A affinity chromatography then serves to capture the product and to remove a majority of process related impurities and is followed by a low pH viral inactivation step. After this step, an additional 1 or 2 chromatography steps are utilized to further purify the protein removing both, process- and product-related impurities. These steps can be a combination of anion exchange, cation exchange, hydrophobic interaction, mixed mode, or ceramic hydroxyapatite chromatography. Chromatographic purification is followed by viral filtration and finally tangential flow filtration, which concentrates and diafilters the product into the desired formulation conditions.

All process steps are scaled down during process characterization studies and are assessed individually to show they are representative of the full scale manufacturing process. In this section, examples will be given for typical scale-down parameters and methodology for 2 of the steps, centrifugation and column chromatography.

Centrifugation is the initial step in the harvest clarification process and is typically used for microbial and mammalian cell culture for separating the cells from the culture broth. A disc stack centrifuge is typically used for this purpose and employs centrifugal force to separate solid cell debris from the liquid or cell culture supernatant. Cell debris and solid impurities are discharged from periphery of the centrifuge bowl using centrifugal force to drive the solids through an orifice while clarified centrate is recovered from the center of the bowl.

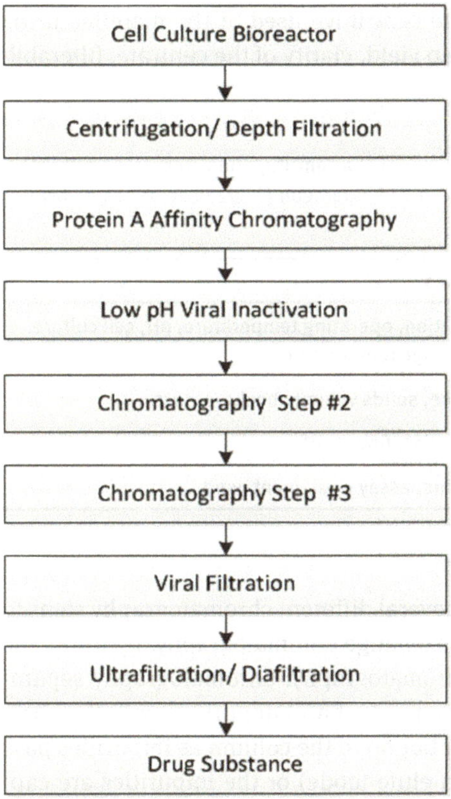

Figure 7.3.8: Sequence of unit operations from cell culture to drug substance including the several purification steps for a protein molecule.

The principle scale-down parameter for a centrifuge is Q/Σ, where Q is the flow rate of cell culture solution through the centrifuge and Σ is the area of a gravity settling tank that has equal sedimentation characteristics as the centrifuge, which is dependent on the mechanical design and geometry of a centrifuge [36]. Other parameters that should be kept constant upon scale up include the ratio of solids collection volume to total shot volume and solution properties of the feed material such as pH and temperature. In addition, cell culture properties play a significant role in centrifuge performance, thus culture variability in respect to cell densities, viability and duration should be minimized. Because of differences in equipment design between scales, it is challenging to replicate the exact process at small scale, where shear differences generate changes in the particle size distribution. Recently, some small-scale models have had success where cell culture is first sheared through a capillary or rotating device and then separated using bench scale centrifugation [37, 38]. However, these approaches have not yet been widely accepted. Instead, as a general practice, a pilot scale centrifuge of similar design is used to scale down the centrifugation process. Shown in Table 7.3.3 are types of scale-down parameters for a Westfalia CSC-6 disc stack centrifuge. At target operating conditions the clarification area of the CSC-6 centrifuge is approximately 31-fold less than the centrifuge used at the manufacturing scale. The primary output parameters are step yield, clarity of the centrate, filterability of the centrate, and amount of cell lysis.

Table 7.3.3: Partial list of parameters for scale-down model of a centrifugation step using a Westfalia CSC-6 disc stack centrifuge, grouped according to type of parameter (* Q/Σ used as scale down criteria for centrifugation RPM).

Type of Parameter	List of Parameters
Scale independent	Cell culture duration, operating temperature, pH, cell culture and centrate storage temperature
Scale dependent linear	Total shot volume, solids volume, backpressure
Scale dependent nonlinear	Centrifugation RPM*, flow rate*
Operational	Turbidity monitors, assay equipment used

Downstream purification processes employ several different chromatography steps to purify the protein of interest, including ion exchange chromatography, hydrophobic interaction chromatography, and affinity chromatography. Chromatography separation is based on differential interaction of the product and impurities with the chromatographic absorbent. The protein is either bound to the column as impurities pass through and is later eluted (called bind-and-elute mode) or the impurities are captured in the column as the protein flows through (called flow-through mode). Scaling down a chromatography step is achieved by keeping scale independent parameters constant. These are the bed height and fluid velocity, which determine residence time in the column, as well as the ratio of process fluid volume to column volumes. This ensures that the column load (g product/L of resin) and the equilibration, wash, and elution volumes (in column volumes) are kept constant across scales. Typically, scale-down is achieved by holding the residence time in the column constant (constant bed height and fluid velocity) and reducing the column diameter and thereby reducing column volumes and amount of product consumed. In addition, other scale independent parameters such as buffer composition, pH and conductivity, and temperature are maintained the same as manufacturing scale. The comparability of column packing is measured by height equivalent of a theoretical plate (HETP) and asymmetry factor. The small scale chromatography needs to meet the same criteria as manufacturing scale to demonstrate packing is comparable [39]. Finally, equipment and operational aspects must be matched as closely as possible. This includes tubing, pumps, controllers, UV monitors, detection equipment, and such. Product pool start and stop criteria must be matched with the manufacturing scale to ensure similar levels of yield and purity in the product streams [40]. Table 7.3.4 summarizes the scale-down parameters for a chromatography step.

Table 7.3.4: Partial list of parameters for scale-down model of a chromatography step, grouped according to type of parameter (* Bed height and fluid velocity used as scale down criteria for column diameter).

Type of Parameter	List of Parameters
Scale independent	Bed height, fluid velocity, equilibration, elution and wash volumes, buffer pH and conductivity, temperature
Scale dependent linear	Product loaded, resin volume
Scale dependent nonlinear	Column diameter*, flow rate*
Operational	Column packing, tubing, pumps, controllers, UV monitors, detection equipment, start-stop criteria, assay equipment used

7.3.4.4 Additional Considerations for Scale-down Modeling

In addition to qualifying a scale-down model at operating setpoints, it increases the credibility of the model if comparison is shown to manufacturing scale at different setpoints, if such at-scale data exists. Ultimately, the main use of scale-down models for process characterization is to test the process, as it would perform in manufacturing, at different operating setpoints.

Product quality comparison across scales can pose several challenges. Sample handling and time of processing can change the quality of harvest material. Preprocessing steps such as low pH harvest can alter certain product attributes, and should be matched across scales for a meaningful comparison. Hold times and multiple thawing and refreezing cycles impact the quality of downstream process intermediates. It is important to obtain the product quality of the manufacturing scale from the appropriate step instead of drug substance for a head-to-head comparison. Retains from intermediate steps must be entered into the nonroutine sampling plan of manufacturing campaigns that will be used for scale-down modeling in the future.

A set of variables that are difficult to quantify or control include raw materials. Lot-to-lot variability in raw materials can complicate scale-down modeling efforts. Program timelines typically do not permit extensive evaluation of multiple lots of different raw materials. As a result, where possible, lots of raw materials already used in previous manufacturing campaigns should be used. If such material is not available, then the same vendor should be used for non-GMP raw material.

7.3.5 DoE for Process Characterization

As outlined in Section 7.3.3, risk assessments help to identify the input parameters that could have the most influence on the process performance and product quality (performance parameters). However, it must be experimentally determined which of these parameters truly have a significant effect and by how much. This exercise is the crux of process characterization, which forms the focus of this section. Figure 7.3.9 outlines the flow chart for this experimental process, which mainly consists of screening, modeling and model verification experiments, before a design space could be built. The prerequisite for these experiments is to have a sound scale-down model that will predict the same outcomes in the manufacturing scale as obtained from the scale-down system (discussed in Section 7.3.4).

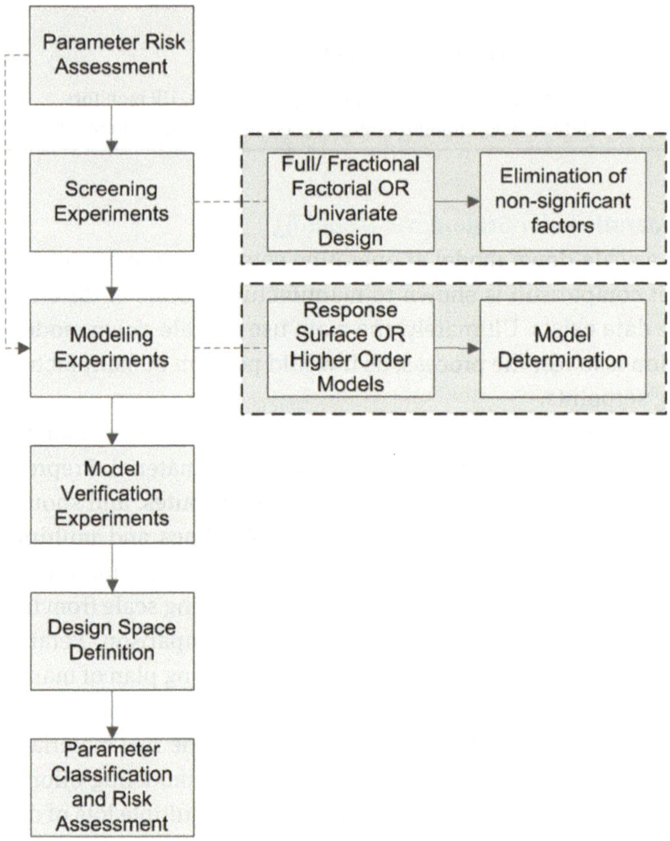

Figure 7.3.9: Flow chart of the sequence of experiments for process characterization studies.

The most commonly used approach for process characterization is a design of experiments (DoE) strategy, since (i) it requires less number of experiments compared to one-factor-at-a-time (OFAT) approach, (ii) it reveals any interactions between the different input parameters, which leads to sound predictions over the entire design space, and (iii) where optimization is concerned, it leads to the global optimum as opposed to a local optimum from OFAT testing [41-44]. There are numerous software programs available for setting up a DoE study and analyzing the results, including Design Expert (Stat-Ease), JMP (SAS Institute), Statgraphics (Statpoint Technologies), MODDE (Umetric), Minitab 16 (Minitab) and *rsm* package in R [45].

The first step in experimental design is to identify factors and responses; the factors are typically input parameters that can be controlled and are short-listed during risk assessment for their potential to impact the process performance and product quality, which form the responses. It may not always be feasible or economical to include all input parameters identified from risk assessments within a DoE mode. Sometimes it is easier to perform an OFAT study for such factors, especially when there is a low likelihood of interaction with other factors, based on prior process knowledge. Examples of such cell culture parameters include medium hold times and feed addition timings [14]. Some other parameters such as pCO_2 levels and culture osmolality, which are outputs dependent on process inputs but can have a direct impact on process performance, are independently studied. It is important to have established analytical assays available for each of the process and product quality responses and have thorough knowledge of the assay variability, prior to embarking on process characterization.

7.3.5.1 Screening and Modeling Experiments

The DoE strategy for process characterization can be broken into 3 broad categories of experiments: screening, modeling, and model verification. The purpose of screening experiments is two-fold:

1. Experimentally "screen" for the factors that impact the responses: Screening experiments are designed to identify significant main effects. This means that screening designs should have enough statistical power to confirm the significant effect of a factor on 1 or more responses without being confounded by interactions between other factors. This is called aliasing and an important aspect of the DoE design [41]. If 1 or more factors fall out due to insignificant effect on the responses, it reduces the resource requirement for the subsequent modeling experiments.
2. Revisit the experimental range of the factors: The testing range selected for the factors is a crucial design element of the DoE experimentation for process characterization. The range should be wider than the operating range in manufacturing and should ideally encompass previous action limits for the parameter. Historical development and manufacturing experience must be thoroughly assessed to assist in the range setting process. This is illustrated by Figure 7.3.10, which is a modified version of illustration by Li et al. [35] of the different ranges associated with a given input parameter. Range 1, which is within the operating limits, may be too narrow to give a detectable signal over noise ratio, and the parameter may not have a statistically significant effect on the response(s). For example Abu-Absi et al. [46] found from screening studies that temperature did not affect

N-glycosylation. However, when they widened the temperature range tested, a quadratic model was built and temperature was identified as a critical process parameter (CPP). Range 2, which goes beyond the known failure points, is too wide and will require repeating experimental runs with a narrower range. Range 3 is the ideal range as it leads to the identification of the acceptable range, which is in between the operating range and the characterization range [46]. Appropriate action limits could then be set at or within the acceptable range for the parameter. Often, the edge of failure is not known and the range set during the screening experiments may be too wide or too narrow. Depending on the results from the screening experiments, the ranges for select input factors may need to be modified in order to capture the appropriate parameter space in the modeling experiments. Other common indicators of range modifications include

(i) identification of a process optimum at the edge of the current experimental space, indicating that the real optimum is outside this space, and
(ii) a product quality attribute is undesirable for a large fraction of the experimental space.

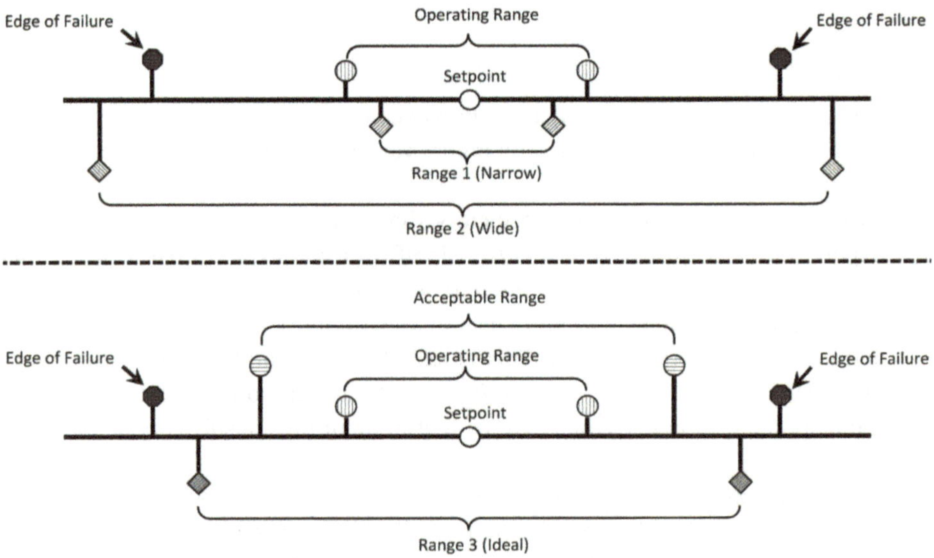

Figure 7.3.10: Ideal range setting process.

Typically screening designs use a partial factorial design with at least resolution IV to be able to identify main effects. For resolution III designs, the main effects are aliased with 2-factor interactions and may confound the results obtained. As a rule of thumb, the power of the design should be at least 80 % for detecting signal/noise ratio of 2 or greater at 5 % alpha risk (i.e., 5 % chance of detecting false positive effects). In order to increase the power of the design, the most straightforward method is by increasing the replicates or number of runs. Additionally, a good design should have adequate degrees of freedom for lack of fit and pure error, which is possible by increasing replicates at the center point [41].

Since screening designs are usually very lean in nature, they cannot model quadratic effects or factor interactions due to aliasing. The next step after screening is modeling, where response surface methodology (RSM) is used to model the output responses (process and PQ parameters) as a function of the significant factors identified from the screening stage. Typically a new design is built after revisiting factor effects and their ranges. However in some cases, the factorial design used in the screening stage could be augmented to complete the modeling stage. To cite an example, an internal project tested the following 5 parameters during the screening stage using a resolution V fractional factorial design: pH, temperature, dissolved oxygen, seed density, and medium concentration. The medium concentration was not significant and the ranges of the factors were appropriate for modeling the responses. Hence, the factorial screening runs were just augmented with axial runs to complete a full central composite design (CCD) with 4 factors. Since the fifth factor was not significant, the 5-factor screening runs were still valid to be used towards the 4-factor CCD, based on the assumption that there was no impact from the fifth factor.

A central composite experimental design (CCD) is one of the most commonly used designs to build response surface models for bioprocess applications [44]. This design models quadratic effects with high quality predictions over the entire design space, when there is no limitation on the number of runs and there are no constraints that deem certain factor settings in the design space infeasible to run. It does not, however, take into account 3-factor interactions, which are aliased with 2-factor interactions. This is acceptable for bioprocessing applications as most interactions in upstream and downstream process development involve only 2 factors at a time. In special circumstances, other designs such as Box-Behnken, D-optimal or Plackett-Burman may be used. The Box-Behnken designs are useful for 3 to 4 factors, especially when the simultaneous extremes of all factors should be avoided for risk of process failure. This is because the design points are such that 1 of the factors is always at the center of its range. The Plackett-Burman designs are most relevant only for robustness testing of an existing model, since the aliasing is very complex and as a result they can model only main effects effectively. D-optimal designs are used when certain factor settings are not feasible to run in the design space because of physical constraints. The reader is referred to Montgomery [41] for detailed description of each of these designs since it is beyond the scope of this text. In the next section, we will focus on statistical and operational considerations encountered in the application of DoE experimental strategy for upstream and downstream process characterization.

Statistical Considerations

Typically, the software used to build the DoE strategy would ensure the design has the appropriate statistical rigor to make the necessary conclusions. However, the experimenter should scrutinize it closely to understand potential limitations of the design. One of the most important design parameters is the standard error of the design as a function of fraction of the design space (FDS), which is a measure of precision. This is then used to calculate the fraction of the design space that has the desired precision based on the maximum allowable half width of the confidence interval of a response and also depends on the actual standard deviation of the response in the design space. For process characterization RSM models, the FDS should be >90 %. The standard error decreases with increase in the number of runs, which is not always possible due to resource and scheduling constraints. As a result, to get a higher FDS, the more common approach is to allow for more variation in the confidence intervals compared to the standard deviation. Alternatively, if the overall variability of the process is minimized leading to lower standard deviations, the FDS becomes larger. Finally if one or more model terms are not significant, it decreases the degrees of freedom required for the model, as a result of which there is an increase in the FDS.

Once the data is generated, it is important to check for normality of the residuals and remove any outliers that might increase the 'lack of fit' of the model. However, the degrees of freedom lost by removal of outliers may lead to inadequate fit for the model itself. If this is the case, such runs must be repeated. Sometimes the data may require a transformation to increase the accuracy of fit. This is especially true if the ratio between the maximum response value and minimum response value is greater than 10. The resulting analysis of variance (ANOVA) shows whether the model is significant and which factors significantly impact the response. A statistical measure for the strength of the model is the value of the adjusted and predicted R-square. As a general rule of thumb, these should be at least greater than 0.5 and not more than 0.2 apart from each other for a good model fit and prediction. The remaining degrees of freedom not explained by the model are lumped together to form the 'lack of fit'. If the 'lack of fit' is significant, the fit of the model may be questionable, since this could happen due to a lot of variability in the data that could not be fit. In some cases, the error associated with repeatability at the center point (called 'pure error') is much less compared to the variability at the other design points, which could also lead to significant 'lack of fit'. In such cases, if the R-square values are high, the model can still be considered valid.

The factorial designs generate linear models as a function of the significant factors, which are valid only at the corners of the design space. An additional statistical parameter called "curvature" is calculated for the overall model and is a function of how far off the actual response is at the center point from the hypothetical response assuming the linear model is valid over the entire design space. If the curvature is significant, the linear model is not sufficient to model the space, and a higher order model is required. Sometimes, if the responses at high and low levels of a factor are very similar to each other, the screening model may not pick up the factor as being significant. However, this may not be true if the curvature is significant, since the curvature can be due to any factor(s).

Process or Operational Elements

Due to inherent biological variability, DoE results from cell culture experiments are often fraught with high standard deviations. This can be somewhat alleviated by implementation of appropriate controls. If the parameter setpoints for the control process are not at the true center of the design, then it is recommended to run a control condition with each experimental run, to account for any inoculum source or raw material variability. Blocking is also helpful in such scenarios, since it will help eliminate the effect of the variation before computing the model.

For cell culture processes, temperature and pH are the 2 most common factors that affect the process and product quality. In addition, other factors such as dissolved oxygen, seed density, temperature shift timing, culture duration, and feed percentages are often characterized. There may be instances where the effect of 1 factor partially overlaps that of a closely interacting or dependent factor, and caution must be exercised when designing and analyzing such experiments. For example, the feed percentage variation results in higher or lower amount of nutrients per cell, while increasing or decreasing the seed density leads to a somewhat similar outcome. Similarly analyzing pH setpoint and deadband in the same experiment may be convoluted. As an example, a process with pH setpoint of 7.0 +/− 0.1 with a variation of 0.2 pH units in setpoint and 0.05 pH units in deadband will result in the following 5 combinations of pH ranges: (i) 6.75 to 6.85, (ii) 6.65 to 6.95, (iii) 7.15 to 7.25, (iv) 7.05 to 7.35, and (v) 6.9 to 7.1. While this covers the entire pH range in 5 different regimes and could provide more resolution, from a strictly DoE perspective, the same could be achieved by just changing the setpoint by 0.25 units resulting in the following experimental combinations: (i) 6.65 to 6.85, (ii) 6.9 to 7.1, and (iii) 7.15 to 7.35. This strategy would give similar information with less experimental input.

Another subject of discussion in setting up cell culture DoEs is whether to evaluate inoculum and seed expansion stages separately than the production bioreactor stage. Since the main purpose of each of the inoculum expansion shake flask stages and seed expansion bioreactor stages is to expand cells to inoculate the production bioreactor, the characterization of these steps could be lumped together as the characterization of a single unit operation to simplify the experimental design. However, the final seed bioreactor stage (N-1) may affect the production bioreactor performance, hence the N-1 stage is typically lumped with the production bioreactor stage. Abu-Absi et al. [46] followed this two-tiered approach.

Often the upstream and downstream groups work in a silo. Communication between the upstream and downstream groups is essential for developing a successful control strategy. It is important to understand which product quality attributes do not change through the downstream process and which ones do. This will determine where the primary control for that attribute should lie. For example, for most monoclonal antibodies, the glycosylation patterns affect the effector function [47] and are important CQAs to control. Typically, they do not change through the downstream process, hence it is very important to understand and quantify the level of control possible through the cell culture process for these glycoforms. On the other hand, host cell proteins (HCPs) are typically cleared downstream by several logs, and the HCP control should be predominantly a downstream effort.

Typically, each step in the downstream purification process is characterized independently of each other with their own dedicated DoE study [48-51]. However, there are times the starting impurity level can be high and impact the capability of the subsequent downstream steps to remove the impurity and are dependent on receiving representative feed streams from cell culture. In the industry, often the impact of incoming feed streams has been ignored. The output from the first column becomes the input for the next column. It can be difficult to control the output from the first column to the second column. If the variability is high in the eluate of the first column, one way to address this is forward linking of parameter, which is described below.

In the forward linkage approach (Figure 7.3.11) [52, 53], critical quality attributes are identified and incorporated as input parameters in each of the purification unit operation statistical designs. Load pools are generated for each unit operation that represent low, mid, and high levels of the critical quality attributes. The screening and modeling experimental design for each purification step is constructed by inserting these attribute load ranges from the preceding step as input factors. The attribute ranges investigated are made wide during the screening design to potentially find the edges of failure and narrowed during the modeling designs, if needed. This approach can help build a better understanding of the impact of one process step on a subsequent step. In cases in which a small number of critical quality attributes are identified, the generation of load pools can be straightforward. However, if a large number of critical attributes are identified, it may not be possible to generate 3 load pools that represent the full range for each attribute. In this case, a ranking of the attributes may have to be used in order to include a full operational range for the most critical of the attributes within these established product quality and/or process performance limits.

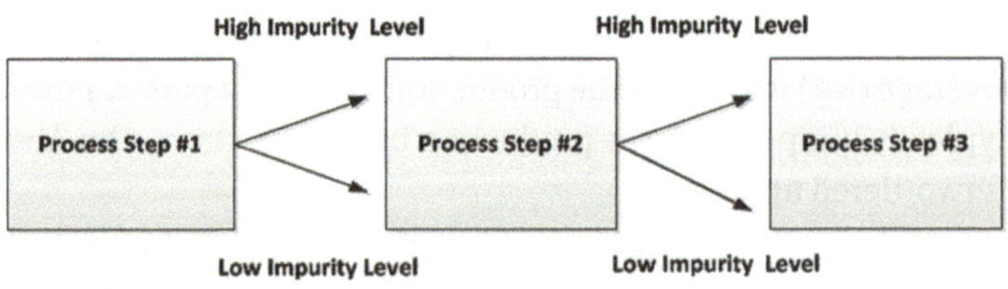

Figure 7.3.11: Forwarding linking approach.

7.3.5.2 Model Verification Experiments

The final stage in DoE modeling is to verify the models created for each of the responses. This is an important step with multiple applications:

1. it verifies the model predictions at different spots in the design space and identifies regimes where the prediction is weaker or stronger,
2. it confirms if the model is systematically under- or over-predicting in the design space, and
3. it confirms the control strategy definition (Section 7.3.7), by additional verification runs at the proposed action limits for the process.

There are 2 aspects to consider when choosing verification points:

1. the points should be well distributed across the entire design space in order to check for model predictability across the entire space,
2. the predicted value of the responses at the points should encompass the entire range of the values of the responses obtained from the DoE experiments.

This checks if the model is equally predictive, irrespective of the actual levels of the responses. If the model verification run results fall within the 99 % prediction intervals from the predicted mean, the model can be considered predictive. However, care must be taken to interpret the prediction intervals, since intervals wider than the assay variability combined with process variability may result from high variability in the experimental data set.

Sometimes, verification runs are also used for robustness testing. In this case, a lower resolution design (such as res III) is created, in which the factors are changed in smaller increments around a point of interest, such as a process optimum or an action limit [44]. This will verify that the response is robust against small changes in the input factors in regimes where it is expected to be.

7.3.5.3 Model Augmentation

Augmenting a model means adding more experimental points to make an existing model stronger. Adding axial points to the design after the factorial runs are completed is an example of model augmentation, where a linear model is augmented to a quadratic model. Model augmentation is typically done when the existing model has limitations significant enough to warrant the investment of resources for additional experiments required for filling the gaps. There are several ways of augmenting a model, and the choice is made based on the particular scenario. It may range from just repeating a few design points that were outliers, or adding a few extra points in a particular regime of importance within the design space to get better predictions, to adding an entire set of extra factorial or axial conditions by expanding the design space in 1 or more directions. A scenario when the latter is typically chosen is when the desired optimum is outside the existing design space. Most DoE software has built in algorithms to augment designs and evaluate the new power and other statistical measures for the design.

7.3.5.4 Design Space Definition

When all the upstream and downstream process steps are fully characterized and predictive models generated, design space establishment can be initiated. The predictive models can define the process parameter ranges for assurance of acceptable quality. Models for each process step are first analyzed separately to determine the capability of the step to perform its intended function. Initial analysis illustrates that only a few process parameters can impact CQAs and confirms critical process parameters. These CQAs are found to be controlling output parameters that constrain the design space for each process step. Process models developed from characterization studies are then analyzed in combination to link the individual process steps and assign acceptable process parameter ranges for each process step when viewed in the context of the entire process. This is accomplished by defining the limits for each CQA in the drug substance and then propagating back through the process using the models to determine CQA limits that can be fed into each subsequent process step. The CQA design space boundaries have been defined in literature based on either the confidence interval [1], prediction interval [16], or the tolerance interval [28], depending on the desired level of control. Once these intermediate quality levels are established, process input parameters can be constrained to ensure that quality levels do not exceed these levels and ensure that final levels do not exceed their limits. The process parameter ranges that ensure each CQA within desired limits may or may not overlap and the most restrictive process parameters will constrain the process parameter limits that will define the process design space for each step.

A design space can be established for one or more unit operations, or a single design space can span multiple unit operations [6]. The choice of unit operations to include in a design space depends on the amount of relevant scientific knowledge and the degree of desired operational flexibility. Thus, a design space linkage through the complete manufacturing process to the drug substance allows the maximum manufacturing flexibility, but could have the most regulatory scrutiny. On the other hand, design space linkage through only 1 or 2 unit operations may potentially allow for a simpler regulatory approval. This can be a great option if these 2 steps provide the greatest opportunity for increased process robustness to the overall process.

7.3.6 Upstream and Downstream Process Characterization Case Study for a Monoclonal Antibody

The following sections provide a case study for the process characterization and design space definition of a humanized monoclonal antibody along the lines of the

Table 7.3.5: Process steps and CQAs evaluated at each upstream and downstream step for a monoclonal antibody (ProA = Protein A, VI = viral inactivation, CEX = cation exchange chromatography, AEX = anion exchange chromatography, VF = viral filtration, TFF = tangential flow filtration, LMW = low molecular weight).

	Production Bioreactor	Centrifugation	ProA	VI	AEX	CEX	VF	TFF
Product-related substance or impurity								
Aggregation	✓	✓	✓	✓	✓	✓	✓	✓
LMW	✓	✓	✓	✓	✓	✓		✓
Low pI isoforms	✓	✓	✓		✓	✓		✓
Galactosylation	✓		✓		✓	✓		✓
Oxidation	✓		✓		✓	✓		✓
Process-related impurities								
HCP	✓	✓	✓		✓	✓		
DNA	✓	✓	✓		✓	✓		
Antifoam C	✓	✓	✓		✓	✓		
Protein A			✓		✓			
Process performance								
Cumulative cell mass	✓							
Yield	✓	✓	✓	✓	✓	✓	✓	✓
Turbidity		✓						
Processing time								✓
Flux								✓
Adventitious agents								
Viruses	✓		✓	✓	✓	✓	✓	

concepts discussed in Sections 7.3.3 to 7.3.5. Prior to performing process characterization studies, product quality and process performance attributes to be tested at each process step were identified. This activity depends on the function of each individual process step within the manufacturing process. Individual process steps do not affect all CQAs [1]. Table 7.3.5 shows a list of the most relevant CQAs and process performance parameters tested at each individual process step based on previous development and manufacturing history for a monoclonal antibody. The purpose of the cell culture inoculum expansion train is to expand cells for inoculating the production bioreactor, and there is no accumulation of product during such steps. Hence, it is not expected to impact product quality, but only process consistency. The exception to this could be the couple of seed bioreactors immediately preceding the production bioreactor, since these stages could directly impact the health of the cells and the initial trajectory of the process in the production bioreactor. The production bioreactor process parameters have the potential to impact all product related substance and impurities as well as most process related impurities. On the other hand, the downstream process parameters' potential impact on quality attributes vary based on each process step's capability and intended function. Multivariate DoE studies are typically performed for all steps except, cell culture inoculum train, viral inactivation (VI), and viral filtration (VF). Univariate studies are performed for these steps because no interaction between process parameters is expected.

7.3.6.1 Process Risk Assessment

A risk ranking and filtering tool, similar to the one used in the CQA criticality analysis was used to focus the characterization effort on the few most important process parameters that may affect product quality and process consistency. The anion exchange chromatography step is used here as an example to illustrate this. The impact score (on a 1 to 5 scale) is multiplied by the uncertainty score (on a 2 to 6 scale) to generate the RPN. Table 7.3.6 shows the definition and scoring system used for the RRF risk assessment. Figure 7.3.12 shows a Pareto plot of the scoring results from the risk assessment. Load HCP is an output parameter from the previous chromatography step, but can have a significant effect on the AEX step capability to remove additional HCP, so it was included in the DoE study. Process parameters with RPN ≥ 10 were

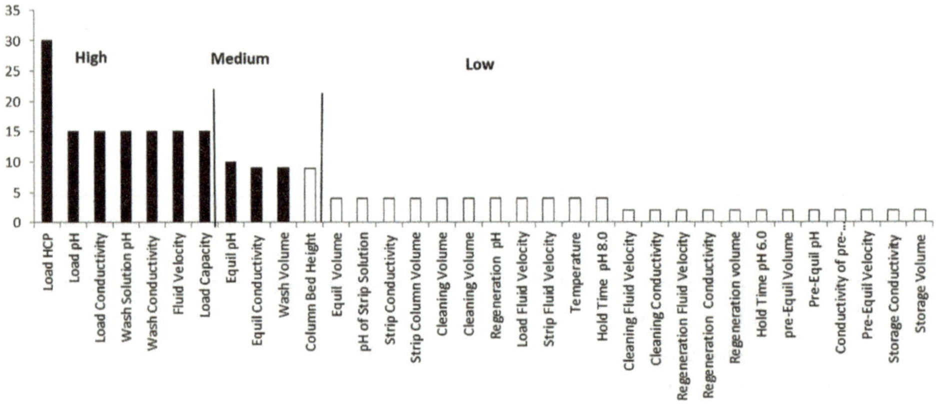

Figure 7.3.12: Pareto plot showing results from risk assessment. The high risk process parameter RPN threshold was set at ≥ 10, medium risk process parameter RPN threshold was 6 to 9; such parameters maybe included in the DoE study and justification is provided if not included, low risk process parameter RPN threshold was <5, for which no further study is required. Process parameters highlighted in black were included in the DoE study.

included in the DoE, along with equilibration conductivity and wash volume with RPN of 6 to 9 in the medium risk range. Column bed height and load fluid velocity together can have an effect on the product residence time in the column. Fluid velocity was given a higher RPN of 15 and included in the DoE study. At the same time, bed height was given a lower RPN of 9 because it can be tightly controlled in manufacturing and hence not included in the DoE study.

Table 7.3.6: Definitions and scoring system for a process risk assessment using rank ranking and filtering tool (CQA = critical quality attribute).

Impact			**Uncertainty**	
Risk	Score	Definition	Score	Definition
High	5	Significant effect on a CQA	6	Effect not known: limited or no knowledge over planned range
Medium	3	Moderate effect on CQA and a significant effect on process consistency	3	Possible effect: theoretical knowledge indicates parameter may have an effect over planned range
Medium/low	2	Minor effect on a CQA or a moderate effect on process consistency	–	NA
Low	1	No effect on CQA and/or process consistency	1	Known effect: experimental and/or historical data shows parameter clearly demonstrates effect on attribute

7.3.6.2 Screening and Modeling Designs

The RRF assessment outlined above was conducted for all process parameters for each unit operation. Results of the assessment summarizing potential CPP and key parameters included in the screening experiments are shown in Table 7.3.7, along with the initial characterization study performed. Both multivariate and univariate studies were performed for the initial screening step for the cell culture bioreactor. Full factorial was performed using pH lower limit, seed density, and feed timing. All other production process parameters were evaluated at center points as well as high and low extreme set points to challenge the potential process action limits. At least 2 replicates were performed for each test condition in all experiments. Less than 5 process parameters were identified as potentially having significant impact on the centrifugation step, hence predictive response surface modeling experiments were performed directly. A central composite design was chosen for centrifugation because curvature was expected in the characterization space being evaluated based on previous experience. Fractional factorial resolution IV designs with at least 4 center points were selected for all other process steps. As explained in Section 7.3.5.1, these are good for identifying main effects and are suitable for the initial screening experiments.

Table 7.3.7: Potential critical and key process parameters evaluated in screening and modeling designs and initial DoE used in process characterization studies.

Process Step	Initial Characteriza-tion study	Screening Process Parameters		Modeling Process Parameters
Cell culture Bioreactor	Full factorial and univariate studies bracketing high and low ranges	Seed density pH Temperature Feed timing Feed volume	Seeding density Cell age DO% set point	pH Temperature Seed density
Centrifuge	Central composite design	Harvest pH Cell culture viability G-force Q/Sigma (flow rate)		Harvest pH Cell culture viability G-force Q/Sigma (flow rate)
Protein A chromatography	Fractional factorial resolution IV design	Harvest pH Load HCP Load capacity Load velocity Load pH Wash I velocity Wash I chaotrope concentration	Wash I NaCl concentration Wash II pH Wash II volume Elution velocity Elution pH Elution volume	Harvest pH Wash I chaotrope concentration Elution velocity Load capacity Load HCP
CEX chromatography	Fractional factorial resolution IV design	Fluid velocity Equilibration pH Equilibration conductivity Load HCP Load capacity Load pH	Load conductivity Wash pH Wash conductivity Elution pH Elution NaCl concentration	Load HCP Load capacity Wash volume Elution pH Elution NaCl concentration

AEX chromatography	Factional factorial resolution IV design	Fluid velocity Equilibration pH Equilibration conductivity Load HCP Load capacity	Load pH Load conductivity Wash pH Wash conductivity Wash volume	Additional study not need adjusted $r^2 = 0.97$ and predicted $r^2 = 0.93$
UF/DF	Central composite design	Mass load Load concentration Cross-flow rate TMP	Load pH Diafiltration pH Diafiltration concent-ration Diavolumes	Mass load Diafiltration pH TMP

Process parameters identified as being significant during screening studies were carried forward into the modeling design phase of development, as shown earlier in Table 7.3.7. An exception to this was the AEX chromatography step, since the resolution IV fractional factorial design was adequate to predict product quality and process performance. For all other process steps, process-modeling experiments were performed using either central composite or D-optimal response surface experimental designs to build predictive models. The results of these studies confirmed or eliminated significant parameters identified in screening designs and identified critical interactions between process parameters using ANOVA analysis. Detailed models for the centrifugation and Protein A steps are described by Cecchini [19]. Table 7.3.8 summarizes the ANOVA for anion exchange (AEX) chromatography to illustrate characterization of a process step that can be used to define a design space for the step. The fractional factorial design was comprised of 40 experiments, divided into 4 blocks with 2 center points per block, to examine the impact of the 10 factors on process responses.

Table 7.3.8: ANOVA of AEX chromatography step. Process parameter considered significant if p-value ≤ 0.05, lack of fit of model considered significant if p-value ≤ 0.05. Square root transformation was required to fit a normal distribution. R^2 values were calculated to examine the fraction of the overall variation that is explained by the model. The adjusted R^2 value is adjusted for the number of terms in the model relative to the number of points in the study. Predicted R^2 values were examined to evaluate how well the model will predict future experimental data (PP = process parameter).

Process Parameter	HCP (ppm) – p-Value	% Relative Contribution (PP Sum of Squares/Total Sum of Squares)
Fluid velocity (cm h^{-1})	–	
Equilibration pH	–	
Equilibration conductivity (mS cm^{-1})	–	
Load capacity (%)	–	
Load HCP (ppm)	<0.0001	34 %
Load pH	*	
Load conductivity (mS cm^{-1})	<0.0001	43 %
Wash pH	*	
Wash conductivity (mS cm^{-1})	0.0217	1 %
Load HCP*load conductivity	<0.0001	22 %
Lack of fit	0.3281	
R^2	0.9657	
Adjusted R^2	0.9595	
Predicted R^2	0.9377	
Transformation	square root	

Load HCP, load conductivity, and the interaction between them were found to significantly influence output HCP levels, as illustrated by Figure 7.3.13. Aggregate percentage, low molecular weight impurities, low pI isoform levels and yield were evaluated during AEX chromatography experimentation, however none of the parameters had significant influence on these CQAs or performance attributes. DNA was below the limit of quantitation for all experiments. In addition to identifying the significant parameters that influence HCP, a mathematical model was developed to predict HCP levels in the AEX flow-through pool. Adjusted and predicted R^2 values illustrate the model's ability to fit the experimental data set and predict response values, which were both >0.9. This model was utilized to help define the design space, as explained in the next section.

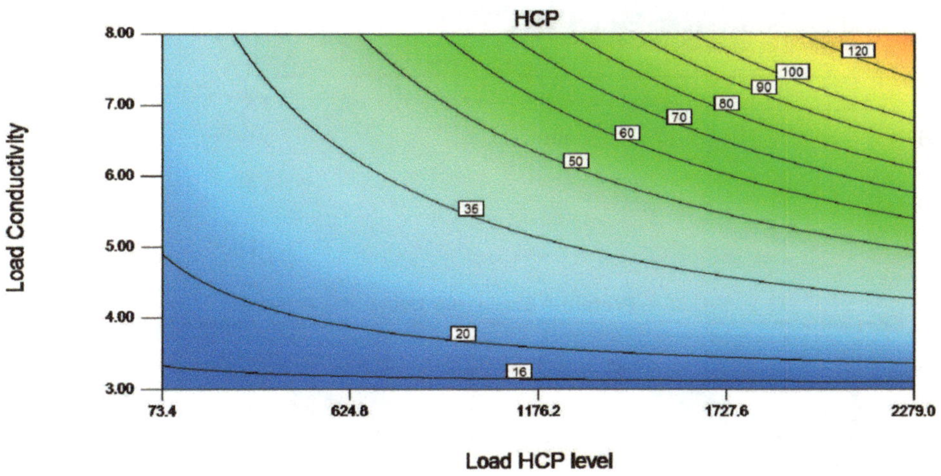

Figure 7.3.13: Contour plot from the process characterization of the AEX chromatography showing the interaction between load HCP and load conductivity.

7.3.6.3 Design Space Establishment

Once the DoE experiments were completed and process models were built, the information was used to establish design space linkage through the complete manufacturing process to drug substance as illustrated in this section. The cell culture production bioreactor process was shown during process characterization studies to be very robust and to consistently produce low levels of product-related impurities over the complete process parameter range evaluated (e.g., aggregate LMW was less than 1 %, charged isoform and various glycoforms were within historical ranges). As a result, downstream purification was not required for reduction of these impurities and a design space for these attributes could be directly established, relating cell culture parameters to levels in the drug substance. In addition, the purification process was capable of removing process-related impurities (HCP, DNA, and cell culture additives)

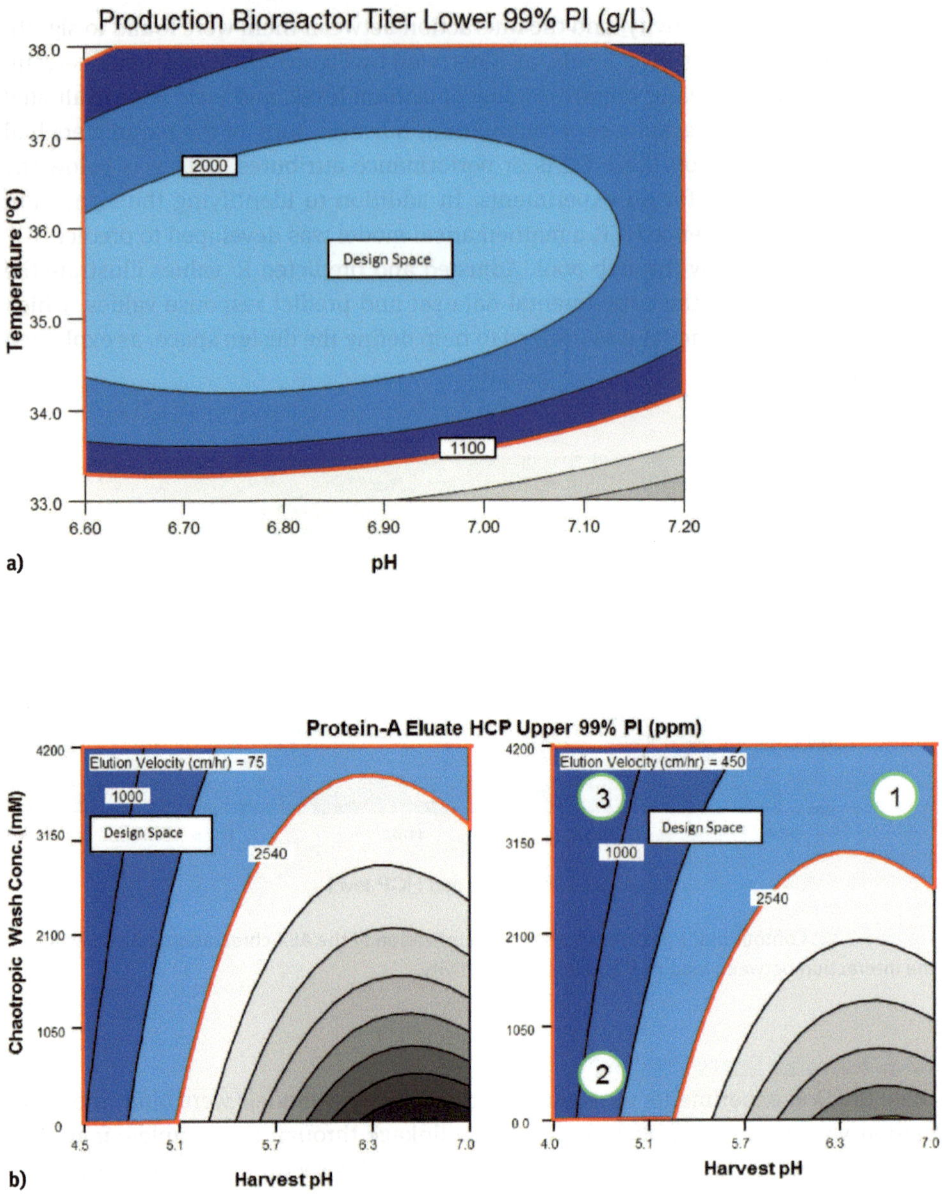

Figure 7.3.14: **a)** Design space for cell culture production bioreactor. The design space is constrained by cell culture titer for process consistency and is depicted within the red border. Temperature and pH were the most significant process parameters that effect product quality and consistency.
b) Centrifuge/Protein A design space. The area within the red box represents the combination of all levels of harvest pH, Protein A wash concentration and elution velocity that produce a Protein A eluate with HCP levels less than 2,540 ppm within 99 % prediction intervals. Numbers represent the following operating regimes: 1. neutral pH operating space, 2. low pH harvest operating space, 3. maximum robustness space.

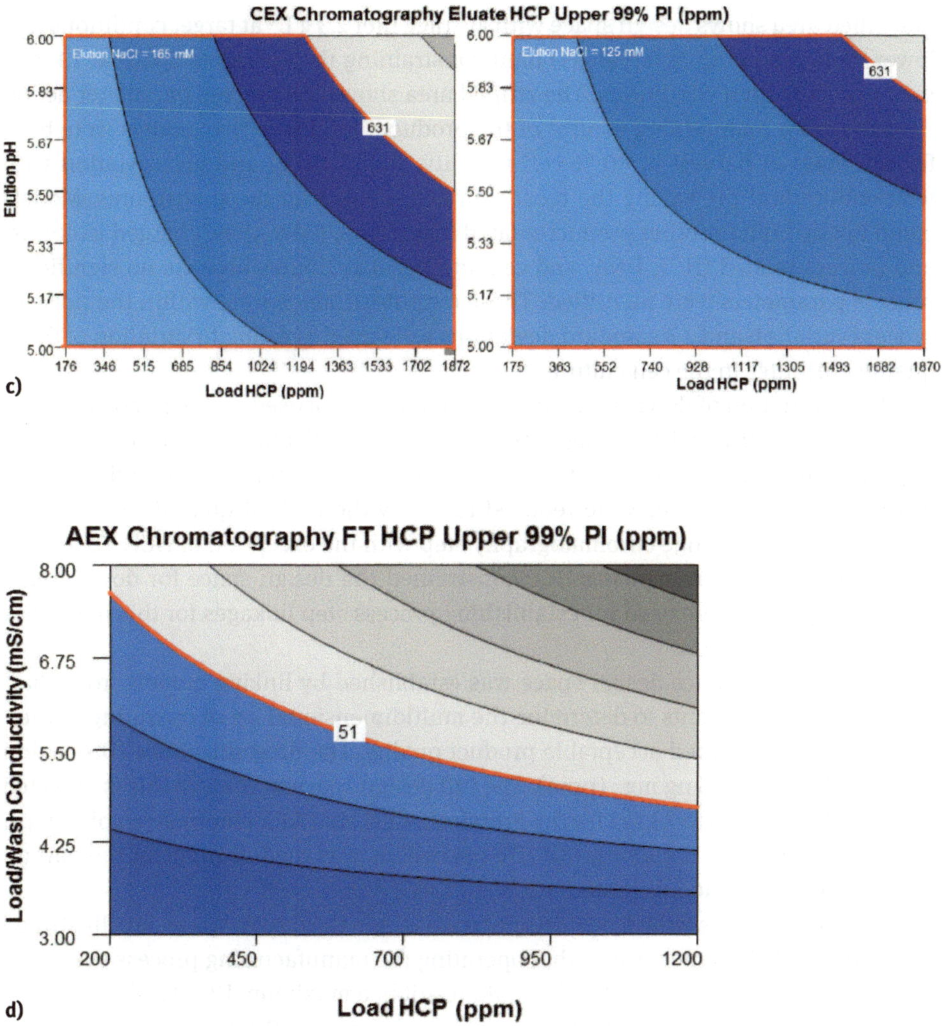

Figure 7.3.14: **c)** CEX design space. The area within the red box represents the combination of all levels of load HCP, pH and NaCl concentration that produce a CEX chromatography eluate with HCP levels less than 1,000 ppm within 99 % prediction intervals. **d)** AEX design space. The area within the red box represents the combination of all levels of load HCP and load/wash conductivity that produce AEX chromatography flow through with HCP levels less than 51 ppm within 99 % prediction intervals.

to below set limits in the drug substance even when they were generated or used in cell culture at maximum levels. Thus, the cell culture design space spanned the complete process parameter range that was evaluated during characterization.

Product titer was the only attribute that constrained the cell culture bioreactor operating space to maintain process consistency and is depicted in Figure 7.3.14a. The blue filled area shows design space with average titer 2.3 g L^{-1} at target conditions and lower 99 % prediction interval (PI) limit constraining the operating space at 1.1 g L^{-1} under process limit conditions. The striped area shows the process parameter action limit range for maintaining control of the product titer. Predictive models were built for cell mass at harvest at 40 % cell viability, product titer and glycosylation with temperature and pH having the most significant effect on these attributes. Models could not be built for other product-related (aggregate, LMW <1 %, charged isoforms) and process-related (HCP, DNA, and cell additives) attributes because no significant process parameters were identified. These output variables were within the noise of process variation and ± 3 standard deviations were used to estimate variation of these quality attributes in the cell culture.

The evaluation of downstream operation linkages focused on the removal of process-related impurities beginning at the centrifuge clarification step. Process related impurities included DNA, HCP, rProtein A leachate, and cell culture additives. All process-related impurities were reduced to below the limit of quantification (LOQ) after the anion exchange chromatography step with the exception of HCP. Therefore, process parameters influencing HCP constrained the design space for downstream process steps and were used for establishing process step linkages for this particular manufacturing process.

The drug substance design space was established by linking models from characterization experiments to determine the multidimensional set of operating parameter ranges that assured acceptable product quality. The drug substance (DS) action limit for HCP was ≤ 25 ng mg^{-1} (ppm). The full design space to achieve this is depicted graphically in 7.3.14b to 7.3.14d for the Protein-A, CEX, and AEX chromatography steps. The design space for the VF and UF/DF operations was the full range of the parameters tested during modeling studies.

Parameter ranges, which ensure operation within the design space, are illustrated in Table 7.3.9. This demonstrates that operating the manufacturing process even with all variables set to worst case levels would result in a maximum HCP level in the drug substance of 25 ppm based on model prediction intervals (99 %).

With the design space defined, alternate operating spaces can be examined that could potentially improve the process. Harvest pH and Protein A wash parameters were found to interact and contribute significantly to HCP removal during the purification process. Figure 7.3.14b illustrates the impact of these parameters on protein-A eluate HCP levels. By lowering the cell culture bioreactor pH prior to harvest from pH 7 (Region 1 in Figure 7.3.14b) to pH 5 (Region 2 in Figure 7.3.14b), cells and cell debris flocculate and particulates are more efficiently removed by centrifugation. Lowering the pH had the added benefit of greatly reducing DNA level to close to assay level of detection level and lowering HCP levels two- to fivefold [19]. This change could eliminate an expensive chaotrope wash, that was difficult to handle, and an environmentally unfavorable component of the waste stream. Alternatively, low pH combined with a chaotropic wash (Region 3 in Figure 7.3.14b) could add increased robustness for

HCP removal for the remainder of the manufacturing process. Movement to this new area demonstrates the harvest and Protein A steps can reduce HCP to similar levels < 1,000 ppm as are seen after the cation exchange (CEX) chromatography and could potentially facilitate removal of this step.

Table 7.3.9: Operating space definition within the design space using HCP as the limiting CQA.

Process Step	Factors that Significantly Influence HCP	Factor Ranges that Assure Operation within the Design Space	Maximum 99 % HCP Prediction throughout the Design Space (ppm)
Production bioreactor	No significant parameters	N/A	1,972,737
Centrifugation	Harvest pH	Interaction: see design space in Figure 7.3.14b	1,380,000
MabSelect chromatography	Harvest pH		2,540
	Chaotropic wash concentration (mmol L^{-1})	Harvest pH = 4.5 to 7.3 if wash concentration = 3,800 to 4,200 mmol L^{-1} or Harvest pH = 4.5 to 5.1 if wash concentration < 3800 mmol L^{-1}	
	Elution velocity (cm h^{-1})	100 to 450	
CEX chromatography	Elution pH	5 to 5.6	1,000
	Elution [NaCl] (mmol L^{-1})	125 to 165	
AEX chromatography	Load conductivity	3.6 to 5.0	51
	Wash conductivity	3.6 to 5.0	
Viral filtration	Load concentration (g L^{-1})	0.8 to 3.2	43
UF/DF	None, UF/DF provides an average reduction factor over all process ranges tested	N/A	25

7.3.7 Manufacturing Control Strategy

As outlined in Figure 7.3.1, the main objective of product and process characterization is to define the manufacturing control strategy for process performance qualification (PPQ) runs and the proposed commercial process. This section will discuss how the control strategy is set once the characterization package is generated. The main steps involved are (i) process parameter classification, (ii) action limit setting, (iii) risk assessment, and (iv) PFD documentation and batch record review.

7.3.7.1 Process Parameter Classification

Process parameters can either be inputs or outputs. Process control is the demonstration that controlled parameters (inputs) can be maintained within their acceptable limits and, as a result, performance parameters (outputs) remain within predicted limits. The inputs are tested during process characterization and any input that significantly impacts one or more critical quality attributes (CQAs) is classified as a critical process parameter (CPP). CPPs have action limits which when exceeded leads to a major process deviation. On the other hand, if an input parameter affects only the process or non-CQAs, it is classified as a key process parameter (KPP). If action limits for KPPs are exceeded it results in a minor process deviation. Sometimes an alternate approach is taken by completely removing action limits for inputs that do not affect process or product quality. This approach avoids having a lot of nuisance deviations, but strong justification must be provided for such classification. In a similar fashion to the inputs, the output parameters are also classified in the order of their criticality. An extremely critical output is one whose specifications have to be met, or the batch is terminated. Examples include microbial, viral, or mycoplasma contamination in the harvest sample and post-use integrity test of filters prior to bottling of drug substance. The less critical ones such as cell densities and viabilities in the upstream expansion and production bioreactor stages have action limits, while the noncritical ones do not. Accurate parameter classification is essential to understand the triggers and levers for process control.

7.3.7.2 Action Limit Setting

Setting of action limits for the parameters is the main step for control strategy definition. Normally, wider action limits are preferred by manufacturing due to operational flexibility; however, this may increase the risk of driving product quality outside of its specifications. Hence, a balance between these 2 aspects has to be taken into consideration while setting the action limits. The collective body of knowledge generated from the SAR studies on the product (product characterization) and CQA response surface models as a function of inputs (process characterization) is used to define the acceptance criteria for the CQAs and determine the levels of input factors that meet the criteria. In a strictly QbD study, the CQA acceptance criteria must be met for all combinations of critical input factors operating at their action limits. This can be done by a global optimization routine within the software used for DoE process characterization. It is prudent to take a conservative approach when setting action limits for CPPs. This means leaving some room outside the action limits where the process and PQ are still acceptable. This becomes harder when more than one input affects a CQA. For example, if high temperature in the production bioreactor impacts aggregation and high pH makes it worse, the operating space for temperature and/or pH may become quite narrow. However, this may not be feasible mainly due to operational limitations of manufacturing equipment. In such instances, action limits could be made wider for a single parameter by considering its effects in isolation, based on low likelihood of multiple parameters running at the action limits at the same time. This is usually considered a better strategy than locking oneself in to a very narrow operating range for an input that is not feasible to control due to equipment limitations. Ranges for the outputs are set based on historical manufacturing data and data from control runs during process characterization. Appropriate statistical measures are used to set the action limits. To summarize, the action limits must be set such that the process is able to consistently maintain the input parameters within the action limits, to generate output parameters within their action limits, which further results in meeting the CQA acceptance criteria and drug substance release specifications. This is the main objective of the PPQ runs.

7.3.7.3 Risk Assessment

After parameter classification and action limit setting, a risk assessment is typically performed to assess the risk imparted to the process by the process parameters. This risk assessment is typically different than the one conducted before process characterization, as additional knowledge from process characterization is used for the assessment. An FMEA based risk assessment could be used where potential modes of failure are identified and additional risk mitigation procedures are prescribed based on the severity of the failure, likelihood of occurrence, and probability of detection. This type of assessment pays more attention to manufacturing equipment capabilities, automation, and measures for detecting deviations such as alarms, which are typically not assessed during process characterization experiments. The objective is to completely derisk the process before initiating the PPQ runs. Since Section 7.3.3 goes into details of the risk assessment types and procedures, we will skip that in this section.

7.3.7.4 Documentation and Batch Record Review

The final step before PPQ runs is to document the process capturing the target values, operating ranges and action limits for all the process parameters. Step by step instructions to execute the process are summarized in master production records and documented in batch records to be filled out by manufacturing operators during the execution of the batch. These records must be reviewed by a cross-functional team for accurately capturing findings from process characterization, parameter ranges, and risk mitigation procedures for successful execution of the PPQ batches. Manufacturing operators should be appropriately trained to carry out the specified tasks using good manufacturing practices (GMP).

7.3.8 Conclusions

A significant section of the biological license application (BLA) for a biopharmaceutical entity deals with the ability to manufacture the drug consistently using the intended commercial process. This requires an in-depth understanding of the quality attributes that affect the safety and efficacy of the drug and how these attributes can be "controlled" by controlling a set of input parameters that significantly affect them. This forms the main objective of a collective set of activities that spans over multiple cross-functional groups and months of effort and is called process characterization. A thorough process characterization effort is imperative for a successful process performance qualification (PPQ) in manufacturing.

We covered several key elements of process characterization in this chapter, starting from CQA criticality analysis and quality risk management, which form the basis of QbD. Significant advances in analytical technology, such as high-throughput methods, are improving the speed and resolution of product characterization. On the flip side, the bar is even higher for product comparability protocols between 2 different processes. Different types of risk assessments and their applications were covered in this chapter. Raw material risk assessment and management will prove to be very important in the next decade. As most of the industry has made the transition to chemically defined raw materials, the first set of challenges due to raw material variability has surfaced. The importance of scale-down modeling for process characterization cannot be stressed enough. Advances in this area are migrating towards further reduction of scales for increased efficiency and more aggressive timelines. For example, traditional upstream scale-down modeling in 2 to 5 L bioreactors may shift to mini- or microbioreactors with the advent of new technology in this area.

One of the several advantages of using DoE for process characterization is the flexibility to expand into a QbD study for design space definition and freedom of moving within the design space. At present, the implementation of QbD for BLA filings is not widespread as most companies adopt a semi-QbD approach. However, advances in process analytical technology (PAT) and advanced process control (APC), have opened the doors for endless possibilities for real-time product quality control. From a regulatory point of view, this will provide the highest level of control for a process to deliver a product with a consistent quality profile.

References

[1] Eon-duval A, Valax P, Solacroup T, et al. Application of the quality by design approach to the drug substance manufacturing process of an Fc fusion protein: towards a global multi-step design space. J Pharm Sci 2012;101:3604-3618.

[2] US Food and Drug Administration .Pharmaceutical cGMPs for the 21st Century – A Risk-Based Approach, 2004. at http://www.fda.gov/downloads/Drugs/DevelopmentApprovalProcess/Manufacturing/QuestionsandAnswersonCurrentGoodManufacturingPracticescGMPforDrugs/UCM176374.pdf, accessed 15th May 2014.

[3] Korakianiti E, Rekkas D. Statistical thinking and knowledge management for quality-driven design and manufacturing in pharmaceuticals. Pharm Res 2011;28:1465-1479.

[4] Yacoub F, Lautens J, Lucisano L, Banh W. Application of Quality by Desig Principles to Legacy Drug Products. J Pharm Innov 2011;6:61-68.

[5] International Conference on Harmonisation of Technical Requirements for Registration of Pharmaceuticals for Human Use. ICH Harmonised Tripartite Guideline: Quality Risk Management Q9, 2005 at http://www.ich.org/fileadmin/Public_Web_Site/ICH_Products/Guidelines/Quality/Q9/Step4/Q9_Guideline.pdf, accessed 15th May 2014.

[6] International Conference on Harmonisation of Technical Requirements for Registration of Pharmaceuticals for Human Use. ICH Harmonised Tripartite Guideline: Pharmaceutical Development Q8(R2), 2009 at http://www.ich.org/fileadmin/Public_Web_Site/ICH_Products/Guidelines/Quality/Q8_R1/Step4/Q8_R2_Guideline.pdf, accessed 15th May 2014.

[7] International Conference on Harmonisation of Technical Requirements for Registration of Pharmaceuticals for Human Use. ICH Harmonised Tripartite Guideline: Pharmaceutical Quality System Q10, 2008 at http://www.ich.org/fileadmin/Public_Web_Site/ICH_Products/Guidelines/Quality/Q10/Step4/Q10_Guideline.pdf, accessed 15th May 2014.

[8] International Conference on Harmonisation of Technical Requirements for Registration of Pharmaceuticals for Human Use. ICH Harmonised Tripartite Guideline: Development and Manufacture of Drug Substances (Chemical Entities and Biotechnological/Biological Entities) Q11, 2012 at http://www.ich.org/fileadmin/Public_Web_Site/ICH_Products/Guidelines/Quality/Q11/Q11_Step_4.pdf, accessed 15th May 2014.

[9] Kozlowski S, Swann P. Considerations for Biotechnology Product Quality by Design. In: Rathore AS, Mhatre R (eds). Quality by Design for Biopharmaceuticals: Principles and Case Studies. 1st ed. New York, NY, USA: John Wiley and Sons; 2009:9-30.

[10] Kozlowski S, Swann P. Current and future issues in the manufacturing and development of monoclonal antibodies. Adv Drug Delivery Rev 2006;58:707-722.

[11] Rathore AS, Winkle H. Quality by design for biopharmaceuticals. Nat Biotechnol 2009;27:26-34.

[12] Rathore AS. Roadmap for implementation of quality by design (QbD) for biotechnology products. Trends Biotechnol 2009;27:546-553.

[13] Arora T, Greene R, Mercer J, et al. Quality by Design for Biotechnology Products – Part 1: A PhRMA Working Group's Advice on Applying QbD to Biotech. Biopharm Int 2009;22:26-36.

[14] Looby M, Ibarra N, Pierce JJ, et al. Application of quality by design principles to the development and technology transfer of a major process improvement for the manufacture of a recombinant protein. Biotechnol Prog 2011;27:1718-1729.

[15] Martin-Moe S, Lim FJ, Wong RL, et al. A new roadmap for biopharmaceutical drug product development: Integrating development, validation, and quality by design. J Pharm Sci 2011;100:3031-3043.

[16] CMC Biotech Working Group. A-Mab: a Case Study in Bioprocess Development, Version 2.1, 2009. at http://www.casss.org/associations/9165/files/A-Mab_Case_Study_Version_2-1.pdf, accessed 15th May 2014.

[17] Cook S, Patton KA, Bazemore LR. Quality by design in the CMO environment. Biopharm Int 2007;20:28-37.
[18] Chhatre S, Farid SS, Coffman J, et al. How implementation of Quality by Design and advances in Biochemical Engineering are enabling efficient bioprocess development and manufacture. J Chem Technol Biot 2011;86:1125-1129.
[19] Cecchini DJ. Applications of Design Space for Biopharmaceutical Purification Processes. In: Rathore AS, Mhatre R (eds). Quality by Design for Biopharmaceuticals: Principles and Case Studies. 1st ed. New York, NY, USA: John Wiley and Sons; 2009:127-142.
[20] Shivhare M, McCreath G. Practical Considerations for DoE Implementation in Quality By Design. BioProcess Int 2010;8:22-30.
[21] PDA technical report no. 42: Process validation of protein manufacturing. Parenteral drug Association. PDA Journal Pharm Sci Technol 2005;59:1-28.
[22] Mire-Sluis A, Ramnarine E, Siemiatkoski J, et al. Practical Applications of Quality Risk Management. Bioproc Int 2010;8:20-32.
[23] Weese D. Presentation: Overview of Risk Assessment Methods and Applications. At: CMC Strategy Forum meeting on "Practical Applications of Quality Risk Management" 27th/28th July 2009, Bethesda, MD, USA.
[24] Charoo NA, Ali AA. Quality risk management in pharmaceutical development. Drug Dev Ind Pharm 2013;39:947-960.
[25] Stark C. Practical considerations to risk assessments for efficient use in pharmaceutical development and production. SBE's 3rd International Conference on Accelerating Biopharmaceutical Development and Quality by Design in Biologics 2013; Coronado Island, CA.
[26] Seely RJ, Haury J. Applications of Failure Modes and Effects Analysis to Biotechnology Manufacturing Processes. In: Rathore AS, Sofer G (eds) Process Validation in Manufacturing of Biopharmaceuticals. 3rd ed. Boca Raton, FL, USA: CRC Press, Taylor and Francis; 2012:51-62.
[27] Schenerman M, Axley MJ, Oliver CN, et al. Using a Risk Assessment Process to Determine Criticality of Product Quality Attributes. In: Rathore AS, Mhatre R (eds). Quality by Design for Biopharmaceuticals: Principles and Case Studies. 1st ed. New York, NY, USA: John Wiley and Sons; 2009:53-84.
[28] vanHoek P, Harms J, Wang X, Rathore AS. Case Study on Definition of Process Design Space for a Microbial Fermentation Step. In: Rathore AS, Mhatre R (eds). Quality by Design for Biopharmaceuticals: Principles and Case Studies. 1st ed. New York, NY, USA: John Wiley and Sons; 2009:85-110.
[29] van der Voet H, Perry JN, Amzal B, Paoletti C. A statistical assessment of differences and equivalences between genetically modified and reference plant varieties. BMC Biotechnol 2011;11:15.
[30] Pieracci J, Yusuf-Makagiansar H. Leveraging Multivariate Analysis Tools to Qualify Scaled-Down Models. In: Rathore AS, Sofer G (eds). Process Validation in Manufacturing of Biopharmaceuticals. 3rd ed. Boca Raton, FL, USA: CRC Press, Taylor and Francis; 2012:411-440.
[31] Heath C, Kiss R. Cell culture process development: advances in process engineering. Biotechnol Prog 2007;23:46-51.
[32] Nienow AW. Reactor engineering in large scale animal cell culture. Cytotechnology 2006;50:9-33.
[33] Tsang VL, Wang AX, Yusuf-Makagiansar H, Ryll T. Development of a scale down cell culture model using multivariate analysis as a qualification tool. Biotechnol Prog 2014;30(1):152-160.
[34] Betts JI, Doig SD, Baganz F. Characterization and application of a miniature 10 mL stirred-tank bioreactor, showing scale-down equivalence with a conventional 7 L reactor. Biotechnol Prog 2006;22:681-688.

[35] Li F, Hashimura Y, Pendleton R, Harms J, Collins E, Lee B. A systematic approach for scale-down model development and characterization of commercial cell culture processes. Biotechnol Prog 2006;22:696-703.
[36] Ambler CM. The theory of scaling up laboratory data for the sedimentation type centrifuge. J Biochem Microbiol Technol Eng 1959;1:185-205.
[37] Westoby M, Rogers JK, Haverstock R, et al. Modeling industrial centrifugation of mammalian cell culture using a capillary based scale-down system. Biotechnol Bioeng 2011;108:989-998.
[38] Boychyn M, Yim SS, Bulmer M, et al. Performance prediction of industrial centrifuges using scale-down models. Bioproc Biosys Eng 2004;26:385-391.
[39] Conley L, McPherson J, Thommes J. Validation of the Zevalin(R) Purification Process: A Case Study. In: Rathore AS, Sofer G (eds). Process Validation in Manufacturing of Biopharmaceuticals. 3rd ed. Boca Raton, FL, USA: CRC Press, Taylor and Francis; 2012:455-490.
[40] Godavarti R, Petrone J, Robinson J, et al. Scaled-Down Models for Purification Processes: Approaches and Applications. In: Rathore AS, Sofer G (eds). Process Validation in Manufacturing of Biopharmaceuticals. 3rd ed. Boca Raton, FL, USA: CRC Press, Taylor and Francis; 2012:89-140.
[41] Montgomery DC. Design and Analysis of Experiments. 8th ed. Mississauga, ON, Canada: John Wiley and Sons; 2012.
[42] Seely RJ, Seely JE. A Rational, Step-Wise Approach to Process Characterization. Biopharm Int 2003;16:24-34.
[43] Frigon NL, Mathews D. Practical Guide to Experimental Design. 1st ed. New York, NY, USA: John Wiley and Sons; 1996.
[44] Mandenius CF, Brundin A. Bioprocess optimization using design-of-experiments methodology. Biotechnol Prog 2008;24:1191-1203.
[45] Lenth RV. Response-Surface Methods in R, Using rsm. J Stat Softw 2009;32:1-17.
[46] Abu-Absi SF, Yang L, Thompson P, et al. Defining process design space for monoclonal antibody cell culture. Biotechnol Bioeng 2010;106:894-905.
[47] Jefferis R. Glycosylation of recombinant antibody therapeutics. Biotechnol Prog 2005;21:11-16.
[48] Follman DK, Fahrner RL. Factorial screening of antibody purification processes using three chromatography steps without protein A. J Chromatogr A 2004;1024:79-85.
[49] Kelley BD. Establishing Process Robustness Using Designed Experiments. In: Sofer G, Zabriskie DW (eds). Process Validation in Manufacturing of Biopharmaceuticals: Guidelines, Current Practices, and Industrial Case Studies. 1st ed. New York, NY, USA: Marcel Dekker; 2000:29-60.
[50] Xu Z, Li J, Zhou JX. Process development for robust removal of aggregates using cation exchange chromatography in monoclonal antibody purification with implementation of quality by design. Prep Biochem Biotechnol 2012;42:183-202.
[51] Jiang C, Flansburg L, Ghose S, et al. Defining process design space for a hydrophobic interaction chromatography (HIC) purification step: application of quality by design (QbD) principles. Biotechnol Bioeng 2010;107:985-997.
[52] Gardner AR, Smith TM. Identification and Establishment of Operating Ranges of Critical Process Variables. In: Sofer G, Zabriskie DW (eds). Process Validation in Manufacturing of Biopharmaceuticals: Guidelines, Current Practices, and Industrial Case Studies. 1st ed. New York, NY, USA: Marcel Dekker; 2000:61-76.
[53] Pieracci J, Perry L, Conley L. Using partition designs to enhance purification process understanding. Biotechnol Bioeng 2010;107:814-824.

8 Pharmaceutical Aspects of Biologics from Animal Cell Culture Processes

8.1 Spatiotemporally Controlled Delivery of Biopharmaceuticals

Balder Rebmann, Peter van Hoogevest, Matias Zurbriggen, and Wilfried Weber

8.1.1 Introduction

The fast growing market of biopharmaceuticals presents current drug delivery approaches with a challenge. The huge molecular size of biologics (compared to most of the small molecule therapeutics), their low stability *in vitro* and *in vivo*, and the need of spatiotemporal control for many therapies have to be considered while developing new delivery strategies.

Current approaches are mainly based on parenteral injections, but new methods that enhance the therapeutic index of the biologics or allow noninvasive administration are needed to further enhance the therapeutic possibilities and allow a broader use. Many promising approaches were developed the last decades to reach the aim of Ehrlich's "magic bullet", a drug delivery system (DDS) that only targets the disease site without further cross harm the patients' health.

This chapter focuses on general, noninvasive and/or targeting strategies, their stage of development and future possibilities in the field of spatiotemporally controlled drug delivery of biopharmaceuticals.

8.1.2 Biologics Delivery

8.1.2.1 Lipid-based

Lipids are a huge class of molecules with a wide spectrum of biophysical properties, providing characteristics from strong lipophilic compounds to amphiphilic surfactants. Many of them are classified as GRAS (generally recognized as safe) components by the United States Food and Drug Administration (FDA), especially in the field of oral application. Their high modular structure and the broad spectrum of possible synthetic modifications allow additional adjustment of the lipid properties for the required application while obtaining good biocompatibility. The high flexibility and the tendency to self-assemble leads to the formation of many different nanostructures which are still subjects of current nanoparticle research [1]. These attributes make them an ideal candidate for drug delivery.

During the last decades the use of lipid-based drug delivery systems (DDS) was investigated with the main focus on poorly water-soluble or lipophilic drugs. A broad spectrum of products reached the market (from parenteral feeding, Intralipid®, to cancer treatment, Doxil®, or local anesthetic, DepoDur®; see Table 8.1.1), which show reduced toxicity and/or enhanced pharmacokinetics compared to the free drugs.

Main lipids used for these formulations are fatty acids, phospholipids, triglycerides, cholesterol, and their derivatives. The most common formulations used for drug delivery are described below (Figure 8.1.1).

Table 8.1.1: Selection of lipid-based drug delivery system on the market or in clinical trial.

Product	Drug	Status
Liposomes		
AmBisome (Gilead)	Amphotericin B	Approved (1990)
Doxil/Caelyx (Johnson & Johnson)	Doxorubicin	Approved (1995)
DepoDur (Pacira)	Morphine sulfate	Approved (2004)
Exparel (Pacira)	Bupivacaine	Approved (2011)
ThermoDox (Celsion)	Thermosensitive doxorubicin	Phase III
MBP-436 (Mebiopharm)	Transferrin-targeted oxaliplatin	Phase II
Solid lipid nanoparticles		
LNP technology (Tekmira Pharmaceuticals)	ALN-TTR02 (RNAi-therpeutic)	Phase II
MC3 (Alnylam)	ALN-PCS (RNAi-therapeutic)	Phase I
(Micro)emulsions		
Diprivan (AstraZeneca)	Propofol	Approved (1986)
Estrasorb (King)	Estrogen	Approved (2003)
Self-emulsifying drug delivery systems		
Sandimmune (Novartis)	Cyclosporin A	Approved (1983)
Neoral (Novartis)	Cyclosporin A	Approved (1995)

Lipid-Drug Complexes

Peptides and proteins have often a very complex surface structure, including different charges, local pH-environments, and lipophilicity. Amphiphilic surfactants like phospholipids or cholesterol derivatives can interact with these surface structures, leading to enhanced stability by preventing aggregation and can further increase encapsulation efficiency [2].

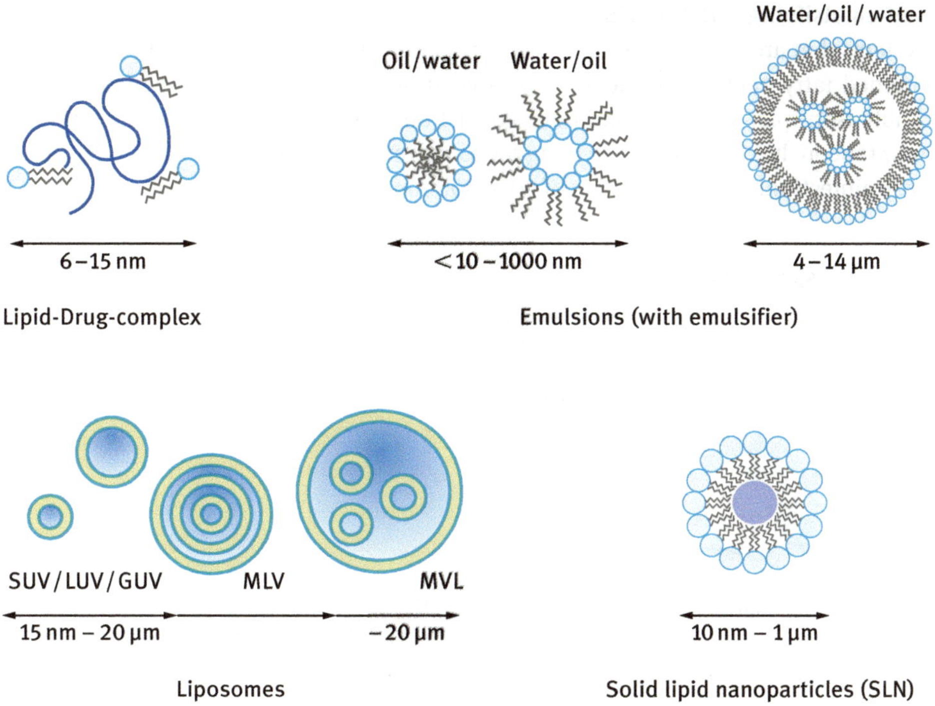

Figure 8.1.1: Examples of lipid-based drug delivery formulations.

(Micro-)Emulsions

Emulsions are mixtures of at least two immiscible liquids whereas one of these is dispersed into small droplets within the other. Basic emulsions contain an aqueous buffer and triglycerides often with an added surfactant for a higher thermodynamic stability [3]. There are different kinds of formulation types (Figure 8.1.1), like aqueous medium dispensed in oil (w/o), oil dispensed in an aqueous medium (o/w), or double emulsions (w/o/w) where a first w/o is further emulsified in an aqueous surfactant solution. Depending on the complexity of the formulation the size of the droplets varies from 100 to 1,000 nm and can be up to 14 µm for w/o/w emulsions [3].

Due to addition of higher amounts of at least one surfactant and/or cosurfactants so-called microemulsions can be formed (Figure 8.1.2). These formulations are thermodynamically more stable than normal emulsions and show other beneficial properties like low viscosity and high transparency [4].

Self-emulsifying DDS (SEDDS)

Systems that are self-emulsifying consist normally of a triglyceride lipid with medium chained fatty acids, a surfactant and a cosurfactant (Figure 8.1.2), which are preferably isotropic. These formulations consist only of lipid components and emulsify spontaneously in the presence of aqueous buffer under gentle agitation (e.g., in the gastrointestinal tract) to colloids of 50 to 300 nm [2].

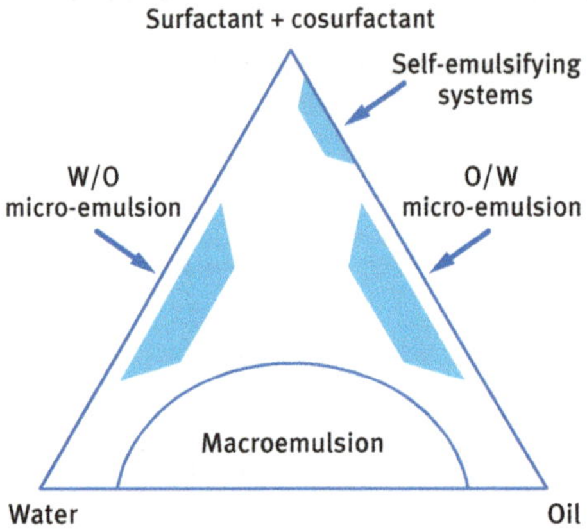

Figure 8.1.2: Hypothetical phase-diagram of oil/water/surfactant + cosurfactant emulsion-based systems.

Solid Lipid Nanoparticles (SLN)/Nanostructured Lipid Carriers (NLC)

In contrast with emulsions, these particles consist of lipids with long, saturated fatty acids, which lead to a solid state at temperatures below 40 °C. In most cases small amounts of surfactants are added to this formulation, leading to an emulsifying monolayer encapsulating a highly organized lipid crystal. These additional moieties allow the modification of the nanoparticles surface (e.g., protection or targeting ligands) [5].

By changing the fatty acid composition of SLN to more unsaturated fatty acids or mixing them with oils, semiliquid core particles form, so-called nanostructured lipid carriers [5]. This new approach differences in the physical properties of the compounds, the lipids form an amorphous or imperfect crystalline solid. The higher liquid state of the lipid core leads to smaller, smoother, and more regular structures with gaps within the crystal structure, which allow including bigger payloads.

Liposomes

Liposomes consist of an aqueous compartment entirely enclosed by a lipid bilayer composed by phospholipids. They can be distinguished by their size and number of bilayers [6]:

1. small unilamellar vesicles (SUV, 25 to 100 nm),
2. large unilamellar vesicles (LUV, 200 to 800 nm),
3. giant unilamellar vesicles (GUV, 1 to 300 µm), and
4. multilaminar vesicles (MLV, 100 to 1,000 nm).

The formation of these particles is driven by the self-assembling behavior of the phospholipids in aqueous solutions. These stabilizing forces protect them from erosion or other destabilizing effects upon administration (in comparison to other marked polymers). Due to the high structural similarities between their surface and the cellular membranes, unique ways of interaction like membrane fusion or disruption are possible.

The use of these nanocarriers as drug delivery systems was investigated over the last decades, resulting in several products on the market or in development (Table 8.1.1). For new approaches, the FDA [7] and also more recently the European Medicines Agency (EMA) [8] published guidelines for manufacturing and monitoring recommendations needed for new registration.

A huge advantage of liposome systems is the high flexibility in composition, including physiological active ingredients, and the ease to modify their surface. This versatile composition flexibility results in a huge variety of liposome types available for drug delivery (Table 8.1.2). Moreover, the use of phospholipids and other natural occurring membrane molecules for the formation of these vesicles lead to excellent biocompatibility and negligible toxic effects during systemic application, even at high concentrations.

Table 8.1.2: Examples of commonly known liposome types.

Type	Characteristics
Archeosomes	Etherlipid-based liposomes
Ethosomes	Flexible liposomes composed of about 30 % ethanol
Immuno-Liposomes	Glycoprotein anchored liposomes for vaccination purposes
Lipoplexes	Cationic lipid-DNA complexes for tranfection purposes
Niosomes	Small liposomes made from nonionic surfactants
Polymerised liposomes	Liposomes bearing phospholipids with cross-linked headgroups
Proliposomes	Dry lipid particles that form liposomes on contact with water
Stealth liposomes	Liposomes with polyethylene-glycol modified headgroups
Transfersomes	Cholate containing, very flexible liposomes
Virosomes	Liposomes with a functional viral envelope for vaccination purposes
Multivesicular liposomes	Nonconcentric, honeycomb-structured multilaminar liposomes

Nowadays, a broad spectrum of lipid formulations are used and investigated for pharmaceutical applications. The lipid-based systems have distinct physical properties and show different behavior *in vivo*, which allows choosing suitable platform for the required pharmaceutical need. Emulsions and other basic drug-lipid formulations allow for good "straightforward" strategies for constant release. More complex approaches like SLN and liposomes give the possibility by functionalization of its surface and/or interior to target specific tissues or trigger the site-specific release of their cargo. This possible spatiotemporal resolution during delivery can improve the pharmacological efficacy and/or reduce toxic side effects.

Archaeosomes

Alternative lipid-based colloidal particles are the so-called "archaeosomes". They have the same fundamental structure as liposomes but consist of polar ether-lipids that are extracted from archaeobacterial membranes. These lipids are characterized by 2 isoprene chains (C20-C40) ether-linked to glycerol. Two basic structures are found in nature:

1. the *archeol* (diether)-lipids, which correlate to the conventional phospholipid, and
2. the *caldarchaeaol* (tetraether)-lipids, which are able to form a monolayer structure due to their 2 polar head groups [9].

This chemical structure gives the archaeosomes reasonable advantages over conventional liposomes in respect to stability against oxidation, high pressure, alkaline or acid pH, high and low temperature, lipases, bile salts, and serum proteins [10]. These characteristics make them an interesting candidate for drug delivery, especially in the fields of oral administration. However, one of the main drawbacks at the moment is the lack of sufficient and affordable archaealipids on the market. This is a result of the complex cultivation and fermentation of archaea and the quantitative insufficient methods of chemical synthesis. First trials with archaeosomes *in vivo* showed their use as new nanocarriers for peptides, vaccine, and *in vitro* DNA delivery [11] as they did not show toxic effects in short-term repeated administration of therapeutic doses [12]. Moreover, first successful attempts to functionalize the ether-lipids with polyethylene glycol (PEG) and other moieties were carried out [13]. However additional studies are needed to estimate the real potential of this new type of nanocarrier.

Manufacturing

A key aspect for successful production of drug delivery systems is the preparation and loading. Due to the low stability of biopharmaceuticals towards detergents, solvents and physical forces (e.g., shear forces), mild conditions during the whole process are needed. Considering the high costs of biopharmaceutical production, the entrapment efficiency has to be as high as possible. Nonentrapped drugs are complex to recover and would need additional quality controls, leading to a considerable cost increase. While covalent modifications ("prodrugs") can increase the encapsulation efficiency [14], they in turn can affect the pharmaceutical properties leading to regulatory issues. Thereby only noncovalent excipient interaction will be reviewed in this chapter. Finally, the upscale potential to industrial quantities and the stable storage of the formulations has to be kept in mind. Nonstable storage will lead to a loss of monodispersity, which brings a huge disadvantage due to possible aggregation and embolism after administration. Due to the high variation of physical properties and stabilities of peptides and proteins drugs the finding of the right method of preparation is case-by-case different.

(Micro-)Emulsions

The preparation of emulsions (w/o, o/w, w/o/w) is a two-stage process:
1. the dispersion of the liquid into small droplets and
2. the homogenization of these droplets to a monodisperse solution.

The cargo of DDS is in general dissolved in the aqueous phase (w/o, w/o/w); however, the oil phase can also be used for lipophilic protein or peptide delivery (e.g., cyclosporine A, Sandimmune Neoral®) [2].

The formation of small droplets is achieved by stirring during the mixing of the two liquids. The shear forces (and resulting heat) can lead to degradation of the payload and should therefore be minimized by addition of a surfactant. The better the surfactant the less shear force is needed, but it has to be kept in mind, that strong surfactants often show toxic effects. The following homogenization proceeds normally under relative mild conditions and should not further destabilize the cargo of the droplets. In the case of double emulsions (w/o/w), the process of emulsification is repeated, whereby the second aqueous solution often includes a surfactant to enhance the stability.

Emulsions are already a widespread DDS for small molecule drugs, thereby efficient methods for up-scaling were established. In addition new technical approaches, like for instance microfluidic systems, provide new opportunities for the production process of emulsion-based systems [15]. Moreover, the formation of emulsions is a fundamental preparation step of many lipidic or polymeric nanocarrier systems, e.g., SLNs or poly(lactic-co-glycolic acid) (PLGA) particles. Alternatively solid nanocapsules can be formed by a polymerization step after (micro-)emulsion formation. These very stable particles provide interesting possibilities for the development of biomedical applications [16].

The disadvantage of long-term storage of many emulsion systems, caused by the metastability of bigger droplets in solutions, can be solved by considering monodispersity, a stable pH, and the avoidance of oxidation of the lipids. Therefore, these paras have to be monitored during the whole preparation process.

Self-emulsifying Drug Delivery Systems (SEDDS)

The ability of SEDDS to spontaneously emulsify gives them several advantages. The whole preparation is carried out under mild conditions (no shear forces, high temperatures, organic solvents), the resulting "presolution" is easy to store and can be loaded into soft shell capsules. The process is easy to scale up and the anhydrous storage of the formulation enhances the stability of the enclosed drug. One limitation of this system is the high hydrophilicity of peptides and proteins, resulting in a low loading efficiency and low total drug-lipid ratios. To increase the lipophilicity of the biopharmaceuticals, an additional preparation step can be included. The interaction of the drug with a lipid surfactant (drug-lipid complex) or lyophilization in the presence of a cosurfactant (anhydrous reverse micelle) show promising encapsulation rates while still providing sufficient bioactivity of the compounds [2]. A high drug-lipid ratio is needed due to a limited dosage by the high amounts of surfactant and cosurfactant present in this formulation. These excipients can show toxic effects above certain concentrations. An agreement has to be made between adequate dosage, appropriate self-emulsifying behavior, and potential toxicity.

Solid Lipid Nanoparticles (SLN)/Nanostructured Lipid Carriers (NLC)

Several ways to produce SLNs and NLCs were established since their discovery (Table 8.1.3), together with cheap and effective upscale strategies. For the encapsulation of peptides and proteins, the double emulsion technique (w/o/w) or high-pressure homogenization (HPH) seem to be the most promising methods. The absence of solvents during the process is helpful while working with sensitive cargo-like peptides and proteins. Nevertheless, relative high temperatures (above 40 °C) and other physical forces (shear forces or high pressure) are needed to melt and disperse the lipids in small droplets during the preparation.

Emulsion-based techniques thereby constitute a good method to overcome the main disadvantage of the SLNs, the low encapsulation efficiency of hydrophilic macromolecules. Due to introduction of an additional aqueous phase by reverse-micelle double emulsion preparation, proteins can be encapsulated into an aqueous phase within the SLN [17]. An alternative approach is the preformulation of drug-lipid complexes or micelles (see also SEDDS) to enhance the lipophilicity of the biopharmaceuticals [2].

The amorphous structure of NLC shows in general higher rates of encapsulation by their bigger gaps within the crystal structure. Additionally the more fluid character allows the use of lower temperatures in case of HPH. Nevertheless, the concentration of the drugs within the formulation has to be considered but can be counteracted by higher doses due to the high lipid concentration possible during preparation and their low toxicity.

The stability during storage is discussed controversial. While showing quite robust characteristics stored as solution up to 1 year, the lyophilization or spray-drying in the presence of cryoprotectants is recommended in most cases [18]. This has an additional advantage, as it is possible to encapsulate the dried powder into capsules or tablets when oral application is desired. However, a drying process has to be adopted for every blend and monitored due to the possible appearance of colloidal side-products (micells, liposomes, and others). The loss of the monodispersity is not

Table 8.1.3: Common preparation techniques for lipid-based DDS.

Method	Process
Emulsion	
Water in oil (w/o) emulsion	Mixing of aqueous solution (with a surfactant) in an excess of oil by stirring, further downsizing by high pressure or membrane-based extrusion
Oil in water (o/w) emulsion	Mixing of oil in an excess of aqueous solution (with a surfactant) by stirring, further downsizing by high pressure or membrane-based extrusion.
Double (w/o/w) emulsion	A first w/o-emulsion is further emulsified in an aqueous-phase containing surfactants by stirring
SLN/NLC	
High-pressure homogenization (hot/cold)	Homogenization of liquid (hot) or preformed (cold) lipid particles to monodispersed predefined nanoparticles by high pressure
Microemulsion	mixture of lipids and surfactants (+/− cosurfactant) dispersed under stirring, formation of solid particles due to an excess of buffer
Solvent emulsification-evaporation	Lipids are dissolved in a water-immiscible/partially water-miscible solvent and emulsified in an aqueous buffer, removal of solvent by reduced pressure/heating
Double emulsion	The drug is encapsulated in an internal water-phase of a liquid w/o/w emulsion, particle formation due to cooling or solvent evaporation
Liposomes	
Mechanical dispersion	Lipid film formation by lipid containing solvent-evaporation, rehydration of lipid film with aqueous buffer, homogenization and further downsizing by ultrasound, high pressure or shear homogenization, water-miscible
Solvent dispersion	Lipid containing solvent is drop-wise added into aqueous buffer, self-assembly of liposomes due to quick solvent dilution
Detergent removal	Formation of mixed micelles by mixing lipids and detergent in an aqueous buffer, formation of liposomes by removal of the detergent due to dilution, column separation, dialysis or absorption

acceptable for many applications and the agglomeration of the particles will lead to reduced stability and loss of encapsulated drugs.

Liposomes

The formation of liposomes is a broad studied field. Most methods include the preparation of multidisperse liposomes and the downsizing to a monodisperse solution. Moreover, methods leading to direct formation of the desired size and lamellarity were established (Table 8.1.3).

The separation of the aqueous core region by the lipid bilayer and the high hydrophilicity of the protein and peptide drugs make the inner phase of the liposomes a promising system for drug encapsulation. As general rule, it is admitted that the size of the inner aqueous phase is proportional to the possible amount of cargo to be loaded. Liposomes should be thereby as small as necessary (delivery and targeting) and as big as possible (payload). Especially at sizes <100 nm, the ratio of core to membrane decreases considerably [19].

A first passive entrapment (e.g., film hydration or ethanolic injection with/into buffer with high drug content) followed by freeze-thaw cycles (or dehydration and rehydration) and additional mechanical downsizing is an often-used method. Depending on the interaction of the protein with the lipid, entrapment efficiency between 30 to 80 % is reached [20, 21]. Techniques that separate the liposome preparation and drug loading include the production of empty liposomes and the subsequent encapsulation of proteins or peptides by adapted free-thaw methods [21]. This gives thereby a preparative advantage by easier process monitoring.

New micro fluidics-based preparation methods were investigated recently for the formation of liposomes and other lipid-based carriers [22]. With these mixing techniques, small, monodisperse, and unilamellar liposomes can be produced in a continuous-flow process. The absence of harsh conditions and the no longer required additional downsizing step make these methods an interesting approach for drug delivery applications. First *in vivo* test with siRNA loaded particles showed auspicious results in terms of encapsulation efficiency and bioavailability [23].

Other techniques try to increase the encapsulation efficiency by formation of multivesicular liposomes (DepoFoam®, Paira Pharmaceuticals). These vesicles are formed by double emulsion preparation and show several compartments within a bigger sphere. The first products based on this technique were recently approved for small molecule delivery by the FDA (DepoDur®), but also the use as peptide DDS was tested with uplifting results (DepoInsulin®, DepoLeuprolid®).

Liposome formulations have to be held under physical and chemical stable conditions to avoid aggregation, hydrolysis, or oxidation. Liposome products on the market used for drug delivery are either in liquid form (e.g., Doxil®) or lyophilized (e.g., Myocet®, AmBisome®) and show a shelf life of at least 1 year. In order to store liposomes in a liquid form, the physical properties (e.g. pH, inert gas phase, osmolarity, light protection, temperature) have to be considered. Moreover, used lipid formulation shows impact while long-term storage. The addition of cholesterol up to 40 mol-% has stabilizing effect on the colloidal structure of the liposomes. Furthermore, the addition of negative charged phospholipids (e.g., phosphatidyl-glycerol/

-serin/-inositol) or modification of the lipids headgroups (e.g., PEG conjugates) lead to repulsion of the liposomes and thereby to enhanced stability.

Lyophilization of the liposomes in the presence of cryoprotectants (e.g., lactose, sucrose) provides a good alternative to liquid storage. The resulting cake or powder can then be reconstituted prior administration and allows thereby higher chemical stability for the lipids and drug. In addition, lyophilization techniques, the selection of lipids, and modification will affect the stability of the colloidal structure [24].

The preparation of DDS is a key point in the development of new therapeutic strategies. Formation and drug loading implicate physical forces, the use of solvent or detergents and other drug degrading conditions. The properties and stability of the used biopharmaceuticals are essential keystones. While many peptides are stable in the presence of solvents or shear forces, complex proteins are much more sensitive and need very mild conditions while preparation.

While emulsion techniques and solid nanoparticles show promising results for peptides and small, stable proteins, various studies showed that the encapsulation into liposomes seems to be a suitable method for complex proteins. However, the established methods still lack on high encapsulation rates for the most proteins. New preparation techniques, like the microfluidics, are needed to achieve sufficient payloads. Lipid-based DDS can partial antagonize this by their very low toxicity and high biocompatibility; thus, they are not limited to low dosages, especially during nonparenteral applications. Anyway, the preparation and loading processes of lipid-based carriers are still an ongoing field in lipid research with a high output of new techniques.

For validation of storage conditions and stability, a broad spectrum of optimization processes can be adopted from the field of small molecule DDS. Considering the instability of many biopharmaceuticals during storage, lyophilization or “in situ preparation methods” approaches seem to be the most promising.

Delivery

The preferred route of administration of biopharmaceuticals is the parenteral injection. Through the direct entrance into the bloodstream, hepatic first pass effects are avoided and reproducible dosing with high bioavailability is ensured. In any event, injections need medical briefing or the presence of trained personal. While the compliance of the patient in the case of serious diseases might be reasonable, this limits the use of biopharmaceutical for minor complaints. However, a safe and easy noninvasive route for application would lower the cost due to reduced production effort and the lack of medical supervision.

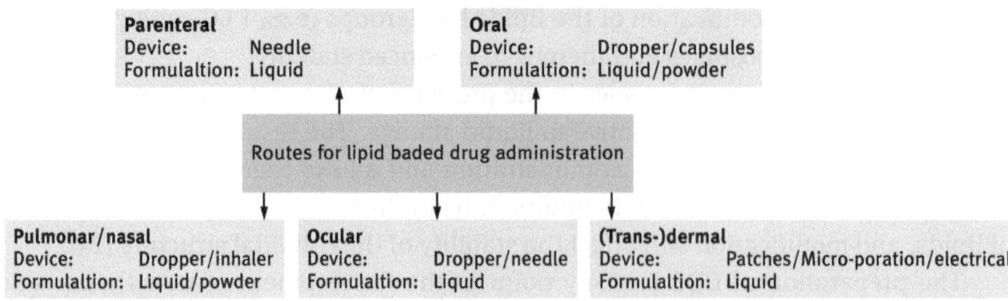

Figure 8.1.3: Overview of administration routes for lipid-based drug delivery systems.

Parenteral Administration

There are many ways to inject a pharmaceutical into the body. While intravenous seems to be the most direct method to enter the bloodstream, also subcutaneous injections showed comparable results while being less painful and easier to handle. In addition, intramuscular injection, for sustained release, and side-specific (e.g., intratumoral) injections are common methods.

After administration, the DDS are exposed to the serum proteins, including lipases, proteases, and opsonins. Most emulsions are thereby quickly digested and directed to the hepatic tissue. The use of small, neutral, droplets with large molecular weight emulsifiers or surfactants provides with prolonged circulation times. However, due to their physicochemical properties, the circulation time is very limited and will lead to a quick accumulation within the liver, with minor fractions in lung, spleen, and bone marrow [25]. Unmodified nanoparticles like liposomes or SLNs are also quickly cleared from the blood by the mononuclear phagocyte system (MPS). The surface loading, shape, and size influences the half-life time of these carriers [26]. In general, small sized particles with a neutral or slightly negative-charged surface show the highest serum stability. The use of specific phospholipids like phosphatidylglycerol (PG) or phosphatidyl-serine (PS) can reduce the elimination by the Mononuclear Phagotic System (MPS) and lowers immunogenicity. Furthermore, the addition of cholesterol increases the stability of the lipid membrane by integrating between the fatty acid chains and increase thereby the circulation time. However, the half-life time of lipid-based carriers barely reaches few hours.

A breakthrough in the field of nanocarrier-based DDS was the surface modification with hydrophilic polymers (Table 8.1.4). The coating of nanocarriers with PEG leads to highly stable, long circulating (>1 day), weakly immunogenic drug carriers [27, 28]. High-density coating with these polymers avoids the interaction with opsonins and other proteins, which prevents the recognition by MPS macrophages. Several DDS using this technology already reached the market (Table 8.1.1) and comprising thereby a general technique for stabilization of nanoparticles.

Table 8.1.4: Common copolymers used in nanocarrier-based drug delivery.

Hydrophilic Polymers [108]
Polyethylene glycol (PEG)
Poly (oxazoline)
Polyvinyl alcohol
Poly(glycerol)
Poly-N-vinylpyrolidone
Poly(amino acids)
Polysaccharides
Mucoadhesive Polymers [109, 2]
Alginate
Amylose
Cellulose
Chitosan
Hyaluronan
Poly (acrylamide)

The size and surface properties of nanocarriers are important to ensure long circulation and avoidance of immunogenic effects. This is sufficient to ensure spatiotemporal resolution by advanced strategies like targeting and inducible release within therapeutic approaches.

Oral administration

The historical most common route for pharmaceutical drug delivery is the oral application. It is not only preferred by the patient due to lack of pain or need of an administration device, but also by the pharmaceutical industry due to reduced cost of production. Oral administration of protein and peptide drugs by classical formulations has serious disadvantages like the low stability in the gastrointestinal (GI) tract and very low permeability across the mucous membrane due to their high molecular weight and hydrophilicity [2]. An appropriate formulation is needed which protects the biopharmaceuticals from the harsh environment in the small intestine, including acid pH, proteases, and peptidases, enhances the permeability through the mucus layer and the intestinal epithelium, and finally releases the drug in a temporally or spatially controlled manner [29].

To protect the protein or peptide drugs from digestion, the coadministration of protease inhibitors seems to be an obvious solution. Studies in which different protease inhibitors were given in addition to an oral dose of insulin showed increased bioavailability and high stability in the GI tract [30]. However, high concentrations are needed due to large quantity and distribution of the enzymes within the small intestine. This is not only a significant drawback in costs, buts also this approach implicates the danger of serious toxic effects during chronic application. New approaches try to overcome these drawbacks by covalently conjugating protease inhibitors to nanoparticles while retaining their bioactivity [31].

A further approach is the physical protection of the biopharmaceuticals by encapsulation into a colloidal system like liposomes, SLNs, or (micro-)emulsions. In these formulations the drug is solved in the inner phase of the carrier, which is protected by the lipid layer against proteases [32]. Moreover, lipidified peptides solved in the lipid phase of SEDDS or SLN show increased stability [33]. A downside of the high biocompatibility of lipid carriers is that they are also targets of the lipases and bile salts in the GI tract. This leads during the application to an increased amount of free fatty acids and bile salt/phospholipid (BS/PL) micells. Together, these components form partially spontaneously uni- and/or multilamellar vesicles within the GI tract [34]. Nevertheless, these new vesicles are still potentially able to protect the encapsulated cargo. On the other hand the digestion products, especially lysophospholipids and medium chain fatty acids, are good absorption enhancer by transiently losing the tight junctions and thereby enhancing the transcellular transport of the remaining carriers. In addition, the high concentration of lipid components at the GI epithelium seems to facilitate transcellular transport by increased membrane disorder, but the exact mechanism for this observation is not well understood [2].

The formation of lipid drug complexes of peptides and proteins with a surfactant provides an interesting straightforward approach. The increased lipophilicity of these complexes enhances their transcellular transport and thereby the bioavailability. For example the use of C-8 fatty acid or its derivate (N-[8-(2-hydroxybenzoyl) amino caprylate (SNAC)) show their potential in clinical trials for oral delivery of insulin, glucagon-like peptide1 (GLP1) analogues (Eligen® Technology, Emisphere), or octreotide (Octreolin®, Chiasma).

Other approaches try to use strategies of mucosa adhesion or penetration, sometimes combined with ligand mediated transport through the GI epithelium. Based on the above-mentioned reduced stability of lipid-based carriers in the GI tract, SLN and NLC are the most promising candidates for this route of administration because of their dense structure compared to emulsions and liposomes. The mucus is a viscous film within the GI tract (and other hollow organs, e.g., respiratory tract) mainly consisting of highly glycosylated proteins, lipids, and water. It is separated in 2 layers, the luminal part, which is rapidly cleared, and the adherent layer with a slow turnover. The function of the mucus is the protection of the epithelium from pathogens. Two different strategies were established to overcome this barrier.

By coating nanoparticles with biocompatible, mucoadhesive polymers like chitosan or polyacrylic acid (Table 8.1.4), the retention time of the carriers is prolonged and thereby the bioavailability increased [35]. Additional, these polymers often show permeability-enhancing effects by interacting with the intracellular space of the epithelium. An opposite strategy is adapted from viruses, which are often able to diffuse easily through the mucus. Their small size and a high density of negative and positive charges on the surface inspired the coating of nanoparticles with hydrophilic, nearly neutral polymers. High density PEGylation of nanoparticles showed up to 3 order of magnitude higher diffusion rates within the mucus, just four- and sixfold lower than in water [36], and thereby an interesting strategy to overcome the mucus barrier.

While the unspecific enhancement of the mucosa permeability is limited to its protectoral function against pathogens, ligand-mediated transport offers an elegant alternative. The use of surface targeting structures like lectins showed enhancement of bioavailability of the encapsulated cargos. Moreover, targeting of the active transport system of vitamin B12, a high affinity ligand-receptor interaction (Kd ≈ 5 fmol L^{-1}), seems to be an interesting approach to overcome the barrier of the intestinal epithelium [37]. However, these approaches need a high mobility of the carriers to reach the surface of the epithelium. Thereby, additional mucus penetrating strategies have to be considered.

Summarizing the actual studies focused on oral delivery of peptides and proteins formulations with sufficient bioavailability for clinical applications might be feasible. But the lack of lipid-based products on the market (excluding the strong lipophilic cyclic peptide cyclosporine, Sandimmune® and Neoral®) and the many discontinued clinical trials during the last decade show the limitation of this route. The need of precise and reproducible dosing independent from the physical state of the GI tracts seems to be the major challenge. A focus to peptides, not sensitive to exact concentration levels, contrary to insulin, could lead to new products on this field.

While classic approaches for enhanced permeability of the GI tract (e.g., mucolytics, membrane destabilization, tight junctions disruption) are leading to limited protectoral function of the mucosa, combined approaches of protective coating (e.g., chitosan, PEGylation) with reversible permeability increasing substances (e.g., fatty acids) or specific targeting (e.g., vitamin B12) show enhanced bioavailability without noticeable toxic effects.

In the field of lipid-based DDS, especially SEDDS and SLNs/NLCs show good potential for at least slightly lipophilic peptides or proteins. The possibility to administrate these formulations as capsule (SEDDS) or tablets (dried SLN/NLC) and their high structural stability give them important advantages. In addition, the lipid drug complex approaches are cheap and straightforward methods for oral delivery of peptides, also giving the opportunity to combine these complexes with further encapsulation into SEDDS or SLPs/NLCs.

Pulmonary and Nasal Administration

The delivery of biopharmaceuticals across the respiratory tract is a promising non-invasive way due to its easy accessibility, proximity to system bloodstream and its large surface (80 to 120 m^2). Furthermore, this thin mucus layer and epithelium with its low proteolytic activity and low acidity, compared to the GI tract, makes it a viable target for drug delivery. However, the mucosa still forms a tight barrier for macromolecules. Additionally, the mucocilliary clearance and the high activity of the lung macrophages are obstacles to overcome, leading to a bioavailability of most peptides and proteins below 1 %. The highest absorption rates are observed in the deeper lung, especially in the alveolar region. To obtain a good distribution within this tissue, the DDS have to be finely dispersed into an aerosol (particles size < 3 μm). Particles smaller than 500 nm show in addition a relative low uptake by the lung macrophages while reaching easily the alveoli [38].

Common applicators for DDS aimed to the lungs are nebulizer (liquid) or dry powder inhaler (dry). While most developed formulation for pulmonary delivery are dry powder-based, also nebulizers have been improved in their functionality constituting now an alternative method [39].

In general strategies to overcome the barriers within the lung are comparable to the GI tract: the use of mucoadhesive (Table 8.1.4) or mucupenetrating polymers combined with permeation enhancers (Table 8.1.5) exhibit increased bioavailability of peptides and proteins (up to 25 %) [17]. However, it has to be considered that these excipients have to be very well tolerated. The lung is a very sensitive organ and irritation or toxicity within would abolish the benefits of this noninvasive route.

Table 8.1.5: Classification of permeation enhancers (adapted from [110]).

Class	Examples
Bile salts	Sodium glycocholate, sodium fusidate
Surfactants	Brij®-35, polysorbate-80, sodium lauryl sulfate, lyso-phosphatidylcholine
Fatty acids and derivates	Sodium carprate, sorbitan laurate, sodium myristate
Glycerides	Phospholipids, medium chain glycerides
Chelators	Ehtylene diamine tetraacetate (EDTA), citric acid
Salicylates	Salicylic acid, sodium methoxysalicylate
Polymers	Chitosan and derivates, polycarbophil, sodium methylcellulose
Other	Azone®, cyclodextrins, benzalkonium chloride, menthol

The use of lipid-based DDS, especially phospholipids, in pulmonary administration is due to their high biocompatibility and their role as surfactant a widely accepted method [40]. Besides, dry mixtures of drugs and lipids, particularly nanocarrier systems, show benefits in the delivery of biopharmaceuticals. The lung epithelium shows high rates of endocytosis and also transcytotic delivery, which allows high rates for systemic delivery. But also, local treatment with sustained release kinetics is wanted for a wide spectrum of diseases like lung cancer, tuberculosis, asthma, and pulmonary hypertension. Nebulized liposomes showed thereby encouraging results for local targeting, either as local depot of peptides [41] or in local protein-based cancer therapy [42], while showing excellent compatibility.

Systemic delivery based on the surfactant effect of the used lipids or surface coating was especially shown for the SLN/NLC DDS. High systemic bioavailability of protein or peptide was achieved by nebulization [17] as well as by dry powder inhalation [43].

Nasal drug delivery shows some special characteristics. The nasal epithelium has relative high molecular weight cut-off (~1 kDa) while only allowing relative small doses (25 to 200 μL). The bigger gaps in the epithelium make this route especially interesting for (micro-)emulsions. While most nanocarrier systems still need additional permeation enhancer to cross the epithelium, this DDS can enter the blood stream and reach bioavailability comparable to subcutaneous injections for peptides like insulin [44]. This route offers additionally a possible targeting to the brain. The local proximity and the possibility to bypass the blood-brain barriers trough the olfactory route can lead to an enhanced accumulation of biologics within the brain [45].

Drug delivery through the airways was the first successful, FDA-approved way of noninvasive insulin delivery (Exubera®, Pfizer). But this product is also a good example for the possible problems and obstacles new products on this field have to pace. It has to be ensured that the delivery is safe for the purpose of reproducibility and toxicity. Exubera® only reached a market share of less than 1 % and was therefore quickly withdrew from the market. While this flop was probably also based on marketing failures and missing long-term studies, other products try to enter the market (e.g., AFREZZA®, MannKind, insulin) or are already successful (Miacalcin®, Novartis, calcitonin-salmon), showing the high potential of this route for protein and peptide delivery.

Ocular Administration

Noninvasive administration of biopharmaceuticals into the eye is not focused on systemic delivery but rather on avoiding the risk of ocular complications and increase the patient compliance. Eye disorders like corneal neovascularization, age-related macular degeneration, retinoblastoma or uveal melanoma are hard to treat by systemic administration of biopharmaceuticals since only about 2 % reach the vitreous cavity. Due to that many therapies include injections within the eyeball [46], a desired type of therapy would rather be the administration of eye drops on the corneal surface, which is an easy and cheap method.

The main barrier for uptake through the corneal epithelium is its very dense structure, mechanical clearance, draining and the high amount of proteases. To overcome this tight barrier, especially methods benefiting from the thin structure of the epithelium show good results, including good compatible permeation enhancers and mucoadhesive copolymers (e.g., chitosan).

Emulsions are often used for the delivery of small molecules into the eye due to their similarity to tears in their composition. Especially the use of nonionic emulsifiers (e.g., polysorbate-80) show high biocompatibility and lead to first FDA approved products (Restasis®, Alllergan) [47].

The use of nanocarriers is due to their enhanced protective effect and possible mucoadhesive properties an interesting approach. Liposomes and SLN showed either as solution with mucoadhesive properties [48] or incorporated into hydrogels [49] encouraging results.

In current therapies that include intravitreal injections of antibodies (Lucentis®, Novartis or Avastin®, Roche), a prolonged residency was shown for liposomal formulation [50]. This reduces the amount of injections needed and thereby possible ocular complications.

Ocular drug delivery is due to the low systemic availability and the risks of complications through invasive methods an interesting field for new drug delivery systems. While many studies show encouraging results, no products are yet on the market. However, especially in this field, lipid-based system seems to be a suitable carrier due to their high biocompatibility and low toxicity which are both needed for a successful local therapy.

Dermal and Transdermal Delivery

While the delivery of small molecules and peptides with lipid-based formulation (especially emulsions and micelles) into the skin is a common and accepted method in pharmaceutics and cosmetics, the system delivery through the cornea layers is more difficult. The skin has no mucus layer and also lacks of high amounts of proteases, it is rather protected by a physical barrier. The 10 to 20 μm thick *Stratum corneum* is a layer of dead corneocytes with additional lipids and waxes forming a "brick and mortar" barrier [51]. To penetrate this layer, an increased residential time and permeation enhancers are the most applied methods.

The use of patches as carriers is an appropriate method due to their high patient compliance and good control over exposure time and dosing. Patches that treat disorders like hormonal deficiencies, restless legs syndrome or Parkinson's disease are already successful on the market. The main mechanism of lipid-based DDS rely on their interaction with the corneum lipid bilayers. This leads to fluidization and thereby to enhanced permeation. Other cosolvents like PEG can additionally enhance this effect by providing a better distribution within the protective layer. The formation of transient drug-lipid conjugates further enhances the permeation of the drugs by increasing their lipophilicity and thereby their possibility to cross the lipidic barriers [52].

While these already established formulations might work for small peptides (smaller than 500 Da with moderate lipophilicity), bigger peptides and proteins need new methods to overcome the lipophilic barrier of the skin.

Liposomes show good results in increased skin delivery and deposition, but their use is limited to topical treatment due to their inability to penetrate the *Stratum corneum* as intact vesicles. To overcome this obstacle, new types of liposomes like Transferosomes®, Niosomes® or Ethosomes® (Table 8.1.2) where developed which are more flexible and thereby able to penetrate the skin barriers [53]. The use of SLN/NLC formulations is wide spread in cosmetic applications and is now also adapted for pharmaceutical approaches. The stable structure, controlled release, and film formation behavior make them an effective delivery system but also limit them to topical application [54].

The targeting of the skin appendages like sweat gland pores or hair follicle is an approach to overcome the *Stratum corneum* and reach the system blood circulation. They constitute just 0.1 % of the skin but show a much higher penetrability for large hydrophobic molecules like for instance insulin. The coadministration of specific peptides allows targeting these structures and enhances thereby the uptake into the bloodstream [55].

Beside these passive approaches, many active penetration methods (Table 8.1.6) were developed in the last decade. They can be separated into 2 main groups: microporation techniques and electrically assisted enhancement techniques [56].

Table 8.1.6: Active penetration methods for transdermal delivery of proteins and peptides [56]

Method	Associated Companies
Microporation	
Microneedles	Zosano Pharma, DermaRoller, Becton Dickinson, Corium, 3M
Thermal microporation	Altea Therapeutics
Radiofrequency ablation	TransPharma Medical
Laser ablation	Biosolutions AG, Norwood Abbey
Electrically assisted techniques	
Iontophoresis	Teikoku Pharma, Empi, Isis Biopolymer, Chattanooga, Vyteris
Electroporation	–
Other	
Sonophoresis	Echo Therapeutics

While the topical application of biopharmaceuticals into the skin is already feasible by passive, formulation-based approaches, the transdermal delivery is still limited to a small group of peptides. For the systemic delivery of bigger peptides and proteins the additional use of active transport mechanisms is needed. First devices like microneedles (Fluzone® Intradermal, Sanofi) or iontophoretic patches (Zecuity®, NuPathe) are already approved by the FDA and now enter the markets. For the successful delivery of biopharmaceuticals combinations of formulations and active transport devices, e. g. liposome filled self-dissolving microneedles [57], appear to be interesting.

Targeting

New therapeutic approaches try to increase their pharmaceutical effects while reducing unwanted or toxic side effects. The ability of many DDS (not only lipid-based) to passively or actively target specific tissues or even cell types give them a big advantage over common delivery approaches. The modification of only the carrier without changing the drug structure itself, compared to drug conjugates, allows modular and exchangeable targeting for the next step in personalized medicine.

It has to be differentiated between passive and active targeting. While passive targeting relies on the physical properties of the DDS like size, shape and loading, active targeting is based on high specific ligand-target interaction. However, to achieve a targeting effect, a long circulation time within the bloodstream is needed to either allow passive accumulation or ligand interaction [58]. The methods described below will focus on lipid-based systems but remain widely applicable on the broad field of nanoparticles, including many polymer-based particles.

Passive targeting can be achieved by choosing the route of delivery (e.g., GI tract, lung, skin, or eye) or by biophysical interaction of the DDS with the tissue. Lipid-based systems will be recognized as metabolites or lipid transport vehicles (lipoprotein complexes) and transported to the liver or spleen. This mechanism allows easy targeting of these tissues or the cells included in this route of transport (e.g., Myocet®, Enzon Pharmaceuticals, targeting mononuclear phagocyte system).

For targeting solid tumors or inflamed tissues the enhanced permeability and retention effect (EPR) plays a role. Based on the phenomena that these pathological sites show increased blood vessel permeability compared to healthy tissues an accumulation of particles smaller than 200 nm (at least <400 nm) can be observed [59]. This leads, in combination with the absence of lymphatic systems for most tumors, to a high local concentration of the nanocarriers. Several anticancer drugs on the market already successfully use this targeting effect (e.g., Doxil® or Marqibo®).

Subcutaneous injection of lipid-based systems or the use of specific fatty acid chains while oral application will lead to accumulation within the lymphatic system and uptake by macrophages of the lymph nodes [60]. This may give interesting opportunities in the field of immune regulation, infections control, or treating of tumor infection within the lymph nodes.

Cationic DDS achieve passive targeting to the epithelial cells and enhanced internalization by ionic interaction with the negatively charged cell surface. Tumor vasculature can be targeted due to their high negative charge during angiogenesis. First trials with classical anti-tumor drugs (EndoTAG-1®, Medigene, phase II) or antiangiogenic siRNA [61] show the high potential of this concept.

Active targeting is an additional approach to further enhance the specificity of the DDS. The surface decoration of nanocarriers or even emulsions [62] with targeting moieties can enhance the therapeutic effect not only by higher accumulation at the target site but also due to receptor-mediated internalization [63]. To ensure optimal interaction possibilities while retaining long circulation time the combination of surface PEGylation with coupling the targeting moiety at the distal end of the PEG-chain (Figure 8.1.4) seems to be most promising [64]. Many strategies for site-specific conjugation of the targeting domains and the incorporation into the DDS where established [65].

The moieties used vary from complex proteins like monoclonal antibodies to small molecules like sugars or aptamers (Table 8.1.7). Thereby has to be considered that the exposed targeting moieties do not reduce the circulation time of the nanocarrier [64]. Also a high cell specific density (minimum 4×10^4 targets per cell) of the targeted surface structure is needed to ensure adequate interaction [66]. Newer approaches focus therefore on at least two different targets. This ensures not only a higher density of possible targets, it also further enhances the specificity and allows targeting more heterogeneous cell populations (e.g., tumor tissue).

Furthermore, continuative targeting of specific organelles is possible. Some metabolic or storage diseases for example need the specific delivery of enzymes to the mitochondria or lysosome. This can be achieved by incorporation of cationic lipophilic compounds (e.g., dequalinium or triphenylphosphonium) or specific surface markers (e.g., mannoside residues, certain cell penetrating peptides) into the DDS [67]. While the nanocarrier devices on the market already use passive targeting effects (EPR: Doxil®, macrophage targeting: Myocet®), just a few small molecule formulation (e.g., MM-302, Merrimack, phase I; MBP-436, Mebiopharm, phase II) and siRNA/RNAi delivery systems (e.g., ALN-VSP, Alnylam, phase I; TKM-PLK1, Tekmira, phase I) with active targeting are in progress into clinical use. This may rely on the increase in cost associated to the production and quality control of the targeting-ligand while not showing for all applications superior benefits. Especially anatomic barriers and accessibility of the target domains can reduce the effect of active targeting considerably. However, due to the many advantages successful targeting offers and the new developments in multitargeted DDS, an increase in promising products reaches the clinic.

Tabelle 8.1.7: Examples for targeting moieties (adapted from [64]).

Type	Ligand
Immuno-based	
Antibody	Anti-CD19 IgG
Fab	Anti-MT1-MMP
scFv	Anti-HER2
Protein and peptide	
Growth factor/interleukins	Interleukin 13
Glycoproteins	Transferrin
Cell signaling	RGD-containing motifs
Small molecules	
Hormones	Estrogen
Sugar	Lactose
Nutrients	Folate
RNA/DNA	
Aptamer	E-selectin aptamer

Induced Release and Endosomal Escape

Tissue-specific targeting is not the only requirement for a successful therapeutic delivery. Upon arrival at its target, the release rate of the transported drug is critical for its therapeutic outcome. Additionally, many peptides and proteins have to be delivered intracellularly to reach their therapeutic effect. Receptor-mediated uptake by the targeted cells leads to encapsulation into endosomes. This cell compartment will develop into a lysosome and thereby digest its content by very low pH and proteases. The escape of the entrapped drug must be guaranteed to assure its pharmacological effect [68].

The coating of DDS with hydrophilic polymers enhances their stability and circulation time, but it also decreases their cell interaction and drug release. To overcome these obstacles, different cleavable linker-based polymer coatings were developed. Specific physiological characteristics of its targeted tissue like pH or redox state (Table 8.1.8) shall lead to deshielding of the carriers and thereby enhance their release behavior. Other approaches focus on the disruption of the colloidal structure of the DDS application by the local appliance of external stimuli like heat or ultrasound (Table 8.1.8) or special physiological conditions like enhanced shear-stress in coronary heart diseases [69]. These methods will lead to extracellular release of their payloads.

Table 8.1.8: Inducible release stimuli (adapted from [101]).

Method	Stimulus	Method
Local triggered		
Enzymes	Altered enzyme expression	Peptide-substrate linker
pH-changes	Shift from pH 7.4 to pH 6.5 (tumor tissue) or pH 5.5 (endosome)	pH sensitive linker
Redox condition	Higher glutathione or reductive enzymes levels in tumors	Redox sensitive linker
Shear forces	Reduced blood vessel diameter in coronary heart diseases	Shear force sensitive membrane
Remote triggered		
Heat	Phase transition from solid (37 °C) to liquid (42 °C)	Lipids with phase transition near physiological temperatures
Ultrasound	Expansion of gas	(Co)encapsulation of gas containing micro-bubbles
Light	Photo-induced crosslinking of	Use of photosensitive lipids
Magnetic field	Magnetic force rapture	Incorporation of magnet particles

For intracellular delivery, endosomal escape strategies are needed. These methods are mainly based on 3 mechanisms: fusion with the endosomal membrane, pore formation, or proton sponge effects, but also other mechanisms like for instance the hijacking of native receptor transport mechanisms were investigated (Table 8.1.9).

Lipid-based DDS have several interaction possibilities with the endosomal membrane like lipid-transfer or membrane fusion. Specific lipids, like for instance dioleoyl-phosphatidyl-ethanolamin (DOPE) show enhanced membrane fusion activity and work thereby as endosomal escape agents [70].

Several polymers show also endosomal disruptive properties due to their buffering capacity at low pH. Their ability to absorb protons leads to swelling of the endosome and finally to rapture of its membrane, the so-called "proton sponge" effect [71].

The biggest group of endosomal escape agents are proteins or peptides. They are either derived from biological systems like viruses and bacteria or are synthetically engineered. Their mechanism is mostly based on (pH change-induced) pore formation or membrane fusion [72]. Other approaches try to hijack natural endosomal escape mechanism of membrane receptors [73].

An interesting group are the cell penetrating peptides (CPP) (Table 8.1.10). They offer an alternative mechanism to enter the cell compared to receptor-mediated approaches. This leads to direct delivery to the cytosol or enhanced release form the endosomes whereat the exact mechanisms are a matter of controversies [74]. Compared to targeting ligands, these peptides lack of specificity to surface structures and

Tabelle 8.1.9: Examples for endosomal escape agents (adapted from [68]).

Proteins and Peptides	Mechanism
Virus-derived agents	
Hemagglutinin	Fusion
gp41	Pore/fusion
L2 from papillomavirus	Fusion
Envelope protein of West Nile virus	Fusion
Bacteria derived agents	
Listeriolysin O	Pore
Diphtheria toxin	Fusion
Pseudomonas aeraginosa exotoxin A	Pore
Shiga toxin	Pore
Plant derived agents	
Ricin	Unclear
Saporin	Unclear
Gelonin	Unclear
Human/animal derived agents	
Human calcitonin-derived peptide (hCT)	Unclear
Fibroblast growth factor receptor (FGFR3)	Unclear
Melittin	Pore
Synthetic peptides	
KALA	Fusion
GALA	Fusion
Poly(L-histidine)	Proton sponge
Chemicals	
Polyethylene Imine (PEI)	Proton sponge
Poly(amidoamine)s (PAAs)	Proton sponge
Ammonium chloride	Proton sponge

will thereby enter every cell they come in contact with. Therefor current nanocarrier approaches focus on combination of cleavable surface coating with specific cell targeting moieties and a second surface layer with CPP (Figure 8.1.4) [75]. After reaching the targeted site the protection coating will be cleaved (by methods mentioned above) and the CCP will be exposed to the cell surface.

Table 8.1.10: Examples of cell-penetrating peptides (adapted from [75]).

Cell-penetrating Peptide	Origin	Mechanism
TAT	HIV-1 transciptional activator	Direct penetration/pore
Penetratin (pAntp)	Drosophila melanogaster	Direct penetration/endocytosis
Polyarginines	Synthetic	Direct penetration/endocytosis
Pep-1	Synthetic	Direct penetration/pore
Transportan	Synthetic	Direct penetration/endocytosis
MAP	Synthetic	Multiple mechanisms

While external release of biopharmaceuticals is interesting for receptor or surface targeting therapeutics, the bypassing of endocytosis or endosomal escape is an important aspect for many protein and peptide drugs. First small molecule DDS using inducible release techniques (e.g., ThermoDox, Celsion, phase III) are already in late clinical trials. Also, the successful use of CCP proteins for enhanced intracellular drug delivery *in vitro* (Chariot™ Delivery Reagent, active motif) and many encouraging *in vivo* studies show the high potential of these helper peptides for the future.

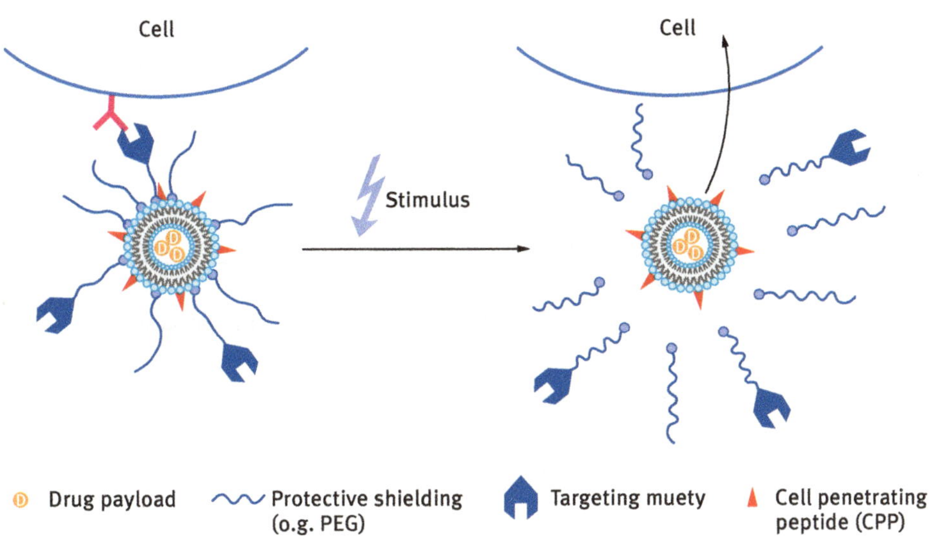

Figure 8.1.4: Illustration of a targeted nanocarrier-based delivery system with induced release mechanism. After accumulation at the targeted tissue/cells the protective coating is removed on command exposing uptake/release enhancing moieties (e.g., CPP) (adapted from [63]).

8.1.2.2 Polymer-based

A wide variety of polymers are established for use in medical application like for instance in the development of implants, patches, and other tissue substitution approaches. Moreover, their use as general DDS for a wide variety of drugs, including proteins and peptides, is topic of many current studies. The used polymers can be separated in 3 major groups (Table 8.1):

1. native derived polymers, extracted or inspired from polymers common in the human body,
2. biodegradable polymers, which contain chemical bonds that are sensitive to hydrolyzation or substrates for human enzymes, and
3. biocompatible/nonbiodegradable polymers, which lack of interaction with most cells and structures within the body.

Table 8.1.11: Overview of polymers used in drug delivery (adapted from [111]).

Natural Polymers	
	Protein-based polymers
	Gelatine, collagen, albumin, fibrin
	Polysaccharides
	Agarose, alginate, hyaluronic acid, dextran, chitosan, cyclodextran
Synthetic Polymers	
Biodegradable	**Polyesters**
	Poly(lactic acid), poly(glycolic acid), poly-D,L-lactide-co-glycolide
	Polyanhydrides
	Poly(sebacic acid), poly(adipic acid)
	Polyamides
	Poly(imino carbonates), polyamino acids
	Phosphorous-based polymers
	Polyphosphates, polyphosphonates
Biocompatible	**Cellulose-derivates**
	Ethyl cellulose, cellulose acetate, carboxymethyl cellulose
	Silicons
	Polydimethylsiloxane, colloidal silica
	Acrylic polymers
	Polymethacrylates and its derivates
	Polyethers
	Polyethylene glycol (PEG), polyglycerol derivates
	Other
	Polyvinyl pryrrolidone, poloxamers, poloxamines

The polymeric nature of this heterogenic group of molecules leads to a high modularity. Their biophysical characteristics do not only depend on their side chains (e.g., anionic, hydrophilic, hydrophobic), but also on the blending of different block polymers or the fusion of single polymer blocks to copolymer (e.g., dimeric or trimeric polymer blocks), which allows further adjustments. Therefore, the amount of possible combinations is nearly unlimited. This is reflected by the broad variety of nanoparticulare structures formed by these polymers, including nanoparticles, -spheres, dendrimers or hydrogels (Figure 8.1.5).

Most of the delivering, targeting, and release strategies mentioned above are also applicable for these polymeric nanoparticles, as long as it is possible to modify their surface. Therefore, this part will focus on the special field of drug conjugates, the interaction of the polymers systems with their surrounding tissue and strategies to further control their spatiotemporal release kinetics.

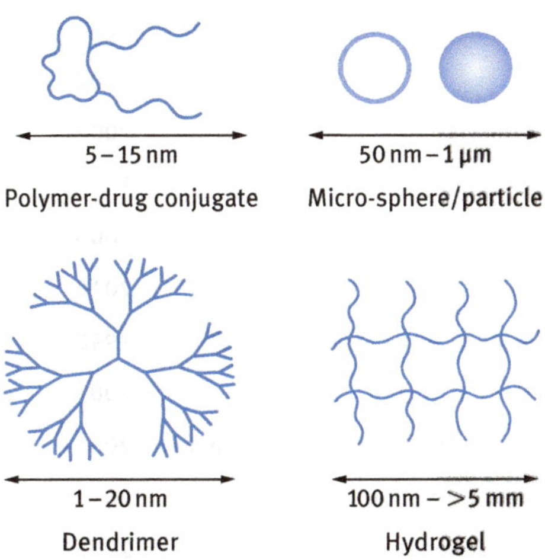

Figure 8.1.5: Examples of polymer-based drug delivery formulations.

Drug conjugates

Besides polymeric nanocarriers, especially drug-conjugates play an important role in the delivery of proteins and peptides and are already present on the market (Table 8.1.12). Fusion approaches mainly focuses on the low half-life time of biologics, but also show positive effects on storage stability, solubility, and *in vivo* immunogenicity [76, 77].

The fate of many small proteins and peptides within the human blood stream is the clearing by renal filtration caused by their small size. The average kidney pore size of 6 nm correlates with an 40 to 50 kDa cut-off (for globular molecules) and is additional influenced by surface charge induced repulsive effects (negative proteoglycans at kidney pores) [78]. Given that most peptides and many proteins have a smaller size, the enhancement of their hydrodynamic volume by fusion will lead to higher half-life times. Additionally, the conjugation of polymers leads to a reduced interaction with proteins or cells. While this is generally a benefit due to reduced protease degradation or recognition by the immune system, enzymes or signaling molecules can be affected in their activity. To avoid strong reduction of activity and to fit the enhanced regulatory requirements that arose the last years, high specific and reproducible conjugation methods are needed.

Table 8.1.12: Fusion-based products in clinical practice (adapted from [77, 80]).

	Name	Drug	Approval
PEGylated	Adagen®	Adenosine deaminase	1990
	Oncaspar®	Asparaginase	1994
	Neulasta®	Granulocyte-colony stimulating factor (G-CSF)	2002
	PegIntron®	Interferon α2b	2000
	PEGASYS®	Interferon α2b	2002
	Somavert®	Human growth hormone mutein (antagonist)	2002
	Mircera®	Erythropoietin (EPO)	2007
	Cimiza®	Anti-TNF Fab'	2008
	Krystexxa®	Uricase	2010
Fc-fusion	Enbrel®	Tumor necrosis factor receptor domain (TNFR2)	1998
	Amevive®	Anti-CD2	2003
	Orencia®	Cytotoxic T lymphocyte antigen 4 (CTLA4) domain	2005
	NPlate®	Thrombopoietin peptide analog	2008
	Arcalyst®	Interleukin-1 receptor domain	2008
	Eylea®	Vascular endothelia growth factor receptor (VEGFR) 1 and 2 derived domain	2011
	Nulojixy®	Cytotoxic T lymphocyte antigen 4 (CTLA4) domain	2011

Polymer-conjugates/PEGylation

The conjugation of many natural or synthetic polymers has been investigated, including hydroxypropyl-methacrylamide (HPMA), hydroxyl-ethyl-starch (HESylation), polyoxazoline (POZylation) and polysialic acid (PolyXen). Most of them show promising results [79], however only conjugation with polyethylenglycol (PEGylation) lead to the approval of marketed products jet (Table 8.1.12). The high flexibility, rate of hydration, low immunogenicity, and biocompatibility make this polymer an optimal candidate for drug conjugation.

Early coupling strategies were based on random coupling PEG (linear or branched) to ε-amino groups of lysine residues (by carboxylates or carbonates). The resulting products had a high disparity in number and location of the PEGylated sides. As a consequence, these proteins exhibit indeed prolonged circulation times in the patients but also reduction in bioactivity [20]. Since then, many new strategies were developed for side-specific PEGylation (Table 8.1.13). Together with optimized reaction conditions and purification processes, conjugates with high homogeneity, reproducibility, and exact pharmacological characteristics can be produced which fit the increased requirements for drug approval [80]. Besides the covalent coupling approaches, also releasable or noncovalent PEGylation strategies are investigated to overcome the drawbacks of reduced activity. The addition of hydrolysable linkers (e.g., bicine linker) or unstable PEGylation sites (e.g.. histidine PEGylation, PegIntron®) are common approaches. Moreover, PEGylation by hydrophobic interaction or by complex formation was shown to be functional but still need to be further investigated for its *in vivo* behavior [76].

Table 8.1.13: Site-specific PEGylation techniques [76]

Name	Method	Status
ε-amino PEGylation	Relative unspecific coupling to primary amino groups (lysine and N-terminus) by carboxylates or carbonates under physiological conditions	Clinical research
N-terminal PEGylation	Different pKs values of ε-amino groups and the N-terminus allow specific N-term PEGylation under weak acidic conditions	Clinical research
Thiole PEGylation	Genetically engineered or naturally unpaired cysteins can be specificity PEGylated by thoroughly buffered alkaline pH (no amino cross-reaction)	Clinical research
Disulfide bridging PEGylation	Agents with 2 thiol-reactive moieties allows specific PEGylation of disulfide bridges due to substitution by a PEGylated linker	Lead identification phase
Unnatural amino acid PEGylation	Substitution of single amino acids by synthetic analogs allows the addition of chemical unique residues, which allows high specific PEGylation	Phase I-II (PEG-hGH)
Glyco-PEGylation	By partial digestion of the glycosylation or the use of substrate unspecific enzymes N- or O-glycosylation sites can be PEGylated	Phase I (PEG-factor IX)
C-terminal PEGylation	Generation of a C-terminal chemical reactive side by internal fusion and following spontaneous splicing	In research

Fusion Proteins

Besides the chemical addition of polymeric structures, the genetically engineering of fusion proteins has shown to be a good strategy to increase plasma half-life of peptide and protein drugs. Different strategies were developed and resulting products can be found on the market. The fusion of therapeutics to long circulating serum proteins, like the Fc domain of immunoglobulin or serum albumin, is one prominent approach. The prolonged plasma half-life is for these fusion constructs is not only based on their slower renal clearance, also salvage processes further enhance the presents in the serum [77]. While albumin fusion products just enters the market (e.g., Albufuse® technology, Novozymes Biopharma) [81], Fc fusion constructs have already shown their potential with several products on the market (Table 8.1.12), including one of the most sold biologics, Enbrel® (Pfizer, sales 2012: 8.4 billion U. S. dollars), and several in late clinical trials [77].

Further approaches are inspired by natural pathogen strategies. Several virulence factors of protozoa and bacteria have highly repetitive and random coiled sequences. It was shown that by manipulating these sequences, the circulation time of the proteins can be altered [82]. Inspired by these native occurring sequences, new synthetic amino acid sequences where developed, either based on highly repetitive (e.g., PASylation [83]) or systematically screened nonrepetitive (e.g., XTEN [84]) sequences.

Glycoengineering/Hyperglycosylation

Posttranslational glycosylation is an essential step in the production of many biologics and one major benefit of mammalian expression systems. The (partial) absence of these complex polysaccharide structures leads to lower stability (*in vitro* and *in vivo*), loss of functionality, and/or immunogenic reactions [85]. Glycoengineering approaches try to control the effect of glycosylation by changing the glycan composition or by adding new glycosylation sites (hyperglycosylation) to enhance thereby their half-life time and bioactivity[86].

The terminal glycans of the N-linked carbohydrate seems the most important factor leading to physiological effects like reduced proteolytic degradation, renal or hepatic clearance. The presence of charged glycans at this position (e.g., sialic acid) shows thereby the best results [86]. By engineering additional N-glycosylation consensus sequences (Asn-X-Ser/Thr) into therapeutic proteins, higher *in vivo* activities could be reached for several proteins, including very successful products like AraNESP® (Amgen) [87].

Other glycoengineering approaches focus on lysosome targeting by receptor-mediated endocytosis. Through exposure of terminal mannose or mannose-6-phosphate domains the proteins will be directed into the lysosome. Products for treatment of lysosomal storage diseases can therefor easily directed to their target location (e.g. Cerezyme®, Genzyme; Naglazyme®, Biomarin Pharmaceuticals) [86].

The conjugation or modification of peptides and proteins is a well-established method for the prolongation of plasma half-life, but has to be included in an early phase of the development of new therapeutics. Modification of the primary structure or posttranslational modifications can have a huge impact on expression levels, stability, and/or activity. Furthermore, "fast forward" approaches of chemical conjugation by unspecific PEGylation are not suitable anymore for drug approval. More specific strategies were developed the last years and have now to be integrated in the development of new biologics.

Passive Polymer Systems

For many therapeutic applications, constant release kinetics are needed to maintain certain drug levels in the patient. Passive polymer systems can offer this behavior by passive release, either based on diffusion out of a depot or biodegradation of the carrier system. Common used materials are biodegradable and/or native-derived polymers (Table 8.1.11) in the form of micro particles or hydrogels.

Micro-particle Systems

The synthetic polymer poly(lactic-Co-glycolic acid)(PLGA) is a FDA- and EMA-approved biodegradable material for parenteral administration and the only marketed micro particle system for delivery of proteins and peptides (Table 8.1.14). After administration, intramuscularly or subcutaneously as immobile depot, the polymer particles will slowly start to hydrolyze into smaller fragments until just lactic and glycolic acid monomers remain. This process takes from several weeks up to years depending on the particle size, polymer length, and ratios of the 2 monomers [88]. Moreover, the use of pure polylactic acid (PLA) particles was also shown to work as constant release depot for peptides (Table 8.1.14). These particles show much slower degradation rates than PLGA (50:50 blends) and can, by mixing them with PLGA, modify the release kinetics. Further surface modification (e.g., with mucoadhesive polymers, PEGylation) can adapt their characteristics, resulting in a manifold platform, also in the field of targeted delivery or noninvasive drug delivery [89].

The use of double or triple block copolymers based on PLA/PLGA and other synthetic or natural polymers (e.g., PEG, poly-ε-caprolactone, chitosan) expand the platform for sustained releasing particles [88-90].

Main drawbacks of these systems are still low drug-polymer ratios (1 %), potential drug degradation while particle formation and high acidity while digestion (acid monomers of PLA and PLGA). In addition, the release kinetics are at least in the early stage not a "zero order" function, burst release with a following lag phase can be observed in many formulations, probably based on desorption and erosion effects on surface bound drugs [89]. While optimized preparation processes like double emulsion solvent evaporation (PLGA) [88] or salting-out techniques (PLA) [90] are suitable for many proteins and peptides, still cheap and more effective methods are needed to establish these systems as popular delivery platforms. Moreover, the fine-tuning by blending or the use of copolymer blocks can help to overcome the remaining obstacles (burst release, acidity) [88].

Table 8.1.14: Examples for peptide and protein containing PLA/PLGA-particle products on the market.

Name	Carrier	Drug	Approval
Sandostatin LAR depot	PLGA-glucose	Octreotide acetate	1988
Lupron depot	PLA/PLGA	Leuprolide acetate	1993
Nutropin depot	PLGA	Recombinant hGH	1999
Decapeptyl	PLA/PLGA	Triptorelin	2000
Somatuline LA	PLGA	Lanreotide	2007
Trelstar depot	PLGA	Triptorelin pamoate	2008

Hydrogel Systems

Hydrogels are attractive systems for the establishment of local drug depots. This water swollen, cross-linked polymer networks offer a good platform for transient, local, and controlled release. Implanted or injected, they release of their payload by diffusion, swelling, or erosion/degradation (Table 8.1.15) to its surrounding tissue. Moreover, their use in topical applications (e.g., patches or contact lenses) shows potential in regenerative medicine [91].

Table 8.1.15: Passive release mechanism (adapted from [93]).

Name	Mechanism
Fickian diffusion	Mesh-size bigger than cargo, concentration gradient leads to burst released followed by a short timespan with lower release
Swelling	Mesh-size smaller than cargo, swelling leads to enhanced pore size which allows than Fickian diffusion of the cargo
Erosion/degradation	Mesh-size smaller than cargo, release due to degradation beginning from the surface, burst effect due to absorption effects followed by „zero order“ release
Absorption	Weak interaction (electrostatic, hydrophobic, hydrogen bonding) leads to retention of the cargo and thereby to slow release
Covalent coupling	Covalent coupling of the cargo due to a biodegradable linker or polymer-backbone leads to slow release dependent on the stability of the crosslinking structures

While most hydrogels are based on *in vitro* preformation, many *in situ* gelling systems were developed. These formulations are easy to inject and form spontaneously hydrogels after administration at the targeted site [92]. The incorporation of therapeutic molecules is thereby either based on physical encapsulation by coincubation while hydrogel formation or on interaction of the payload with the polymers (specific or unspecific).

Diffusion-based hydrogel approaches often use synthetic polymers (e.g., PEG and its copolymer blocks) due to their high biocompatibility and modification possibilities. The duration of the controlled release is thereby limited to several days and can be adjusted by modification of cross-linking density and addition of excipients [93].

The use of natural polymers, polysaccharides, and proteins, is more focused on degradation-based release. They profit from their high biocompatibility, cell adhesion properties, and slow degradation rates within the body (enzymatic or hydrolytic) but suffer from poor mechanical properties and difficult tailoring [93]. While most natural polymers show due to their low cross-linking rates bigger mesh sizes, they often have weak interaction with their protein or peptide payload. Polysaccharides like hyaluronan or chitosan for example show high surface charges (negative or positive) and thereby interact with the charge of their payloads [94] or exhibit additional mucoadhesive behavior [91]. Heparin, a negative charged sulfated polysaccharide, has additional specific interaction with many growth factors. "Heparinization" of natural polymers constitutes therefore a good method for noncovalent immobilization [95]. Further natural occurring interactions of growth factors with ECM binding partners like fibrin or chondroitin sulfate can be used to functionalize these polymer backbones [96]. Besides these natural-based interactions, other approaches like the use of chelating ions, specific peptides interaction, or covalent linkage to the polymer backbone were also shown to be efficient methods for payload immobilization [93].

The variety of passive polymer systems with different release kinetics gives also interesting opportunities in regenerative medicine. Complex healing processes need different growth factors regulated independently in localization, stability, and availability. For example, the recovery of bone fractures requires first immigration of cells, then angiogenesis events, and finally mineralization of osteoblast for solid bone formation [97].

The combination of different passive polymer systems allows separate representation of two or more growth factors, based on distinct diffusion or degradation rates (Figure 8.1.6) [97, 98]. By tailoring mesh-size, accessibility or polymer composition the different release kinetics can be further fine-tuned to reach the desired characteristics [99].

Passive polymer systems already entered the market for spatiotemporal controlled biopharmaceutical delivery. While most systems are based on degradable synthetic polymer particles, also hydrogel approaches either for topical use (Regranex®Gel, Healthpoint Biotherapeutics) or as implants (*OP-1® PUTTY,* Olympus Biotech) are commercialized. These systems allow sustained, local release and lower thereby the need of high doses and/or repeated injection. In addition, more complex delivery systems for tissue regeneration with several payload proteins entered the market (e.g., *Emdogain®*, Straumann) depicting the high potential of these systems as tool for new therapeutic approaches.

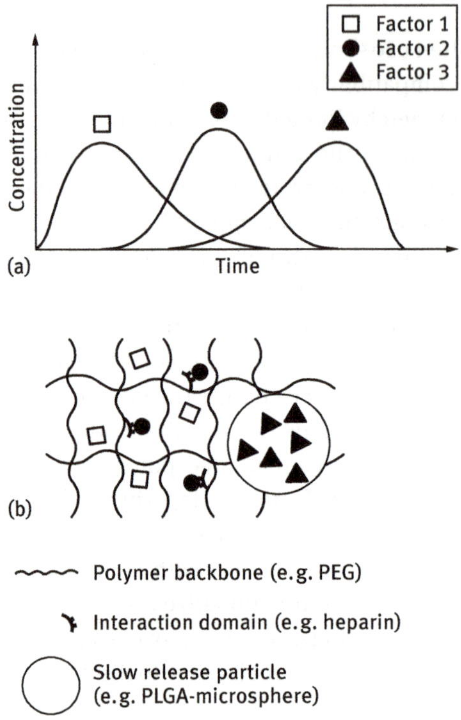

Figure 8.1.6: Schematic multifactor release system for regenerative medicine. **a)** Example of distinct release kinetics for different stimulating factors (e.g., growth factors). **b)** Hydrogel-depot for the release of independent factors. In this hypothetical approach, factor 1 (square) is released by diffusion based on the mesh size, release of factor 2 is also based on diffusion but retained by noncovalent interaction (e.g., heparin), and factor 3 is incorporated into a microparticle (e.g., PLGA-microsphere) and is released by slow degradation of the polymeric particle.

Stimuli-responsive Polymer Systems

The ability to further tune the release behavior of the incorporated cargos by external or physiological triggers gives a dynamic adjustable control for the therapeutic application. While passive systems have predefined release kinetics with sustained release, stimulus sensitive systems work more like biological on/off switches. This may be suitable for therapies that need adjustable and/or pulsative dosing, like for instance in cancer therapy or vaccination schemes [100]. Additional external stimuli like light, hypothermia, or ultrasound (Table 8.1.8) have a good local resolution and allow thereby further targeting in the case of free circulating microspheres or -particles.

Moreover, smart biomaterials that are sensitive to physiological conditions, like for instance pH changes, redox potential, or overexpression of specific proteases (Table 8.1.8), can react directly to changes in condition thereby increasing or reducing the release of its payload [101]. This crosstalk between the drug depot and its target allow very fast and exact adaptation of the drug dose to it needs.

Two major strategies were developed to make polymers reactive to physical stimuli: the use of stimulus sensitive linkers and the ability to switch between hydrophobic and hydrophilic properties [101]. Linker-based methods are well suited for deshielding approaches (see also “inducible release”) or dendrimers. In the case of dendrimers the use of sensitive linkers near the core region results in dissolution of many surface structures. This amplification allows a higher sensitivity to the used stimulus. Switching-based systems mainly focus on amphiphilic polymers that form micelles or hydrogels (Figure 8.1.6). By changing the amphiphilic properties of the polymers these nanostructures will collapse (micelles) or shrink/swell (hydrogels) and thereby release their cargo [101].

Furthermore the use of nanoferro particles that are sensitive to magnetic fields show interesting properties as trigger in DDS. These particles with diameters in the range of 50 to 200 nm consist mostly of iron oxide and can be incorporated into colloidal nanostructures or hydrogels. By applying oscillating magnetic fields these particles will generate local hypothermia due to the Néel relaxation mechanisms [102]. In combination with a thermo sensitive polymer that contains biopharmaceuticals these devices can release the cargo in a controlled manner [103].

An alternative strategy is the functionalization of polymers with biomolecules. The ability of these molecules to sense small molecules like drugs [100, 104] or metabolites [105, 106] in physiological relevant concentrations make them an interesting tool for stimulus sensitive DDSs. Moreover the origin of these switches (Table 8.1.16) ensures optimally functionality under physiological conditions, which is a drawback for many polymer-based switches [101]. Developments in biomolecule coupling chemistry lead to the investigation of efficient and specific coupling methods (see also [107]) thus ensure a predictable and reproducible behavior of these systems.

Table 8.1.16: Biological switches used to control polymer-systems.

Type	Example	Reference
Anti-body <> antigen	scFv <>FITC	[100]
Ligand <> receptor	VEGF <> receptor domain	[112]
Drug <> target	GyrB <> Coumermycin	[104]
DNA <> binding domain	Operator <> Repressor	[106]

A good example for such a stimulus responsive polymer system is the vaccination approach developed by Gübeli et al. [100]. By using a human single chain variable fragment (scFv) against a diagnostic agent (fluorescein) physical coupled to a branched eight-arm PEG as polymer backbone, an *in vivo* functional and biocompatible DDS was constructed. In this system, the payload is physical entrapped within the hydrogel, retained by the small mash size. Upon addition of the diagnostic agent fluorescein, which competitively inhibits the polymer interactions, the hydrogel dissolve, and thereby releasing its cargo (Figure 8.1.7a).

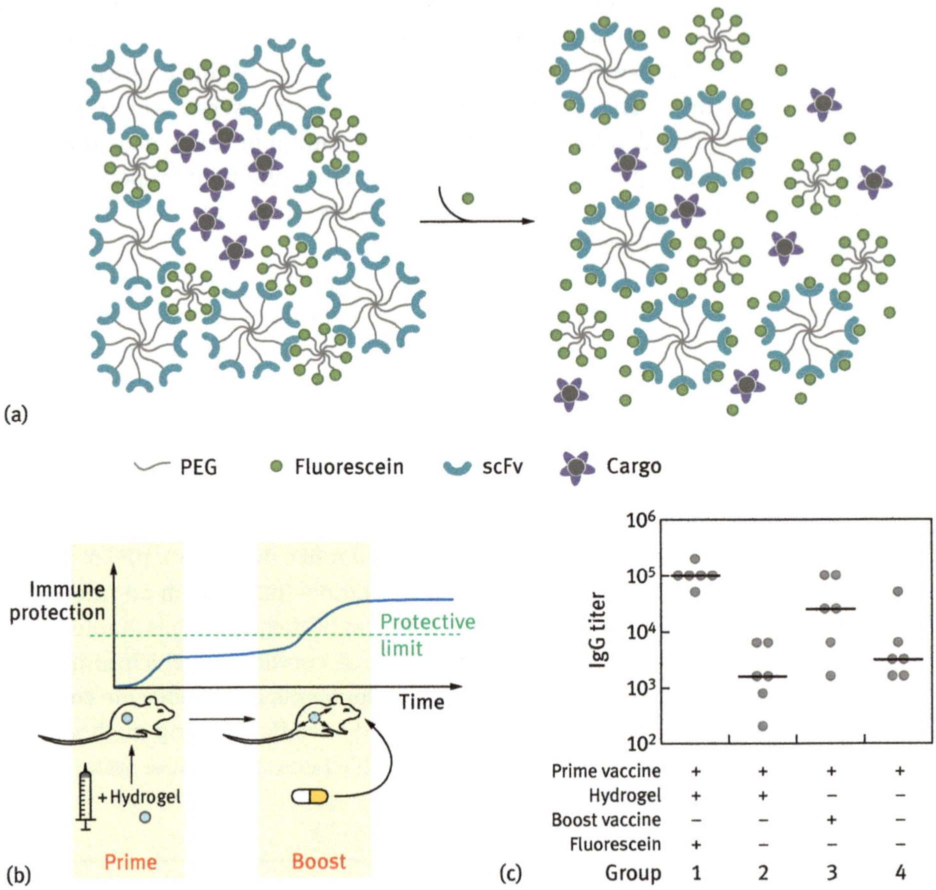

Figure 8.1.7: Stimulus sensitive hydrogel-mediated vaccination approach against HPV16. **a)** Hydrogel design. The hydrogel is based on the interaction of fluorescein and a single chain variable fragment (scFv) against it. To form a hydrogel these components are physically coupled to a star-shaped eight-arm PEG-backbone. The payload is physically caged due to the polymer mesh size and is released by induced dissolution of the hydrogel upon addition of unbound fluorescein. **b)** Vaccination scheme. The hydrogel depot is applied together with the prime vaccination. Instead of a second, boost-injection of the vaccine, fluorescein is orally administered. This leads to dissolution of the hydrogel depot, with a subsequent release of the entrapped vaccine is released. **c)** HPV16-specific IgG titers of different prime-boost vaccinations in mice. All mice received a prime vaccine injection of alum-adsorbed HPV16-capsomeres. Groups 1 and 2 additionally received a hydrogel depot harboring alum-adsorbed HPV16-capsomeres, but group 1 only received an additional oral dose of fluorescein. Group 3 received a classic boost injection of the vaccine. HPV16 specific IgG titers in the serum were measured by ELISA (for details also see [100]).

This approach was used for the development of a single injection vaccination against human papillomavirus type 16. While the first vaccine dose is applied in combination with the hydrogel at day 1, an oral dose of fluorescein at day 7 is sufficient to reach an immunization comparable to classic vaccination with a second "boost" injection at the same time point (Figure 8.1.7b/c). These results show the high potential of biomolecule-polymer copolymers in the field of stimulus responsive DDS.

The ability to gain additional control over the release of DDS can give a huge advantage in therapy. Many promising physical- or small molecule-based systems where developed. While physical inducers seem due to their additional targeting effect very interesting, most of the used polymers show low biocompatibility or lack of *in vivo* studies. Copolymerization with other, biocompatible, polymer blocks can help to overcome these obstacles, but further research is still needed. Systems adapted from biological switches are more complex to develop but already showed their suitability for tissue engineering or medical applications.

8.1.3 Conclusion

The emerging field of biopharmaceuticals presents current drug delivery approaches with a challenge. The low stability and big molecular size, especially for complex peptides or proteins, prevent the use of most established noninvasive strategies such as oral, pulmonary, or transdermal delivery. Therefore, the majority of marketed formulation is administered by injection. By prolonging serum half-life times by simple and effective strategies like drug conjugations or constant release systems, patient discomfort and cost of the therapies can be reduced. Several products based on these approaches are successful in the market.

Further, the constantly growing market of regenerative medicine is desperately looking for reliable peptide and protein delivery systems with high spatiotemporal resolutions and cell-interacting properties. In this field, more complex, multistage delivery system can have a big impact in the near future developments.

Another obstacle for many promising peptide and protein drugs is their need for intracellular delivery. To overcome the barrier of the membrane and the downstream degradation mechanism of the cell the incorporation of the drugs into nanoscaled drug delivery devices is a promising approach. Many strategies were developed the last decade that not only include solid release mechanism but also allow specific targeting of tissues or cell types. While these multifunctional nanocarrier systems show very promising results in preclinical experiments and bring the goal of the "magic bullet" closer.

Beside this optimization of parenteral application, the development of noninvasive strategies in the delivery or peptides and proteins take an important role in current research. Benefits like local targeting (lung, skin, eyes, or gut) and much higher patient compliance can not only reduce the cost and increase the effect of many therapies, but also allow the development of new pharmaceuticals. While current delivery methods can be adapted to small and robust peptides, also approaches for big and sensitive proteins show auspicious results.

Lipids, with their high biocompatibility and good *in vivo* characteristics, as well as polymers, with their high specific properties, showed their possible use in the delivery of proteins and peptides.

Further advances in analytical methods and therapy monitoring combined with more reliable *in vitro* and *in preclinical* models simplify the future developments for the delivery of biopharmaceutical. Encouraged by the success of small molecule loaded drug delivery systems and the high number of possible benefits due to enhanced resolution in time and space, additional and more complex biopharmaceutical delivery systems can be expected in the near future on the market.

References

[1] Mashaghi S, Jadidi T, Koenderink G, Mashaghi A. Lipid nanotechnology. Intl J Mol Sci 2013;14:4242-82.

[2] Li P, Nielsen HM, Mullertz A. Oral delivery of peptides and proteins using lipid-based drug delivery systems. Exp Opin Drug Delivery 2012;9:1289-304.

[3] Bibette J, Calderon FL, Poulin P. Emulsions: basic principles. Rep Prog Phys 1999;62:969.

[4] Mehta SK, Kaur G. Microemulsions as carriers for therapeutic molecules. Rec Pat Drug Delivery Form 2010;4:35-48.

[5] Wissing SA, Kayser O, Muller RH. Solid lipid nanoparticles for parenteral drug delivery. Adv Drug Del Rev 2004;56:1257-72.

[6] Bonacucina G, Cespi M, Misici-Falzi M, Palmieri GF. Colloidal soft matter as drug delivery system. J Pharm Sci 2009;98:1-42.

[7] U. S. Department of Health and Human Services, Food and Drug Administration, Center for Drug Evaluation and Research. Draft Guidance for Industry: Liposome Drug Products - Chemistry, Manufacturing and Controls; Human Pharmacokinetics and Bioavailability; and Labeling Documentation. 2002, accessed August 2002 at http://www.fda.gov/cder/guidance/2191dft.pdf.

[8] Committee for Human Medicinal Products (CHMP). Reflection paper on the data requirements for intravenous liposomal products developed with reference to an innovator liposomal product. 2013, accessed March 2013 at http://www.ema.europa.eu/ema/pages/includes/document/open_document.jsp?webContentId=WC500140351.

[9] Sprott GD. Structures of archaebacterial membrane lipids. J Bioenerg Biomem 1992;24:555-66.

[10] Choquet CG, Patel GB, Beveridge TJ, Sprott GD. Stability of pressure-extruded liposomes made from archaeobacterial ether lipids. App Microbiol Biotechnol 1994;42:375-84.

[11] Benvegnu T, Lemiegre L, Cammas-Marion S. New generation of liposomes called archaeosomes based on natural or synthetic archaeal lipids as innovative formulations for drug delivery. Rec Pat Drug Delivery Form 2009;3:206-20.

[12] Omri A, Agnew BJ, Patel GB. Short-term repeated-dose toxicity profile of archaeosomes administered to mice via intravenous and oral routes. Intl J Toxicol 2003;22:9-23.

[13] Laine C, Mornet E, Lemiegre L, et al. Folate-equipped pegylated archaeal lipid derivatives: synthesis and transfection properties. Chemistry 2008;14:8330-40.

[14] Taneja D, Namdeo A, Mishra PR, et al. High-entrapment liposomes for 6-Mercaptopurine–a prodrug approach. Drug Dev Ind Pharm 2000;26:1315-9.

[15] Okushima S, Nisisako T, Torii T, Higuchi T. Controlled production of monodisperse double emulsions by two-step droplet breakup in microfluidic devices. Langmuir : ACS J Surf Col 2004;20:9905-8.

[16] Kothamasu P, Kanumur H, Ravur N, et al. Nanocapsules: the weapons for novel drug delivery systems. BioImpacts : BI 2012;2:71-81.

[17] Liu J, Gong T, Fu H, et al. Solid lipid nanoparticles for pulmonary delivery of insulin. Intl J Pharm 2008;356:333-44.

[18] Mehnert W, Mader K. Solid lipid nanoparticles: production, characterization and applications. Adv Drug Delivery Rev 2001;47:165-96.

[19] Xu X, Khan MA, Burgess DJ. Predicting hydrophilic drug encapsulation inside unilamellar liposomes. Intl J Pharm 2012;423:410-8.

[20] Pisal DS, Kosloski MP, Balu-Iyer SV. Delivery of therapeutic proteins. J Pharm Sci 2010;99:2557-75.

[21] Xu X, Costa A, Burgess DJ. Protein encapsulation in unilamellar liposomes: high encapsulation efficiency and a novel technique to assess lipid-protein interaction. Pharm Res 2012;29:1919-31.

[22] van Swaay D, deMello A. Microfluidic methods for forming liposomes. Lab Chip 2013;13:752-67.

[23] Belliveau NM, Huft J, Lin PJ, et al. Microfluidic Synthesis of Highly Potent Limit-size Lipid Nanoparticles for In Vivo Delivery of siRNA. Mol Ther Nucleic Acids 2012;1:e37.

[24] Chen C, Han D, Cai C, Tang X. An overview of liposome lyophilization and its future potential. J Controlled Release 2010;142:299-311.

[25] Carrstensen H, Muller RH, Muller BW. Particle size, surface hydrophobicity and interaction with serum of parenteral fat emulsions and model drug carriers as parameters related to RES uptake. Clin Nut 1992;11:289-97.

[26] Albanese A, Tang PS, Chan WC. The effect of nanoparticle size, shape, and surface chemistry on biological systems. Ann Rev Biomed Eng 2012;14:1-16.

[27] Klibanov AL, Maruyama K, Torchilin VP, Huang L. Amphipathic polyethyleneglycols effectively prolong the circulation time of liposomes. FEBS Lett 1990;268:235-7.

[28] Blume G, Cevc G. Liposomes for the sustained drug release in vivo. Biochim Biophys Acta 1990;1029:91-7.

[29] Ensign LM, Cone R, Hanes J. Oral drug delivery with polymeric nanoparticles: the gastrointestinal mucus barriers. Adv Drug Delivery Rev 2012;64:557-70.

[30] Lee VHL. Protease inhibitors and penetration enhancers as approaches to modify peptide absorption. J Controlled Release 1990;13:213-23.

[31] Werle M, Takeuchi H. Chitosan-aprotinin coated liposomes for oral peptide delivery: Development, characterisation and in vivo evaluation. Intl J Pharm 2009;370:26-32.

[32] Dogru ST, Calis S, Oner F. Oral multiple w/o/w emulsion formulation of a peptide salmon calcitonin: in vitro-in vivo evaluation. J Clin Pharm Ther 2000;25:435-43.

[33] Hu FQ, Hong Y, Yuan H. Preparation and characterization of solid lipid nanoparticles containing peptide. Intl J Pharm 2004;273:29-35.

[34] Porter CJ, Trevaskis NL, Charman WN. Lipids and lipid-based formulations: optimizing the oral delivery of lipophilic drugs. Nat Rev Drug Dis 2007;6:231-48.

[35] Takeuchi H, Yamamoto H, Kawashima Y. Mucoadhesive nanoparticulate systems for peptide drug delivery. Adv Drug Delivery Rev 2001;47:39-54.

[36] Lai SK, Wang YY, Hanes J. Mucus-penetrating nanoparticles for drug and gene delivery to mucosal tissues. Adv Drug Delivery Rev 2009;61:158-71.

[37] Petrus AK, Fairchild TJ, Doyle RP. Traveling the vitamin B12 pathway: oral delivery of protein and peptide drugs. Ang Chem 2009;48:1022-8.

[38] Dombu CY, Betbeder D. Airway delivery of peptides and proteins using nanoparticles. Biomaterials 2013;34:516-25.
[39] Nanjwade BK, Adichwal SA, Gaikwad KR, et al. Pulmonary drug delivery: novel pharmaceutical technologies breathe new life into the lungs. J Pharm Sci Technol 2011;65:513-34.
[40] Veldhuizen R, Nag K, Orgeig S, Possmayer F. The role of lipids in pulmonary surfactant. Biochim Biophys Acta 1998;1408:90-108.
[41] Hajos F, Stark B, Hensler S, et al. Inhalable liposomal formulation for vasoactive intestinal peptide. Int J Pharm 2008;357:286-94.
[42] Skubitz KM, Anderson PM. Inhalational interleukin-2 liposomes for pulmonary metastases: a phase I clinical trial. Anti-cancer Drugs 2000;11:555-63.
[43] Grenha A, Remunan-Lopez C, Carvalho EL, Seijo B. Microspheres containing lipid/chitosan nanoparticles complexes for pulmonary delivery of therapeutic proteins. Eur J Pharm Biopharm 2008;69:83-93.
[44] Sintov AC, Levy HV, Botner S. Systemic delivery of insulin via the nasal route using a new microemulsion system: In vitro and in vivo studies. J Controlled Release 2010;148:168-76.
[45] Lochhead JJ, Thorne RG. Intranasal delivery of biologics to the central nervous system. Ad Drug Delivery Rev 2012;64:614-28.
[46] Gaudana R, Ananthula HK, Parenky A, Mitra AK. Ocular drug delivery. AAPS J 2010;12:348-60.
[47] Di Tommaso C, Behar-Cohen F, Gurny R, Moller M. Colloidal systems for the delivery of cyclosporin A to the anterior segment of the eye. Ann Pharm Fr 2011;69:116-23.
[48] Shen J, Wang Y, Ping Q, et al. Mucoadhesive effect of thiolated PEG stearate and its modified NLC for ocular drug delivery. J Controlled Release 2009;137:217-23.
[49] Lajavardi L, Camelo S, Agnely F, et al. New formulation of vasoactive intestinal peptide using liposomes in hyaluronic acid gel for uveitis. J Controlled Release 2009;139:22-30.
[50] Bochot A, Fattal E. Liposomes for intravitreal drug delivery: a state of the art. J Controlled Release 2012;161:628-34.
[51] Prausnitz MR, Langer R. Transdermal drug delivery. Nat Biotechnol 2008;26:1261-8.
[52] Naik A, Kalia YN, Guy RH. Transdermal drug delivery: overcoming the skin's barrier function. Pharm Sci Technol Today 2000;3:318-26.
[53] Pierre MB, Dos Santos Miranda Costa I. Liposomal systems as drug delivery vehicles for dermal and transdermal applications. Arch Dermatol Res 2011;303:607-21.
[54] Puglia C, Bonina F. Lipid nanoparticles as novel delivery systems for cosmetics and dermal pharmaceuticals. Expert Opin Drug Delivery 2012;9:429-41.
[55] Chen Y, Shen Y, Guo X, et al. Transdermal protein delivery by a coadministered peptide identified via phage display. Nat Biotechnol 2006;24:455-60.
[56] Herwadkar A, Banga AK. Peptide and protein transdermal drug delivery. Drug Discovery Today Technol 2012;9:e147-e54.
[57] Guo L, Chen J, Qiu Y, et al. Enhanced transcutaneous immunization via dissolving microneedle array loaded with liposome encapsulated antigen and adjuvant. Intl J Pharm 2013;447:22-30.
[58] Torchilin VP. Affinity liposomes in vivo: factors influencing target accumulation. J Mol Recog 1996;9:335-46.
[59] Choudhary S, Mathew M, Verma RS. Therapeutic potential of anticancer immunotoxins. Drug Discovery Today 2011;16:495-503.
[60] Oussoren C, Storm G. Liposomes to target the lymphatics by subcutaneous administration. Adv Drug Delivery Rev 2001;50:143-56.
[61] Tagami T, Suzuki T, Matsunaga M, et al. Anti-angiogenic therapy via cationic liposome-mediated systemic siRNA delivery. Intl J Pharm 2012;422:280-9.

[62] Tamilvanan S. Formulation of multifunctional oil-in-water nanosized emulsions for active and passive targeting of drugs to otherwise inaccessible internal organs of the human body. Intl J Pharm 2009;381:62-76.
[63] Koshkaryev A, Sawant R, Deshpande M, Torchilin V. Immunoconjugates and long circulating systems: origins, current state of the art and future directions. Ad Drug Delivery Rev 2013;65:24-35.
[64] Perche F, Torchilin VP. Recent trends in multifunctional liposomal nanocarriers for enhanced tumor targeting. J Drug Delivery 2013;2013:705265.
[65] Manjappa AS, Chaudhari KR, Venkataraju MP, et al. Antibody derivatization and conjugation strategies: application in preparation of stealth immunoliposome to target chemotherapeutics to tumor. Journal Controlled Release 2011;150:2-22.
[66] Hosokawa S, Tagawa T, Niki H, et al. Efficacy of immunoliposomes on cancer models in a cell-surface-antigen-density-dependent manner. Brit J Cancer 2003;89:1545-51.
[67] Torchilin VP. Recent approaches to intracellular delivery of drugs and DNA and organelle targeting. Ann Rev Biomed Eng 2006;8:343-75.
[68] Varkouhi AK, Scholte M, Storm G, Haisma HJ. Endosomal escape pathways for delivery of biologicals. Journal Controlled Release 2011;151:220-8.
[69] Holme MN, Fedotenko IA, Abegg D, et al. Shear-stress sensitive lenticular vesicles for targeted drug delivery. Nat Nanotechnol 2012;7:536-43.
[70] Hart SL. Lipid carriers for gene therapy. Cur Drug Delivery 2005;2:423-8.
[71] Midoux P, Pichon C, Yaouanc JJ, Jaffres PA. Chemical vectors for gene delivery: a current review on polymers, peptides and lipids containing histidine or imidazole as nucleic acids carriers. Brit J Pharm 2009;157:166-78.
[72] Li W, Nicol F, Szoka FC, Jr. GALA: a designed synthetic pH-responsive amphipathic peptide with applications in drug and gene delivery. Adv Drug Delivery Rev 2004;56:967-85.
[73] Cho JY, Guo C, Torello M, et al. Defective lysosomal targeting of activated fibroblast growth factor receptor 3 in achondroplasia. Proc Natl Acad Sci USA 2004;101:609-14.
[74] Koren E, Torchilin VP. Cell-penetrating peptides: breaking through to the other side. Trends Mol Med 2012;18:385-93.
[75] Koren E, Apte A, Jani A, Torchilin VP. Multifunctional PEGylated 2C5-immunoliposomes containing pH-sensitive bonds and TAT peptide for enhanced tumor cell internalization and cytotoxicity. Journal Controlled Release 2012;160:264-73.
[76] Pasut G, Veronese FM. State of the art in PEGylation: the great versatility achieved after forty years of research. Journal Controlled Release 2012;161:461-72.
[77] Czajkowsky DM, Hu J, Shao Z, Pleass RJ. Fc-fusion proteins: new developments and future perspectives. EMBO Mol Med 2012;4:1015-28.
[78] Kontermann RE. Strategies for extended serum half-life of protein therapeutics. Cur Opin Biotechnol 2011;22:868-76.
[79] Chen C, Constantinou A, Deonarain M. Modulating antibody pharmacokinetics using hydrophilic polymers. Expert Opin Drug Delivery 2011;8:1221-36.
[80] Veronese FM, Pasut G. PEGylation: Posttranslational bioengineering of protein biotherapeutics. Drug Discovery Today Technol 2008;5:e57-e64.
[81] Fanali G, di Masi A, Trezza V, et al. Human serum albumin: from bench to bedside. Mol Asp Med 2012;33:209-90.
[82] Buscaglia CA, Alfonso J, Campetella O, Frasch AC. Tandem amino acid repeats from Trypanosoma cruzi shed antigens increase the half-life of proteins in blood. Blood 1999;93:2025-32.

[83] Schlapschy M, Binder U, Borger C, et al. PASylation: a biological alternative to PEGylation for extending the plasma half-life of pharmaceutically active proteins. Prot Eng Design Selection 2013;26:489-501.

[84] Schellenberger V, Wang CW, Geething NC, et al. A recombinant polypeptide extends the in vivo half-life of peptides and proteins in a tunable manner. Nat Biotechnol 2009;27:1186-90.

[85] Sola RJ, Griebenow K. Effects of glycosylation on the stability of protein pharmaceuticals. J Pharm Sci 2009;98:1223-45.

[86] Sola RJ, Griebenow K. Glycosylation of therapeutic proteins: an effective strategy to optimize efficacy. BioDrugs 2010;24:9-21.

[87] Elliott S, Lorenzini T, Asher S, et al. Enhancement of therapeutic protein in vivo activities through glycoengineering. Nat Biotechnol 2003;21:414-21.

[88] Mundargi RC, Babu VR, Rangaswamy V, et al. Nano/micro technologies for delivering macromolecular therapeutics using poly(D,L-lactide-co-glycolide) and its derivatives. J Controlled Release 2008;125:193-209.

[89] Danhier F, Ansorena E, Silva JM, et al. PLGA-based nanoparticles: an overview of biomedical applications. J Controlled Release 2012;161:505-22.

[90] Kumari A, Yadav SK, Yadav SC. Biodegradable polymeric nanoparticles based drug delivery systems. Col Surf B Biointerfaces 2010;75:1-18.

[91] Ludwig A. The use of mucoadhesive polymers in ocular drug delivery. Ad Drug Delivery Rev 2005;57:1595-639.

[92] Van Tomme SR, Storm G, Hennink WE. In situ gelling hydrogels for pharmaceutical and biomedical applications. Intl J Pharm 2008;355:1-18.

[93] Censi R, Di Martino P, Vermonden T, Hennink WE. Hydrogels for protein delivery in tissue engineering. J Controlled Release 2012;161:680-92.

[94] Lee F, Chung JE, Kurisawa M. An injectable hyaluronic acid-tyramine hydrogel system for protein delivery. J Controlled Release 2009;134:186-93.

[95] Jeon O, Powell C, Solorio LD, et al. Affinity-based growth factor delivery using biodegradable, photocrosslinked heparin-alginate hydrogels. J Controlled Release 2011;154:258-66.

[96] Perez RA, Won JE, Knowles JC, Kim HW. Naturally and synthetic smart composite biomaterials for tissue regeneration. Adv Drug Delivery Rev 2013;65:471-96.

[97] Lienemann PS, Lutolf MP, Ehrbar M. Biomimetic hydrogels for controlled biomolecule delivery to augment bone regeneration. Adv Drug Delivery Rev 2012;64:1078-89.

[98] Tsurkan MV, Hauser PV, Zieris A, et al. Growth factor delivery from hydrogel particle aggregates to promote tubular regeneration after acute kidney injury. J Controlled Release 2013;167:248-55.

[99] Yilgor P, Hasirci N, Hasirci V. Sequential BMP-2/BMP-7 delivery from polyester nanocapsules. J Biomed Mat Res Part A 2010;93:528-36.

[100] Gübeli RJ, Hövermann D, Seitz H, et al. Remote-Controlled Hydrogel Depots for Time-Scheduled Vaccination. Adv Funct Mat 2013:23(43):5355-62.

[101] Fleige E, Quadir MA, Haag R. Stimuli-responsive polymeric nanocarriers for the controlled transport of active compounds: concepts and applications. Adv Drug Delivery Rev 2012;64:866-84.

[102] Neuberger T, Schöpf B, Hofmann H, et al. Superparamagnetic nanoparticles for biomedical applications: Possibilities and limitations of a new drug delivery system. J Magnet Magn Mat 2005;293:483-96.

[103] Chan A, Orme RP, Fricker RA, Roach P. Remote and local control of stimuli responsive materials for therapeutic applications. Adv Drug Delivery Rev 2013;65:497-514.

[104] Ehrbar M, Schoenmakers R, Christen EH, et al. Drug-sensing hydrogels for the inducible release of biopharmaceuticals. Nat Mat 2008;7:800-4.

[105] Gordijo CR, Koulajian K, Shuhendler AJ, et al. Nanotechnology-Enabled Closed Loop Insulin Delivery Device: In Vitro and In Vivo Evaluation of Glucose-Regulated Insulin Release for Diabetes Control. Adv Funct Mat 2011;21:73-82.
[106] Geraths C, Christen EH, Weber W. A Hydrogel Sensing Pathological Urate Concentrations. Macromol Rapid Commun 2012.
[107] Hotz N, Wilcke L, Weber W. Design, Synthesis, and Application of Stimulus-Sensing Biohybrid Hydrogels. Macromol Rapid Commun 2013:34(20):1594-610.
[108] Cavadas M, Gonzalez-Fernandez A, Franco R. Pathogen-mimetic stealth nanocarriers for drug delivery: a future possibility. Nanomed Nanotechnol Biol Med 2011;7:730-43.
[109] Asane GS, Nirmal SA, Rasal KB, et al. Polymers for mucoadhesive drug delivery system: a current status. Drug Dev Ind Pharm 2008;34:1246-66.
[110] Lee JW, Park JH, Robinson JR. Bioadhesive-based dosage forms: the next generation. J Pharm Sci 2000;89:850-66.
[111] Pillai O, Panchagnula R. Polymers in drug delivery. Cur Opin Chem Biol 2001;5:447-51.
[112] Yamaguchi N, Zhang L, Chae BS, et al. Growth factor mediated assembly of cell receptor-responsive hydrogels. J Am Chem Soc 2007;129:3040-1.

Index

Bei Fragen zur Produktsicherheit wenden Sie sich bitte an:
If you have any questions regarding product safety,
please contact:

Walter de Gruyter GmbH
Genthiner Straße 13
10785 Berlin
productsafety@degruyterbrill.com